QUALITATIVE THEORY OF
SECOND-ORDER DYNAMIC SYSTEMS

# A. A. Andronov, E. A. Leontovich, I. I. Gordon, and A. G. Maier

# QUALITATIVE THEORY OF SECOND-ORDER DYNAMIC SYSTEMS

Translated from Russian by D. Louvish

A HALSTED PRESS BOOK

JOHN WILEY & SONS

New York · Toronto

ISRAEL PROGRAM FOR SCIENTIFIC TRANSLATIONS

Jerusalem · London

© 1973 Israel Program for Scientific Translations Ltd.

Sole distributors for the Western Hemisphere and Japan

HALSTED PRESS, a division of
JOHN WILEY & SONS, INC., NEW YORK

Library of Congress Cataloging in Publication Data
Main entry under title:

Qualitative theory of second-order dynamic systems.

   Translation of *Kachestvennaĩa teoriĩa dinamicheskikh
sistem vtorogo porĩadka*.
  "A Halsted Press book."
  1. Differential equations.  2. Dynamics.
I. Andronov, Aleksandr Aleksandrovich, 1901–1952.
QA372.K14513          531'.11          73-4704
ISBN 0-470-03195-6

Distributors for the U.K., Europe, Africa and
the Middle East

JOHN WILEY & SONS, LTD., CHICHESTER

Distributed in the rest of the world by

KETER PUBLISHING HOUSE JERUSALEM LTD.

ISBN 0 7065 1292 8
IPST cat. no. 22054

This book is a translation from Russian of
KACHESTVENNAYA TEORIYA DINAMICHESKIKH
SISTEM VTOROGO PORYADKA
Izdatel'stvo "Nauka"
Glavnaya Redaktsiya Fiziko-Matematicheskoi Literatury
Moscow 1966

Printed in Israel

## TRANSLATOR'S PREFACE

This is a translation of the first volume of a definitive treatise on dynamic systems and their applications, planned by the celebrated Soviet mathematician A. A. Andronov some thirty years ago, left in an incomplete form owing to Andronov's untimely death in 1952 and subsequently completed by his colleagues and students. The translation of the second volume, published in 1971 (Theory of Bifurcations of Dynamic Systems on a Plane, IPST No. 5438, Israel Program for Scientific Translations, Jerusalem 1971), has sold out, and a second printing is being published simultaneously with this volume in cooperation with John Wiley and Sons, Inc.

The translator has made every effort to adhere to the terminology used in Theory of Bifurcations, at any rate in regard to the specific terminology of the qualitative theory of differential equations. Small differences are inevitable, owing to the two different translators involved, but the only major change has been to replace partition into paths by the more manageable term phase portrait.

The entire two-volume treatise should be of interest as perhaps the only really self-contained account of the Soviet approach to the theory in English. The exposition presupposes very little prior knowledge on the reader's part and proceeds in easy stages.

## PREFACE

This book was begun in 1949 by A. A. Andronov, in collaboration with
E. A. Leontovich and A. G. Maier. After the death of Andronov (in 1952)
and Maier (in 1951) it was completed by Leontovich and I. I. Gordon.
The final version is due to Leontovich.

The book presents the classical results of the qualitative theory of
differential equations on the plane, due primarily to Poincaré and Bendixson,
and also some recent results having a direct bearing on the classical
theory (see Chapters VII through XI).

Though it is independent, the book may be regarded as a realization of
the first volume of Andronov's monograph on second-order dynamic
systems and their applications. Apart from the present book, the
monograph [has a second volume, "Theory of Bifurcation of Dynamic
Systems on a Plane" (IPST cat. no. 5438)], which contains the theory of
structurally stable dynamic systems, Andronov's work on the theory of
bifurcations, and applications of the methods of bifurcation theory to
various problems in oscillation theory.

Also participating in preparation of the book were N. A. Gubar', who
wrote Chapter IX, and R. M. Mints, who assembled and examined the
examples in Chapters IV (Sections 7 and 9) and XII. The book was edited
by Yu. M. Romanovskii.

The chapters are to some extent independent of each other. The
material of Chapters I through VI, IX and XII may serve as an introduction
to the qualitative theory of differential equations. From this standpoint,
the proofs of several propositions in Chapter II (Lemmas I to XIII) may
be omitted and attention concentrated only on their geometric meaning.
This material could be the content of an introductory course in the
qualitative theory of differential equations for third to fifth year university
students of physics, mathematics and mechanics.

Chapters IV, V, VI and IX may be used as references.

Chapters VII, VIII, IX, X and XI contain a complete exposition of original
research.

Chapters VII, VIII, X and XI present results of Leontovich and Maier.
The study of multiple equilibrium states by Bendixson's method in
Chapter IX is due to Gubar'.

An acquaintance with Chapters I, II and III is sufficient for an understanding of Chapters VII, VIII, X and XI.

The appendix summarizes several elementary concepts and propositions
required for the main text, and also presents some additional material.

The book contains numerous figures and examples illustrating the
methods discussed; some of the examples are of independent interest,

as byproducts of applications. Each chapter is preceded by an introduction containing a brief survey of its contents.

Sections, theorems and figures are numbered continuously through the book. Lemmas, formulas and examples are numbered independently within each section. References to a lemma, formula [or subsection] indicate the section number, followed by the number of the lemma, formula [or subsection].

Gor'kii, 1965

E. Leontovich,
I. Gordon

# Contents

1. Introduction (1).  2. Geometric interpretation of the dynamic system (I) in the space $R^3$ (2).  3. Simplest properties of solutions of system (I) (3).  4. Geometric interpretation of a dynamic system on the phase plane (7).  5. Phase portrait in a region $G$ of the phase plane. Elementary discussion (8).  6. Comparison of geometric interpretation in $R^3$ and geometric interpretation on the phase plane (12).  7. Direction on a path. Change of parametrization (13).  8. Terminology and notation (16).  9. Continuity with respect to initial values (17).  10. Change of variables (18).  11. The differential equation corresponding to a dynamic system (20).  12. Isoclines (22).  13. The classical concepts "integral," "integral curve," "general integral" in regard to analytic systems (22).  14. Examples (24).  15. Comments on the examples (38).

1. Introduction (39).  2. Definition of dynamic system on the sphere (40).  3. Dynamic system on the sphere as a vector field on the sphere (43).  4. Solutions and paths of a dynamic system on the sphere (43).  5. Examples of dynamic systems on the sphere (49).

1. Arc without contact (55).  2. Generalized arc without contact (57).  3. Intersection of a path with an arc without contact (57).  4. Configuration of paths in the neighborhood of an arc without contact (58).  5. Some properties of the functions $\Phi(t, s)$, $\Psi(t, s)$ (61).  6. Paths crossing two arcs without contact. Correspondence function (64).  7. Case of a path having several points in common with an arc without contact (69).  8. Succession function (73).  9. Closed curves formed by combining an arc of a path and an arc without contact; regions bounded by such curves (75).  10. Cycle without contact (78).  11. Family of cycles without contact. Paths entering region filled by cycles without contact (79).  12. Single-crossing cycle (80).  13. Differentiation of function along paths of system (I) (81).  14. Cycle without contact between two successive turns of a path crossing an arc without contact (82).

## INTRODUCTION

This is a mathematical book. However, the mathematical questions with which it is concerned are directly and organically pertinent to problems of natural science and technology. A brief discussion of this relationship is in order.

Exact science, whose beginnings may be placed in Newton's era, is intimately bound up with the creation of an adequate mathematical language, the "differential and integral calculus" and the apparatus of differential equations.

Since Newton, the laws of nature have been described by differential equations. Celestial mechanics was the first realm of science whose laws (interactions between bodies obeying the law of universal gravitation) were subjected to this treatment. Consideration of the solutions of these differential equations enabled scientists, starting from information about the positions and velocities of bodies at a given time, to predict with a high degree of accuracy their positions at any other time (e.g., to predict precisely the occurrence of solar and lunar eclipses, the position of the planets over various periods, etc.).

Celestial mechanics has played a crucial part in the creation of the scientific outlook. It was celestial mechanics which first convinced scholars of the very existence of unchanging laws of nature, which could be utilized together with information concerning the present to reveal the past and to predict the future. Celestial mechanics was the solid foundation of determinism, first stated explicitly by Laplace (at the beginning of the nineteenth century) who, after Newton, was one of the creators of celestial mechanics.*

The differential equations of celestial mechanics are ordinary differential equations, which satisfy existence and uniqueness theorems. Laplacian determinism is embodied in the mathematical proposition that the solutions of the equations are uniquely determined by their initial conditions.

With the development of science after Newton, in particular physics, differential equations found ever-increasing use as a tool for the mathematical description of phenomena. In many fields of physics, such as electromagnetic field theory, theory of heat and statistical physics, the fundamental role has passed from ordinary differential equations to partial differential equations, together with their special features. Nevertheless, ordinary differential equations have by no means receded into the background but have even conquered new realms.

---

* Laplacian determinism represents "an ideal form of causal relationships" (Bohr /3/, p.1). Subsequent developments in physics, chiefly quantum mechanics, have led to certain changes in "ideal" Laplacian determinism, though they have left its main ideas intact.

Although differential equations constitute a mathematical tool, one must emphasize that the actual derivation or "setting up" of differential equations for some domain of science obviously transcends the bounds of mathematics and belongs essentially to the domain under consideration. The derivation of differential equations always involves a certain idealization of reality, so that the result is no more than a mathematical description of a certain simplified model of real phenomena. Moreover, even when extremely general principles are available for the derivation of differential equations (for example, in mechanics, where one has a "prescription" for the equations of motion — the Lagrange equations of the first and second kind), the study of particular cases usually involves informal arguments, which are far beyond the scope of mathematics.

We have no intention, however, of discussing questions of idealization in the study of reality. Our primary interest lies in the mathematical questions and problems relevant to the ordinary differential equations of natural science.

The first questions to be considered date from the development of celestial mechanics.

One of the classical problems of celestial mechanics is that of the motion of $n$ material points under the action of Newtonian gravitation, the $n$-body problem, described by a system of differential equations:

$$\dot{x}_i = \Phi_i (x_1, \ x_2, \ \ldots, \ x_N) \qquad (i = 1, \ 2, \ \ldots, \ N), \qquad (1)$$

where $N$ obviously depends on the number of material points. As stated above, the right-hand sides of these equations satisfy conditions guaranteeing the existence and uniqueness of the solution for given initial conditions.

The independent variable in system (1) — time — does not appear explicitly in the right-hand sides. Such systems are said to be autonomous, in contrast to nonautonomous systems,* in whose right-hand sides the time appears explicitly:

$$\dot{x}_i = F_i (x_1, \ x_2, \ \ldots, \ x_N, \ t). \qquad (2)$$

Systems of type (1), sometimes also of type (2), are known as dynamic systems, reflecting their role in mechanics, in particular celestial mechanics.

Besides being autonomous, the equations of motion of celestial mechanics possess certain other specific properties. They are conservative

---

* The classification of systems as autonomous and nonautonomous is somewhat arbitrary. Indeed, by adding to (2) a dependent variable $\tau$ such that

$$\frac{d\tau}{dt} = 1,$$

we can replace (2) by an equivalent autonomous system:

$$\frac{dx_i}{dt} = F_i (x_1, \ x_2, \ \ldots, \ x_N, \ \tau), \qquad \frac{d\tau}{dt} = 1.$$

The classification is nonetheless extremely convenient from certain standpoints.

equations possessing an energy integral, i.e., they can be written in what is known as canonical or Hamiltonian form:

$$\frac{dx_i}{dt} = -\frac{\partial H}{\partial y_i}, \qquad \frac{dy_i}{dt} = \frac{\partial H}{\partial x_i}. \qquad (3)$$

Thus, on the basis of existence and uniqueness theorems, we know that the equations of motion of celestial mechanics have a unique solution when initial conditions are specified relative to any given instant of time.

This means that "knowledge of the state of system (3)," i.e., knowledge of the coordinates of the material points and their velocities at some given time, enables one to determine its state (the coordinates and velocities of the points) at any future or past time. The problem is thus reduced to finding a solution of system (3), in other words, to integrating it.

When $n = 2$, the case of two bodies, equations (3) can be integrated by elementary quadratures. The resulting analytical expressions yield exhaustive information about the type of motion and enable one to find the position of the bodies at any time for given initial conditions. However, the situation is immeasurably complicated when we pass to $n = 3$ — the "three-body problem" — not to speak of $n > 3$. As the French mathematician Borel has stated, "in celestial mechanics, as in counting by primitives, three is many."

In case $n \geqslant 3$ the very words "to integrate the system" or to find its solution are meaningless without further specification. Indeed, if "integration of system (3)" is taken to mean determination of an analytical expression for the solution, one naturally asks: what kind of expression? As early as the beginning of the nineteenth century it was known that only in very special cases can the solution of (3) for $n \geqslant 3$ be expressed in terms of integrals of elementary functions (i.e., solved by quadratures). Moreover, even when this is possible, the expressions obtained may well be so complex that their direct analysis involves enormous difficulties and in fact requires special methods.

Alternatively, we may consider system (3) to have been integrated if the solution has been found in the form of uniformly and absolutely convergent series. Unfortunately, experience shows that such series may converge so slowly that they are of little practical use.*

The three-body problem occupied many famous nineteenth-century mathematicians. As far back as the eighteenth century, the pressing needs of astronomy stimulated the development of approximate and numerical methods of solution. True, these methods only enable one, as a rule, to compute one particular solution over a given interval of the independent variable (the time $t$); however, in a great many questions approximate computation of a few particular solutions is quite adequate. Approximate and numerical determination of particular solutions is meaningful and significant not only in problems of celestial mechanics, in which field it has by now reached a high degree of sophistication, but also for many problems of physics and technology.

---

\* Sundman obtained a solution of the three-body problem by series which converge absolutely and uniformly for any $t$. However, it was subsequently shown that these series are practically useless, since they converge too slowly: in order to achieve sufficient accuracy (say, in computation of ordinary solar eclipses) one would have to take a number of terms exceeding the so-called "number of electrons in the Einstein universe."

There are numerous problems in which approximate computation yields an exhaustive solution.

Powerful tools are available today for numerical determination of solutions — electronic computers. "Integration" of differential equations in the sense of numerical solution is employed extensively in various problems and the precision that can be achieved is considerable.

This situation notwithstanding, approximate computation of particular solutions on finite intervals of the independent variable is inadequate in principle for the solution of many problems in natural science. The availability of modern high-speed computers has not changed anything in this respect. Such problems first arose, naturally, in celestial mechanics itself, in relation to questions of cosmogony. Thus, with regard to the three-body problem alone, examples of such problems were listed by Poincaré in the introduction to one of his classical memoirs /5/: "Will one of the bodies remain always in a certain region of the heavens or can it escape to infinity? Will the distance between two of these bodies decrease indefinitely or, on the contrary, will it always remain between definite limits?" and many other problems of this sort. It is evident that approximate knowledge of a few particular solutions over a finite interval of time cannot provide an answer to these questions. On the contrary, one needs solutions over an arbitrarily large time interval, in other words, one needs to know, as it were, the properties of the solution "as a whole," or "in the large."

The investigation of solutions of differential equations from this standpoint has become known as "qualitative investigation" or "qualitative integration," and the corresponding theory as the qualitative theory of differential equations. Among the questions relevant to the qualitative theory are the following:

Does system (1) possess integral curves which are closed curves (or, in the accepted terminology, does it have closed paths)?

Are the solutions corresponding to a given equilibrium state stable or unstable (i.e., does the point return to this equilibrium state after once leaving it, or not?) For what regions of values of the variables do the points of the integral curves tend to the given equilibrium state with increasing $t$?

The task of the qualitative investigation of differential equations in its nontrivial aspects was first clearly delineated by Poincaré, at the end of the last century.

At approximately the same time, Lyapunov formulated an extremely important particular problem of the qualitative theory — the problem of stability of a motion (solution); he examined this problem for a quite broad range of cases.

Poincaré formulated the problem in its general form for the simplest case, a system of two differential equations:

$$\dot{x} = P(x, y), \quad \dot{y} = Q(x, y). \tag{I}$$

The fundamental constituents of the qualitative theory of system (I) were set forth in Poincaré's classical monograph "Sur les courbes définies par une équation différentielle." Simultaneously, in his three-volume treatise "Les méthodes nouvelles de la mécanique céleste," Poincaré considered many questions of qualitative theory in connection with the three-body problem. Lyapunov's investigations of stability were presented in his book "The General Problem of Stability of Motion." Poincaré's studies of systems of type (I) were subsequently completed by Bendixson, while his investigations of the equations of celestial mechanics were put on a firm basis by Birkhoff, who used set-theoretic methods for this purpose.

Certain investigations of Poincaré, continued by Birkhoff, were the basis of the so-called "metric theory of dynamic systems," which may be regarded as an entirely independent part of the qualitative theory of differential equations with its own specific viewpoints and methods. The applied field most intimately connected with the metric theory of dynamic systems is statistical physics.

Not wishing to dwell upon later publications on the qualitative theory of differential equations, we go on to discuss its penetration into other realms of natural science. At the end of the nineteenth century, the only field enjoying the fruits of the qualitative theory of differential equations was celestial mechanics. Since the beginning of the present century, the situation has changed radically. The study of periodic processes, periodic phenomena in various fields of physics — mechanics, optics, acoustics, etc. — is now subsumed under the heading "oscillation theory." The end of the previous century saw the publication of the first comprehensive exposition of the general theory of oscillations — Rayleigh's celebrated "Treatise on the Theory of Sound."

Developments in technology also involved the need, to a greater or lesser degree, for oscillation theory. With the increasing velocities and sizes of structures, engineers urgently required methods for the elimination of harmful and sometimes even destructive vibrations arising at certain critical velocities or periods (for instance, because of resonance).

In this connection, the prevailing mathematical tool of oscillation theory was the theory of linear differential equations. Many questions of physics and technology are bound up with linear systems. With regard to ordinary differential equations alone, the fundamental equations of the classical theory of oscillations were linear differential equations with constant coefficients and periodic right-hand side. This classical tool was generally capable of coping with numerous problems, such as questions of resonance and ways of overcoming it. It is an extremely simple tool and moreover quite effective.

An essential change in this situation took place at the beginning of the twentieth century with the advent of radiophysics and radio-engineering. It was soon established that the majority of phenomena in radio-engineering cannot be described by linear differential equations; they are governed by essentially nonlinear equations. In addition, the dominant oscillation problems of radio-engineering are in a sense the inverse of those of the classical theory. The fundamental problem of the classical theory as evidenced in early technology was how to eliminate harmful oscillations, whereas one of the basic problems of modern radio-engineering is the generation of oscillations. The devices used in radio-engineering for generation of oscillations employ time-independent energy sources, so that one is concerned with what are known as self-oscillations. In mathematical terms, this means that the systems of differential equations describing these devices are autonomous, of type (I). In accordance with a long-standing tradition in the theory of oscillations, attempts were continually made to force essentially "nonlinear" phenomena into the confines of the linear theory. This not only precluded any correct description of the phenomena common in radio-engineering, but simply led to outright errors.

The question of mathematical tools adequate to deal with the phenomena of radio-engineering was set in sharp relief by Mandel'shtam in the 1920s, and subsequently solved by Andronov.

It transpired that the required mathematical theory was that developed by Poincaré and Lyapunov, of which we spoke above, originally aimed at the needs of celestial mechanics, in other words, the qualitative theory of differential equations. It is true that the systems of differential equations governing radio-engineering phenomena are not Hamiltonian, as is the case in celestial mechanics. Poincaré, however, in his book "Sur les courbes définies par une équation différentielle," in which he laid the foundations of the qualitative theory, did not assume that the equations are necessarily Hamiltonian.

Some of the fundamental questions relevant to radio-engineering are reflected in the mathematical terminology of the qualitative theory of differential equations; for example, the question of the existence or non-existence of oscillations for various values of the parameters corresponds to the question of the existence or nonexistence of a closed path (limit cycle) of a system of differential equations. In this connection, approximate computation of particular solutions over a given, finite time interval, which plays such an important role in astronomy, seems to be far less significant in the differential equations of radio-engineering systems, or, at any rate, is subordinated to qualitative investigation of the equations.

We have been discussing a relatively restricted field — radio-engineering, for which the qualitative theory of differential equations has proved its worth as a mathematical tool. However, even the study of radio-engineering problems made it obvious that the corresponding mathematical problems are amenable to a much broader interpretation. There are numerous problems in various domains of natural science and technology whose mathematical interpretation is analogous, involving the use of the qualitative theory of differential equations. Not to speak of problems of mechanics and acoustics, examples of such problems are the question of variable stars, the so-called cepheids, the question of certain periodic reactions in chemistry, photosynthesis in biology, and many others. The theory of automatic control is yet another important domain of application of the qualitative theory (though in a rather specialized form).

The qualitative theory of differential equations has thus provided an adequate mathematical tool for describing phenomena in a host of domains, both such as can be subsumed under the heading of oscillation theory and others (celestial mechanics).

In a certain sense, the "history" of phenomena which have the same description in the qualitative theory of differential equations is similar: irrespective of the physical nature of these phenomena, they are as it were "behaviorally isomorphic." Moreover, the existence of such an isomorphism by no means requires that the phenomena be governed by the same differential equations. It is sufficient that the equations have the same "qualitative structure" as to partition of space into paths.

In the realm of ideas now known as cybernetics, the question of generality, isomorphic behavior in different scientific and technological disciplines, is all-pervading. In the tremendous arsenal of mathematical tools utilized by cybernetics, differential equations, in particular the qualitative theory of differential equations, occupy a relatively small place. Nevertheless, even today this tool retains its meaning and significance and will undoubtedly continue to penetrate new regions.

This book is devoted to the qualitative theory of second-order dynamic systems, i. e., systems of two autonomous differential equations (I) considered on the $(x, y)$-plane.

The case of second-order systems may naturally be regarded as the first and simplest case, and should be studied both for its own importance and as a preliminary to consideration of the more complex case of three or more autonomous differential equations. In addition, systems of type (I) are still of independent interest for applications, inasmuch as many phenomena and problems in various fields of physics and technology may be described, subject to suitable idealization, by such systems.

For systems of three or more autonomous differential equations, the situation becomes immeasurably more complicated. Hitherto, results in the qualitative theory of such dynamic systems have been extremely sparse, though developments in the past decade have been quite intensive.

The problem of qualitative investigation can be formulated in a natural way not only for autonomous dynamic systems, the main topic of the present discussion, but also for many nonautonomous dynamic systems. Although the nonautonomous formulation has its own specific features, it is nevertheless organically related, as to both content and methods (the method of point mappings), to the problem of qualitative investigation of autonomous dynamic systems and, in particular, to the qualitative theory of second-order dynamic systems.

Even in the simplest case of a system of two nonautonomous differential equations with right-hand sides periodic in $t$,

$$\dot{x} = F(x, y, t), \quad \dot{y} = \varphi(x, y, t),$$

where

$$F(x, y, t+\tau) \equiv F(x, y, t), \quad \varphi(x, y, t+\tau) \equiv \varphi(x, y, t) \quad (\tau - \text{the period}),$$

one encounters difficulties of the same type as in consideration of autonomous systems of order $n = 3$.

Finally, the question of qualitative investigation of distributed systems, i. e., systems of partial differential equations, arises naturally in the context of diverse physical problems. At present the qualitative theory of distributed systems is only in its infancy, but one can undoubtedly expect intensive developments. The theory should naturally be based on the qualitative theory of ordinary differential equations, and in particular the qualitative theory of second-order dynamic systems.

*Chapter I*

## DYNAMIC SYSTEMS IN A PLANE REGION AND ON THE SPHERE

### §1. DYNAMIC SYSTEMS IN A PLANE REGION

**1. INTRODUCTION.** We shall consider systems of differential equations of type

$$\frac{dx}{dt} = P(x, y), \qquad \frac{dy}{dt} = Q(x, y), \qquad \text{(I)}$$

where $P(x, y)$ and $Q(x, y)$ are continuous functions defined in some region $G$ in the euclidean plane ($x, y$ are cartesian coordinates) and at least once continuously differentiable in this region with respect to both variables. The region may be either bounded or unbounded; in particular, it may coincide with the entire ($x, y$)-plane.

Systems of type (I) are a special case of systems of two differential equations in two unknown functions: the independent variable $t$ does not occur explicitly in their right-hand sides. Such systems are said to be autonomous. Autonomous systems of differential equations are also known as dynamic systems.

The system (I) will be referred to as a dynamic system on the plane or in a plane region. We shall also say that the dynamic system is given or defined in the region $G$. Henceforth we shall omit the words "on the plane" or "in a plane region."

A dynamic system (I) defined in a region $G$ is called a system of class $C_n$ if the functions $P(x, y)$ and $Q(x, y)$ are functions of class $C_n$, i. e., they are $n$ times continuously differentiable in $G$.

A dynamic system (I) is called a system of analytic class, or an analytic system, if $P$ and $Q$ are analytic functions in $G$.

It is obvious that any system of class $C_k$ ($k > 1$) is at the same time a system of class $C_{k_1}$ for any $k_1 < k$, in particular, a system of class $C_1$. An analytic system is of class $C_k$ for any natural number $k$.

All dynamic systems considered in this book will be of class $C_1$. Throughout the sequel, therefore, the term dynamic system should always be understood as a system of class $C_1$, even when this is not made explicit.

In this section we shall study the simplest properties of dynamic systems in a plane region. These properties are characteristic for autonomous systems of differential equations.

Nonautonomous systems (i. e., systems in whose right-hand sides $t$ appears explicitly) generally do not possess these properties.*

**2. GEOMETRIC INTERPRETATION OF THE DYNAMIC SYSTEM (I) IN THE SPACE $R^3$.** We now consider the customary geometric interpretation of systems of two differential equations in two unknown functions, i. e., the interpretation in three-dimensional space with cartesian coordinates $x, y, t$.

In this context, $P(x, y)$ and $Q(x, y)$ should be treated as functions of three variables $x, y$ and $t$. But since they are independent of $t$, their domain of definition in three-dimensional space $R^3$ is an infinite cylindrical region $H$, generated by all straight lines parallel to the $t$-axis which cut the $(x, y)$-plane at points of the region $G$.**

The solutions

$$x = \varphi(t), \quad y = \psi(t)$$

of system (I) are represented by curves lying in the region $H$. These curves are known as i n t e g r a l   c u r v e s of the system (I) (see Appendix, §8). Here and below, any solution of a system of differential equations will be assumed to be continued over the m a x i m u m   p o s s i b l e i n t e r v a l   o n   t h e   $t$-a x i s (see Appendix, §8).

Since the functions $P(x, y)$ and $Q(x, y)$ are at least of class $C_1$, system (I) satisfies the conditions of the existence-uniqueness theorem (see Appendix, §8) at all points of $H$; in this special case, the theorem reduces to the following statement:

*T h e o r e m   1. For any point $M_0(x_0, y_0) \in G$ and any $t_0$, $-\infty < t_0 < +\infty$, there is exactly one solution*

$$x = \varphi(t), \quad y = \psi(t)$$

*of the system (I) which satisfies initial conditions*

$$x_0 = \varphi(t_0), \quad y_0 = \psi(t_0),$$

*and is defined for all $t$ in a certain interval $(\tau, T)$ containing $t_0$. In particular, the solution may be defined for all $t$, i. e., $\tau = -\infty$, $T = +\infty$.*

The geometric meaning of Theorem 1 is that through each point of the region $H$ there is exactly one integral curve of the system (I).

For systems of type (I) we also have the following theorem, which will play a significant role in the sequel:

*T h e o r e m   2. Let $\overline{G}_1$ be a closed bounded subregion of $G$ $(\overline{G}_1 \subset G)$,*

$$x = \varphi(t), \quad y = \psi(t) \tag{1}$$

*a solution of system (I) defined in an interval $(\tau, T)$, such that for all $t$ in $(\tau, T)$ the point $N(\varphi(t), \psi(t))$ always remains in $\overline{G}_1$. Then $\tau = -\infty$, $T = +\infty$, i. e., the solution (1) is defined for all $t$.*

P r o o f. Assume that the solution $x = \varphi(t), y = \psi(t)$ is defined for $t = t_0$.

---

* The properties of dynamic systems described in this section remain valid, subject to obvious modifications, for dynamic systems of any order, i. e., systems of the type (1) considered in the Introduction, for any $n$.
** When $G$ is the entire $(x, y)$-plane, $H$ will be the entire space $R^3$.

Let $\tau_1$ and $\tau_2$ be two arbitrary numbers such that $\tau_1 < t_0$, $\tau_2 > t_0$. Let $\bar{H}_1$ denote the bounded cylindrical region in $R^3$ consisting of all points $M(t, x, y)$ such that $\tau_1 \leqslant t \leqslant \tau_2$, $(x, y) \in \bar{G}_1$ (see Figure 1). The integral curve corresponding to the solution (1) passes

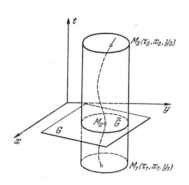

through the point $M_0 (t_0, \varphi(t_0), \psi(t_0))$, which is in the region $H_1$. But then, by Theorem A' of the Appendix (continuation of solution up to the boundary of the domain of definition, see § 8), this integral curve will leave $\bar{H}_1$ for some $t$ greater than $t_0$ and for some $t$ smaller than $t_0$. Now the integral curve cannot leave $\bar{H}_1$ through its lateral surface, since then there would be a point $N(\varphi(t), \psi(t))$ outside the closed region $\bar{G}_1$, contradicting the assumption. Consequently, the integral curve leaves $\bar{H}_1$ through its upper and lower bases (Figure 1). But this means that (1) is defined for $t = \tau_1$ and $t = \tau_2$. Since $\tau_1$ and $\tau_2$ are arbitrary, the solution (1) is defined for all $t$. Q.E.D.

FIGURE 1.

3. **SIMPLEST PROPERTIES OF SOLUTIONS OF SYSTEM (I).** We shall now establish certain properties of solutions of system (I) which follow from the fact that it is autonomous.

*Lemma 1. If*

$$x = \varphi(t), \quad y = \psi(t)$$

*is a solution of system (I) defined on an interval $(\tau, T)$, then*

$$x = \varphi(t+C), \quad y = \psi(t+C), \qquad (2)$$

*where $C$ is any constant, is also a solution of system (I) defined on the interval $(\tau - C, T - C)$.*

Proof. Since (1) is a solution of the system (I), it follows that for all $t \in (\tau, T)$, identically,

$$\frac{d\varphi(t)}{dt} = P(\varphi(t), \psi(t)), \quad \frac{d\psi(t)}{dt} = Q(\varphi(t), \psi(t)).$$

Replacing $t$ by $t+C$ in these equalities, we get the following identities for all $t \in (\tau - C, T - C)$:

$$\frac{d\varphi(t+C)}{d(t+C)} = P(\varphi(t+C), \psi(t+C)),$$
$$\frac{d\varphi(t+C)}{d(t+C)} = Q(\varphi(t+C), \psi(t+C)). \qquad (3)$$

But obviously

$$\frac{d\varphi(t+C)}{d(t+C)} \equiv \frac{d\varphi(t+C)}{dt}, \quad \frac{d\psi(t+C)}{d(t+C)} \equiv \frac{d\psi(t+C)}{dt},$$

and so we can rewrite (3) as

$$\frac{d\varphi\,(t+C)}{dt}=P\,(\varphi\,(t+C),\quad \psi\,(t+C)),\qquad \frac{d\psi\,(t+C)}{dt}=Q\,(\varphi\,(t+C),\quad \psi\,(t+C)).$$

These equalities show that (2) is a solution of system (I). The fact that this solution is defined on $(\tau - C,\ T - C)$ is proved by simple arguments which will be omitted here. Q.E.D.

From the geometric standpoint, Lemma 1 means that the curve obtained from any integral curve by shifting it along the $t$-axis for an arbitrary distance is again an integral curve.

In fact, the integral curve

$$x=\varphi\,(t+C),\qquad y=\psi\,(t+C)$$

is obtained from the integral curve

$$x=\varphi\,(t),\qquad y=\psi\,(t)$$

by shifting it for a distance $C$ along the $t$-axis.*

*Lemma 2. a) Two solutions*

$$x=\varphi\,(t),\qquad\qquad y=\psi\,(t) \tag{1}$$

*and*

$$x=\varphi\,(t+C),\qquad y=\psi\,(t+C) \tag{2}$$

*of the system (I) may be regarded as solutions satisfying initial conditions with identical $x_0$, $y_0$ and different $t$.*

*b) Two solutions satisfying initial conditions with identical $x_0$, $y_0$ and different $t$ may be obtained from each other by substituting $t + C$ for $t$, with a suitably chosen constant $C$.*

Proof. If the solution (1) corresponds to initial values $t_0$, $x_0$, $y_0$, so that

$$\varphi\,(t_0)=x_0,\qquad \psi\,(t_0)=y_0, \tag{3}$$

then, by virtue of the obvious equalities

$$\varphi\,(t_0-C+C)=\varphi\,(t_0)=x_0,\qquad \psi\,(t_0-C+C)=\psi\,(t_0)=y_0,$$

---

* Another elementary fact is worth mentioning here: since the integral curve is given parametrically by equations

$$x=\varphi(t),\ y=\psi\,(t),$$

where $\varphi$ and $\psi$ are single-valued functions, it follows that any $t$ for which the solution is defined determines a unique point on the integral curve, so that the tangent to the latter can never be parallel to the $(x,\,y)$-plane. Indeed, the direction cosines of the tangent to the integral curve are proportional to $\dot\varphi$, $\dot\psi$ and 1, i. e., the angle between the tangent and the $t$-axis cannot be equal to $\pi$. In particular, this implies that an integral curve cannot have maxima or minima.

the solution (2) corresponds to initial values $t_0-C$, $x_0$, $y_0$, so that part a) is proved.

Now let the solution (1) correspond to initial values $t_0$, $x_0$, $y_0$ and consider the solution

$$x = \varphi^*(t), \quad y = \psi^*(t), \tag{4}$$

corresponding to initial values $t_0^*$, $x_0$, $y_0$, where $t_0^* \neq t_0$. Setting the constant $C$ in the solution (2),

$$x = \varphi(t+C), \quad y = \psi(t+C),$$

equal to $t_0-t_0^*$, we see that it obviously corresponds to the same initial values $t_0^*$, $x_0$, $y_0$ as (4). Since the solution is uniquely determined by the initial conditions, we have

$$\varphi^*(t) \equiv \varphi(t+t_0-t_0^*), \quad \psi^*(t) \equiv \psi(t+t_0-t_0^*),$$

which proves part b). Q.E.D.

Henceforth, when considering two solutions of type (1) and (2), we shall often say that the solutions differ as to the choice of initial value for $t$.

The solution of any system of two differential equations corresponding to arbitrary initial values $t_0$, $x_0$, $y_0$ is clearly a function of $t$, $t_0$, $x_0$, $y_0$ (see Appendix, §8), i. e.,

$$x = \Phi(t, t_0, x_0, y_0), \quad y = \Psi(t, t_0, x_0, y_0). \tag{5}$$

It is clear from the definition of the functions $\Phi(t, t_0, x_0, y_0)$ and $\Psi(t, t_0, x_0, y_0)$ that $\Phi(t_0, t_0, x_0, y_0) \equiv x_0$, $\Psi(t_0, t_0, x_0, y_0) \equiv y_0$. In the case of autonomous system (1), however, the functions (5) are essentially functions not of $t$ and $t_0$ but of the difference $t-t_0$. This is shown in the following

Lemma 3. As a function of $t$ and the initial values $t_0$, $x_0$, $y_0$, a solution of system (I) can be written

$$x = \varphi(t-t_0, x_0, y_0), \quad y = \psi(t-t_0, x_0, y_0). \tag{6}$$

Proof. Together with (5), we consider the solution

$$x = \Phi(t, 0, x_0, y_0), \quad y = \Psi(t, 0, x_0, y_0),$$

satisfying the initial conditions: $x = x_0$, $y = y_0$ for $t = 0$.
By Lemma 1,

$$x = \Phi(t-t_0, 0, x_0, y_0), \quad y = \Psi(t-t_0, 0, x_0, y_0) \tag{7}$$

is also a solution of the system (I). Since the solutions (5) and (7) correspond to the same initial values $t_0$, $x_0$, $y_0$, they must coincide:

$$\Phi(t, t_0, x_0, y_0) \equiv \Phi(t-t_0, 0, x_0, y_0),$$
$$\Psi(t, t_0, x_0, y_0) \equiv \Psi(t-t_0, 0, x_0, y_0).$$

Setting

$$\Phi(t-t_0, 0, x_0, y_0) = \varphi(t-t_0, x_0, y_0),$$
$$\Psi(t-t_0, 0, x_0, y_0) = \psi(t-t_0, x_0, y_0)$$

we get the assertion of the lemma. Q.E.D.

Henceforth the solution of system (I) corresponding to initial values $t_0$, $x_0$, $y_0$ will always be written in the form (6).

*Lemma 4. If the solution*

$$x = \varphi(t-t_0, x_0, y_0), \quad y = \psi(t-t_0, x_0, y_0) \tag{8}$$

*is defined for* $t-t_1$ *and*

$$\left.\begin{array}{l} \varphi(t_1-t_0, x_0, y_0) = x_1, \\ \psi(t_1-t_0, x_0, y_0) = y_1, \end{array}\right\} \tag{9}$$

*then*

$$\varphi(t-t_0, x_0, y_0) \equiv \varphi(t-t_1, x_1, y_1),$$
$$\psi(t-t_0, x_0, y_0) \equiv \psi(t-t_1, x_1, y_1). \tag{10}$$

Proof. It follows immediately from (9) that (8) and the solution

$$x = \varphi(t-t_1, x_1, y_1), \quad y = \psi(t-t_1, x_1, y_1)$$

correspond to the same initial values $t_1$, $x_1$, $y_1$. But then they must coincide, and this proves (10). Q.E.D.

Remark. Setting $t = t_0$ in (10), we get

$$x_0 = \varphi(t_0-t_1, x_1, y_1), \quad y_0 = \psi(t_0-t_1, x_1, y_1).$$

This is obviously true for any $t_1$, $x_1$, $y_1$ satisfying (10). Dropping the indices, we obtain

$$x_0 = \varphi(t_0-t, x, y), \quad y_0 = \psi(t_0-t, x, y).$$

*Lemma 5. If (I) is a system of class $C_n$, then the functions* $x = \varphi(t-t_0, x_0, y_0)$, $y = \psi(t-t_0, x_0, y_0)$ *have partial derivatives of the following orders, jointly continuous in all variables throughout their domain of definition:*

*1) up to order $n+1$, inclusive, with respect to $t$ (or $t_0$),*
*2) up to order $n$, inclusive, with respect to $x_0$ and $y_0$*

$$\left(\frac{\partial^k \varphi(t-t_0, x_0, y_0)}{\partial x_0^i \partial y_0^{k-i}}, \quad \frac{\partial^k \psi(t-t_0, x_0, y_0)}{\partial x_0^i \partial y_0^{k-i}} \quad \begin{array}{l} i = 0, 1, \ldots, k \\ k = 1, 2, \ldots, n \end{array}\right),$$

*3) up to order $n+1$ with respect to $t$ (or $t_0$), $x_0$ and $y_0$, provided the functions are differentiated at least once with respect to $t$ (or $t_0$)*

$$\left(\frac{\partial^{k+l}\varphi(t-t_0, x_0, y_0)}{\partial t^l, \partial x_0^i \partial y_0^{k-i}}, \quad \frac{\partial^{k+l}\psi(t-t_0, x_0, y_0)}{\partial t^l \partial x_0^i \partial y_0^{k-i}} \quad \begin{array}{l} i = 0, \ldots, k \\ k+l = 1, 2, \ldots, n \\ l = 1, 2, \ldots, n+1 \end{array}\right).$$

This lemma follows at once from Theorems B' and B" of the Appendix, §8.

**4. GEOMETRIC INTERPRETATION OF A DYNAMIC SYSTEM ON THE PHASE PLANE.** In this book, the geometric interpretation of system (I) in three-dimensional space $(x, y, t)$ is of secondary importance. The fundamental geometric interpretation of an autonomous system (I) refers to the $(x, y)$-plane, which is known as the p h a s e  p l a n e of system (I).

At each point $M(x, y)$ of the region $G$ on the $(x, y)$-plane, we consider a vector $v$ with components $P(x, y)$, $Q(x, y)$. Thus, the dynamic system (I) defines a v e c t o r  f i e l d in the region $G$.*

Since the functions $P(x, y)$ and $Q(x, y)$ are by assumption continuously differentiable, the vector field defined by system (I) is what is known as a c o n t i n u o u s l y  d i f f e r e n t i a b l e  v e c t o r  f i e l d.

Suppose that at least one of the functions $P(x, y)$, $Q(x, y)$ does not vanish at the point $M(x, y)$. Then the length of the vector at this point,

$$r = \sqrt{P^2(x, y) + Q^2(x, y)},$$

is positive, and the sine and cosine of the angle $\theta(x, y)$ between the positive $x$-axis and the direction of the vector are given by

$$\sin \theta = \frac{Q}{\sqrt{P^2 + Q^2}}, \qquad \cos \theta = \frac{P}{\sqrt{P^2 + Q^2}}$$

(see Appendix, §5).

At points where

$$P(x, y) = 0, \qquad Q(x, y) = 0,$$

the length of the vector is zero and no direction is defined. Such points are known as s i n g u l a r  p o i n t s of the vector field (or of system (I)); points at which at least one of $P(x, y)$, $Q(x, y)$ is not zero are r e g u l a r or n o n s i n g u l a r points of the vector field. At every regular point $M$ of a vector field, the angle $\theta(x, y)$ is continuous. At a singular point, $\theta(x, y)$ is undefined and as $x$ and $y$ tend to the coordinates of the point $\lim \theta(x, y)$ may not even exist.

Let

$$x = \varphi(t), \qquad y = \psi(t) \tag{11}$$

be any solution of system (I). The set of points $M(\varphi(t), \psi(t))$, where $t$ runs through all values for which the solution (11) is defined, is called the p a t h corresponding to the solution, or a path of the vector field defined

---

* The mathematical literature makes extensive use of vector notation for systems of differential equations. In this notation system (I) becomes a vector equation

$$x = F(x).$$

Vector notation is an extremely convenient tool for systems of more than two equations. For systems of two equations the use of vector notation would only encumber the presentation and we shall therefore not use it, except in the case of dynamic systems on a sphere.

by the dynamic system (I), or simply a path of the dynamic system (some-times also a phase trajectory or simply trajectory).

Equations (11) are obviously parametric equations of the path. Conversely, the solution defining a given path will be called the solution corresponding to the path.

If a point $M (x, y)$ on a path is not a singular point of the vector field, the vector $(P (x, y), Q (x, y))$ is a tangent vector to the path (Figure 2). Indeed, since $x = \varphi (t), y = \psi (t)$ is a solution of system (I), we have

$$\dot\varphi (t) \equiv P (\varphi, \psi), \quad \dot\psi (t) \equiv Q (\varphi, \psi). \tag{12}$$

But the vector with components $\dot\varphi (t), \dot\psi (t)$ is obviously tangent to the path, and by (12) it is a vector of the field defined by (I).

Treating the parameter $t$ as "time," we have a "kinematic" inter-pretation of system (I): the solution $x = \varphi (t), y = \psi (t)$ describes, as it were, the motion of a point along a path in the phase plane. At each point of the phase plane, the vector defined by (I), i. e., the vector $(P (x, y), Q (x, y))$, is clearly equal to the velocity of the moving point or the "phase velocity." Solutions with the same initial values $x_0$ and $y_0$ and different $t_0$ define motions beginning at the same point but at different "times" ($t_0$ and $t^*$). The point with coordinates $(\varphi (t), \psi (t))$ is also known as a "representative" point.

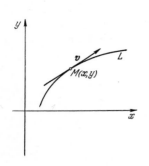

Let $M (a, b)$ be a singular point of system (I),

FIGURE 2.

$$P (a, b) = Q (a, b) = 0. \tag{13}$$

Then obviously $x = a, y = b$ is a solution of (I), and so a singular point of a vector field is itself a path. A path of this type is called an equilibrium state.* The converse is also clearly true: if system (I) has a solution

$$x = a, \quad y = b \tag{14}$$

(where $a$ and $b$ are constants) then the point $(a, b)$ must be an equilibrium state (a singular point of the vector field), i. e., it satisfies (13). Since it does not involve $t$, the solution (14) is obviously defined for all $t$.

In the sequel, a point $(x, y)$ of the region $G$ such that $P (x, y) = Q (x, y) = 0$ will usually be called an equilibrium state (and not a singular point).

An equilibrium state $M (a, b)$ of system (I) is said to be isolated if there exists $\varepsilon_0 > 0$ such that there are no other equilibrium states in the $\varepsilon_0$ -neighborhood of $M$.

## 5. PHASE PORTRAIT IN A REGION $G$ OF THE PHASE PLANE. ELEMENTARY DISCUSSION.

Lemma 6. Any two solutions differing only as to choice of initial value $t_0$ correspond to the same path.

* Other terms used are "critical point" or "equilibrium position."

Proof. By Lemmas 1 and 2, any two solutions differing only as to choice of initial value $t_0$ (but with the same initial values $x_0, y_0$) may be obtained from each other by substituting $t + C$ for $t$. But any two solutions

$$x = \varphi(t), \qquad y = \psi(t) \tag{15}$$

and

$$x = \varphi(t+C), \qquad y = \psi(t+C), \tag{16}$$

the first defined on $(\tau, T)$ and the second on $(\tau - C, T - C)$, obviously correspond to the same path (since substitution of $t + C$ for $t$ in (15) is simply a change of notation for the variable). Q.E.D.

Theorem 3. Through every point of the region $G$ there is exactly one path of the dynamic system (I).

Proof. Let $M_0(x_0, y_0)$ be an arbitrary point in $G$. By Theorem 1 (existence-uniqueness theorem), for every $t$ there exists a solution with initial values $t_0, x_0, y_0$:

$$x = \varphi(t), \qquad y = \psi(t).$$

This clearly means that there is at least one path $L$ through the point $(x_0, y_0)$.

Now suppose that there are two distinct paths $L$ and $L^*$ through some point $M_0(x_0, y_0)$ of $G$.

Let

$$x = \varphi^*(t), \qquad y = \psi^*(t)$$

be the solution corresponding to the path $L^*$. Then there is necessarily some $t = t^*$ such that

$$x = \varphi^*(t^*) = x_0, \qquad y = \psi^*(t^*) = y_0,$$

but then, by Lemma 2, for a suitably chosen constant $C$,

$$\varphi^*(t) = \varphi(t+C), \qquad \psi^*(t) = \psi(t+C),$$

and so (Lemma 6) the paths $L$ and $L^*$ cannot be distinct, contrary to our assumption. Q.E.D.

Remark 1. It follows at once from the reasoning used to prove the theorem that any two different solutions corresponding to the same path can be obtained from each other by substituting $t + C$ for $t$, i. e., they differ only as to choice of initial value $t_0$ (see Lemma 2).

Remark 2. Suppose that, for some choice of a solution corresponding to a path $L$, a point $M_0$ on the path corresponds to $t_0$ and a point $M_1$ to $t_0 + \tau$. Then it follows from Remark 1 that if $M_0$ corresponds to $t^*$ for some other choice of a solution corresponding to $L$, then $M_1$ corresponds to $t^* + \tau$.

Remark 3. If a path lies entirely within a bounded closed region $\overline{G}_1 \subset G$, it follows from Theorem 2 that the corresponding solution is defined for all $t$ $(-\infty < t < +\infty)$.

By virtue of Theorem 3, a dynamic system defined in a region $G$ defines a certain family of paths, known as the p h a s e  p o r t r a i t of the system in $G$.*

The principal topic of the qualitative theory of dynamic systems, with which we are concerned in this book, is the study of these phase portraits. An essential component of this program is to determine the possible features of individual paths. We now indicate a few fundamental properties of paths. We have already considered a special type of path — equilibrium states. Namely,

$$x = a, \quad y = b$$

is an equilibrium state if and only if

$$P(a, b) = Q(a, b) = 0.$$

Suppose now that the path $L$ corresponding to a solution

$$x = \varphi(t), \quad y = \psi(t),$$

is not an equilibrium state. At all points of this path, obviously,

$$[\dot{\varphi}(t)]^2 + [\dot{\psi}(t)]^2 \equiv [P(\varphi, \psi)]^2 + [Q(\varphi, \psi)]^2 \neq 0.$$

Indeed, let $M^*(x^*, y^*)$ be a point on $L$, corresponding to $t^*$, at which

$$\dot{\varphi}^2(t^*) + \dot{\psi}^2(t^*) = [P(\varphi(t^*), \psi(t^*))]^2 + [Q(\varphi(t^*), \psi(t^*))]^2 = 0.$$

Then

$$\dot{\varphi}(t^*) = P(\varphi(t^*), \psi(t^*)) = 0, \quad \dot{\psi}(t^*) = Q(\varphi(t^*), \psi(t^*)) = 0,$$

and this clearly means that $(x^*, y^*)$ is an equilibrium state. But an equilibrium state is itself a path, and by Theorem 3 the point $M^*(x^*, y^*)$ cannot belong to a path $L$ other than the equilibrium state.

Can a path which is not an equilibrium state intersect itself? In other words, can there exist $t_1$ and $t_2$, $t_1 \neq t_2$, which determine the same point on the path?

The answer to this question is given by the following

*L e m m a  7. Let $L$ be a path corresponding to a solution*

$$x = \varphi(t), \quad y = \psi(t) \quad (\tau < t < T), \tag{17}$$

*not an equilibrium state, and suppose that there exist $t_1$ and $t_2$ $(\tau < t_1 < t_2 < T)$ such that*

$$\varphi(t_1) = \varphi(t_2), \quad \psi(t_1) = \psi(t_2).$$

*Then the solution (17) is defined for all $t$ (i. e., $\tau = -\infty$, $T = +\infty$ ), the functions $\varphi(t)$ and $\psi(t)$ are periodic in $t$, and the path is a smooth simple closed curve.*

* [The authors use the more expressive but less elegant and unfamiliar (in English) term "partition of $G$ into paths."]

10

**Proof.** Let

$$\varphi(t_1) = \varphi(t_2) = x_0, \qquad \psi(t_1) = \psi(t_2) = y_0. \tag{18}$$

Together with the solution (17), consider the solution

$$x = \varphi(t + t_2 - t_1), \qquad y = \psi(t + t_2 - t_1), \tag{19}$$

defined on $(\tau - C, T - C)$, where $C = t_2 - t_1$ (see Lemma 1).

It follows from (18) that the solutions (17) and (19) satisfy the same initial conditions ($x = x_0$, $y = y_0$ for $t = t_1$). Hence they are equal, and a fortiori defined on the same interval of the $t$-axis. But if $C \neq 0$ the intervals $(\tau, T)$ and $(\tau - C, T - C)$ cannot coincide unless $\tau = -\infty, T = +\infty$. We have thus shown that the solutions (17) and (19) are defined for all $t$ ($-\infty < t < +\infty$).

Now, since the solutions (17) and (19) coincide, it follows that for all $t$ ($-\infty < t < +\infty$)

$$\varphi(t + C) \equiv \varphi(t), \qquad \psi(t + C) \equiv \psi(t), \tag{20}$$

where $C = t_2 - t_1 > 0$. This clearly means that $\varphi(t)$ and $\psi(t)$ are periodic functions, with common period $\theta = t_2 - t_1$.

Let

$$\theta_0 \ (\theta_0 \leqslant \theta \leqslant t_2 - t_1) \tag{21}$$

be the smallest positive number such that

$$\varphi(t + \theta_0) = \varphi(t), \qquad \psi(t + \theta_0) = \psi(t). \tag{22}$$

Such a number necessarily exists, for otherwise we could find a sequence of positive numbers $\{\theta_n\}$ such that $\lim\limits_{n \to \infty} \theta_n = 0$ and

$$\varphi(t + \theta_n) \equiv \varphi(t), \qquad \psi(t + \theta_n) \equiv \psi(t).$$

But then, for any $n$ and any (positive or negative) integer $k$,

$$\varphi(t + k\theta_n) \equiv \varphi(t), \qquad \psi(t + k\theta_n) \equiv \psi(t)$$

or, for some fixed $t_0$,

$$\varphi(t_0 + k\theta_n) = \varphi(t_0), \qquad \psi(t_0 + k\theta_n) = \psi(t_0).$$

Thus, each of the functions $\varphi(t)$ and $\psi(t)$ takes on the same value $\varphi(t_0)$ and $\psi(t_0)$, respectively, at each point of the sequence

$$\ldots t_0 - N\theta_n, \ \ldots, \ t_0 - \theta_n, \ t_0, \ t_0 + \theta_n, \ \ldots, \ t_0 + N\theta_n, \ \ldots,$$

where $N$ may be any integer and $\theta_n$ is arbitrarily small for sufficiently large $n$. Consequently, for any desired $t^*$ we have either $t^* = t \pm k\theta_n$, and then $\varphi(t^*) = \varphi(t_0), \ \psi(t^*) = \psi(t_0)$, or $t^*$ lies in some (open) interval

$(t_0 + (k-1)\theta_n, t_0 + k\theta_n)$ and, since $\theta_n$ may be made arbitrarily small by choosing $n$ sufficiently large, there exists $t'$ arbitrarily close to $t^*$ such that

$$\varphi(t') = \varphi(t_0), \quad \psi(t') = \psi(t_0).$$

But then, since the functions $\varphi(t), \psi(t)$ are continuous, it is evident that

$$\varphi(t^*) = \varphi(t_0), \quad \psi(t^*) = \psi(t_0).$$

This means that the functions $\varphi(t)$ and $\psi(t)$ are constants, so that the path $L$ is an equilibrium state — contradiction.

Clearly, all points on the path $L$ can be obtained by varying $t$ in equations (17) from $t_0$ to $t_0 + \theta_0$ ($t_0 \leqslant t \leqslant t_0 + \theta_0$), where $t_0$ is any fixed number. Now, by definition, $\theta_0$ is the smallest number for which condition (22) holds; hence any two numbers $t'$ and $t''$, $t_0 \leqslant t' < t'' \leqslant t_0 + \theta_0$, necessarily determine distinct points of $L$. And this means (see Appendix, §5) that $L$ is a simple closed curve. By Lemma 5 this curve is obviously smooth. Q.E.D.

A solution in which $\varphi(t)$ and $\psi(t)$ are periodic functions of $t$ is known as a periodic solution. The smallest number $\theta_0 > 0$ satisfying (22) is the period of the solution.

A path $L$ corresponding to a periodic solution is known as a closed path. It is obvious that all solutions corresponding to the same closed path are periodic solutions with the same period. Any path which is not a closed path or an equilibrium state is called a nonclosed path.

It follows from Lemma 7 that the paths of system (I) cannot intersect themselves, i.e., any portion of a nonclosed path corresponding to values of $t$ in a finite closed interval is a simple smooth arc.

At this point, therefore, we have the following elementary information concerning paths. A path may be 1) an equilibrium state, 2) a closed path, or 3) a nonclosed (nonselfintersecting) path. This information is rather inconclusive, since as yet the possible features of nonclosed paths remain unknown. This question will be discussed in detail in §4.

**6. COMPARISON OF GEOMETRIC INTERPRETATION IN $R^3$ AND GEOMETRIC INTERPRETATION ON THE PHASE PLANE.** As mentioned above, every solution of system (I) determines an integral curve in $R^3$.

A path is obviously the projection of an integral curve on the $(x, y)$-plane. By Lemma 4, the integral curves which project onto a given path are precisely those obtained from a certain integral curve (hence, of course, from each other) by a translation along an arbitrary segment on the $t$-axis. We thus have a natural correspondence between paths of a dynamic system on the phase plane and integral curves in $R^3$. Depending on the properties of the path $L$, the following situations may arise here:

1) $L$ is an equilibrium state $M(a, b)$. The corresponding integral curve in $R^3$ is a straight line $x = a$, $y = b$, parallel to the $t$-axis and passing through $M$. Translation along the $t$-axis maps this line onto itself.

2) $L$ is a closed path, corresponding to a $\theta_0$-periodic solution. The corresponding integral curves are "helical curves" of pitch $\theta_0$ which project onto the path $L$. Translation along a segment $C$ on the $t$-axis maps each integral curve onto another curve, unless $C$ is a multiple of $\theta_0$, in which case the integral curve is mapped onto itself (Figure 3).

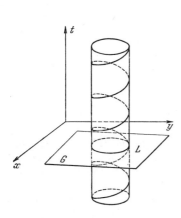

FIGURE 3.

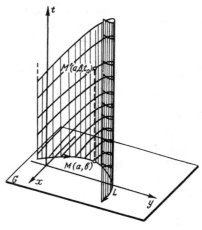

FIGURE 4.

3) $L$ is a nonclosed path. A nontrivial translation along the $t$-axis maps any integral curve corresponding to $L$ onto another integral curve (Figure 4).

The following elementary facts should be emphasized. A point moving along a path which is not an equilibrium state (i. e., a representative point with coordinates $x = \varphi(t)$, $y = \psi(t)$) cannot tend to a point of any other path as $t$ tends to a finite limit. Indeed, otherwise the integral curves would intersect in the space $(x, y, t)$, which is impossible by virtue of Theorem 1. In particular:

*A point moving along a path which is not an equilibrium state can tend to an equilibrium state only as $t$ tends to $+\infty$ or to $-\infty$.*

**7. DIRECTION ON A PATH. CHANGE OF PARAMETRIZATION.** Let $L$ be a path of system (I) and $x = \varphi(t)$, $y = \psi(t)$ any corresponding solution.

We define a direction on $L$ by stipulating that the positive direction on $L$ is that of increasing $t$. With this definition, the positive direction at any point of $L$ coincides with the direction of the vector defined at that point by system (I).

Using the "kinematic" interpretation, we might say that the positive direction on $L$ is the direction in which the point $(\varphi(t), \psi(t))$ moves along the path with increasing $t$, such that the direction of its velocity at each point coincides with that of the phase velocity.

This definition of a direction on $L$ is independent of the specific solution chosen (since all such solutions may be obtained from one another by substituting $t + C$ for $t$).

In the sequel, we shall generally omit the word "positive," i. e., *by the direction on a path $L$ of system (I) we shall mean the positive direction defined (or, as is often said, induced) on $L$ by the system.*

Together with system (I), let us consider the system

$$\frac{dx}{dt} = -P(x, y), \qquad \frac{dy}{dt} = -Q(x, y). \tag{I$'$}$$

The vector field of system (I') is obtained from that of (I) by reversing the direction of all vectors (without changing their lengths).

A direct check shows that for every solution

$$x = \varphi(t), \quad y = \psi(t) \tag{23}$$

of system (I) there is a solution

$$x = \varphi(-t), \quad y = \psi(-t) \tag{24}$$

of system (I'). Hence it is evident that *systems (I) and (I') have the same paths*, but induce opposite directions on the paths. Thus, passage from system (I) to (I') may be interpreted as a change of parametrization on the paths, $t$ being replaced by $-t$.

We now consider a more general change of parametrization on the paths of system (I). Let $f(x, y)$ be a function of class $C_1$ defined in the region $G$. We assume that $f(x, y)$ does not vanish in $G$ except at equilibrium states of system (I), and moreover that it does not change sign in $G$.

Together with system (I), we consider the system

$$\frac{dx}{ds} = P^*(x, y) = P(x, y) f(x, y), \quad \frac{dy}{ds} = Q^*(x, y) = Q(x, y) f(x, y). \tag{I*}$$

By virtue of our assumptions concerning the function $f(x, y)$, it is clear that the equilibrium states of system (I) coincide with those of system (I*).

*L e m m a  8.  If*

$$x = \varphi(t), \quad y = \psi(t) \tag{25}$$

*is a solution of system (1) such that the corresponding path is not an equilibrium state, then there exists a monotone function $t = \beta(s)$ of class $C_1$ such that*

$$x = \varphi(\beta(s)) = \varphi^*(s), \quad y = \psi(\beta(s)) = \psi^*(s) \tag{26}$$

*is a solution of system (I*).*

P r o o f.  For any given initial value $t_0 \in (\tau, T)$, where $(\tau, T)$ is the interval of definition of the solution (25), and an arbitrary number $s_0$, consider the following function $s(t)$:

$$s = s_0 + \int_{t_0}^{t} \frac{dt}{f(\varphi(t), \psi(t))}.$$

Since $f(x, y)$ does not vanish at points which are not equilibrium states, $s(t)$ is a monotone function of class $C_1$ defined on $(\tau, T)$. Thus the inverse function $t = \beta(s)$ is defined on some interval $(\sigma, S)$, also of class $C_1$ and monotone. Obviously,

$$\frac{d\beta(s)}{ds} = f(\varphi(\beta(s)), \psi(\beta(s))).$$

14

Therefore,

$$\frac{d\varphi\,(\beta)}{ds} = \frac{d\varphi\,(\beta)}{d\beta}\cdot\frac{d\beta}{ds} = P\,(\varphi\,(\beta),\ \psi\,(\beta))\cdot f\,(\varphi\,(\beta),\ \psi\,(\beta)),$$
$$\frac{d\psi\,(\beta)}{ds} = \frac{d\psi\,(\beta)}{d\beta}\cdot\frac{d\beta}{ds} = Q\,(\varphi\,(\beta),\ \varphi\,(\beta))\cdot f\,(\varphi\,(\beta),\ \psi\,(\beta)).\qquad(27)$$

These relations show that the functions (26) constitute a solution of system (I*). It is readily seen that $(\sigma,\ S)$ is the maximal interval of definition of the solution (26), since otherwise the interval $(\tau,\ T)$ would not be maximal for the solution (25). Q.E.D.

Equations (25) and (26) are obviously different parametric equations for the same path. It therefore follows from Lemma 8 that *the dynamic systems (I) and (I\*) have the same paths, but with different parametrizations.* Upon passage from system (I) to system (I*), the directions on the paths remain the same if $f\,(x,\ y) > 0$ and are reversed if $f\,(x,\ y) < 0$.

Suppose now that $f\,(x,\ y)$ may vanish at points other than the equilibrium states of system (I) and may also change sign in the region $G$. Then the equilibrium states of system (I*) are obviously those of (I) plus all points of $G$ which are not equilibrium states of (I) but such that $f\,(x,\ y) = 0$.

The curve

$$f\,(x,\ y) = 0$$

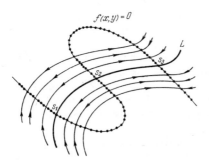

FIGURE 5.

is called a **singular curve** of system (I*) (each point of this curve is an equilibrium state of system (I*)).

Let $L$ be a path of system (I), not an equilibrium state. If $f\,(x,\ y) \neq 0$ on $L$, then, as before, $L$ is a path of system (I*), generally with a different parametrization.

But if there are points of the curve $f\,(x,\ y) = 0$ on $L$, they divide the path into finitely or countably many smooth curves which are paths of system (I*) (Figure 5). The direction on each such path coincides with the direction on $L$ if $f\,(x,\ y) > 0$, and is the opposite direction otherwise.

Thus, every path of system (I) is either a path of system (I*) or the union of finitely or countably many paths of system (I*).*

In the sequel, in several applications and examples, we shall frequently have to deal with dynamic systems of type

$$\frac{dx}{dt} = \frac{P\,(x,\ y)}{f\,(x,\ y)}, \qquad \frac{dy}{dt} = \frac{Q\,(x,\ y)}{f\,(x,\ y)}, \qquad (I**)$$

where the functions $P\,(x,\ y)$, $Q\,(x,\ y)$ are either of class $C_N$ $(N \geqslant 1)$ or analytic and $f\,(x,\ y)$ is either of class $C_N$ or analytic and may vanish in the region $G$

---

* This set may also be uncountable, since the path $L$ may contain uncountably many equilibrium states of system (I*) (i. e., points of the curve $f\,(x,\ y) = 0$). However, the number of paths of system (I*) lying on $L$ and distinct from equilibrium states is always either finite or countable. In Figure 5, the path $L$ of system (I) is the union of seven paths of system (I*), three of which are equilibrium states $S_1$, $S_2$, $S_3$.

(in which the system is being studied). Of course, the right-hand sides of (I**) are undefined at points for which $f(x, y) = 0$. However, by a substitution of the parameter $t$ one can reduce study of system (I**) to consideration of a system of type (I).

Indeed, supposing that $f(x, y)$ does not vanish at $x, y$, we put $dt = f(x, y)\, d\tau$ and obtain the system

$$\frac{dx}{d\tau} = P(x, y), \qquad \frac{dy}{d\tau} = Q(x, y). \qquad (\text{I}^{***})$$

This system is also defined for $x, y$ at which the function $f(x, y)$ vanishes (this corresponds to extending the definition by continuity), so that system (I***) is defined throughout the region $G$. It is clear that, in any subregion of $G$ in which $f(x, y)$ does not vanish, the paths of systems (I**) and (I***) coincide as point sets, although they are differently parametrized. Moreover, wherever $f(x, y) > 0$ the direction with respect to $\tau$ is the same as the direction with respect to $t$, while if $f(x, y) < 0$ the direction is reversed. Points with coordinates $x, y$ such that $f(x, y)$ vanishes, at which the right-hand sides of system (I**) are undefined, must of course be "dropped" from the paths of system (I**) (it is easily seen by consideration of simple examples that these points may be the limits of points on the path as $t$ tends to a finite value).

**8. TERMINOLOGY AND NOTATION.** When we wish to emphasize that solutions corresponding to some path $L$ are defined for all $t\,(-\infty < t < +\infty)$, we shall sometimes say that $L$ is a w h o l e   p a t h. By Theorem 2, any path lying in a bounded portion of the plane such that the distance of any of its points from the boundary of the region $G$ is always greater than some $\varrho_0 > 0$ is necessarily a whole path.

The converse is false. A path on which there are points arbitrarily close to the boundary of $G$ may or may not be a whole path.

Let $M_0$ be a point on a path $L$ corresponding (with respect to some fixed solution) to $t = t_0$. If the solution is defined for all $t \geqslant t_0$, the set of points on $L$ corresponding to $t \geqslant t_0$ is called a p o s i t i v e   s e m i p a t h on $L$, denoted by $L^+$ or $L_{M_0}^+$. Similarly, if the solution is defined for all $t \leqslant t_0$ the set of all points on $L$ corresponding to $t \leqslant t_0$ is called a n e g a t i v e   s e m i - p a t h on $L$, denoted by $L^-$ or $L_{\bar{M}_0}$.

Evidently, if we take another solution corresponding to $L$, for which $M_0$ corresponds to some $t_1 \neq t_0$, the points of the semipath $L_{\dot{M}_0}$ (or $L_{\bar{M}_0}$) will correspond to $t > t_1(t \leqslant t_1)$. $M_0$ will sometimes be called the e n d p o i n t of the semipath. Later we shall frequently have to consider semipaths without specifying whether they are positive or negative. In such cases we shall use the notation $L^{(\,)}$ or $L_{M_0}^{(\,)}$.

If $L$ is an equilibrium state or a closed path, any positive or negative semipath on $L$ is the whole path. A semipath on a nonclosed path will be called a n o n c l o s e d   s e m i p a t h, a semipath on a closed path (which, as stated, coincides with the whole path) — a c l o s e d   s e m i p a t h.

In the [Soviet] mathematical literature, a solution of system (I) is often termed a m o t i o n. This terminology is in accord with the kinematic interpretation of a dynamic system. We shall also make use of this widely employed terminology. Thus, we shall speak of the motion corresponding to given initial values, the path corresponding to a given motion, the motion corresponding to a given path or the motion on a path (i. e., the solution corresponding to the path), a periodic motion, etc.

We shall also say that a path $L$ passes through a point $M_0$ at $t = t_0$, meaning that a certain motion has been fixed on $L$ and the point $M_0$ corresponds to $t = t_0$ for this motion. Similarly, we shall use phrases like "a point $M_1$ on the path $L$ corresponds to $t = t_0$" or "the path intersects (or crosses) an arc $l$ at $t = t_1$" and so on, meaning always that the point $M_1$ or the common point of $L$ and the arc $l$ correspond to $t = t_1$ with respect to a specific motion on $L$.

We shall often use expressions "the path $L$ enters a given region or leaves the region with increasing (or decreasing) $t$," "the path remains in the region for $t > T_0$," and so on, which need no explanation. The following notation will also be used. If

$$x = \varphi(t), \qquad y = \psi(t) \tag{28}$$

is some motion (i. e., solution), the point with coordinates $\varphi(t)$, $\psi(t)$ will be denoted by $M(t)$ and the solution (28) by $M = M(t)$. If initial values are specified, i. e., the motion (solution) is written

$$x = \varphi(t - t_0, x_0, y_0), \qquad y = \psi(t - t_0, x_0, y_0), \tag{29}$$

we shall denote the point with coordinates $\varphi(t - t_0, x_0, y_0)$, $\psi(t - t_0, x_0, y_0)$ by $M(t - t_0, M_0)$ and the solution (29) by $M = M(t - t_0, M_0)$, where $M_0$ denotes the point $(x_0, y_0)$.

**9. CONTINUITY WITH RESPECT TO INITIAL VALUES.** A fundamental theorem in the theory of differential equations, second only to the existence-uniqueness theorem, is the theorem that the solution is continuous in the initial values (see Appendix, §8).

In relation to the autonomous systems (I) considered here, the theorem may be stated as follows.

*Theorem 4. Let $x = \varphi(t - t_0, x_0, y_0)$, $y = \psi(t - t_0, x_0, y_0)$ be a solution of System (I) defined on an interval $(\tau, T)$, and $\tau_1, \tau_2 (\tau_1 < \tau_2)$ two arbitrary numbers in this interval. Then, for any $\varepsilon > 0$, there exists $\eta > 0$ such that if*

$$|x_0 - x_0^*| < \eta, \quad |y_0 - y_0^*| < \eta,$$

*then the solution $x = \varphi(t - t_0, x_0^*, y_0^*)$, $y = \psi(t - t_0, x_0^*, y_0^*)$ is defined for all $t$, $\tau_1 \leqslant t \leqslant \tau_2$, and for all these values of $t$*

$$|\varphi(t - t_0, x_0, y_0) - \varphi(t - t_0, x_0^*, y_0^*)| < \varepsilon,$$
$$|\psi(t - t_0, x_0, y_0) - \psi(t - t_0, x_0^*, y_0^*)| < \varepsilon.$$

Remark. By definition, the functions $\varphi(t - t_0, x_0, y_0)$, $\psi(t - t_0, x_0, y_0)$ are continuous in $t - t_0$. Since according to Theorem 4 they are continuous in $x_0, y_0$ uniformly with respect to $t$ in any finite closed interval on the $t$-axis, it is clear that *they are jointly continuous in all their variables throughout their domain of definition.*

We shall mainly use the following geometric version of Theorem 4.

*Theorem 4'. Let $M_0(x_0, y_0)$ and $M_1(x_1, y_1)$ be two points on an arbitrary path $L$, corresponding to values $t_0$ and $t_1$ of $t$. Then for any $\varepsilon > 0$ there* *exists* $\delta > 0$ *such that if* $M_0' \in U_\delta(M_0)$, *then the path* $L'$ *passing through* $M_0'$ *at* $t = t_0$ *is defined for all* $t$ *in the interval* $t_0 \leqslant t \leqslant t_1$ *(or* $t_0 \geqslant t \geqslant t_1$*) and the point* $M'$ *on* $L'$ *corresponding to any* $t$ *in this interval lies in the* $\varepsilon$*-neighborhood of the point* $M$ *on* $L$ *corresponding to the same* $t$ *(Figure 6).*

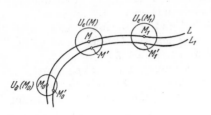

FIGURE 6.

The following lemma follows immediately from Theorem 4.

*L e m m a 9. Let* $K$ *be a closed bounded subset of* $G$. *Then there exists* $h_0 > 0$ *such that for any* $t_0$ *and any point* $M_0(x_0, y_0) \in K$ *the solution*

$$x = \varphi(t - t_0, \ x_0, \ y_0), \qquad y = \psi(t - t_0, \ x_0, \ y_0) \quad (30)$$

*is defined for all* $t$ *in the interval* $t_0 - h \leqslant t \leqslant t_0 + h$.

P r o o f. Suppose the assertion false, i. e., for any $h > 0$ there is a point $\tilde{M} \in K$ such that the solution (30), written briefly as

$$M = M(t - t_0, \ \tilde{M}),$$

is not defined on the entire interval $[t_0 - h, \ t_0 + h]$. Then there exist a sequence $\{h_i\}$ converging to zero and a sequence of points $\{M_i\}$ in $K$ such that the solution

$$M = M(t - t_0, \ M_n)$$

is not defined on the entire interval $[t_0 - h_n, \ t_0 + h_n]$. Since $K$ is closed and bounded, there is a subsequence of $\{M_i\}$ that converges to some point $M^*$ in $K$. We may therefore assume without loss of generality that the sequence $\{M_i\}$ itself converges to $M^* \in K$.

Consider the solution

$$M = M(t - t_0, \ M^*).$$

There always exists $h^* > 0$ such that this solution is defined (at least) for $t$ in the closed interval $[t_0 - h^*, \ t_0 + h^*]$. By Theorem 4, it follows that for sufficiently large $n$ any solution

$$M = M(t - t_0, \ M_n)$$

is defined on $[t_0 - h^*, \ t_0 + h^*]$. But $h_n < h^*$ for sufficiently large $n$ (because $h_n \to 0$), and so the solution $M = M(t - t_0, \ M_n)$ must be defined for all $t$ in $[t_0 - h_n, \ t_0 + h_n]$, contradicting the choice of the points $M_n$. Q.E.D.

**10. CHANGE OF VARIABLES.** Let us assume that the region $G$ in which system (I) is defined is bounded, and consider a r e g u l a r  m a p p i n g of this region onto some region $G^*$ on the $(u, v)$-plane (concerning regular mappings, see Appendix, §6).

Suppose that the mapping is defined by formulas

$$x = f(u, v), \qquad y = g(u, v) \tag{T}$$

or by equivalent formulas

$$u = f^* (x, y), \quad v = g^* (x, y), \tag{T*}$$

where the functions $f$, $g$, $f^*$, $g^*$ are of class $C_2$. We shall assume moreover that $G^*$ is a bounded region; a necessary and sufficient condition for this to be true is that the functions $f^*$ and $g^*$ be bounded in $G$.

The reader will recall that the variables $u$ and $v$ may be treated not only as cartesian coordinates on the $(u, v)$-plane but also as curvilinear coordinates in the region $G$ on the $(x, y)$-plane. Formulas (T) and (T*) then become the formulas for c h a n g e  of  v a r i a b l e s  or  t r a n s - f o r m a t i o n  o f  c o o r d i n a t e s.

Suppose that, after transformation to coordinates $u$, $v$, system (I) becomes

$$\frac{du}{dt} = U (u, v), \quad \frac{dv}{dt} = V (u, v). \tag{31}$$

The functions on the right are obviously given by

$$U (u, v) = \frac{\partial f^*}{\partial x} P (f (u, v), g (u, v)) + \frac{\partial f^*}{\partial y} Q (f (u, v), g (u, v)),$$
$$V (u, v) = \frac{\partial g^*}{\partial x} P (f (u, v), g (u, v)) + \frac{\partial g^*}{\partial y} Q (f (u, v), g (u, v)). \tag{32}$$

Thus, transformation to the new coordinates $u$, $v$ transforms the vector $\boldsymbol{r}$ with coordinates $P (x, y)$, $Q (x, y)$ to a vector $\boldsymbol{r}^*$ with coordinates $U (u, v)$, $V (u, v)$ expressed in terms of $P (x, y)$, $Q (x, y)$ by (32).

The mapping (T) takes every path of the system (I)

$$x = \varphi (t), \ y = \psi (t)$$

onto a path of system (31)

$$u = \varphi^* (t) = f^* (\varphi (t), \psi (t)), \quad v = \psi^* (t) = g^* (\varphi (t), \psi (t)), \tag{33}$$

and, conversely, the mapping (T*) takes paths of system (31) onto paths of system (I). One checks directly that the pair of functions (33) is a solution of system (31).

In the sequel we shall not confine ourselves to r e g u l a r  mappings. In particular, we shall often use the transformation to polar coordinates, which is clearly not regular.* The necessary clarifications regarding transformation to polar coordinates will be given later, when the need arises.

---

* Indeed, transformation to polar coordinates

$$x = \varrho \cos \theta, \quad y = \varrho \sin \theta,$$

is not one-to-one, and moreover the Jacobian

$$\begin{vmatrix} \dfrac{\partial x}{\partial \varrho} & \dfrac{\partial y}{\partial \varrho} \\ \dfrac{\partial x}{\partial \theta} & \dfrac{\partial y}{\partial \theta} \end{vmatrix} = \varrho$$

vanishes at $\varrho = 0$.

## 11. THE DIFFERENTIAL EQUATION CORRESPONDING TO A DYNAMIC SYSTEM.

If we divide one of equations (I) by the other, we get one of the differential equations

$$\frac{dy}{dx} = \frac{Q(x, y)}{P(x, y)}, \qquad (II)$$

or

$$\frac{dx}{dy} = \frac{P(x, y)}{Q(x, y)}. \qquad (II*)$$

Consider equation (II). Let $M_0(x_0, y_0)$ be any point in the region $G$. By the existence-uniqueness theorem, if $P(x_0, y_0) \neq 0$ for $x_0, y_0$, there is a unique solution

$$y = f(x),$$

with initial values $x_0, y_0$, and therefore there is a unique integral curve of equation (II) through the point $M_0(x_0, y_0)$. At each point of this curve the slope of the tangent is given by (II).

Let $x = \varphi(t)$, $y = \psi(t)$ be a solution of system (I) with initial values $t_0, x_0, y_0$. By assumption, $\varphi'(t_0) = P(x_0, y_0) \neq 0$ and therefore $t$ may be expressed as a function of $x$, $t = \gamma(x)$, in the neighborhood of $t_0, x_0, y_0$. Substituting this function into $y = \psi(t)$, we get a solution of equation (II):

$$y = \psi(\gamma(x)) = f(x).$$

Clearly, the integral curve of equation (II) coincides, at points where it is defined, with a path (or part of a path) of system (I).

Suppose that the solution $y = f(x)$ is defined on an interval $(x_1, x_2)$,* and let $x$ tend to an endpoint of this interval, say $x \to x_1$ (the entire argument can be repeated for the case $x \to x_2$). Using standard theorems, one readily sees that if the point $(x, f(x))$ does not tend to the boundary of $G$ as $x \to x_1$, it must tend to a point $M(x_1, f(x_1))$ at which $P(x_1, f(x_1)) = 0$, i.e., to a point at which equation (II) is undefined. If $Q(x_1, f(x_1)) \neq 0$, then $M$ is obviously a point on the path of system (I) at which the tangent is parallel to the $y$-axis (Figure 7). In the neighborhood of a point of this type it is natural to appeal to equation (II*), and to "continue" the integral curve corresponding to the solution $y = f(x)$ of equation (II) as the integral curve of equation (II*) that passes through the point $M(x_1, f(x_1))$. Obviously, at any point for which neither $P(x, y)$ nor $Q(x, y)$ vanishes the solution of equation (II*) can be obtained from the solution $y = f(x)$ of equation (II) by expressing $x$ as a function of $y$, $x = g(y)$; in a sufficiently small neighborhood of such a point, the corresponding parts of the integral curves of equations (II) and (II*) will coincide.

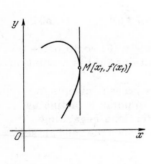

FIGURE 7.

* Recall that the solution of a differential equation is always assumed to be continued over the maximum possible interval (see Appendix, §8).

Similarly, at a point $N\,(g\,(y_1),y_1)$ for which $Q\,(g\,(y_1),\,y_1)=0$ but $P\,(g\,(y_1),\,y_1)\neq 0$, the natural "continuation" of the integral curve of equation (II*) is the integral curve of equation (II) passing through the point.

It is readily seen that the set consisting of all points of the integral curve of equation (II) through a point $M_0\,(x_0,\ y_0)$ of $G$ (not an equilibrium state), together with all its "continuations" in the above sense, is precisely the path through the point $M_0$.

To summarize, *equations (II) and (II\*) in conjunction determine all paths of system (I) other than its equilibrium states.* In contrast, however, whereas examination of system (I) yields the paths in the form of para-metric equations, equations (II) and (II*) determine the paths by equations in $x$ and $y$ (equations in cartesian coordinates). In the sequel, whenever equations (II) and (II*) are considered together we shall not write them both; it will be understood, however, that both are to be considered. We shall also employ the following symmetric notation for equations (II) and (II*):

$$P\,(x,\ y)\,dy - Q\,(x,\ y)\,dx = 0$$

or

$$\frac{dx}{P\,(x,\ y)} = \frac{dy}{Q\,(x,\ y)}\,. \tag{III}$$

Paths of system (I) other than equilibrium states will be called integral curves of equation (III) (and also, not quite rigorously, integral curves of equation (II) or (II*)).

Points at which both

$$P\,(x,\ y)=0 \quad \text{and} \quad Q\,(x,\ y)=0$$

so that both equations (II) and (II*) are undefined, are called singular points of equations (II), (II*) or (III). Thus, the equilibrium states of system (I) correspond to singular points of equations (II), (II*) or (III), and conversely: singular points correspond to equilibrium states.

Whereas system (I) defines a vector field in the region $G$ of the phase plane, whose elements are vectors $v\,(x,\ y)$ with components $P\,(x,\ y)$, $Q\,(x,\ y)$, (see §1.5), equation (III) (or the pair of equations (II) and (II*)) defines a field of directions or field of line elements. A line element is defined to be a point $M$ and an undirected straight-line segment containing $M$ as an interior point. The field of line elements defined by equation (III) is obtained by taking a straight-line segment with slope $\frac{Q\,(x,y)}{P\,(x,y)}$ through each point $M\,(x,\ y)$ (if $P\,(x,\ y)=0$, the segment is taken parallel to the $y$-axis).

The line element corresponding to a point $M\,(x,\ y)$ obviously lies on the tangent to the path passing through $M$.

If $f\,(x,\ y)$ is a function of class $C_1$ which does not vanish in $G$, the system

$$\frac{dx}{dt}=P\,(x,\ y)\,f\,(x,\ y), \qquad \frac{dy}{dt}=Q\,(x,\ y)\,f\,(x,\ y) \tag{I*}$$

(see §1.7) obviously corresponds to the same differential equation (III)

$$\frac{dx}{P(x,\ y)} = \frac{dy}{Q(x,\ y)},$$

as the system

$$\frac{dx}{dt} = P(x,\ y), \qquad \frac{dy}{dt} = Q(x,\ y). \tag{I}$$

Hence an alternative proof of the assertion in §1.7 that systems (I) and (I*) have the same paths.*

**12. ISOCLINES.**   Curves defined in the region $G$ by an equation

$$Q(x,\ y) - C \cdot P(x,\ y) = 0 \tag{34}$$

(where $C$ is a constant), or

$$P(x,\ y) = 0, \tag{35}$$

are known as i s o c l i n e s (curves of equal inclination) of system (I) or equation (III). These curves obviously have the property that the tangents to the paths of system (I) passing through their points (other than equilibrium states) have equal directions at these points. In fact, the slope of the paths at points of an isocline (34) is $C$, while the paths at points of the isocline (35) have slope $\infty$. Thus, the direction of the tangent to a path changes only when the point moves from one isocline to another. The isoclines $Q(x,\ y) = 0$ and $P(x,\ y) = 0$ are known as p r i n c i p a l   i s o c l i n e s. Tangents to paths through points of the first (second) principal isocline are horizontal (vertical); hence an alternative term for the principal isoclines: i s o c l i n e s   o f   h o r i z o n t a l   (v e r t i c a l)   d i r e c t i o n s.

Clearly, all equilibrium states lie on each isocline, and conversely any point common to any two (distinct) isoclines is an equilibrium state of the system. In particular, any intersection point of two principal isoclines is an equilibrium state.

**13. THE CLASSICAL CONCEPTS "INTEGRAL," "INTEGRAL CURVE," "GENERAL INTEGRAL" IN REGARD TO ANALYTIC SYSTEMS.** In this subsection we introduce the concepts "integral," "integral curve," "general integral" of a differential equation or system of equations, as is usually done in the classical literature for analytic equations and systems. In this context, however, the meaning of the term "integral curve" is not quite the same as that explained in the Appendix, §8.

We shall devote some attention to these concepts, for although they play no role in the theory to be presented below they will often figure in the examples given throughout the book.

---

* If the function $f(x,\ y)$ vanishes in $G$, it is clear that by considering the equation $\frac{dy}{dx} = \frac{Q(x,y)}{P(x,y)}$ we are "losing" the singular points of system (I*) (other than equilibrium states of system (I)) for which $f(x,\ y) = 0$.

Suppose that the system (I) under consideration

$$\frac{dx}{dt}=P\,(x,\ y), \qquad \frac{dy}{dt}=Q\,(x,\ y)$$

is analytic in a region $G$. Write the corresponding differential equation in symmetric form (III):

$$\frac{dx}{P\,(x,\ y)}=\frac{dy}{Q\,(x,\ y)}\,.$$

Let $F\,(x,\ y)$ be a function which
  a) is analytic at all points of the curve

$$F\,(x,\ y)=0, \tag{36}$$

  b) satisfies the following identity at all points of the curve (36):

$$F'_x\,(x,\ y)\ P\,(x,\ y)+F'_y\,(x,\ y)\,Q\,(x,\ y)=0. \tag{37}$$

Then equation (36) is called an **integral** or **particular integral** of equation (III) and the curve that it defines is an **integral curve** of equation (III) or system (I).

Let $F\,(x,\ y)=0$ be an integral of system (I), and consider the corresponding integral curve. This curve may contain equilibrium states of system (I), and also points at which $F'_x\,(x,\ y)=F'_y\,(x,\ y)=0$, i. e., singular points of the curve (36).

*Any "piece" of an integral curve which does not contain equilibrium states of system (I) or singular points is either a path or part of a path of system (I).*

In fact, let $M_0\,(x_0,\ y_0)$ be a point on part of the curve (36) satisfying the above condition. Suppose that $F'_y\,(x_0,\ y_0)\neq0$. Then, in some neighborhood of $M_0$ the curve is defined by an equation of type $y=f\,(x)$ such that $\frac{dy}{dx}=f'\,(x)=-\frac{F'_x\,(x,\ y)}{F'_y\,(x,\ y)}$ at all points of the curve in the neighborhood. Since $F'_y\,(x_0,\ y_0)\neq0$, the derivative $F'_y\,(x,\ y)$ does not vanish in a neighborhood of $M_0$. It follows from (37), i. e.,

$$F'_x\,(x,\ y)\,P\,(x,\ y)+F'_y\,(x,\ y)\,Q\,(x,\ y)=0,$$

that $P\,(x,\ y)\neq0$ in the neighborhood of $M_0$, and

$$\frac{dy}{dx}=f'\,(x)=-\frac{F'_x\,(x,\ y)}{F'_y\,(x,\ y)}=\frac{Q\,(x,\ y)}{P\,(x,\ y)}\,.$$

But this means that the function $y=f\,(x)$ satisfies equation (II):

$$\frac{dy}{dx}=\frac{Q\,(x,\ y)}{P\,(x,\ y)}\,.$$

23

The case $F'_x(x_0, y_0) \neq 0$ is treated in similar fashion. Thus, the piece of the curve (36) in question is a part of an integral curve in the sense of §1.11, so that it is (part of) a path of system (I).

We now consider a family of curves

$$F(x, y, C) = 0, \tag{38}$$

defined for $C$ in some region (usually an interval).

*Equation (38) is called a general integral of equation (III) or system (I) if every curve in the family that it defines is an integral curve in the above sense and every point of the region $G$ lies on at least one of the curves (38).*

In particular, it follows from this definition that if $\Phi(x, y)$ is a function defined in $G$, which is analytic at all points of $G$, except possibly at equilibrium states of system (I), and satisfies there the identity

$$\Phi'_x(x, y) \cdot P(x, y) + \Phi'_y(x, y) \cdot Q(x, y) \equiv 0,$$

then

$$\Phi(x, y) = C \tag{39}$$

is a general integral of system (I).

If system (I) (or equation (III)) has a general integral of type (39), with $\Phi(x, y)$ analytic at all points of $G$, we shall say that s y s t e m (I) ( o r e q u a t i o n (III)) h a s a n a n a l y t i c i n t e g r a l i n $G$.*

In certain special cases, knowledge of an analytic integral of system (I) is an aid to qualitative investigation of the system.

**14. EXAMPLES.** In this subsection we shall present a few simple examples of dynamic systems illustrating the material of the preceding subsections. In all these examples, the dynamic systems will be defined throughout the plane.

We first give two simple examples of systems with no equilibrium states.

E x a m p l e 1.

$$\frac{dx}{dt} = 1, \quad \frac{dy}{dt} = 0.$$

The paths are straight lines parallel to the $x$-axis:

$$y = c_1, \quad x = t + c_2.$$

There are clearly no equilibrium states, and all paths (coinciding with the integral curves) are whole paths.

---

* In particular, the so-called Hamiltonian systems discussed in the introduction,

$$\frac{dx}{dt} = \frac{\partial H}{\partial y}, \quad \frac{dy}{dt} = -\frac{\partial H}{\partial y},$$

where $H(x, y)$ is an analytic function, have an analytic integral: $H(x, y) = C$ is an analytic integral (known as the "energy integral") of this system.

Example 2.

$$\frac{dx}{dt} = 1, \qquad \frac{dy}{dt} = 1 + y^2,$$
$$y = \text{tg}\,(t + c_1), \qquad x = t + c_2.$$

There are no equilibrium states, but the paths are not "whole" since they escape to infinity as $t$ tends to a finite limit: $y = \text{tg}(t + c_1) \to \infty$ as $t + c_1 \to$
$\to \frac{\pi}{2}(2k + 1)$.

Example 3.

$$\frac{dx}{dt} = a_1 x, \qquad \frac{dy}{dt} = a_2 y, \tag{40}$$

where $a_1$ and $a_2$ have equal signs.

This system defines a vector field on the $(x, y)$-plane (the phase plane of system (40)), illustrated approximately in Figure 8a for $a_1 < 0$, $a_2 < 0$ and in Figure 8b for $a_1 > 0$, $a_2 > 0$. The straight lines in the figures are isoclines.

System (40) obviously has a single equilibrium state $O(0, 0)$. Solving the system as a linear system with constant coefficients, one easily sees that the solution with initial values $t_0$, $x_0$, $y_0$ is

$$x = x_0 e^{a_1(t - t_0)}, \qquad y = y_0 e^{a_2(t - t_0)}. \tag{41}$$

As expected (Lemma 3), this solution is a function of $t - t_0$.

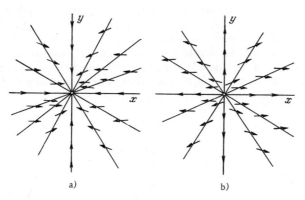

a) b)

FIGURE 8.

The simplest way to find the paths of system (40) is to eliminate $t$ from equation (41) and work in cartesian coordinates.

The result is

$$y - y_0 \frac{x^{a_2/a_1}}{x_0^{a_2/a_1}} = 0.$$

Setting $\frac{y_0}{x_0^{a_2/a_1}} = C$ for $y_0 \neq 0$, we get "parabolas"

$$y - Cx^{a_2/a_1} = 0, \qquad (42)$$

while for $y_0 = 0$

$$x = 0. \qquad (43)$$

Equation (42) gives $y = 0$ when $C = 0$.

It is readily seen that if we replace system (40) by a single equation, say

$$\frac{dy}{dx} = \frac{a_2 y}{a_1 x}$$

or

$$\frac{dy}{a_2 y} = \frac{dx}{a_1 x}, \qquad (44)$$

and integrate it, the integral curves in the sense of §1.13 are the "parabolas" (42) and the two coordinate axes.

As mentioned in §1.13, equation (44) defines a field of line elements, as illustrated in Figure 9.

The paths of system (40) are the parts (halves) of the parabolas (42) and the coordinate axes $x = 0$, $y = 0$ on either side of the equilibrium state $O\,(0,\,0)$. It is evident from (41) that then the points on any path other than $O$ tend to $O$ as $t \to +\infty$ if $a_1 < 0$, $a_2 < 0$, and as $t \to -\infty$ if $a_1 > 0$, $a_2 > 0$. We abbreviate this statement by saying that the paths tend to the equilibrium state $O$ as $t \to +\infty$ or $t \to -\infty$.

Recall that if a "representative" point, moving along any path $L$ other than an equilibrium state, tends to some equilibrium state $A\,(x_0,\,y_0)$, this can only occur if $|t| \to \infty$. Indeed, as shown in §1.6, if $t$ were to tend to some finite $\tau$, two integral curves would pass through the point of space with coordinates $\tau,\,x_0,\,y_0$: a straight line parallel to the $t$-axis (corresponding to the equilibrium state $A\,(x_0,\,y_0)$) and one other, corresponding to the path $L$. This of course contradicts the existence-uniqueness theorem.

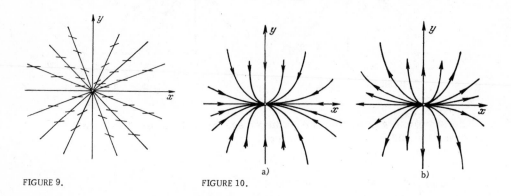

a)

b)

FIGURE 9.                    FIGURE 10.

Thus, the phase portrait of system $(40)$ (with directions indicated on the paths*) has the form illustrated in Figure 10. An equilibrium state of this type is known as a n o d e, s t a b l e if $a_1 < 0$, $a_2 < 0$ (Figure 10a) and u n s t a b l e if $a_1 > 0$, $a_2 > 0$ (Figure 10b).

We now consider the other interpretation of the solutions of $(40)$, by integral curves in three-dimensional space $R^3$ with coordinates $x$, $y$, $t$. It follows from $(41)$ that the integral curves of system $(40)$ in the space $(x, y, t)$ are the following:

1) The $t$-axis, i. e., $x = 0$, $y = 0$ (obtained from equations $(41)$ when $x_0 = y_0 = 0$); the projection of this curve on the phase space is the equilibrium state $O$.

2) Exponential curves

$$x = x_0 e^{a_1 (t-t_0)}, \quad y = 0,$$

situated in the coordinate half-planes $x > 0$, $y = 0$ or $x < 0$, $y = 0$ and tending asymptotically to the $t$-axis as $t \to + \infty$ if $a_1 < 0$ (Figure 11a) and as $t \to - \infty$ if $a_1 > 0$; these curves project onto the positive and negative $x$-axes, which are paths of the system.

3) Exponential curves

$$x = 0, \quad y = y_0 e^{a_2 (t-t_0)},$$

similar to curves of type 2.

4) Curves

$$x = x_0 e^{a_1 (t-t_0)}, \quad y = y_0 e^{a_2 (t-t_0)} \quad (x_0 \neq 0, \; y_0 \neq 0),$$

lying on the parabolic cylinders $y - C x^{a_2/a_1} = 0$ $(C \neq 0)$ with generators parallel to the $t$-axis. The $t$-axis divides each such cylinder into two "halves" and each integral curve of this type lies wholly inside one half of the cylinder and tends asymptotically to the $t$-axis, as $t \to + \infty$ if $a_1 < 0$, $a_2 < 0$ (Figure 11b) and as $t \to - \infty$ if $a_1 > 0$, $a_2 < 0$. Integral curves of type 4 are obtained from each other by translation along the $t$-axis. This is also true of curves of type 2 or 3.

E x a m p l e 4.

$$\frac{dx}{dt} = -y + ax, \qquad \frac{dy}{dt} = x + ay \tag{45}$$

(where $a$ is a nonzero constant).

The vector field defined by this system $(a < 0)$ is illustrated in Figure 12. Solving $(45)$ as a linear system with constant coefficients, we obtain the following solution for initial values $t_0$, $x_0$, $y_0$ (in agreement with Lemma 3, it is again a function of $t - t_0$)

---

* When there are no singular curves, the directions on the paths are completely determined by the direction at a single point, for then the direction at any other point is determined by considerations of continuity. To determine the direction at some point $x_0$, $y_0$ for which $P(x_0, y_0) \neq 0$, simply compute $P(x_0, y_0)$ and observe its sign; if $P(x_0, y_0) > 0$, then $dx/dt > 0$ at $(x_0, y_0)$, so $x$ increases with increasing $t$ near this point and we have thus determined the direction on the path passing through $(x_0, y_0)$. Quite similarly, one determines directions on paths by examining the sign $dy/dt$ at points for which $Q(x_0, y_0) \neq 0$.

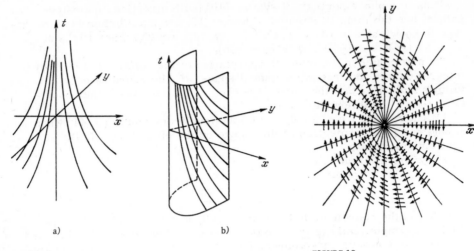

a)                                      b)

FIGURE 11.                                      FIGURE 12.

$$x = e^{a(t-t_0)}[x_0 \cos(t-t_0) - y_0 \sin(t-t_0)], \tag{46}$$
$$y = e^{a(t-t_0)}[x_0 \sin(t-t_0) + y_0 \cos(t-t_0)].$$

It is more convenient to study the paths of this system in polar coordinates. Let $\varrho_0$ and $\theta_0$ be the polar coordinates of a point $M_0(x_0, y_0)$. Setting $x = \varrho \cos\theta$, $y = \varrho \sin\theta$, one readily finds the equations $\theta = \theta(t)$, $\varrho = \varrho(t)$ of the paths in polar coordinates ($\theta(t)$, $\varrho(t)$ are continuous functions of $t$, $\theta(t) \geqslant 0$, $\theta(t_0) = \theta_0$, $\varrho(t_0) = \varrho_0$). Elementary calculations give

$$\varrho = \varrho_0 e^{a(t-t_0)}, \qquad \theta = t - t_0 + \theta_0. \tag{47}$$

Eliminating $t$, we get

$$\varrho = \varrho_0 e^{a(\theta-\theta_0)}. \tag{48}$$

Equation (48) clearly yields all the paths of system (46). If $\varrho_0 \neq 0$ these paths are logarithmic spirals. If $\varrho_0 = 0$ we get an equilibrium state $O(0, 0)$.

The first of equations (47) shows that the paths all tend to the equilibrium state $O$, as $t \to +\infty$ if $\alpha < 0$ (Figure 13a) and as $t \to -\infty$ if $\alpha > 0$ (Figure 13b). An equilibrium state of the type illustrated by this example is known as a focus, stable if $\alpha < 0$ and unstable if $\alpha > 0$ (the rigorous definition of a focus will be given later, in Chapter IV).

The equation

$$\frac{dx}{-y+ax} = \frac{dy}{x+ay}$$

corresponding to system (45) is homogeneous. Integrating it by means of the substitution $\frac{y}{x} = u$ or $\frac{x}{y} = u$, we get

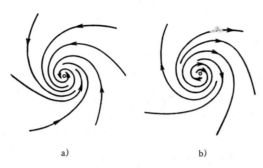

a)                          b)

FIGURE 13.

$$x^2 + y^2 - Ce^{2a \operatorname{arctg} \frac{y}{x}} = 0 \tag{49}$$

or

$$x^2 + y^2 - Ce^{2a \operatorname{arctg} \frac{x}{y}} = 0. \tag{50}$$

The first of these equations is a general integral of the system (in the sense of §1.13) in any region not containing points of the $y$-axis ($x = 0$), the second in any region not containing points of the $x$-axis. But neither of them is a general integral, in the strict sense of the phrase, in a region containing the point $O$. The "full" integral curve for such a region may be constructed by "gluing together" the two partial curves (49) and (50).

We now consider the three-dimensional interpretation. As in the preceding example, the $t$-axis is an integral curve of system (45) in the space $(x, y, t)$. The other integral curves lie on cylindrical surfaces whose directrices are the spirals (48) and whose generators are parallel to the $t$-axis. These integral curves tend asymptotically to the $t$-axis as $t \to +\infty$ if $\alpha < 0$ and as $t \to -\infty$ if $\alpha > 0$.

We note that, although the actual shapes of the paths in Examples 3 and 4 for $a_1 < 0$, $a_2 < 0$ and $\alpha < 0$ ($a_1 > 0$, $a_2 > 0$ and $\alpha > 0$), respectively, are quite different, there is nevertheless a certain similarity in their behavior: in both cases all paths which are not equilibrium states tend to an equilibrium state as $t \to +\infty$ (or $t \to -\infty$).

Accordingly, once we have rigorously defined the "qualitative structure" of a phase portrait, we shall regard the phase portraits of Examples 3 and 4 as possessing the same "qualitative structure."

Example 5.

$$\frac{dx}{dt} = -y, \quad \frac{dy}{dt} = x. \tag{51}$$

This system is the special case of (45) obtained by setting $\alpha = 0$. The solutions with initial values $t_0, x_0, y_0$ are

$$x = x_0 \cos(t - t_0) - y_0 \sin(t - t_0), \tag{52}$$
$$y = x_0 \sin(t - t_0) + y_0 \cos(t - t_0).$$

One easily checks directly (or via (52)) that

$$x^2 + y^2 = C \qquad (53)$$

is a general integral of the system. In this case, therefore, the system has an analytic integral.

The paths of the system are obviously the equilibrium state $O\,(0,\,0)$ and closed paths — concentric circles about the origin (Figure 14). The solutions (52) corresponding to the closed paths (circles) are $2\pi$-periodic functions.

The integral curves in three-dimensional space are the $t$-axis and helical curves lying on circular cylinders with directrices (53). The pitch of each helical curve is $2\pi$ (Figure 15).

Example 6.

$$\frac{dx}{dt} = x, \qquad \frac{dy}{dt} = -y. \qquad (54)$$

The corresponding vector field is illustrated in Figure 16. The solution with initial values $t_0$, $x_0$, $y_0$ is

$$x = x_0 e^{(t-t_0)}, \qquad y = y_0 e^{-(t-t_0)}. \qquad (55)$$

The point $O\,(0,\,0)$ is an equilibrium state.

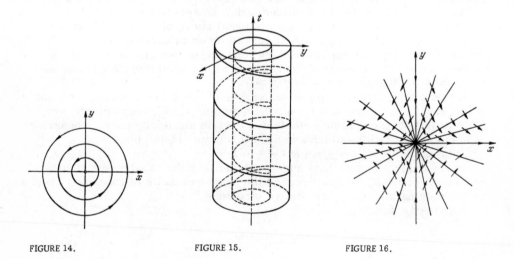

FIGURE 14.                    FIGURE 15.                    FIGURE 16.

The system has an analytic integral

$$xy = C. \qquad (56)$$

When $C \neq 0$, the integral curves are hyperbolas (56), and when $C = 0$ they are the coordinate axes $x = 0$ and $y = 0$. Each hyperbola is the union of two paths (its branches), each of the coordinate axes the union of three paths (the equilibrium state $O$ and two half-axes). The phase portrait is illustrated in Figure 17.

It is evident from (55) that the paths lying on the $x$-axis (obtained from (55) by setting $y_0 = 0$) tend to the equilibrium state as $t \to -\infty$, while those lying on the $y$-axis do so as $t \to +\infty$. The system has no other paths which tend to the equilibrium state $O$.

An equilibrium state of this type is known as a **saddle point**. Paths which tend to a saddle point $O$ (in this case, the half-lines $x = 0$ and $y = 0$) are called **separatrices** of the saddle point.

Any path which is not a separatrix, however close to the separatrix it may be, will recede from the separatrix when $t$ increases without bound. This behavior is obviously not in conflict with Theorem 4 (continuous dependence on initial values), for that theorem considers the behavior of neighboring paths only over a finite $t$-interval. It is readily seen that if the initial path is a separatrix, Theorem 4 is valid for any finite $t$-interval. However, the longer the time interval chosen, the smaller must we take the value of $\delta$ (see Theorem 4').

We shall not consider the integral curves of system (54) in the space $(x, y, t)$, since the reasoning is analogous to that in the preceding examples.

We now consider rather more complicated examples, involving nonlinear systems. Here we shall study only the phase portrait, and not the three-dimensional interpretation as in the linear examples.

Before proceeding to the examples, we offer an elementary remark which is nevertheless of essential importance in understanding certain basic properties of phase portraits: In the neighborhood of any point other than an equilibrium state, the paths behave "in the small" like parallel curves. This is illustrated graphically in Figure 18. The rigorous meaning and proof of this statement will be given in §3.3. We need one more preliminary remark.

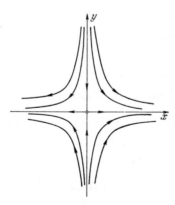

FIGURE 17.

FIGURE 18.

Given the system

$$\frac{dx}{dt} = P(x, y), \qquad \frac{dy}{dt} = Q(x, y), \tag{I}$$

consider the system

$$\frac{dx}{dt} = P(x, y) - f(x, y)Q(x, y), \qquad \frac{dy}{dt} = Q(x, y) + f(x, y)P(x, y), \tag{$I_f$}$$

where $f(x, y)$ is of class $C_N$ or analytic, and defined in the same region as system (I).

It is easy to see that the equilibrium states of system (I) coincide with those of system $(I_f)$. At each point of $G$ we consider the vectors $v$ and $v_f$ defined respectively by systems (I) and $(I_f)$. Denoting by $\theta$ and $\theta_f$ the angles between the positive $x$-axis and the vectors $v$ and $v_f$, respectively, we clearly have

$$\operatorname{tg}\theta = \frac{Q(x, y)}{P(x, y)}, \qquad \operatorname{tg}\theta_f = \frac{Q(x, y) + fP(x, y)}{P(x, y) - fQ(x, y)}.$$

Then the tangent of the angle between the vectors $v$ and $v_f$ is*

$$\operatorname{tg}(\theta_f - \theta) = \frac{\dfrac{Q + fP}{P - fQ} - \dfrac{Q}{P}}{1 + \dfrac{Q}{P}\left(\dfrac{Q + fP}{P - fQ}\right)} = f. \tag{57}$$

Formula (57) means, as we shall say for short, that t h e  v e c t o r  f i e l d of  s y s t e m  $(I_f)$ i s  r o t a t e d  r e l a t i v e  t o  t h e  v e c t o r  f i e l d  o f system (I) t h r o u g h  a n  a c u t e  a n g l e  w h o s e  t a n g e n t  i s  e q u a l to $f$.

E x a m p l e 7.

$$\frac{dx}{dt} = -y - x(x^2 + y^2 - 1),$$

$$\frac{dy}{dt} = x - y(x^2 + y^2 - 1). \tag{58}$$

It is easily seen that this is a system of type $(I_f)$ with $P(x, y) = -y$, $Q(x, y) = x$ and $f(x, y) = x^2 + y^2 - 1$. The corresponding system (I), $\frac{dx}{dt} = P(x, y)$, $\frac{dy}{dt} = Q(x, y)$, is simply system (51) of Example 5. It follows that system (58), like (51), has a unique equilibrium state $O(0, 0)$ and the vector field of system (58) is rotated relative to that of system (51) through an acute angle with tangent $x^2 + y^2 - 1$.

This angle is obviously positive when $(x^2 + y^2 - 1) > 0$, negative when $(x^2 + y^2 - 1) < 0$, and vanishes on the circle $x^2 + y^2 - 1 = 0$.

Observing the sign of $x^2 + y^2 - 1$, one readily checks that when $C > 1$ the paths of system (58) e n t e r  the circles $x^2 + y^2 = C$, while they  l e a v e

---

* Note that the scalar product $(vv_f) = P^2 + Q^2$ is positive at any regular point of the region $G$. Thus the vectors $v$ and $v_f$ are not orthogonal.

them when $C < 1$. Figure 19 illustrates the directions of the vectors defined by system (58) (the vectors in the drawing are not those of system (58), for they are drawn with equal lengths).

One easily checks directly that the circle

$$x^2 + y^2 - 1 = 0$$

is an integral curve of system (58), and therefore a closed path. By virtue of the above relationship between the vector fields of systems (51) and (58), the paths

$$x^2 + y^2 = C \qquad (59)$$

of system (51) are cycles without contact for the paths of system (58), i.e., for $C \neq 1$ the paths of system (58) are never tangent to the circles (59). As for the circle

$$x^2 + y^2 = 1$$

it is a path for both systems (51), (58).

Summarizing the above discussion, it becomes clear that the paths of system (58) have the form illustrated in Figure 20. This can be proved rigorously either by the general theory set forth in §§3.11 − 3.13 or by determining the equations of the paths in polar coordinates.

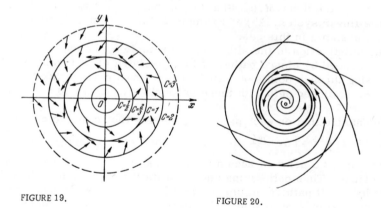

FIGURE 19.                    FIGURE 20.

Setting

$$x = \varrho \cos \theta, \qquad y = \varrho \sin \theta$$

or

$$\varrho^2 = x^2 + y^2, \qquad \theta = \text{arctg } \frac{y}{x},$$

we find

$$\frac{d\varrho^2}{dt} = 2x\frac{dx}{dt} + 2y\frac{dy}{dt} = 2\varrho^2(\varrho^2 - 1),$$

$$\frac{d\theta}{dt} = \frac{x\frac{dy}{dt} - y\frac{dx}{dt}}{x^2 + y^2} = 1$$

(60)

and

$$\frac{d\varrho^2}{d\theta} = 2\varrho^2(\varrho^2 - 1).$$

(61)

Integrating this equation, we get

$$\varrho^2 = \frac{1}{1 - Ce^{-2\theta}}, \qquad \varrho = \frac{1}{\sqrt{1 - Ce^{-2\theta}}}.$$

This is the equation of a path in polar coordinates. We get the path passing through a point $M_0(\varrho_0, \theta_0)$ by setting $C = \frac{(\varrho_0^2 - 1)e^{2\theta_0}}{\varrho_0^2}$. If $\varrho_0 > 1$, then $C > 0$ and $\varrho > 1$; $\varrho \to 1$ as $\theta \to +\infty$, and $\varrho \to +\infty$ as $\theta \to \frac{\lg C}{2} + 0$. (It is clear that in this case $\theta$ varies in the interval $\frac{\lg C}{2} < \theta < +\infty$.) $\varrho = 1$ is a solution of equation (61). If $\varrho_0 < 1$, then $C < 0$ and $\varrho < 1$. Then $\varrho \to 0$ as $\theta \to -\infty$ and $\varrho \to 1$ as $\theta \to +\infty$. Thus the phase portrait of the system is indeed as illustrated in Figure 20. The second of equations (60) shows that if a path passes through the point $M_0(\varrho_0, \theta_0)$ at $t = t_0$, then $\theta = t + (\theta_0 - t_0)$. As in the case of the linear system (45) of Example 4, the equilibrium state $O(0, 0)$ is a focus, unstable in this case.

In contradistinction to the situation in Example 6, the path $x^2 + y^2 - 1 = 0$ is not surrounded by closed paths. It is an isolated closed path and all paths passing through points of a sufficiently small neighborhood tend to it as $t \to +\infty$. A closed path with these properties is known as a limit cycle.

We emphasize that on each path lying outside the limit cycle $t$ varies from the finite value $\frac{\lg C}{2}$ to $+\infty$. Another way of saying this is that with decreasing $t$ a point on such a path escapes to infinity in a finite time. Thus paths lying outside a limit cycle are not whole. On the other hand, all paths lying inside the limit cycle are obviously whole, i. e., $t$ varies from $-\infty$ to $+\infty$. The direction on the paths can be established by direct examination of the system. For example, for $x = 0$ and $y > 0$ we have $\frac{dx}{dt} < 0$, so that $x$ decreases with increasing $t$ along the $y$-axis. This is obviously sufficient to determine the direction on all paths of the system.

Example 8.

$$\frac{dx}{dt} = 2y, \qquad \frac{dy}{dt} = 12x - 3x^2.$$

(62)

The system has two equilibrium states $O$ $(0, 0)$ and $A$ $(4, 0)$. It has an analytic integral

$$y^2 - 6x^2 + x^3 = C. \tag{63}$$

It is not hard to determine the main features of the family of curves (63), starting from the auxiliary family

$$u = 6x^2 - x^3 + C. \tag{64}$$

Since $y = \sqrt{u}$, the curves (64) have the shape illustrated in Figure 21a, and the curves (63) that illustrated in Figure 21b.

The equilibrium state $O$ $(0, 0)$ lies on an integral curve, obtained from (63) by setting $C = 0$. This curve is the union of four paths: the equilibrium state $O$, two nonclosed paths, one of which tends to $O$ as $t \to -\infty$ and the other as $t \to +\infty$, and a "loop" which tends to $O$ both as $t \to -\infty$ and as $t \to +\infty$.

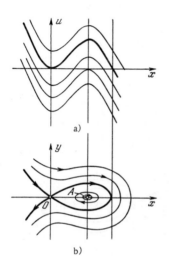

a)

b)

FIGURE 21.

It is readily seen that the equilibrium state $A$ $(4, 0)$ lies on the curve (63) with $C = -32$. This curve has one branch and an isolated point — the equilibrium state. All other integral curves contain no equilibrium states. When $C < -32$ the curve (63) has one branch, situated to the left of the infinite branch of the curve (63) with $C = 0$. When $-32 < C < 0$, the curve (63) is the union of two branches, one a closed curve (oval) enclosing $A$. When $C > 0$ the curve has a single branch (situated to the right of the curve (63) for $C = 0$). Each branch of an integral curve ($C \neq 0$) is a path.

The equilibrium state $A$ is a center (see Example 5). The other equilibrium state $O$ is a saddle point, and the paths that tend to $O$ as $t \to -\infty$ or $t \to +\infty$ are its separatrices (see Example 6).

The rigorous definition of a separatrix will be given in §5.4. Here we merely remark that a separatrix of a saddle point is not a path but a semipath. In this connection, when dicussing separatrices tending to a saddle point we shall not distinguish between separatrices one of which is a part of the other (for example, $C_1O$ and $C_2O$ in Figure 22). With this convention in mind, we see that in the present example there are exactly four separatrices tending to the saddle point. Two of these separatrices lie on the same path — the "loop."

The direction on the paths is determined (for example) by setting $x = 0$, $y > 0$ in the first equation of (62):

$$\frac{dx}{dt} = 2y > 0,$$

so that we can determine all directions on paths (Figure 21b).

Example 9.

$$\frac{dx}{dt} = 2y - \alpha(12x^2 - 3x^3), \qquad \frac{dy}{dt} = 12x - 3x^2 + 2\alpha y. \qquad (65)$$

The field of this system is obtained by rotating the field of system (62) through a constant angle $\varphi$ such that tg $\varphi = \alpha$. Consequently, the paths of system (65) are not tangent to those of system (62) at any point. In particular, the closed paths of system (62) are cycles without contact for the paths of system (65).

Using this fact in conjunction with the general theory of §§3.11 – 3.13, one can determine the position of the paths of system (65). We shall confine ourselves to geometrically intuitive considerations. To fix ideas, let us assume that $\varphi < 0$. Then any path of system (65) crossing some path of system (62) at $t = t_0$ will tend to an equilibrium state $A$ as $t \to +\infty$, and as $t$ increases it will leave the region filled by closed paths.

As in the case of system (62), the equilibrium state $O$ (0, 0) of system (65) is a saddle point, but here the position of the separatrices (Figure 22) is different. One might say that after rotation of the field, i.e., after passage to system (65), each separatrix of system (62) "splits" into two separatrices.

The separatrices $L$ of system (65) situated to the left of the $y$-axis are in positions similar to those of system (62).

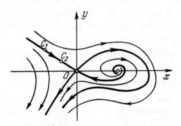
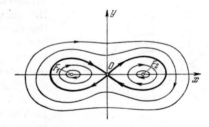

FIGURE 22.            FIGURE 23.

Example 10.

$$\frac{dx}{dt} = 2y, \qquad \frac{dy}{dt} = 4x - 4x^3. \qquad (66)$$

Equating the right-hand sides to zero, we find the equilibrium states to be $O$ (0, 0), $F_1(-1, 0)$, $F_2(1, 0)$.

It is easy to see that

$$y^2 - 2x^2 + x^4 = C \qquad (67)$$

is an analytic general integral of system (66).

Investigation of the family of curves (67) is easy and proceeds along lines analogous to the procedure in Example 8. Using the auxiliary family of curves

$$u = 2x^2 - x^4 + C,$$

(68)

one easily constructs the family (67) (Figure 23). The integral curve $y^2 - 2x^2 + x^4 = 0$ is the union of three paths — two loops and the equilibrium state $O\,(0,\ 0)$. When $C > 0$ each curve (67) is a single closed curve (oval), and when $C < 0$ it is the union of two ovals. Each of these ovals is a path. When $C = -1$ we get two isolated points — the equilibrium states $F_1$ and $F_2$.

The equilibrium state $O$ is a saddle point, while $F_1$ and $F_2$ are centers.

Example 11.

$$\frac{dx}{dt} = 2y - \mu\,(y^2 - 2x^2 + x^4)\,(4x - 4x^3),$$

$$\frac{dy}{dt} = 4x - 4x^3 + \mu\,(y^2 - 2x^2 + x^4)\,2y.$$

(69)

It is easy to see that the vector field of system (69) is rotated relative to that of system (66) (Example 10) through an acute angle with tangent $\mu\,(y^2 - 2x^2 + x^4)$. A direct check shows that

$$y^2 - 2x^2 + x^4 = 0$$

(70)

is an integral of system (69). Therefore the curve (70), which is an integral curve of system (66), is also an integral curve of system (69).

Finally, we observe that inside the curve (70) the expression $y^2 - 2x^2 + x^4$ is negative, while outside it is positive.

Comparing the vector fields of systems (66) and (69), one readily sees that all the closed paths of system (66) (Figure 23) are cycles without contact for the paths of system (69).

Hence, on the basis of the general theory of §§ 3.11 − 3.13, one can show that the phase portrait has the form of Figure 24a for $\mu > 0$ and of Figure 24b for $\mu < 0$. When $\mu > 0$ all paths outside the curve (70) escape to infinity as $t$ increases, while they "wind" onto the curve (70) as $t \to -\infty$. Paths within the curve (70) wind onto one of the simple closed curves comprising (70) as $t \to +\infty$, while as $t \to -\infty$ they tend to one of the equilibrium states $F_1$ and $F_2$, which are foci (see Chapter IV).

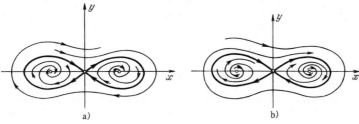

a)　　　　　　　b)

FIGURE 24.

The equilibrium state $O$ is again a saddle point; the curve (70) is a limit continuum (see §4) for paths lying outside it, and each of its loops (together with the equilibrium state $O$) is a limit continuum for paths lying inside the loop. The situation for $\mu < 0$ is similar.

**15. COMMENTS ON THE EXAMPLES.** The above examples, whose main purpose was to illustrate the general theory developed up to now, also illustrate "exhaustive" investigation of the "qualitative structure" of the phase portrait, exhaustive qualitative investigation of a dynamic system.

A precise definition of what we mean by the "qualitative aspect" of the phase portrait and qualitative investigation of a dynamic system will be given in Chapter III. For the moment we wish to dwell only on some geometrically intuitive considerations. From the qualitative standpoint, knowledge of the exact shape of the paths is of little interest: we have already stressed this by pointing to the identical "qualitative behavior" of the paths in the vicinity of nodes and foci.

On the other hand, it is essential to know, e. g., the number of equilibrium states, whether or not there exists an isolated closed path (limit cycle), the behavior of the separatrices, and so on.

In the above cases we were able to study the qualitative structure of the phase portraits "exhaustively," thanks to the extreme simplicity of the dynamic systems in question. The systems of Examples 1 through 6 were linear. Even in the other examples we derived relatively simple analytical expressions for the solutions or integrals. This enabled us fully to determine the structure of the phase portrait. The investigation of Examples 9 and 11 was based on results obtained in Examples 8 and 10, on the concept of cycles and curves without contact (of which more will be said in the sequel) and on the property of rotation of fields.

It is self-evident that arbitrary dynamic systems of type (I) will not always be amenable to such an elementary and exhaustive treatment.

We cannot expect to obtain elementary expressions for the solutions or integrals in the general case. Because of this, even dynamic systems of extremely simple types arising in applications require special techniques for their qualitative investigation. An example is the van der Pol system

$$\dot{x} = y, \quad \dot{y} = k\,(1 - x^2)\,y - x,$$

whose qualitative theory has given birth to a wealth of literature.

It is thus natural to ask whether one can find regular methods for qualitative investigation of dynamic systems, or at least fairly effective techniques to this end.

We reiterate that even in cases where the dynamic system in question has an analytic integral (in the sense of §1.13) for which an analytical expression

$$\Phi\,(x,\ y) = C \tag{71}$$

is available (as was the case in Examples 8 and 9), qualitative investigation of the phase portrait is not thereby rendered trivial. It may indeed be reduced to qualitative investigation of the family of curves (71), but to date there exist no regular methods to this end, even for a single curve

$$F\,(x,\ y) = 0.$$

Such methods are lacking even when the functions $\Phi(x, y)$ and $F(x, y)$ are polynomials.

One should not imagine, therefore, that knowledge of an analytic integral (when it exists) provides the immediate answer to our needs: it merely reduces one problem, direct investigation of the phase portrait of system (I), to the qualitative investigation of a family of curves (71).

It is therefore reasonable to seek out methods or techniques for direct qualitative investigation of system (I), without the intermediary of analytical expressions for its solutions. Such techniques are discussed in Chapters IV, V, VI and IX.

Before proceeding to a description of these techniques, we must establish certain general properties of phase portraits: what can we say in general of the phase portrait defined by a system of type (I)? The first question coming to mind here is: what types of paths can a dynamic system of type (I) have in general?

We have already ascertained (§ 1.5) that paths may be either equilibrium states, closed paths or nonclosed paths. This information is however still too general and vague, especially as regards nonclosed paths.

The examples considered above have already provided instances of various types of nonclosed paths or, more precisely, semipaths: semipaths tending to a closed path as $t \rightarrow +\infty$, semipaths tending to a certain "limit continuum" consisting of two loops of separatrices of a saddle point and the saddle point itself. Do these examples represent all possible types of paths or do there exist paths of a completely different nature? This question will be answered in Chapter II.

Apart from questions pertaining to the possible types of individual paths, there are several questions which bear on the phase portrait as a whole. These questions will be discussed later.

We conclude this chapter with a discussion of dynamic systems on the sphere.

## §2. DYNAMIC SYSTEMS ON THE SPHERE

**1. INTRODUCTION.** In this section we define dynamic systems on the sphere and prove some of their fundamental properties.

Dynamic systems on the sphere are worthy of consideration, because, on the one hand, they retain all essential features of plane systems (I), and on the other they are free of certain complicating properties of the latter. Namely, the study of plane dynamic systems defined in an o p e n region $G$ of the plane is hindered by the fact that an open plane region is not compact. And our problems are not solved by considering the system in the closure $\bar{G}$, for in a closed region there are two different kinds of points — interior and boundary points. The sphere, however, is compact, and its points are all of the same kind.

A dynamic system on the sphere is a special case of a dynamic system on a closed orientable surface of arbitrary genus $k \geqslant 0*$ /15/. The general

---

* Dynamic systems on a surface arise in numerous applied problems. For example, one is led to consider dynamic systems on a cylinder in Zhukovskii's problem of the dynamics of symmetric flight of an aircraft (see /16/), the problem of automatic frequency control and many others. Dynamic systems on a torus arise in consideration of the stability of parallel operation of synchronous machines, etc. /17/. Of course, dynamic systems on a surface must also be studied in connection with the properties of dynamic systems in spaces of three or more dimensions. Concerning dynamic systems on surfaces, see /18/.

definition is entirely analogous to that given below for dynamic systems on the sphere. However, only systems on closed orientable surfaces of genus zero retain the essential properties of plane systems. Only in this case do the individual paths and the phase portrait display the same features as in the plane case. On the other hand, dynamic systems on closed surfaces of a more complex topological structure — orientable surfaces of genus $k \geqslant 1$ and nonorientable surfaces — possess certain properties which set them apart from plane systems.

For example, on a surface of genus 1 (torus) a system may have paths which are dense in the entire surface. As we shall see, this cannot occur on the plane or on surfaces of genus zero.

To summarize: if we wish to consider dynamic systems on surfaces which nevertheless preserve the main features of plane systems, we are necessarily confined to systems on arbitrary surfaces of genus zero. We shall consider only the case of the sphere, since in this case elementary analytical techniques are available.

**2. DEFINITION OF DYNAMIC SYSTEM ON THE SPHERE.** We shall assume for simplicity's sake that the sphere is in three-dimensional space, defined by the equation $x^2 + y^2 + z^2 = 1$. This obviously involves no loss of generality.

Suppose that we have on the sphere some regular atlas of class $r$ (or analytic). This means (see Appendix, §7.3) that the sphere is covered by regions (charts) $g_1, g_2, \ldots, g_N$ homeomorphic to plane regions, and in each chart $g_i$ we have a coordinate system, defined by functions

$$x = \varphi_i(u_i, v_i), \qquad y = \psi_i(u_i, v_i), \qquad z = \chi_i(u_i, v_i), \tag{1}$$

satisfying the following conditions:

a) the functions (1) define a topological mapping of some region $H_i$ on the $(u_i, v_i)$-plane onto the region $g_i$ on the sphere;

b) $\varphi_i, \psi_i, \chi_i$ are of class $C_r$ or analytic;

c) the Jacobians

$$\frac{D(\varphi_i, \psi_i)}{D(u_i, v_i)}, \qquad \frac{D(\varphi_i, \chi_i)}{D(u_i, v_i)}, \qquad \frac{D(\psi_i, \chi_i)}{D(u_i, v_i)}$$

do not vanish together at any point of $H_i$.

Equation (1) may be treated as parametric equations of the spherical region $g_i$, and $u_i, v_i$ as curvilinear coordinates in $g_i$.

If two charts $g$ and $\widetilde{g}$ of the atlas (we are using a different, index-free notation) have common points, and therefore a common region

$$\omega = g \cap \widetilde{g},$$

the region $\omega$ is provided with two coordinate systems: $u, v$ and $\widetilde{u}, \widetilde{v}$. Then the functions

$$\widetilde{u} = f(u, v), \quad \widetilde{v} = h(u, v) \quad \text{or} \quad u = \widetilde{f}(\widetilde{u}, \widetilde{v}), \, v = \widetilde{h}(\widetilde{u}, \widetilde{v}), \tag{2}$$

expressing the coordinates of one system in terms of those of the other define in $\omega$ a regular mapping of class $C_r$ (or analytic mapping) (see Appendix, §6 and §1.10).

*Definition I. A dynamic system of class $C_k$ (an analytic dynamic system) is said to be defined on the sphere $s$ if the following conditions hold for some atlas of class $C_{k+1}$ (or analytic):*
  1) *a dynamic system*

$$\frac{du_i}{dt} = U_i(u_i,\ v_i), \qquad \frac{dv_i}{dt} = V_i(u_i,\ v_i) \tag{3}$$

*of class $C_k$ (or analytic) is given in each chart $g_i$ of the atlas ($u_i$, $v_i$ are the local coordinates in $g_i$), and defined for all points $u_i$, $v_i$ in the region $H_i$;*
  2) *at points common to two charts $g$ and $\widetilde{g}$, i.e., points of $\omega = g \cap \widetilde{g}$, the dynamic systems are transformed into each other by the corresponding transformations of local coordinates; in other words, if*

$$\frac{d\widetilde{u}}{dt} = \widetilde{U}(\widetilde{u},\ \widetilde{v}), \qquad \frac{d\widetilde{v}}{dt} = \widetilde{V}(\widetilde{u},\ \widetilde{v}) \tag{4}$$

*is the dynamic system in $\widetilde{g}$, it is obtained in the region $\omega$ from the system (1) defined in $g$ by the transformation of coordinates (2), so that*

$$\begin{aligned}
\widetilde{U}(\widetilde{u},\ \widetilde{v}) &= \frac{\partial f}{\partial u} U(\widetilde{f},\ \widetilde{g}) + \frac{\partial f}{\partial v} V(\widetilde{f},\ \widetilde{g}), \\
\widetilde{V}(\widetilde{u},\ \widetilde{v}) &= \frac{\partial h}{\partial u} U(\widetilde{f},\ \widetilde{g}) + \frac{\partial h}{\partial v} V(\widetilde{f},\ \widetilde{g})
\end{aligned} \tag{5}$$

*(and conversely system (3) can be obtained from system (4) by the corresponding transformation of coordinates).*

System (3) will be called the system corresponding to the chart $g_i$.

We now define when two dynamic systems on the sphere are identical.

Let $\Sigma_1$ and $\Sigma_2$ be two different atlases of class $C_{k+1}$ (or analytic) on the sphere $S$. Denote the local coordinates of the first atlas by $u_i$, $v_i$, those of the second by $\xi_i$, $\eta_i$. If a chart $g$ of $\Sigma_1$ and a chart $\gamma$ of $\Sigma_2$ overlap, each point of the intersection

$$\lambda = g \cap \gamma$$

has both coordinates $u$, $v$ and coordinates $\xi$, $\eta$ (indices are omitted). The corresponding transformation of coordinates, defined by formulas

$$\begin{aligned}
\xi &= f^*(u,\ v), \qquad u = f^{**}(\xi,\ \eta), \\
\eta &= h^*(u,\ v), \qquad v = h^{**}(\xi,\ \eta),
\end{aligned} \tag{6}$$

is regular of class $C_{k+1}$ or analytic (this follows from Theorem VIII of §6, Appendix, just as in the case of two charts $g$ and $\widetilde{g}$ belonging to the same atlas).

Let $D_1$ and $D_2$ be two dynamic systems defined on the sphere, the first referred to the atlas $\Sigma_1$ and the second to $\Sigma_2$.

*We shall say that these dynamic systems are identical or equivalent if, at points in the intersection of two charts $g$ and $\gamma$ belonging to different atlases, the transformation from local coordinates in $g$ to local coordinates in $\gamma$ converts system $D_1$ into system $D_2$.*

41

Thus, if system $D_1$ is identical to system $D_2$, the former having the form

$$\frac{du}{dt}=U\,(u,\ v), \qquad \frac{dv}{dt}=V\,(u,\ v) \qquad (7)$$

in $g$ and the latter the form

$$\frac{d\xi}{dt}=\Phi\,(\xi,\ \eta), \qquad \frac{d\eta}{dt}=\Psi\,(\xi,\ \eta) \qquad (8)$$

in $\gamma$, then system (8) is obtained from system (7) in the region $\lambda=g\cap\gamma$ by the transformation (6), so that (as in the case of two overlapping charts of the same atlas)

$$\frac{d\xi}{dt}=\frac{\partial f^*}{\partial u}\,U\,(f^{**},\ h^{**})+\frac{\partial f^*}{\partial v}\,V\,(f^{**},\ h^{**})=\Phi\,(\xi,\ \eta),$$

$$\frac{\partial\eta}{\partial t}=\frac{\partial h^*}{\partial u}\,U\,(f^{**},\ h^{**})+\frac{\partial h^*}{\partial v}\,V\,(f^{**},\ h^{**})=\Psi\,(\xi,\ \eta).$$

This definition enables one to "change atlases" when studying dynamic systems on the sphere.

For the most part, we shall use simple atlases, as described in §7 of the Appendix. Here one chart $g$ of the atlas consists of all points on the sphere but one, $N$ say, and the other chart $\widetilde{g}$ of all points other than the point $\widetilde{N}$ antipodal to $N$. Then, if $u$, $v$ are the coordinates in $g$ and $\widetilde{u}$, $\widetilde{v}$ the coordinates in $\widetilde{g}$, we have at all points common to $g$ and $\widetilde{g}$

$$\widetilde{u}=\frac{4u}{u^2+v^2}, \qquad \widetilde{v}=\frac{4v}{u^2+v^2} \qquad (9)$$

and, by symmetry,

$$u=\frac{4\widetilde{u}}{\widetilde{u}^2+\widetilde{v}^2}, \qquad v=\frac{4\widetilde{v}}{\widetilde{u}^2+\widetilde{v}^2}. \qquad (10)$$

If $D$ is a dynamic system on the sphere and $g$ some proper subregion of the sphere, then $D$ is equivalent in $g$ to a dynamic system in a plane region. Indeed, one can always define an atlas of the sphere such that $g$ lies entirely within one chart. To do this, take a point $N$ not in $g$ and consider the simple atlas described above by stereographic projection from $N$ and the antipodal point $\widetilde{N}$. If the closure $\bar{g}$ of the region $g$ is also not the entire sphere, a dynamic system $D$ in the closed region $\bar{g}$ on the sphere is equivalent to a dynamic system in a closed plane region.

We end this subsection with a remark concerning analytic dynamic systems. An analytic dynamic system on the sphere is uniquely and completely determined if it is specified in an arbitrary region $g$ (however small it may be), by virtue of the fact that an analytic function has a unique continuation. However, not every system defined and analytic in some part of the sphere can be continued to the entire sphere (for in the process of continuation the right-hand sides of the equations may have poles or essential singularities).

### 3. DYNAMIC SYSTEM ON THE SPHERE AS A VECTOR FIELD ON THE SPHERE.

In complete analogy with the case of a system in a plane region, specification of a dynamic system on the sphere may be interpreted as specification of a vector field on the sphere.

Let $u$, $v$ be coordinates in a chart $g$ of an atlas on the sphere, and

$$\frac{du}{dt} = U(u, v), \qquad \frac{dv}{dt} = V(u, v)$$

a dynamic system in this chart. Let us write the parametric equations

$$x = \varphi(u, v), \qquad y = \psi(u, v), \qquad z = \chi(u, v) \qquad (11)$$

of $g$ in vector notation:

$$r = \Phi(u, v),$$

where $r = \overrightarrow{OM}$ is the radius-vector of an arbitrary point $M$ of $g$ in the space $(x, y, z)$. At each point $M(u, v)$ of the region $g$, consider the vector

$$v(M) = \frac{\partial \Phi(u, v)}{\partial u} U(u, v) + \frac{\partial \Phi(u, v)}{\partial v} V(u, v).$$

Since it is a linear combination of the vectors $\dfrac{\partial \Phi}{\partial u}$, $\dfrac{\partial \Phi}{\partial v}$, this vector lies in the tangent plane to the sphere at $M$. It is readily seen that the vector is independent of the specific coordinate system introduced on the sphere. Indeed, let $\omega$ be a region on the sphere referred both to coordinates $u$, $v$ and to coordinates $\tilde{u}$, $\tilde{v}$ related to $u$ and $v$ by equations of type (2). By the definition of identical dynamic systems on the sphere, the dynamic system is defined in $\omega$ in terms of coordinates $\tilde{u}$, $\tilde{v}$ by

$$\frac{d\tilde{u}}{dt} = \tilde{U}(\tilde{u}, \tilde{v}), \qquad \frac{d\tilde{u}}{dt} = \tilde{V}(\tilde{u}, \tilde{v}),$$

where $\tilde{U}(\tilde{u}, \tilde{v})$ and $\tilde{V}(\tilde{u}, \tilde{v})$ are expressed in terms of $U(u, v)$ and $V(u, v)$ by (5).

If $r = \tilde{\Phi}(\tilde{u}, \tilde{v})$ is the parametric vector equation of the spherical region $\omega$ in coordinates $\tilde{u}$, $\tilde{v}$ (analogous to equations (11)) and $v$ the vector defined above, then, obviously,

$$v = \frac{\partial \tilde{\Phi}}{\partial \tilde{u}} \tilde{U} + \frac{\partial \tilde{\Phi}}{\partial \tilde{v}} \tilde{V} = \frac{\partial \Phi}{\partial u} U + \frac{\partial \Phi}{\partial v} V. \qquad (12)$$

Thus, at each point of the sphere the dynamic system defines a vector tangent to the sphere, in other words, it defines a vector field on the sphere. The vector $v$ clearly has zero length at points where $V = U = 0$; these are the singular points of the vector field.

### 4. SOLUTIONS AND PATHS OF A DYNAMIC SYSTEM ON THE SPHERE.

We now define the solution of a dynamic system on the sphere. Some preliminary remarks are in order.

As before, let $g$ be a chart in a given atlas on the sphere, with coordinates $u$ and $v$, and

$$x = \varphi(u, v), \qquad y = \psi(u, v), \qquad z = \chi(u, v),$$

the parametric equations of $g$, which we shall again write in vector notation:

$$r = \Phi(u, v).$$

Let

$$\frac{du}{dt} = U(u, v), \qquad \frac{dv}{dt} = V(u, v)$$

be the system corresponding to $g$,

$$u = u(t), \qquad v = v(t) \tag{13}$$

any solution of this system. Consider the functions

$$x = \varphi(u, v) = f_1(t), \qquad y = \psi(u, v) = f_2(t), \qquad z = \chi(u, v) = f_3(t)$$

or the equivalent vector function

$$r = F(t) = \Phi(u(t), v(t)). \tag{14}$$

The function (14) is continuous for all $t$ such that the solution (13) is defined. It is natural to call this vector function a solution or "part of a solution" of the dynamic system on the sphere. It may happen that for some $t_0$ the point $M_0$ with coordinates $u_0 = u(t_0)$, $v_0 = v(t_0)$ on the sphere belongs to a chart $\tilde{g}$ of the atlas, overlapping the chart $g$ (Figure 25). Let

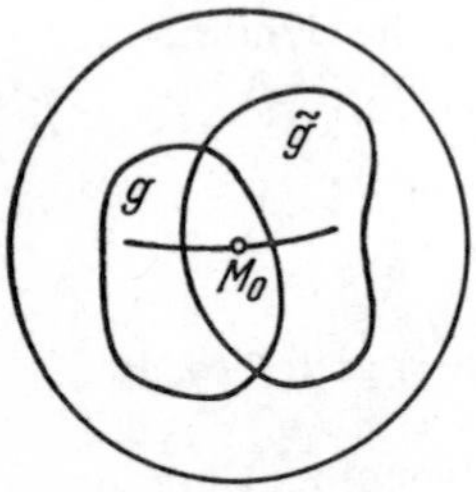

$$\omega = g \cap \tilde{g},$$

and let $\tilde{u}, \tilde{v}$ be the coordinates in $\tilde{g}$, related to $u$ and $v$ in $\omega$ by

FIGURE 25.

$$\tilde{u} = f(u, v), \qquad \tilde{v} = h(u, v) \ (u = \tilde{f}(\tilde{u}, \tilde{v}), \qquad v = \tilde{h}(\tilde{u}, \tilde{v})),$$

and

$$\frac{d\tilde{u}}{dt} = \tilde{U}(\tilde{u}, \tilde{v}), \qquad \frac{d\tilde{v}}{dt} = \tilde{V}(\tilde{u}, \tilde{v}) \tag{15}$$

the dynamic system corresponding to $\tilde{g}$. Consider the solution of system (15) with initial values

$$\tilde{u}_0 = f(u_0, v_0), \qquad \tilde{v}_0 = h(u_0, v_0)$$

at $t = t_0$, denoting it by

$$\tilde{u} = \tilde{u}(t), \qquad \tilde{v} = \tilde{v}(t).$$

In view of the discussion in §1.10, it is obvious that, at points of ω,

$$\widetilde{u}'(t) \equiv f\left(u\left(t\right),\, v\left(t\right)\right), \qquad \widetilde{v}'(t) \equiv h\left(u\left(t\right),\, v\left(t\right)\right). \tag{16}$$

Let

$$x = \widetilde{\varphi}\left(\widetilde{u},\, \widetilde{v}\right), \qquad y = \widetilde{\psi}\left(\widetilde{u},\, \widetilde{v}\right), \qquad z = \widetilde{\chi}\left(\widetilde{u},\, \widetilde{v}\right)$$

be the parametric equations of $\widetilde{g}$, written in vector notation as

$$\boldsymbol{r} = \widetilde{\boldsymbol{\Phi}}\left(\widetilde{u},\, \widetilde{v}\right).$$

Consider the vector function (analogous to the vector function $\boldsymbol{r} = \boldsymbol{\Phi}\left(u,\, v\right)$)

$$\boldsymbol{r} = \widetilde{\boldsymbol{\Phi}}\left(\widetilde{u},\, \widetilde{v}\right). \tag{17}$$

It is readily seen that the following identity holds for all $t$ such that the point with coordinates $u\left(t\right),\, v\left(t\right)$ (or $\widetilde{u}\left(t\right),\, \widetilde{v}\left(t\right)$) belongs to ω :

$$\widetilde{\boldsymbol{\Phi}}\left(\widetilde{u},\, \widetilde{v}\right) \equiv \boldsymbol{\Phi}\left(u,\, v\right). \tag{18}$$

This follows from the fact that by the definition of $\boldsymbol{\Phi}\left(u,\, v\right)$ and $\widetilde{\boldsymbol{\Phi}}\left(\widetilde{u},\, \widetilde{v}\right)$ we have

$$\widetilde{\boldsymbol{\Phi}}\left(f,\, h\right) \equiv \boldsymbol{\Phi}\left(u,\, v\right),$$

at points of ω, and also from (16). In particular, it may happen that the functions

$$\widetilde{u} = \widetilde{u}\left(t\right), \qquad \widetilde{v} = \widetilde{v}\left(t\right)$$

are defined not only in ω but also at points of $\widetilde{g}$ not in ω. We then define a vector function

$$\boldsymbol{r} = \boldsymbol{F}\left(t\right)$$

such that

$$\boldsymbol{F}\left(t\right) \equiv \boldsymbol{\Phi}\left(u\left(t\right),\, v\left(t\right)\right)$$

for all $t$ such that the point $u\left(t\right),\, v\left(t\right)$ belongs to $g$, and

$$\boldsymbol{F}\left(t\right) \equiv \widetilde{\boldsymbol{\Phi}}\left(\widetilde{u}\left(t\right),\, \widetilde{v}\left(t\right)\right)$$

for all $t$ such that the point $\left(\widetilde{u}\left(t\right),\, \widetilde{v}\left(t\right)\right)$ belongs to $\widetilde{g}$. By (18),

$$\boldsymbol{r} = \boldsymbol{F}\left(t\right)$$

is a continuous vector function, which it is natural to treat as the "solution" or part of a solution of the dynamic system on the sphere.

One might say that the vector function $r=\tilde{\Phi}(\tilde{u}(t),\ \tilde{v}(t))$ is a continuation to $\tilde{g}$ of the vector function $r=\Phi(u(t),\ v(t))$ defined in $g$. Similarly, given a point on the sphere with coordinates $\tilde{\tilde{u}}(t), \tilde{\tilde{v}}(t)$, belonging to some chart $\tilde{\tilde{g}}$ overlapping $\tilde{g}$, we can define a continuous vector function in the same way for points of $\tilde{g}$, and so on. After these preliminary remarks we can define the solution of a dynamic system on the sphere.

A *system of continuous functions* $x=f_1(t),\ y=f_2(t),\ z=f_3(t)$, *or the equivalent vector function* $r=F(t)$, *defined on an interval* $(\alpha,\ \beta)$, *is called a solution of the dynamic system* D *on the sphere if:*

*a) in every chart* $g$,

$$F(t) \equiv \Phi(u(t),\ v(t)),$$

*where* $r=\Phi(u,\ v)$ *is the vector parametric equation of* $g$ *relative to the chosen atlas, and the functions* $u(t)$ *and* $v(t)$ *satisfy the system corresponding to* $g$, *so that*

$$\frac{du(t)}{dt} \equiv U(u(t),\ v(t)), \qquad \frac{dv(t)}{dt} = V(u(t),\ v(t));$$

*b) there exists no vector function* $r=F(t)$ *defined on an interval larger than* $(\alpha,\ \beta)^*$ *which satisfies condition a) and coincides with* $F(t)$ *on* $(\alpha,\ \beta)$.

The last condition simply means that the solution is assumed to be defined on the maximal possible interval on the $t$-axis.

In view of the rules for change of variables and identity (18), it is easy to see that *the solution of a dynamic system on the sphere is independent of the choice of an atlas.* Theorem 1 of §1 (existence-uniqueness theorem) and the definition of a solution directly imply the following

Theorem 5. *For any point* $M_0$ *on the sphere and any* $t_0$, *there exists exactly one solution* $r=F(t)$ *satisfying the initial condition*

$$r_0 = F(t_0),$$

*where* $r_0$ *is the radius-vector of* $M_0$.

A solution $r=F(t)$ of a dynamic system on the sphere also satisfies propositions analogous to Lemmas $1-5$ of §1. In particular, if $r=F(t)$ is a solution defined on an interval $(\alpha,\beta)$, then $r=F(t+C)$ is also a solution, defined on $(\alpha-C,\ \beta-C)$.

A path of a dynamic system on the sphere is a set of points defined by equations $x=f_1(t),\ y=f_2(t),\ z=f_3(t)$ or the equivalent vector function $r=F(t)$. Each solution uniquely determines a path $L$. As in the plane case, we shall speak of the solution corresponding to a given path.

The solution will be called a motion corresponding to the path, or a motion on the path.

At each point $M_0$ of the sphere which is not a singular point of the vector field defined by the dynamic system, the corresponding vector is the tangent vector at $M_0$ to the path through this point. Indeed, let

$$r=F(t)$$

---

* I. e., an interval containing $(\alpha,\ \beta)$ as a proper subinterval.

be some solution of the system $D$ and $L$ the corresponding path. Suppose that the point $M_0$ on the sphere with radius-vector $r_0 = F(t_0)$ lies in a chart $g$ with local coordinates $u$, $v$. Let $u_0$, $v_0$ be the local coordinates of $M_0$. Let

$$\frac{du}{dt} = U(u, v), \qquad \frac{dv}{dt} = V(u, v)$$

be the system corresponding to $g$. We have

$$\frac{dr}{dt} = \left( \frac{dF(t)}{dt} \right)_{t=t_0} = \left( \frac{\partial \Phi(u, v)}{\partial u} \frac{d(u)}{dt} + \frac{\partial \Phi(u, v)}{\partial v} \frac{d(v)}{dt} \right).$$

$\left( \dfrac{dF(t)}{dt} \right)_{t=t_0}$ is obviously the tangent vector to the path $L$ at $M_0$. Since $r = F(t)$ is a solution of system $D$, it follows by definition that

$$\left( \frac{du(t)}{dt} \right)_{t=t_0} = U(u_0, v_0), \qquad \left( \frac{dv(t)}{dt} \right)_{t=t_0} = V(u_0, v_0),$$

and so

$$\left( \frac{dF(t)}{dt} \right)_{t=t_0} = \frac{\partial \Phi(u_0, v_0)}{\partial u} U(u_0, v_0) + \frac{\partial \Phi(u_0, v_0)}{\partial v} V(u_0, v_0).$$

But the vector on the right is precisely the vector $v$ defined by the dynamic system at $M_0$.

We thus get the following "kinematic" interpretation of a dynamic system on the sphere, independent of the specific atlas chosen on the sphere. The dynamic system defines on the sphere a field of tangent vectors $v(M)$, known as the phase velocities of the points $M$. A solution of the dynamic system on the sphere is a motion $z = F(t)$ on the sphere, with the property that the velocity of the moving point as it passes through a point $M_0$ is equal to the phase velocity at $M_0$. A path of the dynamic system is the curve traced out by the moving point.

The following theorem is proved along the same lines as the analogous theorem for a plane region, and the proof is omitted:

*Theorem 6. Through each point of the sphere there is exactly one path of the dynamic system.*

We append a few remarks concerning the paths of a dynamic system on the sphere.

Let $L$ be some path of a dynamic system $D$ on the sphere, referred to some given atlas.

There are two possible cases: 1) $L$ lies entirely within one chart; 2) $L$ contains points in different charts $g, \tilde{g}, \tilde{\tilde{g}}, \ldots$.

In the first case, let $u$, $v$ be local coordinates in $g$ and

$$\frac{du}{dt} = U(u, v), \qquad \frac{dv}{dt} = V(u, v) \tag{19}$$

the system corresponding to $g$. Let us regard $u$ and $v$ as cartesian coordinates on the plane and assume that system is defined in a region $h$

of the $(u, v)$-plane. Then by §2.2 the functions (1) define a mapping $T$ of $h$ onto the region $g$ on the sphere, satisfying conditions a), b), c) stated in §2.2. It is evident from the definition of a path of $D$ that the path $L$ on the sphere is the image of some path $L'$ of system (19) on the $(u, v)$-plane.

Now consider case 2), in which the path $L$ contains points in different charts of the atlas. Then these charts must overlap in pairs. To fix ideas, suppose that $L$ has points in two charts $g$ and $\tilde{g}$ (the case of more than two charts is treated similarly).

Let

$$\frac{d\tilde{u}}{dt}=\tilde{U}\,(\tilde{u}, \tilde{v}), \qquad \frac{d\tilde{v}}{dt}=\tilde{V}\,(\tilde{u}, \tilde{v}) \tag{20}$$

be the system corresponding to $\tilde{g}$. As in $g$, we shall regard $\tilde{u}, \tilde{v}$ as cartesian coordinates on the $(\tilde{u}, \tilde{v})$-plane and assume that system (20) is defined in a region $\tilde{h}$ on the $(\tilde{u}, \tilde{v})$-plane. We then have a mapping, defined by functions analogous to (1), of $\tilde{h}$ onto $\tilde{g}$. The path $L$ falls into (one or more) parts, each the image of (one or more) paths of system (19) or of system (20). Those parts of $L$ lying in the intersection $\omega = g \cap \tilde{g}$ are images of parts of paths of both systems.

We now prove another fundamental theorem about paths on the sphere. There is no analog of this theorem for dynamic systems on the plane.

*Theorem 7. Every path on the sphere is a whole path, i.e., every solution of a dynamic system on the sphere is defined for all $t$ from $-\infty$ to $+\infty$.*

Proof. Suppose that on the contrary, some solution $r=F(t)$ is defined only for $t < \tau_0$, where $\tau_0$ is some finite number. Let $L$ denote the path corresponding to $F(t)$.

Let $\{t_i\}$ $(t_i < \tau_0)$ be a sequence converging to $\tau_0$. Let $M_1, M_2, \ldots, M_n \ldots$ denote the corresponding points on $L$. Since the sphere is compact, we may assume that the sequence $M_n$ is convergent (otherwise we need only select a suitable convergent subsequence). Let the sequence $\{M_n\}$ converge to a point $M_0$. Let $g$ be a chart on the sphere containing $M_0$, with local coordinates $u, v$, and let $T$ be the corresponding mapping of $h$ onto $g$ (Figure 26). Let $M_n^*$ be the points of the $(u, v)$-plane mapped onto $M_n$, $M_n^*=T^{-1}(M_n)$ $(n = 0, 1, 2, \ldots)$ and $u_n$, $v_n$ the coordinates of $M_n^*$.

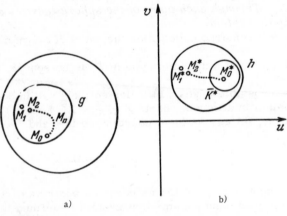

a)                  b)

FIGURE 26.

Consider some closed region $\overline{K}^*$ contained in $h$ and containing $M_0^*$ in its interior (Figure 26b). By Lemma 8 (§1.9), there exists a number $\eta_0 > 0$ such that any solution

$$u = u(t), \qquad v = v(t)$$

of the system (19) (corresponding to $g$) passing through a point of $\overline{K}^*$ at $t = t_0$ is defined for all $t$ in the interval $t_0 - \eta_0 \leqslant t \leqslant t_0 + \eta_0$. Choose $n$ so large that

$$0 < \tau_0 - t_n < \frac{\eta_0}{2}. \tag{21}$$

This is possible since $M_n \to M_0$, $t_n \to \tau_0$. Consider the solution

$$u = u(t - t_n, u_n, v_n), \qquad v = v(t - t_n, u_n, v_n) \tag{22}$$

of system (19). By the choice of $n$, this solution is defined for all $t$ in $t_n \leqslant t \leqslant t_n + \eta_0$. By (21), we have $\tau_0 - \frac{\eta_0}{2} < t_n < \tau_0$. Thus, the solution (22) is also defined in the interval $t_n \leqslant t \leqslant \tau_0 + \frac{\eta_0}{2}$. Now, when "transferred" to the sphere by the mapping $T: h \to g$ this solution must coincide with the solution $r = F(t)$; hence the latter is also defined for $t_n \leqslant t \leqslant \tau_0 + \frac{\eta_0}{2}$. But then it is obvious from the definition of the solution of a dynamic system on the sphere that the solution $r = F(t)$ is defined for all $t$ such that $t_n \leqslant t \leqslant \tau_0 + \frac{\eta_0}{2}$.

This contradicts our assumption that the solution $F(t)$ is defined only for $t < \tau_0$. Q.E.D.

By Theorem 6, a dynamic system on the sphere defines a phase portrait on the sphere, and by Theorem 7 the phase portrait consists solely of whole paths.

In §1 we established certain fundamental properties of paths of a dynamic system on the plane. The same properties hold true for paths on the sphere. As on the plane, paths on the sphere are either equilibrium states, nonclosed paths, or closed paths. The proof is exactly the same as in the plane case (see §1).

Equilibrium states are singular points of the vector field on the sphere, i.e., points at which $v(M) = 0$. Moreover, paths on the sphere are also continuous with respect to the initial conditions. This theorem may be stated in exactly the same geometric version as for the plane (see §1.9, Theorem 4').

In addition, the terms positive semipath, negative semipath, semipath and the relevant notation (see §1.7) will also be employed for the case of the sphere.

**5. EXAMPLES OF DYNAMIC SYSTEMS ON THE SPHERE.** All the examples given below will be considered relative to a simple atlas on the sphere (see §2.2 and Appendix, §7). As in §2.2, we shall denote the charts of this atlas by $g$ and $\tilde{g}$ ($g$ consists of all the points of the sphere except one point $N$, and $\tilde{g}$ of all the points except the point $\tilde{N}$ antipodal to $N$).

The transformation from local coordinates $u$, $v$ in $g$ to local coordinates in $\tilde{g}$ is given by formulas (9) and (10):

$$\tilde{u} = \frac{4u}{u^2+v^2}, \qquad \tilde{v} = \frac{4v}{u^2+v^2},$$

$$u = \frac{4\tilde{u}}{\tilde{u}^2+\tilde{v}^2}, \qquad v = \frac{4\tilde{v}}{\tilde{u}^2+\tilde{v}^2}.$$

Example 1. Let $A$ be the dynamic system defined by equations

$$\frac{du}{dt} = \frac{u}{1+u^2+v^2}, \qquad \frac{dv}{dt} = \frac{v}{1+u^2+v^2} \tag{23}$$

in $g$ and equations

$$\frac{d\tilde{u}}{dt} = \frac{-\tilde{u}\,(\tilde{u}^2+\tilde{v}^2)}{16+\tilde{u}^2+\tilde{v}^2}, \qquad \frac{d\tilde{v}}{dt} = \frac{-\tilde{v}\,(\tilde{u}^2+\tilde{v}^2)}{16+\tilde{u}^2+\tilde{v}^2} \tag{24}$$

in $\tilde{g}$.

Using (9) and (10), one readily checks that the right-hand sides of these equations satisfy condition (5).

The equilibrium points of system $A$ are the point $N$ in $g$ and the point $\tilde{N}$ in $\tilde{g}$. System (23) differs from the system

$$\frac{du}{dt} = u, \qquad \frac{dv}{dt} = v \tag{25}$$

only in the presence of the nonvanishing factor $\frac{1}{1+u^2+v^2}$. Hence the paths of system (23) differ only as to parametrization (see §1.7) from those of system (40) in §1.14 for $a_1 = a_2 = 1$. "Transferring" these paths from the $(u, v)$-plane to the sphere and recalling that $\tilde{N}$ is an equilibrium state, we see that the paths of system $A$ are as illustrated in Figure 27 — the meridians of the sphere.

Example 2. Let $A$ be the dynamic system defined by equations

$$\frac{du}{dt} = \frac{-v+au}{1+u^2+v^2}, \qquad \frac{dv}{dt} = \frac{u+av}{1+u^2+v^2} \tag{26}$$

in $g$ and equations

$$\frac{d\tilde{u}}{dt} = \frac{(-\tilde{v}-a\tilde{u})\,(\tilde{u}^2+\tilde{v}^2)}{16+\tilde{u}^2+\tilde{v}^2}, \qquad \frac{d\tilde{v}}{dt} = \frac{(\tilde{u}-a\tilde{v})\,(\tilde{u}^2+\tilde{v}^2)}{16+\tilde{u}^2+\tilde{v}^2} \tag{27}$$

in $\tilde{g}$. As in the preceding example, one readily checks that the right-hand sides of these equations satisfy (5) of §2.2. The equilibrium states are $N$ and $\tilde{N}$. System (26) differs from system (45) in Example 4 of §1.14 only in notation and the nonvanishing factor $\frac{1}{1+u^2+v^2}$. Its paths on the $(u, v)$-plane are therefore as shown in Figure 13. "Transferring" these paths as before to the sphere, we get the phase portrait of Figure 28 $(a > 0)$.

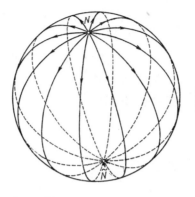

FIGURE 27.

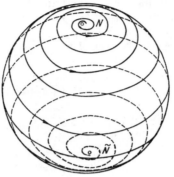

FIGURE 28.

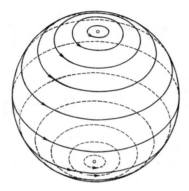

FIGURE 29.

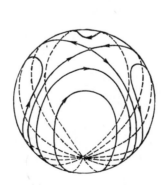

FIGURE 30.

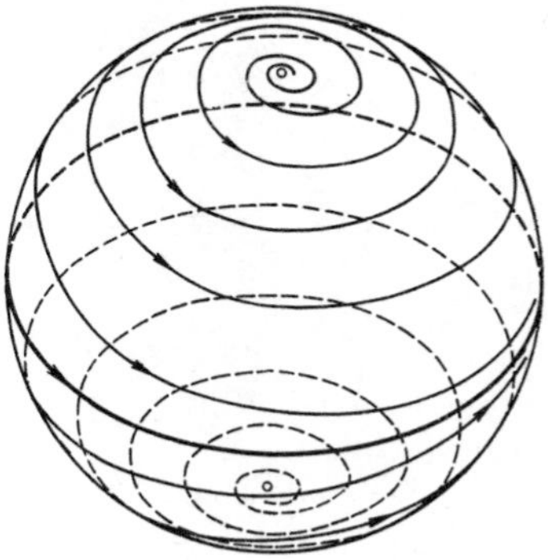

FIGURE 31.

If $a=0$, the phase portrait of system (26) on the $(u, v)$-plane is that shown in Figure 14. The corresponding phase portrait on the sphere is illustrated in Figure 29.

Example 3. Let $A$ be the dynamic system defined by equations

$$\frac{du}{dt}=\frac{u}{1+u^2+v^2}, \qquad \frac{dv}{dt}=\frac{-v}{1+u^2+v^2} \tag{28}$$

in $g$ and equations

$$\frac{d\widetilde{u}}{dt}=\frac{\widetilde{u}(3\widetilde{v}^2-4\widetilde{v}^2)}{16+\widetilde{u}^2+\widetilde{v}^2}, \qquad \frac{d\widetilde{v}}{dt}=\frac{\widetilde{v}(4\widetilde{v}^2-3\widetilde{u}^2)}{16+\widetilde{u}^2+\widetilde{v}^2} \tag{29}$$

in $\widetilde{g}$.

Comparing (28) with the system of Example 6 in §1.14 (Figure 17), one readily shows that the phase portrait on the sphere is as illustrated in Figure 30.

Example 4. Let $A$ be the dynamic system defined by equations

$$\frac{du}{dt}=\frac{-v-u(u^2+v^2-1)}{1+u^2+v^2}, \qquad \frac{dv}{dt}=\frac{u-v(u^2+v^2-1)}{1+u^2+v^2} \tag{30}$$

in $g$ and equations

$$\frac{d\widetilde{u}}{dt}=\frac{16\widetilde{u}-(\widetilde{u}+\widetilde{v})(\widetilde{u}^2+\widetilde{v}^2)}{16+\widetilde{u}^2+\widetilde{v}^2}, \qquad \frac{d\widetilde{v}}{dt}=\frac{16\widetilde{v}+(\widetilde{u}-\widetilde{v})(\widetilde{u}^2+\widetilde{v}^2)}{16+\widetilde{u}^2+\widetilde{v}^2} \tag{31}$$

in $\widetilde{g}$.

Comparing system (30) with system (58) of Example 7, §1.14, we see that the phase portrait of $A$ on the sphere has the form shown in Figure 31.

*Chapter II*

*LIMIT POINTS AND SETS.*
*FUNDAMENTAL PROPERTIES OF PATHS*

INTRODUCTION

This chapter is devoted to a fundamental problem of the qualitative
theory of dynamic systems: the possible behavior and "shape" of an
individual path.

We have already touched upon this question ($1.10). In this chapter we
set forth the classical results, which answer the question exhaustively for
a dynamic system of type (I). These results were outlined by Poincaré /5/,
and were subsequently made rigorous and generalized by Bendixson with
the help of set-theoretic methods.

In §1 we acquired some knowledge about paths: a path may be either an
equilibrium state, a closed path or a nonclosed path. From the standpoint
of the qualitative theory, this is an exhaustive description of the
behavior of the path in the case of an equilibrium point (single point) or a
closed path (simple closed curve). The remaining case, that of a nonclosed
path, clearly requires further investigation. The examples of specific
dynamic systems considered in §1 exhibit several different types of
nonclosed path. One would naturally like to determine what types of
nonclosed path are possible in general.

A nonclosed path is a curve defined by parametric equations

$$x = \varphi(t), \quad y = \psi(t)$$

$(\varphi(t)$ and $\psi(t)$ are at least of class $C_1$ and defined on some interval
$\tau < t < T$) which has no "self-intersections," i. e., for no two distinct $t_1$
and $t_2$ in the interval $(\tau, T)$ is it true that both

$$\varphi(t_1) = \varphi(t_2), \quad \psi(t_1) = \psi(t_2).$$

By no means, however, can any curve satisfying these requirements be
a path of a dynamic system of type (I).

We shall see, for example, that under our assumptions concerning
system (I) no path can have the "shape" of the function $y = \sin \frac{1}{x}$. A path
cannot densely fill out an entire region,* though such curves (satisfying the
conditions stated above for a nonclosed path) exist. The fact that a path

---

* For an example of a "smooth" simple curve (i. e., without self-intersections) densely filling out a region,
see, e.g., /20/.

53

is merely one member of a whole family of curves, the phase portrait of system (I) (in a region $G$), imposes additional restrictions on its possible behavior. The results of Poincaré and Bendixson on the behavior of nonclosed paths of system (I) follow essentially from two theorems (valid in the entire region $G$): the existence-uniqueness theorem and the theory on continuity of solutions with respect to initial values.

The chapter comprises two sections.

§3 is auxiliary in nature, containing a series of lemmas which, though elementary, are fundamental for the exposition of this and the following chapters. We first introduce the concepts "arc without contact" and "cycle without contact," which are the basic auxiliary tool in subsequent propositions. We then establish an elementary, but again fundamental, proposition on the local phase portrait in the neighborhood of a point which is not an equilibrium state (this proposition was anticipated in §1.14): the configuration of paths in the neighborhood of such a point resembles that of parallel segments. Also proved in this section are several lemmas which describe how paths can intersect arcs and cycles without contact.

A few remarks concerning the proof of the auxiliary lemmas in §3. Most of these lemmas are perfectly obvious from the intuitive, geometric viewpoint. It is quite natural to query the very need to prove these propositions rather than to accept them simply as obvious, all the more so as throughout this book certain fundamental geometric theorems are unavoidably taken for granted since their proofs involve special techniques and methods far removed from the subject-matter of the book.*

Moreover, one may well ask whether the proof of many elementary geometrically obvious assertions (for instance, Lemma 2 or 4 of §3) is not in a certain sense tautologous, for the essence of the matter is to reduce one geometrically obvious assertion to another one which is by no means more obvious, often encumbering and complicating the text. This objection, however, may be met as follows: if some assertion were reduced without proof to an appeal to geometric intuition, this would naturally justify omitting the proofs of a great many other no less "obvious" assertions.

The mere appeal to "geometric intuition" provides no genuine criterion as to what assertions may be accepted without proof. The treacherous and vague path of appeal to geometric intuition may well lead us unwittingly to assertions which are no longer obvious, or simply to vagueness and outright error.

Thus, observing the laws of elementary mathematical rigor, we must unavoidably single out certain c l e a r - c u t   a n d   w e l l - d e f i n e d   f a c t s which we accept as known or obvious (the number of these "axioms" should be kept to a minimum). All other assertions must be p r o v e d on the basis of the propositions assumed to be given or known.

---

* For example, the fact that any simple closed curve separates the plane into two regions, of which it is the common boundary. This theorem is known as the Jordan Curve Theorem. The need for a proof is often questioned by beginners. In this connection we recommend the discussion in Courant, R. and Robbins, H. What is Mathematics? Oxford University Press, 1941, Chapter V, §3.

The need to prove intuitively obvious theorems is most convincingly demonstrated by counterexamples showing that apparently self-evident geometric assertions can fail to be true. For an effective example of this type, see, e.g., Aleksandrov, P.S., Kombinatornaya topologiya (Combinatorial Topology), Moskva, 1947, p.68. [English translation: New York, Graylock Press, 1956, Vol.I, p.41.]

The facts accepted as known in this book are collected in the Appendix. In particular, the immediate grounds for the proofs of the lemmas of §3 will be found in §§1 through 6 of the Appendix. These are certain elementary considerations concerning simple arcs, elementary theorems on simple closed curves and the properties of what are known as regular mappings. These facts, in turn, are either proved in the Appendix or suitable references are given to the literature.

Irrespective of their geometric "obviousness" (especially in the case of Lemmas 1 and 3), the lemmas of §3 are all proved in detail. Nevertheless, without damage to his understanding of the sequel the reader may omit the proofs of the most "obvious" lemmas, acquainting himself only with their statements and referring to the proofs only if some uncertainty arises in further reading.

§4 is devoted to the main content of this chapter — determination of the possible features of paths of system (I). We define limit points and limit sets of semipaths, which play a major role not only in the problem considered in this chapter but also in examination of the phase portrait in the large (Chapters VII, VIII, X and XI). At the end of the section we present two classical theorems concerning certain basic properties of phase portraits in the large; we also consider isolated closed paths — limit cycles — and examine the phase portrait in the neighborhood of a limit cycle.

## §3. AUXILIARY PROPOSITIONS ON THE INTERSECTION OF PATHS WITH CYCLES AND ARCS WITHOUT CONTACT

Throughout the sequel the term "path" will refer to a path of a dynamic system

$$\frac{dx}{dt} = P(x, y), \qquad \frac{dy}{dt} = Q(x, y), \tag{I}$$

defined in a region $G$ of the plane and satisfying the conditions of §1.1.

**1. ARC WITHOUT CONTACT.** Let $l$ be some smooth simple arc (see Appendix, §3.5) in the region $G$ and $M$ a point on $l$ which is not an equilibrium state of system (I).

If the path through $M$ is not tangent to $l$ at this point, we shall say that the arc $l$ has no contact at $M$. If the path through $M$ is tangent to $l$, we shall say that $l$ has contact at $M$.

A smooth simple arc $l$ is called an arc without contact of system (I) if: a) there are no equilibrium states on $l$; b) $l$ has no contact at any of its points. We shall say that an arc without contact $l$ passes through a point $M$ if $M$ is an interior point of $l$ (Figure 32). It is clear that through each point $M$ of $G$ which is not an equilibrium state one can draw an arc without contact. An example is any sufficiently small segment on the normal to the path at $M$.

Let $l$ be an arc without contact, defined by parametric equations

$$x = f(s), \qquad y = g(s),$$

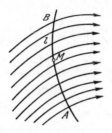

FIGURE 32.

where $f(s)$ and $g(s)$ are of class $C_1$ and defined on an interval $a \leqslant s \leqslant b$. The points $A (f(a), g(a))$ and $B (f(b), g(b))$ are the endpoints of $l$. Since $l$ is a smooth simple arc, we have $[f'(s)]^2 + [g'(s)]^2 \neq 0$ for all $s$, $a \leqslant s \leqslant b$ (see Appendix, §5.5).

It follows from condition a) that for all $s$

$$P^2 (f(s), g(s)) + Q^2 (f(s), g(s)) \neq 0. \tag{1}$$

Condition b), in turn, implies that for all $s$ (see Appendix, §5.5)

$$\Delta(s) = \begin{vmatrix} P(f(s), g(s)) & f'(s) \\ Q(f(s), g(s)) & g'(s) \end{vmatrix} \neq 0. \tag{2}$$

The sign of this determinant determines the sign of the angle between the arc and the path. Since the angle between an arc without contact $l$ and any path crossing it* never vanishes, this angle is clearly of constant sign.

If

$$x = \varphi(t - t_0, x_0, y_0), \qquad y = \psi(t - t_0, x_0, y_0)$$

is a solution of system (I) (see §1.3), then for any $s \in [a, b]$

$$x = \varphi(t - t_0, f(s), g(s)), \qquad y = \psi(t - t_0, f(s), g(s))$$

is the equation of the path of system (I) which passes at $t = t_0$ through the point $(f(s), g(s))$ of the arc $l$. We can write condition (2) as

$$\Delta(s) = \begin{vmatrix} \varphi'(0, f(s), g(s)) & f'(s) \\ \psi'(0, f(s), g(s)) & g'(s) \end{vmatrix} \neq 0. \tag{3}$$

The arc $l$ may be defined in closed form rather than parametrically:

$$F(x, y) = 0,$$

where $x$ (or $y$) varies, say, over an interval $a \leqslant x \leqslant b$. Then it follows from condition a) that at points of $l$

$$P^2 (x, y) + Q^2 (x, y) \neq 0, \tag{4}$$

and from condition b) that (again at points of $l$)

$$F'_x (x, y) P (x, y) + F'_y (x, y) Q (x, y) \neq 0. \tag{4'}$$

A final remark: let $l$ be an arc without contact, $A$ and $B$ its endpoints. Since by definition $l$ has no contact at $A$ and $B$, it can be continued, i. e., there always exists an arc without contact $l_1$, of which $l$ is a subarc, containing $A$ and $B$ as interior points (Figure 32).

---

* Concerning the angle between two vectors or smooth arcs, see Appendix, §5.

**2. GENERALIZED ARC WITHOUT CONTACT.** In many contexts a role similar to that of an arc without contact is allotted to a "generalized arc without contact."

A simple arc $l$ (not necessarily smooth) will be called a g e n e r a l i z e d a r c  w i t h o u t  c o n t a c t for system (I) if: a) there are no equilibrium states on $l$; b) for any path passing at $t = t_0$ through some point $M$ of $l$ (not an endpoint), there exists $\delta > 0$ such that points on the path corresponding to $0 < t - t_0 < \delta$ $(0 < t_0 - t < \delta)$ lie on the positive (negative) side of $l$,* or conversely. In particular, a smooth arc $l$ is a generalized arc without contact for system (I) if at any of its points it has either no contact or contact of even order with the paths of the system (Figure 33).

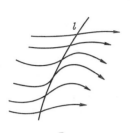

FIGURE 33.

**3. INTERSECTION OF A PATH WITH AN ARC WITHOUT CONTACT.**

*L e m m a  1.  Let $l$ be an arc without contact, $M_0(x_0, y_0)$ one of its interior points, $L$ the path passing through $M_0$ at $t = t_0$.*

*There exists $h > 0$ such that the arc on $L$ defined by $|t - t_0| \leqslant h$ satisfies the following conditions: a) it is a simple arc, b) it has no points other than $M_0$ in common with the arc $l$, c) the subarc corresponding to $t$ such that $t_0 - h \leqslant t < t_0$ lies on one side of $l$, the subarc corresponding to $t_0 < t \leqslant t_0 + h$ on the other side of $l$ (Figure 34).*

Lemma 1 follows directly from Lemma 1 of §5 in the Appendix, and from the properties of paths proved in §1.4. Obviously, if $L$ is a closed path the number $h$ is certainly smaller than half the period of the corresponding periodic solution.

In the next lemma we assume that a prescribed motion on $L$ is defined for all $t, \tau < t < T$ ($T$ and $\tau$ may be equal to $+\infty$ and $-\infty$, respectively).

*L e m m a  2.  No section of a path $L$ corresponding to values of $t$ in an arbitrary closed interval $[\alpha, \beta], \tau < \alpha < \beta < T$, can have more than a finite number of points of intersection with any arc without contact.*

P r o o f.  Suppose that this is false, and let $L$ be a path having infinitely many points of intersection with an arc without contact $l$, corresponding to values of $t$ in $[\alpha, \beta]$. Let $\{t_n\}$ be a convergent sequence of such values of $t$, $\lim_{n \to \infty} t_n = t^*$. It is clear that $t^* \in [\alpha, \beta]$. Let $M^*$ be the corresponding point on the path, $M_n$ the point corresponding to $t_n$. Since $M^*$ is a point of accumulation** of the points $M_n$, which are on $l$, it is also a point of $l$. We may assume without loss of generality that it is not an endpoint of $l$, for otherwise we need only replace $l$ by a larger arc without contact $l'$ (see §3.1).

By Lemma 1, there exists $h > 0$ such that all points corresponding to values of $t$ on the segment $|t - t^*| \leqslant h$, except $M^*$, are not on $l$. But $|t_n - t^*| < h$ for sufficiently large $n$, and so $M_n$ cannot lie on $l$, contrary to assumption. Q.E.D.

---

\* Concerning the positive and negative sides of a simple arc, see Appendix, §3 and §6.5.

\* Throughout this book we shall employ the term "point of accumulation" instead of the more customary [in the Soviet literature] term "limit point," since we shall need the term "limit point of a semipath" in a different connotation from the set-theoretic one.

FIGURE 34.

## 4. CONFIGURATION OF PATHS IN THE NEIGHBORHOOD OF AN ARC WITHOUT CONTACT.

In the preceding subsection we considered the intersection of an arc without contact with one path. We now consider all the paths crossing an arc without contact, in a neighborhood of the arc.

Let

$$x = \varphi\,(t - t_0,\, x_0,\, y_0), \qquad y = \psi\,(t - t_0,\, x_0,\, y_0)$$

be a solution of system (I) and

$$x = f\,(s), \qquad y = g\,(s), \qquad a \leqslant s \leqslant b,$$

the parametric equation of an arc without contact $l$. Then for any fixed $s \in [a,\, b]$

$$\begin{aligned} x &= \varphi\,(t - t_0,\, f\,(s),\, g\,(s)) = \Phi\,(t,\, s), \\ y &= \psi\,(t - t_0,\, f\,(s),\, g\,(s)) = \Psi\,(t,\, s) \end{aligned} \tag{5}$$

is a motion on the path crossing the arc $l$ at $t = t_0$, at the point corresponding to $s$.

The following simple lemma describes the configuration of paths near an arc without contact. Though auxiliary in nature, this lemma is fundamental for what follows.

*Lemma 3. There exists* $h_0 > 0$ *such that for all* $s$ *and* $t$,

$$a \leqslant s \leqslant b, \qquad |t - t_0| \leqslant h_0, \tag{6}$$

*the functions (5) define a regular mapping* $T$ *of the rectangle (6) on the* $(t,\, s)$*-plane onto a certain closed region* $\overline{W}$ *on the* $(x, y)$*-plane satisfying the following conditions: a)* $\overline{W}$ *is bounded by a simple closed curve; b) all points of the arc* $l$ *are in the interior of* $\overline{W}$, *except its endpoints, which are on the boundary of* $\overline{W}$; *c)* $T$ *maps a segment of the straight line* $t = t_0$ *onto the arc* $l$, *and segments of the straight lines* $s = $ const *onto arcs of the paths crossing* $l$ *(Figure 35). \**

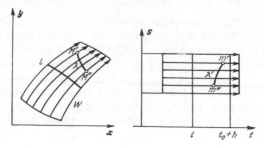

FIGURE 35.

---

\*  An analogous statement is valid for a generalized arc without contact $l$: there exists $h_0 > 0$ such that for all $s$ and $t$, $a \leqslant s \leqslant b$, $|t - t_0| \leqslant h_0$, the functions (5) define a topological (not necessarily regular) mapping $T$ of $R$ onto some closed region $W$ satisfying conditions a), b), c) of Lemma 3.

Proof. By Lemma 6 of §1, the functions $\varphi\,(t - t_0,\, x_0,\, y_0)$, $\psi\,(t - t_0,\, x_0,\, y_0)$ have continuous partial derivatives with respect to $t$, $t_0$, $x_0$, $y_0$. Since moreover the functions $f\,(s)$ and $g\,(s)$ are continuously differentiable with respect to $s$, the functions $\Phi\,(t,\, s)$, $\Psi\,(t,\, s)$ (see (5)) have continuous partial derivatives with respect to $s$ and $t$ throughout their domain of definition, i. e., for all $s \in [a,\, b]$, and, for each fixed $s$, for all $t$ such that the solution $\varphi\,(t - t_0,\, f\,(s),\, g\,(s))$, $\psi\,(t - t_0,\, f\,(s),\, g\,(s))$ is defined. By Lemma 8 of §1, there exists $h^* > 0$ such that for all $s \in [a,\, b]$ the solutions $\varphi\,(t - t_0,\, f\,(s),\, g\,(s)) = \Phi\,(t,\, s)$, $\psi\,(t - t_0,\, f\,(s),\, g\,(s)) = \Psi\,(t,\, s)$ are surely defined for all $t$ on $|t - t_0| \leqslant h^*$. Moreover, since $l$ is an arc without contact, it follows that for $t = t_0$ and all $s \in [a,\, b]$

$$\Delta\,(s) = \begin{vmatrix} \Phi'_t\,(t_0,\, s) & \Phi'_s\,(t_0,\, s) \\ \Psi'_t\,(t_0,\, s) & \Psi'_s\,(t_0,\, s) \end{vmatrix} \neq 0 \tag{7}$$

(see (3) in §3.1). But this means that the functions $\Phi\,(t,\, s)$ and $\Psi\,(t,\, s)$ satisfy all conditions of Lemma 3 in §6 of the Appendix, and so there exists $h_0 > 0$, $h_0 < h^*$, satisfying all conditions of our lemma. Q.E.D.

Remark. In particular, the properties of the functions $\Phi\,(t,\, s)$, $\Psi\,(t,\, s)$ imply the following assertions:

a) There exists $h_0 > 0$ such that all paths crossing an arc without contact $l$ at $t = t_0$ have no other points in common with $l$ for $0 < |t - t_0| \leqslant h_0$.

b) Any path passing through a point $M^*$ in the region $\overline{W}$ at $t = t^*$ must cross the arc $l$ at some time $t'$ such that

$$|t' - t^*| \leqslant h_0.$$

This time $t'$ is unique.

c) Let $M'$ and $M''$ be points on two arcs of paths which cross $l$ at $t = t_0$; assume moreover that $M'$ and $M''$ are in $\overline{W}$, lie on the same side of $l$ and correspond to $t'$ and $t''$ such that $|t_0 - t'| \leqslant h$, $|t_0 - t''| \leqslant h$, where $h$ is any positive number, $h < h_0$. Then $M'$ and $M''$ can be joined by a simple arc $\lambda$ in the region $\overline{W}$ which does not cut the arc $l$, and all points of $\lambda$ correspond on the paths passing through them to values of $t$ such that $0 < |t - t_0| \leqslant h$ (see Figure 35; the points $m'$, $m''$ and arc $\lambda'$ on the $(t,\, s)$-plane are mapped by the functions (5) onto the points $M'$, $M''$ and arc $\lambda$ on the $(x,\, y)$-plane).

The parameters $t$ and $s$ may of course be treated as curvilinear coordinates in the region $\overline{W}$.

The following theorem, which is an immediate corollary of Lemma 3, describes the "local qualitative structure" of a dynamic system in the neighborhood of any point which is not an equilibrium state.

Theorem 8. Let $M$ be a point in $G$ which is not an equilibrium state of system (I). There exist a closed region $\overline{W}$, bounded by a simple closed curve and containing $M$ in its interior, and a topological mapping (see Appendix, §1.12) of $\overline{W}$ onto a rectangle $R$ in the euclidean plane with sides parallel to the coordinate axes, such that arcs of paths lying in $\overline{W}$ are mapped onto straight-line segments parallel to one of the coordinate axes.

To prove Theorem 8, simply draw any arc without contact through the point $M$ and apply Lemma 3.

In connection with Theorem 8, recall that we discussed the features of the phase portrait in the neighborhood of a point other than an equilibrium state in §1.14. Appealing to intuition, we stated there that in the neighborhood of any such point, i. e., "in the small," the paths behave "like a family of parallel straight lines." Theorem 8 gives this statement a rigorous meaning. This is our first use of the concept of a t o p o l o g i c a l m a p p i n g to characterize the qualitative features of the phase portrait (see Appendix, §1.12).

As we shall see in Chapter III, with further specification of the notions "qualitative features of paths," "qualitative investigation of a dynamic system," etc., topological mappings will play a major role.

The next lemma is again quite simple but of fundamental importance.

L e m m a 4. *Let $l$ be an arc without contact, $M_0$ any interior point of $l$. For any $\varepsilon > 0$, $\Delta > 0$, there exists $\delta > 0$ ($\delta = \delta(\varepsilon, \Delta)$) such that any path passing at $t = t_0$ through a point $M^* \in U_\delta(M_0)$ crosses the arc without contact at some $t = t^*$ such that $|t^* - t_0| < \Delta$ and does not leave $U_\varepsilon(M_0)$ as $t$ varies from $t^*$ to $t_0$ (Figure 36).*

P r o o f. Let $M_0$ correspond on $l$ to the parameter value $s = s_0$ ($a < s_0 < b$). Since the functions

$$x = \varphi(t - t_0, \, f(s), \, g(s)) = \Phi(t, s),$$
$$y = \psi(t - t_0, \, f(s), \, g(s)) = \Psi(t, s)$$

are continuous, it follows that for any $\varepsilon > 0$ there exist $h$ and $\sigma$, $\Delta > h > 0$ and $\sigma > 0$, such that for all $s$ and $t$ with

$$|t - t_0| \leqslant h, \quad |s - s_0| \leqslant \sigma \tag{8}$$

the corresponding closed region $\overline{W}$ (see Lemma 3) lies entirely within $U_\varepsilon(M_0)$. Moreover, $M_0$ is an interior point of $\overline{W}$. Hence there exists $\delta > 0$

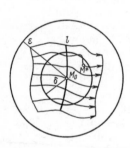

such that $U_\delta(M_0)$ is contained in $\overline{W}$, i. e., the points of $U_\delta(M_0)$ correspond to values of $t$ and $s$ satisfying inequalities (8). But then, by Remark b) to Lemma 3, any path passing through a point of $U_\delta(M_0)$ must cross the arc $l$ for some $t'$ such that $|t' - t_0| < h < \Delta$. Q.E.D.

The next lemma follows at once from Lemma 4 and the continuity of solutions with respect to initial values.

FIGURE 36.

L e m m a 5. *Let $L_0$ be a path passing at $t = t_0$ through a point $M_0$ and crossing an arc without contact $l$ at $t = t_1$, at an interior point $M_1$ of $l$. Then for any $\varepsilon > 0$ and $\Delta > 0$ there exists $\delta > 0$ such that any path $L$ passing at $t = t_0$ through an arbitrary point $M' \in U_\delta(M_0)$ will cross the arc $l$ at some $t = t_1'$ at a point $M_1'$ such that: a) $|t_1' - t_1| < \Delta$; b) $M_1' \in U_\varepsilon(M_1)$; c) any point of $L$ corresponding to $t \in [t_0, t_1]$ lies in the $\varepsilon$-neighborhood of the point defined on $L_0$ by the same value of $t$.*

The proof is obvious.

**5. SOME PROPERTIES OF THE FUNCTIONS $\Phi(t, s)$, $\Psi(t, s)$.** We have been considering the functions

$$x = \varphi(t - t_0, f(s), g(s)) = \Phi(t, s),$$
$$y = \psi(t - t_0, f(s), g(s)) = \Psi(t, s) \tag{9}$$

($x = f(s)$, $y = g(s)$ being parametric equations of an arc without contact $l$, $a \leqslant s \leqslant b$) at times $t$ sufficiently close to $t_0$, $|t - t_0| \leqslant h_0$. Later, however, we shall have to consider them at all $t > t_0$ for which they are defined. We now establish some properties of these functions.

We first prove the following

*Lemma 6. a) The functions* $x = \varphi(t - t_0, x_0, y_0)$, $y = \psi(t - t_0, x_0, y_0)$ *satisfy the partial differential equations*

$$\frac{\partial \varphi}{\partial t} = \frac{\partial \varphi}{\partial x_0} P(x_0, y_0) + \frac{\partial \varphi}{\partial y_0} Q(x_0, y_0), \quad \frac{\partial \psi}{\partial t} = \frac{\partial \psi}{\partial x_0} P(x_0, y_0) + \frac{\partial \psi}{\partial y_0} Q(x_0, y_0); \tag{10}$$

*b)*

$$I = \begin{vmatrix} \dfrac{\partial \varphi}{\partial x_0} & \dfrac{\partial \varphi}{\partial y_0} \\ \dfrac{\partial \psi}{\partial x_0} & \dfrac{\partial \psi}{\partial y_0} \end{vmatrix} = e^{\int_{t_0}^{t} [P_x'(\varphi, \psi) + Q_y'(\varphi, \psi)] \, dt}.$$

Proof. We first prove part a). By the Remark to Lemma 4, §1, if

$$x = \varphi(t - t_0, x_0, y_0), \quad y = \psi(t - t_0, x_0, y_0), \tag{11}$$

then

$$x_0 = \varphi(t_0 - t, x, y), \quad y_0 = \psi(t_0 - t, x, y).$$

But then, by the definition of the functions $\varphi(t - t_0, x_0, y_0)$, $\psi(t - t_0, x_0, y_0)$,

$$\frac{\partial x_0}{\partial t_0} = P(x_0, y_0), \quad \frac{\partial y_0}{\partial t_0} = Q(x_0, y_0). \tag{12}$$

Substituting the expressions for $x_0$ and $y_0$ into (12), we get identities

$$x \equiv \varphi[t - t_0, \varphi(t_0 - t, x, y), \psi(t_0 - t, x, y)],$$
$$y \equiv \psi[t - t_0, \varphi(t_0 - t, x, y), \psi(t_0 - t, x, y)].$$

Differentiating these identities with respect to $t_0$ and using (12), we get

$$\frac{\partial \varphi}{\partial t_0} + \frac{\partial \varphi}{\partial x_0} P(x_0, y_0) + \frac{\partial \varphi}{\partial y_0} Q(x_0, y_0) = 0,$$
$$\frac{\partial \psi}{\partial t_0} + \frac{\partial \psi}{\partial x_0} P(x_0, y_0) + \frac{\partial \psi}{\partial y_0} Q(x_0, y_0) = 0. \tag{13}$$

But it is obvious that

$$\frac{\partial \varphi}{\partial t_0} = -\frac{\partial \varphi}{\partial t}, \quad \frac{\partial \psi}{\partial t_0} = -\frac{\partial \psi}{\partial t}.$$

Substituting these expressions into (13), we get (10).

This proves a). To prove b), we first determine $\frac{dI}{dt}$:

$$
\frac{\partial I(t)}{\partial t} = \frac{\partial}{\partial t}
\begin{vmatrix} \frac{\partial \varphi}{\partial x_0} & \frac{\partial \varphi}{\partial y_0} \\ \frac{\partial \psi}{\partial x_0} & \frac{\partial \psi}{\partial y_0} \end{vmatrix} =
\begin{vmatrix} \frac{\partial}{\partial t} \frac{\partial \varphi}{\partial x_0} & \frac{\partial \varphi}{\partial y_0} \\ \frac{\partial}{\partial t} \frac{\partial \psi}{\partial x_0} & \frac{\partial \psi}{\partial y_0} \end{vmatrix} +
\begin{vmatrix} \frac{\partial \varphi}{\partial x_0} & \frac{\partial}{\partial t} \frac{\partial \varphi}{\partial y_0} \\ \frac{\partial \psi}{\partial x_0} & \frac{\partial}{\partial t} \frac{\partial \psi}{\partial y_0} \end{vmatrix} =
$$

$$
= \begin{vmatrix} \frac{\partial}{\partial x_0} P(x, y) & \frac{\partial \varphi}{\partial y_0} \\ \frac{\partial}{\partial x_0} Q(x, y) & \frac{\partial \psi}{\partial y_0} \end{vmatrix} +
\begin{vmatrix} \frac{\partial \varphi}{\partial x_0} & \frac{\partial}{\partial y_0} P(x, y) \\ \frac{\partial \psi}{\partial x_0} & \frac{\partial}{\partial y_0} Q(x, y) \end{vmatrix} =
\begin{vmatrix} P'_x \frac{\partial \varphi}{\partial x_0} + P'_y \frac{\partial \psi}{\partial x_0} & \frac{\partial \varphi}{\partial y_0} \\ Q'_x \frac{\partial \varphi}{\partial x_0} + Q'_y \frac{\partial \psi}{\partial x_0} & \frac{\partial \psi}{\partial y_0} \end{vmatrix} +
$$

$$
+ \begin{vmatrix} \frac{\partial \varphi}{\partial x_0} & P'_x \frac{\partial \varphi}{\partial y_0} + P'_y \frac{\partial \psi}{\partial y_0} \\ \frac{\partial \psi}{\partial x_0} & Q'_x \frac{\partial \varphi}{\partial y_0} + Q'_y \frac{\partial \psi}{\partial y_0} \end{vmatrix} = (P'_x + Q'_y) I(t).
$$

Thus,

$$
\frac{\partial I}{\partial t} = [P'_x(\varphi(t-t_0, x_0, y_0), \psi(t-t_0, x_0, y_0)) +
$$

$$
+ Q'_y(\varphi(t-t_0, x_0, y_0), \psi(t-t_0, x_0, y_0))] I(t). \tag{14}
$$

We now find $I(t_0)$. At $t = t_0$, by the definition of $\varphi$ and $\psi$,

$$
\varphi(0, x_0, y_0) \equiv x_0, \qquad \psi(0, x_0, y_0) \equiv y_0.
$$

Therefore

$$
\left( \frac{\partial \varphi}{\partial x_0} \right)_{t=t_0} = 1, \qquad \left( \frac{\partial \varphi}{\partial y_0} \right)_{t=t_0} = 0,
$$

$$
\left( \frac{\partial \psi}{\partial x_0} \right)_{t=t_0} = 0, \qquad \left( \frac{\partial \psi}{\partial y_0} \right)_{t=t_0} = 1
$$

and

$$
I(t_0) = \begin{vmatrix} 1 & 0 \\ 0 & 1 \end{vmatrix} = 1. \tag{15}
$$

Treating $t_0$, $x_0$, $y_0$ as fixed and solving the homogeneous linear differential equation (14) with initial condition (15), we get

$$
I = e^{\int_{t_0}^{t} [P'_x(\varphi, \psi) + Q'_y(\varphi, \psi)] dt}. \tag{16}
$$

Q.E.D.

As before, let $l$ be an arc without contact defined by parametric equations $x = f(s)$, $y = g(s)$, $a \leqslant s \leqslant b$.

*Lemma 7. For any $t$ and $s$ at which the functions*

$$
x = \varphi(t-t_0, f(s), g(s)) = \Phi(t, s), \qquad y = \psi(t-t_0, f(s), g(s)) = \Psi(t, s), \tag{9}
$$

*are defined, the Jacobian*

$$\Delta(t, s) = \frac{D(\Phi, \Psi)}{D(t, s)} = \begin{vmatrix} \Phi'_t & \Phi'_s \\ \Psi'_t & \Psi'_s \end{vmatrix}$$

*does not vanish.*

Proof. We have

$$\Delta(t, s) = \begin{vmatrix} \dfrac{\partial \varphi}{\partial t} & \dfrac{\partial \varphi}{\partial x_0} f'(s) + \dfrac{\partial \varphi}{\partial y_0} g'(s) \\[2ex] \dfrac{\partial \psi}{\partial t} & \dfrac{\partial \psi}{\partial x_0} f'(s) + \dfrac{\partial \psi}{\partial y_0} g'(s) \end{vmatrix}.$$

Substituting $\frac{\partial \varphi}{\partial t}$ and $\frac{\partial \psi}{\partial t}$ from (10) and bearing in mind that the roles of $x_0$ and $y_0$ are played here by $f(s)$ and $g(s)$, respectively, we obtain

$$\Delta(t, s) = \begin{vmatrix} \dfrac{\partial \varphi}{\partial x_0} P(f(s), g(s)) + \dfrac{\partial \varphi}{\partial y_0} Q(f(s), g(s)) & \dfrac{\partial \varphi}{\partial x_0} f'(s) + \dfrac{\partial \varphi}{\partial y_0} g'(s) \\[2ex] \dfrac{\partial \psi}{\partial x_0} P(f(s), g(s)) + \dfrac{\partial \psi}{\partial y_0} Q(f(s), g(s)) & \dfrac{\partial \psi}{\partial x_0} f'(s) + \dfrac{\partial \psi}{\partial y_0} g'(s) \end{vmatrix} =$$

$$= \begin{vmatrix} \dfrac{\partial \varphi}{\partial x_0} & \dfrac{\partial \varphi}{\partial y_0} \\[2ex] \dfrac{\partial \psi}{\partial x_0} & \dfrac{\partial \psi}{\partial y_0} \end{vmatrix} \cdot \begin{vmatrix} P(f(s), g(s)) & g'(s) \\ Q(f(s), g(s)) & f'(s) \end{vmatrix} =$$

$$= [P(f(s), g(s)) g'(s) - Q(f(s), g(s)) f'(s)] I(t).$$

The bracketed expression does not vanish, since $l$ is an arc without contact (see (2) in §3.1), and $I(t) \neq 0$ by Lemma 6. Hence $\Delta(t, s) \neq 0$. Q.E.D.

We now assume that the functions (9)

$$x = \varphi(t - t_0, f(s), g(s)) = \Phi(t, s), \qquad y = \psi(t - t_0, f(s), g(s)) = \Psi(t, s)$$

are defined for all $s$ and $t$ such that $a \leqslant s \leqslant b$, $t_0 \leqslant t \leqslant \tau(s)$ (or $t_0 \geqslant t \geqslant \tau(s)$), where $\tau(s)$ is a single-valued continuous function of $s$.

*Lemma 8. Suppose that every path which passes at $t = t_0$ through a point $(f(s), g(s))$ of the arc without contact $l$, $a \leqslant s \leqslant b$, has no other points in common with $l$ for $t_0 < t \leqslant \tau(s)$ (or $t_0 > t \geqslant \tau(s)$). Then equations (9) define a regular mapping $T$ of the region $R$ defined on the $(t, s)$-plane by*

$$a \leqslant s \leqslant b, \qquad t_0 \leqslant t \leqslant \tau(s)$$

*(or $t_0 \geqslant t \geqslant \tau(s)$) onto some closed region $\overline{W}$ on the $(x, y)$-plane (Figure 37).*

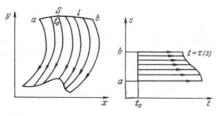

FIGURE 37.

Proof. To fix ideas, we shall assume that $t_0 \leqslant t \leqslant \tau(s)$. By Lemma 7, the Jacobian $\Delta(t, s) = \frac{D(\Phi, \Psi)}{D(t, s)}$ does not vanish in $R$. Hence, to prove that the mapping $T$ defined by (9) is regular it will suffice to show that it is one-to-one, i.e., maps any two distinct points $K_1(t_1, s_1)$ and $K_2(t_2, s_2)$, $t_0 < t_1 < \tau(s_1)$, $t_0 < t_2 < \tau(s_2)$, of $R$ onto distinct points of the region $W$. We suppose first that $s_1 \neq s_2$. Let $L_1$ be the path crossing the arc $l$ at $t = t_0$ at the point $M_1$ corresponding to $s_1$, and $L_2$ the path crossing $l$ at $t = t_0$ at the point $M_2 (\neq M_1)$ corresponding to $s_2$.

By the definition of the functions (9), the image $T(K_1)$ of $K_1(t_1, s_1)$ on the $(x, y)$-plane lies on the path $L_1$, and the image $T(K_2)$ of $K_2(t_2, s_2)$ lies on the path $L_2$. By assumption, the arcs of these paths corresponding to $t_0 \leqslant t \leqslant \tau(s_1)$ and $t_0 \leqslant t \leqslant \tau(s_2)$, respectively, have no points other than $M_1$ and $M_2$ in common with $l$ (and do not intersect each other). Hence it is clear that the points $T(K_1)$ and $T(K_2)$ are distinct.

Now suppose that $s_1 = s_2$. To fix ideas, let $t_0 < t_1 < t_2$. The points $T(K_1)$ and $T(K_2)$ lie on one path $L_1$ crossing the arc $l$ at the point $M_1$ corresponding to $s = s_1$. Were these points equal, the path $L_1$ would be closed and the number $t_2 - t_1$ a multiple of the period $\theta_0$ of the corresponding periodic motion.

The point $M$ of $L$ lying on the arc $l$ would then correspond both to $t_0$ and to $t'_0 = t_0 + (t_2 - t_1)$. But $t'_0 = t_0 + (t_2 - t_1) < t_2 < \tau(s_1)$ and this contradicts the assumption. Q.E.D.

Remark 1. If the system (I) and the functions $f(s)$, $g(s)$ (in the parametric equations of $l$) are of class $C_k$ (analytic), then the regular mapping $T$ defined by (9) is also of class $C_k$ (analytic).

Remark 2. Consider the set of points in $W$ such that $t = h(s)$, $t_0 < h(s) < \tau(s)$, $a \leqslant s \leqslant b$. Let $h(s)$ be a single-valued continuous function. This set is clearly a generalized arc without contact, whose parametric equations in cartesian coordinates are

$$x = \Phi(h(s), s), \qquad y = \Psi(h(s), s).$$

If $h(s)$ has a continuous derivative, this arc is smooth. It is easy to see that if $h(s) \equiv t_0 + h_0$, where $h_0$ is a sufficiently small positive constant $(0 < h_0 < \tau(s))$, then the curve $t = t_0 + h_0$ is an arc without contact.

Remark 3. Any two points $M_1$ and $M_2$ in the region $W$, corresponding to $(t_1, s_1)$ and $(t_2, s_2)$, $s_1 < s_2$, $t_0 < t_1 < \tau(s_1)$, $t_0 < t_2 < \tau(s_2)$, may be joined by an arc without contact $t = h(s)$, where $s_1 \leqslant s \leqslant s_2$ and $h(s)$ is a function of class $C_1$. Moreover, this arc can be chosen in such a way that it has arbitrary prescribed directions at $M_1$ and $M_2$, different from the direction of the paths at these points (see Appendix, §6.7).

## 6. PATHS CROSSING TWO ARCS WITHOUT CONTACT. CORRESPONDENCE FUNCTION.

In the sequel we shall often have occasion to deal with paths crossing two arcs without contact. We now devote some attention to this situation.

Let $l$ and $\tilde{l}$ be two arcs without contact having no common points,

$$x = f(s), \qquad y = g(s), \qquad a \leqslant s \leqslant b,$$

and

$$x = \tilde{f}(\tilde{s}), \qquad y = \tilde{g}(\tilde{s}), \qquad \tilde{a} \leqslant \tilde{s} \leqslant \tilde{b},$$

their parametric equations. Denote the point on $l$ $(\tilde{l})$ corresponding to parameters $s$ $(\tilde{s})$ by $M(s)$ $(\tilde{M}(\tilde{s}))$. Let $L_0$ be a path crossing $l$ at a point $M_0$ and $\tilde{l}$ at a point $\tilde{M}_0$, where $M_0$ and $\tilde{M}_0$ are not endpoints of the arcs. Suppose that

$$x = \varphi(t), \quad y = \psi(t) \tag{17}$$

is a motion on $L_0$ under which $M_0$ corresponds to $t = t_0$ and $\tilde{M}_0$ to $t = \tilde{t}_0$. To fix ideas we assume that $t_0 < \tilde{t}_0$.

Assume moreover that the section of $L_0$ defined by $t \in [t_0, \tilde{t}_0]$ is a simple arc (so that if $L_0$ is a closed path the difference $\tilde{t}_0 - t_0$ must be less than the period) having no points other than $M_0$ and $\tilde{M}_0$ in common with the arcs $l$ and $\tilde{l}$ (Figure 38). Let $M_0$ correspond on $l$ to the parameter values $s = s_0$ and $\tilde{M}_0$ on $\tilde{l}$ to $\tilde{s} = \tilde{s}_0$ (i. e., $M_0 = M(s_0)$, $\tilde{M}_0 = \tilde{M}(\tilde{s}_0)$).

We may assume that $s_0 = \tilde{s}_0 = 0$ (this can always be achieved by reparametrization, replacing $s$ and $\tilde{s}$ by $s - s_0$ and $\tilde{s} - \tilde{s}_0$, respectively). Since $l$ and $\tilde{l}$ are arcs without contact

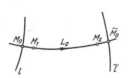

FIGURE 38.

$$\Delta(s) = \begin{vmatrix} P(f(s),\, g(s)) & Q(f(s),\, g(s)) \\ f'(s) & g'(s) \end{vmatrix} \neq 0$$

and

$$\tilde{\Delta}(\tilde{s}) = \begin{vmatrix} P(\tilde{f}(\tilde{s}),\, \tilde{g}(\tilde{s})) & Q(\tilde{f}(\tilde{s}),\, \tilde{g}(\tilde{s})) \\ \tilde{f}'(\tilde{s}) & \tilde{g}'(\tilde{s}) \end{vmatrix} \neq 0$$

(see (2) in § 3.1; both these inequalities hold for $s \in [a, b]$ and $\tilde{s} \in [\tilde{a}, \tilde{b}]$).

Assume that the parameters $s$ and $\tilde{s}$ have been so chosen that the determinants $\Delta(0)$ and $\tilde{\Delta}(0)$, hence also the angles between the path $L_0$ and the arcs $l$ and $\tilde{l}$, have the same sign (see Appendix, § 5). If this is not so we need only replace $s$ by $- s$. To fix ideas, suppose that $\Delta(0) > 0$, $\tilde{\Delta}(0) > 0$.

L e m m a 9. *There exist a constant $\sigma_0 > 0$, $a < -\sigma_0 < \sigma_0 < b$, and functions $t = \chi(s) > t_0$, $\tilde{s} = \Omega(s)$, defined for all $s$ with $|s| < \sigma_0$, such that a path $L$ crossing the arc $l$ at $t = t_0$ at a point $M(s)$ ($|s| < \sigma_0$) will cross the arc $\tilde{l}$ at $t = \chi(s)$ at a point $\tilde{M}(\tilde{s})$, where $\tilde{s} = \Omega(s)$, and the following conditions are satisfied: a) $\chi(s)$ and $\Omega(s)$ are of class $C_1$; b) for $t_0 < t < \chi(s)$ the path $L$ has no points in common with $l$ and $\tilde{l}$; c) $\Omega'(s) > 0$, so that $\tilde{s} = \Omega(s)$ is a monotone increasing function of $s$.*

P r o o f. Consider the functions $\Phi(t, s)$, $\Psi(t, s)$ of (9). For any fixed $s$, $a \leqslant s \leqslant b$, these functions define a motion on the path such that the point $M(s)$ common to the arc $l$ and the path corresponds to $t = t_0$. When $s = 0$,

$$x = \Phi(t, 0) = \varphi(t), \quad y = \Psi(t, 0) = \psi(t),$$

which is the prescribed motion (17) on the path $L_0$. By assumption, this motion is defined for all $t \in [t_0, \tilde{t}_0]$, hence also for all $t \in [t_0, t^*]$, where $t^*$ is slightly greater than $\tilde{t}_0$. But then, by the theorem on continuity with respect to initial values (see § 1), there exists $\sigma^* > 0$ such that all motions on paths crossing the arc $l$ at points $|s| < \sigma^*$ are defined for the same values of $t$, i.e., for all $|s| < \sigma^*$ the functions (9) are defined for all $t \in [t_0,\ t^*]$.

If the paths crossing $l$ cross the arc $\tilde{l}$ at some $t > t_0$, this obviously implies that the system of equations

$$\Phi(t,\ s) = \tilde{f}(s), \qquad \Psi(t,\ s) = \tilde{g}(s) \tag{18}$$

has a solution.

In any case, the functions

$$\Phi_1(t,\ s,\ \tilde{s}) \equiv \Phi(t,\ s) - \tilde{f}(\tilde{s}), \qquad \Psi_1(t,\ s,\ \tilde{s}) \equiv \Psi(t,\ s) - \tilde{g}(\tilde{s})$$

are defined for $|s| < \sigma^*$, $t_0 \leqslant t \leqslant t^*$, $\tilde{a} \leqslant \tilde{s} \leqslant \tilde{b}$, and are of class $C_1$. Since by assumption the path $L_0$ cuts $\tilde{l}$ at $t = \tilde{t}_0$, the system of equations (18) is satisfied by $t = \tilde{t}_0$, $s = \tilde{s} = 0$:

$$\Phi_1(\tilde{t}_0,\ 0,\ 0) = 0, \qquad \Psi_1(\tilde{t}_0,\ 0,\ 0) = 0. \tag{19}$$

Finally, for $t = \tilde{t}_0$, $s = 0$, $\tilde{s} = 0$, the Jacobian

$$\Delta(t,\ s,\ \tilde{s}) = \frac{D(\Phi_1,\ \Psi_1)}{D(t,\ \tilde{s})} = \begin{vmatrix} \Phi'_t(t,\ s) & -\tilde{f}'(\tilde{s}) \\ \Psi'_t(t,\ s) & -\tilde{g}'(\tilde{s}) \end{vmatrix}$$

becomes

$$\Delta(\tilde{t}_0,\ 0,\ 0) = \begin{vmatrix} \varphi'_t(\tilde{t}_0 - t_0,\ f(0),\ g(0)) & -\tilde{f}'(0) \\ \psi'_t(\tilde{t}_0 - t_0,\ f(0),\ g(0)) & -\tilde{g}'(0) \end{vmatrix} = -\begin{vmatrix} P(\tilde{f}(0),\ \tilde{g}(0)) & -\tilde{f}'(0) \\ Q(\tilde{f}(0),\ \tilde{g}(0)) & -\tilde{g}'(0) \end{vmatrix},$$

and therefore, since $\tilde{l}$ is an arc without contact,

$$\Delta(\tilde{t}_0,\ 0,\ 0) = D_1 \neq 0. \tag{20}$$

It follows from (19) and (20) that equations (18) satisfy all the assumptions of the implicit function theorem (Appendix, §4). Thus there exists a unique pair of functions

$$t = \chi(s), \qquad \tilde{s} = \Omega(s),$$

defined for $|s| \leqslant \sigma_1$, where $\sigma_1$ is some positive number, $\sigma_1 < \sigma^*$, which satisfy system (18). Moreover, the functions $\chi(s)$ and $\Omega(s)$ are of class $C_1$ and

$$\chi(0) = \tilde{t}_0, \qquad \Omega(0) = 0.$$

We have thus proved the existence of the functions $\chi(s)$ and $\Omega(s)$ for any $\sigma_0 \leqslant \sigma_1$, and part a) of our lemma is proved.

We now prove part b). Let $\alpha$ be a positive number such that

1) for all $|s| < \sigma_1$ $\quad \alpha < \frac{\chi(s) - t_0}{2}$;

2) for all $|s| < \sigma_1$, $t_0 < t < t_0 + \alpha$, $\chi(s) - \alpha < t < \chi(s)$, the paths

$$x = \varphi(t - t_0, f(s), g(s)), \qquad y = \psi(t - t_0, f(s), g(s))$$

have no points in common with $l$ and $\tilde{l}$. The existence of $\alpha$ follows from Remark a) to Lemma 3 (see §3.4) and from the assumption that $l$ and $\tilde{l}$ have no points in common (Figure 38).

By assumption, the path $L_0$ has no points in common with $l$ and $\tilde{l}$ for $t_0 < t < \tilde{t}_0 = \chi(0)$.

Consequently, the arc of this path corresponding under the motion (17) to $t$ such that $t_0 + \alpha \leqslant t \leqslant \tilde{t}_0 - \alpha$ (the arc $M_1 M_2$ in Figure 38) lies at a positive distance from each of the arcs $l$ and $\tilde{l}$. But then it obviously follows from the continuity of the functions $\varphi$, $\psi$, $f$, $g$ and $\chi$ that, for sufficiently small $\sigma_0$, $0 < \sigma_0 < \sigma_1$, the same is true for all paths crossing $l$ at points with $|s| < \sigma_0$: their arcs defined by $t \in [t_0 + \alpha, \; \chi(s) - \alpha]$ in equations (18) lie at a positive distance from $l$ and $\tilde{l}$. By the choice of $\alpha$, this implies the truth of part b).

To prove part c), we shall calculate $\Omega'(s)$. Since $\chi(s)$ and $\Omega(s)$ are solutions of equations (18), we have

$$\Phi(\chi(s), \; s) \equiv \tilde{f}(\Omega(s)), \qquad \Psi(\chi(s), \; s) \equiv \tilde{g}(\Omega(s))$$

$(|s| \leqslant \sigma)$. Differentiating with respect to $s$, we get

$$\Phi_t'(\chi(s), \; s)\, \chi'(s) + \Phi_s'(\chi(s), \; s) = \tilde{f}'(\Omega(s))\, \Omega'(s),$$
$$\Psi_t'(\chi(s), \; s)\, \chi'(s) + \Psi_s'(\chi(s), \; s) = \tilde{g}'(\Omega(s))\, \Omega'(s),$$

whence

$$\Omega'(s) = \frac{\begin{vmatrix} \Phi_t'(\chi(s), \; s) & \Phi_s'(\chi(s), \; s) \\ \Psi_t'(\chi(s), \; s) & \Psi_s'(\chi(s), \; s) \end{vmatrix}}{\begin{vmatrix} \Phi_t'(\chi(s), \; s) & \tilde{f}'(\Omega(s)) \\ \Psi_t'(\chi(s), \; s) & \tilde{g}'(\Omega(s)) \end{vmatrix}} \tag{21}$$

(the denominator does not vanish because $\tilde{l}$ is an arc without contact).

The numerator of this fraction is clearly the determinant $\Delta(\chi(s), \; s)$ of Lemma 7 (see §3.4), which is equal to

$$\Delta(\chi(s), \; s) = \begin{vmatrix} P(f(s), \; g(s)) & f'(s) \\ Q(f(s), \; g(s)) & g'(s) \end{vmatrix} I(\chi(s)), \tag{22}$$

where $I(\chi(s)) > 0$ by (16). The denominator of (21) is

$$\begin{vmatrix} \varphi_t'(\chi(s) - t_0, \; f(s), \; g(s)) & \tilde{f}'(\Omega(s)) \\ \psi_t'(\chi(s) - t_0, \; f(s), \; g(s)) & \tilde{g}'(\Omega(s)) \end{vmatrix} = \begin{vmatrix} P[\tilde{f}(\Omega(s)), \; \tilde{g}(\Omega(s))] & \tilde{f}'(\Omega(s)) \\ Q[\tilde{f}(\Omega(s)), \; \tilde{g}(\Omega(s))] & \tilde{g}'(\Omega(s)) \end{vmatrix}. \tag{23}$$

Substituting (22) and (23) into (21) and recalling the notation $\Delta(s)$ and $\tilde{\Delta}(\tilde{s})$ introduced at the beginning of the subsection, we obtain

$$\Omega'(s) = \frac{\Delta(s)}{\tilde{\Delta}(\Omega(s))} I(\chi(s)). \tag{24}$$

Since $I(\chi(s)) > 0$ and the signs of the determinants $\Delta(s)$ and $\tilde{\Delta}(\Omega(s))$ are by assumption the same, it follows that $\Omega'(s) > 0$ and so $\Omega(s)$ is indeed a monotone increasing function. Q.E.D.

Remark 1. The assertion that $\tilde{s} = \Omega(s)$ is a monotone function of $s$ has a simple geometric interpretation. Let $\lambda$ denote the arc of the path $L_0$ corresponding to $\tau_0 \leqslant t \leqslant \tau_1$, where $\tau_0 < t_0$, $\tau_1 > \chi(s)$, with $\tau_0 - t_0$ and $\tau_1 - \chi(s)$ so small that $\lambda$ is a simple arc.* The mono-tonicity of $\Omega(s)$ means that any path crossing the arc $l$ at $t = t_0$ on the positive (negative) side of $\lambda$ at a point $M(s^*)$ sufficiently close to $M_0$ will cross the arc $\tilde{l}$ at $t = \chi(s^*)$, again on the positive (negative) side of $\lambda$, at the point $\tilde{M}(s^*)$ (Figure 39).

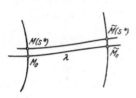

FIGURE 39.

Remark 2. If system (I) and the functions $f(s)$, $g(s)$, $\tilde{f}(s)$, $\tilde{g}(s)$ are all of class $C_k$ (analytic), then the functions $\Omega(s)$ and $\chi(s)$ are also of class $C_k$ (analytic).

Let us assume that the functions $\Omega(s)$ and $\chi(s)$ of Lemma 9 are defined for all $s \in [a, b]$ and

$$\Omega(a) = \tilde{a}, \quad \Omega(b) = \tilde{b}. \tag{25}$$

Let $A$ be the endpoint of the arc $l$ corresponding to $s = a$, $B$ the endpoint of $l$ corresponding to $s = b$. Let $L_A$ and $L_B$ denote the paths passing through $A$ and $B$. By (25), $L_A$ and $L_B$ pass through the endpoints $\tilde{A}$ and $\tilde{B}$, respectively, of the arc $\tilde{l}$, where $\tilde{A}$ and $\tilde{B}$ are the endpoints corresponding to $\tilde{s} = \tilde{a}$ and $\tilde{s} = \tilde{b}$, respectively.

Consider the simple closed curve $\gamma$ consisting of the arcs $A\tilde{A}$ and $B\tilde{B}$ of the paths $L_A$ and $L_B$ and the subarcs $AB$ and $\tilde{A}\tilde{B}$ of $l$ and $\tilde{l}$. Let $\Gamma$ denote the region interior to the curve $\gamma$ and $\bar{\Gamma}$ its closure.

Lemma 10. *Any path passing through a point of the region $\Gamma$ crosses $l$ with decreasing $t$ and $\tilde{l}$ with increasing $t$.*

Proof. We shall use Lemma 8 and the notation introduced therein, with the role of $\tau(s)$ played by $\chi(s)$. Since all assumptions of Lemma 8 hold here, the mapping $T$ defined by equations (9),

$$x = \varphi(t - t_0, f(s), g(s)) = \Phi(t, s), \quad y = \psi(t - t_0, f(s), g(s)) = \Psi(t, s),$$

is a regular mapping of the region $\bar{R}$:

$$a \leqslant s \leqslant b, \quad \tau_0 \leqslant t \leqslant \chi(s)$$

---

* If $L_0$ is a nonclosed path, it is clear (see §1) that any arc corresponding to a finite interval $[\tau_0, \tau_1]$ on the $t$-axis is simple.

    Thus the condition that $\tau_0$ and $\tau_1$ be sufficiently close to $t_0$ and $t_1$ is necessary only when $L_0$ is a closed path.

onto some region in the $(x, y)$-plane. Let $C$ denote the boundary of $R$. It is readily seen that $T(C) = \gamma$.

Consequently, the regular mapping $T$ takes the boundary of the simply connected region $R$ onto the boundary of the simply connected region $\Gamma$. But then, since both regions $R$ and $\Gamma$ are bounded (see Appendix, §1.7), $T(R) = \Gamma$. And this means that all points of $\Gamma$ lie on paths which cross $l$ with decreasing $t$ and $\tilde{l}$ with increasing $t$.

Remark 1. All paths which cross the arc $l$ at its interior points clearly enter $\Gamma$ with increasing $t$, while those crossing $\tilde{l}$ at its interior points leave $\Gamma$ with increasing $t$.

Remark 2. Let $A$ be a fixed point and $B$ be a variable point on the arc $l$. Then for every $\varepsilon > 0$ there exists $\delta > 0$ such that if $B \in U_\delta(A)$ the corresponding region $\Gamma$ is contained in the $\varepsilon$-neighborhood of the arc $A\tilde{A}$ of $L_A$. This follows directly from the continuity of the functions $\chi(s)$ and $\Omega(s)$.

Any closed region of the type $\bar{\Gamma}$, i. e., a simply connected region bounded by two nonintersecting arcs without contact $AB$ and $\tilde{A}\tilde{B}$ and two arcs of paths $A\tilde{A}$ and $B\tilde{B}$ for which the statement of Lemma 10 holds true, will be called an elementary topological quadrangle, or simply an elementary quadrangle (Figure 40).

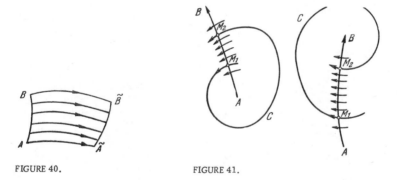

FIGURE 40.　　　　　　　　　FIGURE 41.

## 7. CASE OF A PATH HAVING SEVERAL POINTS IN COMMON WITH AN ARC WITHOUT CONTACT.

The lemmas proved above are valid not only for dynamic systems on the plane and on surfaces of genus zero, but also on surfaces of higher genus, since the proofs make no use of specific properties of the plane. In contrast, the proofs of the lemmas to be established in this subsection depend essentially on a property of the plane or the sphere — the fact that they are simply connected, so that any simple closed curve divides the plane (or sphere) into two regions. Surfaces of higher genus are not simply connected. For this reason, the lemmas and propositions to be proved below are not valid for such surfaces.

As above, let $l$ be an arc without contact, with parametric equations

$$x = f(s), \quad y = g(s), \quad s \in [a, b],$$

where the endpoints $A$ and $B$ of the arc correspond to parameter values $a$ and $b$, respectively. Let $L_0$ be a path which has two distinct points $M_1 (s_1)$ and $M_2 (s_2)$ in common with $l$, not endpoints of $l$, and corresponding under some motion

$$x = \varphi(t), \quad y = \psi(t) \tag{26}$$

on $L_0$ to parameter values $t_1$ and $t_2$ $(t_1 < t_2)$. Assume that $L_0$ and $l$ have no other common points for $t \in [t_1, t_2]$. The arc $M_1 M_2$ of the path $L_0$ will be called a **t u r n  o f  t h e  p a t h**. To fix ideas, let $s_1 < s_2$. It is clear from geometric considerations that under these assumptions there are two possible configurations of $L_0$ and $l$, schematically depicted in Figure 41 (or the mirror images of these configurations). It is clear that the path cannot cross the arc as shown in Figure 42, since then $l$ would not be an arc without contact.

Let $C$ denote the simple piecewise-smooth curve formed by the arc $M_1 M_2$ of $L_0$ (corresponding to parameter values $t_1 \leqslant t \leqslant t_2$) and the subarc $M_1 M_2$ of $l$.

Since $l$ is an arc without contact, the angles between $l$ and the paths that cross it all have the same sign. Hence, by Lemma 4 in §6 of the Appendix, the paths crossing the subarc $M_1 M_2$ of $l$ at its interior points either all enter $C$ or all leave $C$ (Figure 43).

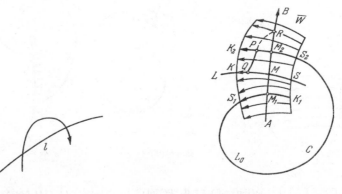

FIGURE 42.                    FIGURE 43.

**L e m m a  11.** *If $C$ is the simple smooth curve formed by the arc $M_1 M_2$ of the path $L_0$ and the subarc $M_1 M_2$ of the arc without contact $l$, then: a) points of $L_0$ corresponding to parameter values $t < t_1$ lie inside (outside) $C$, and points corresponding to $t > t_2$ lie outside (inside) $C$; b) any path $L$ which crosses $l$ at $t = t_0$ between the points $M_1$ and $M_2$ will never cross this subarc of $l$ for other values of $t$; c) the subarc $AM_1$ of $l$ lies inside (outside) the curve $C$, the subarc $BM_2$ lies inside (outside) $C$ (Figure 43).*

P r o o f. The proof is based on the properties of the region $W$ constructed in Lemma 3 about the arc $l$, and on Lemma 4 in §6 of the Appendix.

Consider the paths which cut the arc $l$ at $t = t_0$. To avoid confusion, we shall denote the parameter along these paths ("time") not by $t$, as in the equations (26) of the path $L_0$, but by $T$. Thus, the equations of these paths will have the form

$$x = \varphi(T - t_0, \ f(s), \ g(s)) = \Phi(T, \ s),$$
$$y = \psi(T - t_0, \ f(s), \ g(s)) = \Psi(T, \ s). \tag{27}$$

For sufficiently small $h$, the set $\overline{W}$ of points with coordinates

$$x = \Phi(T, \ s), \qquad y = \Psi(T, \ s)$$
$$(t_0 - h \leqslant T \leqslant t_0 + h, \qquad a \leqslant s \leqslant b)$$

has the properties established in Lemma 3. ($h$ is necessarily such that $t_1 + h < t_2 - h$, for otherwise the mapping defined by equations (27) is readily shown to be nonregular.) The region $\overline{W}$ is illustrated in Figure 43.

If we set $s = s_1$ and $s = s_2$ in (27), we get two motions on the path $L_0$, neither of them identical with (26). One of these motions

$$x = \varphi(T - t_0, \ f(s_1), \ g(s_1)), \ y = \psi(T - t_0, \ f(s_1), \ g(s_1)),$$

is obtained from the motion

$$x = \varphi(t), \qquad y = \psi(t)$$

by the substitution $t = T - T_0 + t_2$, and the other

$$x = \varphi(T - t_0, \ f(s_2), \ g(s_2)), \qquad y = \psi(T - t_0, \ f(s_2), \ g(s_2))$$

by the substitution $t = T - T_0 + t_1$.

Consider the subarc of the arc $M_1 M_2$ on $L_0$ corresponding to parameter values $t_1 + h < t < t_2 - h$ (the open arc $S_1 S_2$ in Figure 43). By assumption, for $t$ in $(t_1, t_2)$ the path $L_0$ has no points in common with $l$, and so the open arc $S_1 S_2$ on $L_0$ cannot have points in common with $\overline{W}$. Indeed, otherwise it would follow from Remark b) to Lemma 3 that $L_0$ must have a point in common with $l$ for some parameter value $t$, $t_1 < t < t_2$.

Thus, the only points of the curve $C$ in $\overline{W}$ are those of the subarc $M_1 M_2$ of $l$ and the arcs $M_2 S_2$, $M_1 S_1$ on $L_0$ (corresponding to parameter values $t_2 - h \leqslant t \leqslant t_2$ and $t_1 \leqslant t \leqslant t_1 + h$, respectively).* The arc $l$ separates $\overline{W}$ into two subregions $\overline{W}^-$ and $\overline{W}^+$, corresponding to parameter values $t_0 - h \leqslant T < t_0$ and $t_0 < T \leqslant t_0 + h$, respectively.

Now let $L$ be any path crossing the arc $l$ at a point $M$ between $M_1$ and $M_2$ (Figure 43). In view of Lemma 4 in §6 of the Appendix, we may assume, say, that the "half-open" arc $MK$ on $L$ corresponding to $t_0 < T \leqslant t_0 + h$ lies outside $C$ (in $\overline{W}^+$) and the half-open arc $MS$ corresponding to $t_0 - h \leqslant T < t_0$ lies inside $C$ (in $\overline{W}^-$).

Now consider the section of the path $L_0$ defined by $t_2 < t \leqslant t_2 + h$** (the half-open arc $M_2 K_2$ in Figure 43). It is obviously in $\overline{W}^+$. By the properties

---

* The arcs $M_2 S_2$ and $M_1 S_1$ on $L_0$ are defined by equations (27) with $t_0 - h \leqslant T < t_0$ and $t_0 + h \geqslant T > t_0$, respectively.

** Or parameter values $t_0 < T \leqslant t_0 + h$ in equations (27).

of $\overline{W}$, any point $P$ on this arc can be joined to a point $Q$ on the arc $MK$ of $L$ by a continuous curve $PQ$ lying entirely within $\overline{W}$ and having no points in common with $C$ (see Remark to Lemma 3). It follows that the half-open arc $M_2K_2$ on $L_0$, like the arc $MK$ on $L$, lies outside $C$. Similarly, one shows that both arcs $M_1K_1$ and $MS$ on the paths $L_0$ and $L$, respectively, lie inside $C$.

Next, it is clear that for $t < t_1$ the path $L_0$ can have no points outside $C$. Otherwise, $L_0$ would have to leave $C$ with decreasing $t$, but this is impossible since $L_0$ cannot cross the arc $M_1M_2$ and cannot cross the arc without contact $l$ with decreasing $t$ while leaving $C$ (indeed, by assumption, the path $L$, hence also all paths crossing $l$, enter $C$ with decreasing $t$).

Entirely analogous arguments show that the points on $L$ corresponding to parameter values $t > t_2$ lie outside $C$. This completes the proof of part a).

Now for part b). Let $L$ be some path passing at $t = t_0$ through a point $M$ between $M_1$ and $M_2$. As $t$ decreases, this path crosses $C$ inward. It cannot cross the curve outward at points between $M_1$ and $M_2$ (since then it would necessarily cross the arc $l$ in the opposite direction, which is impossible).

Obviously, with decreasing $t$ it cannot pass through the points $M_1$ and $M_2$ either.

Indeed, suppose that the path $L$ crosses $C$ outward with decreasing $t$, at the point $M_1$. But there is a path through $M_1$, namely $L_0$, which contains no points near $M_1$ inside $C$.

Thus, except for its initial point $M$, the negative semipath $L_{\overline{M}}$, issuing from an interior point $M$ of the subarc $M_1M_2$ of $l$, remains inside $C$. Similarly one shows that the positive semipath $L_M^+$ remains (except for its initial point $M$) outside $C$. This proves part b).

Finally, we prove part c). Consider the subarc $M_2B$ of $l$. It is readily seen that any point other than $M_2$ on this subarc may be joined to a point $Q$ on the arc $M_2K_2$ of $L_0$ by a continuous curve inside $\overline{W}$ which has no points in common with $C$ (the dashed curve in Figure 43). Consequently, the half-open arc $M_2B$ (corresponding to $s_2 < s \leqslant b$), like the arc $M_2K_2$ on the path $L_0$, lies outside $C$. One proves in similar fashion that the half-open arc $M_1A$ lies inside $C$. This completes the proof of Lemma 11.

Corollary 1. If a path $L$ has two distinct points $M_1$ and $M_2$ in common with an arc without contact $l$, corresponding to parameter values $t_1$ and $t_2$, and no points in common with $l$ for $t_1 < t < t_2$, then there are no other points of $L$ on $l$ between $M_1$ and $M_2$.

This follows directly from part a) of Lemma 11.

Corollary 2. Suppose that a path $L$ crosses an arc without contact $l$ at more than two points, and let $M_1, M_2, M_3, \ldots$ be the (finite or countable) set of all such points. Assume moreover that for some fixed motion on the path these points are $t$-successive, i.e., the corresponding parameter values $t$, $t_1$, $t_2$, $t_3$, ... vary monotonically, say

$$t_1 < t_2 < t_3 < \cdots,$$

and $L$ does not cross $l$ at intermediate $t$, $t_i < t < t_{i+1}$, $i = 1, 2, \ldots$ Then the points $M_i$ are also $s$-successive, i.e., their parameter values $s_i$ on $l$ also vary monotonically: either

$$s_1 < s_2 < s_3 < \cdots$$

or

$$s_1 > s_2 > s_3 > \ldots$$

(Figure 44). Moreover, the path $L$ has no points in common with $l$ which lie (on $l$) between the points $M_i$ and $M_{i+1}$ $(i = 1, 2, \ldots)$.

Corollary 2 follows from parts a) and b) of the lemma. Briefly, this may be stated as follows:

*Points of intersection of a path $L$ and an arc without contact which are $t$-successive are also $s$-successive.*

**L e m m a 12.** *A closed path cannot have more than one point in common with an arc without contact.*

P r o o f. Suppose the statement false, and let $L$ be a closed path having at least two distinct points in common with an arc without contact $l$, say $M_1$ and $M_2$, corresponding to parameter values $t_1$ and $t_2$ $(t_1 < t_2)$. We may assume that the path $L$ does not cut $l$ at intermediate times $t_1 < t < t_2$.

By virtue of Lemma 11, the semipath $L_{\bar{M}_1}$ lies inside the curve $C$ formed from the arc $M_1 M_2$ on $L$ and the subarc $M_1 M_2$ of $l$, and the semipath $L_{\bar{M}_2}$ outside this curve, or conversely. To fix ideas, let $L_{\bar{M}_1}$ lie inside $C$ (Figure 45). Let $M^*$ be a point on $L$ corresponding to some $t = t^* < t_1$. Clearly, $M^* \in L_{\bar{M}_1}$. By assumption, $L$ is a closed path. Let $\theta_0$ be the period of the motion on $L$. Then all points on $L$ corresponding to parameter values $t = t^* + n\theta_0$ ($n$ an integer) coincide with $M^*$. But if $n$ is sufficiently large we have $t^* + n\theta_0 > t_2$ and this means that $M^* \in L_{\bar{M}_2}^+$. This is a contradiction, for the same point $M^*$ cannot be both inside and outside the simple closed curve $C$. Q.E.D.

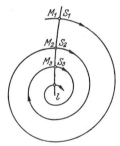

FIGURE 44.

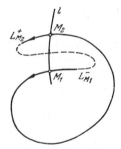

FIGURE 45.

**8. SUCCESSION FUNCTION.** As before, let $L_0$ be a path crossing an arc without contact $l$ at two points $M_1$ and $M_2$, not endpoints of $l$, which are $t$-successive and correspond to times $t_1$ and $t_2$, $t_1 < t_2$. The path $L_0$ may be either nonclosed or closed. In the first case, $M_1$ and $M_2$ are distinct points on $l$, corresponding to different parameter values $s_1$ and $s_2$.

In the second case, the points $M_1$ and $M_2$ coincide, $s_2 = s_1$, but $t_1$ and $t_2$ are distinct: $t_2 = t_1 + \theta_0$, where $\theta_0$ is the period of a motion on the path (Figure 46). Paths which cross the same arc without contact twice may be treated as a limiting case of paths crossing two different arcs without contact $l_1$ and $l_2$ at times $t_0$ and $t = \chi(s)$, obtained when the arcs $l_1$ and $l_2$ approach one another, but in such a way that the length of the interval $t_0 < t < \chi(s)$ does not tend to zero. The following analog of Lemma 9 is valid.

**Lemma 13.** *There exist* $\sigma_0 > 0$ *and functions* $t = \chi(s) > t_1$, $\tilde{s} = \Omega(s)$ *defined for* $s$ *with* $|s - s_1| < \sigma_0$, *such that any path* $L$ *crossing the arc* $l$ *at* $t = t_1$ *at a point* $M(s)$ ($|s - s_1| < \sigma_0$) *will cross* $l$ *at some time* $t = \chi(s) > t_1$ *at the point* $M(\tilde{s})$, $\tilde{s} = \Omega(s)$ *(Figure 47), and the following conditions are satisfied: 1)* $\chi(s)$ *and* $\Omega(s)$ *are functions of class* $C_1$; *2) for* $t_2$, $t_1 < t_2 < \chi(s)$ *the path* $L$ *has no points in common with* $l$; *3)* $\Omega'(s) > 0$, *so that* $\tilde{s} = \Omega(s)$ *is a monotone increasing function of* $s$. *(An analogous statement is true if* $t_1 > t_2$; *in this case* $\chi(s) < t_1$.)

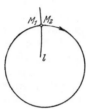

FIGURE 46.

FIGURE 47.

Proof. Consider the equations

$$x = \varphi(t - t_1, \quad f(s), \quad g(s)) = \Phi(t, s),$$
$$y = \psi(t - t_1, \quad f(s), \quad g(s)) = \Psi(t, s),$$

which, for any fixed $s$, define a motion on the path such that at time $t_1$ the path crosses the arc $l$ at a point $M(s)$ ($a \leqslant s \leqslant b$).

If the functions

$$t = \chi(s), \quad \tilde{s} = \Omega(s),$$

of our lemma exist, they must obviously satisfy the system of equations

$$\Phi(t, s) = f(\tilde{s}), \quad \Psi(t, s) = g(\tilde{s}).$$

By assumption, this sytem of equations is certainly satisfied by $t = t_2$, $s = s_1$, $\tilde{s} = s_2$. The proof is now easily completed by arguments along lines entirely similar to the proof of Lemma 9. Q.E.D.

Remark 1. If system (I) and the functions $f(s)$, $g(s)$ are of class $C_h$ (analytic), then $\Omega(s)$ and $\chi(s)$ are also of class $C_h$ (analytic).

Remark 2. It follows from the properties of the function $\tilde{s} = \Omega(s)$ that it has a single-valued inverse

$$s = \Omega^{-1}(\tilde{s}).$$

The function $\tilde{s} = \Omega(s)$ and its inverse $s = \Omega^{-1}(\tilde{s})$ are called s u c c e s s i o n f u n c t i o n s on $l$.

It is clear that a succession function cannot be defined on any arc without contact, but only on an arc crossed more than once by a path.

Let $\tilde{s} = \Omega(s)$ be a succession function on an arc without contact $l$. Let $M(s)$ be the point corresponding to $t = t_1$. Then the point $M(\tilde{s})$, i.e., $M(\Omega(s))$, corresponds to $t = \chi(s)$. If $t_2 > t_1$, then $\chi(s) > t_1$ for all $s$ for which the succession function is defined; but if $t_2 < t_1$ then $\chi(s) < t_1$. In the first case, we say that the succession function $\tilde{s} = \Omega(s)$ is consistent with motion in the sense of increasing $t$; in the second case the function is consistent with motion in the sense of decreasing $t$.

Clearly, of the two succession functions $\tilde{s} = \Omega(s)$ and $s = \Omega^{-1}(\tilde{s})$, one is consistent with motion in the sense of increasing $t$, the other in the sense of decreasing $t$. When $\chi(s) > t_1$, the point $M(\Omega(s))$ is called the s u c c e s s o r of $M(s)$, and $M(s)$ is the p r e d e c e s s o r of $M(\Omega(s))$.

By Lemma 12, a path passing through a point $M(s)$ is closed if and only if $\Omega(s) = s$.

## 9. CLOSED CURVES FORMED BY COMBINING AN ARC OF A PATH AND AN ARC WITHOUT CONTACT; REGIONS BOUNDED BY SUCH CURVES.

Let $\tilde{s} = \Omega(s)$ be a succession function defined on an arc $l$ for all $s \in [a, b]$, and suppose that every path passing at $t = t_0$ through a point $M(s)$, $s \in [a, b]$, passes through the point $M(\Omega(s))$ at $t = \chi(s)$. To fix ideas, we shall assume that $\chi(s) > t_0$, so that the succession function $\Omega(s)$ is consistent with motion in the sense of increasing $t$. Denote $\Omega(a)$ and $\Omega(b)$ by $a'$ and $b'$, respectively, and let $A$, $B$, $A'$, $B'$ be the points on $l$ corresponding to values $a$, $b$, $a'$, $b'$, respectively, of the parameter $s$ (Figure 48). The paths through $A$ and $B$ will be denoted by $L_A$ and $L_B$, respectively. The points $A$ and $A'$ ($B$ and $B'$) may be either distinct or identical. If $A$ and $A'$ coincide, the path $L_A$ is closed and $a' = \Omega(a) = a$ (Figure 49).

Since the succession function $\Omega(s)$ is an increasing function of $s$ and by assumption $a < b$, it follows that

$$a' = \Omega(a) < \Omega(b) = b' \qquad (28)$$

and for any $s \in (a, b)$,

$$a' < \Omega(s) < b'. \qquad (29)$$

The subarcs $AB$ and $A'B'$ may or may not have common points. The case in which $AB$ and $A'B'$ have no common points (Figure 50) can clearly be reduced to the case considered in §3.5 of paths crossing two different arcs without contact.

Suppose that the arcs $AB$ and $A'B'$ have points in common. Then $B$ cannot lie between $A$ and $A'$ (when the latter are distinct), for otherwise it would follow from (28) and (29) that

$$a < b < a' < b',$$

but this implies, contrary to assumption, that the subarcs $AB$ and $A'B'$ have no points in common. Similarly, the point $A$ cannot lie between $B$ and $B'$. Thus, when the subarcs $AB$ and $A'B'$ of $l$ have common points, the subarcs $AA'$ and $BB'$ have either no common points or only one (if the points $A'$ and $B$ coincide; Figure 51). It is not difficult to determine all the possible configurations (one example is shown in Figure 52).

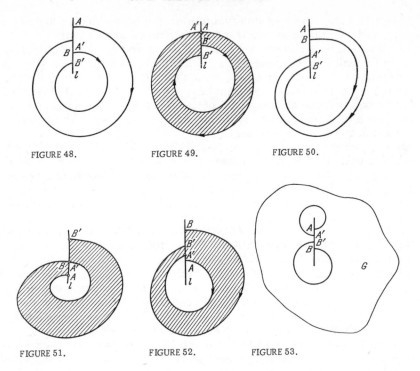

FIGURE 48.   FIGURE 49.   FIGURE 50.

FIGURE 51.   FIGURE 52.   FIGURE 53.

In case $L_A$ is a nonclosed path, we let $C_1$ denote the simple closed curve consisting of the arc $AA'$ of $L_A$ and the subarc $AA'$ of $l$; if $L_A$ is a closed path, $C_1$ will denote the path itself. Similarly, we let $C_2$ denote the simple closed curve consisting of the arc $BB'$ of $L_B$ and the subarc $BB'$ of $l$, or the entire path $L_B$. $C_1$ and $C_2$ have either no common points (Figures 48, 49, 52) or only one common point (Figure 51), depending on whether or not the subarcs $AA'$ and $BB'$ of $l$ have common points. The regions bounded by $C_1$ and $C_2$ will be denoted by $\Gamma_1$ and $\Gamma_2$, respectively.

Lemma 14. *If the subarcs $AB$ and $A'B'$ of $l$ have common points, the regions $\Gamma_1$ and $\Gamma_2$ lie inside one another,* so that the curves $C_1$ and $C_2$ bound a finite region. Any path passing through a point of this region not on $l$ crosses the subarc $AB$ with decreasing $t$ and therefore crosses the subarc $A'B'$ of $l$ with increasing $t$ (the regions bounded between $C_1$ and $C_2$ are hatched in Figures 51 and 52).*

Proof. We start with the first assertion of the lemma. Suppose that it is false, so that the regions $\Gamma_1$ and $\Gamma_2$ are mutually exterior (Figure 53).

Let $\Gamma^*$ be a region whose boundary contains both curves $C_1$ and $C_2$ and no other point of the region $G$ in which the dynamic system is defined. The region $G \smallsetminus (\Gamma_1 \cup \Gamma_2)$ meets these conditions. Apart from $C_1$ and $C_2$, the boundary of this region obviously contains the boundary of $G$. Let $\lambda$ be the union of the subarcs $AB$ and $A'B'$ of $l$; this is clearly an arc without contact.

---

* [Employing a slight abuse of language, we shall often speak of (any number of) curves "lying inside one another," meaning that of any two of these curves, one lies inside the other.]

One endpoint of $\lambda$ is either $A$ (as in Figure 48) or $A'$, the other is either $B$ or $B'$.

In view of Lemma 11 and our assumption concerning the regions $\Gamma_1$ and $\Gamma_2$, the following statements are readily verified:

a) All points on $\lambda$ not on the subarc $AA'$ lie outside $C_1$, and all its points not on the subarc $BB'$ lie outside $C_2$. Consequently, every point of $\lambda$ is either a boundary point or an interior point of $\Gamma^*$.

b) If a path crosses the arc $\lambda$ at $t = t_0$, at a point $M(s)$ between $A$ and $B$ (so that $a < s < b$), all points of this path corresponding to parameter values $t_0 < t < \chi(s)$ lie in the region $\Gamma^*$.

Let $W^*$ denote the (open) set of all points $M(x, y)$ with coordinates

$$x = \varphi(t - t_0, f(s), g(s)), \quad y = \psi(t - t_0, f(s), g(s)),$$

where $t_0 < t < \chi(s)$ with $a < s < b$. (It is obvious that $W^* \subset \Gamma^*$, and $W^*$ has no points in common with $l$ and, therefore, with $\lambda$.) By Lemma 8, these equations define a regular mapping of the region $R$ defined on the $(s, t)$-plane by

$$t_0 < t < \chi(s), \quad a < s < b,$$

onto the set* $W^*$. Consequently (see Appendix, §6), $W^*$ is a region and its boundary is the image of the boundary of $R$. Now the boundary of $W^*$ is the union of the arc $AA'$ on $L_A$, the arc $BB'$ on $L_B$ and the subarc $\lambda$ of $l$, in other words, of the closed curves $C_1$ and $C_2$ and the arc $\lambda$. This is impossible, for if the regions $\Gamma_1$ and $\Gamma_2$ are mutually exterior the curves $C_1$ and $C_2$ together with the arc $\lambda$ "connecting" them cannot form the boundary of any region contained entirely in $G$. (Any such region, con-sisting of points of $\Gamma^*$ not on the arc $\lambda$, would necessarily contain boundary points of $G$.)

This contradiction proves the first part of the lemma, i.e., the regions $\Gamma_1$ and $\Gamma_2$ lie inside one another. Hence there exists a region $\Gamma$ whose boundary consists only of points of $C_1$ and $C_2$.

To prove the second part, we note first that by Lemma 11 all points of the arc $\lambda$ are either in the boundary of $\Gamma$ or in $\Gamma$ itself. Therefore $\Gamma \backslash \lambda$ is a region whose boundary is the union of the curves $C_1, C_2$ and the arc $\lambda$. As demonstrated above, this is precisely the boundary of the region $W^*$, and so these regions must coincide. But then (see Appendix, §1.7) the regions $\Gamma \backslash \lambda$ and $W$ coincide. And in view of the definition of $W^*$ this implies the truth of the second part. Q.E.D.

Remark 1. Let $A$ be a fixed point and let $B$ vary along the arc $l$. Then for any $\varepsilon > 0$ there exists $\delta > 0$ such that if $B \in U_\delta(A)$ the region $\Gamma$ lies entirely in the $\varepsilon$-neighborhood of the arc $AA'$ of $L_A$. This follows from Remark 2 to Lemma 10 and Lemma 14.

Remark 2. Let $L_0$ be a closed path, $A_0$ a point on $L_0$, such that there are closed paths other than $L_0$ through points arbitrarily close to $A_0$. Then there exists $\delta > 0$ such that all closed paths passing through points of $U_\delta(A_0)$ lie inside one another. This follows directly from Lemma 4 and

---

\* Note that the corresponding mapping of $R$ onto $W$ in Lemma 8 is also single-valued on the boundary, while in the present case this is not true.

Lemma 14. Indeed, letting $l$ be an arc without contact through $A_0$ and using Lemmas 4 and 5, we can always choose $\delta$ in such a way that all paths passing through the points of this neighborhood cross the arc twice. And then, by virtue of Lemma 14, the regions bounded between any two closed paths through points of $U_\delta (A_0)$ lie inside one another, and this proves our assertion.

**10. CYCLE WITHOUT CONTACT.** Let $C$ be a smooth simple closed curve in the region $G$, $M$ any point on $C$. As in the case of a simple smooth arc $l$, we shall say that the curve $C$ has or does not have contact (with the paths of system (I)) at the point $M$ according as $C$ is or is not tangent to a path of system (I) at this point.

A smooth simple closed curve $C$ is called a cycle without contact of a dynamic system (I) if a) there are no equilibrium states on $C$, b) $C$ does not have contact at any of its points. If $C$ is a cycle without contact, given by parametric equations

$$x = f_0 (s), \quad y = g_0 (s),$$

where $s \in [s_0, s_0 + \tau]$, then $f_0 (s)$ and $g_0 (s)$ are functions of class $C_1$ on this interval (because $C$ is a smooth closed curve) and

$$f_0 (s_0) = f_0 (s_0 + \tau), \quad g_0 (s_0) = g_0 (s_0 + \tau),$$
$$f_0' (s_0) = f_0' (s_0 + \tau), \quad g_0' (s_0) = g_0' (s_0 + \tau).$$

A cycle without contact $C$ satisfies the following analogs of conditions (4) and (4') (§3.1):

$$P^2 (f_0 (s), \ g_0 (s)) + Q^2 (f_0 (s), \ g_0 (s)) \neq 0, \tag{30}$$

$$\Delta (s) = \begin{vmatrix} P (f_0 (s), \ g_0 (s)) \ f_0' (s) \\ Q (f_0 (s), \ g_0 (s)) \ g_0' (s) \end{vmatrix} \neq 0 \tag{30'}$$

for all $s \in [s_0, s_0 + \tau]$.

If a cycle without contact is defined implicitly,

$$\Phi (x, y) = 0,$$

then at all points of the cycle we have conditions

$$P^2 (x, y) + Q^2 (x, y) \neq 0, \tag{31}$$
$$\Phi_x' (x, y) P (x, y) + \Phi_y' (x, y) Q (x, y) \neq 0, \tag{31'}$$

analogous to conditions (7) and (8) for an arc without contact.

Just as in the case of an arc without contact (see §3.1), the angle between a cycle without contact $C$ and any path crossing it does not vanish, and therefore has the same sign at all points of $C$ (which obviously depends on the sign of $\Delta (s)$). Hence it follows that, with increasing $t$, paths which cross the cycle $C$ at $t = t_0$ either all enter it or all leave it (see Appendix, §6). As a consequence, we see in particular that *any path crossing a cycle without contact has exactly one point in common with it* (Figure 54).

FIGURE 54.

Every arc of a cycle without contact is of course an arc without contact. We encountered cycles without contact in Examples 7 and 9 ($1.14).

**11. FAMILY OF CYCLES WITHOUT CONTACT. PATHS ENTERING REGION FILLED BY CYCLES WITHOUT CONTACT.** Let $C_0$ and $C_1$ be two cycles without contact, $C_0$ lying inside $C_1$. Let $\Gamma$ denote the annular region bounded by $C_0$ and $C_1$, and suppose that $\Gamma$ is covered by cycles without contact having the following properties:

a) there is exactly one cycle $C$ through each point of $\Gamma$; b) the cycles without contact lie inside one another, $C_0$ lies inside each cycle, and each cycle lies inside $C_1$.

Such a family of curves in $\Gamma$ will be called a family of cycles without contact.

Suppose that a family of cycles without contact is defined by an equation

$$F(x, y) = \gamma,$$

where $F(x, y)$ is a function of class $C_k$, $k \geqslant 1$, and the cycles lying in the region $\Gamma$ correspond to values of $\gamma$ between $\gamma_0$ and $\gamma_1$, $\gamma_0 < \gamma_1$. Then it is obvious that the expression

$$F'_x(x, y) P(x, y) + F'_y(x, y) Q(x, y) \tag{32}$$

does not vanish at any point of $\Gamma$, and so is of constant sign. This means that with increasing $t$ the paths either all enter each cycle without contact or all leave each cycle without contact. Assuming that the paths cross each of these cycles inward, we now prove the following.

*L e m m a 15. With increasing $t$, every path crossing the cycle without contact $C_1$ crosses all cycles of the family in $\Gamma$ and leaves $\Gamma$ through the cycle $C_0$.*

P r o o f. Let $L$ be a path crossing the cycle $C_1$ at $t = t_1$. To begin with, we note that if $L$ crosses some cycle $C^*$ of the family at $t = t^*$, it obviously crosses all cycles containing $C$ (at times $t < t^*$).

Suppose that $L$ does not cross all the cycles without contact in $\Gamma$. Divide these cycles into two types, according to whether $L$ crosses them or not. All cycles of the second type lie inside all cycles of the first. Thus there necessarily exists a cycle $C$ which is either the "last" cycle of the first type or the "first" cycle not of the first type, i.e., the path $L$ crosses all cycles containing $C^*$ but no cycle contained inside $C^*$. But there cannot exist a last cycle of the first type. Indeed, on any path crossing a cycle $C^*$ at some $t = \tau$, all points corresponding to $t > \tau$ sufficiently close to $\tau$ lie inside $C^*$, so that the path must cross cycles without contact inside $C^*$ for $t > \tau$.

Suppose, therefore, that $C^*$ is the first cycle not of the first type. Since by the definition of $C^*$ the path $L$ crosses all cycles containing $C^*$, there are points on $L$ arbitrarily close to $C^*$ and so $C^*$ contains at least one point of accumulation of these points. Let $P$ be this point and $l$ some arc of the cycle $C^*$ containing $P$ as an interior point. Then by Lemma 4, $l$ is obviously an arc without contact and the path $L$ must cross $l$, so that it also crosses $C^*$. Thus $C^*$ is a cycle of the first type, contrary to assumption. This contradiction completes the proof.

Suitably modified, the lemma remains valid if the path enters $C_0$ with decreasing $t$.

**Lemma 16.** *Let $C_1$ and $C_2$ be two cycles without contact, of which one, $C_2$ say, lies inside the other, and suppose that all paths that cross one cycle, $C_1$ say, at $t = t_0$ cross the cycle $C_2$ at some $t > t_0$.*

*Then there is a path through each point of the annular region between $C_1$ and $C_2$ which crosses $C_2$ with increasing $t$ and $C_1$ with decreasing $t$.*

P r o o f.  Let $P_1$ and $P_2$ be arbitrary points on $C_1$, $L_1$ and $L_2$ the paths passing through these points, $Q_1$ and $Q_2$ the points at which they cut the cycle $C_2$ (Figure 55). The arcs $P_1Q_1$ and $P_2Q_2$ of $L_1$ and $L_2$ clearly divide the annular region $R$ between $C_1$ and $C_2$ into two elementary quadrangles.

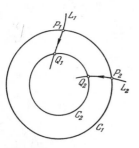

The proof now follows by applying Lemma 10 to each of these quadrangles.  Q.E.D.

In the sequel, we shall often deal with families of cycles without contact not in a region bounded by two cycles without contact, as in the present case, but in a doubly connected domain whose boundary is the union of a cycle without contact and an equilibrium state in its interior.  An example is the family of circles without contact

FIGURE 55.

$$x^2 + y^2 = c, \quad 0 < c \leqslant 1,$$

considered in Example 7 of §1.14.

**12. SINGLE-CROSSING CYCLE.**  In certain cases, cycles without contact may be replaced by "generalized cycles without contact" or "single-crossing cycle." We shall say that a simple (not necessarily smooth) closed curve $C$ is a  s i n g l e - c r o s s i n g  c y c l e  for paths of system (I) if a) there are no equilibrium states on $C$, b) for any path passing through a point of $C$ at $t = t_0$, there exists $\delta > 0$ such that points on the path corresponding to $0 < t - t_0 < \delta$ $(0 < t_0 - t < \delta)$ lie inside (outside) $C$, or conversely.  In particular, a single-crossing cycle is a smooth simple closed curve which is not a cycle without contact is a single-crossing cycle if at any of its points it has either no contact or contact of  e v e n  order with paths of the system* (Figure 56). Obviously, if a single-crossing cycle is smooth and

$$F(x, y) = 0$$

is its equation, then the expression

$$F'_x(x, y)P(x, y) + F'_y(x, y)Q(x, y)$$

may vanish at points of the cycle.  Later, we shall sometimes consider families of single-crossing cycles instead of families of cycles with contact;

---

* Paths cannot be tangent to a smooth closed curve at all its points, since then the curve itself would be an integral curve. This is impossible, since by assumption the existence-uniqueness theorem is valid for our dynamic system.

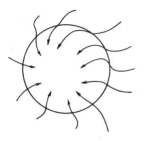

FIGURE 56.

from a certain standpoint, they possess the same properties as families of cycles without contact.

**13. DIFFERENTIATION OF FUNCTION ALONG PATHS OF SYSTEM (I).** Conditions (4'), (30'), (31'), which define arcs and cycles without contact, and also the conditions defining a family of cycles without contact and a family of single-crossing cycles, may be derived using a rather different approach, based on the concept of d i f f e r e n t i a - t i o n  a l o n g  p a t h s  o f  s y s t e m  (I). Let

$$z = F(x, y)$$

be some function of class $C_k$ $(k > 1)$ (or analytic) defined in a region $G_1$ which is either the domain of definition $G$ of the dynamic system or a subregion thereof.

Let $x = \varphi(t)$, $y = \psi(t)$ be a motion on some path of system (I), such that the point corresponding to a fixed $t = t_0$, i.e., the point with coordinates $x_0 = \varphi(t_0)$, $y_0 = \psi(t_0)$, is in $G_1$.

Consider the function

$$z(t) = F(\varphi(t), \psi(t)).$$

This function describes the variation of $F(x, y)$ along the path $L$. At $t = t_0$, we have

$$\left(\frac{dz}{dt}\right)_{t=t_0} = F'_x(x_0, y_0) P(x_0, y_0) + F'_y(x_0, y_0) Q(x_0, y_0). \tag{33}$$

The derivative $\left(\frac{dz(t)}{dt}\right)_{t=t_0}$ is clearly independent of the choice of the motion on the path, depending only on the point $x_0$, $y_0$.

The expression

$$\frac{dz}{dt} = \frac{d}{dt}[F(x, y)] = \frac{\partial F(x, y)}{\partial x} P(x, y) + \frac{\partial F(x, y)}{\partial y} Q(x, y)$$

is known as the d e r i v a t i v e  o f  $F(x, y)$  a l o n g  p a t h s  o f  s y s t e m  (I) [or, when there is no danger of confusion, the p a t h  d e r i v a t i v e  of the function].*

If

$$F(x, y) = 0$$

is the equation of a simple arc $l$, which is an arc without contact, then

$$\frac{dF(\varphi, \psi)}{dt} = F'_x P + F'_y Q \neq 0$$

* [Literally translated, the Russian term is "derivative of the function by virtue of system (I)."]

81

at all points of the arc. But if

$$\frac{dF\,(\varphi,\,\psi)}{dt} = F'_x P + F'_y Q = 0$$

at some point of $l$, then $l$ has contact with a path at this point.

Similar reasoning is valid when $F\,(x,\,y) = 0$ is a simple closed curve $C_1$. In particular, if $C_1$ is not a cycle without contact but a single-crossing cycle, then at some of its points

$$\frac{dF\,(x,\,y)}{dt} = F'_x P + F'_y Q = 0,$$

i. e., $C$ has contact of even order with a path at these points. If the dynamic system and the function $F\,(x,\,y)$ are of class $C_k$, $k \geqslant 2$, then, by the definition of contact of even order, we have at these points

$$\frac{d^2F\,(\varphi,\,\psi)}{dt^2} = 0.$$

Assume now that all the curves

$$F\,(x,\,y) = C\,,$$

where $C$ varies over some interval $C_1 \leqslant C \leqslant C_2$, are simple smooth closed curves, lying inside one another, curves for smaller values of $C$ lying inside curves with larger values of $C$.

The condition

$$\frac{dF\,(x,\,y)}{dt} = F'_x P + F'_y Q \neq 0$$

is obviously necessary and sufficient for the family $F\,(x,\,y) = C$ to be a family of cycles without contact. If moreover $\frac{dF\,(x,\,y)}{dt} < 0$ at all points of the closed region $\Gamma$ bounded by the cycles

$$F\,(x,\,y) = C_1, \qquad F\,(x,\,y) = C_2,$$

this means that every path through a point of $\Gamma$ crosses each cycle of the family inward with increasing $t$.

Conversely, if $\frac{dF}{dt} > 0$ in $\Gamma$, each path crosses the cycles of the family outward with increasing $t$.

If $F\,(x,\,y) = C$ is a family of single-crossing cycles, not necessarily cycles without contact, then the expression

$$F'_x P + F'_y Q$$

does not change sign, but it may vanish.

**14. CYCLE WITHOUT CONTACT BETWEEN TWO SUCCESSIVE TURNS OF A PATH CROSSING AN ARC WITHOUT CONTACT.** Let $L_0$ be a path having more than two points in common with an arc without contact. Let $M_1\,(s_1)$, $M_2\,(s_2)$ and $M_3\,(s_3)$

be $t$-successive common points. By Corollary 2 to Lemma 11, they are also $s$-successive, so that either $s_1 < s_2 < s_3$ or $s_1 > s_2 > s_3$. To fix ideas, we shall assume that $s_1 < s_2 < s_3$ (Figure 44). Assume moreover that every path crossing the subarc $M_1M_2$ of $l$ at $t = t_0$ at a point $M(s)$ crosses the subarc $M_2M_3$ of $l$ at $t = \chi(s) > t_0$ ($s_1 \leqslant s \leqslant s_2$) and has no points in common with $l$ for $t_0 < t < \chi(s)$ (the path that passes through $M_1$ at $t = t_0$ and also through $M_2$ is of course $L_0$). Let $C_1$ be the closed simple curve formed by the turn $M_1M_2$ of $L_0$ and the subarc $M_1M_2$ of $l$, and $C_2$ the analogous curve formed by the turn $M_2M_3$ of $L_0$ and the subarc $M_2M_3$ of $l$. To fix ideas, we shall assume that $C_1$ (except for its point $M_2$) lies inside $C_2$.

Let $\bar{\Gamma}$ be the closed region (see Lemma 14) whose points $M(x, y)$ have coordinates (9)

$$x = \varphi(t-t_0, f(s), g(s)) = \Phi(t, s),$$
$$y = \psi(t-t_0, f(s), g(s)) = \Psi(t, s),$$

where $s_1 \leqslant s \leqslant s_2$, $t_0 \leqslant t \leqslant \chi(s)$. Equations (9) define a mapping of the closed region $\bar{R}$ defined on the $(t, s)$-plane by

$$s_1 \leqslant s \leqslant s_2, \quad t_0 \leqslant t \leqslant \chi(s), \tag{34}$$

onto the closed region $\bar{\Gamma}$ (see Lemma 14). This mapping is clearly regular at all points of $\bar{R}$ except two, $(\chi(s_1), s_1)$ and $(t_0, s_2)$, which are both mapped onto the point $M_2$ on the phase plane.

*L e m m a  17. One can always draw through $M_2$ a cycle without contact lying entirely within $\bar{\Gamma}$ and crossing all paths which pass through $\bar{\Gamma}$ (i.e., all paths which cross the subarc $M_1M_2$, hence also the subarc $M_2M_3$, of the arc $l$).*

P r o o f. Let $k < 0$ and $r < 0$ be two arbitrary negative numbers. By Remark 3 to Lemma 10 (see also §6 of the Appendix), there exists a function $\lambda(s)$ of class $C_1$, defined on $s_1 \leqslant s \leqslant s_2$, such that

a) $\qquad \lambda(s_1) = \chi(s_1), \quad \lambda(s_2) = t_0; \tag{35}$

b) for all $s$ in the interval $(s_1, s_2)$,

$$t_0 < \lambda(s) < \chi(s);$$

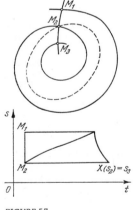

FIGURE 57.

c) $\qquad \lambda'(s_2) = k, \quad \lambda'(s_1) = \chi(s_1) + r. \tag{36}$

Consider the curve defined on the $(t, s)$-plane by

$$t = \lambda(s), \quad s_1 \leqslant s \leqslant s_2.$$

This curve (Figure 57) lies entirely within the region $R$ defined by (34), and is never tangent to the straight lines $s = \text{const}$ (since the function $\lambda(s)$ is differentiable).*

The mapping (9) takes the curve $t = \lambda(s)$ onto some curve $C$ in the $(x, y)$-plane, described by parametric equations

$$x = \Phi(\lambda(s), s) = \varphi(\lambda(s) - t_0, f(s), g(s)),$$
$$y = \Psi(\lambda(s), s) = \psi(\lambda(s) - t_0, f(s), g(s)). \tag{37}$$

$C$ is obviously a simple closed curve through $M_2$, lying entirely within $\overline{\Gamma}$, and having exactly one point in common with every path passing through $\overline{\Gamma}$. It follows from the properties of the functions $\varphi, \psi, f, g$ that $C$ is smooth at all its points, with the possible exception of $M_2$. Since the curve $t = \lambda(s)$ is never tangent to the lines $s = \text{const}$ at any point, the regularity of the mapping (9) implies (see Appendix, §6) that $C$ is not tangent to a path at any point, except possibly $M_2$.

We now show that for arbitrary $k < 0$ we can choose $r < 0$ in such a way that $C$ is smooth at $M_2$. Clearly, if this can be done the curve will not be tangent to the path $L_0$ at $M_2$ either, and will therefore be the required cycle without contact. A necessary and sufficient condition for the curve (37) to be smooth at $M_2$ is

$$\left. \frac{\dfrac{d\Psi(\lambda(s), s)}{ds}}{\dfrac{d\Phi(\lambda(s), s)}{ds}} \right|_{s=s_1+0} = \left. \frac{\dfrac{d\Psi(\lambda(s), s)}{ds}}{\dfrac{d\Phi(\lambda(s), s)}{ds}} \right|_{s=s_2-0}, \tag{38}$$

which we can rewrite, using (35) and (36), as

$$\frac{\Psi_t'(\chi(s_1), s_1)[\chi'(s_1) + r] + \Psi_s'(\chi(s_1), s_1)}{\Phi_t'(\chi(s_1), s_1)[\chi'(s_1) + r] + \Phi_s'(\chi(s_1), s_1)} = \frac{\Psi_t'(t_0, s_2)k + \Psi_s'(t_0, s_2)}{\Phi_t'(t_0, s_2)k + \Phi_s'(t_0, s_2)}. \tag{39}$$

Now the parametric equations of the subarc $M_2M_3$ on $l$ can be written as either

$$x = \Phi(t_0, s), \quad y = \Psi(t_0, s) \quad (s_2 \leqslant s \leqslant s_3), \tag{40}$$

or

$$x = \Phi(\chi(s), s), \quad y = \Psi(\chi(s), s) \quad (s_1 \leqslant s \leqslant s_2). \tag{41}$$

In the first case, $M_2$ corresponds to the parameter value $s_2$, in the second to $s_1$.

Calculating the slope of the tangent to $l$ at $M_2$ first by (41) and then by (40), and equating the resultant expressions, we obtain

$$\frac{\Psi_t'(\chi(s_1), s_1)\chi'(s_1) + \Psi_s'(\chi(s_1), s_1)}{\Phi_t'(\chi(s_1), s_1)\chi'(s_1) + \Phi_s'(\chi(s_1), s_1)} = \frac{\Psi_s'(t_0, s_2)}{\Phi_s'(t_0, s_2)}. \tag{42}$$

---

* Were the curve $t = \lambda(s)$ tangent to some straight line $s = s_0$, it would follow that $\lambda'(s) \to \infty$ as $s \to s_0$, so that $\lambda(s)$ would not be differentiable for all $s$.

Next, the parametric equations of the path $L_0$ can be written as either

$$x = \Phi(t, s_2), \quad y = \Psi(t, s_2),$$

(43)

or

$$x = \Phi(t, s_1), \quad y = \Psi(t, s_1).$$

(44)

In the first case, $M_2$ corresponds to the parameter value $t_0$, in the second to $t = \chi(s_1)$. Calculating the slope of the tangent to $L_0$ at $M_2$ in two ways, we obtain

$$\frac{\Psi_t'(\chi(s_1), s_1)}{\Phi_t'(\chi(s_1), s_1)} = \frac{\Psi_t'(t_0, s_2)}{\Phi_t'(t_0, s_2)}.$$

(45)

One now readily shows by way of (42) and (45) that (39) is equivalent to

$$r = \frac{\begin{vmatrix} \Psi_t'(t_0, s_2) & \Psi_s'(\chi(s_1), s_1) \\ \Phi_t'(t_0, s_2) & \Phi_s'(\chi(s_1), s_1) \end{vmatrix}}{\begin{vmatrix} \Psi_t'(\chi(s_1), s_1) & \Psi_s'(t_0, s_2) \\ \Phi_t'(\chi(s_1), s_1) & \Phi_s'(t_0, s_2) \end{vmatrix}} k,$$

which, in view of (42) and (45), can be written

$$r = \frac{\begin{vmatrix} \Psi_t'(t_0, s_2) & \Psi_t'(\chi(s_1), s_1) \chi'(s_1) + \Psi_s'(\chi(s_1), s_1) \\ \Phi_t'(t_0, s_2) & \Phi_t'(\chi(s_1), s_1) \chi(s_1) + \Phi_s'(\chi(s_1), s_1) \end{vmatrix}}{\begin{vmatrix} \Psi_t'(\chi(s_1), s_1) & \Psi_s'(t_0, s_2) \\ \Phi_t'(\chi(s_1), s_1) & \Phi_s'(t_0, s_2) \end{vmatrix}} k.$$

(46)

We claim that this number $r$ is negative. Indeed, the elements of the first column of each determinant in (46) are components of a vector tangent to $L_0$ at $M_2$ and pointing in the direction of increasing $t$. Thus the first columns in both determinants differ by a positive constant factor. The same holds true for the second columns of the determinants (their elements are the components of a vector tangent to $l$ at $M_2$ and pointing in the direction of increasing $s$).

Hence it follows that the determinants in the numerator and denominator of (46) have the same signs, and since $k < 0$ this shows that $r < 0$. Thus, if $r$ is the negative number defined by (46), the curve $C$ is a cycle without contact. Q.E.D.

## §4. LIMIT POINTS AND SETS. FUNDAMENTAL PROPERTIES OF PATHS

**1. LIMIT POINTS OF A SEMIPATH AND A PATH.** As mentioned in the introduction to this chapter, §4 presents the main subject matter of the chapter — investigation of single paths.

Let

$$\frac{dx}{dt} = P(x, y), \qquad \frac{dy}{dt} = Q(x, y) \qquad\qquad (I)$$

be a dynamic system defined in a plane region $G$.

In this section we shall study single semipaths or whole paths of this system lying in some bounded closed region $\bar{G}_1$, $\bar{G}_1 \subset G$. We shall call such semipaths or whole paths b o u n d e d. Obviously, any semipath taken from a bounded semipath or bounded whole path is also bounded. If $G$ is an unbounded (bounded) region, a bounded semipath or bounded whole path has no points arbitrarily close to infinity (arbitrarily close to the boundary of $G$).

Throughout this section, the terms s e m i p a t h and p a t h will usually mean a b o u n d e d semipath and a b o u n d e d whole path, unless stipulated otherwise. In addition, we shall also consider single paths or semipaths of dynamic systems on the sphere. Recall that every path on the sphere is a whole path (§2, Theorem 7).

*D e f i n i t i o n  II. A point $M$ is called a limit point of a positive (negative) semipath $L^+$ $(L^-)$ if for every $\varepsilon > 0$ and every $T > 0$ the neighborhood $U_\varepsilon(M)$ contains at least one point of $L^+$ $(L^-)$ (which may or may not be distinct from $M$) which corresponds under any motion on the path to a time $t > T$ $(t < -T)$.*

This definition remains meaningful without any modification for semipaths on surfaces of arbitrary genus, and also for dynamic systems in spaces of arbitrary dimension.

By Definition II, if $M$ is a limit point of a positive semipath $L^+$, then either a) there exists a sequence of d i s t i n c t points $M_k$ on $L^+$, corresponding to times $t_k$ ($k = 1, 2, \ldots$), such that $M_k \to M$ and $t_k \to \infty$ as $k \to \infty$, or b) the point $M \in L^+$ corresponds to an infinite set of $t = t_k$ such that $t_k \to +\infty$ as $k \to \infty$. A similar statement holds for a limit point of a negative semipath.

Any two positive (semipaths) taken from the same path have the same limit points. Since a bounded closed region (or a sphere) is compact, the semipaths considered here (bounded on the plane or arbitrary on the sphere) must have at least one limit point. If a semipath lies entirely within $\bar{G}_1 \subset G$, its limit points must also be in $\bar{G}_1$.

A limit point of a semipath may or may not belong to the semipath.

Let $L^+$ be a positive semipath, $M^*$ any point of accumulation of a sequence of points on $L^+$ which is not itself on the semipath. We claim that $M^*$ is a limit point for $L^+$. Indeed, let $M_k$ be a sequence of points on $L^+$ converging to $M^*$, and let $M_k$ correspond to the parameter value $t_k$ ($k = 1, 2, \ldots$). If the sequence $t_k$ is not bounded above, it is clear that $M^*$ is a limit point of $L^+$. If the sequence $t_k$ is bounded, we may assume that it converges to some number $t^*$ (otherwise we need only consider a convergent subsequence). But then the points $M_k$ converge to the point on $L^+$ corresponding to $t = t^*$, and they cannot converge to $M^*$ which by assumption is not on $L^+$. Thus the sequence $t_k$ cannot be bounded. The same holds true for a negative semipath $L^-$.

In the sequel, $M(t)$ will denote the point on the path corresponding to parameter value (time) $t$. In addition, all proofs of statements concerning properties of semipaths (and sometimes the statements themselves) will be phrased for positive semipaths, and we shall not take the trouble to repeat that they remain valid for negative semipaths.

We now consider a whole (bounded) path $L$ of the dynamic system (I) or a path of a dynamic system on the sphere.

*Definition III. A point $M$ is called a limit point of the path $L$ if it is a limit point for a positive semipath $L^+$ or a negative semipath $L^-$ taken from $L$. In the first case $M$ is also called an $\omega$-limit point, in the second case an $\alpha$-limit point of $L$.*

**2. EXAMPLES OF LIMIT POINTS.** A few examples will clarify the meaning of the new concept.

Any equilibrium state obviously has a unique limit point — the equilibrium state itself. In this case the limit point is simultaneously an $\alpha$- and $\omega$-limit point.

Now let $L$ be a closed path, $\theta_0$ the period of a motion on the path. Any point $M$ on $L$ corresponding to some time $t_0$ also corresponds to an infinite sequence

$$\ldots, t_0 - n\theta_0, \ldots, t_0 - 2\theta_0, \quad t_0 - \theta_0, t_0, t_0 + \theta_0, \quad t_0 + 2\theta_0, \ldots, t_0 + n\theta_0, \ldots,$$

where $n \to +\infty$. But then, by Definitions II and III, it is both an $\alpha$-limit point and an $\omega$-limit point of the path $L$. Thus all points of a closed path are $\alpha$- and $\omega$-limit points. It is obvious that a closed path has no other limit points.

If a path $L$ has at least one limit point not on $L$, it is called a s e l f - l i m i t i n g p a t h. Thus:

*Equilibrium states and closed paths are self-limiting paths.*

Now consider the examples of §1.14. Semipaths which tend to an equilibrium state (node, focus, or saddle point; Examples 3, 4, 5)* have a unique limit point — the equilibrium state. In Example 4, all the spiral-shaped paths lying inside the closed path (limit cycle) $x^2 + y^2 = 1$ have a unique $\alpha$-limit point — the equilibrium state $O$; the $\omega$-limit points of these paths are all points of the limit cycle.

The paths lying outside the limit cycle have no $\alpha$-limit points (their points escape to infinity in a finite time); their $\omega$-limit points are again the points of the limit cycle.

The path in Example 8 which is part of the loop (not the equilibrium state) has one limit point — the equilibrium state, which is both an $\alpha$- and an $\omega$-limit point. In Example 11, when $\mu > 0$ (Figure 22a) the $\omega$-limit points of the paths inside the loop are the equilibrium states (foci), and their $\alpha$-limit points are all points of the loop. The limit points of the paths in Example 11 lying outside the "figure of eight" are also easily determined.

**3. FUNDAMENTAL PROPERTIES OF THE SET OF LIMIT POINTS.**

*L e m m a 1. If $M_0$ is a limit point of a semipath $L^+$, then all points of the path passing through $M_0$ are limit points for $L^+$.*

P r o o f. The lemma is obvious if $M_0$ is an equilibrium state of the dynamic system. Assume therefore that $M_0$ is not an equilibrium state, and let $L_0$ denote the path through $M_0$. It will suffice to prove that any point $M^*$ on $L_0$ other than $M_0$ is a limit point of $L^+$.

---

* The reader should remember that semipaths not coinciding with an equilibrium state can tend to it only as $|t| \to +\infty$.

Let $M_0$ correspond on the path to $t = t_0$ and $M^*$ to $t_0 + \tau = t^*$ ($\tau$ may be either positive or negative). Consider any other motion on $L_0$, under which $M_0$ corresponds to a parameter value $\tilde{t}_0$; then $\tilde{t}_0 + \tau$ (the same $\tau$ as before) marks the point $M^*$. Thus it follows from the theorem on continuity of solutions with respect to initial values that for any $\varepsilon > 0$ and a suitably chosen $\delta > 0$ ($\delta = \delta(\varepsilon)$) any path passing at $t = t'$ (where $t'$ is an arbitrary fixed number) through some point $M' \in U_\delta(M_0)$ will pass at $t = t' + \tau$ through a point of $U_\varepsilon(M^*)$. Since $M_0$ is a limit point of the semipath $L^+$, there exists an infinite sequence of points $M_n(t_n)$ on $L^+$ such that $t_n \to +\infty$ and $M_n(t_n) \in U_\delta(M_0)$ ($n = 1, 2, 3, \ldots$). By the choice of $\delta$, the sequence of points $M_n'(t_n + \tau)$ on the path $L$ which contains $L^+$ possesses the following properties: a) $M_n'(t_n + \tau) \in U_\varepsilon(M')$; b) $t_n + \tau \to \infty$; c) $M_n'(t_n + \tau) \in L^+$) for every $n$.* Since $\varepsilon > 0$ is arbitrary the existence of this sequence $M_n'$ implies that $M^*$ is a limit point of $L^+$. Q.E.D.

A path $L_0$ all of whose points are limit points of a semipath $L^{()}$ is called a **limit path** of $L^{()}$.

Lemma 1 may obviously be phrased as follows: any path through a limit point of a semipath $L^{()}$ is a limit path for $L^{()}$.

*Theorem 9. The set $K$ of all limit points of a semipath $L^+$ (or, what is the same, the set of all $\omega$-limit points of a path $L$) is a) closed, b) connected, c) the union of whole paths.*

P r o o f. Part a) follows directly from the definition of a limit point. In fact, let $M_0$ be a point of accumulation of the set $K$. Then for any $\varepsilon > 0$ the neighborhood $U_\varepsilon(M_0)$ contains limit points of $L^+$, so that there are points of the semipath $L^+$ itself corresponding to arbitrarily large values of $t$. But this means that $M_0$ is a limit point of $L^+$, i.e., $M_0 \in K$.

We now prove part b). Suppose that $K$ is not connected. Since it is closed, it may be expressed as the union of two closed disjoint nonempty sets $A$ and $B$ (see Appendix, §1).

Let $\varrho$ denote the distance between the sets $A$ and $B$; clearly, $\varrho > 0$. Set $\delta = \varrho/4$ and consider the neighborhoods $U_\delta(A)$ and $U_\delta(B)$. These neighborhoods are disjoint. Since the elements of $A$ and $B$ are limit points of $L^+$, one readily sees that there exist two infinite sequences $t_n$ and $t_n'$ ($n = 1, 2, \ldots$) satisfying the following conditions:

1) $t_n \to +\infty$, $t_n' \to +\infty$ as $n \to \infty$.

2) The points on $L^+$ corresponding to $t_n$ lie in $U_\delta(A)$, those corresponding to $t_n'$ lie in $U_\delta(B)$.

3) $t_1 < t_1' < t_2 < t_2' < \ldots < t_n < t_n' < \ldots$

But then, for every $n = 1, 2, \ldots$ there exists $\tau_n$, $t_n < \tau_n < t_n'$ such that the point $M_n(\tau_n)$ of $L^+$ is in neither $U_\delta(A)$ nor $U_\delta(B)$.

Consider the set $\{M_n(\tau_n)\}$. Since $L^+$ is a bounded semipath or a semipath on the sphere, this set has at least one point of accumulation $M^*$, either outside or on the boundary of $U_\delta(A)$ and $U_\delta(B)$. $M^*$ is obviously a limit point of $L^+$. But this contradicts the assumption that $K$, which is a subset of the union of $U_\delta(A)$ and $U_\delta(B)$, contains all the limit points of $L^+$. This proves part b).

Part c) follows directly from Lemma 1. Q.E.D.

---

\* Under a suitable motion, the points on the semipath $L^+$ correspond to times $t > t_0$ (where $t_0$ is a fixed number). If $\tau > 0$ it is clear that all the points $M_n'(t_n + \tau)$ are on $L^+$. But if $\tau < 0$, then $M_n'(t_n + \tau)$ lies on $L^+$ only starting from some $t_n$ such that $t_n + \tau > t_0$.

Since $K$ is closed and connected, it is a c o n t i n u u m (see Appendix, §1).

The set $K$ of all limit points of a semipath $L^+$ is known as the limit set or limit continuum of $L^+$. In the case of a positive (negative) semipath the set is also known as the ω-limit (α-limit) set or continuum. Similarly, the set of all α (ω)-limit points of a path $L$ is called the α (ω)-limit continuum of the path $L$. We shall sometimes denote the limit continua of paths or semipaths by $K_\alpha$ and $K_\omega$, or $K_\alpha$ $(L)$ and $K_\omega$ $(L)$.

Instead of saying that "$K$ is the limit set of a semipath $L^{()}$ or the ω (α)-limit set of a path $L$," we shall sometimes say that the semipath $L^{()}$ t e n d s  t o  t h e  s e t $K$ or the path $L$ t e n d s  t o  t h e  s e t $K_\omega$ $(K_\alpha)$ as $t \to +\infty$ $(t \to -\infty)$.*

Note that Lemma 1 and Theorem 9 are valid not only for plane regions or spheres but also for paths on any closed manifold and for paths in a bounded region of a space of arbitrary dimension.

**4. PROPERTIES OF PATHS CHARACTERISTIC FOR DYNAMIC SYSTEMS ON THE PLANE OR THE SPHERE.** From now on, all propositions proved in this section depend essentially on Lemma 11 of §3 and are therefore valid only for a plane region or a sphere. The following theorem is true for any (not necessarily bounded) path of system (I).

*T h e o r e m  10.* a) *Any path of a system (I) defined in a plane region G is nowhere dense in G.* b) *Any path of a dynamic system on the sphere is nowhere dense on the sphere.*

P r o o f. We begin with part a). Suppose the assertion false, and let $L$ be a path of system (I) which is dense in some subregion $G_1$ of $G$. $L$ is of course not an equilibrium state. Let $M$ be an arbitrary point on $L$, $M \in G_1$, and let $l$ be an arc without contact through $M$, lying in $G_1$. All points of $l$, as points of $G_1$, are points of accumulation for the path $L$. But then it follows from Lemma 4 of §3 that points of $L$ are dense on the arc $l$. Let $M_1$ $(t_1)$ and $M_2$ $(t_2)$ be two $t$-successive points of $L$ on the arc $l$, so that $L$ does not cross $l$ for any $t$ between $t_1$ and $t_2$. Such $t$-successive points exist by virtue of Lemma 1 of §3. It follows from Lemma 11 of §3 that there are no points of $L$ between $M_1$ and $M_2$ on the arc $l$. But this contradicts the fact that points of the path are dense on $l$, proving part a).

Part b), relating to paths on the sphere, is proved in exactly the same way as part a).** Q.E.D.

R e m a r k. One result of Theorem 10 is to reduce the investigation of paths or semipaths of the sphere $S$ to investigation of bounded paths or semipaths on the plane. Indeed, let $L$ be any path on the sphere. Since $L$ is nowhere dense on the sphere, there exists a point on the sphere which is neither on $L$ nor a point of accumulation of points on $L$. Thus there exists a region $G_1$ containing no point of $L$. Taking any point $N$ of $G_1$ as the center of stereographic projection, we project the closed region $H = S^2 \setminus G_1$ onto a bounded closed plane region $\tilde{H}$, and the path $L$ onto a path $\tilde{L}$ contained together with its closure in $\tilde{H}$. This means that investigation of the path $L$ on the sphere reduces to investigation of the bounded path $\tilde{L}$ on the plane.

All further propositions are valid for bounded semipaths and paths.

---

\* We have already used the expression "a semipath tends to an equilibrium state" in consideration of the examples.

\*\* Theorem 10 fails to hold for dynamic systems on surfaces [of higher genus], such as the torus. A classical example of a dynamic system on the torus whose paths are dense on the torus is given, e.g., in /23/.

*Lemma 2. Let $M^*$ be a limit point of a nonclosed semipath $L^+$, not an equilibrium state. Let $l$ be an arc without contact passing through $M^*$, and assume that $M^*$ is not an endpoint of $l$. Then the semipath $L^+$ intersects $l$ in a countable sequence of points tending to $M^*$. If these points are indexed in order of increasing parameter values $t$,*

$$M_1(t_1), M_2(t_2). \ldots, M_n(t_n) \rightarrow M^*$$

*(where $\{t_n\}$ is a monotone increasing sequence, $t_n \rightarrow \infty$), then the corresponding sequence $s_n$ of parameter values on the arc $l$ ($s_n$ corresponding to $M_n$) is also monotone, and $\lim s_n = s^*$, where $s^*$ is the $s$ corresponding to $M^*$.*

P r o o f. By the definition of a limit point, any neighborhood of $M^*$ contains points corresponding to arbitrarily large $t$. But then, by Lemma 4 of §3, it follows that there are also points on $l$ corresponding to arbitrarily large $t$ and arbitrarily close to $M^*$. It clearly follows from Lemma 2 of §3 that the set of all such points is countable. The rest of the lemma follows from Corollary 2 to Lemma 11 of §3.

C o r o l l a r y 1. An arc without contact $l$ can contain only one limit point of a semipath $L^+$. Otherwise the sequence $s_n$ could not be monotone.

C o r o l l a r y 2. The points of intersection of a semipath $L^+$ with an arc without contact $l$ lie to one side of the point $M^*$. This assertion also follows directly from the monotonicity of the sequence $s_n$.

The next theorem, whose proof uses Lemma 2, establishes one of the fundamental properties of paths of dynamic systems on the plane and on the sphere.

*Theorem 11. A nonclosed path $L_0$ which has at least one limit point other than an equilibrium state cannot be the limit path of any path.*

P r o o f. Let $L_0$ be a nonclosed path with a limit point $M^*$ which is not an equilibrium point. Let $l$ be an arc without contact through $M^*$. By Lemma 2, there are countably many points $M_n$ of $L_0$ on the arc $l$.

Now suppose that $L_0$ is a limit path of some semipath, $L^+$ say ($L^+$ is obviously part of a nonclosed path, which may or may not be $L_0$ itself). Then, by definition, all points of $L_0$, including all the $M_n$, are limit points of $L^+$. Thus the arc $l$ contains countably many limit points of the semipath $L^+$, but this contradicts Corollary 1 to Lemma 2. Q.E.D.

C o r o l l a r y. No point of a nonclosed path $L$ can be a limit point of $L$.

This follows at once from Theorem 11. An alternative formulation of this corollary is as follows:

*A nonclosed path cannot be a limit path of itself.*

On the basis of this theorem, one can prove the following proposition:

*Every nonclosed path is homeomorphic to a straight line (or an open interval).*

Indeed, let $L$ be a nonclosed path, and

$$x = \varphi(t), \quad y = \psi(t) \tag{1}$$

some motion on $L$. By Theorem 11, a nonclosed path is a single-valued and continuous image of a straight line and equations (1) define a mapping $T$ of this straight line onto $L$; moreover, $T$ is one-to-one. To prove the assertion, we must show that $T^{-1}$ is continuous. In other words, we must prove that if $M_n (t_n) \in L$ is a sequence of points converging to a point $M^* (t^*)$ then (a) if $t^*$ is finite, $t_n \to t^*$; (b) if $t_n \to \infty$, the sequence $M_n$ cannot converge to a point on $L$. (a) follows from the properties of a nonclosed path, and (b) from Theorem 11.

The same proposition is true when the nonclosed path $L$ is not a whole path.

Thus, any i n d i v i d u a l p a t h of system (I) is either a) a point, b) a simple closed curve, or c) homeomorphic to an open interval.

We thus have exhaustive information on the nature of each path on the plane and on the sphere. However, we are primarily interested in properties properties of a path not as an isolated entity but as a set of points in a plane region or on a sphere.

By Theorem 11, a nonclosed path cannot behave as illustrated in Figure 58, i. e., it cannot tend to one of its own points as $t \to + \infty$ (or $t \to - \infty$).

The coming propositions will further characterize paths of system (I). All of them stem essentially from the properties of limit continua of paths and semipaths described in Theorem 11.

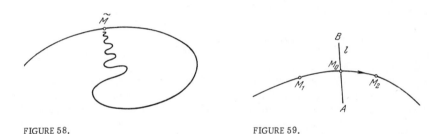

FIGURE 58.                                FIGURE 59.

**5. SOME PROPERTIES OF LIMIT PATHS.** The properties presented below are also characteristic for paths of dynamic systems in a plane region or on the sphere. Before proceeding to these properties, we cite another property of arcs without contact.

Let $L$ be a path, not an equilibrium state, $M_0$ a point on $L$ and $l$ an arc without contact passing through $M_0$.

L e m m a 3. *There exists a subarc of $l$, containing $M_0$ as an interior point, which contains no points of $L$ other than $M_0$.*

P r o o f. If $L$ is a closed path the assertion is obvious, for a closed path cannot have more than one point in common with an arc without contact (Lemma 12 in §3). Let $L$ be a nonclosed path and let the point $M_0$ correspond to a parameter value $t_0$. By Theorem 11, $L$ can contain none of its limit points. Hence there exist $T_2 < t_0$, $T_1 > t_0$, and $\varepsilon > 0$ such that $U_\varepsilon (M_0)$ contains no points of $L$ corresponding to parameter values $t < T_2$ and $t > T_1$. Then the subarc of $l$ lying in $U_\varepsilon (M_0)$ (this subarc contains $M_0$ as an interior point) can contain only points of $L$ corresponding to parameter values $T_2 \leqslant t \leqslant T_1$ (including the point $M_0$). But by Lemma 2 of §3 there can be only finitely many such points. This completes the proof.

Suppose now that an arc without contact $l$ passing through a point $M_0$ of the path $L$ contains no points of this path other than $M_0$. Let $A$ and $B$ be the endpoints of $l$ ($M_0$ is not an endpoint).

Let $M_1M_2$ be some arc on the path $L$, containing $M_0$ as an interior point, and suppose that the positive direction on this arc agrees with the positive direction on $L$ (Figure 59). To fix ideas, we shall assume that the half-open subarc $(M_0A]$ of $l$ lies on the negative side of the arc $M_1M_2$ of $L$, and the subarc $(M_0B]$ on its positive side (see Appendix, §3). We shall then say that $(M_0A]$ lies on the negative side of the path $L$ and $(M_0B]$ on its positive side. It is obvious that the property of the subarcs $(M_0A]$, $(M_0B]$ to lie on a given side of $L$ is independent of the choice of the arc $M_1M_2$ on $L$.

We now proceed to our investigation of the properties of limit paths. Let $L^+$ be a semipath, $L_0$ a limit path of $L^+$ (not an equilibrium state), $M_0$ a point on $L_0$ and $l_0$ an arc without contact passing through $M_0$. By Corollary 1 to Lemma 2, the arc $l_0$ contains no point of $L_0$ other than $M_0$, and by Corollary 2 to the same lemma all points of intersection of the semipath $L^+$ and the arc $l_0$ lie on the same side of $M_0$ on the arc.

Let $M$ be any point on $L_0$ other than $M_0$, and $l$ an arc without contact through $M$, not containing any point of $L_0$ other than $M$. The semipath $L^+$ crosses $l$ at infinitely many points. In this situation:

*Lemma 4. If the points of intersection of $L^+$ and $l_0$ lie on a subarc of $l_0$ on the positive (negative) side of $L_0$, then the points of intersection of $L^+$ with the arc without contact $l$ also lie on a subarc of $l$ which is on the positive (negative) side of $L_0$.*

(In Figure 60, the points of intersection of the semipath $L^+$ with the arcs $l_0$ and $l$ lie on the negative side of $L_0$).

Lemma 4 follows immediately from the fact that $M$ is a limit point of the semipath $L^+$ and from Remark 1 to Lemma 9 of §3.

*Definition IV. We shall say that $L_0$ is a positively (negatively) situated limit path of a semipath $L^{()}$ if the points at which $L^{()}$ crosses arcs without contact through the points of $L_0$ lie on the positive (negative) side of $L_0$. We shall also say that $L_0$ is a positively situated $\omega$- (or $\alpha$-) limit path of the path $L$ if $L_0$ is a positively situated limit path for a semipath $L^+$ ($L^-$) on $L$.*

*Lemma 5. If a path has more than one point in common with some arc without contact, then its $\alpha$- and $\omega$-limit sets are disjoint.*

Proof. It follows obviously from Lemma 12 that any path having more than one point in common with an arc without contact is nonclosed. Let $L$ be a nonclosed path and $l$ an arc without contact having more than one point in common with $L$. Then there exist two $t$-successive common points of $L$ and $l$, say $M_1(t_1)$ and $M_2(t_2)$, $t_1 < t_2$.

Let $C$ be the simple closed arc consisting of the arc $M_1M_2$ on $L$ and the subarc $M_1M_2$ of $l$ (Figure 61). By Lemma 11 of §3, one of the semipaths $L^-_{M_1}$, $L^+_{M_2}$ lies entirely (except for its initial point) inside $C$, the other outside the curve. No point of $C$ can be a limit point of either $L^-_{M_1}$ or $L^+_{M_2}$. Thus the limit set of one of these semipaths is entirely outside $C$, that of the other entirely inside $C$. Consequently, they must be disjoint. Q.E.D.

*Theorem 12. If at least one of the limit points of a nonclosed (whole) path $L$ is not an equilibrium state, then the $\alpha$- and $\omega$-limit sets $K_\alpha(L)$ and $K_\omega(L)$ are disjoint.*

The proof follows directly from Lemmas 2 and 5.

*Theorem 13. If none of the limit points of a semipath $L^{()}$ is an equilibrium state, it has a closed limit path.*

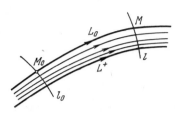

FIGURE 60.

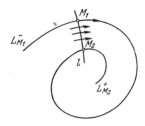

FIGURE 61.

P r o o f. Let $L^+$ be the semipath in question. If it is part of a closed path, the latter is a limit path of $L^+$. Suppose, therefore, that $L^+$ is a semipath of a nonclosed path. Let $L_0$ be one of its limit paths. Suppose that $L_0$ is nonclosed and let $\tilde{M}$ be any of its limit points. By Theorem 11, $\tilde{M}$ is an equilibrium state. But $\tilde{M}$, as a limit point of $L_0$, is clearly also a limit point of $L^+$ (since the limit set is closed), and this contradicts the assumption of the theorem. Thus $L_0$ is a closed path. Q.E.D.

*T h e o r e m   14. If a semipath $L^+$ has a closed limit path $L_0$, then $L_0$ is the unique limit path of the semipath $L^+$, i. e., its limit continuum.*

P r o o f. If $L^+$ is part of the closed path $L_0$, the theorem is obvious. Assume, therefore, that $L^+$ is part of a nonclosed path.

Let $M_0$ be a point on $L_0$ and $l$ an arc without contact passing through $M_0$ (Figure 62).

We claim that for any $\varepsilon > 0$ there exists a number $\tau$ such that all points on the semipath $L^+$ corresponding to $t > \tau$ lie in $U_\varepsilon(L_0)$. Let $\theta_0$ be the period of a motion on $L_0$. If $M_0$ corresponds under this motion to time $t = t_0$, it will also correspond to time $t = t_0 + \theta_0$ (in other words, the closed path, passing through $M_0$ at $t = t_0$, will pass through the same point at $t = t_0 + \theta_0$).

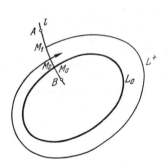

FIGURE 62.

Using Lemma 13 of §3, one readily sees that there exists a subarc $AB$ of $l$, containing $M_0$ as an interior point, with the following property: every path passing through a point of $AB$ will cross $l$ again with increasing $t$. Let $M_1$ be some point on the semipath $L^+$, lying on the subarc $AB$ of $l$ in $U_\delta(M_0)$, where $\delta$ is a positive number, and $M_2$ the point succeeding $M_1$ (on $l$). It is clear that $M_2$ is closer to $M_0$ on the arc than $M_1$. Let $C$ be the closed curve formed by the arc $M_1M_2$ on $L^+$ and the subarc $M_1M_2$ of $l$. By Lemma 14 of §3, one of the curves $C$ and $L_0$ lies inside the other, and the two curves together bound a certain region $\Gamma$. By Remark 1 to Lemma 14, if $\delta$ is sufficiently small (i. e., the point $M_1$ is sufficiently close to $M_0$) the region $\Gamma$ is contained in $U_\varepsilon(L_0)$. Finally, it follows from Lemma 11 of §3 (part a)) that the semipath $L^+$ (beginning from $M_2$) lies entirely in $\Gamma$. But then all the limit points of $L^+$, in other words, the limit continuum $K_\omega(L^+)$, lie in $U_\varepsilon(L_0)$:

93

$$K_\omega(L^+) \subset U_\varepsilon(L_0).$$

This is true for any $\varepsilon > 0$. But then obviously $L^+$ can have no limit point not on $L_0$. Indeed, any point $Q$ not on $L_0$ is at some nonzero distance from $L_0$. For sufficiently small $\varepsilon > 0$ it cannot be in $U_\varepsilon(L_0)$ and so cannot be in $K_\omega(L^+)$. On the other hand, the path $L_0$ is of course a subset of $K_\omega(L^+)$. Thus $K_\omega(L^+) = L_0$. Q.E.D.

**6. LIMIT PATHS OF DYNAMIC SYSTEMS WITH FINITELY MANY EQUILIBRIUM STATES. POSSIBLE TYPES OF PATHS.** An equilibrium state $O$ of a dynamic system is said to be isolated if it has a neighborhood $U_\varepsilon(O)$ containing no other equilibrium state.

Obviously, all equilibrium states of a dynamic system which has only finitely many equilibrium states must be isolated. The converse is also true, provided the region in which the system is considered is compact: If the system is being considered on the sphere or in a closed bounded region $\overline{G}$ on the plane, and has only isolated equilibrium states, the number of the latter is necessarily finite. In fact, otherwise the equilibrium states would have a point of accumulation, and this would be a non-isolated equilibrium state.

*Theorem 15. If $L^{()}$ is a semipath all of whose limit points are isolated equilibrium states, its limit set consists of a single equilibrium state.*

Theorem 15 is a consequence of the fact that the limit set of a semipath is connected.

Remark. Let $L^+$ be a semipath whose limit set consists of a single equilibrium state $O$. Then for any $\varepsilon > 0$ there exists $T = T(\varepsilon)$ such that all points $M(t)$ on $L^+$ with $t > T(\varepsilon)$ are in $U_\varepsilon(O)$. In other words, if $O$ is the only limit point of the semipath $L^+$, then $M(t) \to O$ as $t \to +\infty$. (The proof follows easily by reductio ad absurdum.) The converse is obvious: If $M(t) \to O$ as $t \to +\infty$, then $O$ is a limit point of $L^+$.

In the sequel we shall be concerned almost exclusively with dynamic systems (in a plane region or on the sphere) with finitely many equilibrium states. It follows from Theorems 9, 11, 13, 14 and 16 that the limit set of every semipath $L^{()}$ of such a system may be of one of the following types: a) a single equilibrium state; b) a single closed path; c) a connected set which is the union of whole paths, some of which are equilibrium states and the others nonclosed paths tending to equilibrium states both as $t \to -\infty$ and as $t \to +\infty$ (Figure 63).

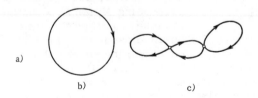

a)

b)

c)

FIGURE 63.

The examples in Chapter I show that all three types of limit set actually exist. In Examples 3 and 4 the equilibrium states are limit sets of nonclosed paths. In Example 7 the closed path is a limit path for the paths both inside and outside it. The limit continuum in Example 11 is a "figure of eight" consisting of three paths (one of which is an equilibrium state).

Every bounded semipath $L^{()}$ tends to a limit set of one of types a), b) and c).

Thus, subject to the assumption that the number of equilibrium states is finite, we have the following possible types of semipath: 1) equilibrium state; 2) closed path; 3) nonclosed semipath tending to an equilibrium state; 4) nonclosed semipath tending to a closed path; 4) nonclosed semipath tending to a limit continuum of type c).

These five types of semipath exhaust all possibilities.

The dynamic system in Example 7 of §1 has semipaths of types 1) to 4) (Figure 20). A semipath of type 5) appears in Example 11.

It is not hard to enumerate all types of bounded w h o l e p a t h s of a dynamic system with finitely many equilibrium states. Namely, a whole path $L$ is one of the following: 1) equilibrium state; 2) closed path; 3) nonclosed path tending to the limit set $K_\omega(L)$ as $t \to +\infty$ and to the limit set $K_\alpha(L)$ as $t \to -\infty$, where each of the sets $K_\alpha$ and $K_\omega$ is of type a), b) or c).

By Theorem 12, the limit sets $K_\alpha$ and $K_\omega$ may have common points only if each of them is an equilibrium state. But then they must coincide, and so the path $L$ is a "loop" (Figure 21b); loop-type paths also figure in Example 11 of §1. In all other cases, the $\alpha$- and $\omega$-limit sets of a nonclosed path are disjoint.

Figure 64 illustrates a few types of nonclosed whole paths $L$ of the dynamic system (I).

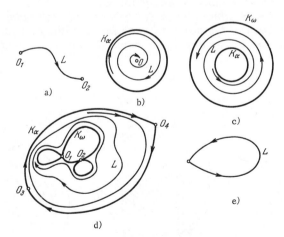

FIGURE 64.

In conclusion, we emphasize that the above characterization of limit sets is valid only provided the number of equilibrium states is finite. It fails if we relax the requirement that the equilibrium states be isolated. For example, consider the system

$$\frac{dx}{dt} = [-y - x(x^2 + y^2 - 1)](x^2 + y^2 - 1),$$

$$\frac{dy}{dt} = [x - y(x^2 + y^2 - 1)](x^2 + y^2 - 1),$$

obtained from the system of Example 7 in §1 by multiplying the right-hand sides by $x^2 + y^2 - 1$. By the results of §1.6 each point of the circle $x^2 + y^2 - 1 = 0$ is an equilibrium state of the new system, but all the other paths coincide with the corresponding paths of the original system (Example 7). The direction on the paths is easily determined according to the sign of the function $x^2 + y^2 - 1$. The configuration and directions of the paths are shown in Figure 65. The limit set of each positive semipath lying outside the circle $x^2 + y^2 - 1 = 0$ and of each negative semipath lying inside the circle is the circle itself, i. e., a continuum all of whose points are equilibrium states. Thus Lemma 15 fails to hold in this case.

**7. THEOREM ON EXISTENCE OF AN EQUILIBRIUM STATE WITHIN A CLOSED PATH.**
In this subsection we present a fundamental theorem, which has a bearing on the phase portrait as a whole: we shall show that there exists at least one equilibrium state within every closed path.

The proof given here is due to Bendixson, and makes essential use of the lemmas and theorems proved above. Later, in §8, we shall present another proof, independent of that given here, based on consideration of the rotation of a vector field along a closed curve.

Let $L_0$ be a closed path of system (I) and $M_0$ a point on this path. We shall assume that any neighborhood of $M_0$ contains points through which pass closed paths other than $L_0$ (either outside or inside $L_0$). Then, by Theorem 3 (continuity with respect to initial values), there are closed paths other than $L_0$ through every neighborhood of any point on $L_0$.

*Lemma 6. If K is some closed set lying inside (outside) $L_0$, there exists a neighborhood of the point $M_0$ such that all closed paths passing through its points lie inside one another and the set K is inside (outside) all these paths.*

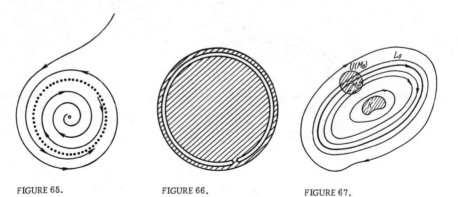

FIGURE 65.                FIGURE 66.                FIGURE 67.

Proof. By Remarks 1 and 2 to Lemma 14 of §3, for any given $\varepsilon > 0$ there exists $\delta > 0$ such that all closed paths through points of $U_\delta (M_0)$ lie inside one another and the region bounded by $L_0$ and any other closed path through $U_\delta (M_0)$ lies entirely within $U_\varepsilon (L_0)$. Obviously (Figure 66), it may happen that, given a family of closed curves $C$, for any arbitrarily small $\delta > 0$ there exists a curve $C$ entirely contained in the $\delta$-neighborhood of a given closed curve $C_0$, but the region bounded by $C$ and $C_0$ always has points outside $U_\varepsilon (C_0)$, where $\varepsilon$ is a prescribed number.

To fix ideas, suppose that the closed set $K$ lies inside $L_0$, and let the distance from $K$ to $L_0$ be $\varrho(K, L_0) = \varrho_0$. By assumption, $\varrho_0 > 0$. Let $\varepsilon$ be a number smaller than $\varrho_0$ and choose $\delta$ as above. Consider $U_\delta (M_0)$ and a closed path $L_1$ passing through a point in $U_\delta (M_0)$.

If $L_0$ is inside $L_1$, the same is true of $K$. Let $L_1$ be inside $L_0$. Then, if $K$ is not inside $L_1$, it must contain points in $\Gamma$, so that $K$ and $U_\varepsilon (L_0)$ have nonempty intersection. Thus $\varrho_0 = \varrho(L_0, K) < \varepsilon$, contradicting the choice of $\varepsilon$. Thus the set $K$ lies entirely within $L_1$ (Figure 67). Q.E.D.

Lemma 7. *Let $L$ be a nonclosed path having limit points which are not equilibrium states, and $M_0$ some point on $L$. Then no closed paths can pass through a sufficiently small neighborhood of $M_0$.*

Proof. Let $\tilde{M}$ be a limit point of a semipath on $L$, not an equilibrium state. Draw through $\tilde{M}$ an arc without contact $l$, and let $M_1$ and $M_2$ be any two (distinct) points of intersection of $L$ and $l$ (by Lemma 2 such points exist). Suppose that the points $M_0, M_1, M_2$ correspond to times $t_0, t_1, t_2$ $(t_1 \neq t_2)$. We may also assume that $M_0$ is distinct from $M_1$ and $M_2$ (Figure 68).

Let $\varepsilon > 0$ be so small that the neighborhoods $U_\varepsilon (M_1)$ and $U_\varepsilon (M_2)$ are disjoint. By Lemma 5 of §3, there exists $\delta > 0$ with the following properties: any path $L'$ which passes at $t = t_0$ through a point of $U_\delta (M_0)$ will cross the arc $l$ at some time $t$ near $t_1$, at a point $M_1'$ lying in $U_\varepsilon (M_1)$, and again at some time $t$ near $t_2$, at a point $M_2'$ in $U_\varepsilon (M_2)$. Since the neighborhoods $U_\varepsilon (M_1)$ and $U_\varepsilon (M_2)$ are disjoint, the points $M_1'$ and $M_2'$ are distinct. But then, by Lemma 12 of §3, the path $L'$ cannot be closed.

Thus all paths passing through points of $U_\delta (M_0)$ are nonclosed. Q.E.D.

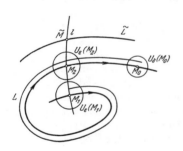

FIGURE 68.

Remark. If all the limit points of a nonclosed path $L$ are equilibrium states, closed paths may approach arbitrarily close to it. Thus, in Example 10 of §1 there are closed paths arbitrarily close to the loop (which is part of the "figure-of-eight" integral curve).

Theorem 16. *If all points inside a closed path $L_0$ are in the domain of definition $G$ of the system, there is at least one equilibrium state inside $L_0$.*

Proof. Suppose the assertion false: the interior of $L_0$ is a subregion of $G$ but contains no equilibrium state of the system. Denote this region by $\Gamma$. Let $L$ be a path passing through some point of $\Gamma$. If $L$ is not closed, it follows from Theorems 12, 13 and 14 that its $\alpha$- and $\omega$-limit continua are distinct closed paths. Consequently, there must be at least one closed path inside $\Gamma$. The same argument shows that within e a c h closed path lying in $\Gamma$ there must be closed paths.

We now define a function $F = F\,(M)$ in the closed region $\bar{\Gamma}$ as follows: if there is a closed path $L$ passing through $M$, we set $F\,(M)$ equal to the area of the interior of $L$; if the path through $M$ is not closed, we put $F\,(M)$ equal to the area $I_0$ of the interior of $L_0$ (i. e., the area of the entire region $\Gamma$). Since by assumption there are no equilibrium points inside $L_0$, the function $F\,(M)$ is defined at all points of $\Gamma$, and its maximum is $I_0$. By definition, $F\,(M) > 0$, and therefore $F$ must have an infimum $m$ in $\bar{\Gamma}$. Obviously, $0 \leqslant m < I_0$. There are two possibilities: 1) $F\,(M)$ assumes the value $m$ at some point $\bar{M}$ of $\Gamma$. 2) $F\,(M)$ does not assume the value $m$ at any point of the region. In the latter case, there exists a sequence of points $\{M_n\}$ in $\Gamma$ such that $F\,(M_n) \to m$ as $n \to \infty$. We may assume that the sequence $M_n$ converges to a point, $\bar{M}$ say. In either case, the infimum of $F\,(M)$ in any neighborhood of $\bar{M}$ is $m$.

Consider the path $\tilde{L}$ passing through $\bar{M}$. This path cannot be nonclosed. For if $\tilde{L}$ is not closed, Lemma 7 implies the existence of a neighborhood of $\bar{M}$ such that all paths crossing this neighborhood are nonclosed. But then the value of $F\,(M)$ at any point of the neighborhood is $I_0$, and the infimum of the function there cannot be equal to $m < I_0$, contradicting the above property of the point $\bar{M}$. Consequently, $\tilde{L}$ is a closed path. Let $L_1$ be a closed path lying inside $L$ and $L_2$ a closed path lying inside $L_1$ (Figure 69) (we know that such paths exist).

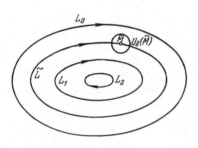

FIGURE 69.

Let $I_1$ and $I_2$ be the areas of the regions bounded by $L_1$ and $L_2$, respectively. Clearly, $I_2 < I_1$. Let $U_\delta\,(\bar{M})$ be a small neighborhood of $\bar{M}$. Any path passing through a point of this neighborhood is either not closed, or it is closed and then, by Lemma 6, the path $L_1$ is inside it. Hence we have $F\,(M) > I_1$ at all points of $U_\delta\,(\bar{M})$, and so at these points $\inf F\,(M) \geqslant I_1 > I_2$. On the other hand, this infimum is equal to $m$. Hence $m > I_2$. But this contradicts the definition of $m$ as the infimum of $F\,(M)$ in the region $\Gamma$.

Thus, our assumption that there are no equilibrium states inside $L_0$ has led to a contradiction. Q.E.D.

Corollary 1. Let $L$ be a nonclosed path crossing an arc without contact $l$ at more than one point; let $M_1\,(t_1)$, $M_2\,(t_2)$ be two $t$-successive points of intersection with $l\,(t_1 < t_2)$ and $C$ the simple closed curve consisting of the arc $M_1 M_2$ on the path $L$ and the subarc $M_1 M_2$ of the arc $l$ (Figure 68). If the region $\Gamma$ interior to $C$ contains no boundary points of $G$, it contains at least one equilibrium state. Indeed, by Lemma 11 of §3, one of the two semipaths $L_{\bar{M}_1}$ or $L_{\bar{M}_2}$, say the former, lies entirely inside $C$ (except for its initial point). Obviously, all limit points of $L_{\bar{M}_1}$ lie in $\Gamma$. If one of these limit points is an equilibrium state, we are done. Otherwise, they form a closed path lying in $\Gamma$. But then, by Theorem 16, there is an equilibrium point inside this closed path, hence in $\Gamma$. Q.E.D.

Corollary 2. If the region $\Gamma$ inside a cycle without contact $C$ contains no boundary points of $G$, it contains at least one equilibrium state (the proof is analogous to that of Corollary 1).

Corollary 3. If $\Gamma$ is a closed simply connected subregion of $G$ containing no equilibrium states, then any path through a point of $\Gamma$ will leave it with both increasing and decreasing $t$. Indeed, otherwise $\Gamma$ would contain the $\alpha$- or $\omega$-limit continuum of the path and therefore at least one equilibrium state.

*Theorem 17. Every dynamic system on the sphere has at least one equilibrium state.*

Proof. Let $L$ be an arbitrary path of a system on the sphere, $K_\omega$ its limit continuum. If $K_\omega$ contains an equilibrium state, we are done. If not, $K_\omega$ is a closed path (Theorem 14). The two regions into which this path separates the sphere are each homeomorphic to a plane region. Thus, by Theorem 16, each of these regions contains at least one equilibrium state. Q.E.D.*

**8. FUNDAMENTAL THEOREM ON EQUILIBRIUM STATES.** The following fundamental theorem on equilibrium states is also due to Bendixson.

*Theorem 18. If $O$ is an isolated equilibrium state, then either any neighborhood of $O$ contains a closed path with $O$ in its interior, or there is a semipath tending to $O$.*

Proof. Suppose there exists a closed neighborhood $\bar{U}$ of the equilibrium state $O$ containing no closed path with $O$ in its interior. We shall show that there must be a semipath tending to $O$.

We may assume without loss of generality that $U$ is a disk with center at $O$, such that neither $U$ nor its boundary contains equilibrium states other than $O$ (since $O$ is an isolated equilibrium state). Denote the circle bounding $U$ by $\sigma$. We shall first show that there exists a positive or negative semipath lying entirely within $\bar{U}$. Suppose that there is no such semipath. Let $\sigma'$ be a circle with center $O$, lying in $U$ (i.e., inside $\sigma$), $M$ an arbitrary point on this circle, $L$ a path passing at $t = t_0$ through $M$ (Figure 70). By assumption, the path $L$ must leave $\bar{U}$ for both increasing and decreasing $t$. Let $A(B)$ be the point nearest $M_0$ (i.e., corresponding to $t$ nearest $t_0$) at which $L$ enters (leaves) $\bar{U}$, and consider the arc $AB$ of the path. (Apart from its endpoints $A$ and $B$, through which $L$ enters and leaves $\bar{U}$, this arc may have interior points lying on the circle $\sigma$. If so, $L$ is tangent to $\sigma$ at such points (Figure 70). Denote the distance from the point $O$ to the arc $AB$ on $L$ by $f(M)$. $f(M)$ is a positive function, defined on the circle $\sigma'$.

Let $\varrho_0$ denote the infimum of the function $f(M)$. Since the circle is compact, it follows easily that there exists a point $M_0 \in \sigma'$ such that inf $\{f(M)\} = \varrho_0$ in each neighborhood of $M_0$ on $\sigma'$.

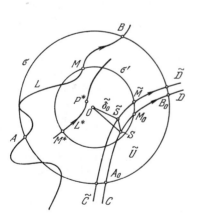

FIGURE 70.

---

* Like Theorem 16, this theorem can be proved by a different method, based on vector fields.

We claim that $f(M_0) = \varrho_0$. Indeed, suppose that $f(M_0) \neq \varrho_0$. Then $f(M_0) = \varrho_0 + \gamma$, $\gamma > 0$. Suppose that the path $L_0$ passing through $M_0$ at $t = t_0$ enters and leaves the disk $\bar{U}$ through points $A_0(t_1)$ and $B_0(t_2)$ $(t_1 < t_0 < t_2)$. If $\tau > 0$ is sufficiently small, there are points $C$ $(t_1 - \tau)$ and $D$ $(t_2 + \tau)$ on $L_0$ such that the arcs $A_0 C$ and $B_0 D$ of the path lie outside $U$.

By the theorem on continuity with respect to initial values, there exists $\delta > 0$ satisfying the following conditions:

a) any path $\tilde{L}$ passing at $t = t_0$ through a point $\tilde{M} \in U_\delta(M_0)$ is defined for all $t$ such that $t_1 - \tau \leqslant t \leqslant t_2 + \tau$; b) the points $\tilde{C}(t_1 - \tau)$, $\tilde{D}(t_2 + \tau)$ lie outside the disk $\bar{U}$; c) the arc $\widetilde{CD}$ of $L$ lies in the $\gamma/2$-neighborhood of the arc $CD$ on $L$.

Now let $\tilde{M}$ be an arbitrary point on the circle $\sigma'$ in $U_\delta(M_0)$. Let $f(\tilde{M}) = \tilde{\delta}_0$ be the distance $\varrho(0, \tilde{S})$, where $\tilde{S} \in \tilde{L}$, and $S$ a point on the arc $CD$ of $L$ such that $\varrho(S, \tilde{S}) < \gamma/2$ (see condition c)). Then $\tilde{\delta}_0 + \varrho(S, \tilde{S}) > \varrho(0, S) > f(M_0) = \tilde{\delta}_0 + \gamma$, and consequently $f(\tilde{M}) = \tilde{\delta}_0 > \tilde{\delta}_0 + \gamma - \varrho(S, \tilde{S}) > \varrho + \gamma/2$. This contradicts the fact that the infimum of $f(M)$ in any neighborhood of $M_0$ is $\varrho_0$.

Thus $f(M_0) = \varrho_0$, and therefore $\varrho_0 > 0$. But this is impossible. Indeed, let $P^*$ be an arbitrary point in $U_{\varrho_0/2}(0)$, $L^*$ the path passing through it, $M^*$ the first point of intersection of $L^*$ and $\sigma'$ for decreasing $t$. Then, obviously, $f(M^*) < \varrho_0/2$, contradicting the definition of $\varrho_0$. This contradiction shows that there exists a semipath $L^{()}$ lying entirely in $\bar{U}$. Since by assumption $\bar{U}$ contains no closed paths, the limit set of $L^{()}$ is either the point $O$ alone or the point $O$ together with paths tending to $O$. In either case the proof is complete.

## 9. ISOLATED CLOSED PATH – LIMIT CYCLE. POSSIBLE CONFIGURATION OF PATHS NEAR A LIMIT CYCLE.

An isolated closed path, i. e., a closed path having a neighborhood in which there are no other closed paths, is known as a limit cycle.

An example of a limit cycle was encountered in Example 7 of §1.

We end this chapter with some basic information about limit cycles. Let $L_0$ be a limit cycle. The behavior of paths in its neighborhood is described by the following

*Theorem 19. All paths through points lying outside (inside) $L_0$ and sufficiently close to $L_0$ tend to $L_0$ either as $t \to +\infty$ (in which case they leave the neighborhood of $L_0$ with decreasing $t$) or as $t \to -\infty$ (in which case they leave the neighborhood of $L_0$ with increasing $t$).*

Proof. Let $\varepsilon_0 > 0$ be so small that the $\varepsilon_0$-neighborhood of the limit cycle $L_0$ contains no equilibrium state and no closed path other than $L_0$.

A number $\varepsilon_0$ meeting these requirements clearly exists.

Let $l$ be an arc without contact through some point $M_0$ of $L_0$, neither of whose endpoints is $M_0$ (Figure 71). Let $A$ and $B$ be points on $l$, on different sides of $M_0$, so close to $M_0$ that the following conditions hold.

1. All paths crossing the subarc $AB$ at $t = t_0$ cross it again at some $t > t_0$. In particular, let the paths $L_A$ and $L_B$ passing through $A$ and $B$, respectively, cross $l$ at

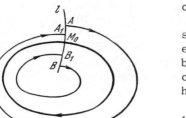

FIGURE 71.

$t > t_0$ at points $A_1$ and $B_1$ (and the arcs $AA_1$ and $BB_1$ on $L_A$ and $L_B$ have no points in common with $l$).

2. If $C_1$ is the simple closed curve consisting of the arc $AA_1$ on $L_A$ and the subarc $AA_1$ of $l$, and $C_2$ the simple closed curve consisting of the arc $BB_1$ on $L_B$ and the subarc $BB_1$ of , then the region $\Gamma$ bounded by $C_1$ and $C_2$ is contained in the $\varepsilon_0$-neighborhood of $L_0$.

That points $A$ and $B$ with these properties can be chosen follows from Lemma 17 of §3 and Remark 1 to that lemma. Moreover, one of the half-open arcs $(M_0A]$ or $(M_0B]$ lies outside the limit cycle $L_0$, the other inside $L_0$. To fix ideas, suppose that the subarc $(M_0A]$ of $l$ is outside $L_0$. Since by assumption there are no closed paths in $U_{\varepsilon_0}(L_0)$ other than $L_0$ itself, $A$ is distinct from $A_1$ and is either closer to $M_0$ than $A_1$ or conversely. Let $t_1$ be the parameter value corresponding to $A_1$, and suppose that the latter is between $M_0$ and $A$. It is easy to see, using Lemma 11 of §3, that in this case $A_1$ is a point of entry for the path $L_A$ into the annular region $\Gamma_1$ bounded by $L_0$ and the curve $C_1$, and as $t$ increases the path cannot leave this region. All limit points of $L_A$ therefore lie in $\Gamma_1$ or on its boundary. By the choice of $\varepsilon_0$ and condition 2, there are no equilibrium states in $\Gamma_1$ or on its boundary; hence, by Theorem 13, the $\omega$-limit set of $L_A$ is a closed path. But the only closed path in the $\varepsilon_0$-neighborhood of $L_0$, hence also in the closed region $\overline{\Gamma}_1$ is the limit cycle $L_0$ itself.

Consequently, the path $L_A$ tends to the limit cycle $L_0$ as $t \to +\infty$. We claim that in general all paths passing through points of the subarc $(M_0A]$ other than $M_0$ tend to $L_0$ as $t \to +\infty$. Indeed, by Lemma 11 of §3 all points of $(A_1M_0]$ on $l$, other than $A_1$ and $M_0$, are in $\Gamma_1$, and all paths crossing the arc $l$ at these points cannot leave $\Gamma_1$ with increasing $t$.

In addition, all paths crossing the subarc $(A_1A)$ of $l$ enter the region $\Gamma_1$ with increasing $t$ and henceforth cannot leave that region. We can thus repeat the reasoning applied above to the path $L_A$ to each of these paths, and so they all tend to the limit cycle $L_0$.

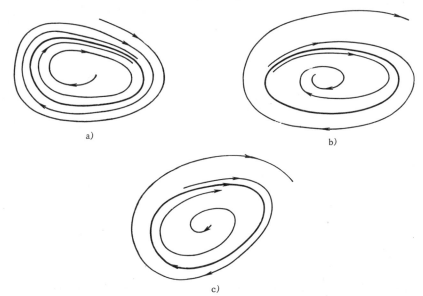

a)

b)

c)

FIGURE 72.

We have thus shown that all paths through points of $(AM_0]$ tend to $L_0$ as $t \to +\infty$. But by Lemma 14 of §3 all paths through points of $\Gamma_1$ must cross the arc $(AM_0]$. Consequently, all paths through points of $\Gamma_1$ tend to the limit cycle $L_0$ as $t \to +\infty$. This of course means that the paths passing through all points sufficiently close to $L_0$ and lying outside $L_0$ tend to $L_0$ as $t \to +\infty$. In view of the fact that no closed paths pass through the points of $\Gamma_1$ and its boundary points, other than $L_0$, we see from Theorem 12 that all paths passing through points of $\Gamma_1$ leave $\Gamma_1$ with increasing $t$ through the subarc $(AA_1)$ of $l$.

The case in which $A$ lies on $l$ between $A_1$ and $M_0$ is dealt with in the same way, with obvious modifications. In this case, all paths through points of the subarc $[AM_0)$ of $l$, hence all paths through points of the corresponding region $\Gamma_1$, tend to the limit cycle $L_0$ as $t \to -\infty$ and leave $\Gamma_1$ with increasing $t$.

The subarc $(M_0B]$ of the arc $l$ is treated similarly.

This completes the proof of Theorem 19.

If all paths passing through points of a neighborhood of a limit cycle $L_0$, outside and inside $L_0$ and distinct from $L_0$, tend to the limit cycle as $t \to +\infty$ $(t \to -\infty)$, $L_0$ is called a s t a b l e (u n s t a b l e) l i m i t  c y c l e (Figure 72a, b).

If all paths passing through points sufficiently close to a limit cycle $L_0$ and lying outside (inside) $L_0$ tend to $L_0$ as $t \to +\infty$, while those lying inside (outside) $L_0$ do so as $t \to -\infty$, then $L_0$ is called a s e m i s t a b l e  l i m i t  c y c l e* (Figure 72c).

---

\* Frequently, especially in applications, the term "unstable" is used for any cycle which is not stable, i.e., either unstable in the sense defined here or semistable.

*Chapter III*

*FUNDAMENTAL CONCEPTS OF THE QUALITATIVE*
*THEORY OF DYNAMIC SYSTEMS*

§5. QUANTITATIVE AND QUALITATIVE
INVESTIGATION OF DYNAMIC SYSTEMS

**1. INTRODUCTION.** The description of individual paths, the main subject of §4, is naturally one of the first questions arising in qualitative investigation of dynamic systems. No less important, however, is information concerning the phase portrait of system (I) as a whole. Some simple and fundamental results of this type are established in Theorems 16, 17 and 18 of §4.

However, in order to proceed to a more detailed study of the "qualitative properties" of the phase portrait we must first rigorously define what we mean by "qualitative property," "qualitative structure of the phase portrait," "qualitative investigation," terms which we have employed up to now only in an intuitive but rather vague sense. We now wish to remedy this situation.

As a preliminary to this goal, the different aspects of investigation of dynamic systems and the role played in this context by qualitative investigation are worthy of attention. The problems arising when one examines the dynamic systems of natural science (celestial and "terrestrial" mechanics, oscillation theory, etc.) fall, roughly speaking, into two categories. On the one hand we have such problems as finding analytical expressions for solutions (e. g., by means of elementary functions or quadratures, or by means of series expansions and various other functions), and also approximate calculation of solutions, for which a formidable arsenal of computational methods is available. This category of problems comprises the "quantitative integration" or quantitative investigation of dynamic systems.*

On the other hand, we have problems relating to the number and properties (in particular, stability or instability) of equilibrium states of a dynamic system, existence of closed paths (periodic solutions), regions of attraction of stable equilibrium states, and so on.

These problems fall into the category of "qualitative investigation" of dynamic systems. Despite the fact that in "qualitative investigation" we

---

* In celestial mechanics, this category includes various methods of series integration in the $n$-body problem (such as Sundman's method); in the classical problem of rigid body motion — methods that yield "totally integrable cases." In these cases, the dynamic systems in question are clearly of higher than second order. Accounts of the many methods for calculation of single paths may be found, e.g., in /43/, /44/, /45/.

may be quite indifferent as to the exact "shape" of the path, the dimensions of closed paths or many other properties characterizing the phase portrait from the quantitative standpoint, the qualitative structure nonetheless reflects highly important features of a dynamic system, which are of both mathematical and applied interest.*

Thus, qualitative investigation differs intrinsically from quantitative integration in regard to its goals and problems, and therefore has its own methods. Quantitative integration in the above-mentioned classical sense cannot supplant qualitative investigation, and in many cases it cannot even render significant assistance. It is frequently easier and more convenient to study the qualitative behavior of paths directly, by examination of the vector field defined by the system, than by means of analytical expressions obtained by integration. We have already pointed out (§1.14) that such analytical expressions yield a complete solution to the problem of qualitative integration only in the simplest cases. In the general case, a knowledge of analytical expressions for integrals is no help to qualitative investigation: it simply reduces the problem of direct qualitative investigation of a dynamic system to "qualitative investigation" of some function $F(x, y, c) = 0$. And this problem may turn out to be no simpler or easier than the original one; plus ça change, plus c'est la même chose. . . Only in special cases do analytical expressions provide a real aid to qualitative investigation.

It must be emphasized that qualitative investigation is by no means, so to speak, a "surrogate" for quantitative investigation; in other words, it is not born of the difficulty or impossibility of quantitative investigation. As we have repeatedly stated, qualitative investigation has its own specific goals, which are not in the scope of quantitative integration. Even more: qualitative investigation is not infrequently an invaluable aid to quantitative integration, as a signpost indicating which paths are worth approximating, and in what region, for the problem at hand.

We are now going to impart to the vague terms "qualitative property," "qualitative structure" a well-defined, rigorous meaning. An adequate mathematical concept has crystallized in the course of mathematical investigation of dynamic systems, the result of careful consideration of what is implied by a "qualitative" property in numerous applications: "qualitative" properties are those properties of paths, sets of paths and phase portraits which are topological invariants, i. e., properties preserved under arbitrary topological mappings of the region or set.

Although topological mappings are considered in the Appendix, a few clarifications are in order here.

---

* The concept of "qualitative investigation" is meaningful and valuable not only for dynamic systems but also for other mathematical objects. As a simple example, consider the case of an algebraic equation of prescribed degree: there are problems which can be solved on the sole basis of the number of real roots of an equation

$$x^n + a_1 x^{n+1} + \ldots + a_n = 0,$$

without any need to determine their true or approximate values. It is natural to call this "qualitative" investigation of the equation. Another example is provided by the theory of curves, e. g., algebraic curves. Disregarding the exact dimensions of the curve given by an equation $F(x, y) = 0$, we may be interested, say, in the number of connected components of the curve and their relative positions. Again, the most natural appellation for a theory of this kind is "qualitative."

A t o p o l o g i c a l   m a p p i n g (of the plane into itself, or of some plane set into another or into itself) is a one-to-one and bicontinuous mapping. In other words, each point $M$ is mapped onto exactly one point $M'$ of the same plane (or set), distinct points $M_1$ and $M_2$ being mapped onto distinct points $M_1'$ and $M_2'$; and [roughly speaking] any two arbitrarily close points $M_1$ and $M_2$ are mapped onto arbitrarily close points $M_1$ and $M_2$. The inverse of a topological mapping is also topological.

An intuitive description of a topological mapping of the plane into itself may be given as follows: imagine the plane to be made from rubber, which is deformed in some way, stretching and squeezing it at various points, but without tearing or folding.

Any topological mapping of the plane into itself is either a deformation of the above type (without tearing and folding) or a mirror reflection of the plane followed by such a deformation (in the former case the topological mapping "preserves orientation," in the latter it "changes orientation" — see Appendix, §2). Clearly, the shape of curves and regions, and of plane sets in general, may undergo drastic changes under a topological mapping, but certain properties will nevertheless be preserved. Thus, a closed curve such as a circle will remain closed after any topological mapping, though its shape may be quite different from that of the original curve. The topological image of a straight-line segment will generally be an arc ("simple arc"), but this arc cannot intersect itself.

Now that we have specified the meanings of the concepts "qualitative property," "qualitative structure," we shall also change our terminology, speaking no longer of "qualitative structure," "qualitative property," etc. but rather using the terms "topological structure," "topological properties," etc.*

We now proceed to a rigorous formulation of the fundamental concepts.

**2. TOPOLOGICAL STRUCTURE OF A DYNAMIC SYSTEM.** Our definition of the topological structure of a dynamic system will relate to an (o p e n) p l a n e   r e g i o n, either the entire domain of definition of the system or a subregion thereof. A similar definition applies to topological structure on a n y   s u b s e t   $M$   o f   t h e   d o m a i n   o f   d e f i n i t i o n, such as a closed bounded region $\overline{G}_1 \subset G$, and also on the s p h e r e. The only change in the definition is to replace the words "in the region $G$" by "in a closed region $\overline{G}_1$," "on the sphere," etc.

The definition given below is in a certain sense i n d i r e c t (definition by abstraction): we do not state exactly what the topological structure is, but specify when two dynamic systems are to have identical [or equivalent] topological structure.** Suppose given two dynamic systems

$$\frac{dx}{dt} = P_1(x, y), \qquad \frac{dy}{dt} = Q_1(x, y) \tag{$A_1$}$$

and

$$\frac{dx}{dt} = P_2(x, y), \qquad \frac{dx}{dt} = Q_2(x, y), \tag{$A_2$}$$

defined in bounded plane regions $G_1$ and $G_2$, respectively.

---

* The term "qualitative" is used in vastly variegated senses, whereas "topological" has a well-defined meaning.
** Examples of indirect definitions are the definitions of a function, the cardinality of a set, etc.

The systems $(A_1)$ and $(A_2)$ may coincide, as may the regions $G_1$ and $G_2$. For example, if $G_1 = G_2$ this means that we are considering two different systems in the same region. It is also possible that some system (I) defined in a region $G$ is being considered in two different (though not necessarily disjoint) regions $G_1 \subset G$ and $G_2 \subset G$.

*Definition V. We shall say that the phase portraits of systems $(A_1)$ and $(A_2)$ have the same or identical topological (or qualitative) structure [or: are topologically equivalent] in regions $G_1$ and $G_2$, respectively, if there exists a mapping $T$ of $G_1$ onto $G_2$ satisfying the following conditions:*

*1) $T$ is a topological mapping;*

*2) if two points of $G_1$ lie on the same path of system $(A_1)$, their images under $T$ lie on the same path of system $(A_2)$;*

*3) if two points of $G_2$ lie on the same path of system $(A_2)$, their images under $T^{-1}$ lie on the same path of system $(A_1)$.*

A mapping satisfying conditions 1), 2), 3) will be called a path-preserving mapping from system $(A_1)$ to system $(A_2)$ or an identifying mapping for systems $(A_1)$ and $(A_2)$. Whenever no confusion can arise we shall say simply "path-preserving mapping" or "identifying mapping."

In the sequel, when considering a given dynamic system

$$\frac{dx}{dt} = P(x, y), \qquad \frac{dy}{dt} = Q(x, y),$$

we shall often speak of all identifying mappings (or path-preserving mappings), meaning the set of all possible topological mappings taking paths of system (I) onto curves which are the paths of some other system (I') (e. g., a system satisfying the conditions of §1.1).

It is obvious that not every topological mapping can be an identifying mapping in the above sense. Indeed, an identifying mapping takes paths of a given dynamic system (I) onto paths of some other system (I') satisfying the conditions of §1. Any path of the system (other than an equilibrium state) is a smooth curve, and so an identifying mapping, though changing the shape of the paths, preserves their smoothness. But it is not hard to construct a topological mapping taking the paths of the system onto non-smooth curves. This being so, the following idea suggests itself: why not define the "qualitative properties" of a dynamic system to be those properties invariant under all possible smooth topological mappings? However (see Appendix, §9), this convention would force us to classify as "qualitative" many properties which disagree completely with the sense of the word as employed in this book, from the standpoint of both purely mathematical considerations and applied questions. We shall discuss this question in the Appendix.

One more comment regarding the concept of an identifying mapping. If there exists an identifying mapping $T$ for two given systems $(A_1)$ and $(A_2)$, this implies the existence of an infinite set of identifying mappings. This set can be constructed as follows. Let

$$x = \varphi(t - t_0, x_0, y_0), \qquad y = \psi(t - t_0, x_0, y_0) \tag{1}$$

be a solution of system (I). It is readily seen that for every fixed $\tau$ the functions

$$x = \varphi\,(\tau,\,x_0,\,y_0), \qquad y = \psi\,(\tau,\,x_0,\,y_0) \tag{2}$$

define a topological mapping $H_\tau$ of the plane into itself which, by virtue of the properties of the functions (1), maps points lying on a path of system (I) onto points on the same path. There are obviously infinitely many (indeed, continuum many) such mappings $H_\tau$, since the number of possible constants $\tau$ is infinite, and if $T$ is an identifying mapping, then $T' = TH_\tau$ (for any $\tau$) is also an identifying mapping.

One can also construct other infinite sets of identifying mappings.

Instead of saying that "the phase portraits of two dynamic systems $(A_1)$ and $(A_2)$ in regions $G_1$ and $G_2$, respectively, have the same topological structure," we shall say more briefly: "the dynamic systems $(A_1)$ and $(A_2)$ have the same topological structure in regions $G_1$ and $G_2$, respectively." We shall also use such abbreviated expressions as "systems $(A_1)$ and $(A_2)$ have the same topological structure" or "the topological structures of the phase portraits in $G_1$ and $G_2$ are identical." In the former case, it should be clear what regions are meant, and in the latter what systems are being studied.

Our definition motivates a few further remarks. Note first that condition 3) of the definition is not a consequence of conditions 1) and 2). This is evident from the following simple example: let the systems $(A_1)$ and $(A_2)$ be

$$\frac{dy}{dt} = 0, \qquad \frac{dx}{dt} = 1, \tag{3}$$

$$\frac{dy}{dt} = 0, \qquad \frac{dx}{dt} = x^2 + y^2, \tag{4}$$

each considered in the entire plane, and let $T$ be the identity mapping. The paths of systems (3) and (4) are illustrated in Figure 73a and b, respectively. The mapping $T$ obviously satisfies conditions 1) and 2), but not condition 3): points lying on the $x$-axis always lie on the same path of system (3), but they may be on different paths of system (4). (The $x$-axis is a single path of system (3), but the union of three different paths of system (4): the two half-axes and an equilibrium point at the origin.)

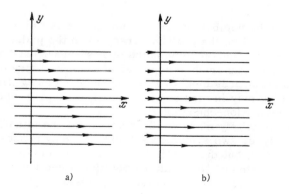

a)                            b)

FIGURE 73.

Another point to be noted here is that the question whether $T$ preserves the positive direction on paths does not arise at all. In other words, it is quite permissible that the positive direction on paths of system $(A_1)$ correspond under an identifying mapping to either positive or negative direction on paths of system $(A_2)$. It is not difficult to cite examples of either situation. In the sequel we shall be concerned mainly with dynamic systems with finitely many equilibrium states (these being the most important for applications). There are then only two possibilities: 1) An identifying mapping $T$ associates the positive direction on paths of system $(A_1)$ with the positive direction on all paths of system $(A_2)$. We then call $T$ a t i m e - n o n r e v e r s i n g identifying mapping. This is the case, for example, when both systems $(A_1)$ and $(A_2)$ are linear and both have stable nodes or foci (one system may have a stable node, the other a stable focus) (compare §6.3). 2) An identifying mapping $T$ associates the positive direction on paths of system $(A_1)$ with the negative direction on all paths of system $(A_2)$. We shall then call $T$ a t i m e - r e v e r s i n g identifying mapping. This happens, for example, when both systems $(A_1)$ and $(A_2)$ are linear, one of them has a stable node or focus and the other an unstable one. In some cases there may exist identifying mappings of both types (for example, when all paths are closed; see Figure 74).

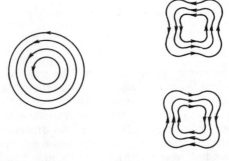

FIGURE 74.

Note that when the number of equilibrium states is finite an identifying mapping cannot associate the positive direction on the paths of system $(A_1)$ with the positive direction on some paths of system $(A_2)$ and the negative direction on others. Inasmuch as topological mappings may either preserve or change orientation, it is clear that a time-nonreversing (time-reversing) identifying mapping may in turn be of either of two types: orientation-preserving or orientation-changing.

In the case of infinitely many equilibrium states the situation may be of another, third type: two systems $(A_1)$ and $(A_2)$ may possess the same topological structure, but the identifying mapping associates the positive direction on some paths of $(A_1)$ with the positive direction on paths of $(A_2)$, and on others with the negative direction on paths of $(A_2)$.

For example, consider the systems

$$\frac{dx}{dy} = -y\,(x^2+y^2-1)^2, \qquad \frac{dy}{dt} = x\,(x^2+y^2-1)^2, \tag{5}$$

$$\frac{dx}{dt} = -y\,(x^2+y^2-1), \qquad \frac{dy}{dt} = x\,(x^2+y^2-1). \tag{6}$$

The circle $x^2 + y^2 = 1$ is a singular curve for both systems. The directions on paths of systems (5) and (6) lying inside the circle $x^2 + y^2 = 1$ are opposed. The identity mapping is clearly an identifying mapping for these systems. The positive directions on the paths lying outside the singular circle is associated with the positive direction, whereas on paths lying inside the circle it is associated with the negative direction. It can be shown that there is no identifying mapping conserving (or reversing) directions with respect to $t$ on all paths.

The relation between dynamic systems defined above is easily seen to be an e q u i v a l e n c e   r e l a t i o n : it is reflexive, symmetric and transitive. It therefore induces a partition of all dynamic systems into disjoint classes of systems possessing the same topological structure. Every dynamic system belongs to one and only one of these classes.

Employing our new notion of identifying (path-preserving) mapping, we can also define t o p o l o g i c a l   p r o p e r t i e s and t o p o l o g i c a l i n v a r i a n t s of a phase portrait (or set of paths).

*D e f i n i t i o n   VI. A topological (qualitative) property of a phase portrait (or set of paths) in a region $G_1$ or topological invariant of the phase portrait is a property or magnitude which is invariant under all identifying mappings.*

Instead of "topological property of a path, set of paths or phase portrait" we shall say briefly "topological property of a dynamic system" and also "topological invariant of a dynamic system." Obviously, the property of a path to be an equilibrium state, or the property of being closed, is a topological property.*

To fix ideas, let us confine the discussion to dynamic systems with finitely many equilibrium states and finitely many limit cycles. Then, clearly, the number of equilibrium states and the number of limit cycles are topological invariants (in other words, they remain invariant under any identifying mapping). Other examples of topological properties are the relative positions of closed paths (when they exist), the existence (or nonexistence) of annular regions filled by closed paths, the existence of a prescribed number of foci or nodes, and so on. On the other hand, the distance between equilibrium states and limit cycles, the exact "shape" of closed paths, etc. are not topological invariants (properties) — they need not be preserved by identifying mappings.

If some collection of properties of a phase portrait is such that two dynamic systems possessing these properties have the same topological structure, we shall call it a collection of defining properties or a c o m p l e t e   s y s t e m   o f   t o p o l o g i c a l   i n v a r i a n t s.

Q u a l i t a t i v e   i n v e s t i g a t i o n of a dynamic system consists in determining its topological properties and invariants (such as the number

---

* The property of being a point or the property of a curve to be closed, like all subsequently mentioned properties of curves which are paths, is invariant not only under identifying mappings but under arbitrary topological mappings.

and type of equilibrium states, number and relative positions of limit cycles, and so on).

Complete qualitative investigation of a dynamic system consists in determining the topological structure of its phase portrait.

This last sentence is quite general and indeterminate. We must of course strive for further specification: what must be done to establish the topological structure of the phase portrait of a dynamic system? There is no universal answer to this question, valid for all possible dynamic systems. Nevertheless, provided one confines attention to certain narrower classes of dynamic systems, the concept "determination of topological structure" can be given a rigorous and well-defined meaning.

This book is concerned with a certain, well-defined class of dynamic systems ("systems with finitely many singular paths"), which is of course the most interesting from the standpoint of both applications and theory. Later (Chapters VII, X, XI) we shall rigorously define the scheme of a phase portrait or of a dynamic system. The scheme will play the role of a complete system of topological properties (invariants). For the moment we shall not go into particulars, leaving the detailed discussion to Chapters VII, X and XI.

To illustrate the concept of phase portraits with the same topological structure, we present a few simple geometric examples. Consider the phase portraits of system (40) of Example 3 and system (45) of Example 4 in §1 (Figures 10 and 13). The origin is an equilibrium state of both systems, a node for system (40) and a focus for system (45). Although the paths of the systems look quite different, it is readily shown that the phase portraits of systems (40) and (45) in the unit disk $C$ have the same topological structure. In other words, one can find a topological mapping of the disk into itself under which paths of system (40) go into paths of system (45) (we shall not describe the construction, for this question will be examined in detail in Chapter VIII). The phase portraits illustrated in Figure 74 are also topologically equivalent* (this can be proved rigorously using the material of Chapter VIII, §18). Figures 75a and 75b show phase portraits which, though sharply differing as to shape, nevertheless have the same topological structure. Figures 76a and 76b illustrate "similar" phase portraits with different topological structure. The same holds true for the phase portraits in Figure 77, which at first sight look "similar."

a)                                                    b)

FIGURE 75.

* The phase portraits of Figure 74 are defined, for example, by the systems

$$\frac{dx}{dt}=y, \qquad \frac{dy}{dt}=-x \qquad \text{and} \qquad \frac{dx}{dt}=y^3, \qquad \frac{dy}{dt}=-x^3.$$

a)                                      b)

FIGURE 76.

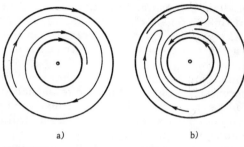

a)                    b)

FIGURE 77.

**3. LOCAL TOPOLOGICAL STRUCTURE.** The question of characterizing the phase portrait "in the small" has already been touched upon (§1.14 and §3.2). We shall now consider this question more closely and introduce the natural concept of local topological structure.

Let $P$ be an interior point of the region $G$ in which a dynamic system (I) is defined. $P$ may be either a regular point or an equilibrium state.

*Definition VII. We shall say that system (I) has local topological structure at the point P if there exists a region $w_0$ containing P and satisfying the following condition: for any $\varepsilon > 0$, there exist a region $w_0'$ and a mapping T such that a) $w_0'$ contains P and is contained in $U_\varepsilon(P)$; b) T is an identifying mapping taking $w_0$ onto $w_0'$ and P onto itself.*

A region $w_0$ satisfying the condition of the definition will be called a region (or neighborhood) of local topological structure. In particular, if $\varepsilon_0 > 0$ is sufficiently small, the neighborhood $U_{\varepsilon_0}(P)$ of a point $P$ (at which the system has local topological structure) will be a neighborhood of local topological structure for $P$.

Before going on to examples demonstrating the existence or nonexistence of a local topological structure, we define identical [equivalent] local topological structures.

*Definition VIII. Let $P_1$ and $P_2$ be points at which two dynamic systems, $(A_1)$ and $(A_2)$, respectively, have local topological structures (the systems $(A_1)$ and $(A_2)$ or points $P_1$ and $P_2$ may be the same). We shall say that the local topological structures at these points are identical [or*

---

* The mapping $T$ takes paths (more precisely, arcs of paths) of system (I) into (arcs of) paths of the same system (see beginning of §5.2).

*equivalent] if $P_1$ has a neighborhood which can be mapped onto a neighbor-hood of $P_2$ in such a way that $P_1$ is mapped onto $P_2$ and paths onto paths.*

It is easy to see that we need only assume the existence of a local topological structure at one of the points $P_1$ and $P_2$ in this definition. If there exists a mapping as specified in the definition, of the neighborhood of one of the points onto a neighborhood of the other, then the second point will "automatically" possess local topological structure, identical with that of the first. It follows from Theorem 8 in §3 and from elementary arguments that any dynamic system of the type considered here (satisfying the conditions of §1) has local topological structure, indeed the same local topological structure, at any point which is not an equilibrium state: the local structure is that of a family of parallel lines.

Local topological structure can exist even at an equilibrium state. It is readily shown that this is true for all the examples of §1.14.

If local topological structure exists at a point which is an equilibrium state, we shall call it the topological structure of the equilibrium state.

The following geometric example will show that local topological structure may fail to exist.* Consider an infinite sequence of circles $\{C_i\}$ of decreasing radii centered at a point $O$, indexed in order of decreasing radii. We construct a family of curves as follows: insert closed curves between every pair of circles $C_1$ and $C_2$, $C_3$ and $C_4$, . . ., $C_{2i-1}$ and $C_{2i}$, . . . and between any two circles $C_{2i}, C_{2i+1}$ exactly $i$ limit cycles lying inside one another and no other closed paths. Then there is no local topological structure at $O$ in the above sense.

In this book we shall confine ourselves in the main to dynamic systems which possess local topological structure at all points. Since this is always the case at regular points, and moreover the local topological structure at any two such points is the same, it is obvious that the crucial question is the existence of local topological structure at each equilibrium state.

The following remark, though elementary, is of fundamental importance. Knowledge of the local topological structure at all points of the region in which a dynamic system is defined does not yield exhaustive information on the behavior of the paths in the large, though it does imply certain conclusions in this respect. This is evident from simple examples. Thus, for example, the local topological structure is the same at all points of the phase portraits illustrated in Figure 78. Nevertheless, the topological structures of these phase portraits "in the large" are obviously quite different: the phase portrait of Figure 78a contains a limit cycle, while there is no limit cycle in that of Figure 78b.

**4. PROPERTIES OF PHASE PORTRAITS IN THE LARGE AND EFFECTIVE METHODS OF QUALITATIVE INVESTIGATION.** Qualitative study of the phase portrait in the large, in other words, study of the topological structure of the phase portrait, is evidently the fundamental problem of the qualitative theory of dynamic systems. This problem possesses two distinct aspects.

---

* We are not concerned here with the construction of an actual dynamic system having this phase portrait (such a system undoubtedly exists), since in the sequel we shall consider systems which have local topological structure at any point.

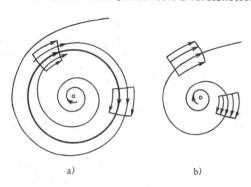

a)                                b)

FIGURE 78.

The first is to determine what properties the phase portrait may possess in general (subject to various restrictions on the right-hand sides of the system). The range of questions arising here borders directly, as to both content and methods, upon the material of Chapter II — investigation of possible types of paths taken singly and the simplest properties of phase portraits in the large (§4). Further investigation of the properties of the phase portrait naturally raises a host of new questions. Elementary examples of phase portraits (§1) shows that not all paths display equivalent behavior: some of them are exceptional or, as we have said, "singular." Examples of such paths are equilibrium states and closed paths. One would naturally like to impart a rigorous sense to the concept of a "singular" path, to determine in general all possible types of "singular" paths and ascertain their roles in the phase portrait, and so on. Finally, the question arises as to what information about paths, in particular, singular paths, is necessary to determine the topological structure of the phase portrait, at least in certain relatively narrow classes of dynamic systems. This last question is directly and organically related to the need, mentioned in §5.2, to specify what is meant by "determining the topological structure of the phase portrait."

The other aspect of qualitative investigation of the phase portrait in the large is the quest for effective techniques or methods to this end, i.e., effective methods to determine the topological structure of the phase portrait or various topologically invariant properties thereof, for a given dynamic system.*

The first problem arising in this context is to find effective methods to study the topological structure of equilibrium states, since the latter clearly play a major role in the phase portrait. In addition, the investigation of equilibrium states is of great importance for applications.

---

\* There is a far-reaching analogy between these two aspects of qualitative investigation of dynamic systems and the same two aspects of "qualitative investigation" of other mathematical objects, such as equations, curves, etc. For example, in qualitative investigation of algebraic curves of fixed degree one asks first what types of curves can occur. On the other hand, here too one is interested in effective methods or techniques for qualitative investigation of an algebraic curve on the basis of the equation $F(x, y) = 0$ defining it.

The next goal, which is of paramount importance, is methods or techniques to ascertain the existence or nonexistence of limit cycles. This question too is crucial for applications. It is far more difficult than determination of the topological structure of equilibrium states and far less attention has been devoted to it. In order to formulate further questions concerning effective methods for qualitative investigation of a given dynamic system, we must first know just what information about paths is necessary to determine the topological structure of the phase portrait. And an obvious prerequisite to this knowledge is a complete answer to the questions comprising the first aspect.

Thus, the logical approach to investigation of the phase portrait would seem to dictate that one should first attack in depth the questions of the first category — the general properties of the phase portrait, the factors conditioning its topological structure, etc., and only then embark on the search for effective methods relating to the phase portrait in the large. However, the sequence of exposition adopted in the present book does not follow this so to speak "formally consistent" order: we first present (in the first three chapters) the basic and classical effective methods, by whose means one can examine numerous examples and accumulate certain "intuitive" material; only then do we proceed to a more detailed examination of the properties of the phase portrait as a whole. Moreover, historically speaking, the development of the qualitative theory of dynamic systems did not follow the above "formally consistent" approach. Classical examples of effective investigation are the methods for investigation of simple equilibrium states presented in the next chapter.

Later we shall also indicate techniques whereby one can establish the existence or nonexistence of limit cycles. We shall then employ them for the (complete or incomplete) qualitative investigation of a great number of specific examples.

*Chapter IV*

*SIMPLE EQUILIBRIUM STATES*

INTRODUCTION

This chapter is devoted to the study of simple equilibrium states.*
If

$$\frac{dx}{dt} = P(x, y), \qquad \frac{dy}{dt} = Q(x, y) \tag{I}$$

is a given dynamic system then, as we have seen (§1), its equilibrium
states are the common points of the curves

$$P(x, y) = 0, \qquad Q(x, y) = 0. \tag{1}$$

An equilibrium state $M(x_0, y_0)$ is said to be s i m p l e if it is a point of
intersection of the curves (1), i.e., a common point at which the curves
have no singularities and their tangents at the point are distinct. A
simple equilibrium state is isolated.

Chapter IV contains four sections (§§6 − 10), given over primarily to
investigation of the topological structure of simple equilibrium states.

The basic tools to this end are the c h a r a c t e r i s t i c  e q u a t i o n
and its roots, the c h a r a c t e r i s t i c  r o o t s (v a l u e s) of an equilibrium
state. The characteristic roots of a simple equilibrium state are not zero.
Depending on the roots of the characteristic equation (on whether they are
real or complex, distinct or equal), system (I) can be reduced by a suitable
linear transformation of variables in the neighborhood of an equilibrium
state to a particularly simple form, known as the c a n o n i c a l  f o r m.
Reduction to canonical form is described in §6.

Sections 7 and 8 discuss investigation proper of the topological structure
of a simple equilibrium state. The investigation makes use of the canonical
form of the system. In §7 we consider all possible cases of characteristic
roots, except for the case in which they are pure imaginary. It is shown
that in these cases the signs of the characteristic roots or of their real
parts completely determine the topological structure of the equilibrium
state: node (characteristic roots $\lambda_1$ and $\lambda_2$ real and of the same sign),
focus (characteristic roots complex but not pure imaginary) or saddle point

---

* Concerning the role of equilibrium states in the context of complete qualitative investigation, see
  Chapter III. The material of §6 may also be found in Stepanov /22/ and Pontryagin /11/. Its inclusion
  here is dictated by the desire to present an exhaustive exposition of the basic classical techniques for
  qualitative investigation of system (I).

(characteristic roots real and of opposite signs). Moreover, the node and focus possess the same topological structure.* In this case, therefore, we have at our disposal an effective method for determining the topological structure of an equilibrium state: compute the characteristic roots and determine their signs (or the signs of their real parts).

Section 8 is devoted to the considerably more difficult case of pure imaginary characteristic roots. In this case the characteristic roots do not fully determine the topological structure of the equilibrium states and additional investigation is necessary. Here the possible types of equilibrium state are focus, center, and what we call centrofocus.

In §9 we consider a question which, though outside the framework of purely topological considerations, arises quite naturally in the context of simple equilibrium states — the question of whether the paths tend to an equilibrium state in a definite direction or not (for the rigorous definitions, see §9.1). Consideration of this subject reveals a "non-topological" distinction between nodes and foci: paths tending to a node do so in a definite direction, whereas paths tending to a focus are spiral-shaped.**

In §10, we present examples of investigation of simple equilibrium states.

An equilibrium state which is not simple is said to be multiple. In contrast to a simple equilibrium state (simple point of intersection of the curves (1)), a multiple equilibrium state is either a point of contact (tangency) of these curves, or a point at which one or both curves have a singularity. At least one characteristic root of a multiple equilibrium state is zero. A few basic types of multiple equilibrium states will be considered in Chapter IX. A multiple equilibrium state may be either isolated or nonisolated. We have already stated (§1.4) that an equilibrium state $O$ is isolated if it has a neighborhood containing no other equilibrium state. Any neighborhood, however small, of a nonisolated equilibrium state $O$ contains other equilibrium states, so that a nonisolated equilibrium state is a point of accumulation of equilibrium states. The converse is also true: any point of accumulation of equilibrium states is an equilibrium state.

We append a brief discussion of equilibrium states of analytic dynamic systems, since in this respect there is an essential difference between analytic and nonanalytic systems. Suppose that the right-hand sides $P(x, y)$ and $Q(x, y)$ of an analytic dynamic system have no common factor which can vanish, i. e., they cannot be expressed as

$$P(x, y) = P_1(x, y) f(x, y), \quad Q(x, y) = Q_1(x, y) f(x, y) \tag{2}$$

where $P$, $Q$, $P_1$, $Q_1$ are analytic functions and $f(x, y)$ is a function which can vanish. Then the system can have only finitely many equilibrium states in any bounded plane region.

---

\*  All these types of equilibrium state were met with in the examples of §1. The procedure requires a knowledge of the coordinates of the equilibrium state, i. e., we must know how to find the common roots of equations (1). Derivation of effective methods to this end is of course far from trivial and involves special considerations. However, we shall not dwell on this subject here but assume throughout the chapter that the coordinates of the equilibrium state are known.

\*\*  In applied problems, the distinction between a node and a focus is important in regard to the study of "transients," the question of how a system settles down to the steady-state conditions corresponding to an equilibrium state (in an oscillatory manner or not).

This follows from the assertion: two analytic functions $P(x, y)$ and $Q(x, y)$, defined in a closed bounded region $\overline{G}$, neither of which is identically zero and which cannot be expressed in the form (2), cannot vanish together at infinitely many points of $\overline{G}$. Indeed, suppose that $P(x, y)$ and $Q(x, y)$ vanish at infinitely many points $M_i(x_i, y_i)$ in $\overline{G}$:

$$P(x_i, y_i) = Q(x_i, y_i) = 0.$$

Let $M_0(a, b) \in \overline{G}$ be a point of accumulation of the points $M_i$. Clearly, $P(a, b) = 0$, $Q(a, b) = 0$. Suppose that at least one of the partial derivatives $P'_x, P'_y, Q'_x, Q'_y$ does not vanish at $M_0(a, b)$. To fix ideas, we assume that $P'_y(a, b) \neq 0$. Since $P(a, b) = 0$, it follows from the implicit function theorem that we can solve the equation $P(x, y) = 0$ for $y$ as a function of $x$ in a neighborhood of the point $M_0(a, b)$. The function $y = \varphi(x)$ is analytic and $\varphi(x_i) = y_i$. Consider the function $F(x) = Q(x, \varphi(x))$. This function vanishes at infinitely many points,

$$F(x_i) = Q(x_i, \varphi(x_i)) = 0,$$

but it is not identically zero by virtue of the assumed properties of $P(x, y)$ and $Q(x, y)$.

Hence (by Rolle's theorem) there exists an infinite set of numbers $\xi_i$ such that $\xi_i \to a$ and $F'(\xi_i) = 0$. Since $F'(x)$ is continuous, this implies that $F'(a) = 0$. Similarly, one proves that $F''(a) = F'''(a) = \ldots = 0$. But then $F(x)$, as an analytic function, must vanish identically, contradicting the assumption. This assertion may also be proved without assuming that one of the derivatives $P'_x, P'_y, Q'_x, Q'_y$ does not vanish, but the proof is a little more difficult.

Thus, in any bounded plane region an analytic dynamic system has either only finitely many equilibrium states or singular curves all of whose points are equilibrium states (points of a curve $f(x, y) = 0$). An infinite set of isolated equilibrium states with a point of accumulation, which may occur in nonanalytic systems, is impossible in the analytic case.

## §6. REDUCTION OF A DYNAMIC SYSTEM TO CANONICAL FORM IN THE NEIGHBORHOOD OF A SIMPLE EQUILIBRIUM STATE

### 1. ANALYTICAL CONDITIONS CHARACTERIZING A SIMPLE EQUILIBRIUM STATE.

$$\frac{dx}{dt} = P(x, y), \qquad \frac{dy}{dt} = Q(x, y) \tag{1}$$

be a dynamic system defined in some plane region $G$ and $M(x_0, y_0)$ an equilibrium state of system (1) ($M \in G$).

$M$ is a simple equilibrium state if it is a point of intersection of the curves

$$P(x, y) = 0, \qquad Q(x, y) = 0, \tag{2}$$

i. e., the tangents to these curves at their common point exist and are distinct. Consequently, $M (x_0, y_0)$ is a simple equilibrium state if

$$\Delta (x_0, y_0) = \begin{vmatrix} P_x' (x_0, y_0) & P_y' (x_0, y_0) \\ Q_x' (x_0, y_0) & Q_y' (x_0, y_0) \end{vmatrix} \neq 0, \tag{3}$$

that is to say, if the Jacobian

$$\frac{D (P, Q)}{D (x, y)} = \Delta (x, y)$$

does not vanish at $M (x_0, y_0)$. A simple equilibrium state is thus charac-terized by the conditions

$$P (x_0, y_0) = 0, \quad Q (x_0, y_0) = 0, \quad \Delta (x_0, y_0) \neq 0$$

and so, by the implicit function theorem (see Appendix, §4.3, Theorem VI, Remark II) a simple equilibrium state $M$ is isolated, i. e., there exists a neighborhood of $M$ containing no equilibrium states other than $M$.

In the neighborhood of $M (x_0, y_0)$, we can express the functions $P (x, y)$ and $Q (x, y)$ as

$$P (x, y) = P_x' (x_0, y_0) (x - x_0) + P_y' (x_0, y_0) (y - y_0) + \varphi (x, y),$$
$$Q (x, y) = Q_x' (x_0, y_0) (x - x_0) + Q_y' (x_0, y_0) (y - y_0) + \psi (x, y).$$

The functions $\varphi (x, y)$ and $\psi (x, y)$ are obviously of class $C_1$ and

$$\varphi (x_0, y_0) = \psi (x_0, y_0) = 0,$$
$$\varphi_x' (x_0, y_0) = \varphi_y' (x_0, y_0) = \psi_x' (x_0, y_0) = \psi_y' (x_0, y_0) = 0.$$

To simplify matters, we shall assume that the equilibrium state under consideration is at the origin, $x_0 = y_0 = 0$ (this can be ensured by the change of variables $x' = x - x_0$, $y' = y - y_0$, followed by a return to the old notation; our assumption therefore involves no loss of generality). Denote the numbers $P_x' (0, 0)$, $P_y' (0, 0)$, $Q_x' (0, 0)$, $Q_y' (0, 0)$ by $a, b, c, d$, respectively. System (1) then becomes

$$\frac{dx}{dt} = ax + by + \varphi (x, y), \quad \frac{dy}{dt} = cx + dy + \psi (x, y), \tag{4}$$

where

$$\varphi (0, 0) = \psi (0, 0) = 0,$$
$$\varphi_x' (0, 0) = \varphi_y' (0, 0) = \psi_x' (0, 0) = \psi_y' (0, 0) = 0. \tag{5}$$

By assumption,

$$\Delta = \frac{D (P, Q)}{D (x, y)} \Big|_{\substack{x=0 \\ y=0}} = \begin{vmatrix} a & b \\ c & d \end{vmatrix} \neq 0, \tag{6}$$

since the point $O (0, 0)$ is a simple equilibrium state. It follows from (5) that

$$\varphi(x,\ y) = O(\varrho), \qquad \psi(x,\ y) = O(\varrho), \tag{7}$$

where $\varrho = \sqrt{x^2 + y^2}$ .

**2. REDUCTION OF A DYNAMIC SYSTEM TO CANONICAL FORM IN THE NEIGHBOR-HOOD OF A SIMPLE EQUILIBRIUM STATE.** We shall first show that system (1) can be reduced by a nonsingular linear transformation to a canonical form which is more convenient to investigate.

Let

$$X = p_{11}x + p_{12}y, \qquad Y = p_{21}x + p_{22}y \tag{8}$$

be a nonsingular linear transformation,* i. e., a transformation such that

$$D = \begin{vmatrix} p_{11} & p_{12} \\ p_{21} & p_{22} \end{vmatrix} \neq 0.$$

The inverse transformation

$$x = q_{11}X + q_{12}Y, \qquad y = q_{21}X + q_{22}Y, \tag{9}$$

exists and is also nonsingular.

Applying the transformation (8) to system (4), we get a system

$$\frac{dX}{dt} = a_1X + b_1Y + \varphi_1(X,\ Y), \qquad \frac{dY}{dt} = c_1X + d_1Y + \psi_1(X,\ Y), \tag{10}$$

where the coefficients $a_1$, $b_1$, $c_1$, $d_1$ are expressed in terms of $a$, $b$, $c$, $d$ and the $p_{ik}$. The exact expressions are easily found by direct computation. If we denote the matrix of the linear terms in system (4) by $A$,

$$A = \begin{pmatrix} a & b \\ c & d \end{pmatrix},$$

the matrix of the transformation (8) by

$$S = \begin{pmatrix} p_{11} & p_{12} \\ p_{21} & p_{22} \end{pmatrix},$$

the matrix of the inverse transformation (9) by $S^{-1}$,

$$S^{-1} = \begin{pmatrix} q_{11} & q_{12} \\ q_{21} & q_{22} \end{pmatrix}$$

---

* We are interested only in nonsingular transformations, i. e., transformations which map the plane onto itself. A transformation such that not all the $p_{ik}$ vanish but their determinant $D$ vanishes is readily seen to map the entire plane onto a straight line. Indeed, if $D = 0$, then clearly $p_{21} = \mu p_{11}$, $p_{22} = \mu p_{12}$, and the transformation is

$$X = p_{11}x + p_{12}y, \qquad Y = \mu(p_{11}x + p_{12}y),$$

so that it maps the entire $(x,\ y)$-plane onto the straight line $Y - \mu X = 0$.

and the coefficient matrix of the linear terms of system $(10)$ by $B$,

$$B = \begin{pmatrix} a_1 & b_1 \\ c_1 & d_1 \end{pmatrix},$$

then a simple computation shows that

$$B = SAS^{-1}, \tag{11}$$

where the multiplication in $(11)$ is matrix multiplication. The functions $\varphi_1(X, Y)$ and $\psi_1(X, Y)$ are obviously expressed in terms of $\varphi(x, y)$ and $\psi(x, y)$ by

$$
\begin{aligned}
\varphi_1(X, Y) &= p_{11}\varphi(x, y) + p_{12}\psi(x, y) = \\
&= p_{11}\varphi(q_{11}X + q_{12}Y, \quad q_{21}X + q_{22}Y) + p_{12}\psi(q_{11}X + q_{12}Y, \quad q_{21}X + q_{22}Y), \quad (12) \\
\psi_1(X, Y) &= p_{21}\varphi(x, y) + p_{22}\psi(x, y) = p_{21}\psi(q_{11}X + q_{12}Y, \quad q_{21}X + q_{22}Y) + \\
&\qquad + p_{22}\psi(q_{11}X + q_{12}Y, \quad q_{21}X + q_{22}Y).
\end{aligned}
$$

It follows from $(5)$ and $(12)$ that

$$
\begin{aligned}
\varphi_1(0,\ 0) &= \psi_1(0,\ 0,) = 0, \\
\varphi_{1x}'(0,\ 0) &= \varphi_{1y}'(0,\ 0) = \psi_{1x}'(0,\ 0) = \psi_{1y}'(0,\ 0) = 0.
\end{aligned} \tag{13}
$$

Thus the functions $\varphi_1$ and $\psi_1$ play the same role for system $(10)$ as the functions $\varphi$ and $\psi$ for system $(4)$.

Consider the quadratic equation

$$\begin{vmatrix} a - \lambda & b \\ c & d - \lambda \end{vmatrix} = \lambda^2 - \sigma\lambda + \Delta = 0. \tag{14}$$

Here $\sigma = a + d$ and $\Delta = ac - bd$ ($\Delta$ has the same meaning as in $(3)$). Equation $(14)$ is known as the characteristic equation of the equilibrium state $O$ and its roots as the characteristic roots (or characteristic values) of the equilibrium state $O$. The characteristic equation and its roots play a major role in investigation of the topological structure of an equilibrium state. Equations of type $(14)$ appear in a great variety of problems. It is also known as the "secular equation." The numbers $\lambda_1$ and $\lambda_2$ satisfying the equation are the characteristic values or eigenvalues of the matrix $A$.

Lemma 1.* Let $\lambda_1$ and $\lambda_2$ be the characteristic roots of an equilibrium state $O$. Then: 1) If $\lambda_1$ and $\lambda_2$ are real and distinct, there exists a non-singular real transformation $(8)$ reducing system $(4)$ to the form

$$\frac{dX}{dt} = \lambda_1 X + \varphi_1(X,\ Y), \qquad \frac{dY}{dt} = \lambda_2 Y + \psi_1(X,\ Y). \tag{15}$$

---

* This lemma follows directly from a standard theorem of linear algebra — reduction of a square matrix of order 2 to Jordan form (see, e.g., /21/). We give an independent proof here, without appeal to matrix algebra or notation. The informed reader may skip the proof of Lemma 1 and take note only of its statement.

2) *If $\lambda_1 = \lambda_2$, there exists a nonsingular real transformation reducing system* (4) *to the form*

$$\frac{dX}{dt} = \lambda_1 X + \varphi_1(X, Y), \qquad \frac{dY}{dt} = \mu X + \lambda_1 Y + \psi_1(X, Y), \tag{16}$$

*where* $\mu = 0$ *if the coefficients $b$ and $c$ in system* (4) *vanish, and $\mu \neq 0$ otherwise (in which case $\mu$ may be any prescribed number).*

3) *If $\lambda_1$ and $\lambda_2$ are complex, $\lambda_1 = \alpha + i\beta$, $\lambda_2 = \alpha - i\beta$, $\beta \neq 0$, $\alpha$ arbitrary, there exists a real nonsingular transformation* (8) *reducing system* (4) *to the form*

$$\frac{dX}{dt} = \alpha X - \beta Y + \varphi_1(X, Y), \qquad \frac{dY}{dt} = \beta X + \alpha Y + \psi_1(X, Y). \tag{17}$$

Proof. We shall try to choose the coefficients $p_{ik}$ of the transformation (8) in such a way that in terms of the variables $X, Y$ system (10) will have the following, "canonical" form:

$$\frac{dX}{dt} = \lambda_1 X + \varphi_1(X, Y), \qquad \frac{dY}{dt} = \lambda_2 Y + \psi_1(X, Y). \tag{18}$$

We must first determine conditions for system (4) to be reducible to the canonical form (18). We have

$$\begin{aligned}
\dot{X} &= p_{11}\dot{x} + p_{12}\dot{y} = p_{11}P(x, y) + p_{12}Q(x, y) = \\
&= p_{11}(ax + by) + p_{12}(cx + dy) + p_{11}\varphi(x, y) + p_{12}\psi(x, y), \\
\dot{Y} &= p_{21}\dot{x} + p_{22}\dot{y} = p_{21}P(x, y) + p_{22}Q(x, y) = \\
&= p_{21}(ax + by) + p_{22}(cx + dy) + p_{21}\varphi(x, y) + p_{22}\psi(x, y).
\end{aligned} \tag{19}$$

On the other hand, transforming system (4) to the form (10) by means of (8), we must have

$$\begin{aligned}
\dot{X} &= \lambda_1 X + \varphi_1(X, Y) = \lambda_1(p_{11}x + p_{12}y) + \varphi_1(p_{11}x + p_{12}y, \ p_{21}x + p_{22}y), \\
\dot{Y} &= \lambda_2 Y + \psi_1(X, Y) = \lambda_2(p_{21}x + p_{22}y) + \psi_1(p_{11}x + p_{12}y, \ p_{21}x + p_{22}y),
\end{aligned} \tag{20}$$

and then the right-hand sides of (19) and (20) must be identically equal. Equating the linear and nonlinear terms separately, we get the following identities in $x$ and $y$:

$$\begin{aligned}
p_{11}(ax + by) + p_{12}(cx + dy) &\equiv \lambda_1(p_{11}x + p_{12}y), \\
p_{21}(ax + by) + p_{22}(cx + dy) &\equiv \lambda_2(p_{21}x + p_{22}y), \\
p_{11}\varphi(x, y) + p_{12}\psi(x, y) &\equiv \varphi_1(p_{11}x + p_{12}y, \ p_{21}x + p_{22}y), \\
p_{21}\varphi(x, y) + p_{22}\psi(x, y) &\equiv \psi_1(p_{11}x + p_{12}y, \ p_{21}x + p_{22}y).
\end{aligned} \tag{21}$$

It follows from the second pair of identities that

$$\begin{aligned}
\varphi_1(X, Y) &= p_{11}\varphi(q_{11}X + q_{12}Y, \ q_{21}X + q_{22}Y) + p_{12}\psi(q_{11}X + q_{12}Y, \ q_{21}X + q_{22}Y), \\
\psi_1(X, Y) &= p_{21}\varphi(q_{11}X + q_{12}Y, \ q_{21}X + q_{22}Y) + p_{22}\psi(q_{11}X + q_{12}Y, \ q_{21}X + q_{22}Y).
\end{aligned}$$

Hence, by (5), it is obvious that the functions $\varphi_1(X, Y)$ and $\psi_1(X, Y)$ satisfy conditions (13). Now consider the first pair of identities (21). Collecting like terms, we get

$$[p_{11}(a-\lambda_1)+p_{12}c]\,x+[p_{11}b+p_{12}(d-\lambda_1)]\,y \equiv 0,$$
$$[p_{21}(a-\lambda_2)+p_{22}c]\,x+[p_{21}b+p_{22}(d-\lambda_2)]\,y \equiv 0.$$

Since these equations must hold identically, the coefficients of $x$ and $y$ vanish and we obtain the following conditions, which $p_{ik}, \lambda_1, \lambda_2$ must satisfy for system (4) to be reducible to the form (18):

$$p_{11}(a-\lambda_1)+p_{12}c=0, \qquad p_{11}b+p_{12}(d-\lambda_1)=0. \tag{22}$$

Similarly, we obtain the following conditions for $p_{21}$ and $p_{22}$:

$$p_{21}(a-\lambda_2)+p_{22}c=0, \qquad p_{21}b+p_{22}(d-\lambda_2)=0. \tag{23}$$

We are of course interested only in nontrivial solutions of these (homogeneous) systems (since the transformation (8) is to be nonsingular). Nontrivial solutions exist only if the determinants of these systems vanish, i. e., if $\lambda_1$ and $\lambda_2$ are roots of the characteristic equation*

$$\begin{vmatrix} a-\lambda & b \\ c & d-\lambda \end{vmatrix} = \lambda^2-\sigma\lambda+\Delta = 0. \tag{24}$$

Let $\lambda_1$ and $\lambda_2$ be roots of this equation.

We now consider in order the three cases figuring in the statement of the lemma.

1) $\lambda_1 \neq \lambda_2$. Since the determinants of systems (22) and (23) vanish, only one equation in each system can be independent. Suppose first that $b = c = 0$, so that the characteristic equation is

$$(a-\lambda)(d-\lambda)=0,$$

and hence $\lambda_1 = a$, $\lambda_2 = d$ (by assumption, $\lambda_1 \neq \lambda_2$). But this clearly implies that the original system is already in canonical form (18), so that the required transformation is in this case the identity.

Now suppose that $b$ and $c$ are not both zero, say $c \neq 0$. Then, by the first equations of systems (22) and (23),

$$\frac{p_{12}}{p_{11}} = \frac{\lambda_1-a}{c}, \qquad \frac{p_{22}}{p_{21}} = \frac{\lambda_2-a}{c}.$$

With this choice of elements $p_{ik}$, it is obvious that $D \neq 0$. Indeed, it is readily seen that if $D$ vanishes then

---

Equation (24) is exactly the characteristic equation of the linear system

$$\dot{x}= ax+by, \qquad \dot{y}=cx+dy$$

(obtained from system (4) by dropping nonlinear terms). Indeed, substituting $x = \alpha e^{\lambda t}$, $y = \beta e^{\lambda t}$ in this linear system, we get equation (24) for $\lambda$ and systems of the type (22), (23) for $\alpha$ and $\beta$.

$$\frac{\lambda_1 - a}{c} = \frac{\lambda_2 - a}{c},$$

i. e., $\lambda_1 = \lambda_2$, contradicting the assumption. We have thus shown that there exists a nonsingular linear transformation reducing system (4) to the form (18).

2) $\lambda_1 = \lambda_2$. In this case the systems (22) and (23) defining the $p_{ik}$ coincide, so that the previous arguments are inapplicable. Suppose first, as in case 1), that $b = c = 0$. Then the characteristic equation is

$$(a-\lambda)(d-\lambda) = 0,$$

and since $\lambda_1 = \lambda_2$ we have $a = d$; thus the original system is already in canonical form (18) with $\mu = 0$.

Now suppose that $b$ and $c$ do not both vanish, say $c \neq 0$. Then not all coefficients in the first equation of (22) are zero. Now, in this case reduction of system (4) to the canonical form (18) is impossible, since there are no solutions of system (22) such that $D \neq 0$. We shall therefore reduce system (4) to the form (16) instead.

As before, the first equation of system (22) gives

$$\frac{p_{12}}{p_{11}} = \frac{\lambda_1 - a}{c}.$$

Since $\lambda_1 = \lambda_2$, we have

$$\sigma^2 - 4\Delta = 0 \quad \text{and} \quad \lambda_1 = \frac{a+d}{2}.$$

Setting $p_{11} = c$, we get $p_{12} = \frac{d-a}{2}$. Thus the first part of the linear transformation of (8) is

$$X = \frac{d-a}{2} y + cx.$$

As mentioned above, there exist no coefficients $p_{ik}$ satisfying system (22) such that $D \neq 0$. As the second part of (8) we therefore take $Y = \mu y$. The whole transformation is thus

$$X = \frac{d-a}{2} y + cx, \qquad Y = \mu y.$$

It is readily seen that this transformation reduces system (4) to the form (16), which we call the canonical form when the characteristic equation has multiple roots.

3) The characteristic roots are complex conjugates, $\lambda_1 = \alpha + i\beta$, $\lambda_2 = \alpha - i\beta$ ( $\beta \neq 0$, $\alpha$ may or may not vanish). Using arguments similar to those in case 1), we can reduce system (4) to the canonical form (18) with complex conjugate $\lambda_1$ and $\lambda_2$. The coefficients of the linear transformation (8) effecting this reduction are also complex conjugates:

$$\frac{p_{12}}{p_{11}} = \frac{a+i\beta-a}{c}, \qquad \frac{p_{22}}{p_{21}} = \frac{a-i\beta-a}{c} \tag{25}$$

(we may assume that $p_{11} = p_{21} = c$). But we are concerned with a dynamic system whose right-hand sides are real functions of real variables. A canonical form in which $X$ and $Y$ assume complex values for real $x$ and $y$ is therefore useless. We shall therefore derive the canonical form (17), as follows. Since the coefficients of (8) are complex conjugates, the variables $X$ and $Y$ in (8) are also complex conjugates. Setting $p_{11} = p_{21} = c$, we get

$$X = (a+i\beta-a)\,y + cx = u + iv, \qquad Y = (a-i\beta-a)\,y + cx = u - iv,$$

whence

$$u = \frac{X+Y}{2} = (a-a)\,y + cx, \qquad v = \frac{X-Y}{2i} = \beta y. \tag{26}$$

The transformation (26) is obviously nonsingular, since its determinant is

$$\begin{vmatrix} c & a-a \\ 0 & \beta \end{vmatrix} = \beta c \neq 0,$$

since $\beta \neq 0$ and by assumption $c \neq 0$. The coefficients of the transformation are real. In terms of the variables $u$ and $v$, the linear terms of the dynamic system obtained from (4) are most simply determined as follows:

$$\frac{du}{dt} = \frac{1}{2}\left[\frac{dX}{dt} + \frac{dY}{dt}\right] = \frac{1}{2}[\lambda_1 X + \varphi_1 + \lambda_2 Y + \psi_1],$$

$$\frac{dv}{dt} = \frac{1}{2i}\left[\frac{dX}{dt} - \frac{dY}{dt}\right] = \frac{1}{2i}[\lambda_1 X - \lambda_2 Y + \varphi_1(X,\ Y) - \psi_1(X,\ Y)],$$

whence, replacing $X$, $Y$ by their expressions in terms of $u$ and $v$, we obtain

$$\frac{du}{dt} = au - \beta v + \varphi_2(u,\ v), \qquad \frac{dv}{dt} = \beta u + av + \psi_2(u,\ v), \tag{27}$$

which is precisely (17) except for notation. This completes the proof of Lemma 1.

Remark. Our proof of Lemma 1 made no use of the fact that $\Delta \neq 0$, and therefore the statement remains valid when $\Delta = 0$.

Throughout the above reasoning we assumed that the equilibrium state is at the origin. In the general case of an equilibrium state $M(x_0, y_0)$, the characteristic equation is

$$\begin{vmatrix} P'_x(x_0,\ y_0) - \lambda & P'_y(x_0,\ y_0) \\ Q'_x(x,\ y_0) & Q'_y(x_0,\ y_0) - \lambda \end{vmatrix} = \lambda^2 - \sigma\lambda + \Delta = 0.$$

In this case, obviously,

$$\sigma = P'_x(x_0,\ y_0) + Q'_y(x_0,\ y_0), \qquad \Delta = \begin{vmatrix} P'_x(x_0,\ y_0) & P'_y(x_0,\ y_0) \\ Q'_x(x_0,\ y_0) & Q'_y(x_0,\ y_0) \end{vmatrix}.$$

### 3. INVARIANCE OF THE CHARACTERISTIC EQUATION UNDER A REGULAR MAPPING.

Suppose that system (4) is reduced by some transformation (8) to the form (10) (not necessarily canonical form). The characteristic equation of the new system is

$$\begin{vmatrix} a_1 - \lambda & b_1 \\ c_1 & d_1 - \lambda \end{vmatrix} = 0.$$

It is natural to expect the roots of this equation to coincide with those of the characteristic equation of system (4). This is indeed, the case:

Lemma 2. The characteristic roots of an equilibrium state of system (4) are invariant under nonsingular linear transformations.

Proof. The proof follows directly from the fact that the matrix $B$ of the linear terms of the transformed system is related to the matrix $A$ of linear terms of the original system by

$$B = SAS^{-1},$$

where $S$ is the matrix of the transformation and $S^{-1}$ that of its inverse. By standard theorems of linear algebra, the matrix $A$ and any similar matrix $B$ have the same characteristic roots.*

Remark. Under any regular transformation of variables (see Appendix, §5) $u = f(x, y)$, $v = g(x, y)$, the characteristic equation of an equilibrium state $M_0(x_0, y_0)$ behaves as under the linear transformation with coefficients

$$p_{11} = f'_x(x_0, y_0), \qquad p_{12} = f'_y(x_0, y_0),$$
$$p_{21} = g'_x(x_0, y_0), \qquad p_{22} = g'_y(x_0, y_0).$$

Hence it clearly follows that the characteristic roots of an equilibrium state are invariant under any regular transformation of variables. This in turn implies that the quantities $\sigma$ and $\Delta$ are also preserved under any regular transformation, so that they are invariants of regular transformations.

### 4. PRELIMINARY REMARKS ON THE POSSIBLE TOPOLOGICAL STRUCTURE OF SIMPLE EQUILIBRIUM STATES.

Consider system (1), assuming as before that the point $O(0, 0)$ is a simple equilibrium state ($\Delta \neq 0$, so that neither of the roots $\lambda_1$ and $\lambda_2$ is zero).

By Lemma 1, we may always reduce the system by a suitable non-singular linear transformation to one of three forms, which we call canonical forms.

Our goal in this chapter is to investigate the topological structure of a simple equilibrium state. It is sufficient to do this for systems in canonical form, for by Lemma 1 any system (4) can be reduced to canonical form by a nonsingular linear transformation of type (8). Since such a trans-formation is certainly a topological mapping, the equilibrium state $O(0, 0)$ of the new system will clearly have the same topological structure as the equilibrium state $O(0, 0)$ of system (4).

---

* A matrix $A$ and any matrix $SAS^{-1}$ (where $S$ is a nonsingular matrix) are said to be similar.

We shall see that in this context it is natural to divide the discussion into cases as follows: 1) characteristic roots $\lambda_1$ and $\lambda_2$ real and of the same sign (node); 2) characteristic roots complex conjugate with nonzero real parts (focus); 3) characteristic roots real and of opposite sign (saddle point); 4) characteristic roots pure imaginary.

It will turn out that in the first three cases *the topological structure of the equilibrium state is completely determined by the linear terms of system (4)*. In other words, it is the same as the topological structure of the equilibrium state of the linear system

$$\frac{dx}{dt} = ax + by, \qquad \frac{dy}{dt} = cx + dy, \tag{28}$$

obtained from (4) by dropping the nonlinear terms.

Common to cases 1), 2), 3) is the circumstance that the characteristic roots have nonzero real parts.* In addition, we shall see that the first two types of equilibrium state have the same topological structure. These cases are sometimes combined and referred to as the case of characteristic roots with nonzero real parts of the same sign.

Case 4) — pure imaginary characteristic roots — is more difficult. Here the topological structure of the equilibrium state is not determined by the characteristic roots, i. e., does not depend on the linear terms alone. It depends on terms of higher degree and may vary accordingly. This case is examined in §8.

In the sequel we shall find it convenient to express the functions $\varphi(x, y)$ and $\psi(x, y)$ on the right of system (4) as

$$\varphi(x, y) = xg_1(x, y) + yg_2(x, y),$$
$$\psi(x, y) = xf_1(x, y) + yf_2(x, y),$$

where $g_1(x, y)$, $g_2(x, y)$, $f_1(x, y)$, $f_2(x, y)$ are continuous functions such that $g_1(0, 0) = g_2(0, 0) = f_1(0, 0) = f_2(0, 0) = 0$. That this can always be done follows readily from Taylor's formula (or Hadamard's lemma).

In this case, Hadamard's lemma can be proved quite easily (e. g., for the function $\varphi$):

$$\varphi(x, y) = \varphi(x, y) - \varphi(0, 0) = \int_0^1 \varphi_t(tx, ty)\, dt =$$

$$= \int_0^1 \varphi'_x(tx, ty)\, x\, dt + \int_0^1 \varphi'_y(tx, ty)\, y\, dt = xg_1(x, y) + yg_2(x, y),$$

where $g_1$ and $g_2$ are defined respectively as $\int_0^1 \varphi'_x(tx, ty)\, dt$ and $\int_0^1 \varphi'_y(tx, ty)\, dt$.

---

* System (28) is often called the linearized system (relative to system (4)); the procedure of dropping the nonlinear terms and studying the equilibrium state of the nonlinear system on the basis of the linearized system is sometimes called "linearization." Our goal in this section is therefore to prove that linearization is legitimate when the characteristic roots have nonzero real parts.

The continuity of $g_1$ and $g_2$ follows from that of the partial derivatives $\varphi'_x$ and $\varphi'_y$, and $g_1(0, 0) = g_2(0, 0) = 0$ since $\varphi'_x(0, 0) = \varphi'_y(0, 0) = 0$.

## §7. CONFIGURATION OF PATHS IN THE NEIGHBORHOOD OF SIMPLE EQUILIBRIUM POINTS: CHARACTERISTIC ROOTS WITH NONZERO REAL PARTS*

### 1. CASE 1: CHARACTERISTIC ROOTS $\lambda_1$ AND $\lambda_2$ REAL AND OF THE SAME SIGN (NODE).

In the notation of formulas (24) of §6, an equilibrium state of the node type is characterized by

$$\sigma^2 - 4\Delta > 0 \quad \text{and} \quad \Delta > 0.$$

The cases of distinct and equal characteristic roots will be considered separately.

1a) If $\lambda_1 \neq \lambda_2$, then by Lemma 1 of §6 the canonical form of the system is

$$\frac{dx}{dt} = \lambda_1 x + \varphi(x, y) = \lambda_1 x + x g_1(x, y) + y g_2(x, y),$$
$$\frac{dy}{dt} = \lambda_2 y + \psi(x, y) = \lambda_2 y + x f_1(x, y) + y f_2(x, y). \tag{1}$$

(Here, as always throughout this section, we are using the special representation of the functions $\varphi$ and $\psi$ cited at the end of §6.)

To fix ideas, let $\lambda_1$ and $\lambda_2$ be negative (the case of positive roots is reduced to this case by substituting $-t$ for $t$). We shall prove the following proposition:

*Any path of the dynamic system (not an equilibrium state) which passes through a point sufficiently close to the origin tends to the origin as $t \to +\infty$ and leaves the neighborhood of the origin with decreasing $t$.*

Consider a neighborhood of the equilibrium state $O$ containing no other equilibrium state, and let

$$x = x(t), \quad y = y(t)$$

be the equation of a path $L$ (not an equilibrium state) through a point of this neighborhood.

Following §3.13, we set up the expression

$$\frac{d}{dt}(x^2 + y^2) = 2x[\lambda_1 x + x g_1(x, y) + y g_2(x, y)] + 2y[\lambda_2 y + x f_1(x, y) + y f_2(x, y)]. \tag{2}$$

Transforming to polar coordinates, $x = \varrho \cos\theta$, $y = \varrho \sin\theta$, we consider the equation of the path in the form

---

* It is worth pointing out that our subsequent investigation of the topological structure of equilibrium states goes through under slightly less restrictive assumptions concerning the functions $\varphi(x, y)$ and $\psi(x, y)$. Namely, instead of the existence of continuous partial derivatives, it is sufficient to require (see /22/) that

$$\lim_{\rho \to 0} \frac{\varphi(x, y)}{\varrho} = 0, \quad \lim_{\rho \to 0} \frac{\psi(x, y)}{\varrho} = 0, \quad \text{where} \quad \varrho = \sqrt{x^2 + y^2}.$$

$$\varrho = \varrho(t), \qquad \theta = \theta(t).$$

We have

$$\varrho^2(t) = x^2(t) + y^2(t).$$

The polar angle $\theta(t)$ obviously varies continuously along the path. In terms of polar coordinates, expression (2) becomes

$$\frac{d\varrho^2}{dt} = 2\varrho^2 \{\lambda_1 \cos^2\theta + \lambda_2 \sin^2\theta + \cos^2\theta\, g_1(\varrho\cos\theta,\ \varrho\sin\theta) +$$

$$+ \cos\theta\sin\theta\, [g_2(\varrho\cos\theta, \varrho\sin\theta) + f_1(\varrho\cos\theta,\ \varrho\sin\theta)] + \qquad (3)$$

$$+ \sin^2\theta f_2(\varrho\cos\theta,\ \varrho\sin\theta)\}.$$

The function $\lambda_1\cos^2\theta + \lambda_2\sin^2\theta$ is bounded in absolute value, periodic in $\theta$ and negative for all (real) values of $\theta$. It therefore has a maximum $-m$ and a minimum $-M$, where $M > 0$ and $m > 0$, $M > m$.

Since the functions $g_1$, $g_2$, $f_1$, $f_2$ are continuous and vanish at $O(0, 0)$, it follows that for any $\varepsilon > 0$ there exists $\varrho_0 > 0$ with the property: if $\varrho(t) < \varrho_0$, then

$$|\cos^2\theta g_1 + \cos\theta\sin\theta\,(g_2 + f_1) + \sin^2\theta f_2| < \varepsilon.$$

Let $\varepsilon < \frac{m}{2}$. Then if $\varrho(t) < \varrho_0$ the expression in braces in (3) is smaller than $-\frac{m}{2}$. Consequently,

$$\frac{d\varrho^2(t)}{dt} < 2\varrho^2(t)\left(-\frac{m}{2}\right) < -m\varrho^2(t). \qquad (4)$$

On the other hand, since $\varepsilon < \frac{m}{2} < \frac{M}{2}$, it follows that when $\varrho(t) < \varrho_0$ the expression in braces in (3) is greater than $-\frac{3M}{2}$, and consequently

$$\frac{d\varrho^2(t)}{dt} > -3M\varrho^2(t), \qquad (5)$$

so that

$$-3M\varrho^2(t) < \frac{d\varrho^2(t)}{dt} < -m\varrho^2(t). \qquad (6)$$

Suppose that the point of the path corresponding to $t = t_0$ is on a circle $C^*$ of radius $\varrho^* < \varrho_0$ centered at $O$, i. e.,

$$x^2(t_0) + y^2(t_0) = \varrho^{*2}.$$

By (4), the function $\varrho^2(t)$, and hence also $\varrho(t)$, is decreasing at $t = t_0$, so that at $t > t_0$ the path crosses the circle $C^*$ inward. In other words, condition (4) means that any circle

$$x^2 + y^2 = C \tag{7}$$

of radius less than $\varrho_0$, in particular, the circle $C^*$, is a cycle without contact (§ 3.10) for the paths of the system. (The condition for a circle (7) to be tangent to the paths is $\frac{d\varrho^2}{dt} = 0$, which contradicts (4).) Hence, with increasing $t$ the path $L$ cannot leave the region bounded by the circle (cycle without contact) $C^*$, and consequently the path $L$ is surely defined for all $t > t_0$ (by Theorem 2 of § 1*) and $\varrho\,(t) < \varrho^* < \varrho_0$ for all such $t$. But then inequality (4) is satisfied for all $t > t_0$. Separating variables in this inequality and integrating from $t_0$ to $t$, we get

$$\lg \frac{\varrho^2\,(t)}{(\varrho^*)^2} < -m\,(t-t_0),$$

i. e.,

$$\varrho^2\,(t) < (\varrho^*)^2 e^{-m(t-t_0)}.$$

It follows that $\varrho\,(t) \to 0$ as $t \to +\infty$, so that the path $L$ tends to the equilibrium state $O$.

We now consider what happens to the path $L$ when $t$ decreases $(t < t_0)$. Since $C^*$ is a cycle without contact, the path cannot cross it inward with decreasing $t$. Moreover, as long as $\varrho^*\,(t) < \varrho_0$, we have (5)

$$\frac{d\varrho^2}{dt} > -3M\varrho^2$$

and so

$$\varrho^2 > \varrho^{*2} e^{-3M\,(t-t_0)}.$$

But this clearly means that $\varrho^2\,(t)$ increases with decreasing $t$ $(t < t_0)$, at any rate, as long as inequality (5) remains true, i. e., until $\varrho\,(t)$ becomes equal to $\varrho_0$. When $\varrho\,(t) = \varrho_0$ inequality (5) is still valid, by assumption. Thus the circle $x^2 + y^2 = \varrho_0^2$ is a cycle without contact. With d e c r e a s i n g $t$ the path $L$ crosses this circle outward and will never enter it again with further decrease in $t$. This proves our assertion in the case of distinct characteristic roots (Figure 79).**

1b) $\lambda_1 = \lambda_2$ (multiple root).

The canonical form of the system is

$$\frac{dx}{dt} = \lambda x + xg_1\,(x,\ y) + yg_2\,(x,\ y),$$
$$\frac{dy}{dt} = \lambda y + \mu x + xf_1\,(x,\ y) + yf_2\,(x,\ y), \tag{8}$$

where $\lambda$ is assumed, as in case 1a), to be negative.

---

* We stated and proved Theorem 2 for paths. However, a suitably modified version is easily seen to be valid for semipaths. At this point we are indeed using the semipath version.
** Note that we have not determined whether the paths tend to the equilibrium state in a definite direction (as in Example 3 of § 1) or along a spiral.

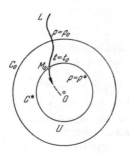

FIGURE 79.

If $\mu = 0$ the discussion is exactly the same as in the previous case. Let $\mu \neq 0$. We shall show that the circles $x^2 + y^2 = C$ are now not necessarily cycles without contact. Indeed, the first term in the expansion of $\frac{d\varrho^2}{dt}$ is now

$$\varrho^2 \, (\lambda_1 \cos^2 \theta + \lambda_1 \sin^2 \theta + \mu \cos \theta \sin \theta),$$

and the expression in parentheses may change sign (when $\mu^2 - 4\lambda_1^2 > 0$). Nevertheless, the proof may be carried through as before, except that the expression

$$\varrho^2 \, (t) = x^2 \, (t) + y^2 \, (t)$$

(square of the distance of points on $L$ from the origin) is replaced by

$$\sigma \, (t) = x^2 \, (t) + ky^2 \, (t),$$

where $k > 0$ will be specified later. $\sigma(t)$ tends to zero if and only if $x^2 \, (t) + y_2 \, (t) \to 0$. It follows from equations (8) that

$$\frac{d\sigma}{dt} = \frac{d \, (x^2 + ky^2)}{dt} = 2 \, \{\lambda x^2 + k\mu xy + \lambda ky^2 + x^2 g_1 \, (x, \ y) +$$

$$+ xy \, [g_2 \, (x, \ y) + kf_1 \, (x, \ y)] + ky^2 f_2 \, (x, \ y)\}.$$

Multiplying and dividing the right-hand side by $x^2 + ky^2$ and transforming to polar coordinates, we get

$$\frac{d\sigma \, (t)}{dt} = 2\sigma \, (t) \left[ \frac{\lambda \cos^2 \theta + k\mu \cos \theta \sin \theta + \lambda k \sin^2 \theta}{\cos^2 \theta + k \sin^2 \theta} + \right.$$
$$\left. + \frac{g_1 \cos^2 \theta + (f_1 k + g_2) \cos \theta \sin \theta + f_2 k \sin^2 \theta}{\cos^2 \theta + k \sin^2 \theta} \right] \qquad (9)$$

Now choose $k > 0$ so that $k^2\mu^2 - 4\lambda k < 0$. Then the numerator of the first fraction in the bracketed expression is a negative definite quadratic form in $\cos \theta$ and $\sin \theta$, and therefore assumes only negative values. The denominators of both fractions assume only positive values and are bounded above and below by positive numbers. The numerator of the second fraction tends to zero as $x \to 0$ and $y \to 0$. By the periodicity of the trigonometric functions, the first fraction has a maximum $-m$ and a minimum $-M$, where $m > 0$, $M > 0$ ($m < M$). The second fraction tends to 0 as $\varrho \, (t) \to 0$. Therefore, continuing exactly as before, one shows that if

$$\sigma \, (t) = x^2 \, (t) + ky^2 \, (t) < \varrho_0^2,$$

where $\varrho_0$ is a sufficiently small positive number, then

$$\frac{d\sigma \, (t)}{dt} < -m\sigma \, (t). \qquad (10)$$

130

On the other hand, it is readily seen that for a suitable choice of $\varrho_0$

$$\frac{d\sigma}{dt} > -3M\sigma,$$

so that for $\varrho < \varrho_0$

$$-3M\sigma < \frac{d\sigma}{dt} < -m\sigma. \tag{11}$$

Thus, the family of ellipses

$$x^2 + ky^2 = C < \varrho_0^2$$

is a family of cycles without contact for the paths of the system, and with increasing $t$ the function

$$\sigma(t) = x^2(t) + ky^2(t)$$

decreases monotonically and tends to zero, i.e., all paths passing through points of the region $u$ inside the ellipse $x^2 + ky^2 = \varrho_0^2$ and distinct from the point $O$ tend to the origin with increasing $t$, never leaving $u$. With decreasing $t$, all paths leave the region $u$. This is proved along the same lines as for distinct $\lambda_1$ and $\lambda_2$, using inequality (11) (the analog of inequality (6)).*

This completes the proof of our assertion.

The case $\lambda_1 > 0$, $\lambda_2 > 0$ is treated in similar fashion.

Thus, when the characteristic roots are both negative, all paths through a sufficiently small neighborhood of the equilibrium state $O$ tend to $O$ as $t \to +\infty$ and leave the neighborhood with decreasing $t$. When the characteristic roots are both positive, all paths passing through a sufficiently small neighborhood of the equilibrium state tend to it as $t \to -\infty$ and leave the neighborhood with increasing $t$.

In the first case, the equilibrium state is known as a s t a b l e  n o d e, in the second case as an  u n s t a b l e  n o d e. When the characteristic roots

---

* The proof for multiple characteristic roots may be carried out in a slightly different way. It is not hard to see that, by a substitution

$$x = a\bar{x}$$

where $a$ is a suitably chosen constant, the constant $\mu$ may first be made so small that

$$\mu^2 - 4\lambda^2$$

is negative. Then, in the expansion of $\frac{d\varrho^2}{dt}$ the coefficient of the first term (coefficient of $\varrho^2$)

$$\lambda_1 \cos^2\theta + \lambda \sin^2\theta + \mu \sin\theta \cos\theta,$$

can obviously no longer change sign, and the proof can be continued as in case 1a).

are multiple, the node is said to be **d e g e n e r a t e** if $\mu \neq 0$ and **d i c r i t i c a l** if $\mu = 0$.*

**2. CASE 2: COMPLEX CONJUGATE CHARACTERISTIC ROOTS** $\lambda_1 = \alpha + i\beta$, $\lambda_2 = \alpha - i\beta$, $\beta \neq 0$, $\alpha \neq 0$ **(FOCUS)**. In this case, obviously, $\sigma^2 - 4\Delta < 0$. The canonical form of the system is

$$\frac{dx}{dt} = \alpha x - \beta y + \varphi(x, y) = \alpha x - \beta y + x g_1(x, y) + y g_2(x, y),$$

$$\frac{dy}{dt} = \beta x + \alpha y + \psi(x, y) = \beta x + \alpha y + x f_1(x, y) + y f_2(x, y). \tag{12}$$

To fix ideas, let $\alpha < 0$ (the case $\alpha > 0$ is reduced to this case by substituting $-t$ for $t$). We shall show that, as in the case of two negative real characteristic roots, all paths passing through a sufficiently small neighborhood of the equilibrium state $O$ tend to $O$ as $t \to +\infty$ and leave the neighborhood with decreasing $t$. To this end, as in case 1a), we consider the expression

$$\frac{d(x^2+y^2)}{dt} = 2\{\alpha(x^2+y^2) + x^2 g_1(x, y) + xy[g_2(x, y) + f_1(x, y)] + y^2 f_2(x, y)\},$$

or in polar coordinates

$$\frac{d\varrho^2}{dt} = 2\varrho^2 \{\alpha + \cos^2\theta\, g_1(\varrho\cos\theta, \quad \varrho\sin\theta) + \cos\theta\sin\theta\,[g_2(\varrho\cos\theta, \varrho\sin\theta) +$$

$$+ f_1(\varrho\cos\theta, \quad \varrho\sin\theta)] + \sin^2\theta f_2(\varrho\cos\theta, \quad \varrho\sin\theta)\}. \tag{13}$$

Since by assumption $\alpha \neq 0$, this expression is of the same form as (3) in §7.1. We may thus employ the same method, almost without modification, as in the case of a node. The final conclusion is the same as in §7.1. If $\alpha < 0$ $(\alpha > 0)$, then every path distinct from $O$ passing at $t = t_0$ through a point of a sufficiently small disk $C_0$ about $O$ is defined for all $t > t_0$ $(t < t_0)$, tends with increasing (decreasing) $t$ to the equilibrium state $O$ without leaving the disk $C_0$, and leaves $C_0$ with decreasing (increasing) $t$.

The equilibrium state is known as a **s t a b l e  f o c u s** if all paths tend to $O$ as $t \to +\infty$, as an **u n s t a b l e  f o c u s** if the paths tend to $O$ as $t \to -\infty$.

We have already remarked (§1.14) that in a certain sense the behavior of the paths near a node or a focus is the same. The definitions of §5 enable us to give this statement a rigorous meaning: the topological structures of a node and a focus are identical.

This makes it natural, as remarked above, to combine cases 1) and 2). We mention here that the properties of the equilibrium state in cases 1) and 2) may be investigated using arguments of a somewhat different type, employing Lemmas 15 and 16 of §3. To elucidate: it follows from (6) or (11) that (as mentioned above) the circle $x^2 + y^2 = C$ or (in case 1b)) the ellipse $x^2 + ky^2 = C$, $C \leqslant \varrho_0^2$, is a cycle without contact for the paths of system (1). Moreover, when $\lambda_1 < 0$ and $\lambda_2 < 0$ (or $\alpha < 0$, as the case may be) the paths cross each of these cycles inward with increasing $t$ (since the right-hand sides of (3) and (13) are negative).

---

\* [In the western literature (e. g., /12/, Chapter 15) it is more customary to call a dicritical node a proper node, both other types being termed improper nodes. The distinction refers to the mode of approach of the paths to the equilibrium state (see below, §9).]

Choosing a cycle $C'_0$ of sufficiently small radius $\varrho'_0 < \varrho_0$ and using Lemmas 15 and 16 of § 3, one readily proves the truth of our assertion concerning the behavior of paths in the neighborhood of the equilibrium points in cases 1), 2).

There is a close connection between the present discussion and the "Lyapunov function." In his investigation of the stability of an equilibrium state of an autonomous dynamic system of arbitrary order, Lyapunov derived sufficient conditions for stability which may be formulated as follows: consider a system

$$\dot{x}_i = p_i\,(x_1,\,\ldots,\,x_n), \qquad i = 1,\,2,\,\ldots,\,n,$$

for which the origin is an equilibrium state. If there exists a function $v\,(x_1,\,\ldots,\,x_n)$, defined in some neighborhood $u$ of the origin and vanishing at the origin, assuming only positive values elsewhere in $u$ (a function satisfying these conditions is said to be positive definite), and if the path derivative of the function

$$\dot{x}_i = p_i\,(x_1,\,\ldots,\,x_n), \text{ i. e., } \sum \frac{\partial v}{\partial x_i} p_i,$$

is negative definite, then the equilibrium state at the origin is stable. A function satisfying the conditions listed above is called a Lyapunov function. In cases 1) and 2) above there exist Lyapunov functions. In cases 1a) and 2) $v\,(x,\,y) = x^2 + y^2$ is a Lyapunov function, in case 1b) $v\,(x,\,y) = x^2 + y^2 k$.

We shall not go into details here concerning the topological equivalence of nodes and foci, since this problem will be considered in Chapter VIII. As an example we shall describe the relevant topological mapping for a system in the $(x,\,y)$-plane

$$\frac{dx}{dt} = P_1\,(x,\,y), \qquad \frac{dy}{dt} = Q_1\,(x,\,y), \tag{14}$$

for which the origin $O_1$ is a node with distinct characteristic roots, and a system in the $(\xi,\,\eta)$-plane

$$\frac{d\xi}{d\tau} = P_2\,(\xi,\,\eta), \qquad \frac{d\eta}{d\tau} = Q_2\,(\xi,\,\eta) \tag{15}$$

(the $(\xi,\,\eta)$-plane may coincide with the $(x,\,y)$-plane) with a focus at the origin $O_2$.

As we have seen, all circles in the $(x,\,y)$-plane and $(\xi,\,\eta)$-plane with centers at $O_1$ and $O_2$, respectively, are cycles without contact, provided their radii are sufficiently small.

Let $C_1$ and $C_2$ be circles in the $(x,\,y)$-plane and $(\xi,\,\eta)$-plane, respectively, of the same radius, which are cycles without contact. Consider motions on the paths of the systems under which the representative points cross the circles $C_1$ and $C_2$ at $t = t_0$ and $\tau = \tau_0$, respectively. We define a mapping of the closed disks bounded by these circles, as follows. Each path of system (14) through a point of $C_1$ is mapped onto the path of system $(15)$ through the point of $C_2$ with the same coordinates, in such a way that corresponding points correspond to equal parameter values $t$ and $\tau$. Now map the point $O_1$ onto $O_2$. We thus obtain a mapping of the closed disk bounded by $C_1$ onto the closed disk bounded by $C_2$, which is readily seen to be a topological mapping.

As stated at the beginning of the chapter, the distinction between a node and a focus is not topological (and is therefore not essential from the purely qualitative standpoint): paths tending to a node do so along definite directions (for the rigorous definition, see §9), whereas paths tending to a focus do so along spirals (see §9.7).

**3. CASE 3 : CHARACTERISTIC ROOTS $\lambda_1$ AND $\lambda_2$ REAL AND OF OPPOSITE SIGN $(\lambda_1\lambda_2 < 0)$ (SADDLE POINT).*** In this case, obviously, $\Delta < 0$ and the system can be reduced to canonical form

$$\frac{dx}{dt} = \lambda_1 x + \varphi\,(x,\ y), \qquad \frac{dy}{dt} = \lambda_2 y + \psi\,(x,\ y), \tag{16}$$

where, say, $\lambda_1 < 0,\ \lambda_2 > 0$. As before, the functions $\varphi\,(x,\ y)$ and $\psi\,(x,\ y)$ can be expressed as

$$\varphi\,(x,\ y) = x g_1\,(x,\ y) + y g_2\,(x,\ y), \qquad \psi\,(x,\ y) = x f_1\,(x,\ y) + y f_2\,(x,\ y), \tag{17}$$

where

$$g_1\,(0,\ 0) = g_2\,(0,\ 0) = f_1\,(0,\ 0) = f_2\,(0,\ 0).$$

The method we shall use here to establish the topological structure of the equilibrium state is quite different from that used in the previous two cases. We shall consider a neighborhood $U_\varepsilon\,(O)$ of the point $O$, containing no equilibrium states other than $O$, and find in this neighborhood arcs without contact which will enable us to elucidate the behavior of paths near $O$.

Let $k_0 > 0$ be arbitrary but fixed, and consider the straight lines $y = \pm\,k_0 x$.

Suppose that the path $L$ corresponding to a solution

$$x = x\,(t), \qquad y = y\,(t),$$

cuts the straight line $y = k_0 x$ (or $y = -k_0 x$) at $t = t_0$, at a point other than the origin.

We have**

$$\left[\frac{d\left(\frac{y(t)}{x(t)}\right)}{dt}\right]_{t=t_0} = \left\{\frac{x\,(t)\,[\lambda_2 y\,(t) + x\,(t)\,f_1\,(x\,(t),\ y\,(t)) + y\,(t)\,f_2\,(x\,(t),\ y\,(t))]}{[x\,(t)]^2} - \right.$$
$$\left. - \frac{y\,(t)\,[\lambda_1 x\,(t) + x\,(t)\,g_1\,(x\,(t),\ y\,(t)) + y\,(t)\,g_2\,(x\,(t),\ y\,(t))]}{[x\,(t)]^2}\right\}_{t=t_0}. \tag{18}$$

---

\* Our investigation of the topological structure of a saddle point follows /22/. A somewhat different approach, though also geometric, is adopted in /11/. See also §7.5.

\*\* The expression in braces is obviously the derivative of the function $z = y/x$ along paths of system (1) (see §3.13). Its sign indicates whether $z\,(t) = y\,(t)/x\,(t)$ increases or decreases along the path with increasing $t$, and thus provides an indication of the way in which the lines $y = \pm k_0 x$ cut the paths.

Since $\frac{y(t_0)}{x(t_0)} = \pm k_0$, it follows that

$$\left[\frac{d\left(\frac{y(t)}{x(t)}\right)}{dt}\right]_{t=t_0} = \{\pm k_0(\lambda_2 - \lambda_1) + f_1(x(t), y(t)) \pm$$
$$\pm k_0[f_2(x(t), y(t)) - g_1(x(t), y(t))] - g_2(x(t), y(t)) k_0^2\}_{t=t_0}. \quad (19)$$

Now consider the straight-line segments $x = \text{const} \neq 0$ defined by the condition $\left|\frac{y}{x}\right| \leqslant k_0$, i.e., the segments cut from vertical straight lines by the straight lines $y = \pm k_0 x$ (Figure 80).

If a path $x = x(t)$, $y = y(t)$ crosses one of these segments at $t = t_0$, then

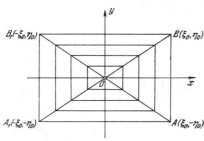

$$\left[\frac{dx(t)}{dt}\right]_{t=t_0} = \left\{x(t)\left[\lambda_1 + \right.\right.$$
$$+ g_1(x(t), y(t)) + \quad (20)$$
$$\left.\left. + \frac{y(t)}{x(t)} g_2(x(t), y(t))\right]\right\}_{t=t_0},$$

and by assumption

FIGURE 80.

$$\left|\frac{y(t_0)}{x(t_0)}\right| \leqslant k_0. \quad (21)$$

Finally, consider the segments defined on the straight lines $y = \text{const} \neq 0$ by the condition $\left|\frac{y}{x}\right| > k_0$, i.e., segments cut from horizontal straight lines by the straight lines $y = \pm k_0 x$ (Figure 80). If the path $x = x(t)$, $y = y(t)$ crosses one of these segments at $t = t_0$, then

$$\left[\frac{dy(t)}{dt}\right]_{t=t_0} = \left\{y(t)\left[\lambda_2 + \frac{x(t)}{y(t)} f_1(x(t), y(t)) + f_2(x(t), y(t))\right]\right\}_{t=t_0}, \quad (22)$$

and

$$\left|\frac{x(t_0)}{y(t_0)}\right| \leqslant \frac{1}{k_0}. \quad (23)$$

Now let $\xi_0$ be so small that the following three conditions hold:
1) if $|x| \leqslant \xi_0$, $|y| \leqslant k_0 \xi_0$, then

$$|f_1(x, y) \pm k_0[f_2(x, y) - g_1(x, y)] - g_2(x, y) k_0^2| < \frac{k_0}{2}(\lambda_2 - \lambda_1); \quad (24)$$

2) if $|x| \leqslant \xi_0$, $|y| \leqslant k_0 \xi$ and $\left|\frac{y}{x}\right| \leqslant k_0$, then

$$\left|g_1(x, y) + \frac{y}{x} g_2(x, y)\right| < \frac{1}{2}|\lambda_1|; \quad (25)$$

3) if $|x| \leqslant \xi_0$, $|y| \leqslant k_0\xi_0$ and $\left|\frac{x}{y}\right| \leqslant \frac{1}{k_0}$, then

$$\left|f_1(x, y)\frac{x}{y} + f_2(x, y)\right| \leqslant \frac{1}{2}|\lambda_2|. \tag{26}$$

That $\xi_0$ exists follows from the properties of the functions $f_1$, $f_2$, $g_1$, $g_2$.

Let $\bar{H}$ be the rectangle, centered at the origin, defined by the inequalities $|x| \leqslant \xi_0$, $|y| \leqslant \eta_0$, where $\eta_0 = k_0\xi_0$ (Figure 80), and denote its vertices by $A(\xi_0, -\eta_0)$, $B(\xi_0, \eta_0)$, $A_1(-\xi_0, -\eta_0)$ and $B_1(-\xi_0, \eta_0)$.* It follows from (19) and (24) that if a point $(x(t), y(t))$ of the path lies on the diagonal $A_1B$ of $\bar{H}$, i. e., on the straight line $y = k_0x$ (but is not the point $O$), then

$$\frac{d\left(\frac{y(t)}{x(t)}\right)}{dt} > \frac{k_0}{2}(\lambda_2 - \lambda_1) > 0. \tag{27}$$

If it lies on the diagonal $AB_1$ (the line $y = -k_0x$), then

$$\frac{d\left(\frac{y(t)}{x(t)}\right)}{dt} < -\frac{k_0(\lambda_1 - \lambda_2)}{2} < 0. \tag{28}$$

It follows from (20) and (25) that if a point $(x(t), y(t))$ of the path lies on one of the vertical segments in $\bar{H}$ and $x(t) > 0$, then

$$\frac{3\lambda_1}{2}x < \frac{dx}{dt} < \frac{\lambda_1}{2}x < 0. \tag{29}$$

It follows from (22) and (26) that if $x(t)$, $y(t)$ lies on one of the horizontal segments in $\bar{H}$ and $y(t) > 0$, then

$$\frac{3\lambda_2}{2}y > \frac{dy}{dt} > \frac{\lambda_2}{2}y > 0. \tag{30}$$

Analogous inequalities hold true if $x(t) < 0$ and $y(t) < 0$ (they are obtained from inequalities (29) and (30) by reversing the inequality signs).

It follows from (27) and (28) that the diagonals $A_1B$ and $AB_1$ of the rectangle $ABB_1A_1$ are not tangent at any of their points other than $O$ to the paths of the system.**

Similarly, it follows from (29) and (30) that all the designated horizontal and vertical segments in the rectangle $\bar{H}$ are segments without contact for our dynamic system.

The diagonals $A_1B$ and $AB_1$ divide the rectangle $\bar{H}$ into four triangles. Consider one of them, say $OAB$. Everything stated and proved for this triangle carries over with suitable modifications to the other three. It clearly follows from (28), (29) and (30) that:

•   The diagonals of $H$ are segments of the straight lines $y = \pm k_0x$.

**   The tangency condition is $\frac{y'(t)}{x'(t)} = \frac{y(t)}{x(t)}$, i. e., $\frac{d\left(\frac{y(t)}{x(t)}\right)}{dt} = 0$. Thus, every closed segment on the diagonals $A_1B$ and $AB_1$, not containing the point $O$, is a segment without contact.

1) any path that crosses the side $OA$ ($OB$) of the triangle $OAB$ at $t = t_0$, at a point other than $O$ and $A$ ($O$ and $B$), enters the triangle with decreasing $t$ and leaves it with increasing $t$;

2) any path that crosses the side $AB$ of the triangle at $t = t_0$, at a point other than $A$ or $B$, enters the triangle with increasing $t$ and leaves it with decreasing $t$;

3) all points on a path passing through the vertex $A$ ($B$) at $t=t_0$, corresponding to parameter values sufficiently close to $t_0$, lie outside the triangle $OAB$ (Figure 81).

Let $M_0$ be an arbitrary point inside $OAB$. Consider a path $L$, $x = x(t)$, $y = y(t)$, passing, say at $t = t_0$ through the point $M_0$. We claim that with decreasing $t$ the path $L$ leaves the triangle $OAB$ through the side $AB$ (at a point other than $A$ or $B$).

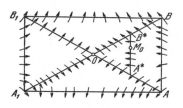

FIGURE 81.

Indeed, the path cannot leave $OAB$ with decreasing $t$ through a $OA$ and $OB$, for by (27) and (28) all paths crossing these sides enter the triangle $OAB$ with decreasing $t$ (it cannot pass through $A$ and $B$ by virtue of condition 3)). Consequently, with decreasing $t$ the path $L$ either leaves $OAB$ through $AB$ or remains inside the triangle. Suppose that $L$ does not cross $AB$, so that it remains inside $OAB$. Now, inside the triangle the abscissa $x(t)$ of the path, which satisfies the differential inequality (29),

is monotone increasing. Integrating (29), we get $x(t) > x_0 e^{\frac{3\lambda_1}{2}(t-t_0)}$.

Since $\lambda_1 < 0$, the expression $\frac{3\lambda_1}{2}(t-t_0)$ is positive for $t < t_0$ and consequently $x(t)$ increases with decreasing $t$, at any rate, until $x(t)$ becomes equal to $\xi_0$ (inequality (29) is valid for $|x| < \xi_0$). But this means that $L$ cannot remain within the triangle $OAB$ for all $t < t_0$, and so necessarily crosses the side $AB$ (which is a segment on the straight line $x = \xi_0$). Since paths crossing $OA$ and $OB$ at their interior points (see condition 1)) enter the triangle $OAB$ with decreasing $t$, it is clear that with further decrease in $t$ they will also cross $AB$.

We have thus shown that any path through a point inside the triangle $OAB$ or on its sides $OA, OB$ (as an interior point) will leave $OAB$ with decreasing $t$ through an interior point of its side $AB$, and the abscissa $x(t)$ of the path is monotone increasing.

We now consider paths passing through interior points of the segment $AB$ and examine their behavior as $t$ increases. Let $M$ be an [interior] point of $AB$ and $L$ a path passing through $M$ at $t = t_0$. With increasing $t$, the path $L$ enters the triangle $OAB$ and subsequently may either a) leave $OAB$ through an interior point of $OA$ (remaining till then inside the triangle) or b) leave $OAB$ through an interior point of $OB$ (remaining till then inside the triangle), or c) remain inside $OAB$.

If case c) obtains, i.e., the path remains inside the triangle, then by Theorem 2 it is defined for all $t > t_0$. We claim that in this case $L$ tends to the equilibrium state $O$ as $t \to +\infty$. Indeed, since it remains in the triangle for all $t > t_0$, it satisfies inequality (29) for all $t > t_0$, and hence

$$x(t) < x_0 e^{\frac{\lambda_1}{2}(t-t_0)},$$

whence it follows that $x(t) \to 0$ and $y(t) \to 0$ as $t \to \infty$.

Both cases a) and b) may actually occur. To verify this statement, it suffices to consider, as before, paths passing through interior points of the sides $OA$ and $OB$. It is obvious that if a path $L$ ($L'$), passing through an interior point $M$ ($M'$) of $AB$, crosses the side $OA$ ($OB$) with increasing $t$ at a point $N$ ($N'$), having remained till then inside the triangle, then any path passing at $t = t_0$ through an interior point of the segment $AM$ ($BM$) will enter the triangle $OAB$ with increasing $t$ and at some time $t > t_0$ will first cross the segment $AN$ ($BN'$) of the side $AO$ ($BO$) outward (Figure 82).

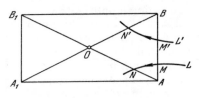

FIGURE 82.

We claim that case c) can also actually occur, but only for one point of the segment $AB$. We shall show that there exists exactly one point $C_0$ on $AB$ such that the path passing through it at $t = t_0$ enters the triangle $OAB$ with increasing $t$ and remains inside, therefore tending to the equilibrium state $O$ as $t \to +\infty$. To prove this, we note first that if the path through a point $M$ of $AB$ crosses $OA$ ($OB$) with increasing $t$, then, by the theorem on continuity of solutions with respect to initial values, paths passing through points of $AB$ sufficiently close to $M$ will have the same property. Let us divide the interior points of $AB$ into two classes: those with the property that the paths passing through them leave $OAB$ through $OA$ with increasing $t$, and all other points. It is clear that neither class is empty, and each point of the first class lies (on $AB$) below all points of the second class. Hence there exists a boundary point $C_0$ such that all lower points are in the first class, all higher points in the second. Let $L_0$ be the path passing at $t = t_0$ through the point $C_0$. If $L_0$ leaves the triangle $OAB$ with increasing $t$ through the side $OA$ ($OB$), then by the above remark all points of $AB$ sufficiently near $C_0$ are of the first (second) class, which contradicts the boundary property of $C_0$. It follows that $L_0$ enters $OAB$ at $t = t_0$ and remains within the triangle with increasing $t$, i. e., as shown above, tends to the equilibrium state $O$. We now show that $C_0$ is the only point on $AB$ with the property that the path passing through it tends to the equilibrium state $O$ with increasing $t$, remaining inside the triangle $OAB$. Suppose that there is another point with this property, $C_1$ say. Now, if $x = x(t)$, $y = y(t)$ is the equation of a path passing through $OAB$, it follows from condition (29) that

$\frac{dx}{dt} < 0$ so that $x(t)$ is a monotone function of $t$ and the equation of the path

may be written $y = y(x)$, with the function $y(x)$ satisfying the differential equation

$$\frac{dy}{dx} = \frac{\lambda_2 y + \psi(x, y)}{\lambda_1 x + \varphi(x, y)}. \tag{31}$$

Let $y = y_0(x)$ and $y = y_1(x)$ be the equations of the paths passing through $C_0$ and $C_1$, respectively, and therefore remaining within the triangle $OAB$

with increasing $t$. $y_0(x)$ and $y_1(x)$ obviously satisfy the inequalities

$$\left|\frac{y_0(x)}{x}\right| \leqslant k_0, \quad \left|\frac{y_1(x)}{x}\right| \leqslant k_0$$

and moreover, by assumption, $y_0(x) \to 0$ and $y_1(x) \to 0$ as $x \to 0$. Since different paths cannot intersect, the sign of the difference $y_0(x) - y_1(x)$ cannot change for all sufficiently large $t$ (i.e., for all sufficiently small $x$). Denote this difference by $z(x)$:

$$z(x) = y_0(x) - y_1(x)$$

and assume, to fix ideas, that $z(x) > 0$. By (31),

$$z'(x) = \frac{dy_0(x)}{dx} - \frac{dy_1(x)}{dx} = \frac{\lambda_2 y_0 + \psi(x, y_0)}{\lambda_1 x + \varphi(x, y_0)} - \frac{\lambda_2 y_1 + \psi(x, y_1)}{\lambda_1 x + \varphi(x, y_1)}. \tag{32}$$

Set

$$\varphi(x, y_0) = \varphi_0, \quad \varphi(x, y_1) = \varphi_1,$$
$$\psi(x, y_0) = \psi_0, \quad \psi(x, y_1) = \psi_1.$$

It is readily shown that equation (32) may be written as

$$\frac{dz}{dx} = \frac{\lambda_2 z + \psi_0 - \psi_1}{\lambda_1 x + \varphi_0} + \frac{(\lambda_2 y_1 + \psi_1)(\varphi_1 - \varphi_0)}{(\lambda_1 x + \varphi_0)(\lambda_1 x + \varphi_1)}$$

or

$$\frac{dz}{dx} = \frac{z}{x}\left[\frac{\lambda_2 + \frac{\psi_0 - \psi_1}{z}}{\lambda_1 + \frac{\varphi_0}{x}} + \frac{\left(\lambda_2 \frac{y_1}{x} + \frac{\psi_1}{x}\right)\left(\frac{\varphi_1 - \varphi_0}{z}\right)}{\left(\lambda_1 + \frac{\varphi_0}{x}\right)\left(\lambda_1 + \frac{\varphi_1}{x}\right)}\right]. \tag{33}$$

By the mean value theorem,

$$\begin{aligned}\varphi_0 - \varphi_1 &= \varphi(x, y_0(x)) - \varphi(x, y_1(x)) = \varphi'_y(x, \xi) z(x), \\ \psi_0 - \psi_1 &= \psi(x, y_0(x)) - \psi(x, y_1(x)) = \psi'_y(x, \eta) z(x),\end{aligned} \tag{34}$$

where $\xi$ and $\eta$ are numbers between $y_0(x)$ and $y_1(x)$. Now $y_1/x$ is bounded $\left(\left|\frac{y_1}{x}\right| \leqslant k_0\right)$ and $\varphi'_y(x, \xi)$, $\psi'_y(x, \eta)$ tend to zero as $x \to 0$ by virtue of the properties of $\varphi$ and $\psi$. Finally, $\frac{\varphi_0}{x}$, $\frac{\varphi_1}{x}$, $\frac{\psi_1}{x}$ tend to 0 as $x \to 0$. In fact,

$$\frac{\varphi_0}{x} = \frac{\varphi(x, y_0(x))}{x} = \frac{x g_1(x, y_0(x)) + y_0 g_2(x, y_0(x))}{x} = g_1(x_1, y_0(x)) + \frac{y_0}{x} g_2(x_1, y_0(x)),$$

and the last expression tends to 0 by virtue of the properties of $g_1(x, y)$, $g_2(x, y)$ and the inequality $\left|\frac{y_0}{x}\right| < k_0$. It follows from (34) and the above remarks that the bracketed expression in (33) tends to $\frac{\lambda_2}{\lambda_1}$. Consequently,

$$\frac{dz(x)}{dx} = \frac{z(x)}{x}\left[\frac{\lambda_2}{\lambda_1} + f(x)\right],\tag{35}$$

where $f(x) \to 0$ as $x \to 0$.

Setting $\frac{\lambda_2}{\lambda_1} = -\gamma$ ($\gamma > 0$ since $\lambda_2 > 0$, $\lambda_1 < 0$) and writing (35) as

$$\frac{dz}{z} = \frac{dx}{x}(-\gamma + f(x)),$$

integrate from $x$ to $x_0 > x$, where $x_0$ will be specified below. The result is

$$z = z(x_0)\, e^{\int_{x_0}^{x}\frac{[-\gamma + f(x)]\,dx}{x}} = z(x_0)\, e^{\int_{x}^{x_0}\frac{[\gamma - f(x)]\,dx}{x}}.$$

Since $f(x) \to 0$ as $x \to 0$, it follows that $x_0$ can always be chosen in such a way that for $x < x_0$ ($x > 0$) we have $|f(x)| < \frac{\gamma}{2}$, and so $\gamma - f(x) > \frac{\gamma}{2}$. Then obviously

$$z > z(x_0)\, e^{\frac{1}{2}\int_{x}^{x_0}\frac{\gamma}{x}\,dx} = z(x_0)\left(\frac{x_0}{x}\right)^{\frac{1}{2}\gamma}.\tag{36}$$

Since $\gamma > 0$, it follows from (36) that $z(x) \to \infty$ as $x \to 0$. But this contradicts our assumption that $z(x) \to 0$ as $x \to 0$.

This contradiction shows that there is only one point $C_0$ on the segment $AB$ such that the path $L_{C_0}$ passing through it tends to the equilibrium state $O$ with increasing $t$, remaining inside the triangle $OAB$ (Figure 83). This path clearly crosses each vertical segment between the straight lines $OA$ and $OB$ at exactly one point.

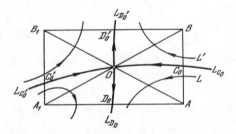

FIGURE 83.

The results proved for the triangle $OAB$ carry over with suitable modifications to the other three triangles in the rectangle $\bar{H}$, and we thus arrive at the following conclusion: there exist points $C_0$ and $C_0'$ on the sides $AB$ and $A_1B_1$, respectively, such that the paths $L_{C_0}$ and $L_{C_0'}$ passing through them do not leave the rectangle $\bar{H}$ with increasing $t$ and tend to the equilibrium state $O$ as $t \to +\infty$. There exist points $D_0$ and $D_0'$ on the sides $A_1A$ and $B_1B$, respectively, such that the paths $L_{D_0}$ and $L_{D_0'}$ do not leave the rectangle $\bar{H}$ with d e c r e a s i n g $t$ and tend to the equilibrium state $O$ as $t \to -\infty$.

Finally, if $M$ is an arbitrary point of $\bar{H}$, distinct from $O$ and not lying on the semipaths $L_{C_0}^+$, $L_{C_0'}^+$, $L_{D_0}^-$, $L_{D_0'}^-$, then the path $L_M$ passing through $M$ at $t = t_0$ leaves $\bar{H}$ with increasing $t$ through one of the horizontal sides of the rectangle, $A_1A$ or $BB_1$, and leaves $\bar{H}$ with decreasing $t$ through one of its vertical sides, $A_1B_1$ or $AB$ (Figure 83).

We have considered the case $\lambda_1 < 0$, $\lambda_2 > 0$. If $\lambda_1 > 0$, $\lambda_2 < 0$, one clearly obtains an analogous pattern, except that the paths through the vertical sides of $\bar{H}$ leave $\bar{H}$, the paths through the horizontal sides enter $\bar{H}$. $\bar{H}$ may be any rectangle centered at $O$ with sides parallel to the coordinate axes, provided it is sufficiently small.

An equilibrium state of this type is known as a s a d d l e p o i n t (Figure 83). The paths $L_{C_0}$, $L_{C_0'}$, $L_{D_0}$, $L_{D_0'}$ are known as the s e p a r a t r i c e s of the saddle point $O$. The term separatrix is also used for any positive semipath on one of the paths $L_{C_0}$ or $L_{C_0'}$ (these semipaths are called $\omega$ -separatrices) and any negative semipath on one of the paths $L_{D_0}$ or $L_{D_0'}$ ($\alpha$ -separatrices). As a rule, no distinction is made between $\omega$ -separatrices (or $\alpha$ -separatrices) on the same path (e. g., all $\alpha$-separatrices on $L_{D_0}$ are treated as the same). With this convention, each saddle point has exactly four separatrices — two $\alpha$- and two $\omega$ -separatrices.*

The separatrices (as semipaths) are also known as the "whiskers of the saddle point."** The topological structure of a saddle point is different from that of a node or a focus.

**4. STABLE AND UNSTABLE EQUILIBRIUM STATES.** We have been studying equilibrium points from the standpoint of their topological structure. In this context, of course, if one disregards the direction of motion in terms of $t$, stable and unstable nodes (foci) have the same topological structure. On the other hand, the topological structure of a saddle point is clearly different from that of a node or focus. In applied problems, however, one often examines equilibrium states exclusively from the standpoint of their stability or instability, not going into the details of their topological structure.† From this point of view, equilibrium states fall into two categories.

a) S t a b l e   e q u i l i b r i u m   s t a t e s. An obvious s u f f i c i e n t condition for an equilibrium state to be stable is that all its characteristic roots have negative real parts. Equilibrium states whose characteristic roots have negative real parts are stable nodes and stable foci.

b) U n s t a b l e   e q u i l i b r i u m   s t a t e s. Of the equilibrium states considered hitherto, those having at least one characteristic root with positive real part are unstable. Such, obviously, are unstable nodes, unstable foci and saddle points.

**5. OTHER METHODS FOR INVESTIGATING EQUILIBRIUM STATES WHOSE CHARACTERISTIC ROOTS HAVE NONZERO REAL PARTS.** The nature of equilibrium states in the cases considered above may be determined by methods other than those set forth in this section.

---

* If the separatrices are understood to be not semipaths on $L_{C_0}$, $L_{C_0'}$, $L_{D_0}$, $L_{D_0'}$, but the paths themselves, then the saddle may have two, three or four separatrices (see Examples 1, 6, 8 of §1; the saddle points in these examples have two, three and four separatrices, respectively).

** [There seems to be no comparable term in English, though some authors speak of the "arcs" of a saddle point.]

† This approach is typical of Lyapunov's work.

One method, developed in the classical work of Poincaré, Picard and Lyapunov for analytic systems, is based on construction of analytic functions

$$u = f(x, y), \quad v = g(x, y),$$

which reduce the system

$$\dot{x} = \lambda_1 x + P(x, y), \quad \dot{y} = \lambda_2 y + Q(x, y)$$

in the neighborhood of the origin to a linear system of type

$$\dot{u} = \lambda_1 u, \quad \dot{v} = \lambda_2 v.$$

Under certain conditions on the characteristic roots $\lambda_1$ and $\lambda_2$, such analytic functions $u$ and $v$ always exist. (One of these conditions is that the expression $p_1\lambda_1 + p_2\lambda_2$ not vanish for any positive integers $p_1$ and $p_2$; this condition is certainly satisfied if $\lambda_1$ and $\lambda_2$ are real and of equal signs, or if they are complex and have nonzero real part.) We shall not dwell here on the techniques for investigation of the equilibrium state when these conditions are satisfied; the interested reader is referred to Dulac /26/.

Brief mention should be made of a completely different method, applicable to dynamic systems of a wider class — systems of class $C_1$. According to this method, expounded in the work of Perron, Petrovskii and others /23/, one considers certain integral equations derived from the original system of differential equations. For example, when the characteristic roots $\lambda_1$ and $\lambda_2$ are real and distinct, the system of integral equations

$$x = x_0 e^{\lambda_1 t} + \int_0^t e^{\lambda_1 (t-\tau)} \varphi(x, y)\, d\tau,$$

$$y = y_0 e^{\lambda_2 t} + \int_0^t e^{\lambda_2 (t-\tau)} \psi(x, y)\, d\tau,$$

is equivalent to system (I) (it can be derived by formulas for solution of an inhomogeneous system of linear equations). One then shows by successive approximations that any path passing through a point sufficiently close to the origin tends to the origin, as $t \to +\infty$ if $\lambda_1$ and $\lambda_2$ are negative, as $t \to -\infty$ if $\lambda_1$ and $\lambda_2$ are positive (paths tending to the origin as $t \to \infty$ are called $0^+$-paths, those tending to the origin as $t \to -\infty$ are $0^-$-paths).

If $\lambda_1$ and $\lambda_2$ have different signs, e. g., $\lambda_1 < 0$, $\lambda_2 > 0$, then it is readily seen that all solutions corresponding to paths that tend to the equilibrium state $O$ as $t \to +\infty$ (i. e., $0^+$-paths) satisfy the system of integral equations

$$x = x_0 e^{\lambda_1 t} \int_0^t e^{\lambda_1 (t-\tau)} \varphi(x, y)\, d\tau, \quad y = -\int_t^\infty e^{\lambda_2 (t-\tau)} \psi(x, y)\, d\tau, \qquad (37)$$

while all solutions corresponding to $0^-$-paths satisfy another, though analogous system. Next, using successive approximations, one can show that these integral equations have solutions corresponding to two $0^+$-paths and two $0^-$-paths, and there are no other $0^-$-paths (see /23/ and /29/).

**6. EXAMPLES.** A few examples will illustrate investigation of equilibrium states of the types considered above.

Example 1.

$$\dot{x} = y \equiv P(x, y), \quad \dot{y} = -(1+x^2+x^4)\, y - x \equiv Q(x, y). \tag{38}$$

The equilibrium states of system (38) are the solutions of the equations

$$y = 0, \quad (1+x^2+x^4)\, y + x = 0.$$

This system has a unique solution $x = y = 0$. Thus $O\,(0,\, 0)$ is the only equilibrium state of system (38). The characteristic equation is

$$\begin{vmatrix} -\lambda & 1 \\ -1 & -1-\lambda \end{vmatrix} \equiv \lambda^2 + \lambda + 1 = 0.$$

Hence $\lambda_{1,2} = -\frac{1}{2} \pm \sqrt{\frac{1}{4} - 1}$. Thus the equilibrium state is a stable focus.

Example 2 /61/.

$$\dot{x} = y \equiv P(x, y), \quad \dot{y} = 2\,(1 - xy) \equiv Q(x, y).$$

The equations $y = 0$, $2(1 - xy) = 0$ have no common solution, and so the system has no equilibrium states.

Example 3 /66/.

$$\dot{x} = y \equiv P(x, y), \quad \dot{y} = x + x^2 - (\varepsilon_1 + \varepsilon_2 x)\, y \equiv Q(x, y),$$

where $\varepsilon_1$ and $\varepsilon_2$ are arbitrary real parameters. The equilibrium states are $O\,(0,\, 0)$ and $A\,(-1,\, 0)$. The partial derivatives of the right-hand sides of the system (at $y = 0$) are

$$P_x' = 0, \quad P_y' = 1, \quad Q_x' = 1 + 2x, \quad Q_y' = -(\varepsilon_1 + \varepsilon_2 x).$$

For the equilibrium state $O\,(0,\, 0)$ we have $P_x' Q_y' - P_y' Q_x' = -1 < 0$, so that $O\,(0,\, 0)$ is a saddle point. The characteristic equation of the equilibrium state $A\,(-1,\, 0)$ is $\lambda^2 + (\varepsilon_1 - \varepsilon_2)\, \lambda + 1 = 0$, so that $\lambda_{1,2} = \frac{\varepsilon_2 - \varepsilon_1}{2} \pm \sqrt{\frac{(\varepsilon_1 - \varepsilon_2)^2}{4} - 1}$.
The equilibrium state $A\,(-1,\, 0)$ is thus of the following type, depending on the parameters $\varepsilon_1$ and $\varepsilon_2$:
1) $0 < \varepsilon_1 - \varepsilon_2 < 2$: stable focus,
2) $\varepsilon_1 - \varepsilon_2 \geqslant 2$: stable node,
3) $0 < \varepsilon_2 - \varepsilon_1 < 2$: unstable focus,
4) $\varepsilon_2 - \varepsilon_1 \geqslant 2$: unstable node.
When $\varepsilon_1 - \varepsilon_2 = 0$ the equilibrium state $A$ has pure imaginary roots; this case has not been studied yet.

In the next example the right-hand sides are not analytic functions.
Example 4 /74/.

$$\dot{x} = y \equiv P(x, y), \quad \dot{y} = 7y + x\,(x^{\frac{1}{2}} - 12) \equiv Q(x, y),$$

where $x \geqslant 0$. We complete the definition of the system, extending it to $x < 0$ in such a way that it possesses central symmetry:

$$\dot{x} = y, \quad \dot{y} = 7y + x(|x|^{\frac{1}{2}} - 12).$$

The equilibrium states are $O(0, 0)$ and $A(144, 0)$.  To determine their types, we find

$$P'_x = 0, \quad P'_y = 1, \quad Q'_x = \frac{3}{2} x^{\frac{1}{2}} - 12, \quad Q'_y = 7.$$

The characteristic equation of the equilibrium state $O$ is

$$\lambda^2 - 7\lambda + 12 = 0,$$

whence $\lambda_1 = 4$, $\lambda_2 = 3$ and $O$ is an unstable node.  The characteristic equation of the equilibrium state $A$ is

$$\lambda^2 - 7\lambda - 6 = 0,$$

whence $\lambda_{1,2} = \frac{7}{2} \pm \sqrt{\frac{49}{4} + 6}$.  Thus the roots $\lambda_1$ and $\lambda_2$ are of opposite sign and the equilibrium state $A$ is a saddle point.

Example 5  /83/.

$$\dot{x} = -[x + (4 + x) y + \delta] \equiv P(x, y), \quad \dot{y} = [-y + (4 + x) x] \equiv Q(x, y), \quad (39)$$

where $\delta > 0$ is a parameter.

The coordinates of the equilibrium states are the solutions of the system

$$x + (4 + x) y + \delta = 0, \quad -y + (4 + x) x = 0.$$

Substituting $y = (4 + x) x$ from the second equation into the first, we get $x^3 + 8x^2 + 17x + \delta = 0$.  Thus we obtain a cubic equation for the abscissas of the equilibrium states.  It is extremely inconvenient to use the explicit solution formulas for cubic equations, but even when the coordinates of the equilibrium states are not explicitly known one can determine their number and types, depending on the parameter $\delta$.  We construct on the $(x, \delta)$-plane a curve

$$x^3 + 8x^2 + 17x + \delta = 0$$

or

$$\delta = -x(x^2 + 8x + 17). \quad (40)$$

This curve cuts the $x$-axis at exactly one point, since the trinomial $x^2 + 8x + 17$ is of fixed sign.  The abscissas of the extremum points of the curve (40) are

$$x_{1,2} = \frac{-8 \pm \sqrt{13}}{3};$$

$$\delta_1 = -\frac{-8+\sqrt{13}}{3}\left[\left(\frac{-8+\sqrt{13}}{3}\right)^2 + 8\left(\frac{-8+\sqrt{13}}{3}\right) + 17\right],$$

$$\delta_2 = -\frac{8-\sqrt{13}}{3}\left[\left(\frac{-8-\sqrt{13}}{3}\right)^2 + 8\left(\frac{-8-\sqrt{13}}{3}\right) + 17\right]$$

are the corresponding values of $\delta$. For given $\delta$, the abscissas of the equi-librium states correspond to the points at which the curve cuts the straight line $\delta = \text{const}$. It is obvious that when $\delta > \delta_1$ and $\delta < \delta_2$ system (39) has one equilibrium state, and when $\delta_2 < \delta < \delta_1$ it has three. When $\delta = \delta_1$ and $\delta = \delta_2$ the system has two equilibrium states. We have thus determined how the number of equilibrium states depends on the value of $\delta$. We now determine their types. We have

$$P'_x = -(1+y), \quad P'_y = -(4+x), \quad Q'_x = 4+2x, \quad Q'_y = -1.$$

Hence

$$\sigma \equiv +(P'_x + Q'_y) = -(x^2 + 4x + 2), \quad \Delta \equiv P'_x Q'_y - P'_y Q'_x = 3x^2 + 16x + 17.$$

The type of equilibrium state depends on the signs of $\sigma$ and $\Delta$. Obviously, $\Delta = 0$ when $x_{1,2} = \frac{-8 \pm \sqrt{13}}{3}$, that is, at the extremum points of the curve (40). It is easy to see that the equilibrium states with abscissas $x > x_1$ and $x < x_2$ have $\Delta > 0$, so that these are nodes or foci, while those with abscissas $x_2 < x < x_1$ have $\Delta < 0$, so that they are saddle points. We now take up the question of stability for the nodes and foci. Obviously, $\sigma = 0$ for $\bar{x}_{1,2} = -2 \pm \sqrt{2}$; the corresponding values of $\delta$ are

$$\bar{\delta}_{1,2} = -(-2 \pm \sqrt{2})\left[(-2 \pm \sqrt{2})^2 + 8(-2 \pm \sqrt{2}) + 17\right].$$

Thus the equilibrium states with $x > \bar{x}_1$ and $x < \bar{x}_2$ have $\sigma < 0$, so that, if not saddle points, they are stable. The equilibrium states with $\bar{x}_2 < x < \bar{x}_1$ have $\sigma > 0$, so that the corresponding nodes or foci are stable. To summarize:

1) $\delta < \delta_2$: one equilibrium state — stable node or focus;
2) $\delta_2 < \delta < \delta_1$: three equilibrium states — two stable nodes or foci and a saddle point;
3) $\bar{\delta}_1 < \delta < \delta_1$: three equilibrium states — a stable node or focus, saddle point, and unstable node or focus;
4) $\delta > \delta_1$: one equilibrium state — stable node or focus.

The values $\delta = \delta_1$ and $\delta = \delta_2$ are not included here, because the corresponding equilibrium states are not simple.

**7. SIMPLE EXAMPLES OF MULTIPLE EQUILIBRIUM STATES.** Before proceeding to the last and most difficult case of a simple equilibrium state (pure imaginary characteristic roots) in §8, we give a few elementary examples of "nonsimple" or multiple equilibrium states. If $M(x_0, y_0)$ is a multiple equilibrium state, then

$$\Delta \equiv \begin{vmatrix} P'_x(x_0, y_0) & P'_y(x_0, y_0) \\ Q'_x(x_0, y_0) & Q'_y(x_0, y_0) \end{vmatrix} = 0,$$

i. e., at least one characteristic root is zero.

In the examples given below the dynamic systems can be integrated explicitly, which facilitates investigation of the structure of their equilibrium states.*

Example 6.

$$\dot{x} = x, \quad \dot{y} = y^2. \tag{41}$$

The roots of the characteristic equation for the equilibrium state $O(0, 0)$ are $\lambda_1 = 1$, $\lambda_2 = 0$, so that $O$ is a multiple equilibrium state. It is easy to see that system $(41)$ can be integrated; the solution is $xe^{\frac{1}{y}} = C$. The behavior of the paths in the neighborhood of the origin is illustrated in Figure 84. An equilibrium state of this type is known as a "saddle node."

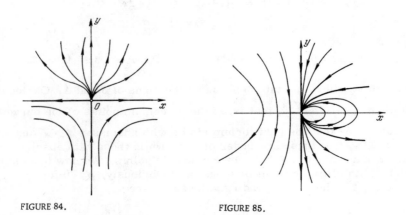

FIGURE 84.                    FIGURE 85.

Example 7.

$$\dot{x} = -xy, \quad \dot{y} = \frac{1}{2}x - y^2.$$

The characteristic equation of the equilibrium state $O(0, 0)$ is $\lambda^2 = 0$, so that both roots $\lambda_1$ and $\lambda_2$ vanish. It is readily seen that $\frac{dy}{dx} = \frac{x - y^2}{-xy}$ is a Bernoulli equation, and its general solution is $\frac{y^2}{x^2} - \frac{1}{x} = C$. The configuration of paths in the neighborhood of the origin is illustrated in Figure 85. When $C < 0$ one obtains ellipses, and when $C > 0$ hyperbolas; when $C = 0$ we have the parabola $x = y^2$.

The behavior of the paths in the neighborhood of a multiple equilibrium state may be topologically equivalent to that in the simple case, as evidenced by the following examples.

Example 8.

$$\dot{x} = x^3, \quad \dot{y} = y.$$

* Chapter VIII will present general methods for investigating certain types of multiple equilibrium state.

146

One of the characteristic roots of the equilibrium state $O\,(0,\,0)$ is zero.
The general solution is $ye^{\frac{1}{2x^2}} = C$. The paths behave in the neighborhood of the origin as in the neighborhood of a simple node. The equilibrium state is known as a "multiple node."

Example 9.

$$\dot{x} = x^3, \quad \dot{y} = -y.$$

One of the characteristic roots of the equilibrium state $O$ is zero, the general solution is $ye^{-1/2\,x^2} = C$. The configuration of paths near the origin is the same as near a simple saddle point. The equilibrium state is therefore known as a "multiple saddle point."

## §8. EQUILIBRIUM STATE WITH PURE IMAGINARY CHARACTERISTIC ROOTS

**1. INTRODUCTORY REMARKS.** In this section we discuss the last of the cases of simple equilibrium states listed in §6, case 4), in which the characteristic roots are pure imaginary. As stated above, this case is considerably more complicated than cases 1)−3), as examination of the linearized system is no longer sufficient. The method to be used here is applicable to arbitrary complex roots, whether pure imaginary or with nonzero real parts. We shall therefore carry the discussion through for the most general case

$$\lambda_1 = a + \beta i, \quad \lambda_2 = a - \beta i \quad (\beta \neq 0),$$

although the case $a \neq 0$ has already been dealt with. We are thus assuming that the canonical form of the system is

$$\frac{dx}{dt} = P\,(x,\,y) = ax + \beta y + \varphi\,(x,\,y) = ax + \beta y + xg_1 + yg_2,$$

$$\frac{dy}{dt} = Q\,(x,\,y) = \beta x - ay + \psi\,(x,\,y) = \beta x - ay + xf_1 + yf_2,$$

(1)

where the functions $\varphi,\,\psi,\,g_1,\,g_2,\,f_1,\,f_2$ satisfy conditions 1 and 2 of §6. To fix ideas, we shall also assume that

$$\beta > 0.$$

(2)

The real number $a$ may or may not vanish.

**2. TRANSFORMATION TO POLAR COORDINATES.** The phase portrait in the neighborhood of an equilibrium state of this type is more conveniently studied in polar coordinates

$$x = \varrho \cos \theta, \quad y = \varrho \sin \theta.$$

(3)

We preface the discussion with a few remarks on transformation from cartesian to polar coordinates. We shall allow the radius-vector $\varrho$ to take on both nonnegative and negative values. In addition, $\varrho$ and $\theta$ will be treated not only as polar coordinates of a point $M(x, y)$ on the $(x, y)$-plane, but also as cartesian coordinates on the $(\varrho, \theta)$-plane. Thus formulas (3) define a single-valued continuous mapping of the $(\varrho, \theta)$-plane onto the $(x, y)$-plane. This mapping is not one-to-one and has the following properties:

a) Any strip $0 \leqslant \varrho < \varrho^*$, where $\varrho^*$ is an arbitrary positive number, is mapped onto the interior of the disk $C^*$ on the $(x, y)$-plane with center at the origin and radius $\varrho^*$.

b) The entire axis $\varrho = 0$ on the $(\varrho, \theta)$-plane is mapped onto a single point, the origin of the $(x, y)$-plane.

c) All points $(\varrho, \theta)$ with equal values of $\varrho$ and values of $\theta$ differing by multiples of $2\pi$, i.e., all points $(\varrho, \theta + 2k\pi)$, $k = 0, \pm 1, \pm 2, \ldots$, are mapped onto the same point of the $(x, y)$-plane.

d) Any two points $(\varrho, \theta)$ and $(-\varrho, \theta + \pi)$ are mapped onto the same point of the $(x, y)$-plane.

e) Any straight line $\theta = \text{const}$ on the $(\varrho, \theta)$-plane is mapped onto a straight line on the $(x, y)$-plane, through the origin $O$, divided by the origin into two rays; one of these rays corresponds to $\varrho > 0$, the other to $\varrho < 0$. Any straight line $\varrho = \text{const}$ is mapped onto a circle about the origin on the $(x, y)$-plane (Figure 86a, b).

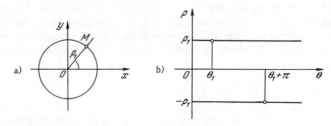

FIGURE 86.

Note that for any fixed $\theta_0$ the mapping (3) is one-to-one on the "half-open" rectangle $W$: $0 < \varrho \leqslant \varrho^*$, $\theta_0 \leqslant \theta < \theta_0 + 2\pi$, or on the rectangle $W'$: $0 > \varrho \geqslant -\varrho^*$, $\theta_0 \leqslant \theta < \theta_0 + 2\pi$ (Figure 87a, b). Each of these rectangles is mapped onto a disk $C^*$ about $O$ of radius $\varrho^*$, punctured at its center. Each point of this punctured disk is the image of exactly one point of the rectangle $W$ (or $W'$). Our mapping thus has an inverse defined on the punctured disk $C^*$ and mapping it [onto] $W$ (or $W'$). This mapping is not continuous. However, it is easy to see that *the mapping (3) is regular in any region $E$ in the $(\varrho, \theta)$-plane of sufficiently small diameter which does not contain points of the axis $\varrho = 0$* (since in any such region the mapping is one-to-one and continuous, and moreover the Jacobian

$$\frac{D(x, y)}{D(\varrho, \theta)} = \varrho$$

does not vanish for $\varrho \neq 0$). In other words, the mapping (3) is locally regular.

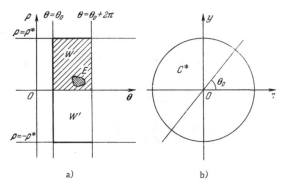

FIGURE 87.

We now apply the transformation $(3)$ to system $(1)$. The result is

$$\frac{dx}{dt} = \frac{d\varrho}{dt}\cos\theta - \varrho\sin\theta\,\frac{d\theta}{dt}\,; \qquad \frac{dy}{dt} = \frac{d\varrho}{dt}\sin\theta + \varrho\cos\theta\,\frac{d\theta}{dt}\,.$$

Assuming that $\varrho \neq 0$ and solving these equations for $\frac{d\varrho}{dt}$, $\frac{d\theta}{dt}$, we get

$$\frac{d\varrho}{dt} = \frac{dx}{dt}\cos\theta + \frac{dy}{dt}\sin\theta, \qquad \frac{d\theta}{dt} = \frac{1}{\varrho}\left(-\frac{dx}{dt}\sin\theta + \frac{dy}{dt}\cos\theta\right).$$

A few simple manipulations yield the system

$$\frac{d\varrho}{dt} = \alpha\varrho + \varphi\,(\varrho\cos\theta,\ \varrho\sin\theta)\cos\theta + \psi\,(\varrho\cos\theta,\ \varrho\sin\theta)\sin\theta,$$

$$\frac{d\theta}{dt} = \beta + \frac{\psi\,(\varrho\cos\theta,\ \varrho\sin\theta)}{\varrho}\cos\theta - \frac{\varphi\,(\varrho\cos\theta,\ \varrho\sin\theta)}{\varrho}\sin\theta. \tag{4}$$

Denoting

$$F\,(\varrho,\ \theta) = \alpha\varrho + \varphi\,(\varrho\cos\theta,\ \varrho\sin\theta)\cos\theta + \psi\,(\varrho\cos\theta,\ \varrho\sin\theta)\sin\theta,$$

$$\Phi\,(\varrho,\ \theta) = \frac{\psi\,(\varrho\cos\theta,\ \varrho\sin\theta)}{\varrho}\cos\theta - \frac{\varphi\,(\varrho\cos\theta,\ \varrho\sin\theta)}{\varrho}\sin\theta, \tag{5}$$

we write system $(4)$ as

$$\frac{d\varrho}{dt} = F\,(\varrho,\ \theta),$$

$$\frac{d\theta}{dt} = \beta + \Phi\,(\varrho,\ \theta). \tag{4'}$$

In view of $(5)$ and the properties of $\varphi\,(x, y)$ and $\psi\,(x, y)$, it is easy to see that if $\varrho^* > 0$ is sufficiently small the functions $F\,(\varrho, \theta)$ and $\Phi\,(\varrho, \theta)$ are defined in the region $\Omega$: $0 < |\varrho| < \varrho^*$, $-\infty < \theta < +\infty$,* and are moreover of class

---

* This region is the union of two open infinite horizontal strips having the axis $\varrho = 0$ as a common boundary component. [ $\Omega$ is not strictly speaking a region, since it is not connected.]

$C_1$ in this region. Furthermore, these functions are $2\pi$-periodic in $\theta$ and tend to zero uniformly in $\theta$ as $\varrho \to 0$.* It follows, in particular, from the last-mentioned property that if $\varrho^*$ is sufficiently small we have

$|\Phi(\varrho, \theta)| < \frac{\beta}{2}$ in the region $\Omega$ and consequently

$$\frac{\beta}{2} < \beta + \Phi(\varrho, \theta) < \frac{3}{2}\beta. \tag{6}$$

We shall assume $\varrho^*$ so small that this condition is indeed satisfied. Then $\beta + \Phi(\varrho, \theta) \neq 0$ and the paths of system (1) in the region $\Omega$ are the integral curves of the equation

$$\frac{d\varrho}{d\theta} = \frac{F(\varrho, \theta)}{\beta + \Phi(\varrho, \theta)} = R(\varrho, \theta), \tag{7}$$

obtained by dividing the first equation of system (4') by the second (see §1.7).

It will be convenient to investigate equation (7) not only for $0 < |\varrho| < \varrho^*$ but also for $\varrho = 0$. To this end, we extend the definition of $R(\varrho, \theta)$ by setting $R(0, \theta) = 0$ for any $\theta$.

Since $\lim_{\rho \to 0} F(\varrho, \theta) = \lim_{\rho \to 0} \Phi(\varrho, \theta) = 0$ and $\beta \neq 0$, it follows that $\lim_{\rho \to 0} R(\varrho, \theta) = 0$, and so the extended function $R(\varrho, \theta)$ is obviously periodic in $\theta$ and continuous throughout the strip $-\varrho^* < \varrho < \varrho^*$. Moreover, in this strip $R(\varrho, \theta)$ has a continuous partial derivative with respect to $\varrho$. Indeed, this is true for $\varrho \neq 0$ by virtue of the fact that the functions $\Phi(\varrho, \theta)$, $\Psi(\varrho, \theta)$ are of class $C_1$ for $\varrho \neq 0$.

Thus, it remains to show that $R(\varrho, \theta)$ is also continuously differentiable at $\varrho = 0$. Multiplying the numerator and denominator of the right-hand side of equation (7) by $\varrho \neq 0$, we write $R(\varrho, \theta)$ as

$$R(\varrho, \theta) = \frac{\alpha\varrho^2 + \varrho\varphi(\varrho\cos\theta, \varrho\sin\theta)\cos\theta + \varrho\psi(\varrho\cos\theta, \varrho\sin\theta)\sin\theta}{\varrho\beta + \psi(\varrho\cos\theta, \varrho\sin\theta)\cos\theta - \varphi(\varrho\cos\theta, \varrho\sin\theta)\sin\theta}.$$

Differentiating the right-hand side with respect to $\varrho$ and letting $\varrho \to 0$, we see, in view of the properties of $\varphi$ and $\psi$, that

$$\lim_{\rho \to 0} \frac{\partial R(\varrho, \theta)}{\partial \varrho} = \frac{\alpha}{\beta}. \tag{8}$$

On the other hand, direct computation of the derivative $\frac{\partial R(\varrho, \theta)}{\partial \varrho}$ at $\varrho = 0$ gives

---

* Indeed,

$$\frac{\varphi(\varrho\cos\theta, \varrho\sin\theta)}{\varrho} = \frac{\varrho\cos\theta g_1(\varrho\cos\theta, \varrho\sin\theta)}{\varrho} + \frac{\varrho\sin\theta g_2(\varrho\cos\theta, \varrho\sin\theta)}{\varrho} =$$

$$= \cos\theta g_1(\varrho\cos\theta, \varrho\sin\theta) + \sin\theta g_2(\varrho\cos\theta, \varrho\sin\theta).$$

By virtue of the properties of $g_1$ and $g_2$, this expression tends to zero as $\varrho \to 0$. The expression $\frac{\psi(\varrho\cos\theta, \varrho\sin\theta)}{\varrho}$ is treated in similar fashion.

$$\frac{\partial R\,(0,\,\theta)}{\partial \varrho} = \lim \frac{R\,(\varrho,\,\theta) - R\,(0,\,\theta)}{\varrho} = \lim_{\rho \to 0} \frac{R\,(\varrho,\,\theta)}{\varrho} = \frac{\alpha}{\beta}\,. \qquad (9)$$

It follows from (8) and (9) that the function $R\,(\varrho,\,\theta)$ is continuously differentiable with respect to $\varrho$ at $\varrho = 0$ as well, hence also in the strip $\Omega^*$: $-\varrho^* < \varrho < \varrho^*$. Thus equation (7), as extended by (8), satisfies in the strip $-\varrho^* < \varrho < \varrho^*$ the assumptions of both the existence-uniqueness theorem and the theorem on continuity of solutions with respect to initial values (Appendix, § 8).

Hence for any $\theta_0$ and $\varrho_0$ equation (7) has a unique solution

$$\varrho = f\,(\theta,\,\theta_0,\,\varrho_0)\ (f\,(\theta_0,\,\theta_0,\,\varrho_0) \equiv \varrho_0),$$

defined on some (maximal) interval $(\theta_1,\,\theta_2)$ containing $\theta_0$. Moreover, since $R\,(0,\,\theta) \equiv 0$ it is clear that $\varrho = 0$ is a solution, i.e., the $\theta$-axis in the $(\varrho,\theta)$-plane is an integral curve of equation (7).

All integral curves of equation (7) in the strip $\Omega^*$ are paths of system (4). If

$$\varrho = \varrho\,(t), \qquad \theta = \theta\,(t)$$

are the equations of some path $L$ of system (4) and $\theta_0,\,\varrho_0$ a point on this path, then the solution

$$\varrho = f\,(\theta,\,\theta_0,\,\varrho_0)$$

of equation (7) is the equation of this path in terms of $\varrho$ and $\theta$. Since $\frac{d\theta}{dt} = \beta + \Phi\,(\varrho,\,\theta) > 0$ for $|\varrho| < \varrho^*$, it follows that $\theta$ is a monotone increasing function of $t$ along any path of system (4). It follows from (6) that for motion along the path in the sense of increasing (decreasing) $t$ the quantities $t$ and $\theta\,(t)$ are either both bounded above (below) or both increase to $+\infty$ (decrease to $-\infty$). Indeed, if, say, $t$ is bounded above, $t < T$ and $\theta\,(t_0) = \theta_0$, then by (6) we have

$$\theta\,(t) = \theta_0 + \int_{t_0}^{t} \frac{d\theta}{dt}\,dt < \theta_0 + \frac{3}{2}\,\beta\,(t - t_0),$$

so that $\theta$ is also bounded above. Similarly one considers the cases in which $\theta$ is bounded above and $t$ (or $\theta$) bounded below.

### 3. COMPARISON OF PATHS OF SYSTEM (I) AND INTEGRAL CURVES OF EQUATION (7).

We shall first interpret the transformation to polar coordinates as mappings of the $(x,\,y)$- and $(\varrho,\,\theta)$-planes onto each other. Consider the integral curves of equation (7) (or, equivalently, the paths of system (4)) lying in the strip defined in the $(\varrho,\,\theta)$-plane by inequalities

$$0 < \varrho \leqslant \varrho^*, \qquad -\infty < \theta < +\infty \qquad (10)$$

(discussion of the strip defined by $-\varrho^* \leqslant \varrho < 0$, $-\infty < \theta < +\infty$ is entirely analogous).

Let $\hat{L}$ be some path in this strip, $\varrho = \varrho(t)$, $\theta = \theta(t)$ the corresponding solution of system (4), defined on an interval $(t, t_0)$, and

$$\varrho = f(\theta, \theta_0, \varrho_0) \tag{11}$$

the solution of equation (7) which yields the equation of $\hat{L}$ in terms of $\varrho$, $\theta$. The mapping (3) takes the path $\hat{L}$ onto a curve

$$x = \varrho(t) \cos \theta(t), \qquad y = \varrho(t) \sin \theta(t), \tag{12}$$

and since the mapping is locally regular this curve is a path of system (1) in the interior of a disk $C^*$ of radius $\varrho^*$. Denote this path by $L$. It is readily seen that since the right-hand sides of system (4) are periodic in $\theta$ the curves

$$\varrho = \varrho(t), \qquad \theta = \theta(t) + 2k\pi \qquad (k = \pm 1, \pm 2, \pm 3, \ldots) \tag{13}$$

are also paths of system (4), lying in the strip (10) and mapped onto $L$ by (3), and they exhaust all such paths. The paths (13) are obviously derived from each other by translation by a multiple of $2\pi$ along the $\theta$-axis. The corresponding solutions of equation (7) are derived from (11) by substituting $\theta_0 + 2k\pi$ for $\theta_0$:

$$\varrho = f(\theta, \theta_0 + 2k\pi, \varrho_0).$$

Conversely, let $x = x(t)$, $y = y(t)$ be a given path of system (1) lying inside the disk $C^*$ and defined on some interval $(t_1, t_2)$. Since the mapping (3) is locally regular, any such path corresponds to at least one path $x = \varrho(t)$, $y = \theta(t)$ of system (4) and with it to all paths of type (13)* (Figure 88a, b).

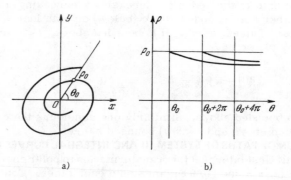

a)  b)

FIGURE 88.

---

* It is readily shown that if $L$ is a nonclosed path of system (1) all the paths (13) mapped onto it by (3) are distinct, while if $L$ is a closed path the paths (13) ($k = 0, 1, 2, \ldots$) all coincide (i.e., in the latter case $L$ is the image of exactly one path of system (4)).

We shall now treat the transformation to polar coordinates $\varrho$, $\theta$ not as a mapping of the $(x, y)$-plane onto the $(\varrho, \theta)$-plane but in the usual sense, as introduction of polar coordinates in the $(x, y)$-plane. Then, obviously, equations (12), and also equations (13), are parametric equations in polar coordinates of a path $L$ of system (1) on the $(x, y)$-plane. A solution $\varrho = f(\theta, \theta_0, \varrho_0)$ of equation (7) is the equation in polar coordinates of the path $L$. Henceforth we shall confine ourselves almost exclusively to this interpretation.

**4. CONSTRUCTION OF THE SUCCESSION FUNCTION ON A RAY** $\theta =$ const.
The following lemma is a direct consequence of the fact that for all $\theta$ and all sufficiently small $\varrho$ $(0 < \varrho \leqslant \varrho^*)$

$$\frac{d\theta}{dt} = \beta + \Phi(\varrho, \theta) \neq 0$$

(since $\beta \neq 0$), in other words, of the fact that $R(\varrho, \theta)$ is bounded for all sufficiently small $\varrho$.

*Lemma 1.\* If $\varrho^* > 0$ is sufficiently small, then each straight line $\theta =$ const on the $(x, y)$-plane has no contact with paths of system (1) at points with $0 < |\varrho| < \varrho^*$ (and each straight line $\theta =$ const on the $(\varrho, \theta)$-plane has no contact with paths of system (4)).*

Thus, any sufficiently small segment of a ray $\theta =$ const on the $(x, y)$-plane, one of whose endpoints is $O$, is "almost" a segment without contact (only "almost" for one of its endpoints is an equilibrium state) (Figure 89). We shall now construct a succession function on a segment of this type, using the equation of the path in polar coordinates — a solution of equation (7):

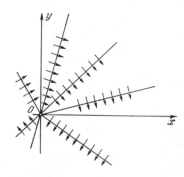

FIGURE 89.

$$\varrho = f(\theta, \theta_0, \varrho_0) \qquad (f(\theta_0, \theta_0, \varrho_0) \equiv \varrho_0).$$

We first establish certain properties of this solution. As stated, above, since $R(0, \theta) = 0$ the function

$$\varrho \equiv 0$$

is a solution of equation (7) (the corresponding integral curve on the $(\varrho, \theta)$-plane is the $\theta$-axis). The solution is of course defined for all $\theta$, i.e., on the interval $-\infty < \theta < +\infty$. Therefore, for all $\theta$ and $\theta_0$,

$$f(\theta, \theta_0, 0) = 0. \tag{14}$$

By the uniqueness theorem it follows that if $\varrho_0 > 0$ $(\varrho_0 < 0)$ then $f(\theta, \theta_0, \varrho_0) > 0$ $(< 0)$ for all $\theta$ in the interval of definition of the solution.

---

\* An alternative proof may be based on the following proposition: The condition for a path to be tangent to a straight line $\theta =$ const, which is a straight line $y = kx$, is obviously $x\dot{y} - y\dot{x} = 0$, or in polar coordinates $\varrho^2 \dfrac{d\theta}{dt} = 0$. But for sufficiently small $\varrho \neq 0$ we have $\dfrac{d\theta}{dt} \neq 0$, so that this condition cannot hold.

Our subsequent reasoning will be based on a lemma which follows from the fact that $\varrho = 0$ is a solution of equation (7) defined for all $\theta$.

*Lemma 2.  Let $\theta_1'$, $\theta_1$, $\theta_2'$, $\theta_2$ be arbitrary numbers such that*
$$\theta_1' \leqslant \theta_1 < \theta_2 \leqslant \theta_2'.$$

*For any $\varepsilon > 0$, there exists $\delta > 0$ such that any solution $\varrho = f(\theta, \theta_0, \varrho_0)$ of equation (7), where $\theta_1 \leqslant \theta_0 < \theta_2$ and $|\varrho_0| < \delta$, is defined on the closed interval $[\theta_1', \theta_2']$ and $|f(\theta, \theta_0, \varrho_0)| < \varepsilon$ at any point of this interval.*

The proof of Lemma 2 is elementary, using the continuity of solutions with respect to initial values and the compactness of the closed interval $[\theta_1, \theta_2]$ (Figure 90).

In the next lemma, which follows at once from Lemma 2, we return to paths of the original system (1).

*Lemma 3.  For any $\varepsilon > 0$ there exists $\delta > 0$ such that any path of system (1) passing at $t = t_0$ through a point $M$ of $U_\delta(O)$ other than $O$ crosses every ray $\theta = $ const with both increasing and decreasing $t$, without leaving the neighborhood $U_\varepsilon(O)$ (Figure 91).*

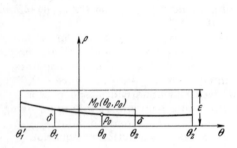

FIGURE 90.

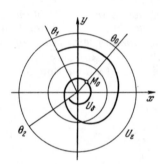

FIGURE 91.

Proof.  We use Lemma 2 with $\theta_1 = 0$, $\theta_2 = 2\pi$ and $\theta_1'$, $\theta_2'$ arbitrary numbers such that $\theta_1' \leqslant -2\pi$, $\theta_2' > 4\pi$.  By Lemma 2, there exists $\delta > 0$ such that any solution $\varrho = f(\theta, \theta_0, \varrho_0)$, where $0 \leqslant \theta_0 < 2\pi$, $0 < \varrho_0 < \delta$, is defined for all $\theta$, $\theta_1' \leqslant \theta < \theta_2'$, and for all these $\theta$ we have $0 < f(\theta, \theta_0, \varrho_0) < \varepsilon$. We may assume without loss of generality that $\varepsilon \leqslant \varrho^*$.  Now let $M_0$ be an arbitrary point in $U_\delta(O)$, $\theta_0$, $\varrho_0$, $0 \leqslant \theta_0 \leqslant 2\pi$, its polar coordinates, $L$ the path passing through this point at $t = t_0$.  As we have shown, the integral curve $\varrho = f(\theta, \theta_0, \varrho_0)$ of equation (7) is a path $\hat{L}$ of system (4), mapped by (3) onto the path $L$ of system (1).  Let

$$\varrho = \varrho(t), \qquad \theta = \theta(t)$$

be a motion on $\hat{L}$, so chosen that

$$\varrho(t_0) = \varrho_0 \quad \theta(t_0) = \theta_0.$$

Suppose that under this motion the path $\hat{L}$ passes through the points

$$\hat{M}_1(\theta_1', f(\theta_1', \theta_0, \varrho_0)) \text{ and } \hat{M}_2(\theta_2', f(\theta_2', \theta_0, \varrho_0))$$

154

respectively, at times $t_1$ and $t_2$. Then $t_0 \in (t_1, t_2)$ (since the function $\theta(t)$ is monotone), and, when $t$ increases from $t_1$ to $t_2$, $\theta(t)$ takes on every value between $\theta_1'$ and $\theta_2'$ exactly once. In particular, $\theta(t)$ takes on all values from $-2\pi$ to $4\pi$. But this means that the path $L$ of system (1) passing at $t = t_0$ through $M_0 \in U_\delta(O)$ crosses every ray $\theta = \text{const}$ with both increasing and decreasing $t$. That $L$ does this without leaving $U_\varepsilon(O)$ follows from the inequality $0 < f(\theta, \theta_0, \varrho_0) < \varepsilon$. Q.E.D.

Remark. It follows from Lemma 3 that if $\theta_0$ is an arbitrary number then the paths of system (1) passing through points of the segment $\theta = \theta_0$, $0 < \varrho < \varepsilon$, on the $(x, y)$-plane include all paths other than $O$ passing through points of the neighborhood $U_\delta(O)$.

Now fix some ray $\theta = \theta_0$, $\varrho > 0$ (or $\varrho < 0$).

By Lemma 3, if $\varrho_0$ is sufficiently small, say $|\varrho_0| < \varepsilon$, the solution

$$\varrho = f(\theta, \theta_0, \varrho_0) \qquad (f(\theta_0, \theta_0, \varrho_0) \equiv \varrho_0)$$

is defined for all $\theta$, $\theta_0 \leqslant \theta \leqslant \theta_0 + 2\pi$ and is the equation (in polar coordinates) of the paths crossing the straight line $\theta = \theta_0$ at points for which $|\varrho_0| < \varepsilon$, $\varrho_0 > 0$ (or $\varrho_0 < 0$). Moreover, by the Remark following Lemma 3 one exhausts all paths passing through points of a sufficiently small neighborhood of $O$ by considering the paths crossing a segment of any fixed ray $\theta = \theta_0$ with $\varrho > 0$ (or $\varrho < 0$), $|\varrho| < \varepsilon$ (where $\varepsilon$ is a suitably chosen constant). Consider the expression

$$\varrho_1 = f(\theta_0 + 2\pi, \theta_0, \varrho_0) = f_{\theta_0}(\varrho_0). \tag{15}$$

It is clear that the points $M_0$ and $M_1$ with polar coordinates $(\varrho_0, \theta_0)$ and $(\varrho_1, \theta_0)$, respectively, are $t$-successive points (for $\beta > 0$) at which the path of system (1) crosses the ray $\theta = \theta_0$, so that (15) is a succession function on a segment of the ray $\theta = \theta_0$.

Since the function $R(\varrho, \theta)$ on the right of equation (7) has a continuous partial derivative with respect to $\varrho$, it follows that the solution (11) is continuously differentiable with respect to the initial value $\varrho_0$, and consequently the succession function $f_{\theta_0}(\varrho_0)$ is also differentiable.

This succession function will be used to investigate the equilibrium state.

**5. DESCRIPTION OF THE PATH THROUGH A POINT IN A SUFFICIENTLY SMALL NEIGHBORHOOD OF THE EQUILIBRIUM STATE.** Lemma 3 yields quite complete information on the possible behavior of a single path through a point sufficiently close to the equilibrium state. This is summarized in the following lemma.

Lemma 4. *A path passing through a sufficiently small neighborhood $U_\delta(O)$ of an equilibrium state $O$ with complex characteristic roots may be of one of the following types 1): a closed path containing $O$ in its interior; 2) a nonclosed path tending to $O$ as $t \to +\infty$ ($t \to -\infty$) but not tending to $O$ as $t \to -\infty$ ($t \to +\infty$); 3) a nonclosed path tending as either $t \to +\infty$ or $t \to -\infty$ to a closed path containing the equilibrium state $O$ (and no other equilibrium state) in its interior.*

Proof. Let $\varepsilon > 0$ be so small that $U_\varepsilon(O)$ contains no equilibrium states other than $O$, and let $\delta > 0$ be a number as indicated in Lemma 3. Let $M_1$ be an arbitrary point of $U_\delta(O)$ other than $O$. Denote some value of the polar angle of $M_1$ by $\theta_1$, and let $L_0$ be the path of system (1) passing through $M_1$ at $t = t_1$. By Lemma 3 there exists $t = t_2 > t_1$ at which the path crosses the ray $\theta = \theta_1$ at a point $M_2$, without crossing the ray at intermediate values of $t$, $t_1 < t < t_2$, and remaining inside $U_\varepsilon(O)$ for all $t$, $t_1 \leqslant t \leqslant t_2$.

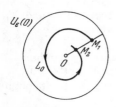

FIGURE 92.

It is clear that as $t$ varies in the interval $(t_1, t_2)$ the path $L_0$ crosses every ray other than $\theta = \theta_1$ exactly once (Figure 92).

The following cases are possible.

a) The points $M_1$ and $M_2$ coincide. Then $L_0$ is a closed path lying entirely within $U_\varepsilon(O)$. By Theorem 16 of §4, there is at least one equilibrium state inside $L_0$. This equilibrium state can only be $O$, since $U_\varepsilon(O)$ contains no other equilibrium states. Hence $L_0$ is a path of type 1).

b) The points $M_1$ and $M_2$ are distinct. Then $L_0$ is a nonclosed path (see Lemma 12, §3). To fix ideas, suppose that $M_2$ lies on the ray $\theta = \theta_1$ between $O$ and $M_1$ (the case in which $M_1$ is between $O$ and $M_2$ is treated similarly). Let $C$ be the simple closed curve formed from the arc $M_1 M_2$ on the path $L$ and the segment $M_1 M_2$ on the ray $\theta = \theta_1$. By Lemma 3, each segment of a ray $\theta = \text{const}$ contained in $U_\varepsilon(O)$ and not containing the point $O$ is a segment without contact. But then we may apply Lemma 11 of §3, according to which all points of the path $L_0$ corresponding to times $t < t_1$ lie outside $C$, and all those corresponding to $t > t_2$ lie inside $C$. The equilibrium state $O$ is inside $C$ — this is evident either directly or as a consequence of part c) of Lemma 11, §3. Hence it follows that the path $L_0$ passing through $M_1$ cannot tend to $O$ both as $t \to +\infty$ and as $t \to -\infty$. And since $M_1$ is an arbitrary point of $U_\delta(O)$ other than $O$, it follows that no path through this neighborhood can tend to $O$ both as $t \to +\infty$ and as $t \to -\infty$.

Let us examine the possible behavior of the path $L_0$ as $t \to +\infty$. Since it remains inside $C$, hence also inside $U_\varepsilon(O)$, with increasing $t$, it may either tend to the equilibrium state $O$ as $t \to +\infty$, in which case it is a path of type 2), or it may tend to a limit continuum $K_\omega$ contained in $U_\varepsilon(O)$. This continuum can only be a closed path containing $O$ in its interior. Indeed, otherwise it follows from Theorem 11 (§4) that the limit continuum must consist of paths which tend both as $t \to -\infty$ and as $t \to +\infty$ to the equilibrium state $O$ (since there are no other equilibrium states in $U_\delta(O)$). But we have already shown that no path can tend to $O$ both as $t \to -\infty$ and as $t \to +\infty$. Thus the limit continuum is necessarily a closed path. As in case a), one shows that this closed path, $L_\omega$ say, contains $O$ in its interior. It is easy to see that the path $L_0$ cannot lie inside $L_\omega$. Consequently, as $t \to +\infty$ the path $L_0$ tends to $L_\omega$, so that $L_0$ is a path of type 3). Q.E.D.*

Similarly, if the path $L_0$ does not leave $U_\varepsilon(O)$ for $t < t_1$, it must tend as $t \to -\infty$ to a closed path lying entirely within $U_\varepsilon(O)$ and containing the equilibrium state $O$ in its interior.

Remark. At some time $t_3 > t_2$ the nonclosed path $L_0$ will cross the segment $M_2 O$ of the ray $\theta = \theta_1$ at some point $M_3$; then, at some time $t_4 > t_3$, it will cross the segment $M_3 O$ at a point $M_4$, and so on. Thus the path $L_0$ crosses the ray $\theta = \theta_1$ at infinitely many points. Similarly, it will cross every ray $\theta = \text{const}$ issuing from $O$ at infinitely many points. Thus it follows that the path is a spiral.

* The lemma may also be proved without appeal to §4, by direct consideration of the properties of solutions of (7).

## 6. PHASE PORTRAIT IN A SUFFICIENTLY SMALL NEIGHBORHOOD OF AN EQUILIBRIUM STATE $O$.

We have determined the possible behavior of a single path passing near an equilibrium state with complex characteristic roots. We now take up the question of the structure of the entire phase portrait in a small neighborhood of the equilibrium state $O$.

Let $\varepsilon, \delta, \theta_1$ be as in Lemma 4. There are two possibilities: 1) There exists a point $M_1$ of $U_\delta(O)$ such that the path $L_0$ passing through it tends to the equilibrium state $O$ as $t \to +\infty$ (or $t \to -\infty$), without leaving $U_\varepsilon(O)$. 2) Through no point of $U_\delta(O)$ is there a path tending to the equilibrium state $O$ (as $t \to -\infty$ or as $t \to +\infty$).

Consider case 1). To fix ideas, suppose that the path $L_0$ passing at $t = t_1$ through $M_1$ tends to $O$ as $t \to +\infty$. By the Remark following Lemma 4, $L_0$ crosses the ray $\theta = \theta_1$ at countably many points $M_1, M_2, \ldots, M_n, \ldots$, corresponding to a monotone increasing sequence of parameter values $t_1 < t_2 < t_3 < \ldots$. Obviously, $t_n \to +\infty$ as $n \to +\infty$. For if the sequence $t_n$ were bounded above, this would imply that $\lim_{n\to\infty} t_n = t^*$, where $t^*$ is some finite number. The point $M^*$ on $L_0$ corresponding to $t = t^*$ is also on the ray $\theta = \theta_1$ and is a point of accumulation of the points $M_n$. But this contradicts Lemma 2 of §3 (since the ray $\theta = \theta_1$ has no contact with paths of the system in the neighborhood of $M^*$). Thus $\lim_{n\to\infty} t_n = +\infty$. Since $t_n \to +\infty$ and the path tends to $O$ as $t_n \to +\infty$, it follows that $O$ is a point of accumulation of the points $M_n$: $\lim_{n\to\infty} M_n = O$. Therefore, each point of the segment $OM$ (except $O$) on the ray $\theta = \theta_1$ lies in one of the segments $M_{n-1}M_n$. But by Lemma 3 every path crossing the segment $M_{n-1}M_n$ will cross the segment $OM$ again with increasing $t$, and this will occur, as is easily seen, on the segment $M_n M_{n+1}$. Hence it clearly follows that any path $\hat{L}_n$ crossing the segment $M_{n-1}M_n$ at some point $\hat{M}_{n-1}$ will cross the segments $M_n M_{n+1}, M_{n+1}M_{n+2}, \ldots$ at points $\hat{M}_n, \hat{M}_{n+1}, \ldots$, respectively, corresponding to times

$$\hat{t}_n < \hat{t}_{n+1} < \ldots,$$

(Figure 93), so that

$$\lim_{n\to\infty} \hat{t}_n = +\infty, \qquad \lim_{n\to\infty} \hat{M}_n = 0.$$

By the Remark to Lemma 3, this implies that all paths passing through a point of a sufficiently small neighborhood $U_\delta(O)$ tend to $O$ as $t \to +\infty$. With decreasing $t$ all paths leave the neighborhood of the equilibrium point $O$.

In this case the equilibrium state $O$ is a stable focus. One shows similarly that if the path tends to $O$ as $t \to -\infty$, then all paths through points close to $O$ are spirals tending to $O$ as $t \to -\infty$, and $O$ is an unstable focus.

Note, moreover, that if the equilibrium state is a stable focus and $\beta > 0$, then $\frac{d\theta}{dt} > 0$ and with increasing $t$ the representative point moves along the path, spiraling around the point $O$ in the counterclockwise sense; but if $\beta < 0$, so that $\frac{d\theta}{dt} < 0$, the motion of the point on the path is clockwise (Figure 94a, b). Analogous statements hold when the focus is unstable.

As mentioned previously, in this case (focus) the topological structure of the equilibrium state is the same as that of a node.

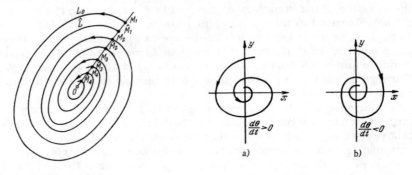

FIGURE 93.                    FIGURE 94.

We now take up case 2): there are no paths tending to the equilibrium state $O$. By Lemmas 3 and 4, there exist closed paths in any neighborhood of $O$ which contain $O$ in their interior.* These paths lie inside one another. Let $L_0$ be a closed path in a sufficiently small neighborhood of $O$, $U_\varepsilon(0)$ say, and let $\Gamma_0$ denote its interior. Consider an arbitrary ray $\theta = \theta_0$ $(\varrho > 0)$ issuing from $O$. Suppose that $L_0$ crosses this ray at a point $M_0$ with radius-vector $\varrho_0$. By the Remark to Lemma 3, each path passing through a point of $\Gamma_0$ will cross the segment $OM_0$ of the ray $\theta = \theta_0$, and we can exhaust all paths passing within $\Gamma_0$ by considering the paths passing through points of the segment $OM_0$. To investigate these paths we use the succession function on some ray $\theta = \theta_0$ (see §8.4), say $\theta_0 = 0$. Let $M$ be an arbitrary point on the segment $OM_0$, with radius-vector $\varrho$ $(0 < \varrho \leqslant \varrho_0)$ (Figure 95). Then, as we have seen, the succession function

$$\bar{\varrho} = f(2\pi, 0, \varrho) = f_0(\varrho)$$

is defined for all $\varrho$, $0 < \varrho < \varrho_0$, continuous and continuously differentiable on this interval (see §8.4).

Since $\varrho \equiv 0$ is a solution of equation (7), so that $\bar{f}(\theta, 0, 0) = 0$, it clearly follows that

$$f_0(0) = 0. \tag{16}$$

Consider the auxiliary function

$$d(\varrho) = f_0(\varrho) - \varrho,$$

defined for the same $\varrho$ as the function $f_0(\varrho)$. By (16), we have $d(0) = 0$. The path passing through the point $M(0, \varrho)$ is closed if and only if $d(\varrho) = 0$.

_____
* This also follows from Theorem 16, §3.

158

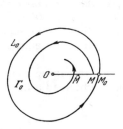

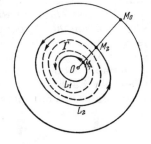

FIGURE 95.                              FIGURE 96.

Let $F$ be the set of all points $M(0, \varrho)$ on the segment $OM_0$ such that $d(\varrho) = 0$. This set is closed (because $d(\varrho)$ is continuous) and contains, apart from $O$, all points on $OM_0$ through which pass closed paths. Now it is obvious that in the case now under consideration (closed paths through points arbitrarily close to $O$) the set $F$ contains points distinct from but arbitrarily close to $O$. Denote the complement of the set $F$ in the segment $[OM]$ by $H$. Paths through points of $H$ are nonclosed. Since $H$ is the complement of a closed set [relative to a closed set], it is open, and therefore is either empty or the union of finitely or countably many open intervals $J_1, J_2,...$ whose endpoints are points of $F$.

Consider any one of these intervals $J$, with endpoints $M_1(0, \varrho_1)$ and $M_2(0, \varrho_2)$ and denote the closed paths through these points by $L_1$ and $L_2$, respectively. To fix ideas, let $\varrho_1 < \varrho_2$ (Figure 96). Then $d(\varrho) \neq 0$ for $\varrho_1 < \varrho < \varrho_2$, and so $d(\varrho)$ does not change sign throughout this interval. Any path $L$ passing through a point of the annulus $\Gamma$ bounded by $L_1$ and $L_2$ crosses the ray $\theta = 0$ between the points $M_1$ and $M_2$ and is therefore a spiral. It follows from Theorems 12 and 13 (§4) that $L$ is a whole path, tending to one of the paths $L_1, L_2$ as $t \to -\infty$ and to the other as $t \to +\infty$. It is clear that if $d(\varrho) < 0$ for $\varrho \in (\varrho_1, \varrho_2)$, then all paths passing through the annulus $\Gamma$ tend to $L_1$ as $t \to +\infty$ and to $L_2$ as $t \to -\infty$, while if $d(\varrho) > 0$ they tend to $L_2$ as $t \to +\infty$ and to $L_1$ as $t \to -\infty$.

Case 1) may be subdivided into two subcases:

a) $d(\varrho) = 0$ for all sufficiently small $\varrho > 0$. Then all paths passing through points sufficiently close to $O$ are closed and contain $O$ in their interior. Thus the equilibrium state $O$ is a c e n t e r.

b) There exist arbitrarily small values of $\varrho > 0$ such that $d(\varrho) \neq 0$. In other words, there are nonclosed paths arbitrarily near $O$, so that in any neighborhood of the equilibrium state there exist annular regions filled by spirals, as described above. In this case we shall call $O$ as c e n t r o - f o c u s.

The exact phase portrait in the neighborhood of a centrofocus depends on whether or not there exist annular regions filled by closed paths, on the nature of the annular regions filled by nonclosed paths, and on the relative positions of annular regions of the two types. The behavior of the paths in the neighborhood of a centrofocus is obviously determined by the distribution of the roots of the function $d(\varrho)$ in the neighborhood of $\varrho = 0$ and by the signs of $d(\varrho)$ on the intervals $J_1, J_2, \ldots$ in which it does not vanish.

We have thus proved the following proposition:

*A simple equilibrium state O with complex characteristic roots $\alpha \pm \beta i$ is either a (stable or unstable) focus, a center, or a centrofocus.*

In the case $\alpha \neq 0$ considered in §7, all paths tend to the equilibrium state $O$ as $t \to +\infty \, (\alpha < 0)$ or as $t \to -\infty \, (\alpha > 0)$, so that the equilibrium state is a stable or unstable focus.

**7. EXAMPLES.** We have already encountered an example of an equilibrium state with pure imaginary characteristic roots which is a center — the linear system

$$\frac{dx}{dt} = -y, \qquad \frac{dy}{dt} = x$$

(see Example 5, §1).

The following examples will show that an equilibrium state with pure imaginary characteristic roots may also be a focus or a centrofocus.

Example 1.

$$\frac{dx}{dt} = -y - x\sqrt{x^2 + y^2}, \qquad \frac{dy}{dt} = x - y\sqrt{x^2 + y^2}. \tag{17}$$

A direct check shows that $O\,(0, 0)$ is the only equilibrium state of system (17), system (17) is a system of class $C_1$ and the characteristic roots of the point $O$ are $\lambda_{1, 2} = \pm i$. To determine the behavior of the paths, we transform the system to polar coordinates. The computations show that in this case system (4) and equation (7) are, respectively

$$\frac{d\varrho}{dt} = -\varrho^2, \tag{18}$$

$$\frac{d\theta}{dt} = 1,$$

$$\frac{d\varrho}{d\theta} = -\varrho^2. \tag{19}$$

Integrating the last equation for initial values $\varrho = \varrho_0$, $\theta = \theta_0$, we get

$$\varrho = \frac{1}{\theta + \dfrac{1}{\varrho_0} - \theta_0}. \tag{20}$$

Equation (20) is the polar equation of the paths of system (17). Since we are concerned only with $\varrho > 0$, the angle $\theta$ varies in the interval $\left( \theta_0 - \dfrac{1}{\varrho_0}, \ +\infty \right)$.

The curve (20) is a hyperbolic spiral. $\varrho \to 0$ as $\theta \to +\infty$, and $\varrho \to \infty$ as $\theta \to \theta_0 - \dfrac{1}{\varrho_0}$. Thus, all paths of system (17) are hyperbolic spirals, which tend to $O$ as $t \to +\infty$ and escape to infinity with decreasing $t$ (in a finite time; see Remark at the end of Example 7 of §1). Thus $O$ is a stable focus (Figure 97).

Consider the function $d\,(\varrho)$ corresponding to this example. Setting $\theta_0 = 0$ and bearing in mind that (20) is a solution of equation (19) (which is an equation of type (7)), we see that

FIGURE 97.

$$d(\varrho) = f(2\pi, 0, \varrho) - \varrho = \frac{1}{2\pi + 1/\varrho} - \varrho =$$

$$= \varrho\left(\frac{1}{1+2\pi\varrho} - 1\right).$$

Since $d(\varrho) < 0$ for $\varrho > 0$, it follows anew from the above previous results that $O$ is a stable focus.

Example 2. Consider the system

$$\frac{dx}{dt} = -y + x(x^2 + y^2)\sin\frac{\pi}{\sqrt{x^2+y^2}},$$

$$\frac{dy}{dt} = x + y(x^2 + y^2)\sin\frac{\pi}{\sqrt{x^2+y^2}},$$

$$(21)$$

defining the right-hand sides to be zero at $O(0, 0)$. Under this assumption, system (21) (like system (17)) is of class $C_1$ and has a unique equilibrium state $O$ with characteristic roots $\pm i$ (this is verified by a direct check). Transforming as in the preceding case to polar coordinates, we see that system (4) and equation (7) are, respectively,

$$\frac{d\varrho}{dt} = \varrho^3\sin\frac{\pi}{\varrho}, \quad \frac{d\theta}{dt} = 1 \tag{22}$$

and

$$\frac{d\varrho}{d\theta} = \varrho^3\sin\frac{\pi}{\varrho}. \tag{23}$$

It is at once evident from equation (23) that the curves $\varrho = \frac{1}{n}$, $n = 1, 2, 3, \ldots$ are closed paths of system (21). Furthermore,

$$\text{if} \quad \varrho > 1, \quad \text{then} \quad \frac{d\varrho}{d\theta} > 0,$$

$$\text{if} \quad \frac{1}{2k} < \varrho < \frac{1}{2k-1}, \quad \text{then} \quad \frac{d\varrho}{d\theta} < 0, \tag{24}$$

$$\text{if} \quad \frac{1}{2k+1} < \varrho < \frac{1}{2k}, \quad \text{then} \quad \frac{d\varrho}{d\theta} > 0 \quad (k = 1, 2, \ldots).$$

The closed paths $\varrho = \frac{1}{n}$ partition the $(x, y)$-plane into countably many annuli (one of which, $\varrho > 1$, contains the point at infinity). We claim that paths passing through interior points of these annuli cannot be closed. Let $M_0(\theta_0, \varrho_0)$, $\varrho_0 \neq 0$, $\varrho_0 \neq \frac{1}{n}$ be such a point and $L_0$ the path passing through it at $t = t_0$. This path clearly lies entirely within one annulus, and therefore $\frac{d\varrho}{dt}$ has the same sign at all its points, say $\frac{d\varrho}{dt} > 0$. If $L_0$ is a closed path with period $T > 0$, then the point corresponding to time $t_0 + T > t_0$ coincides with the point $M_0$, i. e., $\varrho(t_0 + T) = \varrho_0 = \varrho(t_0)$. But this contradicts the condition $\frac{d\varrho}{dt} > 0$.

Thus system (21) has no closed paths other than the circles $\varrho = \frac{1}{n}$ ($n = 1, 2, \ldots$). But then it follows from the general theorems of §4 that paths lying interior to one of the annuli (except for the annulus $\varrho > 1$) tend to one of the circles bounding the annulus at $t \to -\infty$ and to the other as $t \to +\infty$. It follows from (24) that all paths in the annulus $\frac{1}{2k} < \varrho < \frac{1}{2k-1}$ ($k = 1, 2, \ldots$) are spirals tending to the closed path $\varrho = \frac{1}{2k-1}$ as $t \to -\infty$ and to the closed path $\varrho = \frac{1}{2k}$ as $t \to +\infty$. The paths inside the annulus $\frac{1}{2k+1} < \varrho < \frac{1}{2k}$ are spirals which "unwind" from the circle $\varrho = \frac{1}{2k+1}$ and "wind" onto the circle $\varrho = \frac{1}{2k}$. Finally, the paths passing through points of the region $\varrho > 1$ are spirals which tend to the circle $\varrho = 1$ as $t \to -\infty$ and to infinity with increasing $t$. The circles $\varrho = \frac{1}{n}$ are clearly stable limit cycles (see §3.9) of the system (21) for even $n$ and unstable limit cycles for odd $n$.*

The above examples demonstrate that when the characteristic roots are pure imaginary they do not determine the behavior of paths in the neighborhood of the equilibrium state; in other words, the local phase portrait is not uniquely determined by the linearized system.

In this case, therefore, a full characterization of the equilibrium state requires additional investigation.

When the dynamic system is analytic, so that the functions $P(x, y)$ and $Q(x, y)$ can be expanded in series in powers of $x$ and $y$, there are certain cases in which the investigation can be carried through to a successful completion. One then obtains a criterion for an equilibrium state to be a focus. The investigation involves terms of the expansion of degree higher than first.**

## §9. MODE OF APPROACH OF PATHS TO A SIMPLE EQUILIBRIUM STATE

### 1. FUNDAMENTAL DEFINITION. Let

$$\frac{dx}{dt} = P(x, y), \quad \frac{dy}{dt} = Q(x, y) \tag{I}$$

---

* It is not difficult to construct an example of an equilibrium state with pure imaginary roots whose phase portrait consists of annuli completely filled by closed paths alternating with annuli filled by nonclosed paths. In this case one can adduce a geometric example of an equilibrium state which has no definite topological structure in the sense of §3.5; for example, this is the case when we have an infinite sequence of annuli $\Gamma_1, \Gamma_2, \ldots, \Gamma_n, \ldots$, filled by closed paths, alternating with annuli filled by nonclosed paths, which shrink to the origin in such a way that the number of annuli filled by nonclosed paths lying between $\Gamma_i$ and $\Gamma_{i+1}$ increases with increasing $i$.

** The study of these multiple foci is of particular interest in examination of dynamic systems which involve parameters. Suppose that a dynamic system has a focus with pure imaginary characteristic roots for some value of the parameter. Then, as the parameter varies, one or more limit cycles may arise (be "created") from the focus (this problem is considered in /6/).

be the dynamic system under consideration, having an isolated equilibrium state $O(0, 0)$. $O$ may be either simple or multiple, so that the determinant

$$\Delta = \begin{vmatrix} P'_x(0, 0) & P'_y(0, 0) \\ Q'_x(0, 0) & Q'_y(0, 0) \end{vmatrix}$$

may or may not vanish.

Let

$$x = x(t), \quad y = y(t) \tag{1}$$

be a path of system (I) which tends to $O$ as $t \to +\infty$ or as $t \to -\infty$. Since the treatment of either case $(t \to +\infty \text{ or } t \to -\infty)$ is analogous, we shall consider only one, say $t \to +\infty$.

We are thus assuming that $x(t) \to 0$, $y(t) \to 0$ as $t \to +\infty$; $x(t)$ and $y(t)$ never vanish together.

*Definition IX. Let OM be a ray issuing from O and passing through a point M'(t) of the path. If the ray OM tends to a definite limiting position OM\* as $t \to +\infty$, we shall say that the path (1) tends to the equilibrium state O in the direction θ\*, where θ\* is the angle between the positive direction of the x-axis and the ray OM\* (Figure 98).*

The angle $\theta^*$ is of course defined only up to a multiple of $2\pi$.

R e m a r k 1. In the sequel we shall find the following equivalent definition more convenient:

*A path L tends to the equilibrium state O in the direction θ\* if for any ε > 0 all points on L corresponding to sufficiently large t lie in the region bounded by the rays θ = θ\* − ε and θ = θ\* + ε and containing the ray θ = θ\* (Figure 99).*

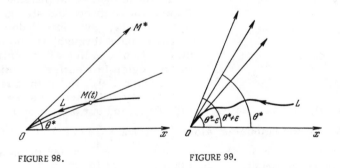

FIGURE 98.                    FIGURE 99.

R e m a r k 2. Instead of saying that the path $L$ tends to $O$ as $t \to +\infty$ in a given direction, we shall say: the semipath $L^+$ tends to $O$ in a given direction.

As stated in the introduction, the question of whether the paths of a system tend to an equilibrium state in definite directions, and if so — in what directions [in other words, the mode of approach to the equilibrium state], oversteps the limits of purely topological investigation of a dynamic system. Nevertheless, identification of these directions provides a more accurate picture of the phase portrait in the vicinity of an equilibrium state. Moreover, as we shall see later (Chapter IX), determination of the directions in which paths can tend to multiple equilibrium states $(\Delta = 0)$ is crucial for investigation of their topological structure.

It follows immediately from the definition that if a semipath $L^+$ tends to an equilibrium state in the direction $\theta^*$, then the quotient $\frac{y(t)}{x(t)}$ has a (finite or infinite) limit, and moreover

$$\lim_{t \to +\infty} \frac{y(t)}{x(t)} = \operatorname{tg} \theta^*.$$

We assert that the converse is also valid: if the semipath $L^+$ tends to an equilibrium state $O$ in such a way that the quotient $\frac{y(t)}{x(t)}$ has a finite or infinite limit $k^*$, then $L^+$ tends to $O$ in a definite direction $\theta^*$ such that $\operatorname{tg} \theta^* = k^*$.

To prove the assertion, suppose that $\lim_{t \to +\infty} \frac{y(t)}{x(t)} = k^*$, and draw through $O$ a straight line $M_1^* M_2^*$ with slope $k^*$ (Figure 100). Next, draw through $O$ two straight lines $A_1 A_2$ and $B_1 B_2$ forming angles $+ \varepsilon$ and $- \varepsilon$ respectively, with $M_1^* M_2^*$, where $\varepsilon$ is a sufficiently small

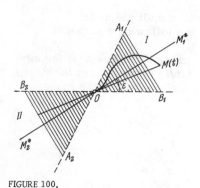

FIGURE 100.

positive number. Since $\lim_{t \to +\infty} \frac{y(t)}{x(t)} = k^*$, it follows that the straight line $OM(t)$ tends as $t \to +\infty$ to the straight line $M_1^* M_2^*$. Consequently, a section of the path $L$ corresponding to sufficiently large $t$ will lie in the region (hatched in Figure 100) between the straight lines $A_1 A_2$ and $B_1 B_2$. But since $L$ does not actually "pass through" the point $O$, this section of $L$ lies either entirely within region I or entirely within region II (see Figure 100). But then it is obvious that the ray $OM(t)$ tends as $t \to +\infty$ to one (and only one!) of the rays $OM_1^*$, $OM_2^*$, and our assertion is proved.

Knowledge of the number $k^*$ alone $\left( k^* = \lim_{t \to +\infty} \frac{y(t)}{x(t)} \right)$ is not yet sufficient to determine the direction in which the path $L$ tends to $O$, for there are two (mutually opposite) directions such that $\operatorname{tg} \theta^* = k^*$ $(0 \leqslant \theta^* < 2\pi)$.

In the sequel, in consideration of multiple equilibrium states, we shall often speak of "paths tending to the equilibrium state with slope $k^*$." This turn of speech will refer both to paths tending to the equilibrium state in the direction $\theta^*$ and to those doing so in the direction $\pi + \theta^*$ $(0 \leqslant \theta^* < \pi, \operatorname{tg} \theta^* = k^*)$.

The question of whether the ray $OM$ has a limiting position $OM^*$ may be formulated slightly differently: does the curve formed by adding the point

$O$ to the path $L$ possess a tangent at $O$?* Another, related question is whether the tangent to the path $L$ at a point $M\,(t)$ has a limiting position as $t \to +\infty$. We shall show in § 9.2 that in the case of a simple equilibrium state the tangent to $L$ at $O$ exists if and only if the tangent to $L$ at $M\,(t)$ has a limiting position as $t \to +\infty$, and when they exist they coincide.

In more general cases, this assertion is usually false. For example, consider the curve defined by

$$y = x^2 \sin \frac{1}{x} \quad \text{for} \quad x \neq 0,$$
$$y = 0 \qquad\qquad \text{for} \quad x = 0.$$

This curve has a tangent at every point, including $x = 0$, but the tangent at a point $M$ with abscissa $x$ does not have a limiting position as $x \to 0$.

This section is devoted to the study of simple equilibrium states, generally of systems of class $C_1$. However, in the case of a multiple characteristic root ($\lambda_1 = \lambda_2 = \lambda$, $\mu \neq 0$: degenerate node, and $\lambda_1 = \lambda_2 = \lambda$, $\mu = 0$: dicritical node), we shall have to make more stringent assumptions and assume the system to be of class $C_2$.

The question of the mode of approach to multiple equilibrium states will be considered in § 20, for analytic systems.

**2. SLOPE AT WHICH A PATH CAN TEND TO A SIMPLE EQUILIBRIUM STATE.** As usual, we write our system as

$$\frac{dx}{dt} = ax + by + \varphi\,(x,\,y), \qquad \frac{dy}{dt} = cx + dy + \psi\,(x,\,y), \tag{2}$$

where $\varphi\,(x,\,y)$ and $\psi\,(x,\,y)$ are functions of class $C_1$, which vanish together with their first derivatives at $O\,(0,\,0)$ (see § 6). The equilibrium state $O$ is assumed to be simple:

$$\Delta = \begin{vmatrix} a & b \\ c & d \end{vmatrix} \neq 0.$$

*Lemma 1.* Let $(I)$ be a dynamic system of class $C_1$, $O\,(0,\,0)$ a simple equilibrium state and $x = x\,(t)$, $y = y\,(t)$ a semipath of the system tending to $0$. Then $\frac{dy}{dx}$ has a (finite or infinite) limit as $t \to +\infty$ if an only if the quotient $\frac{y(t)}{x(t)}$ has a limit, and when they exist the limits are equal:

$$\lim_{t \to +\infty} \frac{dy}{dx} = \lim_{t \to +\infty} \frac{y\,(t)}{x\,(t)}.$$

Proof. Since the functions $\varphi$ and $\psi$ vanish together with their first derivatives at $(0,\,0)$, it follows that $\varphi\,(x,\,y) = o\,(\varrho)$, $\psi\,(x,\,y) = o\,(\varrho)$, where $\varrho = \sqrt{x^2 + y^2}$.

---

* In this connection the tangent is defined as the limiting position of the secant $OM$ as $t \to +\infty$. This remark is necessary in view of the fact that the curve obtained by adjoining the point $O$ to $L$ has no parametric description (the point $O$ does not correspond to any time $t$) and the usual definition of the tangent is valid only for parametrically (or explicitly) defined curves.

Suppose first that $\lim\limits_{t\to+\infty} \frac{y(t)}{x(t)}$ exists and is equal to a finite number $k$. This is possible only if $x(t) \neq 0$ for all sufficiently large $t$, i. e., $x(t)$ is ultimately of fixed sign. We have

$$\frac{dy}{dx} = \frac{\frac{dy}{dt}}{\frac{dx}{dt}} = \frac{c + d\frac{y}{x} + \frac{\psi}{\varrho}\sqrt{1 + \left(\frac{y}{x}\right)^2}}{a + b\frac{y}{x} + \frac{\varphi}{\varrho}\sqrt{1 + \left(\frac{y}{x}\right)^2}}. \tag{3}$$

Since $\Delta \neq 0$, it follows that $c + dk$ and $a + bk$ cannot vanish together, and by (3) we see that $\lim\limits_{t\to+\infty} \frac{dy}{dx}$ exists and is equal to $\frac{c+dk}{a+bk}$ $\left(\frac{\psi}{\varrho} \to 0 \text{ and } \frac{\varphi}{\varrho} \to 0 \text{ as } \varrho \to 0\right)$.

If $\lim\limits_{t\to+\infty} \frac{y(t)}{x(t)} = \infty$, then $\lim\limits_{t\to+\infty} \frac{x(t)}{y(t)} = 0$. It then follows from arguments similar to the foregoing that $\lim \frac{dx}{dt}$ exists, and hence so does $\lim \frac{dy}{dt}$. We have thus proved that the existence of $\lim \frac{y(t)}{x(t)}$ implies that of $\lim \frac{dy}{dx}$.

Now for the converse. Suppose that $\lim\limits_{t\to+\infty} \frac{dy}{dx}$ exists and is equal to $k$. Suppose first that $k$ is finite. This is clearly possible only if $x'(t) \neq 0$, i. e., $x'(t)$ does not change sign, for sufficiently large $t$. But then $x(t)$ is a monotone function for sufficiently large $t$, so that the inverse function $t = t(x)$ exists and the equation of the semipath $L$ near the origin may be written $y = y(x)$. Since $x(t)$ cannot vanish for any $t$ $(x(t) \to 0$ as $t \to +\infty)$, the function $y(x)$ is not defined at $x = 0$. We extend the definition, setting $y(x) = 0$ for $x = 0$. Then

$$\frac{y}{x} = \frac{y(x)}{x} = \frac{y(x) - y(0)}{x - 0} = y'(\gamma x), \text{ where } 0 < \gamma < 1.$$

Hence it follows that

$$\lim\limits_{t\to\infty} \frac{y}{x} = \lim\limits_{t\to\infty} y'(\gamma x) = k = \lim\limits_{t\to+\infty} \frac{dy}{dx}.$$

If $k = \infty$, then $\lim\limits_{t\to\infty} \frac{dx}{dy} = 0$ and it follows from the above arguments that $\lim\limits_{t\to\infty} \frac{x(t)}{y(t)} = 0$, i. e., $\lim \frac{y}{x} = \infty$. We have thus proved that, first, if $\lim\limits_{t\to\infty} \frac{y(t)}{x(t)}$ exists then so does $\lim\limits_{t\to\infty} \frac{dy}{dx}$, and second, if $\lim\limits_{t\to\infty} \frac{dy}{dx}$ exists then so does $\lim\limits_{t\to\infty} \frac{y(t)}{x(t)}$, and the limits are equal. Q.E.D.

Corollary. The slope at which a semipath can tend to an equilibrium state $O$ satisfies the quadratic equation

$$bk^2 + (a - d)k - c = 0 \tag{4}$$

(if $b = 0$, one usually adopts the convention that one root of equation (4) is $\infty$).

In fact, if $k$ is finite we see, letting $t \to +\infty$ in (3), that

$$k = \frac{c+dk}{a+bk},$$

and this is equivalent to (4). But if $k = \infty$, then $\lim \frac{dx}{dy} = \lim \frac{x}{y} = 0$. It is clear that then $y(t) \neq 0$ for sufficiently large $t$. Therefore

$$\frac{dx}{dy} = \frac{a\,\frac{x}{y} + b + \frac{\varphi(x,\,y)}{y}}{c\,\frac{x}{y} + d + \frac{\psi(x,\,y)}{y}}.$$

Letting $t \to +\infty$, we get $0 = b/d$, i. e., $b = 0$, completing the proof.

Note that the discriminant of equation (4) is equal to the discriminant of the characteristic equation. Hence, when the latter is negative (simple or multiple focus) there are no directions along which the paths can tend to the equilibrium state. It is readily shown that the roots $k_1$ and $k_2$ of equation (4) are expressed in terms of the characteristic roots $\lambda_1$ and $\lambda_2$ by

$$k_1 = \frac{d-\lambda_2}{b}, \qquad k_2 = \frac{d-\lambda_1}{b}.$$

All further considerations in this section are valid on the assumption that system (I) is reducible to canonical form in the neighborhood of the equilibrium state $(0,\,0)$.

**3. NODE WITH DISTINCT CHARACTERISTIC ROOTS.** In this case the system can be reduced to the form

$$\frac{dx}{dt} = \lambda_1 x + \varphi(x,\,y), \qquad \frac{dy}{dt} = \lambda_2 y + \psi(x,\,y). \tag{5}$$

To fix ideas, let $\lambda_2 < \lambda_1 < 0$. As shown in §7, there exists $\delta_0 > 0$ such that any path through a point of the neighborhood $U_{\delta_0}(O)$ of the node tends to $O$ with increasing $t$, in such a way that $\varrho(t)$ tends monotonically to zero, i. e., every circle $\varrho = \delta$ of radius $\delta < \delta_0$ is a cycle without contact. Let $\varrho = \varrho(t)$, $\theta = \theta(t)$ be the equations of a path of system (5) in polar coordinates. Then, transforming system (5) to polar coordinates, we get*

$$\frac{d\theta}{dt} = \frac{\lambda_2 - \lambda_1}{2} \sin 2\theta + \frac{\psi(\varrho \cos\theta,\, \varrho \sin\theta) \cos\theta - \varphi(\varrho \cos\theta,\, \varrho \sin\theta) \sin\theta}{\varrho}. \tag{6}$$

The second term on the right tends to zero as $\varrho \to 0$, uniformly in $\theta$. We may therefore write

$$\frac{d\theta}{dt} = \frac{\lambda_2 - \lambda_1}{2} \sin 2\theta + \varrho\Phi(\varrho,\,\theta),$$

---

\* When we did this in the case of complex characteristic roots, the result was a system such that $d\theta/dt \neq 0$ for all sufficiently small $\varrho$ ($|\varrho| < \varrho^*$), that is to say, a system with no equilibrium states in the strip $|\varrho| < \varrho^*$ (the axis $\varrho = 0$ was an integral curve). In the case considered here the situation is different: $d\theta/dt$ vanishes at points of the axis $\varrho = 0$, so that the resulting system in polar coordinates has an equilibrium state on $\varrho = 0$.

where $\Phi(\varrho, \theta)$ is bounded for sufficiently small $\varrho$ (we may assume that this is so for $|\varrho| < \delta_0$) and all $\theta$.

Given any $\varepsilon$, $0 < \varepsilon < \frac{\pi}{2}$, we consider four sectors $T_i^\varepsilon$ $(i = 1, 2, 3, 4)$ of the disk $\varrho \leqslant \delta_0$, defined by the following inequalities (Figure 101):

$$|\theta| \leqslant \varepsilon \quad (i=1), \qquad |\theta - \pi| \leqslant \varepsilon \quad (i=2),$$
$$\left|\theta - \frac{\pi}{2}\right| \leqslant \varepsilon \quad (i=3), \qquad \left|\theta - \frac{3}{2}\pi\right| \leqslant \varepsilon \quad (i=4).$$

On the rays

$$\theta = \varepsilon, \quad \theta = \frac{\pi}{2} - \varepsilon, \quad \theta = \pi + \varepsilon, \quad \theta = \frac{3}{2}\pi - \varepsilon \tag{7}$$

we have $\sin 2\theta > 0$, whereas on the other boundary rays of the sectors $T_i^\varepsilon$ we have $\sin 2\theta < 0$. Therefore, for sufficiently small $\delta(\varepsilon) < \delta_0$ we have $d\theta/dt < 0$ on the rays (7), since $\lambda_2 - \lambda_1 < 0$, and on the other four boundary rays of the $T_i^\varepsilon$, we have $d\theta/dt > 0$.

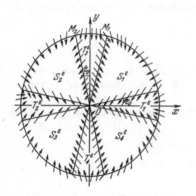

FIGURE 101.

It follows obviously that paths through the boundaries of the sectors $T_1^\varepsilon$ and $T_2^\varepsilon$ enter these sectors with increasing $t$, so that all paths passing through points of $T_1^\varepsilon$, $T_2^\varepsilon$ and their boundaries can never leave them again. Similarly, all paths through the boundaries of the sectors $T_3^\varepsilon$ and $T_4^\varepsilon$ leave these sectors with increasing $t$, so that no path, having once left $T_3^\varepsilon$, $T_4^\varepsilon$, can ever enter them again.

Let $S_1^\varepsilon$, $S_2^\varepsilon$, $S_3^\varepsilon$, $S_4^\varepsilon$ denote the other sectors of the disk $\varrho \leqslant \delta_0$ (Figure 101).

Lemma 2. *Any path passing through a point in one of the regions $S_i^\varepsilon$ will enter either $T_1^\varepsilon$ or $T_2^\varepsilon$ with increasing $t$.*

Proof. To fix ideas, let us consider a semipath $L^+$, whose parametric equations (in polar coordinates) are $\varrho = \varrho(t)$, $\theta = \theta(t)$, passing through a point of $S_1^\varepsilon$. Suppose that it does not enter $T_1^\varepsilon$ and hence never leaves the region $S_1^\varepsilon$. Then, for all sufficiently large $t$,

$$\varepsilon < \theta(t) < \frac{\pi}{2} - \varepsilon, \quad \text{i.e.,} \quad 2\varepsilon < 2\theta < \pi - 2\varepsilon,$$

and so $\sin 2\theta > \sin 2\varepsilon$. It then follows from (6), since $\lambda_2 - \lambda_1 < 0$, that

$$\frac{d\theta}{dt} < \frac{\lambda_2 - \lambda_1}{2} \sin 2\varepsilon + \varrho\Phi(\varrho, \theta).$$

Since $\varrho \to 0$ as $t \to +\infty$ (see §7), it follows that $\frac{d\theta}{dt} < \frac{\lambda_2 - \lambda_1}{4} \sin 2\varepsilon$ for sufficiently large $t$.

Let $\frac{\lambda_2 - \lambda_1}{4} \sin 2\varepsilon = -A$, where $A < 0$. Suppose also that $\theta(t_0) = \theta_0$ at $t = t_0$

$\left( \text{we may obviously assume that } 0 < \theta_0 < \frac{\pi}{2} \right)$. Then for $t > t_0$

$$\theta(t) = \theta_0 + \int_{t_0}^{t} \theta'(t)\, dt < \theta_0 - A(t - t_0) < \frac{\pi}{2} - A(t - t_0),$$

and if $A(t - t_0) > \frac{\pi}{2}$ we have $\theta(t) < 0$. This contradicts the assumption that the semipath $L^+$ never leaves $S_1^\varepsilon$. Q.E.D.

**Theorem 20.** *If $\lambda_2 < \lambda_1 < 0$, all paths tending to the node do so in definite directions. If the dynamic system has canonical form (5) in the neighborhood of the node 0, these directions are $\pi/2$ and $\frac{3}{2}\pi$, 0 and $\pi$. Only one semipath tends to 0 in each of the directions $\pi/2$ and $\frac{3}{2}\pi$; all others tend to 0 in the direction 0 or $\pi$, an infinite number of semipaths in each.*

*A suitably modified statement is valid for the case $\lambda_1 < \lambda_2 < 0$, and also for the case of an unstable node 0 ($0 < \lambda_1 < \lambda_2$ or $0 < \lambda_2 < \lambda_1$).*

Proof. Given $\varepsilon > 0$, we consider any semipath $L^+$ passing through an interior point of the disk $\varrho < \delta(\varepsilon)$ (as before, $\delta(\varepsilon)$ is so chosen that the rays bounding the sectors $T_i^\varepsilon$ have no contact with the paths at points of the disk other than $O$). Suppose that $L^+$ passes through a point of one of the sectors $S_1^\varepsilon$, $S_4^\varepsilon$ or $T_1^\varepsilon$. Then by Lemma 1 the semipath $L^+$ will ultimately be in $T_1^\varepsilon$ and, by the properties of this sector, henceforth never leave it. Since all the circles $\delta < \delta_0$ are cycles without contact, it is readily seen that this is true for any arbitrarily small $\varepsilon > 0$. And this means, by Remark 1 of §9.1, that the semipath $L^+$ tends to the equilibrium state $O$ in the direction $\theta = 0$.

Similarly one shows that if $L^+$ passes through a point of $S_2^\varepsilon$, $S_3^\varepsilon$ or $T_2^\varepsilon$, then it tends to $O$ in the direction $\theta = \pi$. In addition, it is obvious that infinitely many paths tend to $O$ in each of the directions $\theta = 0$ and $\theta = \pi$.

Now consider the sectors $T_3^\varepsilon$ and $T_4^\varepsilon$. Suppose that for some fixed $\varepsilon_0$ the semipath $L^+$ lies entirely inside $T_3^\varepsilon$. We claim that then $L^+$ tends to $O$ in the direction $\pi/2$. Indeed, there are two possibilities: a) For any $\varepsilon$, the path $L^+$ enters the sector $T_3^\varepsilon$ at some time and remains there. And this means that $L^+$ tends to $O$ in the direction $\pi/2$. b) For some $\varepsilon$, a semipath $L^+$ passing through the disk of radius $\delta(\varepsilon)$ corresponding to this $\varepsilon$ has points in the disk outside the sector $T_3^\varepsilon$. But then it follows from Lemma 1 that $L^+$ must enter the sector $T_1^\varepsilon$ or $T_2^\varepsilon$ and so, contrary to assumption, is not entirely inside $T_3^\varepsilon$. Similarly one shows that if $L^+$ lies entirely in some sector $T_4^\varepsilon$ ($\varepsilon < \varepsilon_0$) then it tends to $O$ in the direction $\frac{3}{2}\pi$.

We now show that each of the sectors $T_3^\varepsilon$, $T_4^\varepsilon$ contains exactly one semipath that does not leave the sector with increasing $t$. This will be proved for $T_3^\varepsilon$. Let $\varepsilon$ be fixed, and $M_1 M_2$ the arc of the circle $\varrho = \delta(\varepsilon)$ in the boundary of $T_3^\varepsilon$. The sector $T_3^\varepsilon$ is analogous to the triangle $OAB$ studied in connection with the phase portrait in the neighborhood of a saddle point [see §7.3].

It is not hard to see that all paths entering the sector $T_3^\varepsilon$ through the arc $M_1 M_2$ at points sufficiently close to $M_1$ leave the sector through the segment $OM_1$, while all those entering the sector through this arc at points sufficiently close to $M_2$ leave through the segment $OM_2$. Hence, by the same arguments as in the case of a saddle point (dividing the points of $M_1 M_2$ into two classes), one can show that at least one path entering $T_3^\varepsilon$ through the arc $M_1 M_2$ cannot leave this sector. The proof that this path is unique proceeds along the same lines as for a saddle point. Indeed, write the second equation of (5) as

$$\frac{dy}{dt} = y\left(\lambda_2 + \frac{\psi(x, y)}{y}\right) = y\left(\lambda_2 + \frac{x}{y} g_1 + g_2\right)$$

where $g_1$ and $g_2$ are the same functions as in §7 (see formula (1) of §7). At points of $T_3^\varepsilon$ we have $\frac{x}{y} < \operatorname{tg} \varepsilon$, i. e., $\frac{x}{y}$ is bounded there. If a semipath $L^+$ with parametric equations $x = x(t)$, $y = y(t)$, which does not leave $T_3^\varepsilon$, tends to the node $O$, then it follows from (8) that $\frac{dy}{dt} < 0$ for sufficiently large $t$ (since $\lambda_2 < 0$ and $y > 0$). This means that $y = y(t)$ is a monotone function, so that $t$ may be expressed as a function of $x$ and the equation of $L^+$ written as $x = x(y)$ ($x \to 0$ as $y \to 0$).

The rest of the proof is exactly the same as the proof that the separatrix of a saddle point is unique (§7), with suitable modifications due to the fact that here the equation of the paths has the form $x = x(y)$, unlike the case of §7. Q. E. D.

R e m a r k. It follows from Theorem 20 that the behavior of the semipaths of system (5)

$$\frac{dx}{dt} = \lambda_1 x + \varphi(x, y), \qquad \frac{dy}{dt} = \lambda_2 y + \psi(x, y)$$

($\lambda_1$, $\lambda_2$ distinct real numbers of the same sign), in the sense of their mode of approach to the equilibrium state, is precisely that of the linearized system

$$\frac{dx}{dt} = \lambda_1 x, \qquad \frac{dy}{dt} = \lambda_2 y$$

The same remark is valid for a system not reduced to canonical form

$$\frac{dx}{dt} = ax + by + \varphi(x, y), \qquad \frac{dy}{dt} = cx + dy + \psi(x, y)$$

when the point $O\,(0, 0)$ is a node. Thus the terms $\varphi(x, y)$ and $\psi(x, y)$ exert no influence on the directions in which the semipaths of the system tend to the node.

Using the above results (which, as mentioned, are no longer purely topological) concerning the mode of approach of the paths to a node (with distinct characteristic roots), one sees that the configuration of paths is that schematically depicted in Figure 104a, b (p. 180) (Figure 104a illustrates the case of a system in canonical form, with the paths tending to the node in the directions of the coordinate axes; Figure 104b illustrates the general case).

**4. DICRITICAL NODE.** We now consider the case $\lambda_1 = \lambda_2 = \lambda$, with $\mu = 0$ (see §7.2, case 1b with $\mu = 0$). The canonical form of the system is

$$\frac{dx}{dt} = \lambda x + \varphi\,(x,\,y), \qquad \frac{dy}{dt} = \lambda y + \psi\,(x,\,y). \tag{8}$$

To fix ideas, suppose that $\lambda < 0$, so that the node is stable. We first let $\varphi$ and $\psi$ be arbitrary functions of the class $C_1$ which vanish at the point $O\,(0,\,0)$ together with their first derivatives. We shall show that then, in general, the existence of directions in which the positive semipaths tend to $O$ depends on the functions $\varphi$ and $\psi$, so that system (7) and the linearized system

$$\frac{dx}{dt} = \lambda x, \qquad \frac{dy}{dt} = \lambda y \tag{9}$$

behave in this respect quite differently (compare with the Remark following Theorem 20). To demonstrate this, we consider the following example of Perron:

$$\frac{dx}{dt} = -x - \frac{y}{\lg \sqrt{x^2+y^2}}, \qquad \frac{dy}{dt} = -y + \frac{x}{\lg \sqrt{x^2+y^2}}. \tag{10}$$

The right-hand sides of this system are undefined at $(0,\,0)$. We define them there by setting $\varphi\,(0,\,0) = \psi\,(0,\,0) = 0$.

With this definition, $\varphi$ and $\psi$ are both continuous in the neighborhood of the origin $O$, which is an isolated equilibrium state. A direct check shows that the functions $\varphi\,(x,\,y)$ and $\psi\,(x,\,y)$ have continuous first derivatives in the neighborhood of $O$ (including $O$ itself) and the values of the derivatives there are

$$\varphi_x'\,(0,\,0) = \varphi_y'\,(0,\,0) = \psi_x'\,(0,\,0) = \psi_y'\,(0,\,0) = 0.$$

Let us determine the paths of system (10). We have

$$\varrho^2 = x^2 + y^2,$$
$$\varrho\varrho' = xx' + yy' = x\left(-x - \frac{y}{\lg \sqrt{x^2+y^2}}\right) + y\left(-y + \frac{x}{\lg \sqrt{x^2+y^2}}\right) = -\varrho^2,$$
$$\varrho' = -\varrho; \quad \varrho = Ce^{-t}$$

($C$ may be assumed positive). Now, letting the polar angle on the paths be $\theta = \theta\,(t)$, we get

$$\frac{d\theta}{dt} = \frac{d}{dt}\left(\operatorname{arctg} \frac{y}{x}\right) = \frac{xy' - yx'}{x^2 + y^2}.$$

Expressing $x'$ and $y'$ in terms of $\varrho$ and $\theta$, we get

$$\frac{d\theta}{dt} = \frac{1}{\lg \varrho} = \frac{1}{\lg Ce^{-t}},$$

i. e.,

$$d\theta = -\frac{dt}{t - \lg C}.$$

Integrating with respect to $t$ from $t_0$ to $t > t_0$, we obtain

$$\theta(t) - \theta(t_0) = -\lg(t - \lg C) + \lg(t_0 - \lg C).$$

Hence it is clear that $\theta(t) \to -\infty$ as $t \to +\infty$. But this means that none of the semipaths of our system can tend to $O$ in a definite direction, and they spiral infinitely many times about $O$, "winding" onto the point.

On the other hand, the paths of the linearized system

$$\frac{dx}{dt} = \lambda x, \qquad \frac{dy}{dt} = \lambda y,$$

are the rays

$$x = C_1 e^{\lambda t}, \qquad y = C_2 e^{\lambda t}.$$

Thus each path tends to $O$ as $t \to \infty$ in a definite direction, and moreover there is exactly one path tending to the origin in each direction. Perron's example shows that in the general case ($\varphi$ and $\psi$ arbitrary functions of class $C_1$) the directions of the semipaths are not determined by the linearized system.

Nevertheless, it proves that if more stringent restrictions are imposed on the "perturbations" $\varphi$ and $\psi$ of the linear terms, they lose their influence on the directions of the paths. We shall prove this for functions $\varphi$ and $\psi$ of class $C_2$.

$T h e o r e m$ 21. *If the functions* $\varphi(x, y)$ *and* $\psi(x, y)$ *in system (7)*

$$\frac{dx}{dt} = \lambda x + \varphi(x, y), \qquad \frac{dy}{dt} = \lambda y + \psi(x, y)$$

$(\lambda < 0)$ *are of class* $C_2$ *in the neighborhood of* $O$, *where*

$$\varphi(0, 0) = \varphi'_x(0, 0) = \varphi'_y(0, 0) = 0, \qquad \psi(0, 0) = \psi'_x(0, 0) = \psi'_y(0, 0) = 0,$$

*then each semipath tending to* $O$ *does so in a definite direction, and to each direction corresponds exactly one such path.*

P r o o f. Expand the functions $\varphi$ and $\psi$ by Taylor's formula:

$$\begin{aligned}
\varphi(x, y) &= A_1 x^2 + 2B_1 xy + C_1 y^2 + \alpha_1 x^2 + 2\beta_1 xy + \gamma_1 y^2, \\
\psi(x, y) &= A_2 x^2 + 2B_2 xy + C_2 y^2 + \alpha_2 x^2 + 2\beta_2 xy + \gamma_2 y^2,
\end{aligned} \tag{11}$$

where each of the constants $\alpha_1, \beta_1, \ldots, \gamma_2$ tends to zero as $\varrho \to 0$. Transform the system to polar coordinates, assuming that $\varrho > 0$. We get

$$\begin{aligned}
\frac{d\varrho}{dt} &= \lambda \varrho + \varphi(\varrho \cos\theta, \varrho \sin\theta) \cos\theta + \psi(\varrho \cos\theta, \varrho \sin\theta) \sin\theta, \\
\frac{d\theta}{dt} &= \frac{\psi(\varrho \cos\theta, \varrho \sin\theta)}{\varrho} \cos\theta - \frac{\varphi(\varrho \cos\theta, \varrho \sin\theta)}{\varrho} \sin\theta.
\end{aligned} \tag{12}$$

As we know (see §8), each path $\varrho = \varrho(t)$, $\theta = \theta(t)$ of system (11) in the strip $\Omega^+$ $[-\infty < \theta < +\infty,\ 0 < \varrho < \varrho^*]$ on the $(\varrho, \theta)$-plane (where $\varrho^*$ is sufficiently small) corresponds to a definite path of system (8) (other than the equilibrium state $O$) lying in $U_{\rho^*}(O)$, with parametric equations

$$x = \varrho(t)\cos\theta(t), \qquad y = \varrho(t)\sin\theta(t).$$

Conversely, to each path of system (8) in $U_{\rho^*}(O)$ corresponds an infinite set of paths of system (11) in the strip $\Omega^+$, each obtained from another path of the set by translation along a segment of length $2k\pi$ on the $\theta$-axis. For small $\varrho$, we have

$$\frac{d\varrho}{dt} = \varrho\left[\lambda + \frac{\varphi}{\varrho}\cos\theta + \frac{\psi}{\varrho}\sin\theta\right] \neq 0,$$

and so the paths of system (12) are integral curves of the equation

$$\frac{d\theta}{d\varrho} = \frac{\dfrac{\psi}{\varrho^2}\cos\theta - \dfrac{\varphi}{\varrho^2}\sin\theta}{\lambda + \dfrac{\varphi}{\varrho}\cos\theta + \dfrac{\psi}{\varrho}\sin\theta}. \tag{13}$$

Equation (13) is clearly defined for small negative $\varrho$ as well as positive ones. As in §8.2, we extend the functions $\varphi/\varrho$, $\psi/\varrho$, $\varphi/\varrho^2$ and $\psi/\varrho^2$, defining them at arbitrary points of the $\theta$-axis $(-\infty < \theta < +\infty,\ \varrho = 0)$ by

$$\left.\begin{array}{l}
\left[\dfrac{\varphi(\varrho\cos\theta, \varrho\sin\theta)}{\varrho}\right]_{\rho=0} = \left[\dfrac{\psi(\varrho\cos\theta, \varrho\sin\theta)}{\varrho}\right]_{\rho=0} = 0, \\[2mm]
\left[\dfrac{\varphi(\varrho\cos\theta, \varrho\sin\theta)}{\varrho^2}\right]_{\rho=0} = A_1\cos^2\theta + 2B_1\cos\theta\sin\theta + C_1\sin^2\theta, \\[2mm]
\left[\dfrac{\psi(\varrho\cos\theta, \varrho\sin\theta)}{\varrho^2}\right]_{\rho=0} = A_2\cos^2\theta + 2B_2\cos\theta\sin\theta + C_2\sin^2\theta.
\end{array}\right\} \tag{14}$$

Denote the extended right-hand side of equation (13) by $\Phi(\varrho, \theta)$. The equation

$$\frac{d\theta}{d\varrho} = \Phi(\varrho, \theta) \tag{15}$$

is now defined in the entire strip $\Omega$ $[|\varrho| < \varrho^*]$ (where $\varrho^*$ is sufficiently small) and coincides for $\varrho \neq 0$ with equation (13). We claim that $\Phi(\varrho, \theta)$ is continuous, and continuously differentiable with respect to $\theta$, in the strip $\Omega$. It will suffice to show that the functions $\varphi/\varrho$, $\varphi/\varrho^2$, $\psi/\varrho$, $\psi/\varrho^2$ are continuous and continuously differentiable with respect to $\theta$ on the $\theta$-axis (i. e., for $\varrho = 0$). Continuity follows directly from (11) and (14). We prove continuity of the partial derivatives for $\varphi/\varrho$, $\varphi/\varrho^2$. We have

$$\frac{\varphi}{\varrho} = \frac{\varphi(\varrho\cos\theta, \varrho\sin\theta)}{\varrho} \qquad \text{for} \qquad \varrho \neq 0,$$

$$\frac{\varphi}{\varrho} = 0 \qquad \text{for} \qquad \varrho = 0.$$

When $\varrho \neq 0$,

$$\frac{\partial(\varphi/\varrho)}{\partial\theta} = -\varphi'_x(\varrho\cos\theta, \varrho\sin\theta)\sin\theta + \varphi'_y(\varrho\cos\theta, \varrho\sin\theta)\cos\theta, \tag{16}$$

173

while for $\varrho = 0$ we have $\frac{\varphi}{\varrho} = 0$ for any $\theta$. Therefore $\frac{\partial(\varphi/\varrho)}{\partial\theta} = 0$ at the point $(0, \theta)$, by the definition of the partial derivative.

On the other hand, it follows from (16) that

$$\lim_{\rho \to 0} \frac{\partial(\varphi/\varrho)}{\partial\theta} = 0.$$

This proves that $\frac{\partial(\varphi/\varrho)}{\partial\theta}$ is continuous.

We now consider the function $\varphi/\varrho^2$. By definition (see (14)) it follows that if $\varrho = 0$

$$\frac{\partial(\varphi/\varrho^2)}{\partial\theta} = 2\left[(B_1 \cos\theta + C_1 \sin\theta)\cos\theta - (A_1 \cos\theta + B_1 \sin\theta)\sin\theta\right] \tag{17}$$

for arbitrary $\theta$. But if $\varrho \neq 0$, then

$$\frac{\partial(\varphi/\varrho^2)}{\partial\theta} = \frac{\partial}{\partial\theta}\left[\frac{\varphi(\varrho\cos\theta, \varrho\sin\theta)}{\varrho^2}\right] = \tag{18}$$

$$= \frac{1}{\varrho}\left[-\varphi'_x(\varrho\cos\theta, \varrho\sin\theta)\sin\theta + \varphi'_y(\varrho\cos\theta, \varrho\sin\theta)\cos\theta\right].$$

Now,

$$\varphi'_x(x, y) = \varphi''_{xx}(0, 0)x + \varphi''_{xy}(0, 0)y + \alpha x + \beta y, \tag{19}$$
$$\psi'_y(x, y) = \varphi''_{xy}(0, 0)x + \varphi''_{yy}(0, 0)y + \gamma x + \delta y,$$

where $\alpha$, $\beta$, $\gamma$, $\delta$ tend to zero with $\varrho$, and

$$\varphi''_{xx}(0, 0) = 2A_1, \qquad \varphi''_{xy}(0, 0) = 2B_1, \qquad \varphi''_{yy}(0, 0) = 2C_1.$$

Therefore

$$\frac{\partial(\varphi/\varrho^2)}{\partial\theta} = 2[(-A_1 \cos\theta - B_1 \sin\theta)\sin\theta +$$
$$+ (B_1 \cos\theta + C_1 \sin\theta)\cos\theta] + o(1) \tag{20}$$

(where $o(1)$ has the usual meaning of an infinitesimal quantity).

Letting $\varrho \to 0$ in (20), we get (17), so that the derivative $\frac{\partial(\varphi/\varrho^2)}{\partial\theta}$ is indeed continuous. The functions $\psi/\varrho$, $\psi/\varrho^2$ are treated in exactly the same way.

Thus the function $\Phi(\varrho, \theta)$ is continuous in both variables and continuously differentiable with respect to $\theta$ in the strip $\Omega [|\varrho| < \varrho^*]$. Hence the existence-uniqueness theorem is valid for equation (15) in this strip. Any segment on the straight line $\varrho = 0$ is a segment without contact for the integral curves of equation (15), and it follows that the integral curves behave in the strip $\Omega$ (for small $\varrho^*$) as illustrated in Figure 102. In view of the relation between the paths of system (7) and sections of the integral curves of equation (15), this means that the angle $\theta$ tends to a definite limit as $t \to +\infty$ along any semipath, so that each semipath tends to $O$ in a definite direction. Examination of Figure 102 verifies the truth of the second part of the theorem.

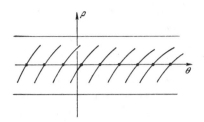

FIGURE 102.

This completes the proof of Theorem 21.

The configuration of the paths in the vicinity of a dicritical node is illustrated in Figure 106 (see p. 181).

**5. DEGENERATE NODE.** When the point $O(0, 0)$ is a degenerate node of a dynamic system, we can always reduce the system by a "scale transformation" to the form*

$$\frac{dx}{dt} = x + \varphi(x, y), \qquad \frac{dy}{dt} = x + y + \psi(x, y). \tag{21}$$

We may therefore confine ourselves to this system. As in the case of a dicritical node, we assume that $\varphi$ and $\psi$ are of class $C_2$ and vanish at $O(0, 0)$ together with their first derivatives. As shown in §7, the point $O(0, 0)$ is an unstable node of system (1): all paths passing through a sufficiently small neighborhood of $O$ tend to $O$ as $t \to -\infty$. By the Corollary to Lemma 1 (see §9.2) the slopes along which the paths may tend to $O$ are determined by the equation

$$bk^2 + (a - d)k - c = 0. \tag{22}$$

In our case, $b = 0$, $a - d = 0$, $c = 1$, both roots of equation (22) are infinite, and we have two possible directions, $\theta = \frac{\pi}{2}$ and $\theta = \frac{3}{2}\pi$.

*Theorem 22. Under the above assumptions concerning the functions $\varphi$ and $\psi$, each path of system (21) tending to the equilibrium state $O$ does so in a definite direction, either $\pi/2$ or $\frac{3}{2}\pi$. For each of these directions there is an infinite set of paths tending to $O$.*

Proof. Assuming that $\varrho \neq 0$ and transforming as usual to polar coordinates, we get the system

$$\frac{d\varrho}{dt} = \varrho\left(\frac{1}{2}\sin 2\theta + 1 + \frac{\varphi(\varrho\cos\theta, \varrho\sin\theta)}{\varrho}\cos\theta + \right.$$
$$\left. + \frac{\psi(\varrho\cos\theta, \varrho\sin\theta)}{\varrho}\sin\theta\right) = R(\theta, \varrho), \tag{23}$$
$$\frac{d\theta}{dt} = \cos^2\theta + \frac{\psi(\varrho\cos\theta, \varrho\sin\theta)}{\varrho}\cos\theta - \frac{\varphi(\varrho\cos\theta, \varrho\sin\theta)}{\varrho}\sin\theta = \Omega(\theta, \varrho).$$

---

* As we have seen, when the characteristic equation has multiple roots the system may be reduced to the form

$$\frac{dx}{dt} = \lambda x + \ldots, \quad \frac{dy}{dt} = \lambda y + \mu x + \ldots$$

Setting $x = \alpha\xi$, $y = \eta$, $t = \beta\tau$, one readily sees that

$$\frac{d\xi}{d\tau} = \beta\lambda\xi + \ldots, \quad \frac{d\eta}{d\tau} = \lambda\beta\eta + \alpha\beta\mu\xi + \ldots$$

$\alpha$ and $\beta$ may always be chosen in such a way that $\beta\lambda = 1$, $\alpha\beta\mu = 1$.

This system is defined for all nonzero $\varrho$ of sufficiently small absolute value. We extend the definition of the system at points $\varrho = 0$ by setting

$$\frac{\varphi\,(\varrho \cos\theta,\ \varrho \sin\theta)}{\varrho} = 0 \quad \text{and} \quad \frac{\psi\,(\varrho \cos\theta,\ \varrho \sin\theta)}{\varrho} = 0.$$

It was shown in §9.4 that, extended in this way, the functions $\varphi/\varrho$, $\psi/\varrho$ have continuous first partial derivatives with respect to $\theta$ in the strip $\Omega$ [$|\varrho| < \varrho^*$]. One readily sees by a direct check that they also have continuous first partial derivatives with respect to $\varrho$.

To solve our problem, it will suffice to consider a negative semipath

$$\varrho = \varrho\,(t), \qquad \theta = \theta\,(t),$$

in the strip $\Omega$, with regard to the behavior of the function $\theta = \theta\,(t)$ as $t \to -\infty$.* Note first that if $\varrho^*$ is sufficiently small the equilibrium states of system (23) lying in the strip $\Omega$ must be on the axis $\varrho = 0$, and thus for any equilibrium state $\cos^2\theta = 0$, i.e., $\theta = \frac{\pi}{2} + k\pi$ ($k = 0, \pm1, \pm2, \ldots$). It follows from the first equation of system (23) that every segment on the axis $\varrho = 0$ bounded by points $\theta = \frac{\pi}{2} + k\pi$, $\theta = \frac{\pi}{2} + (k+1)\,\pi$, is a path of system (23), increasing $t$ on this path corresponding to increase in the parameter $\theta$. These equilibrium states on the $(\varrho,\ \theta)$-plane are not simple. Indeed, consider, say, the equilibrium state $\left(0,\ \frac{\pi}{2}\right)$. Since the functions $\psi/\varrho$ and $\varphi/\varrho$ are continuously differentiable with respect to both $\theta$ and $\varrho$, system (23) admits the following representation in the neighborhood of this equilibrium state: **

$$\frac{d\varrho}{dt} = \varrho + \left(\theta - \frac{\pi}{2}\right)\varphi_1\,(\varrho,\ \theta) + \varrho\psi_1\,(\varrho,\ \theta),$$

$$\frac{d\theta}{dt} = \left(\theta - \frac{\pi}{2}\right)^2 + a\varrho + \left(\theta - \frac{\pi}{2}\right)^2\varphi_2\,(\theta) + \varrho\psi_2\,(\varrho,\ \theta),$$

(24)

where

$$\lim_{\substack{\rho \to 0 \\ \theta \to \frac{\pi}{2}}} \varphi_1\,(\varrho,\ \theta) = 0, \qquad \lim_{\substack{\rho \to 0 \\ \theta \to \frac{\pi}{2}}} \psi_1\,(\varrho,\ \theta) \to 0,$$

$$\lim_{\substack{\rho \to 0 \\ \theta \to \frac{\pi}{2}}} \varphi_2\,(\varrho,\ \theta) \to 0, \qquad \lim_{\substack{\rho \to 0 \\ \theta \to \frac{\pi}{2}}} \psi_2\,(\varrho,\ \theta) \to 0,$$

* Theorem 20 (node with distinct characteristic roots) may also be proved along these lines.
** Denoting the right-hand sides of the first and second equations of (23) by $R\,(\varrho,\ \theta)$ and $\Omega\,(\varrho,\ \theta)$, respectively, we clearly have

$$R'_\theta\left(\frac{\pi}{2},\ 0\right) = 0, \; R'_\rho\left(\frac{\pi}{2},\ 0\right) = 1, \; \Omega'_\theta\left(\frac{\pi}{2},\ 0\right) = 0$$

and

$$\Omega'_\rho\left(\frac{\pi}{2},\ 0\right) = \left[\frac{\partial}{\partial\varrho}\left\{\frac{\psi\,(\varrho \cos\theta,\ \varrho \sin\theta)}{\varrho}\cos\theta - \frac{\varphi\,(\varrho \cos\theta,\ \varrho \sin\theta)}{\varrho}\sin\theta\right\}\right]_{\theta=\frac{\pi}{2},\ \rho=0} = a.$$

and $\alpha$ is a constant, which may or may not vanish. Thus, at the equilibrium state,

$$\Delta = \begin{vmatrix} 0 & 1 \\ 0 & \alpha \end{vmatrix} = 0.$$

And this means that the equilibrium state is not simple.

It follows that the methods of §§6 and 7 are not applicable in this case. Instead one can employ the methods of §§11 and 22. Here, however, we proceed along direct lines.

As mentioned previously, we are interested in the behavior of a negative semipath $\varrho = \varrho(t)$, $\theta = \theta(t)$ $(t < t_0)$ of system (23) in the strip $\Omega$. It clearly follows from the first of equations (23) that if $\varrho^*$ is sufficiently small then $\frac{d\varrho}{dt} \neq 0$ for $\varrho < \varrho^*(\varrho > 0)$, so that the straight lines $\varrho = \text{const}$ through the strip $\Omega$ are arcs without contact for the paths of system (23). Thus $\varrho(t)$ tends monotonically to zero in the strip $\Omega$ as $t \to -\infty$.

As for $\theta(t)$, there are three possibilities: 1) $\theta(t) \to +\infty$ or $\theta(t) \to -\infty$ as $t \to -\infty$; 2) $\theta(t)$ remains bounded as $t \to -\infty$; 3) $\theta(t)$ is not bounded as $t \to -\infty$ but tends neither to $+\infty$ nor to $-\infty$ (example: $t \sin t$).

It is readily seen that the third case cannot occur. Indeed, consider, say, the straight lines $\theta = k\pi$. When $\varrho^*$ is sufficiently small, we have $\frac{d\theta}{dt} > 0$, $\frac{d\varrho}{dt} \neq 0$ on the segments of these lines in the strip $\Omega$, which are therefore segments without contact. Were the third case possible, a path would have to cross the segments $\theta = n\pi$ in opposite directions, which is impossible. In case 2), the semipath remains in a bounded region of the plane and therefore has an $\alpha$-limit continuum on the axis $\varrho = 0$. This continuum is obviously an equilibrium state, i.e., one of the points $\left( 0, \frac{\pi}{2} + k\pi \right)$. We shall now show that the first case is also impossible. To this end, we first prove that there is at least one semipath of system (23) in $\Omega$ tending to each equilibrium state $\left( 0, \frac{\pi}{2} + k\pi \right)$ as $t \to -\infty$. To fix ideas, we consider the point $\left( 0, \frac{\pi}{2} \right)$, in whose neighborhood the right-hand sides of system (23) are given by (24).

Consider the straight line

$$\varrho = \varkappa \left( \theta - \frac{\pi}{2} \right) \tag{25}$$

through the point $\left( 0, \frac{\pi}{2} \right)$, where $\varkappa$ is some positive constant, to be specified later. Let us examine the behavior of the path at points of the straight line (25). Assuming that $\varrho(t_0) = \varkappa(\theta_0 - \pi/2)$, we find an expression (see §7.3) for

$$\frac{d}{dt} \left( \frac{\varrho(t)}{\theta(t) - \frac{\pi}{2}} \right) \Bigg|_{t=t_0}.$$

A simple computation shows that

$$\frac{d}{dt}\left(\frac{\varrho(t)}{\theta(t)-\frac{\pi}{2}}\right)\Bigg|_{t=t_0} = \left(\frac{\frac{d\varrho}{dt}\left(\theta-\frac{\pi}{2}\right)-\frac{d\theta}{dt}\varrho}{\left(\theta-\frac{\pi}{2}\right)^2}\right)\Bigg|_{t=t_0} =$$

$$= \left\{\frac{\left[\varrho+\left(\theta-\frac{\pi}{2}\right)\varphi_1+\varrho\psi_1\right]\left(\theta-\frac{\pi}{2}\right)-\varrho\left[\left(\theta-\frac{\pi}{2}\right)^2+d\varrho+\dots\right]}{(\theta-\pi/2)^2}\right\}\Bigg|_{t=t_0} =$$

$$= \varkappa - \varkappa^2 a + \left(\theta(t_0)-\frac{\pi}{2}\right)\Phi(\theta(t_0)).$$

For any $a$, we can choose a fixed $\varkappa$ so small that $\varkappa - \varkappa^2 a > 0$, and then take $\eta > 0$ so small that for all $\theta$, $\frac{\pi}{2} < \theta \leqslant \frac{\pi}{2}+\eta$,

$$\frac{d}{dt}\left(\frac{\varrho}{\theta-\frac{\pi}{2}}\right)\Bigg|_{t=t_0} > 0.$$

Consider the triangle $CAB$ whose sides are the segment $CA$ on the line (25) and a segment on the straight line $\theta = \frac{\pi}{2}+\eta$ (Figure 103). Choose $\varrho^* > 0$ so small that for all $|\varrho| < \varrho^*$ on the straight line $\theta = \frac{\pi}{2}+\eta$

$$\left[\frac{d\theta}{dt}\right] > 0.$$

Examining Figure 103, one readily sees that every path crossing the segment $CE$ on (25), the segment $ED$ on the straight line $\varrho = \varrho^*$ and

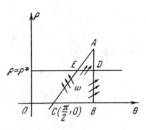

FIGURE 103.

the segment $DB$ on $\theta = \frac{\pi}{2}+\eta$ enters the region $w$ bounded by these segments and the axis $\varrho = 0$ (the latter is a path) with decreasing $t$, and with further decrease in $t$ cannot leave $w$. It is further obvious that every path passing through points of $w$ will never leave this region with decreasing $t$. Thus any such path necessarily tends to the equilibrium state

$$C\left(0, \frac{\pi}{2}\right) \text{ as } t \to -\infty.$$

But this means that there exists an infinite set of paths of system (21) that tend to the equilibrium state $O(0, 0)$ as $t \to -\infty$ in the direction $\frac{\pi}{2}$. Similarly, one shows that there are infinitely many paths of system (21) that tend to $O(0, 0)$ in the direction $\frac{3}{2}\pi$. Hence it follows that for every equilibrium state $\left(0, \frac{\pi}{2}+k\pi\right)$ of system (23) on the $(\varrho, \theta)$-plane there exist semipaths of the system lying above the $\theta$-axis and tending to the equilibrium state as $t \to -\infty$. But then system (23) cannot have a path $\varrho = \varrho(t)$, $\theta = \theta(t)$ such that $\varrho(t) \to 0$ and $\theta(t) \to \infty$ as $t \to -\infty$. Consequently, every semipath of system (21) tending to the equilibrium state $O$ does so in a definite direction — one of the directions $\frac{\pi}{2}, \frac{3}{2}\pi$. Q.E.D.

Thus we see that the paths of a system of class $C_2$ behave in the neighborhood of a degenerate node in the same way (as regards the mode of approach to the node) as those of the linearized system.

An illustration of a degenerate node for a system in canonical form is given in Figure 105a, while Figure 105b illustrates the general case of a degenerate node.

**6. SADDLE POINT AND FOCUS.** We first consider the case of a s a d d l e p o i n t. The canonical form of the system is

$$\frac{dx}{dt} = \lambda_1 x + \varphi(x, y), \qquad \frac{dy}{dt} = \lambda_2 y + \psi(x, y), \qquad (26)$$

where $\lambda_1 \lambda_2 < 0$. It is sufficient to assume that the functions $\varphi$ and $\psi$ are of class $C_1$.

In §7.3 we showed that one of the separatrices, $L_1$ say, of the saddle point $O$, tending to $O$ as $t \to +\infty$, possesses the following property: for any $K_0 > 0$, all points of $L_1$ corresponding to sufficiently large $t$ lie in the region bounded by the straight lines $y = +K_0 x$ and $y = -K_0 x$ and containing the positive $x$-axis (i.e., the ray $\theta = 0$). But this means (see Remark 1 at the beginning of §9.1) that $L_1$ tends to $O$ in the direction $\theta = 0$. Similarly, the other three separatrices tend to $O$ in the directions $\pi$, $\frac{\pi}{2}$ and $\frac{3}{2}\pi$. We thus have the following

*Theorem 23. When the point $O(0, 0)$ is a saddle point, all semipaths of system (26) that tend to $O$, i.e., the separatrices, tend to $O$ in definite directions. Two separatrices tend to $O$ in the directions $0$ and $\pi$, the other two in the directions $\pi/2$ and $\frac{3}{2}\pi$.*

It follows from Theorem 23 that the behavior of the paths of system (26) tending to the saddle point is the same (as regards mode of approach) as in the case of the linearized system

$$\frac{dx}{dt} = \lambda_1 x, \qquad \frac{dy}{dt} = \lambda_2 y,$$

i.e., the terms $\varphi(x, y)$ and $\psi(x, y)$ play no role in this respect.

When the system is not in canonical form,

$$\frac{dx}{dt} = ax + by + \varphi(x, y), \qquad \frac{dy}{dt} = cx + dy + \psi(x, y),$$

the directions $k_1$ and $k_2$ along which the separatrices tend to the equilibrium state $O$ are determined from the quadratic equation

$$bk^2 - (d-a)k + c = 0.$$

The phase portrait near a saddle point is illustrated for canonical form in Figure 107a and for the general case in Figure 107b.

179

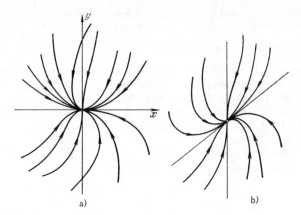

a)                                    b)

FIGURE 104.

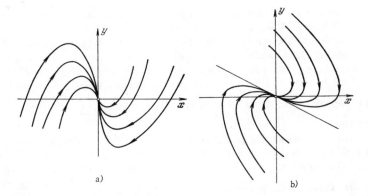

a)                                    b)

FIGURE 105.

We now take up the case of a simple focus or a multiple focus which is not a center or a centrofocus.

The canonical form of the system is

$$\frac{dx}{dt} = \alpha x - \beta y + \varphi(x, y), \qquad \frac{dy}{dt} = \beta x + \alpha y + \psi(x, y).$$

We shall assume that $\alpha \leqslant 0$, $\beta > 0$. Then, as we know from §8, all semipaths passing through a neighborhood of the equilibrium state are spirals which tend to $O$ as $t \to +\infty$, and along each semipath we have $\theta(t) \to +\infty$ as $t \to +\infty$.

Thus no path tending to the focus does so in a definite direction. By Lemma 1, the tangent to a semipath does not tend to any limiting position. We shall show that the tangent as it were "rotates infinitely many times" in a fixed sense, a statement which we shall first make precise.

Let $L^+$ be the semipath in question (a spiral), $M(t)$ the point on it corresponding to time $t$. We define a function $\omega(t)$, continuous for all sufficiently large $t$, as the angle between the positive direction on the abscissa axis and the positive direction on the tangent to $L^+$ at the point $M(t)$ (there are infinitely many such functions, differing from one another by multiples of $2\pi$ (see §8.1); we choose any one of these).

We claim that $\lim\limits_{t\to+\infty} \omega(t) = +\infty$. To prove this, note that in a sufficiently small neighborhood of $O$ segments on the ray $\theta = \text{const}$, $\varrho > 0$ have no contact with the paths of the system (§8.5, Lemma 3). It follows that for no point of this neighborhood is the radius-vector collinear with the tangent to the path at that point.

Consider the difference $d(t) = \omega(t) - \theta(t)$. This is clearly a continuous function which, by virtue of the preceding remark, cannot be equal to multiples of $\pi$. Thus all values of the function $d(t)$ lie in one of the intervals $(k\pi, (k+1)\pi)$ (where $k$ is an integer), so that $d(t)$ is bounded. But then

$$\lim\limits_{t\to+\infty} \omega(t) = \lim\limits_{t\to+\infty} [\theta(t) + d(t)] = +\infty.$$

When $\beta < 0$, we have $\lim\limits_{t\to+\infty} \theta(t) = \lim\limits_{t\to+\infty} \omega(t) = -\infty$. The configuration of paths near a focus is illustrated in Figure 108a, b.

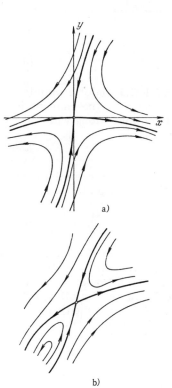

a)

b)

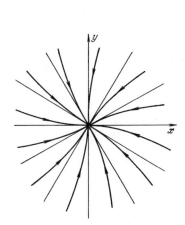

FIGURE 106.

FIGURE 107.

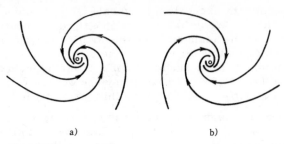

a)                                                    b)

FIGURE 108.

### 7. SURVEY OF SIMPLE EQUILIBRIUM STATES WHOSE CHARACTERISTIC ROOTS HAVE NONZERO REAL PARTS.*

The survey presented below should be useful in qualitative investigation of concrete dynamic systems.

If the equilibrium state in question is at the origin, the system admits the following representation in the neighborhood of the equilibrium state (§ 7):

$$\dot{x} = ax + by + \varphi(x, y), \quad \dot{y} = cx + dy + \psi(x, y).$$

The characteristic equation of the equilibrium state $O$ is

$$\begin{vmatrix} a - \lambda & b \\ c & d - \lambda \end{vmatrix} = \lambda^2 - \sigma\lambda + \Delta = 0 \tag{27}$$

$$[\sigma = a + d, \ \Delta = ad - bc].$$

By the definition of a simple equilibrium state $\Delta \neq 0$, so that the roots of equation (27) — the characteristic roots — do not vanish. The equation of the directions along which the paths tend to the equilibrium state is

$$bk^2 - (d - a)k + c = 0. \tag{28}$$

The relation between the characteristic roots $\lambda_1$ and $\lambda_2$ and the roots $k_1$ and $k_2$ of equation (28) is

$$k_1 = \frac{d - \lambda_1}{b}, \quad k_2 = \frac{d - \lambda_2}{b}.$$

Obviously, $k_1$ and $k_2$ are real if and only if $\lambda_1$ and $\lambda_2$ are real.

Depending on the nature of the characteristic roots, the system can be reduced by a linear transformation in the neighborhood of the equilibrium state to one of the following canonical forms (the original notation of variables is retained):

1. Characteristic roots real and distinct $(\lambda_1 \neq \lambda_2)$:

$$\dot{x} = \lambda_1 x + \varphi(x, y), \quad \dot{y} = \lambda_2 y + \psi(x, y).$$

---

* Equilibrium states with pure imaginary roots (§8) are of course omitted from this survey.

182

2. Characteristic roots equal $(\lambda_1 = \lambda_2 = \lambda)$:

$$\dot{x} = \lambda x + \varphi\,(x,\,y),\quad \dot{y} = \lambda y + \mu x + \psi\,(x,\,y)$$

($\mu$ may or may not vanish).

3. Characteristic roots complex conjugates

$$(\lambda_1 = a + i\beta,\quad \lambda_2 = a - i\beta,\quad \beta \neq 0).$$

The canonical form is

$$\dot{x} = ax + \beta y + \varphi\,(x,\,y),\quad \dot{y} = \beta x - ay + \psi\,(x,\,y).$$

Below we list all possible types of equilibrium state whose characteristic roots have nonzero real parts and illustrate the phase portrait in the neighborhood of the equilibrium state, utilizing the results of §§8 and 9 on the mode of approach of paths to the equilibrium state.* For nodes and foci we illustrate only the stable case. To obtain the phase portrait for unstable nodes or foci, simply reverse the direction of the arrows.

In addition, for all nodes except dicritical nodes, and for saddle points, the figures illustrate two cases: canonical form, where the paths tending to the equilibrium state do so in the directions of the coordinate axes, and the general case, where the directions $k_1$ and $k_2$ may be arbitrary.

I. Node (characteristic roots real and of equal signs):

$a_1$) Nondegenerate stable node $\lambda_1 \neq \lambda_2$ and $\lambda_1 < 0$, $\lambda_2 < 0$ ($\Delta > 0$, $\sigma^2 - 4\Delta > 0$, $\sigma < 0$).

$a_2$) Nondegenerate unstable node $\lambda_1 \neq \lambda_2$, $\lambda_1 > 0$, $\lambda_2 > 0$.

Figures 104a and b illustrate stable nodes, the former for a system in canonical form, the latter in the general case.

$b_1$) Stable degenerate (not dicritical) node $\lambda_1 = \lambda_2 = \lambda < 0$, $\mu \neq 0$ (i. e., $\Delta > 0$, $\sigma^2 - 4\Delta = 0$, $\sigma < 0$).

$b_2$) Unstable degenerate node $\lambda_1 = \lambda_2 = \lambda > 0$, $\mu \neq 0$.

Figure 105a illustrates this case for canonical form, Figure 105b for the general case.

$c_1$) Stable dicritical node (Figure 106) $\lambda_1 = \lambda_2 = \lambda < 0$, $\mu = 0$.

$c_2$) Unstable dicritical node $\lambda_1 = \lambda_2 = \lambda < 0$, $\mu = 0$.

II. Saddle point (characteristic roots $\lambda_1$ and $\lambda_2$ real and of opposite signs) $\lambda_1 < 0$, $\lambda_2 > 0$ (or $\lambda_1 > 0$, $\lambda_2 < 0$), i. e., $\Delta < 0$. Figure 107a corresponds to the case of a system in canonical form. Figure 107b to the general case. (When $\lambda_1 > 0$, $\lambda_2 < 0$, the directions on the paths are reversed.)

III. Focus (characteristic roots complex conjugates)

$$\lambda_1 = a + i\beta,\quad \lambda_2 = a - i\beta,\quad \beta \neq 0$$

(i. e., $\Delta > 0$, $\sigma^2 - 4\Delta < 0$).

$d_1$) Stable focus $\alpha < 0$.

$d_2$) Unstable focus $\alpha > 0$.

---

* The reader should bear in mind that the results for degenerate and dicritical nodes are valid only when the system is (at least) of class $C_2$.

Figure 108a illustrates a stable focus with $\beta > 0$, Figure 108b for $\beta < 0$.

Recall that the topological structure of the phase portraits in the vicinity of a node and a focus is the same.

**8. EXAMPLES.**

Example 1.

$$\dot{x} = y, \quad \dot{y} = x(a^2 - x^2) + by.$$

The equilibrium state $O(0, 0)$ is a saddle point. To determine the directions of the separatrices, we solve the equation $k^2 - bk - a^2$ for the slopes of

the separatrices, obtaining $k_{1,2} = \frac{b}{2} \pm \sqrt{\frac{b^2}{4} + a^2}$.

Example 2 /72/.

$$\dot{x} = y, \quad \dot{y} = 7y + x(|x|^{1/2} - 12).$$

The equilibrium state $O(0, 0)$ is an unstable node. The equation for the directions of the paths at $O$ is $k^2 - 7k + 12 = 0$, whence $k_1 = 3$, $k_2 = 4$.

Example 3 /73/.

$$\dot{x} = -x(2 + y), \quad \dot{y} = x + \beta y.$$

The equilibrium state $O(0, 0)$ is a saddle point. The equation for the directions of the separatrices at the saddle point is $(\beta + 2)k + 1 = 0$, so

that $k = -\frac{1}{2+\beta}$. It is readily seen that the second solution for $k$ is $\infty$,

and the separatrix with slope $k = \infty$ is the straight line $x = 0$.

*Chapter V*

*THEORY OF THE INDEX AND ITS APPLICATIONS*
*TO DYNAMIC SYSTEMS*

INTRODUCTION

Poincaré's concept of the index (and its generalizations) belongs to the theory of vector fields and is now playing a major role not only in the qualitative theory of dynamic systems, but also in many other fields (topology, functional analysis and their applications). The theory of the index yields invaluable information about certain fundamental properties of dynamic systems.

The index concept is based on the concept of the r o t a t i o n of a vector field. Given a continuous vector field on a simple closed curve, the rotation of the field along the curve is, roughly speaking, the number of full revolutions performed by the field vector when the curve is described once in the positive sense (the rigorous definition was given in §6.2). The Poincaré index of an isolated equilibrium state $O$ of a dynamic system is the rotation of the vector field defined by the system along any sufficiently small closed curve containing $O$ in its interior.

Many applications of the theory of the index are based on the fact that the index of a closed curve is equal to the sum of indices of the equilibrium states interior to the curve (Theorem 27), and that the index of a closed path or of a cycle without contact is unity (Theorems 28 and 29). These theorems imply important sufficient conditions for the simultaneous existence of closed paths and equilibrium states of various types.

At the end of §11 we calculate the indices of simple equilibrium states.

§10. THE POINCARÉ INDEX

**1. ROTATION OF A VECTOR FIELD.** We begin with fundamental definitions of the theory of the index.

The concepts we are going to study are meaningful and valuable for vector fields of a more general type than those defined by dynamic systems (the latter are continuously differentiable vector fields), namely, for arbitrary c o n t i n u o u s f i e l d s. Since certain basic facts concerning dynamic systems (the index of a closed path) will be established for continuous but not necessarily differentiable fields, we shall give all the basic definitions on the assumption that the field is continuous but need not be differentiable.

We shall say that a vector field is defined on a set $K$ in the $(x, y)$-plane (in particular, in a plane region) if at each point $M(x, y)$ of $K$ we have a vector vector $v(M)$ whose components $X(x, y)$ and $Y(x, y)$ are continuous functions of the point. The vector field is said to be regular (or without singularities) if it contains no zero vectors, i.e., vectors such that both $X$ and $Y$ vanish. The length and direction of the vector are defined at each regular point. The angle $\omega$ between the positive $x$-axis and the direction of the vector is given by (see Appendix, §5)

$$\cos \omega = \frac{X}{\sqrt{X^2 + Y^2}}, \qquad \sin \omega = \frac{Y}{\sqrt{X^2 + Y^2}}.$$

It is clear that under the above assumptions the length of the vector and the angle $\omega$ are uniquely defined and are continuous functions of the point.*

To define the basic concepts of the theory of the index we need only vector fields defined on a curve (not necessarily in a region of the plane).

Throughout the sequel, all vector fields defined on curves will be assumed continuous and regular, no explicit mention being made of these qualifications.

We first consider the case of a field $v$ defined on a simple arc $l$. The angle function of the field is defined as follows. Assume the $(x, y)$-plane referred to some coordinate system. We define the polar angle of a nonzero vector $v$ as the angle between the positive $x$-axis and the vector $v$, measured counterclockwise. The polar angle is defined not uniquely but only up to a multiple of $2\pi$. An angle function of a vector field $v$ defined on an arc $l$, relative to a prescribed coordinate system, is any function $\alpha = F(M)$ satisfying the following two conditions: 1) $F(M)$ is a single-valued function defined and continuous for all points $M \in l$; 2) for any point $M \in l$, $F(M)$ is the polar angle of the vector $v(M)$ (more precisely, some well-defined value of the polar angle).

We first prove that angle functions exist. Recall that the angle between two nonzero vectors $v_0$ and $v_1$ on the plane is defined as the smallest angle through which the vector $v_0$ must be rotated till it coincides in direction with $v_1$ (see Appendix, §5).

Denote the angle between vectors $v_0$ and $v_1$ by

$$\theta(v_0, v_1).$$

We have (see Appendix, §5)

$$-\pi < \theta(v_0, v_1) \leq \pi.$$

Now the arc $l$ is a closed point set, and therefore the vector-function $v(M)$ is not only continuous but also uniformly continuous on the arc. Hence it follows easily that for any $\varepsilon > 0$ there exists a partition of $l$ by points $M_0, M_1, \ldots, M_{n-1}, M_n$ ($M_0, M_n$ are the endpoints of the arc, and increase in $n$ corresponds to motion along the arc in a definite direction; see Figure 109), satisfying the following condition: for any two points $M', M''$ in the same subarc $M_{k-1}M_k$ of $l$,

---

* The direction of the vector is undefined at a singular point.

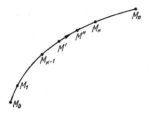

FIGURE 109.

$$|\theta\,(\boldsymbol{v}\,(M'),\ \boldsymbol{v}\,(M''))|<\varepsilon.$$

An angle function $\alpha = F\,(M)$ of the vector field on $l$ may be constructed as follows. Let $\varepsilon$ be less than $\pi$, and take any partition $M_0,\,M_1,\,\ldots,\,M_n$ of $l$ corresponding to this $\varepsilon$. Set $F\,(M_0) = \alpha_0$, where $\alpha_0$ is an arbitrary but fixed value of the polar angle of $\boldsymbol{v}\,(M_0)$. For each point $M$ in the subarc $M_0 M_1$ of $l$, define $F\,(M) = \alpha_0 + \theta\,(\boldsymbol{v}\,(M_0),\ \boldsymbol{v}\,(M))$. Supposing that $F\,(M)$ has already been defined for all points of the subarcs $M_0 M_1, M_1 M_2,\,\ldots,\,M_{k-2}M_{k-1}$, we stipulate now that if $M$ is a point on the subarc $M_{k-1}M_k$, then

$$F\,(M) = F\,(M_{k-1}) + \theta\,(\boldsymbol{v}\,(M_{k-1}),\ \boldsymbol{v}\,(M))$$
$$(k = 2,\ 3,\ \ldots,\ n).$$

We have thus defined a function $F\,(M)$ for all points $M \in l$. It is easy to see that it possesses properties 1) and 2) and so is an angle function of the field $v$.

A vector field may clearly have different angle functions; however:

*L e m m a 1. Let $F\,(M)$ and $F_1\,(M)$ be two angle functions of a vector field $v$ defined on an arc $l$ (relative to the same coordinate system on the plane). Then for all $M \in l$ we have $F_1\,(M) = F\,(M) + 2\pi r$, where $r$ is a constant integer. If $F\,(M)$ and $\Phi\,(M)$ are angle functions of a field $v$ on $l$ relative to different coordinate systems on the plane, then*

$$\Phi\,(M) = F\,(M) + \text{const}.$$

P r o o f. If $F\,(M)$ and $F_1\,(M)$ are two angle functions relative to the same coordinate system on the $(x,\,y)$-plane, it follows from condition 2) that for any point $M \in l$ we have $F_1\,(M) - F\,(M) = 2\pi r\,(M)$, where $r$ is an integer. But the continuity of the functions implies that $r\,(M)$ must be a constant.

Now let $F\,(M)$ be an angle function of the field $v$ relative to a coordinate system $xOy$, and $\Phi\,(M)$ an angle function relative to another coordinate system $x'O'y'$. Let $\alpha_0$ be the angle between the axes $Ox$ and $O'x'$. Then it is obvious that $\Phi\,(M) + \alpha_0$ is an angle function for $v$ relative to the system $xOy$, and by what we have proved $\Phi\,(M) + \alpha_0 = F\,(M) + 2\pi r$. Consequently,

$$\Phi\,(M) = F\,(M) + (2\pi r - \alpha_0) = F\,(M) + \text{const}.$$

Q.E.D.

We can now define the rotation of a vector field along a simple arc.

*D e f i n i t i o n X. Let $l$ be a simple arc, $v\,(M)$ a vector field defined on this arc, $M_1 M_2$ some subarc of $l$ considered with the direction from $M_1$ to $M_2$, and $F\,(M)$ any angle function of the field $v$ on $l$. The rotation of the field $v$ along the subarc $M_1 M_2$ of $l$ is defined as the number*

$$w\,(v,\ M_1 M_2) = [F\,(M_2) - F\,(M_1)]. \tag{1}$$

In particular, if $M_1$ and $M_2$ are the endpoints of the arc $l$ we shall speak of the rotation of the vector field along the arc $l$, denoting it by $w(v, l)$. It follows from Lemma 1 and Definition X that the rotation of a vector field along a simple arc $l$ is independent of the choice of coordinate system on the $(x, y)$-plane or of the choice of the angle function $F(M)$.

The following properties follow at once from (1):

a) if the direction on the arc is reversed, the rotation of any vector field on the arc changes sign:

$$w(v, M_1 M_2) = -w(v, M_2 M_1);$$

b) if $M_1, M_2, M_3$ are arbitrary points on the arc, then

$$w(v, M_1 M_3) = w(v, M_1 M_2) + w(v, M_2 M_3)$$

(additivity).

## 2. INDEX OF A SIMPLE CLOSED CURVE RELATIVE TO A VECTOR FIELD.

Throughout the sequel we shall assume that a positive traversal is defined on the closed curve under consideration.

Let $C$ be some closed curve, $M_1$ and $M_2$ points on $C$. The notation $M_1 M_2$ ($M_2 M_1$) will stand for the arc on $C$ with initial point $M_1$ and terminal point $M_2$ (initial point $M_2$ and terminal point $M_1$), on which the direction from $M_1$ to $M_2$ (from $M_2$ to $M_1$) is that induced by the positive traversal on $C$ (see Appendix, §2). Thus $M_1 M_2$ and $M_2 M_1$ are well-defined and different arcs on $C$ with common endpoints (Figure 110).

*Definition XI. The index of the simple closed curve $C$ on the $(x, y)$-plane relative to a vector field on the curve is defined as the number*

$$I(C, v) = \frac{1}{2\pi}[w(v, M_1 M_2) + w(v, M_2 M_1)]. \tag{2}$$

We shall denote the index of a simple closed curve $C$ by $I(C)$. It is evident that this definition of the index is independent of the specific points $M_1$ and $M_2$ chosen on $C$; this follows easily from property b) (additivity) of the rotation of a vector field.

Let $C$ be described by a parametric equation $M = M(u)$, $\alpha \leqslant u \leqslant \beta$, where $M(\alpha) = M(\beta)$ and increasing $u$ corresponds to the positive traversal of the curve (Figure 111). We define an angle function of a vector field $v$ on $C$ to be any function $F(u)$, defined and continuous for all $u$, $\alpha \leqslant u \leqslant \beta$, such that, for any $u$, $F(u)$ is a polar angle of the vector $v(u)$. The existence of an angle function $F(u)$ is proved exactly as in the case of the function $F(M)$ of §10.1. Here, however, $F(u)$ is defined relative to a certain parametrization of the curve $C$.

It is easy to see that

$$I(C) = \frac{1}{2\pi}[F(\beta) - F(\alpha)]. \tag{3}$$

It follows from this equality, in particular, that the number $F(\beta) - F(\alpha)$ is independent of the choice of parametrization on $C$. Thus, in order to calculate the index of a closed curve $C$ (relative to a given field) we may take any parametrization of the curve, construct an arbitrary angle function

$F(u)$ and use formula (3). It follows from (3) that the index of a closed curve is always an integer, since $F(\beta)$ and $F(\alpha)$ are polar angles of the same vector $v(M(\alpha)) = v(M(\beta))$, and so $[F(\beta) - F(\alpha)]$ is a multiple of $2\pi$.

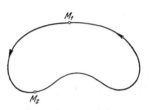

FIGURE 110.

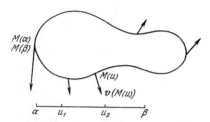

FIGURE 111.

We now prove a number of important propositions.

**Lemma 2.** *If $v(M)$ and $v^*(M)$ are two vector fields defined on a simple closed curve $C$, such that at no point $M \in C$ do the vectors $v(M)$ and $v^*(M)$ have opposite directions, then $I(C, v) = I(C, v^*)$.*

Proof. Consider the curve $C$ relative to some parametrization $M = M(u)$, and let $F(u)$ be the corresponding angle function of the field $v$. Let $\theta(u) = \theta[v(M(u)), v^*(M(u))]$ be the smallest angle between the vectors $v(M(u))$ and $v^*(M(u))$. By assumption, $|\theta(u)| < \pi$. It is easy to see that $\theta(u)$ is continuous in $u$ on the interval $\alpha \leqslant u \leqslant \beta$, and moreover $\theta(\alpha) = \theta(\beta)$. Consider the function $F^*(u) = F(u) + \theta(u)$. This is clearly an angle function of the field $v^*$ (Figure 112). Therefore

$$w(v^*, C) = \frac{1}{2\pi}[F^*(\beta) - F^*(\alpha)] =$$

$$= \frac{1}{2\pi}[F(\beta) - F(\alpha) + \theta(\beta) - \theta(\alpha)] = \frac{1}{2\pi}[F(\beta) - F(\alpha)] =$$

$$= w(v, C).$$

Q.E.D.

**Definition XII.** *Let $v$ and $v^*$ be two fields defined on a curve $C$. We shall say that the field $v$ can be deformed into $v^*$ if there exists a family of fields joining them, i.e., a family of vector fields $v_\tau$ ($0 \leqslant \tau \leqslant 1$) defined on $C$ such that $v = v_0$, $v^* = v_1$, and $v_\tau(M)$, $M \in C$, $\tau \in [0, 1]$, is a continuous function of both arguments $\tau$ and $M$. (All the fields are assumed to have no singularities.)*

**Lemma 3.** *If $v$ can be deformed into $v^*$, the indices of $C$ with respect to these fields are equal:*

$$I(C, v) = I(C, v^*).$$

Proof. Let $v_\tau(M)$ ($0 \leqslant \tau \leqslant 1$) be a family of fields joining $v$ and $v^*$. The vector function $v_\tau(M)$ is defined and continuous on a compact set (the topological product of the curve $C$ and the interval $[0, 1]$), and is consequently uniformly continuous. Hence, for every $\varepsilon > 0$ there exists $\delta > 0$ such that if $|\tau' - \tau''| < \delta$, $\tau' \in [0, 1]$, $\tau'' \in [0, 1]$, then $|v_{\tau'}(M) - v_{\tau''}(M)| < \varepsilon$

for any $M \in C$. Now, since by assumption $v_\tau$ has no singular points, it follows that $v_\tau (M) \neq 0$ and $|v_\tau (M)|$ attains its minimum, $m$ say:

$$m = \min \{|v_\tau (M)|\}; \ M \in C, \ 0 \leqslant \tau \leqslant 1.$$

Letting $\varepsilon$ be this number $m$ and considering the corresponding $\delta$, partition the interval $0 \leqslant \tau \leqslant 1$ by points

$$\tau_0 = 0 < \tau_1 < \tau_2 < \ldots < \tau_{h-1} < \ldots <$$
$$< \tau_{n-1} < \tau_n = 1$$

such that the differences $\tau_k - \tau_{k-1}$ are smaller than $\delta$ ($k = 1, 2, \ldots, n$). Then it is clear that for any point $M \in C$ the vectors $v_{\tau_{k-1}} (M)$ and $v_{\tau_k}(M)$ are not opposite in direction (Figure 113), and so, by the preceding lemma,

$$I (C, v_{\tau_{k-1}}) = I (C, v_{\tau_k}) \qquad (k = 1, 2, \ldots, n).$$

Hence it follows that

$$I (C, v_0) = I (C, v_1),$$

i. e.,

$$I (C, v) = I (C, v^*).$$

Q. E. D.

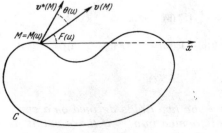

FIGURE 112.

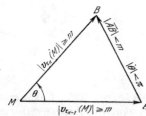

FIGURE 113.

In the next lemma we consider a vector field defined not on a curve but in the region bounded by a simple closed curve.

*Lemma 4. Let $C$ be a simple closed curve, $\Gamma$ the region bounded by the curve and $\bar{\Gamma}$ its closure. If $v$ is a vector field without singularities defined on $\bar{\Gamma}$, then the index of $C$ relative to this field\* is zero:*

$$I (C, v) = 0.$$

---

\* More precisely: relative to the restriction of the field $v$ to the curve $C$, or relative to the field induced by $v$ on $C$.

190

Proof. a) Let us first assume that $C$ is a circle of unit radius about a point $S$, so that $\bar{\Gamma}$ is a disk. Introduce in $\bar{\Gamma}$ polar coordinates $\varrho$ and $\theta$ ($\varrho$ is the distance from the center $S$, $0 \leqslant \varrho \leqslant 1$, and $\theta$ the polar angle; see Figure 114). Let $v(\varrho, \theta)$ be the vector of the field $v$ at the point $(\varrho, \theta)$ of the disk. We define on $C$ a family of fields $v_\tau(\theta)$, $0 \leqslant \tau \leqslant 1$, $0 \leqslant \theta \leqslant 2\pi$, by

$$v_\tau(\theta) = v(\tau, \theta).$$

It is clear that this family joins the field $v_0(\theta)$ to $v_1(\theta)$. Consequently, by Lemma 3, $I(C, v_0) = I(C_1, v_1)$. But it is clear from the construction that the field $v_1$ is simply $v$ (restricted to $C$), while $v_0(\theta)$ consists of equal vectors (all equal to the vector of the field $v$ at $S, v(0, \theta)$). Thus any angle function $F(\varrho)$ of the field $v_0$ is a constant, and $I(C, v_0) = 0$. But then also $I(C, v) = 0$.

b) Now let $C$ be an arbitrary simple closed curve bounding a region $\Gamma$ on the $(x, y)$-plane. Let $K$ be the unit disk on the plane $R^2$, with boundary $C_0$ (Figure 115), and $T$ an orientation-preserving topological mapping of $\bar{\Gamma}$ onto $K$ (see Appendix, §2).

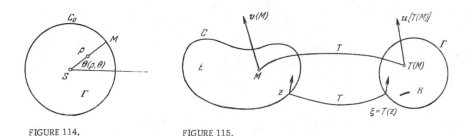

FIGURE 114.                    FIGURE 115.

By assumption, the field $\Gamma$ is defined in $v$. "Transfer" the field by means of the mapping $T$ to the disk $K$, constructing on $K$ a field $u^*$ which coincides with $v^*$ by setting $v^*(T(M)) = v(M)$. It is clear that $v^*$ is a field without singularities, and by part a) of the proof we have $I(C_0, v^*) = 0$. It is not hard to see, however, that $I(C, v) = I(C_0, v^*)$. In fact, to calculate $I(C, v)$ we must parametrize the curve $C$, setting $M = M(u)$ ($M \in C, \alpha \leqslant u \leqslant \beta$), construct an angle function $F(u)$ for the field $v$ and calculate $\frac{1}{2\pi}[F(\beta) - F(\alpha)]$.

Using the mapping $T$ to transfer the parametrization from $C$ to the circle $C_0$:

$$\xi(u) = T(M(u)) \qquad (\xi \in C_0; \ M \in C),$$

and noting that $v^*(\xi) = v(M)$, we at once conclude that the angle function $F(u)$ constructed for the field $v$ on $C$ is also an angle function for the field $v^*$ on $C_0$. Therefore $I(C, v) = I(C_0, v^*) = 0$. Q.E.D.

The next theorem, based on Lemma 4, is essential in consideration of dynamic systems. The proof presented here is due to Hopf.

### 3. FIELD OF TANGENTS TO A CLOSED CURVE.

*Theorem 24 (Poincaré). The index of a smooth simple closed curve relative to its field of tangents is equal to +1.*

Proof. Let $C$ be a smooth simple closed curve. Since $C$ is smooth, it is rectifiable. To fix ideas, we shall assume that the length of $C$ is 1 and that the direction of the tangent vector $v$ at every point of $C$ agrees with the positive traversal on $C$ (the lengths of the vectors $v$ are immaterial, as long as they do not vanish: one may even assume that $v$ is a field of unit tangent vectors).

Let $M_0$ be a point on $C$ with minimum ordinate (at least one such point exists). Then the tangent to $C$ at $M_0$ is horizontal, and the entire curve $C$ lies above it (Figure 116). We shall reckon arc length on $C$ from the point $M_0$, in the positive direction. Then each point $M$ on $C$ other than $M_0$ is uniquely determined by a definite value of $s$ $(0 < s < 1)$ — the coordinate of the point on the curve $C$. To the point $M_0$ correspond two parameter values: $s = 0$ and $s = 1$.

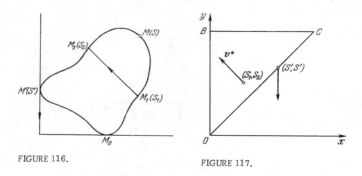

FIGURE 116.

FIGURE 117.

Let $OBC$ be the triangle on the $(x, y)$-plane bounded by the straight lines $x = 0$, $y = 1$, $x = y$ (Figure 117). With each point $(s_1, s_2)$ of this triangle (obviously, $0 \leqslant s_1 \leqslant s_2 \leqslant 1$) we can associate a unit vector $v^*(s_1, s_2)$ in the direction of the vector $\overline{M_1 M_2}$, where $M_1(s_1)$, $M_2(s_2)$ are the points on $C$ with coordinates $s_1$, $s_2$. To make the resulting field continuous, we stipulate that the vector associated with any point $(s', s')$ on the segment $OC$ is associated with the unit tangent vector $v(M')$, i. e., $v^*(s', s') = v(M')$.

The result is a continuous vector field $v^*$ defined on the closed triangle $OCB$ and having no critical points. Therefore, by Lemma 4, the index of the closed curve $OCBO$ relative to $v^*$ is zero:

$$I(OCBO, v^*) = 0.$$

But the index of the polygonal line $OCBO$ is the sum of rotations of the field $v^*$ along the segments $OC$, $CB$, $BO$. Therefore

$$w(v^*, OC) + w(v^*, CB) + w(v^*, BO) = 0.$$

It is easy to see that $w(v^*, OC) = I(C, v)$, since, for a suitable parametrization, the relevant angle functions (for $v^*$ on $OC$ and for $v$ on the curve $C$) are identical.

Now consider the field $v^*$ on the segment $OB$. The vectors of this field cannot point downward, since their directions are those of the vectors joining $M_0$ to points of $C$ (by assumption, the entire curve $C$ lies above $M_0$).

Furthermore, the direction of the vector $v$ $(B) = v^*$ $(0, 1)$ is that of the negative $x$-axis, while that of $v^*$ $(0) = v^*(0, 0)$ is the direction of the positive $x$-axis. Let $F$ $(M)$ be the angle function of the field $v^*$ on $BO$ such that $F$ $(B) = +\pi$. Then $F$ $(0) = 2\pi r$, where $r$ is an integer. If $r \neq 0$, it follows from the continuity of $F$ $(M)$ that there exist points $M$ on $BO$ at which the vectors $v^*$ $(M)$ are directed downward, which is impossible. Therefore

$$r = 0, \quad F \ (0) = 0, \quad F \ (0) - F \ (B) = -\pi \text{ and } w \ (v^*, BO) = \frac{1}{2\pi}(-\pi) = -\frac{1}{2}. \text{ One}$$

shows similarly that $w(v^*, \ CB) = -\frac{1}{2}$. Thus, $I \ (C, v) - \frac{1}{2} - \frac{1}{2} = 0; \ I \ ,C, \ v) = 1.$

Q.E.D.

Corollary. Let $v$ be a field defined on the curve $C$ such that at no point $M \in C$ does the vector $v$ $(M)$ have the direction of the tangent to $C$ at $M$ (i.e., either all vectors of $v$ point into the interior of $C$, or they all point outward). Then $I$ $(C, v) = 1$. The proof follows immediately from Lemma 2 by setting $v^*$ equal to the field of tangents to $C$.

**4. POINCARÉ'S DEFINITION OF THE INDEX.** In this subsection we give Poincaré's definition of the index (see /5/, Chapters III and XIV), in a slightly modified form. In many cases this definition is convenient for calculation of the index of a closed curve.

Let $C$ be a simple closed curve, $v$ a field defined on the curve and $d$ some straight line on the $(x, y)$-plane. Suppose that there exist only finitely many points $M_h$ $(k = 1, 2, \ldots, n)$ on $C$ at which the vector $v$ $(M)$ is parallel to $d$. Let $M$ be a point describing the curve in the positive sense, and let $p$ be the number of points $M_h$ at which the vector $v$ $(M)$ passes through the direction of $d$ in the counterclockwise sense. Let $q$ be the number of points $M_h$ at which $v$ $(M)$ passes through the direction of $d$ in the clockwise sense. Points $M_h$ at which the vector $v$ $(M)$ assumes the direction of $d$ while moving, say, in the clockwise sense and then begins to move in the opposite sense (or vice versa) are not counted. Then

$$I \ (C, v) = \frac{p - q}{2}. \tag{4}$$

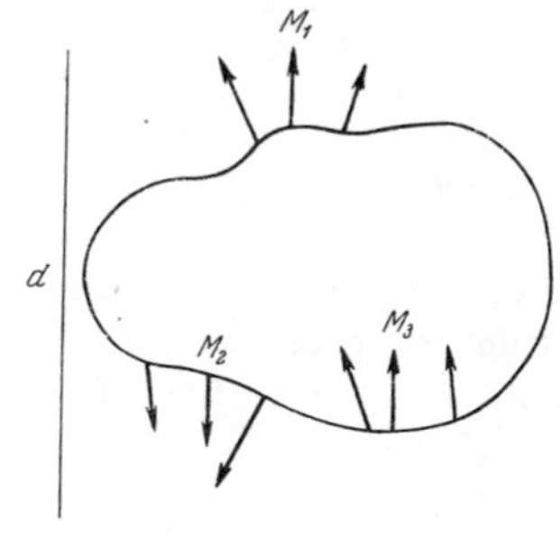

FIGURE 118.

Figure 118 illustrates three points $M_h$, each of one of the above types. The truth of (4) follows from the fact that passage through the direction of $d$ registers an increase or decrease of $\pi$ in the value of the angle function. Hence the total variation of the angle function along the curve $C$ is $(p - q) \pi$, and so the index of $C$ is $\frac{(p-q) \pi}{2\pi} = \frac{p-q}{2}$. A rigorous proof proceeds by consideration of the intersection of the angle function $\alpha = F$ $(M)$ and the straight lines $\alpha = r\pi$ where $r$ is an integer.

## §11. APPLICATION OF THE THEORY OF THE INDEX TO DYNAMIC SYSTEMS

**1. TWO FUNDAMENTAL THEOREMS.** As we know, any dynamic system

$$\frac{dx}{dt} = P(x, y), \quad \frac{dy}{dt} = Q(x, y) \tag{I}$$

defines a vector field (see §1). If system (I) is defined in a region $G$ on the $(x, y)$-plane, we associate with each point $M(x, y) \in G$ the vector $v(M)$ or $v(x, y)$ with components $P(x, y), Q(x, y)$; the result is a vector field $v(M)$ or $v(x, y)$.

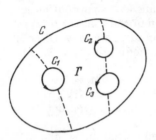

FIGURE 119.

The points of $G$ at which $P(x, y) = Q(x, y) = 0$, i. e., the equilibrium states of system (I), are the singular points of this vector field.

Let $C$ be a simple closed curve lying in $G$. At each point of this curve the field defined by system (I) specifies a certain vector, i. e., a vector field is "induced" on the curve. In the sequel, we shall speak of the index of the closed curve $C$, meaning its index relative to the field induced by the field $v(M)$ defined by the dynamic system. Lemma 4 may be reformulated as follows for this special case:

*Theorem 25. Let $C$ be a simple closed curve in the region $G$, $\Gamma$ the region interior to this curve. If all points of $\Gamma$ belong to $G$ and the closure $\bar{\Gamma}$ contains no singular point of the dynamic system (I), then the index of the curve $C$ is zero: $I(C) = 0$.*

This theorem is easily generalized as follows:

*Theorem 26. If $\Gamma$ is a subregion of $G$ bounded by simple closed curves $C, C_1, C_2, \ldots, C_n$ (Figure 119), with $C$ the outer boundary of $\bar{\Gamma}$, and the region $\bar{\Gamma}$ contains no singular points of system (I), then*

$$I(C) = I(C_1) + I(C_2) + \ldots + I(C_n).$$

The proof follows immediately from Theorem 15 by subdividing the region as indicated in Figure 119 and using the fact that the rotation is additive.

**2. INDEX OF AN ISOLATED SINGULAR POINT.** In a plane region $G$, consider a simple closed curve $C$ whose interior contains a singular point $O$ of system (I), and suppose that there are no other singular points of the field either on or inside $C$.

We claim that the indices of any two such curves are equal.

Let $C$ and $C'$ be two curves of the above description (Figure 120). Suppose first that the curves are disjoint. Then, by Theorem 26, $I(C') = I(C)$ and we are done. If $C$ and $C'$ have common points, let $C''$ be an auxiliary curve having no points in common with either $C$ or $C'$ (e. g., a circle of sufficiently small radius about $O$). Then $I(C) = I(C'')$ and $I(C') = I(C'')$, so that $I(C') = I(C)$.

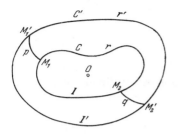

FIGURE 120.

*Definition XIII. The index (or Poincaré index) of an isolated singular point O of the vector field v defined by a dynamic system, or the index of an equilibrium state of system (I), is defined as the index of any closed curve C containing O in its interior such that there are no other singular points of the field v either inside or on C.*

We denote the index of an isolated singular point O of system (I) by $I(O)$.

*Lemma 1. Let Γ be a subregion of G bounded by a simple closed curve C, containing n singular points $O_1, O_2, \ldots, O_n$ of the dynamic system, all in the interior of C. Then the index of C is equal to the sum of indices of these singular points:*

$$I(C) = \sum_{k=1}^{n} I(O_k).$$

The lemma follows at once from Theorem 26 and the definitions of the index of a closed curve and a singular point.

Lemma 1 generalizes in an obvious manner to the case of a multiply connected region $\bar{Γ}$. The lemma yields a series of propositions concerning dynamic systems.

*Theorem 27. Let C be a simple closed curve in the region G and Γ its interior. If all points of Γ belong to G and it contains a finite number of equilibrium states, and there are no equilibrium states on C itself, then the index of C is equal to the sum of indices of all equilibrium states in the interior of C (i.e., in Γ).*

Theorem 27 is easily generalized to the case of a multiply connected subregion Γ of G.

Theorem 24 and its corollary imply the following:

*Theorem 28. The index of any closed path of a dynamic system is equal to +1.*

*Theorem 29. The index of any cycle without contact for a dynamic system is +1.*

Corollary 1. If $L$ is a closed path of a dynamic system whose interior is a subregion of $G$, the sum of indices of the equilibrium states inside $L$ is +1.

Corollary 2. If the interior of a closed path $L$ is a subregion of $G$, the system has at least one equilibrium state inside $L$.*

Corollary 3. If the interior of a cycle without contact $C$ is a subregion of $G$, then the sum of indices of the equilibrium states inside $C$ is +1.

Corollary 4. If the interior of a cycle without contact $C$ is a subregion of $G$, it contains at least one equilibrium state of the system.

**3. THE INDEX AS A LINE INTEGRAL.** When $C$ is a simple smooth closed curve on which there are no singular points of system (I), the index of $C$ may be expressed as a line integral, as we shall now demonstrate.

---

* Corollary 2 is the same as Theorem 17 (§4.7). This is the promised alternative proof of that theorem, based on general properties of vector fields.

Let

$$x = x(\sigma), \qquad y = y(\sigma)$$

be the parametric equations of $C$.

We shall assume that the functions $x(\sigma)$ and $y(\sigma)$ are defined and continuously differentiable for $a \leqslant \sigma \leqslant a + T$,

$$x(a+T) = x(a), \qquad y(a+T) = y(a),$$

with increase in $\sigma$ corresponding to the positive traversal of the curve $C$.

As usual, we are assuming that the functions $P(x, y)$ and $Q(x, y)$ are continuously differentiable. Moreover, since by assumption there are no equilibrium states on $C$, we have $P^2(x, y) + Q^2(x, y) \neq 0$ for all points of $C$.

Let $F(\sigma)$ be any angle function of the vector field induced on $C$ by system (I).

Consider any value of $\sigma^*$, $\sigma^* \in [a, a + T]$. If

$$P(x(\sigma^*), y(\sigma^*)) \neq 0,$$

then this is also true for all $\sigma$ sufficiently close to $\sigma^*$; it is obvious by the definition of an angle function that for all such $\sigma$

$$F(\sigma) = \operatorname{arctg} \frac{Q(x(\sigma),\, y(\sigma))}{P(x(\sigma),\, y(\sigma))} + 2\pi r,$$

where $r$ is a constant integer. Consequently,

$$F'(\sigma) = \frac{P(x(\sigma),\, y(\sigma)) \dfrac{dQ(x(\sigma),\, y(\sigma))}{d\sigma} - Q(x(\sigma),\, y(\sigma)) \dfrac{dP(x(\sigma),\, y(\sigma))}{d\sigma}}{P^2(x(\sigma),\, y(\sigma)) + Q^2(x(\sigma),\, y(\sigma))}. \qquad (1)$$

If $P(x(\sigma^*),\, y(\sigma^*)) = 0$, then $Q(x(\sigma^*), y(\sigma^*)) \neq 0$, and in the neighborhood of $\sigma^*$ the function $F(\sigma)$ differs by an additive constant from

$$\operatorname{arctg} \frac{P(x(\sigma),\, y(\sigma))}{Q(x(\sigma),\, y(\sigma))}.$$

Consequently, formula (1) is true in this case too, so that it holds for all $\sigma$, $a \leqslant \sigma \leqslant a + T$. Then

$$F(\sigma) = \int_a^\sigma \frac{P(x(\sigma),\, y(\sigma)) \dfrac{dQ(x(\sigma),\, y(\sigma))}{d\sigma} - Q(x(\sigma),\, y(\sigma)) \dfrac{dP(x(\sigma),\, y(\sigma))}{d\sigma}}{P^2(x(\sigma),\, y(x)) + Q^2(x(\sigma),\, y(\sigma))} \, d\sigma + F(a),$$

and so the index of the curve $C$ relative to the field $v$ is

$$I(c, v) = \frac{1}{2\pi} \int_a^{a+T} \frac{P(x(\sigma),\, y(\sigma)) \dfrac{dQ(x(\sigma),\, y(\sigma))}{d\sigma} - Q(x(\sigma),\, y(\sigma)) \dfrac{dP(x(\sigma),\, y(\sigma))}{d\sigma}}{P^2(x(\sigma),\, y(\sigma)) + Q^2(x(\sigma),\, y(\sigma))} \, d\sigma. \qquad (2)$$

The integral on the right of (2) is clearly equal to the line integral

$$\int_C \frac{P(x, y)\left(\frac{\partial Q(x, y)}{\partial x} dx + \frac{\partial Q(x, y)}{\partial y} dy\right) - Q(x, y)\left(\frac{\partial P(x, y)}{\partial x} dx + \frac{\partial P(x, y)}{\partial y} dy\right)}{P^2(x, y) + Q^2(x, y)},$$

over the curve $C$, evaluated in the positive sense; we shall abbreviate this integral by

$$\int_C \frac{P\, dQ - Q\, dP}{P^2 + Q^2}. \tag{3}$$

Thus, we have the formula

$$I(C) = \frac{1}{2\pi} \int_C \frac{P\, dQ - Q\, dP}{P^2 + Q^2}. \tag{4}$$

**4. EVALUATION OF INDICES OF SIMPLE EQUILIBRIUM STATES OF A SYSTEM.**
Let (I) be a dynamic system of class $C_1$. We shall assume that the equilibrium state in question is at the origin, so that system (I) can be written

$$\frac{dx}{dt} = ax + by + \varphi(x, y), \qquad \frac{dy}{dt} = cx + dy + \psi(x, y), \tag{5}$$

where $a, b, c, d$ are the values of the first partial derivatives of $P(x, y)$ and $Q(x, y)$ at $(0, 0)$. Since $O$ is a simple equilibrium state, $\Delta = \begin{vmatrix} a & b \\ c & d \end{vmatrix} \neq 0$. As we know, $\varphi(x, y) = 0(\varrho)$ and $\psi(x, y) = 0(\varrho)$ (see §5).

Let $v$ denote the vector field defined by system (5). Consider the system

$$\frac{dx}{dt} = ax + by, \qquad \frac{dy}{dt} = cx + dy \tag{6}$$

and let $v^*$ denote the vector field defined by system (6). Let $I(0, v)$ denote the index of the equilibrium state of system (5) and $I(0, v^*)$ that of the equilibrium state of system (6).

We prove first that

$$I(0, v) = I(0, v^*).$$

The field $v^*$ is clearly defined throughout the plane. Since $\Delta \neq 0$, the vector $v^*(ax + by, cx + dy)$ vanishes only when $x = y = 0$. Therefore $v^*$ does not vanish at points of the unit circle $\varrho = 1$. Let $m$ denote the minimum of $|v^*|$ on this circle. At any other point of the plane,

$$|v^*(x, y)| = \sqrt{(ax + by)^2 + (cx + dy)^2} =$$
$$= \varrho \sqrt{\left(a\frac{x}{\varrho} + b\frac{y}{\varrho}\right)^2 + \left(c\frac{x}{\varrho} + d\frac{y}{\varrho}\right)^2} > \varrho m,$$

since the point $\left(\frac{x}{\varrho}, \frac{y}{\varrho}\right)$ is on the unit circle $(\varrho^2 = x^2 + y^2)$. On the other hand,

$$|v^*(x, y) - v(x, y)| = \sqrt{[\varphi(x, y)]^2 + [\psi(x, y)]^2} = 0(\varrho).$$

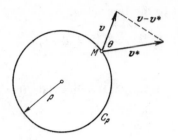

FIGURE 121.

Let $C_\rho$ be a circle of radius $\varrho$ about the origin. Let $(x, y) \in C_\rho$ and let $\theta$ be the smallest (in absolute value) angle between vectors $v(x, y)$ and $v^*(x, y)$. It is clear from Figure 121 that

$$|\sin \theta| \leqslant \frac{|v-v^*|}{|v^*|} < \frac{O(\varrho)}{\varrho m},$$

i. e., $|\sin \theta| \to 0$ as $\varrho \to 0$. Now, for small $\varrho$ the angle $\theta$ cannot be obtuse, since it is the angle opposite the shortest side of a triangle. It follows that $\theta \to 0$ as $\varrho \to 0$. Therefore, if $\varrho$ is sufficiently small, the vectors $v$ and $v^*$ do not have opposite directions at any point of the circle $C_\rho$.

Note now that if $\varrho$ is sufficiently small, then

$$I(O, v) = I(C_\rho, v), \quad I(O, v^*) = I(C_\rho, v^*),$$

and it follows at once from Lemma 2 that

$$I(O, v) = I(O, v^*).$$

Thus, in order to evaluate the index of a simple equilibrium state $O$ of system (5), we need only do this for the linearized system (6).

R e m a r k. The above arguments remain in force when the form of system (I) is

$$\frac{dx}{dt} = P_k(x, y) + O(\varrho^k), \quad \frac{dy}{dt} = Q_k(x, y) + O(\varrho^k),$$

where $P_k(x, y)$, $Q_k(x, y)$ are homogeneous polynomials of degree $k$ and $\sqrt{P_k^2(x, y) + Q_k^2(x, y)}$ vanishes only at $O(0, 0)$. The index of the equilibrium state $O$ of this system is equal to that of the equilibrium state of the system

$$\frac{dx}{dt} = P_k(x, y) \quad \text{and} \quad \frac{dy}{dt} = Q_k(x, y).$$

We now return to the original problem: to evaluate the index of the equilibrium state $O$ of system (6). To this end we use (4), taking the closed curve $C$ enclosing the equilibrium state to be the ellipse

$$(ax + by)^2 + (cx + dy)^2 = 1. \tag{7}$$

Thus

$$I(O) = \frac{1}{2\pi} \int_C \frac{(ax+by)\, d\, (cx+dy) - (cx+dy)\, d\, (ax+by)}{(ax+by)^2 + (cx+dy)^2} =$$

$$= \frac{1}{2\pi} \int_C [(ax+by)\, d\, (cx+dy) - (cx+dy)\, d\, (ax+by)],$$

where $C$ is the ellipse $(7)$ described in the positive sense. If we set

$$ax + by = \xi, \qquad cx + dy = \eta, \tag{8}$$

the ellipse $C$ becomes a circle $\xi^2 + \eta^2 = 1$ on the $(\xi, \eta)$-plane. We para-metrize this circle in the standard way, setting $\xi = \cos \vartheta$, $\eta = \sin \vartheta$. Note that as $\vartheta$ increases from 0 to 1 the circle $\xi^2 + \eta^2 = 1$ is described once in the positive sense. The ellipse $C$ is described in the positive sense if $\Delta = ad - bc > 0$, in the negative sense if $\Delta < 0$. Hence, if $\Delta > 0$,

$$I(O) = \frac{1}{2\pi} \int_0^{2\pi} [\cos \vartheta \, d(\sin \vartheta) - \sin \vartheta \, d(\cos \vartheta)] = +1,$$

and if $\Delta < 0$,

$$I(O) = \frac{1}{2\pi} \int_{2\pi}^{0} [\cos \vartheta \, d(\sin \vartheta) - \sin \vartheta \, d(\cos \vartheta)] = -1.$$

We state our result as a theorem:

*Theorem 30. The index of a simple equilibrium state is +1 in the case of a node or focus, − 1 in the case of a saddle point.*

*Chapter VI*

SPECIAL METHODS FOR QUALITATIVE
INVESTIGATION OF DYNAMIC SYSTEMS

INTRODUCTION

We have already stated (§3.1) that one of the fundamental questions in the qualitative theory of dynamic systems is just what information concerning the paths is necessary to ensure maximal completeness of qualitative investigation.

Although an exhaustive answer will be given only in Chapters VII, VIII, X and XI, we can nevertheless indicate certain basic elements whose necessity in this respect is not open to question.

A first prerequisite for any investigation is of course the number and type of the equilibrium states. When the coordinates of the equilibrium states are known, effective methods for determining their types are available, as shown in Chapter IV, for simple equilibrium states (characteristic roots with nonzero real parts). Moreover, such methods also exist for many types of multiple equilibrium states (some of these will be studied in Chapter IX). True, determination of the coordinates of equilibrium states, even of the total number of equilibrium states, is in itself a far from easy task. In certain cases, however, and in particular when the right-hand sides of the system are polynomials, one can find general methods to determine the number of equilibrium states, which amount to determining the number of points common to two polynomials.*

Apart from information concerning the number and types of equilibrium states, one is also interested in closed paths and in whether they fill out entire regions or are isolated (i. e., limit cycles); one needs the number of limit cycles, their relative positions, and also information concerning their stability.

It is obvious that a knowledge of the types of equilibrium states is far from sufficient to this end; this is easily corroborated by simple examples. For example, the dynamic system of Example 1, §7, which has a single equilibrium state (focus) may have either no limit cycles or an arbitrary finite number thereof. Clarification of the situation requires additional study.

_____

* Determination of the number of points common to two curves, when one is not interested in actually computing their coordinates, is of course a "qualitative" problem. It is intimately connected with the simpler problem of the number of real roots of a function $F(x)$. When the function $F(x)$ is a polynomial, a complete solution is provided by Sturm's test (in this connection, see also Example 5, §7.6, Chapter IV.)

Unfortunately, there are no regular methods with whose aid one can determine whether or not a given dynamic system possesses limit cycles. Moreover, the very question as to what one should expect such methods to accomplish and how is quite vague; the quest for such methods has always been and remains one of the most difficult and important problems in the qualitative theory of dynamic systems.

Because of the lack of general methods, great importance attaches even to partial tests which provide an indication of the existence or nonexistence of limit cycles for particular classes of dynamic systems.

In §12 we present a few comparatively simple tests of this kind, which sometimes yield successful results. The simplest of these is B e n d i x s o n' s t e s t, which states that if the expression

$$\frac{\partial P}{\partial x} + \frac{\partial Q}{\partial y}$$

does not change sign in a simply connected region, the system has no closed paths in this region. This criterion is a special case of the more sophisti- cated D u l a c t e s t (see Theorem 28, §11). At the end of §12 we describe Poincaré's method for determining limit cycles, employing t o p o g r a p h i c s y s t e m s o f c u r v e s a n d c o n t a c t c u r v e s. This method is not regular, but in certain special cases it can be applied successfully. It will be illustrated by examples.

It is readily understood that in qualitative investigation of dynamic systems information concerning the equilibrium states and limit cycles must be supplemented by information concerning the behavior (or position) of separatrices of saddle points. Thus, in Example 4 (§7) there are two equilibrium states, a node and a saddle point. The separatrices of the saddle point may behave in different ways. Two of the logically possible cases which can arise are indicated in Figure 122a and b (the reader may convince himself that there are also other possibilities). Again, no regular methods are available, and one must be satisfied with various specialized devices. Some of these are employed in Examples 4, 9, 12, 13, 14 of Chapter XII. Note that information about the behavior of separatrices may be invaluable in determining whether limit cycles exist (see Examples 16, 17 in Chapter XII). In addition, it is not hard to see that our investigation of qualitative structure will not be complete without a knowledge of the behavior of the separatrices when $x$ and $y$ increase indefinitely.

Consider a simple example — a dynamic system on the plane with no equilibrium states. Even here there are various possible qualitative structures. Thus, it is readily seen that the phase portraits illustrated in Figure 123a and b are distinct. And when there are equilibrium states the topological structure may differ even more, depending on the behavior of the paths as $x$ and $y$ go to infinity.

Thus, when the system is defined throughout the plane, it is natural to study the behavior of the paths "in the neighborhood of infinity." When the right-hand sides of the system are polynomials, it is natural and extremely convenient not to consider the system on the plane but to project it onto a sphere in some way (the Bendixson sphere or Poincaré sphere). When this is done, those points of the sphere (the north pole of the Bendixson sphere and the equator of the Poincaré sphere) which are not images of any points of the plane are interpreted as corresponding to the points at infinity on the plane.* This treatment of the points at infinity (by means of a suitable transformation of variables) is studied in §14. The results of the investigation contribute to investigation of the dynamic system in the finite portion of the plane, for instance, with regard to the behavior of the separatrices or the existence of a limit cycle.

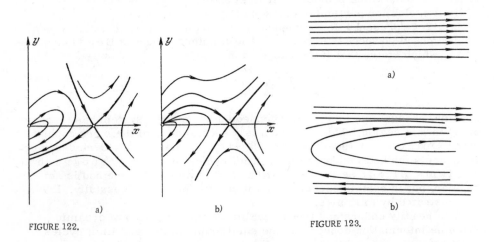

FIGURE 122.                                              FIGURE 123.

The methods for qualitative investigation described in this chapter do not exhaust the arsenal of tools available in the mathematical literature. In particular, there is an extensive literature on certain types of dynamic systems, namely the so-called Liénard equation and its generalizations, whose right-hand sides satisfy special conditions.** Utilizing arguments specific for this type of equation, one can verify the existence or nonexistence of limit cycles, and moreover, when limit cycles exist, their uniqueness. We shall not dwell on the Liénard equation here, referring the reader to the literature (see /37/, /38/, /39/).†

* The reader should note that this approach does not lead to a dynamic system on the sphere in the sense of §2;  certain fundamental conditions may fail to hold.

** The dynamic system corresponding to the Liénard equation is

$$\dot{x}=y, \quad \dot{y}=-\varphi\,(y)-x.$$

† One method widely employed in the technical literature to determine periodic solutions is the so-called "method of harmonic balance." For the second-order dynamic systems treated in this book, the procedure is as follows.  One assumes the required periodic solution to have the form

$$x=A\cos\omega t, \quad y=A\sin\omega t \qquad\qquad\text{(a)}$$

(i. e., if the system has a limit cycle, it must be a circle).  One now substitutes the expressions (a) for $x$ and $y$ into the system

$$\dot{x}=P\,(x,\,y), \quad \dot{y}=Q\,(x,\,y)$$

In §15 we shall discuss the application of numerical methods to determination of qualitative structure.

## §12. TESTS FOR CLOSED PATHS

### 1. GENERAL REMARKS ON ANNULAR REGIONS FILLED BY CLOSED PATHS.

One of the first important questions to be considered in connection with the existence of closed paths is whether there exist entire regions filled by closed paths or whether the closed paths are isolated (i. e., limit cycles).

When the dynamic system is analytic, the following lemma plays an essential role.

*Lemma 1. If $L_0$ is a closed path of an analytic dynamic system, it is either isolated or all paths in its neighborhood are closed.*

Proof. Let $l$ be an analytic arc without contact (e. g., segment without contact) through some point of $L_0$, $s$ the parameter on the arc and $\bar{s} = \omega(s)$ a succession function on $l$, defined for $a \leqslant s \leqslant b$. The values of $s$ corresponding to the points at which closed paths, among these $L_0$, cross $l$ are the zeros of the function $s - \omega(s)$; conversely, the zeros of the function $s - \omega(s)$ correspond to closed paths (see §3.8). Since both the system and the arc $l$ are analytic, $\omega(s)$ is also analytic (see Remark 1 to Lemma 13 in §3). By the properties of analytic functions, if $s - \omega(s)$ is not identically zero it has only finitely many zeroes. And this clearly implies that if $L_0$ is not isolated all paths in its neighborhood are closed, proving the lemma.

Thus, when the system is analytic it cannot have closed paths near which there are both closed and nonclosed paths. In particular, there cannot exist an infinite set of limit cycles tending to a closed path (from either one or both sides). Similarly, it cannot have a closed path approached from its interior (exterior) by closed paths and from its exterior (interior) by nonclosed paths.

On the basis of Lemma 1, one can show that if an analytic dynamic system has an "annular" region filled by closed paths, the boundary of this region is a union of equilibrium states and paths tending to equilibrium states. If all the equilibrium states of the system are simple, the paths making up the boundary of an annular region (other than equilibrium states) are necessarily separatrices of saddle points. Geometric examples of such annular regions are given in Figures 21 and 24 (Chapter I).

One test for the existence of regions filled by closed paths is the existence of an analytic integral in a region containing an equilibrium

expands the functions $P(A \cos \omega t, A \sin \omega t)$ and $Q(A \cos \omega t, A \sin \omega t)$ in Fourier series and drops all higher harmonics. This gives two equations for $A$ and $\omega$. It is clear that unless additional restrictions are imposed on the functions $P(x, y)$ and $Q(x, y)$ the method in the primitive form described here is not valid and its application may (and indeed does) lead to absurd conclusions. However, when employing the method of harmonic balance one usually assumes implicitly that the system under consideration is close to a conservative linear system, i. e.,

$$P(x, y) = y + \mu f_1(x, y), \qquad Q(x, y) = -x + \mu f_2(x, y),$$

where $\mu$ is a small parameter. In this case we have simply the so-called "Poincaré method" (or method of small parameters) /6/.

state with pure imaginary characteristic roots (which in this case is a center). This situation arises in many of the examples considered below (see Examples 4, 7, 8, 11 of this section).

**2. CASE IN WHICH THE CONFIGURATION OF ISOCLINES AND THE PROPERTIES OF THE FIELD BETWEEN THEM DIRECTLY IMPLY THE ABSENCE OF LIMIT CYCLES.** It sometimes happens that the specific "layout" of the isoclines and the field between them directly implies that there are no closed paths. We shall illustrate an elementary example of this type.

Let $O$ be a simple equilibrium state of the system, which is defined in a region $G$ of the $(x, y)$-plane containing $O$. Suppose that the isoclines $P(x, y) = 0$ and $Q(x, y) = 0$ divide $G$ into four subregions:

$$\text{I) } \dot{x} > 0,\ \dot{y} > 0, \quad \text{II) } \dot{x} < 0,\ \dot{y} > 0,$$
$$\text{III) } \dot{x} < 0,\ \dot{y} < 0, \quad \text{IV) } \dot{x} > 0,\ \dot{y} < 0$$

(Figure 124). It is readily seen that in regions I and III the segments $y = C$ have no contact with paths, and the same holds in regions II and IV for the segments $x = C$. In fact, no path which passes at some time $t = t_0$ through any point $M$ of the isocline $P(x, y) = 0$ can return to $M$ with increasing $t$: it can only approach the point $O$.

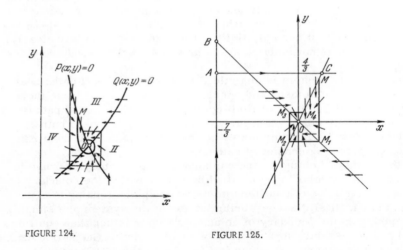

FIGURE 124.                    FIGURE 125.

Example 1.

$$\dot{x} = \left(x + \frac{7}{3}\right)\left(\frac{1}{2}y - x\right), \quad \dot{y} = \left(y - \frac{4}{3}\right)(x + y).$$

Equilibrium states: $O\,(0, 0)$, $A\left(-\frac{7}{3}, \frac{4}{3}\right)$, $B\left(-\frac{7}{3}, \frac{7}{3}\right)$ and $C\left(\frac{2}{3}, \frac{4}{3}\right)$. It is easy to see that $O$ is a stable focus. The integral curves $x = -\frac{7}{3}$ and $y = \frac{4}{3}$ pass through the equilibrium states $A$, $B$ and $C$. Thus, if there is a limit cycle it will contain the equilibrium state $O$ in its interior.

Consider the isocline of vertical directions $y = 2x$ and isocline of horizontal directions $y = -x$ through the point $O$. A limit cycle must cross the segment $OC$ of the former isocline. Let $M$ be an arbitrary point on the segment $OC$; construct a polygonal line $MM_1M_2M_3M_4$ from the corresponding vertical and horizontal segments (Figure 125). All these segments have no contact with paths of the system, and moreover paths crossing these segments enter the polygon $MM_1M_2M_3M_4M$ with increasing $t$. Consequently, a path passing through $M$ at $t = t_M$, after completing one revolution about the point $O$, cannot return to the point $M$ but crosses the isocline at a point closer to $O$; thus no limit cycle can exist. We leave it to the reader to show in analogous fashion that the same holds for the system

$$\frac{dy}{dt} = -\frac{axy}{x+y} + vy^2, \qquad \frac{dx}{dt} = \frac{axy}{x+y} - a,$$

$$x + y > 0.$$

The case discussed in this subsection, in which the absence of limit cycles can be inferred directly from the properties of the field, is quite rare.

In the next subsections we shall present a few other tests for the absence of closed paths.

**3. TESTS OF DULAC AND BENDIXSON.** The two tests presented below are sufficient conditions for nonexistence of closed paths and closed curves which are unions of paths.

*Theorem 31 (Dulac's test for simply connected regions).* *Let*

$$\frac{dx}{dt} = P(x, y), \qquad \frac{dy}{dt} = Q(x, y) \qquad \text{(I)}$$

*be an analytic dynamic system, $G$ a simply connected subregion of the domain of definition of system (I). Let $B(x, y)$ be a continuously differentiable function defined in $G$ such that the function*

$$\frac{\partial}{\partial x}(BP) + \frac{\partial}{\partial y}(BQ) \qquad \text{(1)}$$

*does not change sign. Then there are no simple closed curves in G which are unions of paths of system (I).**

(The above condition is understood to mean that the function (1) may also vanish on finitely many isolated points and smooth curves, but has the same sign at all other points of $G$.)

Proof. We shall first prove that under the assumptions of the theorem the region $G$ contains no closed paths. Suppose on the contrary that there is a closed path $L$ in $G$. Consider the line integral

$$I = \oint_{(L)} (-BQ\,dx + BP\,dy).$$

---

* A simple closed curve of this type is either a closed path, or the union of alternating nonclosed whole paths and equilibrium states (see, e.g., Figures 21 and 24 of Chapter I).

It is easy to see that $I = 0$. Indeed, let

$$x = \varphi(t), \qquad y = \psi(t)$$

be the equations of the path $L$, $0 \leqslant t \leqslant T$. Then

$$I = \int_0^T B(\varphi(t), \psi(t)) \, [-Q(\varphi(t), \psi(t)) \, \varphi'(t) + P(\varphi(t), \psi(t)) \, \psi'(t)] \, dt,$$

and moreover

$$\varphi'(t) = P(\varphi(t), \psi(t)), \quad \psi'(t) = Q(\varphi(t), \psi(t)),$$

since $(\varphi(t), \psi(t))$ is a solution of system (I). It follows that the bracketed expression in the integrand must vanish identically, so that $I = 0$.

On the other hand, by Green's theorem

$$I = \iint_\Omega \left[ \frac{\partial}{\partial x}(BP) + \frac{\partial}{\partial y}(BQ) \right] dx \, dy,$$

where $\Omega$ is the region bounded by the closed path $L$ ($\Omega \subset G$). But this integral cannot vanish, since the integrand is of fixed sign. This contradiction completes the proof for closed paths.

We now show that $G$ cannot contain a simple closed "loop," i. e., union of an equilibrium state $O$ and a closed path $L$ which tends to $O$ both as

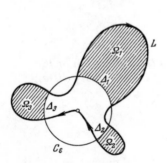

FIGURE 126.

$t \to -\infty$ and as $t \to +\infty$. Suppose that such a loop exists. Let $C_\varepsilon$ be a circle about $O$ with radius $\varepsilon$ so small that there are points of $L$ outside the circle. Since (I) is an analytic system, $L$ is an analytic curve and therefore cuts $C_\varepsilon$ in only finitely many points, which divide $L$ and $C_\varepsilon$ into finitely many subarcs. Let $\Omega$ denote the region bounded by the loop and $\Omega_\varepsilon$ the set of points of $\Omega$ whose distance from $O$ is greater than $\varepsilon$. $\Omega_\varepsilon$ is an open set (see Appendix, §1), and the boundary of each of its components is the union of certain of the above-mentioned subarcs of $L$ and $C_\varepsilon$. It follows that $\Omega_\varepsilon$ is the union of finitely many components, $\Omega_1$, $\Omega_2$, ..., $\Omega_n$, say; denote their positively oriented boundaries by $\gamma_1$, $\gamma_2$, ..., $\gamma_n$. These boundaries are piecewise-smooth curves (Figure 126).

Consider the integral

$$I_\varepsilon = \iint_{\Omega_\varepsilon} \left[ \frac{\partial}{\partial x}(BP) + \frac{\partial}{\partial y}(BQ) \right] dx \, dy = \sum_{k=1}^{n} \iint_{\Omega_k} \left[ \frac{\partial}{\partial x}(BP) + \frac{\partial}{\partial y}(BQ) \right] dx \, dy.$$

Since the integrand is of fixed sign, this integral does not vanish, and moreover $|I_\varepsilon|$ increases with decreasing $\varepsilon$ (since the region $\Omega_\varepsilon$ becomes larger). Therefore, for all $\varepsilon < \varepsilon_0$,

$$|I_\varepsilon| > |I_{\varepsilon_0}| > 0. \tag{2}$$

On the other hand, by Green's theorem

$$I_{\varepsilon_0} = \sum_{k=1}^{n} \oint_{\gamma_k} (-BQ\, dx + BP\, dy).$$

Since as shown above the line integral of $-BQ\, dx + BP\, dy$ over any part of a path $L$ of system (I) vanishes, it follows that $I_\varepsilon$ is equal to the sum of the integrals over the arcs of $C_\varepsilon$ appearing in the boundaries of the regions $\Omega_k$. Denoting these arcs by $\Delta_1, \Delta_2, \ldots, \Delta_n$, we have

$$I_\varepsilon = \sum_{k=1}^{n} \int_{\Delta_k} (-BQ\, dx + BP\, dy).$$

Expressing the terms as line integrals with respect to arc length, we get

$$I_\varepsilon = \sum_{k=1}^{n} \int_{\Delta_k} (-BQ \cos \varphi + BP \sin \varphi)\, ds,$$

where $\varphi$ is the angle between the tangent to the circle $C_\varepsilon$ and the positive $x$-axis. Let $M$ denote the maximum of the expression $|BQ| + |BP|$ in $\Omega$. Estimating the above expression for $I_\varepsilon$ and noting that the sum of lengths of all the arcs $\Delta_k$ is less than the length $2\pi\varepsilon$ of $C_\varepsilon$, we get

$$|I_\varepsilon| < 2\pi M\varepsilon.$$

For sufficiently small $\varepsilon$, this inequality contradicts (2). Thus there cannot be loops in the region $G$.

The proof that $G$ cannot contain a closed path which is the union of several paths and equilibrium states $O_1, O_2, \ldots, O_m$ is entirely analogous, except that here one considers circles $C_{1\varepsilon}, C_{2\varepsilon}, \ldots, C_{m\varepsilon}$ about all the equilibrium states. This completes the proof.

Remark. As far as the nonexistence of closed paths is concerned, the above proof makes no use of the analyticity of the system; it suffices to assume that system (I) is of class $C_1$.

Corollary (Bendixson's test for simply connected regions). If system (I) is analytic and the function

$$\frac{\partial P}{\partial x} + \frac{\partial Q}{\partial y}$$

is of fixed sign in a simply connected region $G$, then $G$ contains no simple closed curves which are unions of paths of the system.

Bendixson's test is simply Dulac's test with $B(x, y) = 1$. It is clear that the above remark is also valid for Bendixson's test.

If $G$ is a multiply connected region and there exists a function $B(x, y)$ such that $\frac{\partial}{\partial x}(BP) + \frac{\partial}{\partial y}(BQ)$ is of fixed sign in $G$, it follows at once that there are no closed paths (or closed curves which are unions of paths) whose interiors are subregions of $G$. Other closed paths may exist, but even concerning these one can draw certain conclusions. We confine ourselves to the following assertion; it is valid for the most frequently arising case — doubly connected regions, which we shall call annular regions.

*T h e o r e m 32 (Dulac's test for annular regions). Let G be a doubly connected subregion of the domain of definition of system (I). Let $B(x, y)$ be a function of class $C_1$, defined in G, such that the function $\frac{\partial}{\partial x}(BP) + \frac{\partial}{\partial y}(BQ)$*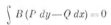

*is of fixed sign. Then G cannot contain more than one simple closed curve which is the union of paths of the system and whose interior contains the inner boundary of G.*

P r o o f. Suppose on the contrary that system (I) has two closed paths $L_1$ and $L_2$ in the annular region $G$. Let $g_1$ be the region bounded by $L_1$ and $L_2$.

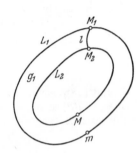

Join some point $M_1$ on $L_1$ to a point $M_2$ on $L_2$ by a simple smooth arc $l$ all of whose interior points lie in $g_1$.

Consider the closed contour $M_1 m M_1 M_2 M M_2 M_1$ (Figure 127), consisting of the points of the paths $L_1$, $L_2$ and the arc $l$.

Along this closed contour, we have

$$\int B(P\,dy - Q\,dx) = 0$$

(since the integrand vanishes identically on $L_1$ and $L_2$ and the arc $l$ is described once in each direction). By Green's theorem,

FIGURE 127.

$$\int B(P\,dy - Q\,dx) = \iint_{g_1} \left[\frac{\partial}{\partial x}(BP) + \frac{\partial}{\partial y}(BQ)\right]dx\,dy.$$

Thus, the integral

$$\iint_{g_1} \left[\frac{\partial}{\partial x}(BP) + \frac{\partial}{\partial y}(BQ)\right]dx\,dy$$

must vanish. But the domain of integration here is the region $g_1$ between $L_1$ and $L_2$, and by assumption the integrand is of fixed sign throughout $g_1$, which is a subregion of $G$. Consequently, there cannot be more than one closed path in the region $G$.

One proves in analogous fashion that $G$ cannot contain more than one simple closed curve consisting of paths.

Note that Dulac's test provides no prescription for the choice of $B(x, y)$ in specific dynamic systems. In any concrete case the search for such a function may or may not succeed, depending on the form of the system and the skill of the investigator.

**4. USE OF POINCARÉ INDICES AND SINGLE-CROSSING CYCLES IN RELATION TO THE EXISTENCE OF LIMIT CYCLES.** In this subsection we consider smooth single-crossing cycles (see §3.12). Recall that a smooth single-crossing cycle is a smooth simple closed curve $C$ having the following properties:

1) There are no equilibrium states on $C$.

2) At all points of $C$ except (possibly) a finite number, the paths have no contact with the curve and they cross $C$ either all inward or all outward.

Paths tangent to $C$ (if such exist) enter or leave the region bounded by $C$ together with all the other paths.

Any cycle without contact is of course a single-crossing cycle (the converse is of course false). By Theorem 29 (Chapter V) the index of a cycle without contact is +1.

In many cases, examination of cycles without contact or single-crossing cycles yields significant information concerning the existence of closed paths or limit cycles. We first describe a few simple tests for nonexistence of closed paths based on the properties of the Poincaré index. These tests are summarized in the following theorem.

*Theorem 33. Let $G$ be a simply connected subregion of the domain of definition of system (I). Then: 1) If $G$ contains no equilibrium states, it contains no closed paths. 2) If $G$ contains finitely many equilibrium states, none of which has index +1 and no combination of which has sum of indices equal to +1, then there are no closed paths in $G$. In particular, if there is only one equilibrium state in the region and its index is not +1 (e.g., a saddle point), then there are no closed paths in $G$. 3) If $G$ contains finitely many equilibrium states, such that for each equilibrium state $O$ with positive index there exists a path tending to $O$ which goes to infinity or has points outside $G$, then there are no closed paths in $G$.*

The proof follows at once from Theorems 25 and 26 of §11.

The next tests are based on the properties of single-crossing cycles. Single-crossing cycles (including cycles without contact) are frequently employed to investigate limit cycles.

*Theorem 34. Let $C$ be a single-crossing cycle bounding a subregion $G$ of the domain of definition of system (I). Suppose that the following conditions are satisfied: 1) all paths crossing $C$ enter $G$ with increasing $t$; 2) $G$ contains a single equilibrium state $O$, which is an unstable node or focus; 3) $G$ contains only finitely many closed paths of the system. Then the number of stable limit cycles in $G$ is greater by one than the number of unstable ones. (Consequently, there exists at least one stable limit cycle.)*

P r o o f. Since by assumption all paths crossing the single-crossing cycle $C$ enter the region $G$ and $O$ is an unstable node or focus, there must be at least one limit cycle in $G$ (see Theorem 13 of §4).

Let $L_1, L_2, \ldots, L_s$ be all limit cycles of the system in $G$, so indexed that $L_{i+1}$ contains $L_i$ in its interior. It is easy to see that these cycles cannot all be semistable. Let $L_{i_1}, L_{i_2}, \ldots, L_{i_k}$ ($i_1 < i_2 < \ldots < i_k$) be those of them which are not semistable. Since $O$ is an unstable equilibrium state, it follows that $L_{i_1}$ is a stable limit cycle, and with increasing $i_j$ unstable and stable limit cycles alternate; the cycle $L_{i_k}$ is obviously stable. But then the number of stable cycles among $L_{i_1}, L_{i_2}, \ldots, L_{i_k}$ must be greater by one than the number of unstable ones. (The number of semistable limit cycles which may exist in $G$ is subject to no restriction other than finiteness.)

An analogous statement holds if one assumes that the paths cross a cycle without contact $G$ o u t w a r d (leaving $G$) and the equilibrium state $O$ is a stable node or focus. But if the paths through $C$ enter $G$ (leave $G$) and $O$ is a stable (unstable) node or focus, the number of stable limit cycles in $G$ is equal to the number of unstable ones; in particular, the region may contain no limit cycles at all.

The next theorem establishes an analogous statement for annular regions.

*Theorem 35.* *Let G be a doubly connected region bounded by two cycles without contact $C_1$ and $C_2$, which contains no equilibrium states and a finite number of closed paths. If all paths crossing $C_1$ and $C_2$ enter G (leave G) with increasing $t$, then the number of stable limit cycles in G is greater (less) by one than the number of unstable ones (Figure 128).*

Hence it follows, in particular, that under these assumptions there is at least one stable (unstable) limit cycle in $G$. But if the paths crossing one of the boundary cycles enter $G$ and those crossing the other leave $G$, the numbers of stable and unstable limit cycles in $G$ are equal, and there may be none of either type.

The proof of Theorem 35 is entirely analogous to that of Theorem 34.

**5. TOPOGRAPHIC SYSTEM OF CURVES AND CONTACT CURVE.** To be able to use Theorems 31 and 32 we must have suitable regions bounded by cycles without contact. No regular methods are known for construction of cycles without contact. In certain special cases, such cycles can be found by judicious choice of a topographic system of curves.

Following Poincaré, we define a topographic system to be a family of nonintersecting simple smooth closed curves

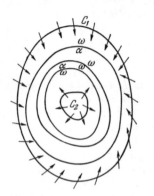

FIGURE 128.

$$F(x, y) = C, \tag{3}$$

lying interior to one another and filling out some doubly connected region (for example, concentric circles about the origin or a family of confocal ellipses). We shall assume that each curve of the system corresponds to a unique $C$. Moreover, to fix ideas we shall assume that curves with smaller $C$ lie interior to curves with greater $C$, so that the "size" of the curves (3) increases with increasing $C$. Finally, we shall assume that there are no equilibrium states of the system on any of the curves (3).

Suppose that a topographic system of curves (3) fills a region $G$. Let

$$x = x(t), \qquad y = y(t)$$

be a path of system (I) in $G$. Consider the function $F(x(t), y(t))$.

The derivative of this system along paths of system (I) (see §3.13) is

$$\frac{dF}{dt} = \frac{\partial F(x, y)}{\partial x} P(x, y) + \frac{\partial F}{\partial y} Q(x, y) = \Phi(x, y).$$

It is clear that a path $L$ is tangent to a curve of the topographic system at their common point $M(x, y)$ if and only if $\Phi(x, y) = 0$. Therefore, any curve of the topographic system on which the function $\Phi(x, y)$ is of fixed sign* is a single-crossing cycle. If moreover $\Phi(x, y) \leqslant 0$ at all

---

* That is, at all of whose points (except possibly a finite set) $\Phi(x, y)$ has the same sign; the function may vanish at finitely many points.

points of such a curve, then $\frac{dF}{dt} \leqslant 0$. $F$ is decreasing, and with increasing $t$ all paths cross the cycle inward. On the other hand, if $\Phi(x, y) \geqslant 0$ on a curve of the topographic system, then with increasing time all paths cross the curve outward. It follows that if the function $\Phi(x, y)$ is of fixed sign in some annular region $G$ whose boundary is a union of curves of the topographic system, then there are no closed paths (or, consequently, limit cycles) in the region.

The curve $\Phi(x, y) = 0$, i. e.,

$$P(x, y) \cdot F'_x(x, y) + Q(x, y) \cdot F'_y(x, y) = 0$$

is known as a **contact curve**. The contact curve consists of all points at which the paths of system (I) are tangent to curves of the topographic system. If the latter is so chosen that the contact curve is closed, it

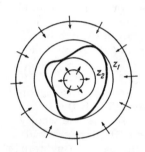

is clear that there exist a "largest" and a "smallest" curve of the topographic system intersecting the contact curve. Denote these curves by $Z_1$ and $Z_2$, respectively (Figure 129). The contact curve obviously lies entirely in the closed annular region $W_1$ bounded by $Z_1$ and $Z_2$, and is tangent to these curves. All curves of the topographic system lying outside $Z_1$ or inside $Z_2$ are single-crossing cycles. It follows that if there exist closed paths in the region $G$ (the region covered by the curves of the topographic system), it must lie between $Z_1$ and $Z_2$. If the function $\Phi(x, y)$ does not change sign but nevertheless vanishes (at points of the contact curve), we shall say that the contact at the points of the contact curve is "false."

FIGURE 129.

## 6. EXAMPLES.

Example 2.

$$\dot{x} = y \equiv P(x, y), \qquad \dot{y} = -(1 + x^2 + x^4) y - x \equiv Q(x, y).$$

The system has a single equilibrium state $O(0, 0)$, a stable focus (see § 7.5, Example 1).

By Bendixson's test this system has no limit cycles, for the function $\frac{\partial P}{\partial x} + \frac{\partial Q}{\partial y} = -(1 + x^2 + x^4)$ does not change sign on the plane.

Example 3 /60/.

$$\dot{x} = y + x(1 + \beta y)(x^2 + y^2 + 1), \qquad \dot{y} = -x + (y - \beta x^2)(x^2 + y^2 + 1).$$

We claim that the system has no limit cycles. Consider a topographic system consisting of the family of circles $x^2 + y^2 = C$. Calculating the derivative $\frac{d(x^2 + y^2)}{dt}$ along paths of the system (see § 3.13), we get

$\frac{d\,(x^2+y^2)}{dt} = 2\,(x^2+y^2+1)(x^2+y^2)$. This function is positive for all points of the $(x,\ y)$-plane except $O\,(0,\ 0)$, which is an equilibrium state. Thus the family $x^2+y^2=C$ is a family of closed curves without contact for the paths of the system. $C$ increases with $t$ along any path. It follows that there are no limit cycles.

Example 4.

$$\dot{x}=1+x^2-y^2+2xy, \qquad \dot{y}=1-x^2+y^2+2xy. \tag{4}$$

Instead of this system, we consider a family of systems depending on a parameter $\alpha$:

$$\dot{x}=-\alpha\,(1-x^2+y^2)+2xy, \qquad \dot{y}=1-x^2+y^2+2\alpha xy. \tag{5}$$

When $\alpha=1$ we get system (4). When $\alpha=0$, we have the system

$$\dot{x}=2xy, \qquad \dot{y}=1-x^2+y^2. \tag{6}$$

It is readily seen that system (5) is obtained from (6) by rotation of the vector field through an angle whose tangent is $\alpha$. The common equilibrium states of the two systems are $A\,(1,\ 0)$ and $B\,(-1,\ 0)$. System (6) can be integrated, its general integral being

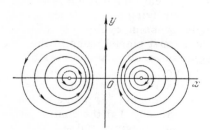

$$\left(x-\frac{C}{2}\right)^2 + y^2 = \frac{C^2}{4}-1 \ \text{(Figure 130)}.$$

System (6) may serve as a topographic system for system (5). If the latter has a limit cycle, it must lie wholly within one of the half-planes $x>0$ or $x<0$, since the paths of system (5) cross the $y$-axis in one direction. But the half-planes $x>0$ and $x<0$ are filled by closed curves having no contact with paths of system (5),

FIGURE 130.

and so the latter can have no limit cycles.

Example 5 /61/.

$$\dot{x}=y, \qquad \dot{y}=2\,(1-xy).$$

This system has no equilibrium states in the finite part of the plane. Since a closed path must contain in its interior at least one equilibrium state, the system has no limit cycles.

Example 6 /62/.

$$\dot{x}=y\equiv P\,(x,\ y), \qquad \dot{y}=-ax-by+ax^2+\beta y^2\equiv Q\,(x,\ y).$$

This system has no limit cycles. This can be verified by Dulac's test, a suitable Dulac function being $B\,(x,\ y)=be^{-2\beta x}$. In fact, the function $\frac{\partial\,(PB)}{\partial x}+\frac{\partial\,(QB)}{\partial y}=-b^2e^{-2\beta x}$ does not change sign on the $(x,\ y)$-plane.

In some cases, the expression in Dulac's test may change sign on certain curves; nevertheless, using specific properties of the system one can sometimes show that no limit cycles can cut these curves.

Example 7 /63/.

$$\dot{x} = x\,(a_1 x + b_1 y + c_1), \qquad \dot{y} = y\,(a_2 x + b_2 y + c_2).$$

It can be shown that this function has no limit cycles. The Dulac function is $B(x, y) = x^{k-1} y^{h-1}$, where $k = \dfrac{b_2\,(a_2 - a_1)}{\Delta}$, $h = \dfrac{a_1\,(b_1 - b_2)}{\Delta}$ $(\Delta = a_1 b_2 - b_1 a_2 \neq 0)$. Then

$$\frac{\partial\,(PB)}{\partial x} + \frac{\partial\,(QB)}{\partial y} = \left( \frac{a_1 c_2\,(b_1 - b_2)}{\Delta} + \frac{b_2 c_1\,(a_2 - a_1)}{\Delta} \right) x^{k-1} y^{h-1}.$$

When $\sigma \equiv a_1 c_2\,(b_1 - b_2) + b_2 c_1\,(a_2 - a_1) \neq 0$, this function vanishes only along the integral curves $x = 0$ and $y = 0$. It follows that there can be no limit cycles, since paths do not intersect. It can be shown that in this case there cannot even exist closed contours which are unions of paths. For suppose such a contour exists, its boundary containing segments of the $x$- and $y$-axes. The contour lies entirely in one of the quadrants of the $(x, y)$-plane. Though the expression $\frac{\partial\,(PB)}{\partial x} + \frac{\partial\,(QB)}{\partial x}$ vanishes on the $x$- and $y$-axes, it does not change sign in any quadrant. Thus $\iint \left( \frac{\partial\,(PB)}{\partial x} + \frac{\partial\,(QB)}{\partial y} \right) dx\,dy \neq 0$ and no such closed contour can exist. When $\sigma = 0$ the system has an analytic integral and one of the equilibrium states is a center, so that there is a whole region of the plane filled by closed paths.

Example 8 /64/.

$$\dot{x} = (y - k)\,x, \qquad \dot{y} = \gamma x + \beta y + q y^2.$$

We shall show, using Dulac's test, that this system has no limit cycles. The Dulac function will be $B(x, y) = x^{-(2q \div 1)}$. The function

$$D(x, y) \equiv \frac{\partial\,(PB)}{\partial x} + \frac{\partial\,(QB)}{\partial y} = (2qk + \beta)\,x^{-(2q+1)} \text{ for } 2qk + \beta \neq 0$$

does not vanish on the half-planes $x > 0$ and $x < 0$. The axis $x = 0$ is an integral curve. Consequently, if $2qk + \beta \neq 0$ the system does not even have closed contours consisting of paths, since any such contour would have to be contained entirely in one of the regions $x \geqslant 0$ or $x \leqslant 0$, but the function $D(x, y)$ does not change sign inside these regions and vanishes only on $x = 0$. If $2qk + \beta = 0$ the system has an analytic integral and there is a whole region filled by closed paths.

Example 9 /61/.

$$\dot{x} = 3xy, \qquad \dot{y} = y^2 - 2x^2 - 4xy + 2x.$$

This system has no limit cycles.

We first determine the possible location of a cycle. It is easy to see that $\bar{x} = 0$ is an integral curve, on which $\dot{y} > 0$ for $y \neq 0$. The equilibrium

213

states are $O(0, 0)$ and $A(1, 0)$. The characteristic roots of $O(0, 0)$ are both zero, so that $O$ is a multiple equilibrium state. Since there is an integral curve through $O(0, 0)$, this point cannot lie inside any limit cycle. The equilibrium state $A(1, 0)$ is a stable focus. Thus, if there exists a limit cycle, it must be in the region $x > 0$, enclosing the equilibrium state $A(1, 0)$. Since $O(0, 0)$ is a multiple equilibrium state, there may also be a closed limit contour which is the union of $O$ and a path issuing from $O$ and returning to it, enclosing the equilibrium state $A$. We now use Dulac's test,

with $B(x, y) = \dfrac{1}{x(x+y)^2}$. The function $\dfrac{\partial(PB)}{\partial x} + \dfrac{\partial(QB)}{\partial y} = -\dfrac{4x^2(x+y)}{x^2(x+y)^4}$ changes sign on the straight line $x + y = 0$. Thus, any closed limit contour consisting of paths must cross this straight line. Let us determine the slopes of the paths on this straight line. Setting $y = -x$ in the equation, we find that

$\dfrac{dy}{dx} = -\dfrac{x(3x+2)}{3x^2}$. Hence it is clear that when $x > 0$ we have $\dfrac{dy}{dx} < -1$, so

that all paths cross the line $x + y = 0$ in the same direction. Consequently, there cannot exist a limit cycle or a closed limit contour which is the union of paths.

In some systems, a limit cycle exists for some parameter values and not for others. One can sometimes verify that there are no limit cycles for a certain range of parameter values. The following examples illustrate this.

Example 10 /65/.

$$\dot{x} = y^2 - (x+1)[(x-1)^2 + \lambda], \quad \dot{y} = -xy.$$

We shall show that when $\lambda \leqslant -1$ the system has no limit cycles. It is easily seen that when $\lambda \leqslant -1$ the system has three equilibrium states in the finite plane: $A(-1, 0)$, $B(1 + \sqrt{|\lambda|}, 0)$ and $C(1 - \sqrt{|\lambda|}, 0)$. All these points lie on the integral curve $y = 0$. Consequently, when $\lambda \leqslant -1$ the system has no limit cycles. We can also show that when $\lambda > 1$ the system has no limit cycles. In this case the equilibrium states are $A(-1, 0)$ (saddle point), $D(0, \sqrt{1+\lambda})$ and $E(0, -\sqrt{1+\lambda})$ (stable foci or nodes). If there are cycles, they lie entirely within the half-planes $y > 0$ and $y < 0$, since $y = 0$ is an integral curve. We use Dulac's test with $B(x, y) = y$. The function

$\dfrac{\partial(PB)}{\partial x} + \dfrac{\partial(QB)}{\partial y} = -y(3x^2 + \lambda - 1)$ does not change sign in the half-planes $y > 0$

and $y < 0$ when $\lambda > 1$.

Example 11 /66/.

$$\dot{x} = y, \quad \dot{y} = x + x^2 - (\varepsilon_1 + \varepsilon_2 x)y. \tag{7}$$

As shown in a previous chapter (§7.5, Example 3), this system has two equilibrium states: $O(0, 0)$ (saddle point) and $A(-1, 0)$ (focus or node). The system is easily seen to be integrable when $\varepsilon_1 = \varepsilon_2 = 0$, the general

solution being $y^2 - x^2 - \dfrac{2x^3}{3} = C$. When $C = 0$ we get a folium of Descartes,

cutting the $x$-axis at the points $x = 0$ and $x = -\dfrac{3}{2}$ (Figure 131). We shall

use this last system as a topographic system. We first find the contact curve for the paths of system (7) and the topographic system

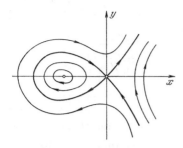

$$(\dot{y})_{\varepsilon_1=\varepsilon_2=0} \cdot \dot{x} - (\dot{x})_{\varepsilon_1=\varepsilon_2=0} \cdot \dot{y} = y^2 \, (\varepsilon_1 + \varepsilon_2 x).$$

It is readily seen that the contact along the curve $y=0$ is "false." $y=0$ is an isocline of vertical directions for both systems. The difference of directions changes sign on the straight line $\varepsilon_1 + \varepsilon_2 x = 0$.

We have the following three possibilities:

1) $\varepsilon_1\varepsilon_2 \leqslant 0$. If system (7) has a limit cycle, it lies in the half-plane $x<0$, for it cannot enclose both focus and saddle point and the paths cross each half of the $y$-axis in one direction. Consequently, the limit cycle cannot cut the straight line $x = -\frac{\varepsilon_1}{\varepsilon_2} \left( -\frac{\varepsilon_1}{\varepsilon_2} \geqslant 0 \right)$. Thus there are no limit cycles.

2) $\varepsilon_1\varepsilon_2 > 0$; $\frac{\varepsilon_1}{\varepsilon_2} > \frac{3}{2}$. In this case the contact curve, the straight line $x = -\frac{\varepsilon_1}{\varepsilon_2}$, passes to the left of the separatrix loop without cutting it (when $\frac{\varepsilon_1}{\varepsilon_2} = \frac{3}{2}$ these curves touch). Thus the separatrix loop and all closed curves within it are closed curves without contact for the paths of system (7). Were there a limit cycle, it would either cut the separatrix loop or lie entirely within it, which is impossible. Thus there are again no limit cycles.

3) $\varepsilon_1\varepsilon_2 > 0$; $\frac{\varepsilon_1}{\varepsilon_2} < \frac{3}{2}$. In this case the contact curve cuts the loop. The existence of limit cycles remains an open question.

In the next example we use Dulac's test for an annulus to prove that the limit cycle is unique.

Example 12.

$$\dot{x} = -y + ax + (3x^2 + 2y^2)\,x \equiv P\,(x,\,y),$$
$$\dot{y} = \quad x + ay + (3x^2 + 2y^2)\,y \equiv Q\,(x,\,y).$$

It is not hard to see that the unique equilibrium state $O\,(0,\,0)$ is an unstable focus if $\alpha > 0$ and a stable focus if $\alpha < 0$. When $\alpha = 0$ it is an equilibrium state with pure imaginary characteristic roots.

We take the family of circles $x^2 + y^2 = C$ as a topographic system. Differentiating the function $x^2 + y^2$ along the paths (§3.13), we get

$$\frac{d\,(x^2 + y^2)}{dt} = 2\,(x^2 + y^2)\,(3x^2 + 2y^2 + \alpha).$$

When $\alpha > 0$ this expression does not change sign, so that there are no limit cycles. When $\alpha < 0$ we have $\frac{d\,(x^2 + y^2)}{dt} > 0$ outside the ellipse $3x^2 + 2y^2 + a = 0$ and $\frac{d\,(x^2 + y^2)}{dt} < 0$ inside this ellipse. Thus there are no limit cycles outside the circle $x^2 + y^2 + \frac{a}{2} = 0$ or inside the circle $x^2 + y^2 + \frac{a}{3} = 0$.

215

FIGURE 132.

All paths passing through points of these two circles enter the annular region between them. Hence there must be an odd number of cycles in the annulus, and the number of unstable cycles must exceed the number of stable ones by unity (Figure 132).

We claim that there is exactly one limit cycle in the annulus. By Dulac's test, the limit cycle in the annulus will be unique if the function

$\frac{\partial P}{\partial x} + \frac{\partial Q}{\partial y}$ does not change sign there. We have

$$\frac{\partial P}{\partial x} + \frac{\partial Q}{\partial y} = 2(6x^2 + 4y^2 + a).$$ It is readily seen that

the ellipse $6x^2 + 4y^2 + a = 0$ lies interior to the smaller circle $x^2 + y^2 + \frac{a}{3} = 0$

bounding the annulus, and so $\frac{\partial P}{\partial x} + \frac{\partial Q}{\partial y}$ does not change sign in the annulus.

Therefore the limit cycle is indeed unique.

## §13. BEHAVIOR OF PATHS AT INFINITY

**1. GENERAL REMARKS. BENDIXSON TRANSFORMATION.** When one is investigating the qualitative structure of the phase portrait of specific dynamic systems defined on the whole plane, it may be considerably important to know how the paths behave as $x$ and $y$ increase indefinitely, or, as is usually said, "at infinity." A knowledge of the behavior of paths at infinity is often also invaluable in the study of dynamic systems in a bounded portion of the plane.

For example, analysis of the situation "at infinity" will sometimes yield unequivocal identification of the equilibrium state to which a separatrix issuing from a saddle point tends (Chapter XII, §30, Examples 5, 6, 9 and others), or enable one to prove the existence of a limit cycle. Thus, investigation of the behavior of the path "at infinity," to which we devote this section, is of essential consequence.

We begin with a special case, which can be solved quite simply. Consider a system

$$\frac{dx}{dt} = P(x, y), \qquad \frac{dy}{dt} = Q(x, y) \tag{I}$$

with the property that, for all $x, y$ such that $\varrho = \sqrt{x^2 + y^2}$ is sufficiently large, say greater than some positive $r_0$, the expression

$$J = P(x, y) \cdot x + Q(x, y) \cdot y$$

does not vanish and is therefore of constant sign. This clearly implies that all circles about the origin of radius greater than $r_0$ are cycles without contact for the paths of system (I). If $x = x(t)$, $y = y(t)$ are the equations of a path, then

$$\frac{1}{2}\frac{d\,(x^2+y^2)}{dt} = \frac{1}{2}\frac{d\varrho^2\,(t)}{dt} = J.$$

Therefore, if $J > 0$ for $\varrho > r_0$, then all paths crossing these circles do so outward with increasing $t$, while if $J < 0$ the paths cross the circles inward. In the first case we shall say that "infinity is stable," and in the second that "infinity is unstable." Sometimes knowledge of the stability or instability of infinity yields a proof of the existence of a limit cycle. For example, if the system is analytic, its only equilibrium state is a stable focus and infinity is stable, then it must have at least one unstable limit cycle.

However, only in special cases does $J$ maintain its sign for large values of $\varrho$. We must therefore evolve methods suitable for investigating a system at infinity in more general cases. Such methods are usually based on the idea of completing the plane by "elements at infinity," thus converting it into a compact manifold, and then examining the behavior of the paths in the neighborhood of the elements at infinity.

The reader probably knows that there are various means to this end. In this subsection we shall study the so-called Bendixson sphere and Bendixson transformation.

Let

$$X^2 + Y^2 + Z^2 = 1 \tag{1}$$

be the unit sphere in $R^3$, $N\,(0,\,0,\,1)$ and $S\,(0,\,0,\,-1)$ its north and south poles, respectively. We shall identify the phase plane $(x,\,y)$ with the plane $\alpha$ tangent to the sphere $\Sigma$ at the pole $S$ (its equation is $z = -1$). Simultaneously, we shall consider the $(u,\,v)$-plane, tangent to $\Sigma$ at $N$, which we denote by $\alpha'$ (Figure 133). The origins on the planes $\alpha$ and $\alpha'$ are $S$ and $N$, respectively, the $y$- and $v$-axes coincide in direction with the $Y$-axis, and the $x$- and $u$-axes with the $X$-axis.

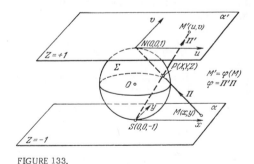

FIGURE 133.

Let $\Pi$ be the stereographic projection of the plane $\alpha$ onto the sphere $\Sigma$ from $N$. $\Pi$ is obviously a topological mapping of $\alpha$ onto the sphere $\Sigma$ punctured at $N$. Any path of system (I) is mapped onto a curve on $\Sigma$,

which we shall call a path on the sphere, though we shall not regard
such paths as paths of some dynamic system defined on the sphere but
simply as images of paths of system (I). To make the projection a mapping
of the plane onto the sphere, we complete the former by adding one "point
at infinity," corresponding to the pole. We shall then refer to the sphere as
the "Bendixson sphere," and investigation of the behavior of paths of
system (I) at infinity is thus reduced to investigation of paths on the sphere
in the neighborhood of the pole $N$.

To this end, we again consider a stereographic projection, but now as a
mapping $\Pi'$ projecting the sphere $\Sigma$ onto the plane $\alpha'$ from $S$ as center.
Consider the mapping $\varphi = \Pi'\Pi$. This is a topological mapping of the plane
$\alpha$ punctured at $S$ onto the plane $\alpha'$ punctured at $N$. The mapping $\varphi$ takes
a point $M(x, y)$ on $\alpha$ onto the point $\varphi(M) = \Pi'(\Pi(M)) = \Pi'(P) = M'(u, v)$
on $\alpha'$ (Figure 133).

Using the obvious relations

$$\frac{X}{x} = \frac{Y}{y} = \frac{Z-1}{-2}, \qquad \frac{u}{X} = \frac{v}{Y} = \frac{2}{Z+1}$$

and equation (1), we obtain, after simple manipulations,

$$x = \frac{4u}{u^2+v^2}, \qquad y = \frac{4v}{u^2+v^2}. \tag{2}$$

$$u = \frac{4x}{x^2+y^2}, \qquad v = \frac{4y}{x^2+y^2}. \tag{3}$$

The transformation (2) (or the equivalent transformation (3)) is known as
the Bendixson transformation. Applying it to system (I), we get

$$\frac{du}{dt} = \frac{1}{4}(v^2-u^2)\cdot P\left(\frac{4u}{u^2+v^2}, \frac{4v}{u^2+v^2}\right) - \frac{1}{2}uvQ\left(\frac{4u}{u^2+v^2}, \frac{4v}{u^2+v^2}\right), \tag{A}$$

$$\frac{dv}{dt} = -\frac{1}{2}uvP\left(\frac{4u}{u^2+v^2}, \frac{4v}{u^2+v^2}\right) + \frac{1}{4}(u^2-v^2)\cdot Q\left(\frac{4u}{u^2+v^2}, \frac{4v}{u^2+v^2}\right).$$

Suppose first that (A) is algebraic, i. e., $P$ and $Q$ are polynomials. Then
equations (A) have the form

$$\frac{du}{dt} = \frac{\Phi(u, v)}{(u^2+v^2)^m}, \qquad \frac{dv}{dt} = \frac{\Psi(u, v)}{(u^2+v^2)^m},$$

where $\Phi$ and $\Psi$ are polynomials, not both divisible by $u^2+v^2$, and $m$ is a
nonnegative integer.

System (A) is generally undefined at the point $(0, 0)$. Consider instead
the system

$$\frac{du}{dt} = \Phi(u, v), \qquad \frac{dv}{dt} = \Psi(u, v). \tag{B}$$

On the punctured plane $\alpha'$, the paths of system (A) coincide with those of
system (B). But system (B) is now defined throughout the plane $\alpha'$, including the origin. Hence, analyzing the configuration of the paths of system (B)
in some neighborhood of the origin and applying the mapping $\varphi^{-1} = \Pi^{-1}(\Pi')^{-1}$,
we shall be able to draw certain conclusions about the behavior of the
paths of system (A) at infinity.

We append a few remarks concerning the Bendixson transformation.

R e m a r k  1.  We have completed the $(x, y)$-plane by a  s i n g l e  "improper" point (point at infinity), whose image under the stereographic projection Π is the north pole $N$ of the sphere. The model of the completed plane is the sphere Σ itself; one then says that system (I) is being studied on the Bendixson sphere.*

R e m a r k  2.  Do the systems (I) and (A) define a dynamic system on the sphere in the sense of §2? Here we are dealing with a simple atlas on the sphere, containing two charts (see Appendix, §7). Equations (I) refer to the first chart (the sphere minus the north pole), equations (A) to the other. It is clear that the result is a dynamic system on the sphere if and only if the definition of the functions $\frac{\Phi(u,v)}{(u^2+v^2)^m}$, $\frac{\Psi(u,v)}{(u^2+v^2)^m}$ can be extended at $u = 0$, $v = 0$ so that they become functions of class $C_1$. This is always possible if $m = 0$, but simple examples show that if $m > 0$ this is generally impossible. Thus systems (I) and (A) do not always define a dynamic system on the sphere.

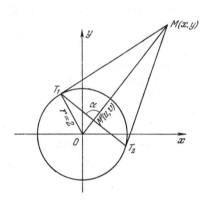

FIGURE 134.

R e m a r k  3.  The Bendixson trans-formation may be derived in a purely plane setting. Indeed, consider the  i n v e r s i o n  o f  t h e  $(x, y)$-p l a n e  in the circle $x^2+y^2=r^2$. Under this mapping, each point $M(x, y)$ on the plane goes into a point $M'(u, v)$ on the same plane, such that $M$ and $M'$ lie on a ray issuing from the origin $O$ and

$$OM \cdot OM' = r^2$$

(Figure 134). It is readily seen that when $r = 2$ the relation between $x, y$ and $u, v$ is given precisely by (2), so that this inversion is the Bendixson transformation.

R e m a r k  4.  Both the Bendixson transformation and the Poincaré transformation to be considered in the next subsection are convenient only when system (1) is algebraic ($P$ and $Q$ polynomials). The origin $(0, 0)$ is a regular point of the system (B) obtained by the transformation. However, since every path of system (I) which goes to infinity as $t \to + \infty$ or $t \to - \infty$ becomes a path of system (B) tending to $(0, 0)$, the Bendixson transformation usually gives rise to an equilibrium state, of quite complicated structure, which is not easy to investigate. The transformation thus yields good results only rarely, and the  P o i n c a r é  t r a n s f o r m a -t i o n, which we are now going to describe, is more convenient.

**2. CONSIDERATION OF ALGEBRAIC DYNAMIC SYSTEMS ON THE POINCARÉ SPHERE.** We now describe another device for completing the plane by elements at infinity, based on a different projection (no longer one-to-one) of the plane onto the sphere. This new device is also applicable when $P$ and $Q$ are polynomials.**

---

* In the theory of analytic functions the plane is also completed by a single point at infinity, and the model is usually called the  R i e m a n n  s p h e r e.

** The same device may also be used for the study of algebraic curves on the plane.

As before, we consider the unit sphere

$$X^2 + Y^2 + Z^2 = 1$$

about the origin $O$ and identify the phase plane $(x, y)$ with the tangent plane $\alpha$ to the sphere at its south pole $S$.

Let $M$ be any point of the $(x, y)$-plane. Draw a straight line through $M$ and the center $O$ of the sphere, cutting the sphere at two antipodal points $M'$ and $M''$. Conversely, any pair of antipodal points, with the sole exception of points lying on the great circle in the plane $Z = 0$ (parallel to the $(x, y)$-plane) defines exactly one point on the $(x, y)$-plane (Figure 135). The great circle in the plane $Z = 0$ is known as the equator and the plane $Z = 0$ as the equatorial plane. The points of the equator represent the "points at infinity" of the plane.

Any two antipodal points not on the equator correspond to the same point of the plane. It is therefore only natural to regard any two antipodal points of the equator as corresponding to the same "point at infinity."*

Related to the plane in this way, the sphere will be called the P o i n c a r é s p h e r e. Note that whereas consideration of the Bendixson sphere was equivalent to addition of one element at infinity to the plane (corresponding to the north pole $N$), in this case we have added an infinite set of elements at infinity, corresponding to the pairwise identified points of the equator.

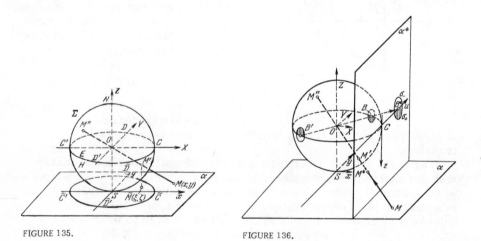

FIGURE 135.                    FIGURE 136.

The above mapping of the plane onto the sphere takes straight lines on the $(x, y)$-plane onto great circles of the sphere, straight lines through the origin being mapped onto great circles perpendicular to the equator. The paths of a dynamic system (I) are mapped onto certain curves on the sphere, which we shall again call "paths of the dynamic system (I)" or simply "paths on the sphere."

---

* This extension of the plane by "points at infinity" leads us to identify every two antipodal points on the sphere (since they correspond to the same point of the plane). It is well known that this yields a manifold which is one of the models for the projective plane.

In order to determine the behavior of the paths in the neighborhood of the equator and on the equator itself, we shall employ the following geometric construction.

Let $C$ and $C'$ be the points at which the equator $E$ cuts the $X$-axis, and $D$ and $D'$ the points at which it cuts the $Y$-axis (Figure 135).

Consider the plane $X = 1$ touching the sphere $\Sigma$ at $C$; denote this plane by $\alpha^*$. We introduce coordinates $u$ and $z$ on this plane, with origin at $C$ and $u$- and $z$-axes parallel to the $Y$- and $Z$-axes, respectively. The $u$-axis is assumed to point in the same direction as the $Y$-axis, the $z$-axis in the opposite direction to the $Z$-axis (i.e., downward; see Figure 136). Let $M'$ and $M''$ be two antipodal points, not in the plane $X = 0$ parallel to the plane $\alpha^*$. (This plane lies in the same position relative to $\alpha^*$ as does the equatorial plane relative to the original plane $\alpha$.)

Using a central projection, map $M'$ and $M''$ onto a certain point $M^*$ of the plane $\alpha^*$.

If $M'$ and $M''$ are the images under central projection of a point $M$ of $\alpha$, not lying on the $y$-axis, it is clear that the point $M^*$ is always defined; in this case $M'$ and $M''$ are not on the equator of the sphere. But if $M'$ and $M''$ are antipodal points on the equator $E$ of the sphere, other than $D$ and $D'$, they also determine a well-defined point of $\alpha^*$, on the axis $z = 0$.

This central projection of the sphere onto the plane $\alpha^*$ is a natural aid to investigation of paths in the neighborhood of any point of the equator other than $D$ and $D'$.

To study the neighborhood of $D$ and $D'$, we consider instead of $\alpha^*$ a plane $\hat{\alpha}$ tangent to the sphere $\Sigma$ at $D$, treating it in exactly the same way as the plane $\alpha^*$. The coordinates in $\hat{\alpha}$ will be denoted by $v$ and $z$, the $v$-axis coinciding in direction with the $X$-axis and the $z$-axis pointing downward.

We now return to the plane $\alpha^*$.

Let $M$ be a point on the plane $\alpha$, not on the $y$-axis, and $M^*$ the corresponding point on $\alpha^*$. Since $M$ and $M^*$ have space coordinates $(x, y, -1)$ and $(1, u, -z)$, respectively, and lie on the same straight line through $O(0, 0)$, we have

$$\frac{x}{y} = \frac{y}{u} = \frac{1}{z}, \tag{4}$$

whence

$$x = \frac{1}{z}, \qquad y = \frac{u}{z} \tag{5}$$

or

$$u = \frac{y}{x}, \qquad z = \frac{1}{x}.$$

For the plane $\hat{\alpha}$ (tangent to the sphere $\Sigma$ at $D$) we get analogous relations between the coordinates $x$, $y$ and $z$, $v$:

$$x = \frac{v}{z}, \qquad y = \frac{1}{z}$$

or

$$v = \frac{x}{y}, \qquad z = \frac{1}{y}. \tag{6}$$

The transformations (5) and (6) are known as Poincaré transformations. Applying transformation (5) to system (I), we obtain

$$\frac{du}{dt} = -uzP\left(\frac{1}{z}, \frac{u}{z}\right) + zQ\left(\frac{1}{z}, \frac{u}{z}\right), \qquad \frac{dz}{dt} = -z^2 P\left(\frac{1}{z}, \frac{u}{z}\right). \tag{$A_1$}$$

For $z \neq 0$ (i.e., points not on the equator), the paths of this system are
obviously projections of paths on the sphere or, what is the same,
projections (through the center $O$) of the paths of system (I) on the plane $\alpha$.

By assumption, the functions $P(x, y)$ and $Q(x, y)$ on the right of system (I)
are polynomials. Let $n$ be the higher of their degrees. Then we can write
system $(A_1)$ as

$$\frac{du}{dt} = \frac{P^*(u, z)}{z^n}, \qquad \frac{dz}{dt} = \frac{Q^*(u, z)}{z^n}. \tag{$A_2$}$$

System $(A_1)$ (or, equivalently, $(A_2)$), is undefined at $z = 0$. We consider
instead the system

$$\frac{du}{d\tau} = P^*(u, z), \qquad \frac{dv}{d\tau} = Q^*(u, z), \tag{$A_3$}$$

obtained from system $(A_2)$ by the substitution

$$\frac{dt}{z^n} = d\tau.$$

Note that the functions $P^*(u, z)$ and $Q^*(u, z)$ on the right of system $(A_3)$ are
also polynomials. If $z \neq 0$, the paths of system $(A_3)$ coincide with those of
system $(A_2)$. Moreover, if $n$ is even, the direction along paths remains
unchanged upon passage from system $(A_2)$ to system $(A_3)$. If $n$ is odd the
direction on paths lying in the half-plane $z < 0$ is reversed.

System $(A_3)$ is now defined for $z = 0$, i.e., throughout the plane $\alpha^*$. It is
readily shown that if the polynomials $P$ and $Q$ are relatively prime, the
same is true of $P^*$ and $Q^*$. In this case the number of singular points of
system (I) on the sphere is finite.

We shall adopt the convention that in the neighborhood of points of the
equator other than $D$ and $D'$ the paths on the sphere (or the paths of
system (I)) are the projections (through $O$) of the corresponding paths of
system $(A_3)$ on the plane $\alpha^*$.

In particular, the singular points on the equator (other than $D$ and $D'$)
are the projections of the singular points (equilibrium states) of system $(A_3)$
on the axis $z = 0$.

Thus, the singular points of system (I) on the equator (other than $D$ and
$D'$) are the projections of the points on the $(u, z)$-plane satisfying the
equations

$$P^*(0, u) = 0, \qquad Q^*(0, u) = 0. \tag{7}$$

In the case under consideration ($P$ ($x$, $y$) and $Q$ ($x$, $y$) are polynomials), the isoclines $P$ ($x$, $y$) $= 0$, $Q$ ($x$, $y$) $= 0$ can be "transferred" in a natural way to the sphere. All common points of these isoclines, both on and off the equator, are singular points of system (I) on the sphere.

However, it is readily seen by direct examination of the expressions for $P^*$ ($z$, $u$), $Q^*$ ($z$, $u$) that there may be singular points of system (I) on the equator which are not common points of isoclines. To illustrate this at first sight paradoxical assertion, we consider the dynamic system

$$\frac{dx}{dt} = 0, \qquad \frac{dy}{dt} = b. \tag{8}$$

where $b \neq 0$ is a constant. In this case, of course, the functions $P$ ($x$, $y$) $= 0$ and $Q$ ($x$, $y$) $= b$ never vanish together. The paths of the system are the straight lines $x = C$. These correspond on the sphere to great circles intersecting at two antipodal points of the equator, which are singular points of system (8). The same conclusions follow from consideration of the functions $P^*$ ($z$, $u$) and $Q^*$ ($z$, $u$) corresponding to system (8).

In order to examine the neighborhoods of the points $D$ and $D'$, we must use transformation (6). Applying this transformation to system (I) we get a system similar to ($A_1$):

$$\frac{dv}{dt} = zP\left(\frac{v}{z}, \frac{1}{z}\right) - zvQ\left(\frac{v}{z}, \frac{1}{z}\right), \qquad \frac{dz}{dt} = -z^2Q\left(\frac{v}{z}, \frac{1}{z}\right). \tag{$\hat{A}_1$}$$

Reducing the expressions on the right to a common denominator, we write this system as

$$\frac{dv}{dt} = \frac{\hat{P}(v, z)}{z^n}, \qquad \frac{dz}{dt} = \frac{\hat{Q}(v, z)}{z^n}. \tag{$\hat{A}_2$}$$

Finally, the substitution $\frac{dt}{z^n} = d\tau$ yields

$$\frac{dv}{d\tau} = \hat{P}(z, v), \qquad \frac{dz}{d\tau} = \hat{Q}(z, v), \tag{$\hat{A}_3$}$$

where $\hat{P}$ ($z$, $v$) and $\hat{Q}$ ($z$, $v$) are polynomials. Whereas system ($\hat{A}_1$) is undefined at $z = 0$, this new system is defined there (hence throughout the plane $\hat{a}$). Using system ($\hat{A}_3$), one can define the paths on the sphere in the neighborhood of $D$ and $D'$.

To summarize: by studying systems ($A_3$) and ($\hat{A}_3$) in the neighborhood of $z = 0$ one can determine the behavior of the paths in the neighborhood of all points of the equator.

As a rule, the polynomials $Q^*$ ($u$, $z$) and $\hat{Q}$ ($z$, $v$) (in systems ($A_3$) and ($\hat{A}_3$)) contain $z$ as a factor, so that the axis $z = 0$ is a union of equilibrium states and paths. Therefore, as a rule, the equator is also a union of equilibrium states and paths.*

---

* The reader will readily convince himself that, as in the case of the Bendixson sphere, a dynamic system considered on the Poincaré sphere is not a dynamic system on the sphere in the sense of §2. In particular, a point moving along a path may tend to a singular point on the equator as $t$ tends to a finite value.

The central projection of the plane $\alpha$ onto the Poincaré sphere which we are considering here associates each point of $\alpha$ with two antipodal points of the sphere. In order to avoid this duality, we can confine attention to a hemisphere (together with the equator), say the lower hemisphere, which we denote by $\overline{H}$. The mapping of $\alpha$ onto the open hemisphere $H$ is topological; antipodal points on the equator are identified. Let us now project the hemisphere $\overline{H}$ orthogonally onto the plane. We get a mapping (clearly topological) of the hemisphere onto the unit disk $\overline{K}$ on $\alpha$ with center at $S$. The equator of the sphere is mapped onto the circle $\Gamma$ bounding $\overline{K}$, and paths on the sphere are mapped onto certain curves which we shall call paths in the disk $\overline{K}$. In particular, singular points on $\overline{H}$ are mapped onto singular points in $\overline{K}$.

The disk $\overline{K}$, together with the corresponding phase portrait, constitutes an extremely convenient model of the dynamic system (I) defined on the plane (extended by the addition of elements at infinity). To each point $M$ on the plane corresponds (in a one-to-one manner) a point $M'$ interior to the circle $\Gamma$. The points of $\Gamma$ (the images of the points on the equator of the sphere) correspond to the points "at infinity" on the plane. Antipodal points on the circle $\Gamma$ correspond to antipodal points of the equator (Figures 135 and 137).

Some remarks are in order concerning investigation of the behavior of paths in the vicinity of points of $\Gamma$.

Let $\widetilde{B}$ and $\widetilde{B'}$ be any two antipodal points of $\Gamma$, which are the images of two antipodal points $B$ and $B'$ on the equator (to fix ideas, assume that $B$ and $B'$ are not $D$ and $D'$) (Figures 136 and 137).

Let $B^*$ be the point of the plane $\alpha^*$ (on the axis $z = 0$) corresponding to $B$ and $B'$, and $\sigma$ some neighborhood of $B^*$. The axis $z = 0$ divides this neighborhood into two half-neighborhoods $\sigma_+$ (in the half-plane $z > 0$) and $\sigma_-$ (in the half-plane $z < 0$). The image of $\sigma_+$ on the lower hemisphere is a half-neighborhood of the point $B$, say, and that of $\sigma_-$ a half-neighborhood of $B'$ (Figure 136).

Since $x = \frac{1}{z}$, when the hemisphere is mapped onto the disk $\overline{K}$ the image of the half-neighborhood $\sigma_+$ is a half-neighborhood of $\widetilde{B}$ corresponding to $x > 0$, and that of $\sigma_-$ a half-neighborhood of $\widetilde{B'}$ (corresponding to $x < 0$) (Figure 137).

Thus, when we investigate the phase portrait in the vicinity of points of the circle $\Gamma$, which is equivalent to examining the neighborhood of points on the equator (i.e., the neighborhood of points on the $z$-axis in the plane $\alpha^*$ or $\alpha$), we are actually identifying antipodal points of the circle $\Gamma$.

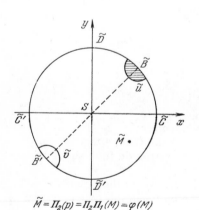

$$\widetilde{M} = \Pi_2(p) = \Pi_2\Pi_1(M) = \varphi(M)$$

FIGURE 137.

A disk with antipodal points on the circumference identified (in the same way as on the sphere) is a well-known model of the projective plane. Each pair of antipodal points on the circumference, once identified, may be interpreted as the point at infinity corresponding to the direction of a straight line on the $(x, y)$-plane. However, if we use the disk $\overline{K}$ as a model of the plane, without identifying antipodal points of $\Gamma$, we can interpret each point of the circumference $\Gamma$ as a point at infinity corresponding to the direction of a ray on the $(x, y)$-plane. Since there are two directions on any straight line on the plane, this interpretation associates two points at infinity with each straight line.

Throughout the sequel, when considering dynamic systems defined on the plane whose right-hand sides are polynomials, we shall use the representation by paths in the disk $\overline{K}$.

We can summarize the foregoing discussion in a scheme (or algorithm) for investigation of dynamic systems at infinity.

Investigation scheme. Given: a dynamic system (A):

$$\frac{dx}{dt} = P(x, y), \qquad \frac{dy}{dt} = Q(x, y),$$

where $P$ and $Q$ are relatively prime polynomials.

1) Apply the Poincaré transformation

$$x = \frac{1}{z}, \qquad y = \frac{u}{z}.$$

The result is a system

$$\frac{du}{dt} = -uzP\left(\frac{1}{z}, \frac{u}{z}\right) + zQ\left(\frac{1}{z}, \frac{u}{z}\right) = \frac{P^*(u, z)}{z^n},$$

$$\frac{dz}{dt} = -z^2 P\left(\frac{1}{z}, \frac{u}{z}\right) = \frac{Q^*(u, z)}{z^n},$$

where $n$ is the smallest nonnegative integer for which this representation is possible.

2) Consider system (A$_3$):

$$\frac{du}{dt} = P^*(u, z), \qquad \frac{dz}{dt} = Q^*(u, z).$$

Find all equilibrium states of this system on the axis $z = 0$ (there is only a finite number). Investigate the local phase portrait of system (A$_3$) in the neighborhood of each such equilibrium state $B^*(u, 0)$.

3) For each point $B^*(u, 0)$, construct the corresponding two points $\tilde{B}$ and $\tilde{B}'$ on the circle $\Gamma$; these are the points at which the straight line $y/x = u$ cuts the circle.

Suppose that the abscissa of $\tilde{B}$ is positive, that of $\tilde{B}'$ negative.

Map a half-neighborhood $\sigma_+$ of the point $B^*$ and the paths therein onto a half-neighborhood $\tilde{\sigma}_+$ of the point $\tilde{B}$ in the disk $K$, as illustrated in Figure 138, and the half-neighborhood $\sigma_-$ onto a half-neighborhood $\tilde{\sigma}_-$ of the point $\tilde{B}'$.

4) If $n$ is odd, the directions on the paths are reversed when the half-neighborhood $\sigma_-$ is mapped onto $\tilde{\sigma}_-$.

5) Apply the Poincaré transformation

$$x = \frac{v}{z}, \qquad y = \frac{1}{z}$$

obtaining the system

$$\frac{dv}{dt} = zP\left(\frac{v}{z}, \frac{1}{z}\right) - zvQ\left(\frac{v}{z}, \frac{1}{z}\right) = \frac{\hat{P}(v, z)}{z^m},$$

$$\frac{dz}{dt} = -z^2Q\left(\frac{v}{z}, \frac{1}{z}\right) = \frac{\hat{Q}(v, z)}{z^m}$$

and then consider the system $(\hat{A}_3)$:

$$\frac{dv}{dt} = \hat{P}(v, z), \qquad \frac{dz}{dt} = \hat{Q}(v, z).$$

6) Investigate the phase portrait of system $(\hat{A}_3)$ in the neighborhood of $D(0, 0)$. Map a half-neighborhood $\hat{\sigma}_+$ of $D$ onto a half-neighborhood $\tilde{\sigma}_+$ of the point $\tilde{D}$ in the disk $K$, as illustrated in Figure 139, and the half-neighborhood $\hat{\sigma}_-$ onto a half-neighborhood $\tilde{\sigma}_-$ of the point $\tilde{D}'$.

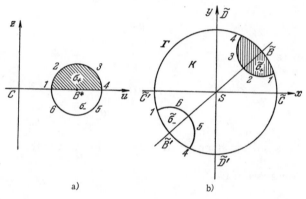

a)                    b)

FIGURE 138.

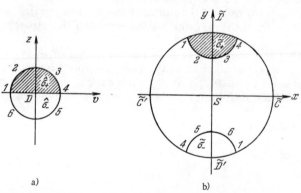

a)                    b)

FIGURE 139.

7) If $m$ is odd, the directions on the paths are reversed when $\hat{\sigma}_-$ is mapped onto $\tilde{\sigma}_-$.

R e m a r k. As we have already indicated, the axis $z = 0$ usually consists of paths of system $(A_3)$, and the corresponding arcs on the circle $\Gamma$ may be treated as paths in the disk $\overline{K}$. In certain exceptional cases, however, this may not be true. For example, it is easily seen that the system

$$\frac{dx}{dt} = x^2 y + x + y, \qquad \frac{dy}{dt} = xy^2 + x - y$$

corresponds to the system

$$\frac{du}{dt} = z - zu^2 - 2uz, \qquad \frac{dz}{dt} = -u - z^2 - uz^2.$$

This system has a single equilibrium state $(0, 0)$ on $z = 0$. At all other points its paths cross the axis at right angles.

**3. SAMPLE INVESTIGATION OF THE EQUATOR.** Consider as an example the behavior at infinity of the paths of the system

$$\frac{dx}{dt} = x(3 - x - ny), \qquad \frac{dy}{dt} = y(-1 + x + y), \tag{9}$$

where $n > 3$.

Applying the Poincaré transformation $x = \frac{1}{z}$, $y = \frac{u}{z}$, we get

$$\frac{du}{dt} = \frac{2u - 4uz + (n+1)u^2}{z}, \qquad \frac{dz}{dt} = 1 + n \cdot u - 3z. \tag{10}$$

Multiplying the right-hand sides by $z$, we have

$$\frac{du}{dt} = 2u - 4uz + (n+1)u^2, \qquad \frac{dz}{dt} = z + nuz - 3z^2. \tag{11}$$

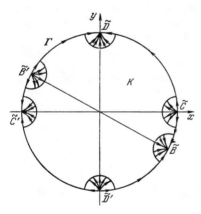

FIGURE 140.

The system has two equilibrium states on the axis $z = 0$ (corresponding to the equator of the Poincaré sphere): $C(0, 0)$ (unstable node) and $B(-2/n + 1, 0)$ (stable node). The axis $z = 0$ is a union of paths.

To investigate the points $D$ and $D'$, we use the second Poincaré transformation $x = v/z$, $y = 1/z$, obtaining the system

$$\frac{dv}{dt} = \frac{4vz - 2v^2 - (n+1)v}{z}, \qquad \frac{dz}{dt} = z - v - 1, \tag{12}$$

and then, multiplying by $z$, the system

227

$$\frac{dv}{dt} = 4vz - 2v^2 - (n-1)\,v,$$

$$\frac{dz}{dt} = z\,(z-v-1). \tag{13}$$

The point $D(0,\,0)$ is a stable node for this system. In view of the fact that the passage from systems $(10)$ and $(12)$ to systems $(11)$ and $(13)$ involved multiplication of the right-hand sides by $z$ (see clauses 4 and 7 of the investigation scheme), the phase portrait near the circle $\Gamma$ (or as one usually says, near the equator) is as illustrated in Figure 140.

## § 14. USE OF NUMERICAL METHODS TO DETERMINE THE QUALITATIVE STRUCTURE OF THE PHASE PORTRAIT

**1. GENERAL REMARKS.** In view of the lack of regular, effective methods applicable to determination of topological structure in all cases, and the rather poor practical performance of the various devices discussed in §12, it is natural to appeal to numerical (approximate) methods.

Of course, the use of numerical methods of integration requires prior knowledge of numerical values for all parameters in the system or, at least, for combinations of these parameters. At the same time, the dynamic systems occurring in applications always involve certain parameters which may take on different values. This need for specification of parameters obstructs one's view of the system as a whole. For this reason, analytical methods, however complicated they may be, are always preferable (when available) to the approximate methods of numerical integration. Nevertheless, there are cases in which the use of numerical integration is the only possible way of acquiring information about the topological structure of the phase portrait of a dynamic system. It should be emphasized that in this context one is not interested in computing the paths over some time interval as an end in itself (though this is of course the ultimate goal in many problems), but rather as a means toward elucidating the qualitative structure of the phase portrait or, at least, deriving certain qualitative characteristics of the phase portrait.

We shall first describe here in brief a quite widespread method of graphical integration, known as the "isocline method," which consists in approximate construction of a "network" of paths.*

**2. METHOD OF ISOCLINES.** The construction proceeds as follows. Given a system (in which all parameters have definite numerical values)

$$\frac{dx}{dt} = P\,(x,\,y), \qquad \frac{dy}{dt} = Q\,(x,\,y), \tag{I}$$

we construct a network of isoclines

$$P\,(x,\,y) + CQ\,(x,\,y) = 0, \qquad C'Q\,(x,\,y) + P\,(x,\,y) = 0,$$

---

* A more detailed exposition may be found, e. g., in /6/.

228

giving the constant $C$ (or $C'$) a sufficiently dense sequence of different numerical values. As we know, the integral curves of system (I) have the same slope at all points of an isocline. In §1 we constructed the family of isoclines for a few simple examples (Figures 8, 12, 16). The points of intersection of any two isoclines are obviously points of intersections of all isoclines — equilibrium states of system (I).

Having constructed a sufficiently fine network of isoclines, we can approximate the paths of the system. We begin the construction from some point $P$ (which should not, of course, be an equilibrium state) (Figure 141). To fix ideas, suppose that $P$ is on the isocline $C = 0$. Draw two segments from $P$, one in the direction of the tangent corresponding to $C = 0$ (i.e., parallel to the $x$-axis) and the other in the direction of the tangent corresponding to the "next" value of $C$, say $C = -0.2$. Extend both segments till they cut the isocline $C = -0.2$, at points $a_1$ and $b_1$ say. On the isocline $C = -0.2$, select a point $P_1$ equidistant from $a_1$ and $b_1$. This is taken as the "next" point of the path through $P$. Now draw two straight lines through $P_1$, at angles corresponding to the isoclines $C = -0.2$ and $C = -0.4$, cutting the isocline $C = -0.4$ at points $a_2$ and $b_2$. The point $P_2$ on the isocline $C = -0.4$ equidistant from $a_2$, $b_2$ will be the "next" point of our path. Continuing in this way, we get a sequence of points $P$, $P_1$, $P_2$, $P_3$, . . . Joining these points by straight-line segments, we get a polygonal line which approximates part of the path through $P$. We can thus continue the construction of the path and also construct other paths, finally obtaining a "network" of paths.

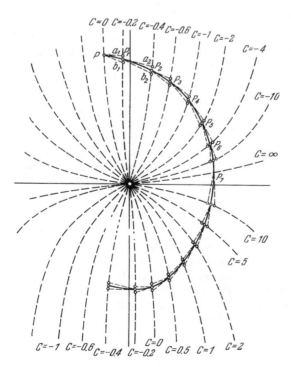

FIGURE 141.

Construction of such a network will sometimes enable one to "hunt" for limit cycles and also to "guess" at the position of separatrices.

If the system involves parameters, we assign them various values and construct for each value an approximate phase portrait. One can thus obtain a veritable "gallery" of phase portraits of the system.

The above construction of "networks" of paths provides no error estimates. Such estimates can clearly be given, but we shall not touch upon this subject here. Suffice it to say that instead of the above geometric method (isocline method) one can apply some method of numerical integration for which error estimates are available, such as the Adams or Runge methods. These methods are beyond the scope of our book; there is an extensive literature on the subject (see /42/, /43/, /44/). Also worthy of mention is the construction of an " ɛ-net" of paths (see /40/ and /41/), which provides error estimates. This construction will not be described here, since qualitative investigation of dynamic systems is more conveniently tackled by the alternative approach presented below.

**3. SPECIFIC CHARACTER OF NUMERICAL METHODS APPLIED TO DETERMINATION OF QUALITATIVE STRUCTURE OF PHASE PORTRAITS.** In none of the approximate constructions described above (networks of paths, ɛ-nets, and so on) were the elements required for knowledge of the qualitative structure determined directly. In particular, the coordinates and types of equilibrium states were not determined directly, but by a rather indirect approach, via the intersection points of the isoclines and the network of paths. Now we know that when the equilibrium states are simple and their characteristic roots have nonzero real parts, their coordinates and types may be determined quite simply by direct computation rather than construction of a network of paths. To this end, one first computes the coordinates of the equilibrium states by finding the (approximate) common points of the curves

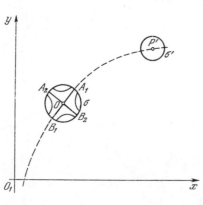

$$P(x, y) = 0, \qquad Q(x, y) = 0,$$

and then determines the characteristic roots. This at once relieves us of the need to construct a network of paths in the neighborhood of an equilibrium state (a construction whose precision is far from satisfactory). Neither is there any need to construct such networks in order to gain information about the behavior of separatrices — it suffices to approximate the separatrices themselves.

Let us indicate in general outline just how this is done.

Let $O$ be a saddle point (whose coordinates have been computed).

Using formula (4) of §9.2, we can compute the directions $\varkappa_1$ and $\varkappa_2$ along which the separatrices tend to the saddle point $O$. Considering a sufficiently small neighborhood of the saddle point, say a disk σ of sufficiently small radius about $O$, we can replace the portions of the separatrices inside σ by segments of straight lines through $O$ with slopes $\varkappa_1$ and $\varkappa_2$. Let $A_1$, $A_2$, $B_1$ and $B_2$ be the points at which these segments cut the circumference of σ (Figure 142).

Starting from the points $A_1$, $A_2$, $B_1$, $B_2$, we approximately construct (compute) the paths through these points and thus obtain an approximation to the true behavior of the separatrices.

In particular, suppose that one of the separatrices, $L_1$ say, which has been replaced inside σ by the segment $OA_1$, tends to a node or a focus. We can always single out a region σ' which acts as a "region of attraction" for the node (focus). Then, by approximate computation of the arc of the path (separatrix) beginning from $A_1$, we can "lead" it right up to σ' and thus fully determine the position of the separatrix $L_1$.

There are methods for estimating the errors involved in this construction of separatrices, but we shall not dwell upon them here. A few more words, concerning the possible construction of "annuli" bounded by turns of paths and containing limit cycles: in the next subsection we shall consider a simple but special case in which approximate construction of paths demonstrates the existence of a limit cycle.

**4. CASES IN WHICH EXISTENCE OF A LIMIT CYCLE CAN BE PROVED BY APPROXI-MATION OF ARCS OF PATHS.** We shall adopt the following assumption, which is not too restrictive: there exists a curve passing through all the equilibrium states which has no contact with paths at points other than equilibrium states. Denote this curve by $\Lambda$. In particular, it may be an isocline, and then the condition that it have no contact with paths at regular points admits the following analytical expression.

Let

$$P(x, y) + CQ(x, y) = 0 \tag{1}$$

be the isocline of slope $C$. If the tangent to this isocline never has the slope $C$, then, obviously, the expression

$$(P'_x + CQ'_x) P + (P'_y + CQ'_y) Q$$

cannot vanish.

Suppose we are considering a portion of the curve $\Lambda$ containing no equilibrium states, which is moreover a simple arc. We may assume this arc parametrized in one-to-one fashion by some parameter $s$.

Suppose that approximation of arcs of paths through points of $\Lambda$ over some sufficiently long portion of the latter shows that the paths cross $\Lambda$ twice. In addition, assume that

1) the approximated arc of some path $L_1$ passing through a point $A_1$ on $\Lambda$ (corresponding to parameter value $s_1$) crosses $\Lambda$ a second time at some point $A_2$ corresponding to $s_2$, such that

$$s_2 > s_1; \tag{2}$$

2) the approximated arc of some path $L'_1$ passing through a point $A'_1$ on $\Lambda$, corresponding to parameter values $s'_1 > s_2$, crosses $\Lambda$ a second time at some point $A'_2$ with parameter $s'_2$, such that

$$s'_2 < s'_1. \tag{3}$$

It is easy to see that under these assumptions the points $A_2$ and $A'_2$ lie on $\Lambda$ between the points $A_1$ and $A'_1$ (Figure 143).

Now denote the parameter value at which the true (not approximate) arc of the path $L_1$ crosses $\Lambda$ for a second time by $s^*_2$, and the analogous parameter value for the true path $L'_1$ by $s^{**}_2$.

Suppose that the precision of the computation is such that inequalities (2) and (3) guarantee the "true inequalities"

$$s^*_2 > s_1, \qquad s^{**}_2 < s'_1. \tag{4}$$

Suppose moreover that the region $\Sigma$ bounded by the simple closed curve which is the union of the arc $A_1A_2$ of $L_1$ and the arc $A_1A_2$ of $\Lambda$, and by the simple closed curve which is the union of the arc $A'_1A'_2$ of $L'_1$ and the arc $A'_1A'_2$ of $\Lambda$, is a bounded annular region. If there are no equilibrium states within this region (as can be verified directly from the computed coordinates of equilibrium states and the approximate paths), it follows from (4), via Theorem 13, that there is at least one limit cycle in the annular region $\Sigma$. Even more: if we assume that none of these limit cycles are semistable their total number must be odd.

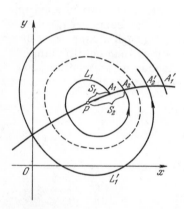

FIGURE 143.

If in addition the point $A_2$ corresponds on the path to a parameter value $t$ greater than that of $A_1$, and the same relation holds for $A'_2$ and $A'_1$, the number of stable limit cycles is greater by 1 than the number of unstable ones.

Thus we see that under the above assumptions approximate computation of arcs of paths enables one actually to prove the existence of at least one limit cycle. Moreover, the construction of a fine network of isoclines and a network of paths may prove superfluous, all information following from the construction of only two arcs of paths.

**5. CASE IN WHICH THE TOPOLOGICAL STRUCTURE OF THE PHASE PORTRAIT IS INDETERMINABLE IN PRINCIPLE BY NUMERICAL METHODS.** As an example, suppose that all paths are closed. In such a case we can never determine by approximation of paths (to within any accuracy) whether these paths are closed or are "slowly winding" spirals.

Other cases may be indicated, which are in a sense more simple. Such is the case, for example, when the system has a semistable limit cycle — a limit cycle with paths winding onto it from one side (i. e., tending to it as $t \to +\infty$ ) and winding off it from the other (tending to it as $t \to -\infty$ )

(Figure 72b). However small the admissible error, one can never establish by approximation of paths whether the system has a semistable limit cycle or whether the paths simply "crowd together" more densely.*

As another example, consider a dynamic system with a saddle point one separatrix of which forms a loop, i.e., issues from the saddle point and returns there (Figure 144a). No degree of precision in approximate construction of the paths, including the separatrix itself, will enable one to distinguish between a true "loop" and the configurations of Figure 144b and c. In other words, the separatrices may "diverge" from each other by a quantity which cannot be detected by any preassigned degree of precision.

The situation is entirely analogous in the configuration of Figure 145 — a "saddle-to-saddle separatrix." No direct approximate computation is capable of differentiating the three configurations illustrated in Figure 145.

a)               b)               c)

FIGURE 144.

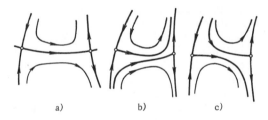

a)               b)               c)

FIGURE 145.

The reader should bear in mind that all the foregoing applies to determination of topological structure by d i r e c t approximation of paths. This by no means implies that there are no i n d i r e c t methods, involving approximations by which one can establish the existence of a double cycle, separatrix loop, and so on.

---

* In the same way as approximate construction of a curve cannot establish whether it is ever tangent to the $x$-axis.

*Chapter VII*

## "SINGULAR" PATHS AND CELLS OF A DYNAMIC SYSTEM*

### INTRODUCTION

Chapter II was devoted to an exhaustive discussion of the possible nature of single paths of a dynamic system of type (I).

In qualitative investigation of dynamic systems, however, one requires information concerning the properties of the phase portrait as a whole, not merely of the paths taken singly. Some propositions of this type were proved in §4 (Theorems 16, 17, 18, 19), but they afford only a rather rough and incomplete idea of the phase portrait as a whole.

We need a more far-reaching and exhaustive study of the possible properties of the entire phase portrait. This is the subject of the present chapter.

Examples (such as the phase portraits of systems (9) and (11), §1.14) indicate that not all paths are "equivalent," that in any phase portrait there are certain paths which are natural candidates for the designation "singular," in contradistinction to the others which are "nonsingular." The paths of this kind encountered hitherto were equilibrium states, limit cycles, and separatrices of saddle points. It is self-evident that a knowledge of the number and configuration of these "singular" paths necessarily plays a fundamental role in establishing the topological structure of the phase portrait.

In this context it is natural to ask: do the above-listed types of singular path exhaust all possibilities? Is there a general definition of "singular" paths, valid for all types? What is their significance vis-à-vis the phase portrait?

The first two questions will be considered in §15. We shall give a general definition of singular and nonsingular paths, valid for paths of any type. According to this definition, a path is singular or nonsingular by virtue not of its own properties but of its behavior in relation to neighboring paths. We shall also determine in §15 all possible types of singular path.

In §16 we adopt the assumption that the number of singular paths of the system is finite. We show that in this case the singular paths divide the phase portrait into finitely many regions, which we call "cells." Each cell

---

* The concepts of orbitally unstable (singular) and orbitally stable (nonsingular) paths were introduced in /46/; they generalize analogous concepts introduced previously by Andronov and Pontryagin /6/ to the case of arbitrary dynamic systems of type (I).

is filled by nonsingular paths whose behavior is "identical," in a sense to be made rigorous below. It is shown that cells may be either simply or doubly connected.

Identification of the singular paths and classification of cells lead to a complete description of the possible types of phase portrait. Among other things, one gains a clearer idea of what paths of a dynamic system play an essential role in determining the topological structure of the phase portrait.

In this chapter we shall confine ourselves to dynamic systems on the plane. However, the entire discussion, definitions and propositions included, carries over to dynamic systems on the sphere.

## §15. ORBITALLY STABLE AND ORBITALLY UNSTABLE PATHS AND SEMIPATHS

**1. FUNDAMENTAL DEFINITIONS.** Assuming that the dynamic system (I) is defined in a region $G$ on the $(x, y)$-plane (possibly the entire plane), we shall consider only bounded semipaths and paths (see §4.1), i. e., semipaths and paths situated in a bounded closed region $\bar{G}_1 \subset G$. This restriction will not be reiterated.

Let $L$ be a path with a bounded positive semipath. Choose some motion on $L$. Let $M$ be the point on $L$ corresponding under this motion to some parameter value $\tau$ and $L_M^+$ the positive semipath on $L$ defined by $t \geqslant \tau$.

*Definition XIV. The path $L$ is said to be $\omega$-orbitally stable at $M$ if for any $\varepsilon > 0$ there exists $\delta > 0$ such that, for any path $L'$ passing at $t = \tau$ through a point $M'$ in the neighborhood $U_\delta(M)$, the positive semipath $L_M^+$ (corresponding to $t \geqslant \tau$) remains within the $\varepsilon$-neighborhood of $L_M^+$.* *

Clearly, if the path $L$ is not $\omega$-orbitally stable at $M$, there exists $\varepsilon_0 > 0$ such that for any $\delta > 0$ there is a path $L'$, passing at $t = \tau$ through a point of $U_\delta(M)$, on which there are points corresponding to $t > \tau$ lying outside the $\varepsilon_0$-neighborhood of $L_M^+$. The definition of $\alpha$-orbital stability of a path at a point $M$ is analogous.

*Definition XV. A path $L$ (with bounded positive semipath) is said to be $\omega$-orbitally stable, or orbitally stable as $t \to +\infty$, if it is $\omega$-orbitally stable at any of its points.*

The definition of $\alpha$-orbital stability (or orbital stability as $t \to -\infty$) is analogous.

*Lemma 1. If a path $L$ is $\omega$-orbitally stable at one point, it is $\omega$-orbitally stable at any other point, i. e., $\omega$-orbitally stable.*

Proof. Suppose that $L$ is $\omega$-orbitally stable at a point $M_1$ which corresponds under some motion on $L$ to time $t = \tau_1$. Let $L_{M_1}^+$ denote the semipath corresponding to $t \geqslant \tau_1$.

Let $M_2$ be any point of $L$, corresponding (under the motion fixed on $L$) to $t = \tau_2$, and $L_{M_2}^+$ the corresponding semipath, corresponding to $t \geqslant \tau_2$. We shall show that $L$ is orbitally stable at $M_2$ as well.

---

* Note that the $\varepsilon$-neighborhood of the semipath $L_M^+$ necessarily contains the $\varepsilon$-neighborhood of the limit points of $L_M^+$.

There are two possibilities.

1. $\tau_2 < \tau_1$. In this case the semipath $L_{M_1}^+$ is part of $L_{M_2}^+$ (Figure 146), and so, by the theorem on continuity of the solution with respect to initial values, the path $L$ is $\omega$-orbitally stable at $M_2$.

2. $\tau_2 > \tau_1$. In this case the semipath $L_{M_2}^+$ is part of $L_{M_1}^+$ (Figure 147).

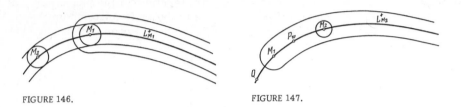

FIGURE 146.  FIGURE 147.

Suppose on the contrary that the path $L$ is not $\omega$-orbitally stable at $M_2$. Then there exists $\varepsilon_0 > 0$ such that for any neighborhood $U_\delta (M_2)$ there is a path passing at $t = \tau_2$ through this neighborhood and leaving the $\varepsilon_0$-neighborhood of $L_{M_2}^+$ at some $t > \tau_2$.

Thus, we can find a sequence of numbers $\{\delta_i\} \underset{i \to \infty}{\to} 0$ and a sequence of paths $\{L_i\}$ with the following property: each path $L_n$ passes at $t = \tau_2$ through a point of the $\delta_n$-neighborhood of $M_2$ and at some time $t > \tau_2$ leaves the $\varepsilon_0$-neighborhood of $L_{M_2}^+$. Consequently, on each path $L_n$ there is a point $P_n$, corresponding to time $t = t_n > \tau_2$, outside the $\varepsilon_0$-neighborhood of the semipath $L_{M_2}^+$.

Note that no infinite subsequence of $\{t_i\}$ can tend to a finite value of $t$ as $i \to \infty$, since otherwise we would have a contradiction to the continuity of the solutions with respect to initial values. Moreover, we may always assume that the sequence of points

$$P_1, \ P_2, \ \ldots, \ P_n, \ldots \tag{1}$$

has one point of accumulation $P_\omega$. Now $P_\omega$ must obviously be distant at least $\varepsilon_0$ from the semipath $L_{M_2}^+$, hence also from the limit points of this semipath, which are the same as those of $L_{M_1}^+$. In addition, $P_\omega$ must be on $L_{M_1}^+$.

Indeed, suppose that $P_\omega$ is not on the semipath $L_{M_1}^+$, and so is at some distance $\varrho_0 > 0$ from $L_{M_1}^+$. Let $\varepsilon < \varrho_0$. The point $P_\omega$ is certainly outside the $\varepsilon$-neighborhood of $L_{M_1}^+$. It is readily seen that, by the theorem on continuity of solutions with respect to initial values, all paths $L_n$ with sufficiently large $n$ pass at $t = \tau_1 < \tau_2$ through points of an arbitrarily small neighborhood of $M_1$. But for sufficiently large $n$ the points $P_n$ on the paths $L_n$, corresponding to $t_n > \tau_2 > \tau_1$, are arbitrarily close to $P_\omega$, hence outside $U_\varepsilon (L_{M_1}^+)$. Consequently, through any neighborhood of $U_{\delta_1}(M_1)$ there are paths, namely the paths $L_n$ with sufficiently large $n$, which leave $U_\varepsilon (L_{M_1}^+)$ with increasing $t$. Hence the path $L$ is not $\omega$-orbitally stable at $M_1$ — contradiction. Thus $P_\omega$ must be on the semipath $L_{M_1}^+$ or, more precisely, on the arc of $L$ defined by times $t$, $\tau_1 \leqslant t < \tau_2$.

Suppose that under the motion chosen on $L$ the point $P_\omega$ corresponds to time $t = T_1$, and let $Q$ be a point on $L$ corresponding to $T_2 < \tau_1$ (Figure 147).

On each path $L_n$, let $Q_n$ be the point corresponding to $t'_n = t_n - (T_1 - T_2)$. Since $t_n \to \infty$ as $n \to \infty$, it follows that also $t'_n \to \infty$ as $n \to \infty$. By continuity with respect to initial values, the sequence of points $\{Q_i\}$ tends to $Q$ as $i \to \infty$. Now $Q$ cannot be an $\omega$-limit point of the path $L$, since then (by Theorem 11, §4) all points of $L$ would be limit points for $L$, including the point $P_\omega$, contradicting the fact that $P_\omega$ lies at a distance greater than $\varepsilon_0$ from $L^+_{M_2}$. But then it is readily seen (since the point $Q$ corresponds to time $T_2 < \tau_1$) that $Q$ is at a finite distance from the semipath $L^+_{M_1}$. And since there are always paths $L_n$ passing through an arbitrarily small neighborhood of $M_1$, this implies that the path $L$ is not $\omega$-orbitally stable at $M_1$. This contradiction completes the proof.

*Definition XVI. A path which is not $\omega$ ($\alpha$)-orbitally stable is said to be $\omega$ ($\alpha$)-orbitally unstable.*\*

Obviously, an $\omega$ ($\alpha$)-orbitally unstable path $L$ is not orbitally stable at any of its points; i.e., if $M$ is a point on the path, there exists $\varepsilon_0 > 0$ such that for any $\delta > 0$ there is a path $L'$ through $U_\delta (M)$ which leaves the $\varepsilon_0$-neighborhood of $L^+_M$ at some time $t > \tau$.

A positive (negative) semipath is said to be orbitally stable if it is a semipath of an $\omega$ ($\alpha$)-orbitally stable path.

A (positive or negative) semipath which is not orbitally stable is said to be orbitally unstable.

*Definition XVII. A bounded path is said to be orbitally stable or nonsingular if it is both $\omega$- and $\alpha$-orbitally stable. A path which is not orbitally stable (therefore either $\omega$-, $\alpha$- or both $\omega$- and $\alpha$-orbitally unstable) is said to be orbitally unstable or singular.*

*Theorem 36. Consider two dynamic systems defined either both on the sphere or both in bounded plane regions $G$ and $G'$. If the topological structure of their phase portraits is the same and $T$ is an identifying mapping (i. e., a path-preserving mapping of $G$ onto $G'$), then orbitally stable semipaths of the one system are mapped onto orbitally stable semipaths of the other, orbitally unstable semipaths onto orbitally unstable semipaths.*

Proof. For dynamic systems on the sphere, the theorem follows at once from the fact that since the sphere is compact the mapping $T$ is uniformly continuous (and moreover all points of the sphere are interior points), and from the definition of orbital stability and instability of semipaths.

For dynamic systems in (bounded) plane regions, consider a bounded semipath $L^+$ of one of the systems. Let $G$ be the domain of definition of the system. The semipath $L^+$ lies in some closed region $\bar{G}_1$ contained together with its boundary in $G$ ($\bar{G}_1 \subset G$).

Let $\bar{G}_2$ be some region, contained in $G_1$ together with its boundary and containing the closure of $G_1$ (i. e., $G \supset \bar{G}_2 \supset G_2 \supset \bar{G}_1 \supset G_1 \supset L^+$). The mapping $T$ is uniformly continuous in $\bar{G}_2$ and the points of $L^+$ cannot be boundary points of $\bar{G}_2$. The theorem now follows readily from the definition of orbital stability and instability of semipaths.

Remark. If the topological mapping $T$ preserves both orientation and direction of time $t$, then $\omega$-orbitally unstable paths are mapped onto $\omega$-orbitally unstable paths, and $\alpha$-orbitally unstable paths onto $\alpha$-orbitally unstable ones.

---

\* Note that all the above definitions, Lemma 1 and also Theorem 36 below (and their proofs) are valid not only for paths on the plane but also for paths of dynamic systems in any number of dimensions.

## 2. SIMPLE EXAMPLES OF ORBITALLY STABLE AND ORBITALLY UNSTABLE PATHS.

The concepts introduced above will now be clarified by examples of paths from dynamic systems considered in previous chapters.

Any semipath tending to a node or focus is orbitally stable.

Indeed, let $O$ be a node or focus. Then, as we know (§7), for any $\varepsilon > 0$ there exists a cycle without contact $C_\varepsilon$ (a circle or ellipse) containing $O$ in its interior and lying entirely in the $\varepsilon$-neighborhood of $O$. It is clear that the $\varepsilon$-neighborhood of any semipath tending to $O$ contains the $\varepsilon$-neighborhood of $O$, and therefore all points inside the cycle $C_\varepsilon$ are in the $\varepsilon$-neighborhood of any semipath tending to $O$.

Let $L^+$ be one of these semipaths. If $M$ is a point of this semipath within $C_\varepsilon$, then all paths through a sufficiently small neighborhood of $M$ remain within $C_\varepsilon$ (hence also within the $\varepsilon$-neighborhood of $L_M^+$) with increasing $t$. If $M$ is outside $C_\varepsilon$ (or on $C_\varepsilon$), then by the theorem on continuity of solutions with respect to initial values (or Lemma 4, §3) one can always find $\delta > 0$ such that all paths passing at $t = t_0$ through $U_\delta(M)$ reach the cycle without contact in a finite time, without having left the $\varepsilon$-neighborhood of $L_M^+$ (or, more precisely, the $\varepsilon$-neighborhood of the arc of $L_M^+$ preceding its intersection with $C_\varepsilon$); once it crosses $C_\varepsilon$ inward, it never leaves the interior of the cycle again, hence never leaves the $\varepsilon$-neighborhood of $L_M^+$. Since the above statement is valid for any $\varepsilon > 0$, it obviously follows that $L_M^+$ is orbitally stable (Figure 148).

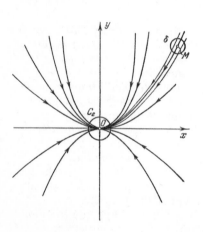

FIGURE 148.

Similarly, all semipaths tending to a limit cycle are also orbitally stable. We shall not dwell on the proof of this geometrically obvious fact, since it follows immediately from a more general proposition (Lemma 8) to be proved below.

Other examples of orbitally stable (i.e., nonsingular) paths are paths that tend to nodes or foci as $t \to +\infty$ and $t \to -\infty$, paths that tend to a node as $t \to +\infty$ ($t \to -\infty$) and to a limit cycle as $t \to -\infty$ ($t \to +\infty$); and also paths that tend to limit cycles both as $t \to +\infty$ and as $t \to -\infty$ (all such paths are both $\omega$- and $\alpha$-orbitally stable).

It is readily seen from these examples that all paths near an orbitally stable (nonsingular) path behave in a highly "similar" manner (this statement will be made rigorous in the sequel). Now this is surely false for all those paths already designated as "singular" paths. We begin with equilibrium states. Stable nodes and foci are orbitally stable as $t \to +\infty$, unstable nodes and foci as $t \to -\infty$. But they are orbitally unstable as $t \to -\infty$ and $t \to +\infty$, respectively. Indeed, in the case, say, of a stable node or focus, paths passing arbitrarily near the point clearly leave any of its neighborhoods with decreasing $t$ (Figure 149).

A saddle point is orbitally unstable both as $t \to +\infty$ and as $t \to -\infty$.

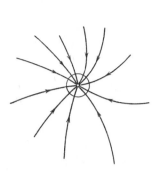

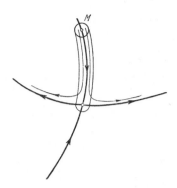

FIGURE 149.                                   FIGURE 150.

In regard to orbital stability, stable and unstable limit cycles exhibit the same behavior as nodes and foci: they may be orbitally stable either only as $t \to +\infty$, or only as $t \to -\infty$, but orbitally unstable as $t \to -\infty$ or as $t \to +\infty$, respectively; consequently, limit cycles, like nodes, foci and saddle points, are singular paths.

Semipaths tending to a saddle point (separatrices) are orbitally unstable. Indeed, if $L^+$ is a semipath tending to a saddle point and $M$ a point on this semipath, one can always find $\varepsilon_0 > 0$ such that for any $\delta > 0$ semipaths other than $L_M^+$ passing through points of $U_\delta(M)$ necessarily leave the $\varepsilon_0$-neighborhood of $L_M^+$ with increasing $t$ (Figure 150).

### 3. POSSIBLE TYPES OF ORBITALLY UNSTABLE SEMIPATHS AND PATHS.

Our next goal is to ascertain which paths of the second-order dynamic systems considered in this book are orbitally stable and which are orbitally unstable.

In so doing we shall rely essentially on Theorem 11 of §4, which concerns limit cycles and is valid only for paths of dynamic systems in plane regions.

We begin with a fundamental theorem:

*T h e o r e m 37. Any path which is a limit path for at least one other path is either ω- or α-orbitally unstable (or both).*

P r o o f. Let $L$ be an ω-limit path for another path $L'$. If $L$ is an equilibrium state, there is of course a point $M_1$ on $L'$ not "on $L$" and hence at a positive distance from $L$.

If $L$ is not an equilibrium state, it follows from Theorem 11 of §4 that $L'$ cannot be a limit path for $L$. Consequently, there is again a point $M_1$ on $L'$ at a positive distance $d$ from $L$. Let $\varepsilon_0 < d$. In both cases, the point $M_1$ is outside the $\varepsilon_0$-neighborhood of $L$. But by assumption $L$ is an ω-limit path for $L'$; consequently, for arbitrarily small $\delta > 0$ the $\delta$-neighborhood of any point of $L$ will contain points on $L'$ corresponding to arbitrarily large times $t$ (in particular, larger than the $t$ corresponding to $M_1$). And since $M_1$ is a point on $L'$ lying outside the $\varepsilon_0$-neighborhood of $L$, this obviously implies that in either case $L$ is α-orbitally unstable.

In exactly the same way one proves that if $L$ is an α-limit path for some other path $L'$ it must be ω-orbitally unstable. Q.E.D.

As shown in Chapter II, a bounded semipath of a dynamic system (i.e., a semipath contained in a closed bounded region $\bar{G}_1 \subset G$) is of one of the following types:

1) Equilibrium state. 2) Semipath tending to an equilibrium state.
3) Semipath of a closed path (hence coinciding with the whole closed path).
4) Semipath tending to a closed path. 5) Semipath tending to some limit contour made up of equilibrium states and paths which are not equilibrium states but tend to equilibrium states both as $t \to +\infty$ and as $t \to -\infty$.*

We are going to examine all possible types of semipath and determine which of them may be orbitally stable and which orbitally unstable.

Theorem 18 (§4) solves the problem immediately for an isolated equilibrium state $O$. Indeed, by that theorem there are two cases: either 1) in any, arbitrarily small neighborhood of $O$ there is a closed path with $O$ in its interior, or 2) there exist paths tending to $O$ (as $t \to +\infty$ or $t \to -\infty$).

In the first case the equilibrium state $O$ is clearly both $\omega$- and $\alpha$-orbitally stable. In the second case, it follows from Theorem 37 that $O$ is orbitally unstable. The problem of orbital stability and instability is thus completely solved for equilibrium states. We now proceed to consider orbitally unstable semipaths tending to an isolated equilibrium state. We first prove a number of auxiliary propositions.

**4. LEMMAS ON THE BEHAVIOR OF SEMIPATHS IN THE NEIGHBORHOOD OF AN EQUILIBRIUM STATE.** Throughout the sequel, we shall assume the plane to be oriented, a positive traversal being fixed for all simple closed curves (e.g., counterclockwise).

Let $O$ be an isolated equilibrium state and $C$ a simple closed curve (not necessarily smooth) containing $O$ in its interior, and such that n e i t h e r inside nor on $C$ is there any equilibrium state other than $O$.

Suppose that there exists a positive semipath $L^+$ tending to $O$ and having points in common with the curve $C$.

Choose some motion on $L^+$. A point $M$ common to $L^+$ and $C$, corresponding to time $t = \tau$ say, is called the l a s t p o i n t of $L^+$ in c o m m o n with $C$ if all points on $L^+$ corresponding to $t > \tau$ lie inside $C$ (Figure 151). As usual, we shall denote the part $MO$ of $L^+$ corresponding to times $t \geqslant \tau$ by $L_M^+$.

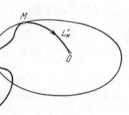

Similarly, if a negative semipath $L^-$ tends to the equilibrium state $O$ and there are points common to the semipath and the curve $C$, we shall speak of the l a s t p o i n t of $L$ in c o m m o n with $C$.

If $L$ is the path of which $L^+(L^-)$ is a semipath, the last point of $L^+(L^-)$ in common with $C$ will sometimes be called the l a s t p o i n t w i t h i n c r e a s i n g (d e c r e a s i n g) $t$ of the path $L$ in c o m m o n with the curve $C$.

FIGURE 151.

We now stipulate that the simple closed curve $C$ with the above properties is a circle, the semipath tending to the equilibrium state $O$ has points outside $C$ and is a positive semipath $L^+$. ($M$ is the last point of $L^+$ in common with $C$, $L_M^+$ the part $OM$ of $L^+$.)

---

* If it is not assumed that the number of equilibrium states is finite, the limit set of a semipath of type 5) may be a continuum of equilibrium states.

We shall prove three lemmas.

*L e m m a  2.  If λ is an arc without contact in the interior of C, on which there is no point of $L_M^+$, then no path crossing λ can cross it again without leaving the circle C.*

P r o o f.  Suppose on the contrary that a path L′ through a point $P_1$ on λ does not leave C but crosses λ again at a point $P_2$ (Figure 152). Thus

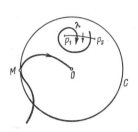

FIGURE 152.

none of the points on the arc $P_1P_2$ of L′ lie outside C (though some may be on C itself). In addition, we may assume that there are no other points of L′ on λ between $P_1$ and $P_2$. Let D be the simple closed curve formed by the arc $P_1P_2$ of L′ and the subarc $P_1P_2$ of λ. The interior of this curve is obviously a subregion of the interior of C. By Corollary 1 to Theorem 16 (§4) there must be at least one equilibrium state inside D.

The semipath $L_M^+$, which by assumption has points outside C, cannot have points inside D, since it does not cross λ (by the choice of λ). Therefore $L_M^+$ cannot tend to a singular point inside D. Consequently, the singular point in question is not O, and this contradicts the choice of C, proving the lemma.

*L e m m a  3.  Any semipath $L'^0$ other than $L_M^+$ which has no points outside C tends to the equilibrium state O.*

P r o o f.  To fix ideas, suppose that $L'^0$ is a positive semipath $L'^+$. Since by assumption it lies within the circle C, all its limit points are either inside or on C. We first prove that $L'^+$ cannot have limit points inside C other than the equilibrium state O. For if $L'^+$ has a limit point N other than O lying inside C, then N cannot be a point of $L_M^+$ because otherwise all points of $L^+$ (of which $L_M^+$ is a part) would be limit points of $L'^+$, including the points of $L^+$ outside the circle C. This is impossible, since by assumption $L'^+$ has no points outside C. In addition, N cannot be a limit point of $L^+$, since this semipath has only one limit point O. Thus we can draw through N an arc without contact S which does not cut the semipath $L^+$ and must contain infinitely many points of $L'^+$. But by Lemma 2 this is clearly impossible. Consequently, $L'^+$ cannot have limit points other than O inside C. Thus, if $L'^+$ has any limit points other than O, they must be points of the circle C.

We claim that this is impossible. Indeed, let N be a point of C which is a limit point of $L'^+$. Then all points of the path $L_0$ passing through N must be limit points for $L'^+$. By the choice of C, the path $L_0$ is certainly not an equilibrium state. Moreover, $L_0$ cannot have points lying outside C, or (by what was proved above) points inside C, so that all points of $L_0$ must lie on C. But this is obviously impossible, since by assumption the semipath $L^+$ tending to the equilibrium state O has points outside C.

*L e m m a  4.  Let Q be a point on $L_M^+$, different from M, and l an arc without contact, one of whose endpoints is Q, lying entirely within the circle C and having no points other than Q in common with $L_M^+$. Then either all paths crossing l at $t = t_0$ at points other than Q leave the circle C with increasing t, or there exists a subarc $QQ_1$ of l such that all paths crossing it tend to the equilibrium state O as $t \to \infty$ without leaving C.*

Proof. Suppose that not all paths crossing $l$ at points other than $Q$ leave $C$ with increasing $t$. In other words, there exists a point $Q_1$ on $l$ through which passes, at $t = T_0$ say, a semipath $L_1^+$ which does not leave $C$ (Figure 153). By Lemma 2, the semipath $L_1^+$ cannot cross $l$ again, and by Lemma 3 it tends to the equilibrium state $O$.

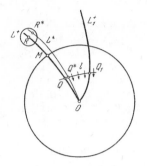

FIGURE 153.

Consider the simple closed curve $S$ formed from the semipath $L_Q^+$ (i. e., the part $QO$ of $L^+$), the semipath $L_{1Q_1}^+$ (the part $Q_1O$ of $L_1^+$), the point $O$ and the subarc $QQ_1$ of $l$. The curve $S$ is in the interior of $C$. Since $QQ_1$ is an arc without contact, it follows that with increasing $t$ the paths that cross it either all enter the interior of $S$ or all leave it. Moreover, each of these paths clearly crosses $QQ_1$ only once (otherwise it would have to cross $QQ_1$ in the opposite direction, which is impossible).

We claim that all paths crossing the arc $QQ_1$ (at points other than $Q$ and $Q_1$) enter the interior of $S$ with increasing $t$. Let $R$ be a point on the semipath $L^+$ (of which $L_M^+$ is a part) outside $C$ — by assumption such a point exists. Suppose that with respect to the motion chosen on $L^+$ the point $Q$ corresponds to time $t = t_0$; then $R$ will correspond to some $\tau < t_0$.

Let $Q^*$ be a point close to $Q$ on the arc $QQ_1$, $L^*$ the path passing through this point. Choosing a motion on $L^*$ under which the point $Q^*$ is reached at $t = t_0$, we see that the point $R^*$ on $L^*$ corresponding to $\tau$ can be made to lie arbitrarily close to $R$ (provided $Q^*$ is sufficiently close to $Q$), and therefore outside the circle $C$ (Figure 153). Since $\tau < t_0$, it is obvious that at $Q^*$ the path $L^*$ crosses $S$ inward with increasing $t$. And then all paths crossing the arc $QQ_1$ at points other than $Q$ and $Q_1$ cross $S$ inward with increasing $t$. Since these paths cannot leave the interior of $C$ again, it follows from Lemma 3 that they tend to the equilibrium state $O$ as $t \to \infty$. Q.E.D.

**5. ORBITALLY UNSTABLE PATHS TENDING TO AN EQUILIBRIUM STATE.** Let the circle $C$, arc without contact $l$ and semipath $L^+$ be as before, $Q$ again being the common point of $l$ and $L^+$; we introduce the following definition:

*Definition XVIII. If the arc without contact $l$ is on the positive side of the semipath $L^+$ and all paths crossing the arc at $t = t_0$ (at points other than $Q$) leave the circle $C$ with increasing $t$, we shall say that $L^+$ is continuable (on the positive side) relative to the circle $C$.*

*If all paths crossing a subarc $QQ_1$ of $l$ at $t = t_0$ remain within $C$ with increasing $t$ (and therefore tend to an equilibrium state as $t \to +\infty$), we shall say that $L^+$ is noncontinuable (on the positive side) relative to $C$ (or: has no continuation relative to the circle $C$ on the positive side).*

A similar definition applies when the arc $l$ is on the negative side of $L^+$, and also when the latter is a negative semipath tending to an equilibrium state $O$.

A semipath $L^+$ may be continuable relative to a circle $C$ from only one side, say positive, or from both sides — both positive and negative.

---

* This definition is due to Bendixson /33/.

Figures 154 and 155 give simple geometric examples of semipaths continuable relative to a circle $C$. The semipath $L^+$ in Figure 154 is continuable relative to $C$ only on the positive side; in Figure 155 the semipath is continuable on both positive and negative sides. Examples of semipaths which are noncontinuable relative to a circle $C$ are given in Figures 156 and 157.

FIGURE 154.

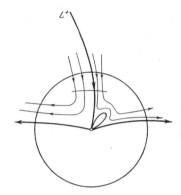

FIGURE 155.

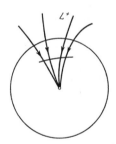

FIGURE 156.

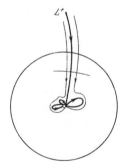

FIGURE 157.

**T h e o r e m  38.** *A semipath $L^+$ tending to an equilibrium state $O$ is orbitally unstable if and only if there exists a circle about $O$ relative to which it is continuable on at least one side.*

P r o o f. Suppose that all paths crossing an arc without contact $l$ (with an endpoint $Q$ on $L^+$) lying inside a circle $C$ leave the circle with increasing $t$.

We can always select $\varepsilon_0 > 0$ so small that the $\varepsilon_0$-neighborhood of the semipath $L_Q^+$ (the part $QO$ of $L^+$) lies entirely inside $C$. And then any path crossing $l$ which leaves the circle with increasing $t$ will a fortiori leave the $\varepsilon_0$-neighborhood of $L_Q^+$. Since we can always find a path crossing $l$ arbitrarily near $Q$, this obviously means that $L_Q^+$ (hence, by Lemma 1, also $L^+$) is orbitally unstable.

243

Conversely, let $L^+$ be a semipath tending to an equilibrium state $O$, and suppose it is orbitally unstable. Then there exists $\varepsilon_0 > 0$ such that, for any small neighborhood of an arbitrary point $R$ on $L^+$, there are paths through this neighborhood which leave $U_{\varepsilon_0}(L^+)$ with increasing $t$.

Consider a circle $C$ (with no equilibrium states other than $O$ either on it or in its interior), of radius $r$ less than $\varepsilon_0$, and let $M$ be the last point of $L^+$ in common with $C$. Through a point $Q$ on $L_M^+$, distinct from $M$, draw an arc without contact $\lambda$ containing $Q$ as an interior point, having no points other than $Q$ in common with $L_M^+$, and lying inside $C$. It is readily seen that there is no subarc of $\lambda$ containing $Q$ as an interior point such that all paths crossing it remain within $C$ (therefore tending to $O$). Indeed, otherwise we could find a neighborhood of each point of $L^+$ such that all paths passing through it remain in the $\varepsilon_0$-neighborhood of $L^+$, and this is precluded by the choice of $\varepsilon_0$. It follows that since the semipath $L^+$ is orbitally unstable all paths crossing either the subarc of $\lambda$ on the positive side of $L_M^+$ or its subarc on the negative side of $L_M^+$ (or both) leave the circle $C$ with increasing $t$. Q.E.D.

We now devote attention to semipaths continuable relative to a circle.

Suppose that apart from the semipath $L^0$ under consideration, which tends to the equilibrium state $O$ and has points outside the circle $C$, there is another semipath $L'^0$ tending to the equilibrium state and having points in common with $C$ (but not necessarily outside $C$).

Let $M$ and $M'$ be the last points of $L^0$ and $L'^0$ in common with the circle $C$, and $L_M^0$, $L'^0_{M'}$ the corresponding semipaths (arcs $MO$ and $M'O$) on $L^0$ and $L'^0$.

Let $\sigma$ be the simple closed curve formed from the semipaths $L_M^0$ and $L'^0_{M'}$, the point $O$ and the arc $MM'$ on $C$ (i.e., the arc of the circle on which the direction from $M$ to $M'$ is that induced by the positive traversal of the circle; see §11.2). We shall call the region $g$ bounded by $\sigma$, which is a subregion of the interior of $\sigma$, a curvilinear sector or simply a sector.

If both semipaths $L_M^0$ and $L'^0_{M'}$ figuring in the boundary of the sector $g$ are positive (negative), the positive traversal of the curve induces on one of them a direction agreeing with that of $t$, and on the other the direction opposite to $t$ (Figure 158).

If one of the semipaths, $L_M^0$ say, is positive, and the other $L'^0_{M'}$ negative, the positive traversal of $\sigma$ induces directions which either coincide on both semipaths with the direction of $t$, or oppose it on both (Figure 159).

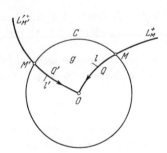

FIGURE 158.

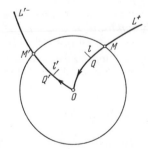

FIGURE 159.

Choose points on $L_M^0$ and $L_{M'}^{'0}$, say $Q$ and $Q'$ respectively, distinct from $M$ and $M'$. Let $l$ be an arc without contact with endpoint $Q$ and no points other than $Q$ in common with $L_M^0$, and $l'$ an arc without contact with endpoint $Q'$ and no other points in common with $L_{M'}^{'0}$.

We assume that the arcs $l$ and $l'$ lie interior to $C$, have no common points, and moreover $l$ has no points other than $Q$ in common with $L_M^0$, $l'$ no points other than $Q'$ in common with $L_{M'}^{'0}$.

Suppose that both the semipaths in question are positive, $L_M^+$ and $L_{M'}^{'+}$. If the arcs $l$ and $l'$ lie on the positive sides of $L_M^+$ and $L_{M'}^{'+}$, respectively, then all points of $l$ other than $Q$ are inside $g$ and all points of $l'$ other than $Q'$ outside $g$ (and vice versa, in case $l$ and $l'$ lie on the negative sides of $L_M^+$ and $L_{M'}^{'+}$, respectively) (Figure 158).

But if one of the semipaths is positive, $L_M^+$ say, and the other negative $(L_{M'}^{'-})$ the situation is different: If the arcs $l$ and $l'$ lie on the positive sides of $L_M^+$ and $L_{M'}^{'-}$, respectively, then all points of both arcs (except $Q$ and $Q'$) are points of $g$. If the arcs both lie on the negative sides of $L_M^+$ and $L_{M'}^{'-}$, the points of both arcs lie outside $g$.

Suppose now that the semipaths $L_M^+$ and $L_{M'}^{'-}$ satisfy the following condition. All paths crossing some subarc $QQ_1$ of $l$ (at points other than $Q$) cross $l'$ with increasing $t$, without leaving $C$; moreover, their points of intersection with $l'$ tend to $Q'$ as their points of intersection with $l$ tend to $Q$.

It is not hard to see (using Lemma 10 of §3) that there then exists a subarc $Q'Q_1'$ on $l'$ such that all paths crossing this subarc (at points other than $Q'$) will cross $l$ with decreasing $t$, without leaving $C$. Indeed, let $Q_2$ be some point on the subarc $QQ_1$ of $l$, and suppose that all paths crossing the subarc $Q_1'Q_2'$ of $l$ cross a subarc $Q_1Q_2$ of $l'$ with increasing $t$, without leaving $C$, in such a way that the paths through $Q_1$, $Q_2$ cross $l'$ at the points $Q_1'$ and $Q_2'$, respectively. Then, by Lemma 10 of §3 and Remark 1 following it, all paths crossing the subarc $Q_1'Q_2'$ of $l'$ will cross the subarc $Q_1Q_2$ of $l$ with decreasing $t$ (again, clearly, without leaving $C$). But by assumption the point $Q_2'$ tends to $Q'$ as $Q_2$ tends to $Q$. Consequently, all paths crossing the subarc $Q'Q_1'$ of $l'$ cross the subarc $QQ_1$ of $l$ with decreasing $t$.

In this case the semipaths $L_M^+$ and $L_{M'}^{'-}$ must be continuable relative to some circle. Indeed, by assumption the semipath $L_M^+$, which has points outside $C$, is obviously continuable relative to $C$. If $L_{M'}^{'-}$ also has points outside $C$, we see directly from Definition XVIII that it is also continuable relative to $C$.* If it has no points outside $C$, but only points in common with $C$ (Figure 160), we can always find a circle $C'$ of radius slightly smaller than $C$ relative to which it is continuable.

To fix ideas, let us assume that the arcs $l$ and $l'$ lie on the positive sides of the semipaths $L_M^+$ and $L_{M'}^{'-}$, respectively.

*Definition XIX. If all paths crossing a subarc $Q_1Q$ of $l$ at points other than $Q$ cross the arc $l'$ with increasing $t$, without leaving the circle $C$, and their points of intersection with $l'$ tend to $Q'$ as their points of intersection with $l$ tend to $Q$, we shall call $L_{M'}^{'-}$ the continuation of $L_M^+$ on the positive side relative to $C$.*

*When the arcs without contact $l$ and $l'$ are on the negative side of $L_M^+$ and $L_{M'}^{'-}$, respectively, we shall call $L_{M'}^{'-}$ the continuation of $L_M^+$ on the negative side relative to $C$.*

---

* By Definition XVIII, a semipath continuable relative to a circle $C$ must have points outside $C$.

A similar definition applies for a negative semipath $L_M^-$.

We shall apply the term "continuation of $L^+$ relative to $C$" not only to the semipath $L_M^+$, but also to the path $L'$ of which it is a part. It is clear that a semipath tending to an equilibrium state cannot have more than two continuations relative to a given circle $C$ — one on the positive and one on the negative side.

The next lemma is valid for a positive semipath $L^+$. An analogous proposition holds for negative semipaths.

The circle $C$, semipaths $L^+$, $L_M^+$, arc without contact $l$, etc. have the same meaning as before.

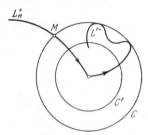

FIGURE 160.

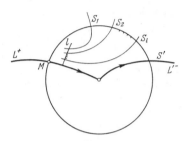

FIGURE 161.

**Lemma 5.** *Let $l$ be an arc without contact on the positive (negative) side of the semipath $L_M^+$. If all paths crossing $l$ leave the circle $C$ with increasing $t$, i.e., $L^+$ is continuable on the positive side relative to $C$, then there exists a negative semipath $L_M^-$ which is the continuation of $L^+$ on the positive (negative) side relative to $C$.*

Proof. To prove the lemma we must show that there exists a negative semipath $L'^-$, tending to the equilibrium state $O$, which satisfies Definition XIX. To fix ideas, let the arc $l$, whose endpoint $Q$ is on $L_M^+$, be on the positive side of the semipath $L_M^+$. Let $\{Q_i\}$ be a sequence of points on $l$ tending to $Q$. By assumption, the path $L_i$ passing at $t = t_0$ through any of the points $Q_i$ must leave the circle C at some $t > t_0$. Let $S_i$ be the first point of this path in common with $C$ for $t > t_0$, so that the arc $Q_i S_i$ (minus its endpoint $S_i$) is inside $C$. It is readily seen, via Lemma 1, that the points $S_i$ are distinct for different $i$. Consequently, there are infinitely many points $S_i$ on $C$, and they must have at least one point of accumulation (Figure 161). Without loss of generality, we may assume that they have exactly one point of accumulation $S'$.

Let $L'$ be the path through $S'$.

Suppose that with respect to a fixed motion on $L'$ the point $S'$ corresponds to time $t = \tau$. We claim that there are no points on $L'$ for $t < \tau$ which lie outside $C$. In fact, suppose on the contrary that the path $L'$ leaves the circle $C$ at $t < \tau$, without previously crossing the arc without contact $l$. Then all paths $L_i$ passing through points $S_i$ sufficiently close to $S'$ leave $C$ with decreasing $t$, without previously crossing $l$. But this is obviously impossible by the definition of the points $S_i$.

Suppose now that $L'$ leaves $C$ with decreasing $t$, after crossing the arc $l$ at some point corresponding to time $t = \tau_1\ (\tau_1 < \tau)$. Then, by the continuity of solutions with respect to initial values, this point must be a point of accumulation of the sequence $Q_i$, so that it coincides with $Q$. Thus $L'$ must coincide with the path $L$ (of which $L_M^+$ is a semipath). But this is impossible, for by assumption all points of $L$ corresponding to times $t > \tau_1$ (i. e., points of $L_Q^+$) lie within $C$, and $L^+$ cannot have in common with $C$ a point $S'$ corresponding to $\tau > \tau_1$.

Thus the semipath $L_{S'}^-$, corresponding to times $t < \tau$, has no points outside outside the circle $C$, and must therefore tend to the equilibrium state $O$.

Let $M'$ be the last point of $L_{S'}^-$ in common with the circle $C$,* and $Q'$ some point of the semipath $L_{M'}^-$ (which is a part or all of $L_{S'}^-$). Let $l'$ be an arc without contact with endpoint $Q'$, on the positive side of $L_{M'}^-$, having no points in common with either $L_M^+$ or $l$ and lying entirely inside $C$ (by Lemma 2, the arc $l'$ has no points other than $Q'$ in common with $L_{M'}^-$). The simple arc formed by the semipath $L_M^+$, the point $O$ and the semipath $L_{M'}^-$ divides the interior of $C$ into two regions, one of which, $g'$ say, contains the arc $l$ minus the point $Q$. Then all points of $l'$ except $Q'$ are also in $g'$ (Figure 159). Moreover, it is obvious that $g'$ also contains all points (other than $S_i$) of the arcs $Q_iS_i$ of the paths $L_i$. For sufficiently large $i$, each path $L_i$ necessarily crosses $l'$ at some point $Q_i'$ (by Lemma 3 of §3 and the fact that the points of the arcs $Q_iS_i$ are in $g'$). The points $Q_i'$ tend to $Q'$ as $i \to \infty$.

We claim that not only the paths $L_i$ but all paths crossing the subarc $Q_iQ$ of $l$ will cross the subarc $Q_i'Q'$ of $l'$.

Consider the interior $D$ of the curvilinear quadrangle $QQ_iQ_i'Q'$ (Figure 162), i. e., the region interior to the simple closed curve made

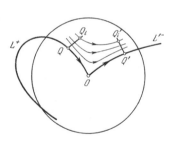

FIGURE 162.

up of the subarcs $QQ_i$ and $Q'Q_i'$ of $l$ and $l'$, respectively, the arc $Q_iQ_i'$ of $L_i$, the semipath $L_Q^+$, the point $O$ and the semipath $L_{Q'}^-$ (the semipath $L_Q^+$ is simply the arc $QO$ of the semipath $L_M^+$, and $L_{Q'}^-$ the arc $Q'O$ of $L_{M'}^-$). It is clear that $D$ is a subregion of the interior of $C$.

All paths crossing the arc without contact $Q_iQ$ can do so only once without leaving $C$ (Lemma 2). Moreover, it is readily seen that with increasing $t$ all these paths enter the region $D$. Indeed, by assumption the semipath $L^+$ passing through the point $Q$ leaves the circle $C$ with decreasing $t$. Thus, by the continuity of solutions with respect to initial values, all paths crossing the arc $Q_iQ$ sufficiently near $Q$ will also leave $C$ with decreasing $t$. Before doing this they cannot cross the arc $l'$, for the latter is situated at a positive distance from $L_M^+$, and thus they cannot cross the boundary of $g'$ a second time before leaving $C$. Consequently, all paths crossing the arc $QQ_i$ sufficiently near $Q$ leave the region $D$ with decreasing $t$, while they enter this region with increasing $t$. But then all paths crossing the arc $Q_iQ$ enter the region $D$ with increasing $t$ (see Appendix, §6, and also Lemma 10 of §3).

---

\* By what we have proved, the semipath $L_{S'}^-$ cannot have points outside $C$, but it may have points on $C$, so that the last point of this semipath in common with $C$ need not be $S'$ but a point corresponding to some smaller parameter value $\tau' < \tau$.

Since by assumption every path crossing $QQ_i$ leaves $C$ with increasing $t$, it of course leaves $D$, which is a subregion of the interior of $C$. But such a path can leave $D$ only across the subarc $Q'Q_i'$ of $l'$. Thus, all paths crossing the subarc $QQ_i$ of $l$ with increasing $t$ also cross the subarc $Q'Q_i'$ of $l'$ prior to leaving the circle $C$. In addition, as $i \to \infty$ the points $Q_i$ tend to $Q$ and the points $Q_i'$ to $Q'$. Hence it clearly follows that the semipath $L_M^+$ satisfies Definition XIX. Q.E.D.

The following propositions will be stated only for positive semipaths continuable on the positive side relative to some circle. Analogous propositions are valid for semipaths (positive or negative) continuable on the negative side relative to a circle.

Let $L^+$ be a semipath, tending to the equilibrium state $O$, which is noncontinuable relative to the circle $C$. It may happen that this semipath is continuable relative to a circle $C'$ about $O$, of radius smaller than that of $C$. This case is illustrated in Figure 163.

Let $L^+$ be continuable on the positive side relative to a circle $C$. Then it is clearly also continuable relative to any smaller circle (about $O$). However, the continuations of $L^+$ on the positive side relative to different circles may be distinct semipaths; a situation of this kind is shown in Figure 164.

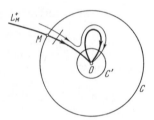

FIGURE 163.

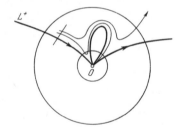

FIGURE 164.

The following lemma employs the same notation (circle $C$, semipaths $L^+$ and $L_M^+$, point $Q_1$, arc $l$, etc.) as before.

Lemma 5'. a) If a semipath $L^+$ tending to an equilibrium state $O$ has no continuation on the positive side relative to a circle $C$, but has such a continuation relative to a smaller circle $C'$, then the path $L'$ which is the continuation of $L^+$ relative to $C'$ cannot have points outside $C$.

b) If a path $L'$ is the continuation of a semipath $L^+$ on the positive side relative to a circle $C$, then the continuation of $L^+$ on the positive side relative to a smaller circle $C'$ is either $L'$ itself or a path lying entirely interior to $C$.

Proof. a) By assumption, the semipath $L^+$ has no continuation on the positive side relative to $C$. Letting $l$ be an arc without contact with endpoint $Q$ on $L_M^+$, we can find a subarc of this arc on the positive side of $L_M^+$, such that all paths through points of this subarc tend to $O$ without leaving $C$. Let $\tilde{L}^+$ denote the semipath through some point $Q^*$ of this subarc of $l$, and $\tilde{g}$ the region interior to the simple closed arc formed from the arc

$QO$ of $L^+$, the arc $Q^*O$ of $\widetilde{L}^+$, the point $O$ and the arc without contact $Q^*Q$. The region $\widetilde{g}$ is clearly inside $C$. In addition, any path crossing the arc $Q^*Q$ at a point other than its endpoints $Q$ and $Q^*$ enters the region $\widetilde{g}$ with increasing $t$ and never leaves it again (Figure 165).

Let $N$ be the last point of $L^+$ in common with the smaller circle $C'$, $P$ some other point on the arc $NO$ of this semipath and $l_1$ an arc without contact with endpoint $P$, on the positive side of $L^+$, lying inside $C'$. Let $L_1$ denote the continuation of $L^+$ on the positive side relative to the circle $C'$; by assumption such a path exists. By definition, the path $L_1$ tends to the equilibrium state $O$ as $t \to -\infty$. Let $S'$ be the last point in common with $C'$ on a semipath $L_1^-$ (so chosen as to have points in common with $C'$) of the path $L_1$, $R$ a point distinct from $S'$ on the arc $S'O$ of $L_1^-$, and $l_1'$ an arc without contact with endpoint $R$, on the positive side of $L_1$.

Assume that the statement of the lemma is false, so that there are points on $L_1$ outside the circle $C$. Let $M_0$ be such a point. All paths crossing the arc without contact $Q^*Q$ sufficiently nearly $Q$ will obviously cross the arc $l_1$ arbitrarily near $P$. Since $L_1^-$ is the continuation of $L^+$ on the positive side relative to $C'$, all paths crossing $l_1$ sufficiently near $P$ will cross $l_1'$ arbitrarily near $R$. With further increase in $t$, these paths must approach arbitrarily near the point $M_0$, therefore leaving the circle $C$. But this implies that all paths crossing the arc $QQ^*$ sufficiently near $Q$ must leave $C$ with increasing $t$. And this is impossible, since all these paths enter $\widetilde{g}$, which is a subregion of the interior of $C$, with increasing $t$, and cannot leave this region again. This contradiction proves part a).

b) By assumption, the semipath $L^+$ has a continuation on the positive side relative to $C$. Let the semipaths $L_M^+$, $L_{M'}^-$ and arcs $l$, $l'$ be as in Definition XIX. The simple arc made up of $L_M^+$, $L_{M'}^-$ and the point $O$ divides the interior of the circle into two sectors. Let $g'$ denote the sector containing the arcs $l$ and $l'$ (Figure 166). To fix ideas, we suppose that the arc of $C$ bounding this sector is $M'M$.

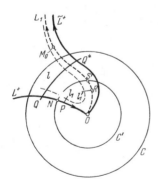

FIGURE 165.

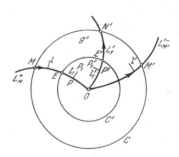

FIGURE 166.

249

Now let $L'_1$ be the continuation of $L^+$ on the positive side relative to the circle $C'$. Let $E$ and $E'$ denote the last points of $L^+$ and $L'^-_1$, respectively, in common with $C'$. Let $P$ and $P'$ be points, distinct from $E$ and $E'$, on the arcs $EO$ and $E'O$ of $L^+$ and $L'^-_1$, respectively, and $l_1$ and $l'_1$ arcs without contact with endpoints $P$ and $P'$, lying inside $C'$ on the positive sides of $L^+$ and $L'^-_1$, respectively.

It is clear that all points other than $P$ on the arc $l_1$ are points of $g'$. Since $L'^-_1$ is a continuation of $L^+$ on the positive side relative to $C'$, all paths crossing a subarc $PP_1$ of $l_1$ will cross some subarc $P'P'_1$ of $l'_1$ with increasing $t$, without first leaving the circle $C'$, and clearly, without leaving $g'$. Moreover, their points of intersection with $l'_1$ tend to $P'$ as their points of intersection with $l_1$ tend to $P$. Hence it obviously follows that all points (other than $P'_1$) on the subarc $P'P'_1$ of $l'_1$ are points of $g'$, and the point $P'$ is an interior or boundary point of $g'$.

If $P'$ is on the boundary of $g'$, we easily see that it can lie only on the semipath $L'_{M'}$, and then the paths $L'$ and $L'_1$ coincide. But if $L'$ and $L'_1$ are different, $P'$ must be a point of $g'$.

Now suppose the assertion of part b) false, and let the path $L'_1$ contain points outside $C$. Let $N'$ be the last point of $L'_1$ in common with $C$.

The simple arc formed from the arc $N'O$ of $L'_1$ and the point $O$, all points of which apart from $O$ and $N'$ are points of $g'$, divides the latter into two disjoint regions $g'_1$ and $g'_2$, and is part of the boundary of both regions. Apart from the semipath $L'_{N'}$, the boundary of one of the regions, say of $g'_1$, includes the path $L^+_M$ and the arc $N'M$ of $C$, and that of the other regions $g'_2$ includes the semipath $L^-_{M'}$ and the arc $M'N'$ of $C$. By the definition of the sectors $g'$, $g'_1$ and $g'_2$, it is obvious that the points of any sufficiently small arc without contact with endpoint on $L^+_M$, lying on the positive side of the latter, are all in $g'_1$, while the points of any sufficiently small arc without contact with endpoint on $L^-_{M'}$, on the positive side of the latter, are in $g'_2$. But this implies that the paths crossing $l$ cannot cross $l'$ without leaving $g'_1$, since they cannot leave $g'_1$ across any component of its boundary except the arc of $C$ (the other components being semipaths). This contradicts the fact that $L^-_{M'}$ is the continuation of $L^+_M$ relative to the circle $C$. Q.E.D.

Corollary. The continuations of a semipath $L^+$ on the positive side relative to circles $C$ and $C'$ coincide if and only if the continuation of $L^+$ on the positive side relative to $C'$ has points outside $C$. (Necessity follows from the lemma, sufficiency directly from the definition of the continuation.)

*Theorem 39 (Bendixson). There exist only finitely many paths which tend to an equilibrium state $O$ and are continuable relative to a given circle $C$.*

Proof. Suppose on the contrary that there is an infinite set of paths $\{L_n\}$ continuable relative to $C$. We may assume without loss of generality that these paths tend to $O$ as $t \to +\infty$. Let $\{A_n\}$ be their last points of intersection (with increasing $t$) with the circle $C$. The sequence $\{A_n\}$ has at least one point of accumulation, $A$ say. It is easy to see that the path $L$ passing at $t = \tau$ through $A$ has no points outside $C$ for any $t > \tau$. And then, by Lemma 2, a semipath $L^+$ of this path also tends to the equilibrium state $O$.

Now let $B$ be any point inside $C$ on the semipath $L^+_A$ (the arc $AO$ of $L^+$), and let $B_1B_2$ be an arc without contact of which $B$ is an interior point. Then at least one of the arcs $B_2B$ and $BB_1$ is cut by infinitely many paths $L_n$.

Suppose, say, that paths $L_{n_1}$ and $L_{n_2}$ cross the arc $BB_1$ at points $S_1$ and $S_2$, respectively, the latter point lying closer to $B$ than the former. By Lemma 4, all paths crossing the arc $BB_1$ tend to the point $O$ with increasing $t$, without leaving $C$. But this contradicts the assumption that $L_{n_1}$ is continuable relative to $C$. Q.E.D.

Note that as the radius of the circle $C$ decreases the number of paths which are continuable relative to $C$ may increase without bound.

**6. SEPARATRICES OF AN EQUILIBRIUM STATE.** The following definition is basic for the sequel:

*Definition XX. If a semipath $L'^-$ is the continuation of a semipath $L^+$ on the positive (negative) side relative to all circles of radius less than some $\varepsilon_0 > 0$, we shall say that $L^+$ is continuable on the positive (negative) side and call $L'^-$ its continuation on the positive (negative) side.*

A similar definition applies for negative semipaths. If the semipath $L^+$ is continuable on the positive (negative) side and its continuation is $L'^-$, then the semipath $L'^-$ is also continuable on the positive (negative) side and the continuation is precisely $L^+$. The path $L'$ of which $L'^-$ (the continuation of $L^+$ on the positive side) is a part will also be called a continuation of $L^+$ on the positive (negative) side.

The following two assertions are easily proved:

1) Any semipath tending to an equilibrium state $O$ can have at most two continuations — one on the positive side and one on the negative side.

2) Any semipath tending to an equilibrium state $O$ may be the continuation of at most two semipaths, on the positive side for one and the negative side for the other.

If a semipath $L^+$ of a path $L$ tends to an equilibrium state $O$ and is continuable on the positive (negative) side, we shall say that $L$ is $\omega$-continuable on the positive (negative) side; both the continuation $L'^-$ of $L^+$ and the path $L'$ of which $L'^-$ is a semipath will be called the $\omega$-continuation of $L^+$ on the positive (negative) side (or the $\omega$-continuation of the path $L$). The definition of an $\alpha$-continuable path and its $\alpha$-continuation is entirely analogous.

We shall sometimes also say that the path $L$ is $\omega$-continuable on the positive (negative) side relative to the equilibrium state $O$, meaning thereby that the path $L$ tends to $O$ and is $\omega$-continuable on the positive (negative) side.

Note that all the orbitally unstable semipaths to be considered below which tend to an equilibrium state are continuable; this is a result of the additional assumption that the number of orbitally unstable paths is finite.

If a semipath $L^+$ is noncontinuable on the positive side, there are two possibilities: 1) $L^+$ has no continuation relative to any circle; 2) $L^+$ has a continuation relative to a certain circle and consequently relative to any smaller circle. But for any circle $C_1$, however small, we can always find a smaller one $C_2$ relative to which the continuation of $L^+$ is distinct from its continuation relative to $C_1$.

Cases 1) and 2) are illustrated in Figures 167a and b (Figure 167b shows an infinite set of decreasing loops).

Any orbitally unstable semipath tending to an equilibrium state $O$ will be called a s e p a r a t r i x of this equilibrium state. Specifically, we shall call a positive semipath an $\omega$-separatrix of $O$ and a negative semipath an $\alpha$-separatrix of $O$. We shall also use the term s e p a r a t r i x for a path of which at least one semipath is orbitally unstable and tends to an equilibrium state (i. e., is a separatrix of an equilibrium state).

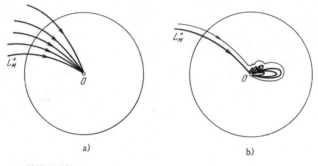

a)                                                       b)

FIGURE 167.

**7. AUXILIARY PROPOSITIONS.** The auxiliary propositions presented below will be used in the subsequent discussion of semipaths some of whose limit points are not equilibrium states.

Let $\bar{g}$ be a bounded closed plane region, in which we have a sequence of simple closed curves $\{C_i\}$ such that either

1) the interior of $C_i$ contains the interior of $C_{i+1}$ as a subregion, or

2) the interior of $C_i$ is a subregion of the interior of $C_{i+1}$.

In addition, we assume one of the following conditions satisfied:

a) the curves $C_i$ have no common points, or

b) each curve $C_i$ has only one point in common with $C_{i+1}$, and the common point of $C_{i-1}$ and $C_i$ is distinct from that of $C_i$ and $C_{i+1}$.

Obviously, if condition b) holds we can extract a subsequence of $\{C_i\}$ whose curves have no points in common (for example, $C_1, C_3, C_5, \ldots$).

Let $\{P_i\}$ be a sequence of points on the curves $C_j$, such that $P_k \in C_{i_k}$, where $i_k$ are natural numbers with $i_1 < i_2 < \ldots$

These points clearly lie on an infinite set of curves $C_j$ (see condition b)).

Let $K$ be the set of all points of accumulation of sequences of this kind. In other words, there are points of infinitely many curves $C_j$ in any neighborhood, however, small, of each point of $K$. But then any such neighborhood will contain points of all curves $C_j$ for sufficiently large $j$.

Indeed, suppose that a given neighborhood of a point of $K$* contains points $P_n$ and $P_m$, $P_n \in C_{i_n}$, $P_m \in C_{i_m}$. Then the straight-line segment joining these points is also in the neighborhood. Since by assumption the interiors of the curves $C_j$ contain one another, this segment cuts all intermediate curves $C_j$, i. e., curves $C_j$ with $j$ between $i_n$ and $i_m$. Hence it follows easily that every neighborhood of a point of $K$ contains points of all curves $C_j$, from some sufficiently large $j$ on. This means that the sequence ($\alpha$) of curves $C_j$ is topologically convergent, and the set $K$ is its topological limit (see Appendix, §1.10).

It is clear that in case 1) the set $K$ is interior to all the $C_i$, and in case 2) exterior to all of them.

---

* We may assume without loss of generality that the neighborhood is a disk about the point.

*Lemma 6. The set K is closed and connected (i. e., it is a continuum).*

Proof. That $K$ is closed is obvious.

If $K$ is not connected, it follows, since it is closed, that $K$ is the union of two closed sets $K_1$ and $K_2$, which are disjoint and therefore at a positive distance $\varrho_0$ from each other. Take $\varepsilon < \frac{\varrho_0}{3}$ and let $K_{1\varepsilon}$ and $K_{2\varepsilon}$ be the $\varepsilon$-neighborhoods of $K_1$ and $K_2$.

It follows from the definition of $K$ that for sufficiently large $j$ each curve $C_j$ has points both in $K_{1\varepsilon}$ and in $K_{2\varepsilon}$. But a closed curve is a connected set, and therefore there is a point on each $C_j$ which is neither in $K_{1\varepsilon}$ nor in $K_{2\varepsilon}$. Take one such point $P_j$ on each curve $C_j$, and let $P$ be a point of accumulation of the points $P_j$. The point $P$ is obviously in $K$, but it belongs to neither $K_1$ nor $K_2$, which is absurd. Q.E.D.

Let $C_i$ and $C_k$ be curves in the sequence $\{C_i\}$, $k > i$. Let $h_{ik}$ denote the region bounded by $C_i$ and $C_k$ (Figure 170). There is exactly one such region. When $j > k$ the region $h_{ij}$ contains $h_{ik}$. Moreover, when $j \geqslant k + 2$ (and if condition a) holds, even when $j \geqslant k + 1$) the curve $C_k$ lies interior to $h_{ij}$.

Let $i$ be fixed, and denote by $\gamma_i$ the region which is the union of all the regions $h_{ij}$ (see Appendix, §7.6):

$$\gamma_i = h_{ii+1} \bigcup h_{ii+2} \bigcup \cdots$$

The region $\gamma_i$ contains all curves $C_j$ for $j \geqslant i + 2$, and all points of $C_{i+1}$ except possibly one (the common point of $C_i$ and $C_{i+1}$, in case b)).

Note that any point of $\gamma_i$ is in some region $h_{ik}$ $(k > i)$. When $k > i$ we have $\gamma_k \subset \gamma_i$. By the definition of $\gamma_i$, all points of $\gamma_i$ lie inside $C_j$ $(j < i)$ in case 1), outside it in case 2).

*Lemma 7. The boundary of the region $\gamma_i$ is the union of $C_i$ and the continuum K.*

Proof. We prove the lemma for case 1): the region bounded by $C_i$ contains that bounded by $C_{i+1}$ (the proof for case 2) is analogous).

The points of $C_i$ are boundary points of $\gamma_i$ by the definition of the latter. The points of the continuum $K$ are not in $\gamma_i$, since $K$ lies inside all curves $C_k$, hence outside all regions $h_{ik}$; however, these points are boundary points of $\gamma_i$, since each point of $K$ is a point of accumulation for a sequence of points of the curves $C_j$, and all these are points of $\gamma_i$.

We now show that these curves make up the entire boundary, i. e., every boundary point of $\gamma_i$ is either on $C_i$ or in $K$.

Let $N$ be a boundary point of $\gamma_i$ which is not on $C_i$. $N$ lies inside all the curves $C_j$. Indeed, if it were outside some $C_{j_0}$ $(j_0 > i)$, since outside $C_i$ there are neither points of $\gamma_i$ nor boundary points of $\gamma_i$), it would be in the region $h_{ij_0}$, therefore in $\gamma_i$, so that it could not be a boundary point of the latter.

Let $\{M_i\}$ be a sequence of points of $\gamma_i$ tending to $N$, and consider an arbitrary neighborhood $U_\delta(N)$.

Let $M_l$ be a point of the sequence $\{M_i\}$ in this neighborhood. Since $M_l$ is in $\gamma_i$, it must be a point of some region $h_{ij_0}$ and therefore outside the curve $C_{j_0}$. But then it must be outside all curves $C_j$ with $j > j_0$.

Let $\lambda$ be the straight-line segment joining $N$ and $M_l$. This segment is inside the neighborhood $U_\delta(N)$ and obviously cuts all curves $C_j$ with $j > j_0$.

Thus, each neighborhood of $N$ contains points of all curves $C_j$ from some $j$ on. This means that $N$ is a point of the continuum $K$. Thus $C_i$ and $K$ make up the entire boundary of the region $\gamma_i$. Q.E.D.

Remark. Apart from the region $\gamma_i$ there is no other region whose boundary is the union of $C_i$ and $K$. Indeed, since $K$ lies entirely inside $C_i$, any such region must contain all points lying inside and sufficiently near the curve  . But any such region has points in common with the region $\gamma_i$, and therefore coincides with it (see Appendix, §1.7).

Lemma 8. *For any* $\varepsilon > 0$ *there exists an integer* $I$ *such that, for all* $i > I$: *a) all the curves* $C_i$ *lie in the* $\varepsilon$*-neighborhood of the continuum* $K$; *b) all the regions* $\gamma_i$ *are contained in the* $\varepsilon$*-neighborhood of* $K$.

Proof. Assuming as usual that condition 1) holds, suppose that part a) of the lemma is false: there exists $\varepsilon_0 > 0$ such that for any natural number $I$ one can find a point $M_j$ on a curve $C_j$, $j > I$, which is not in the $\varepsilon_0$-neighborhood $U_{\varepsilon_0}(K)$ of the continuum $K$. Then there exists a sequence of points $M_{i_1}, M_{i_2}, \ldots$ such that 1) $M_{i_k} \in C_{i_k}$, $i_{k+1} > i_k$; 2) all points $M_{i_k}$ are outside $U_{\varepsilon_0}(K)$. The points of accumulation of this sequence are outside or on the boundary of $U_{\varepsilon_0}(K)$, hence distinct from the points of $K$. But this is impossible, since by the definition of $K$ the points of accumulation of the sequence $M_{i_k}$ must be in $K$. This contradiction proves part a).

To prove part b), suppose on the contrary that there exists $\varepsilon_0 > 0$ such that for arbitrarily large $I$ there exists a region $\gamma_j$, $j > I$, containing a point outside $U_{\varepsilon_0}(K)$.

Then there clearly exist a sequence of regions $\{\gamma_{i_k}\}$ and a sequence of points $\{M_{i_k}\}$ ($i_{k+1} > i_k$) such that 1) $M_{i_k} \in \gamma_{i_k}$; 2) $M_{i_k}$ is outside $U_{\varepsilon_0}(K)$. Moreover, we may assume without loss of generality that the sequence $\{M_{i_k}\}$ has a single point of accumulation $N$, which is of course not in $U_{\varepsilon_0}(K)$. On the other hand, it is readily shown that $N$ is a point of the continuum $K$, and so belongs to $U_{\varepsilon_0}(K)$. This contradiction completes the proof.

## 8. SEMIPATHS WHOSE LIMIT POINTS ARE NOT ALL EQUILIBRIUM STATES.

Let $L^+$ be a nonclosed bounded semipath, whose limit points are not all equilibrium states. Then the limit continuum $K$ of $L^+$ must contain at least one path $L_0$ which is not an equilibrium state. Let $P$ be some point of $L_0$ (of course, $P \in K$) and $l$ an arc without contact through $P$. By Lemma 2 of §3, the semipath $L^+$ crosses $l$ at infinitely many points $\{P_i\}$, corresponding to an indefinitely increasing sequence of parameter values $t_i$, lying on $l$ in order of increasing $t$ and tending to $P$. In this situation (Corollary 1 to Lemma 2, §3) there are no points of the continuum $K$ on $l$ other than $P$. The points $P_i$ divide $L^+$ into arcs with respective endpoints $P_1$ and $P_2$, $P_2$ and $P_3$, . . ., $P_i$ and $P_{i+1}$, . . . As in §3, we shall call these arcs "turns" of the semipath $L^+$. Each turn has in common with $l$ only the points $P_i$ and $P_{i+1}$, and together with the subarc $P_i P_{i+1}$ of $l$ forms a simple closed piecewise-smooth curve which we shall denote by $C_i$ (§3.9).

Lemma 9. *a) From some* $i$ *on, the simple closed curves* $C_i$ *bound regions contained inside one another. b) The topological limit of the sequence of closed curves* $C_i$ *is the limit continuum* $K$ *of the semipath* $L^+$.

Proof. To prove part a), suppose first that there exists $i$ for which the interior of the curve $C_i$ contains $P$ (Figure 168) and therefore also the entire continuum $K$ (by virtue of the fact that the latter is connected). Then all points of the semipath $L^+$ corresponding to $t > t_{i+1}$ must obviously be inside $C_i$ (since if they were outside $C_i$ the limit continuum $K$ of $L^+$ could not lie inside $C_i$). Consequently, the region bounded by $C_{i+1}$ is a subregion of that bounded by $C_i$. To prove a), it remains to show that the point $P$, lying inside $C_{i+2}$, must also be inside $C_{i+1}$. But since the interior of $C_i$ contains the interior of $C_{i+1}$, it follows that the points of $L^+$ corresponding to times $t < t_{i+1}$ are outside $C_{i+1}$. And then, by Lemma 11 of §3, the points of $L^+$ corresponding to $t > t_{i+2}$, including also $P$, must lie inside $C_{i+1}$. Thus, we can repeat for $C_{i+1}$ and $C_{i+2}$ the entire argument given above for $C_i$ and $C_{i+1}$, so that the assertion is proved in this case.

Suppose now that for no $i$ does the curve $C_i$ contain $P$ in its interior, so that $P$ lies outside all the curves $C_i$ (Figure 169). Then, for any curve $C_i$, the points of the semipath $L^+$ corresponding to $t > t_{i+1}$ must lie outside $C_i$. Suppose that the region bounded by $C_{i+1}$ does not contain that bounded by $C_i$, and so does not contain points of $L^+$ for $t < t_{i+1}$. Then, by Lemma 11 of §3, the points of $L^+$ corresponding to times $t > t_{i+2}$ must lie inside $C_{i+1}$, hence the point $P$ must also lie inside $C_{i+1}$. But this contradicts our assumption, and the proof of part a) is complete.

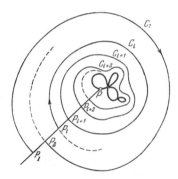

FIGURE 168.

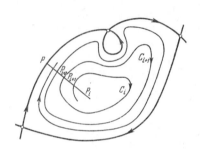

FIGURE 169.

Before we prove part b), note that by what we have already proved, if $i$ is sufficiently large the curves $C_i$ satisfy the conditions laid down in the previous subsection. Let $K^*$ denote the topological limit of this sequence of curves.

Let $N$ be some point of the limit continuum $K$ of $L^+$. Then $N$ is clearly a point of accumulation for some sequence $\{M_i\}$ of points of $L^+$ corresponding to indefinitely increasing times $t$. But $L^+$ is divided by the points $P_i$ into turns $P_1P_2$, $P_2P_3$, . . ., each turn $P_iP_{i+1}$ being an arc of the curve $C_i$. Since both sequences $M_n$ and $P_i$ contain points corresponding to arbitrarily large parameter values $t$, we readily see that for any $i$ there is a turn $P_jP_{j+1}$, $j > i$, which contains a point $M_n$. Since $P_jP_{j+1}$ is an arc of $C_j$, this clearly implies that $N$ is a point of the continuum $K^*$ (the topological limit of the sequence $C_i$). Consequently,

$$K \subset K^*.$$

Now let $Q$ be a point of $K^*$. Then there exists a sequence of points $\{M_i\}$ converging to $Q$ and situated on distinct curves $C_i$. The points $M_i$

consequently lie either on the semipath $L^+$ or on the arc without contact $l$. If there are infinitely many points $M_i$ in the sequence which are points of $L^+$, then $Q$ is a limit point of $L^+$ and

$$Q \in K.$$

But if only finitely many of the points $M_i$ are on $L^+$, then almost all points of the sequence $\{M_i\}$ lie on different subarcs $P_k P_{k+1}$ of $l$. And then it is obvious that $Q$ coincides with $P$, so that again $Q \in K$. Hence it follows that $K^* \subset K$. Since we have already proved the converse inclusion, it follows that

$$K^* = K,$$

proving the lemma.

Part a) of the above lemma implies that if we restrict the sequence of curves $C_i$ (each the union of a turn $P_i P_{i+1}$ of the semipath $L^+$ and a subarc $P_i P_{i+1}$ of the arc without contact $l$) to sufficiently large $i$, all the assertions proved in §15.7 will be valid.

As in §15.7, let $h_{ij}$ denote the region bounded by $C_i$ and $C_j$, and $\gamma_i$ the region bounded by $C_i$ and the continuum $K$ (as shown in the Remark after Lemma 6, this region is uniquely determined) (Figures 170 and 171). The following theorem is fundamental.

*T h e o r e m   40. Any nonclosed semipath $L^+ (L^-)$ whose $\omega (\alpha)$-limit points are not all equilibrium states is orbitally stable.*

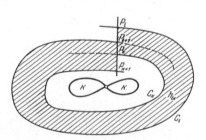 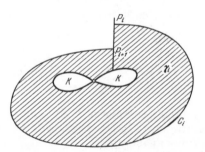

FIGURE 170.                    FIGURE 171.

P r o o f. Retaining the notation of the preceding lemmas, let $\varepsilon > 0$ be arbitrary and consider the neighborhood $U_\varepsilon (K)$. This neighborhood is of course contained in the $\varepsilon$-neighborhood of any positive semipath of $L$. To prove the theorem, therefore, it will suffice to show that all paths passing at $t = t_0$ through a sufficiently small neighborhood of some points of $L^+$ will enter $U_\varepsilon (K)$ within a finite time and with increasing $t$ will never leave $U_\varepsilon (K)$.

By Lemma 7, for any $\varepsilon > 0$ there exists an integer $I$ such that for all $j > I$ the closed curve $C_j$ and region $\gamma_j$ lie entirely within $U_\varepsilon (K)$.

Let $M$ be some point of $L^+$ and $j$ a natural number greater than $I$. By Lemma 5 of §3, there exists a neighborhood of $M$ such that any path passing at $t = t_0$ through its points will cross the subarc $P_j P_{j+2}$ of $l$ at some time

$t = T$, without previously leaving the $\varepsilon$-neighborhood of the arc $MP_j$ of $L^+$. Consequently, at some $t > T$ this path will reach the region $\gamma_j$, But it cannot leave $\gamma_j$ with increasing $t$. Indeed, otherwise it would have points corresponding to $t > T$ in common with either $C_i$ or the continuum $K$. The former alternative is impossible by virtue of Lemma 1 of §3. As for the latter, had the path points in common with $K$, it would have to be one of the paths making up $K$, and this is also impossible because such paths can never have points in common with the subarc $P_jP_{j+2}$ (the continuum $K$ is either entirely outside or entirely inside all the curves $C_i$). This completes the proof of the theorem.

Corollary. Any nonclosed orbitally unstable path is a separatrix of an equilibrium state.

The next theorem establishes when a c l o s e d path is orbitally stable.

*Theorem 41. A closed path which is not a limit path of a nonclosed path is orbitally stable.*

Proof. Let $L$ be a closed path which is not a limit path of any nonclosed path. Let $\varepsilon > 0$ be so small that there are no equilibrium states in $U_\varepsilon(L)$. Through some point $Q$ on $L$ draw an arc without contact $l$, containing $Q$ as an interior point and lying entirely within $U_\varepsilon(L)$.

We shall first show that there are points on $l$, on either side of $Q$ and arbitrarily near $Q$, through which pass closed paths. Indeed, suppose that there is a subarc of $l$ with endpoint $Q$, all paths crossing which are non-closed. To fix ideas, assume that this subarc of $l$ lies outside $L$ (consideration of the other case is analogous). Let $P$ be a point on the subarc such that all paths crossing $l$ between $Q$ and $P$ cross $l$ again with increasing $t$, without leaving $U_\varepsilon(L)$ (see Lemma 5, §3).

Let $L'$ be the path through a point $P'$ of $l$ between $P$ and $Q$, and $P''$ the next point (with respect to $t$) after $P'$ at which $L'$ crosses $l$.

Let $C'$ be the simple closed curve formed from the turn $P'P''$ of $L'$ and the subarc $P'P''$ of $l$. Clearly, $C'$ lies entirely within $U_\varepsilon(L)$. By Lemma 14 of §3.9, the closed path $L$ lies inside the curve $C'$ and the region $\gamma$ bounded by $C'$ and $L$ is a subregion of $U_\varepsilon(L)$ (see Remark 1 to Lemma 14 of §3).

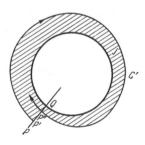

To fix ideas, assume that $P''$, which corresponds to a time $t$ greater than the point $P'$, is closer to $Q$ than $P$ (Figure 172). (The case in which $P'$ is closer to $Q$ than $P''$ is treated similarly.) Then by Lemma 11 of §3 the path $L'$ crosses $C'$ inward at $P''$ with increasing $t$, therefore entering the region $\gamma$ with increasing $t$ and never leaving it again. Thus all $\omega$-limit points of $L'$ are either in $\gamma$ or on its boundary, hence in any case inside $U_\varepsilon(L)$. But by the choice of $\varepsilon$ there are no equilibrium states in

FIGURE 172.

$U_\varepsilon(L)$, and so the set of all limit points of $L'$ must be a closed path $L^*$. Since by assumption $L$ is not a limit path for any nonclosed path, it follows that $L^*$ is distinct from $L$ and so lies entirely within $\gamma$. Moreover, by Lemma 14 of §3 $L^*$ necessarily crosses the subarc $QP$ of $l$. But this contradicts the assumption that no closed paths cross this subarc.

It follows that there are points on $l$, arbitrarily near $Q$ and on both sides of $Q$, through which pass closed paths other than $L$. We can now prove that $L$ is orbitally stable. By what we have proved and by Lemma 14 of §3, for any $\varepsilon > 0$ we can find closed paths $L^*$ and $L^{**}$, one inside and one outside $L$, so close together that the region $\Gamma$ enclosed between $L^*$ and $L^{**}$ is entirely contained in $U_\varepsilon(L)$. The path $L$ will obviously lie inside $\Gamma$.

Let $Q$ be any point on $L$ and $\delta > 0$ so small that $U_\delta(Q)$ is in $\gamma$. Then it is obvious that any path through a point of $U_\delta(Q)$ will remain inside $\gamma$, therefore also inside $U_\varepsilon(L)$, both as $t \to +\infty$ and as $t \to -\infty$. This means that $L$ is orbitally stable. Q.E.D.

### 9. POSSIBLE TYPES OF SINGULAR AND NONSINGULAR PATHS IN CASE OF FINITELY MANY EQUILIBRIUM STATES. CASE OF FINITELY MANY SINGULAR PATHS.

Assume that our dynamic system has only finitely many equilibrium states. In this case the results proved in this chapter yield an exhaustive classification of all types of orbitally unstable or singular paths.

Indeed, we have already listed all possible types of path when there are finitely many equilibrium states (§4.6). It is then readily seen, on the basis of the theorems proved hitherto, that the only possible orbitally unstable paths are of the following three types: 1) Equilibrium state, to which tends at least one other path. 2) Closed path which is the limit path of a nonclosed path. 3) Path of which at least one semipath is a separatrix of an equilibrium state.

Note that even when the number of equilibrium states is finite the number of orbitally unstable paths may be either finite or infinite. There are even geometric examples with infinitely many orbitally unstable paths dense in a certain region of the plane (see Appendix, §9).

We define a s i n g u l a r  p a t h to be any bounded orbitally unstable path or orbitally stable equilibrium state (center).

In the sequel, we shall consider only dynamic systems, defined in a plane region, with the property that the number of singular paths in any bounded subregion is finite.

The case of finitely many singular paths, owing to its simplicity, is obviously the most important for applications. But it is also interesting from the purely mathematical standpoint.

It follows from the classical papers of Bendixson /33/ and Dulac /37/ that every analytic dynamic system whose right-hand sides have no non-constant common divisor has only finitely many orbitally unstable paths in any bounded region of the plane.*

The concept of orbital stability and instability carries over to dynamic systems on the sphere. We shall not dwell upon the details, which are quite obvious.

§16. CELLS OF A DYNAMIC SYSTEM WHICH
HAS FINITELY MANY SINGULAR PATHS

**1. INTRODUCTORY REMARKS.** We shall assume throughout this section (and also in Chapters VIII, X and XI to come) that a l l  t h e  d y n a m i c

---

* The number of singular paths is also finite in the case of the so-called "structurally stable" systems, which are in a sense more general than the systems considered here /6/.

systems under consideration have finitely many singular paths.

In addition, when studying a dynamic system defined in a plane region we shall consider the system not in its entire domain of definition $G$ but in some bounded closed subregion $G^*$ of $G$, with a comparatively simple boundary — a "normal" boundary. This assumption aims at avoiding certain unnecessary complications.

In this context, the set of singular paths in the interior of the closed region $\bar{G}^*$ (i. e., the set of orbitally unstable paths in $\bar{G}^*$ plus all orbitally stable equilibrium states) is supplemented by a finite number of arcs without contact, arcs of paths and certain semipaths, characterizing the normal boundary of the region $G^*$ in which the dynamic system is being considered.

All the singular paths, and also the arcs and semipaths characterizing the boundary, will be called "singular elements."

For dynamic systems on the sphere, the set of singular elements is simply the set of singular paths.

The singular elements partition the sphere or, as the case may be, the plane region $\bar{G}^*$, into finitely many regions, which we shall cell c e l l s. Each cell is filled by orbitally stable or nonsingular paths. For dynamic systems on the plane, there also exist cells filled not by whole orbitally stable paths but by orbitally stable semipaths and arcs of paths.

We shall try to characterize the various types of cell and examine the possible behavior of the nonsingular elements, nonsingular paths, semi-paths and arcs of paths within each cell. In addition, we shall show that each cell is either simply or doubly connected and give a detailed characterization of cells filled by nonclosed paths and cells filled by closed paths.

**2. NORMAL BOUNDARY OF A BOUNDED SUBREGION $\bar{G}^*$ OF THE DOMAIN OF DEFINITION.** Assume that our dynamic system (I), defined in a plane region $G$ (which may also be the entire plane), has only finitely many singular paths in any bounded subregion of $G$. Let $\bar{G}^*$ be a bounded closed subregion of $G$ $(\bar{G}^* \subset G)$.

The boundary of $G^*$ is said to be n o r m a l (for the dynamic system in question) if it satisfies the following conditions:

1. It is the union of finitely many simple closed curves.

2. Each of these simple closed curves is either a cycle without contact for the paths of the system, a closed path of the system, or a closed curve which is the union of a finite (even) number of alternating arcs without contact and arcs of paths.

3. None of the (arcs of) paths figuring in the boundary may be (part of) orbitally unstable paths or semipaths lying entirely in $\bar{G}^*$. In particular, no equilibrium state can be a point of the boundary.

Figures 173 and 174 show examples of regions with normal boundary.

The paths, arcs without contact, arcs of paths and cycles without contact figuring in the boundary will be referred to as boundary paths, boundary arcs, etc.

Any point common to a boundary arc without contact and a boundary arc of a path will be called a c o r n e r p o i n t of the boundary (e. g., the points $M_1$ and $M_2$ in Figure 173).

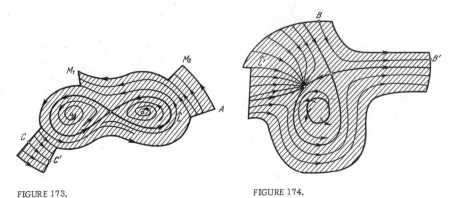

FIGURE 173.                                    FIGURE 174.

Consider an arc of a path whose endpoints are points of boundary arcs without contact, and suppose that all its interior points are points of $\overline{G}^*$. We call this arc a c o r n e r  a r c if at least one of its endpoints is a corner point (e.g., the arc $AM_2$ in Figure 173) and a w h o l e  n o n s i n g u l a r  a r c if neither of its endpoints is a corner point of the boundary (the arcs $CC'$ in Figure 173 and $BB'$ in Figure 174). A semipath whose endpoint is a corner point, all its other points being in $\overline{G}^*$, is a c o r n e r  s e m i p a t h (the semipath $\hat{L}$ in Figure 173 and $\hat{L}'$ in Figure 174). By condition 3), any corner semipath is orbitally stable. It is clear that under our assumptions the number of corner arcs and corner semipaths is always finite.*

Consider the set of singular paths in $\overline{G}^*$.

It follows from the definition of a normal boundary that any singular path lies wholly in the (open) region $G^*$ (see condition 3)).

A path which is not singular is said to be n o n s i n g u l a r. We shall also use the term s i n g u l a r  s e m i p a t h for any orbitally unstable semipath, corner semipath or semipath of a singular path, lying in $\overline{G}^*$. A semipath which is not singular will be called nonsingular. A nonsingular semipath is always orbitally stable, but a singular semipath may be either orbitally unstable or orbitally stable. A singular semipath is always orbitally stable if it is a corner semipath (by condition 3)), but it may be orbitally stable even when it is a semipath of a singular path (since one of the semipaths of an orbitally unstable path may be orbitally stable).

We shall use the term s i n g u l a r  a r c for a boundary arc without contact, boundary or corner arc of a path.

Singular paths, corner semipaths, singular semipaths whose endpoints are on boundary arcs without contact and singular arcs will be called s i n g u l a r  e l e m e n t s. Under the assumption that the number of orbitally unstable paths is finite and the boundary of the region $\overline{G}^*$ in which the dynamic system is being considered is normal, the number of singular elements is finite.

**3. LEMMA ON THE SET OF POINTS OF SINGULAR ELEMENTS.** Let $E \subset \overline{G}^*$ be the set of all points of singular elements in the closed region $\overline{G}^*$.

---

* It can be shown that when the region $G$ is bounded, there exists for any $\varepsilon > 0$ a closed region $\overline{G}^* \subset G$ with normal boundary, contained in the $\varepsilon$-neighborhood of the boundary of $G$.

*Lemma 1. E is a closed set.*

Proof. For any infinite sequence of points of $E$, all points of the sequence can belong only to finitely many singular elements (since there are only finitely many of the latter). To prove the lemma, therefore, we need only show that a point of accumulation of any sequence of points in one singular element is again a member of $E$. But any point of accumulation of a sequence of points on a singular path or semipath either lies on the (semi)path itself or is a limit point thereof, and so must be on a singular path — thus an element of $E$. As for a point of accumulation of points of a singular arc, this must be a point of the arc itself. Q.E.D.

The complement of $E$ in $\bar{G}^*$ is an open set, which may be the union of a finite or countable set of disjoint regions. We shall show later that the number of such regions, which we call c e l l s  o f  t h e  p h a s e  p o r t r a i t, or simply cells, is finite.

Obviously, the points of the cells belong to either  a) whole nonsingular (orbitally stable) paths,  b) nonsingular (orbitally stable) semipaths with endpoints on boundary arcs without contact, or  c) whole nonsingular arcs of paths, i. e., arcs with endpoints on boundary arcs without contact.

The points of a ny nonsingular path or semipath or the points of a nonsingular whole arc must be in some cell.

We now cite a few simple auxiliary propositions pertaining to cells and their boundaries. In particular, it will follow from these propositions that the number of cells is finite.

The first lemma is valid for arbitrary regions consisting of whole paths, i. e., regions which, together with any point, contain the whole path through this point.

*Lemma 2. If $P$ is a boundary point of a region consisting of whole paths (in particular, of a cell consisting of whole paths), then all points of the path $L_P$ through $P$ are boundary points of the region.*

An alternative formulation is as follows: The boundary of a region consisting of whole paths also consists of whole paths.

*Lemma 3. If at least one non-corner point of a boundary arc of a path or a corner arc is a boundary point of a cell, then all points of the arc are boundary points of the cell.*

*Lemma 4. If at least one non-corner point of a singular semipath crossing a boundary arc without contact (i. e., a point of a corner semipath or orbitally unstable semipath crossing a boundary arc without contact) is a boundary point of a cell, then all points of the arc or semipath are also boundary points of the cell.*

The truth of Lemmas 2, 3 and 4 follows directly from the continuity of solutions with respect to initial values.

**4. PROOF THAT THE NUMBER OF CELLS IS FINITE (CASE OF FINITELY MANY SINGULAR ELEMENTS).**

*Lemma 5. A singular path, not an equilibrium state, which is not a limit path for any other path or semipath in $\bar{G}^*$, may be part of the boundary of at most two cells.*

Proof. Let $L$ be a singular path which is not a limit path, $Q$ a point on this path. Let $l$ be an arc without contact with midpoint $Q$ on which there are no points of singular arcs (it is clear that such an arc always exists).

By Lemma 3 of §3, all paths passing through points of a sufficiently small neighborhood of $Q$ must cross $l$. Thus this arc contains points of all cells for which $Q$ is a boundary point. We claim that there is a subarc $Q_1 Q_2$ of $l$, containing $Q$ as an interior point, on which there are no points of a singular path or semipath except $Q$. Indeed, suppose that no such subarc can be found. Then we can obviously find on $l$ a sequence of points, tending to $Q$, belonging to singular paths or semipaths. Since by assumption there are only finitely many singular elements, infinitely many points of this sequence lie on the same singular semipath. But then $L$ is a limit path for another semipath in $\bar{G}^*$, contrary to assumption. Consequently, there is a subarc $Q_1 Q_2$ of $l$, containing $Q$, on which there are no points of singular elements except $Q$. And then all points other than $Q$ on each of the arcs $Q_1 Q$ and $Q_2 Q$ lie in the same cells, so that $Q$ is a boundary point for at most two cells. By Lemma 2, this is true of all other points of the path $L$. Q.E.D.

R e m a r k. All points of $L$ are obviously accessible boundary points of the cell.

L e m m a 6. *All points of a corner arc or corner path, or of an orbitally unstable semipath crossing a boundary arc without contact, may be boundary points of at most two cells.*

P r o o f. We first prove that a corner arc, corner semipath or orbitally unstable semipath crossing a boundary arc cannot consist of limit points of any path lying entirely in $\bar{G}^*$, other than the path of which it is a part. For corner arcs and corner semipaths this follows at once from the definition of normal boundary (condition 3)).

Now consider an orbitally unstable semipath $L_0^{()}$ crossing a boundary arc without contact. By the definition of a normal boundary, the semipath $L_0^{()}$ crosses the boundary arc without contact at one of its interior points. Moreover, if $L_0^{()}$ is a limit semipath for some other semipath, the latter will obviously cross the same arc without contact at infinitely many points. But no semipath can cross a boundary arc without contact more than once without leaving (with increasing a n d decreasing $t$) the region $\bar{G}^*$. It follows that $L_0^{()}$ cannot consist of limit points of any semipath in $\bar{G}^*$. The truth of our lemma now follows as in the proof of Lemma 5.

L e m m a 7. *A singular path $L_0$, not an equilibrium state, which is a limit path for another path, may be part of the boundary of at most finitely many cells.*

P r o o f. Suppose on the contrary that $L_0$ is a boundary path for infinitely many cells. It follows by Lemma 2 that each of its points is a boundary point of infinitely many cells. Through some point $Q$ of $L_0$, draw an arc without contact $l$ which has no points in common with boundary and corner arcs (this is clearly possible). Then (see Lemma 3 of §3) the arc $l$ will contain points of each of the cells of which $Q$ is a boundary point, arbitrarily close to $Q$. Let $\{A_i\}$ be a sequence of points on $l$ tending to the point $Q$, belonging to different cells. Between any two of these points there are boundary points $B_1$, $B_2$ of d i f f e r e n t cells, which are points of singular elements. Since there are only finitely many singular elements, an infinite number of these points must lie on the same singular element, hence on the same singular path $L_1$ (or singular semipath $L_1^{()}$). This (semi)path is thus a boundary (semi)path for infinitely many different cells. The path $L_1$ (semipath $L_1^{()}$) is not closed, since it must have more than one point in common with the arc $l$ and has a limit point $Q$ which is not an equilibrium state. Thus, by Theorem 11, the path $L_1$ itself (or semipath $L'_1$) cannot be a limit (semi)path. But then by Lemma 3 it cannot consist of boundary points of more than two cells. This contradiction completes the proof.

*Theorem 42.* *If a dynamic system defined in a plane region G has only finitely many orbitally unstable paths in any bounded subregion of G, then the number of cells in any closed bounded region $\overline{G}{}^* \subset G$ with normal boundary is finite.*

Proof. Suppose the assertion false, so that some region $\overline{G}{}^* \subset G$ with normal boundary contains a countable set of cells $\{H_i\}$. Take one point in each cell, $A_1, A_2, \ldots$, and join these points by simple arcs lying in $\overline{G}{}^*$ and containing no equilibrium states. Let $B_i$ be a boundary point of the cell $H_i$ on the arc $A_i A_{i+1}$. All the points $B_i$ are in singular elements. If only finitely many of the points $B_i$ are distinct, at least one of them must be a boundary point for infinitely many cells, but this is impossible by Lemmas 5, 6 and 7. Consequently, there are infinitely many distinct points among the $B_i$. But since these are points of singular elements, of which there is only a finite number, an infinite set of points $B_i$ must lie on the same singular element. This element is therefore a boundary element for infinitely many cells, and this contradicts Lemmas 5, 6 and 7. Q.E.D.

**5. DYNAMIC SYSTEMS ON THE SPHERE.** We now digress briefly to dynamic systems on the sphere. As in the plane case, a singular path or singular element is an orbitally unstable path or orbitally stable equilibrium state. A path which is not singular (i. e., orbitally stable path) is said to be nonsingular. A semipath of a singular path will also be called a singular semipath. Let $E$ be the set of points of all singular paths. The following analog of Lemma 1 is valid (the proof proceeds along the same lines):

*Lemma 1'.* *E is a closed set.*

Let $S$ denote the set of points of the sphere. The complement of $E$ in the sphere, i. e., the set $S \setminus E$, is obviously an open set and is therefore the union of a finite or countable number of regions. As in the case of dynamic systems on the sphere, these regions will be called cells of the phase portrait. It is clear that the points of cells belong to whole orbitally stable paths.

Lemmas 2, 5 and 7 remain valid for dynamic systems on the sphere. The proof of the following theorem is analogous to that of Theorem 42.

*Theorem 43.* *If the number of singular paths of a dynamic system on the sphere is finite, the number of cells is also finite.*

**6. BEHAVIOR OF PATHS NEAR ORBITALLY STABLE PATHS.** As before, all propositions concerning semipaths will be stated only for positive semipaths.

*Lemma 8.* *Let $L^+$ be a nonsingular semipath tending to an equilibrium state O. Each point of the semipath has a neighborhood, all paths passing through points of which tend to O as $t \to \infty$ and are nonsingular paths.*

Proof. Let $C$ be a circle about $O$ containing no other equilibrium states, $P$ the last point of $L^+$ in common with this circle. Let $M_0$ be a point on the arc $PO$ of $L^+$; consider an arc without contact through $M_0$, containing $M_0$ as an interior point.

Since $L^+$ is orbitally stable, it is continuable relative to the circle $C$ and so (see Theorem 38) there exists a subarc $AB$ of $l$, containing $M_0$ as an interior point, such that every path crossing $AB$ tends to $O$ with increasing $t$ without leaving the circle $C$ (see Definition XVIII).

Since the number of singular paths is finite and the path $L$, being nonsingular, is not a limit path for any path, it follows easily that there may be only finitely many points of singular paths and semipaths on the subarc $AB$ of $l$. Consequently, there exists a subarc $A_1B_1$ of $AB$, containing $M_0$ as an interior point, all of whose points are on nonsingular paths tending to $O$ as $t \to \infty$. But by Lemma 5 of §3 (Chapter 2) there exists a neighborhood of any point of $L^+$, all paths through the points of which cross the subarc $A_1B_1$ of $l$. This proves our lemma.

Now consider a semipath $L^+$ of a nonclosed path, whose limit points are not all equilibrium states.

Let $K$ be the limit continuum of $L^+$ and $P$ any point of the continuum lying on a path $L_0 \subset K$ which is not an equilibrium state. Draw through $P$ an arc without contact $l$, containing $P$ as an interior point. Let $\{P_i\}$ be a sequence of points of intersection of $l$ and $L$, tending to the point $P$.

As before (§15.8), we let $C_i$ denote the simple closed curve formed from the turn $P_iP_{i+1}$ of the semipath $L^+$ and the subarc $P_iP_{i+1}$ of $l$, $h_{ik}$ the region bounded by $C_i$ and $C_k$ (Figure 170), and $\gamma_i$ the region (see Lemma 7 of §15) whose boundary is the union of $C_i$ and $K$ (Figure 171).

By Lemma 8 of §15, if $i$ is sufficiently large either $C_{i+1}$ lies (except for the point $P_i$) in the interior of $C_i$ and all these curves contain $K$ in their interior, or $C_i$ lies (except for $P_i$) in the interior of $C_{i+1}$ and the continuum $K$ is exterior to all these curves. To fix ideas we shall assume that the first alternative holds.

Lemma 9. *For all sufficiently large $j$ and $k$, $j > k$, any path through a point of the region $h_{jk}$ leaves this region with both increasing and decreasing $t$; with increasing $t$ it leaves across the curve $C_j$ on the subarc $P_jP_{j+1}$ of $l$; with decreasing $t$ it leaves across $C_k$ on the subarc $P_kP_{k+1}$ of $l$.*

Proof. Since there are finitely many singular elements, there exists $\varepsilon_0 > 0$ such that, apart from the singular paths making up $K$, the neighborhood $U_{\varepsilon_0}(K)$ contains no other singular paths.

By Lemma 7 of §15, for all sufficiently large $i$, say $i > I$, all curves $C_i$ and regions $\gamma_i$, hence also all regions $h_{jk}$ $(j > k > I)$, are contained in $U_{\varepsilon_0}(K)$. Hence it follows that the region $h_{jk}$ does not contain any whole singular path. Indeed, the continuum $K$ is interior to all the $C_i$ $(i > I)$ and the neighborhood $U_{\varepsilon_0}(K)$, which contains the region $h_{jk}$, contains no whole singular path. This means that the path $L'$ passing through a point of $h_{jk}$ must leave this region with both increasing and decreasing $t$; otherwise $h_{jk}$ would contain some whole limit path of $L'$, which we have just stated to be impossible. But no path $L'$ can leave $h_{jk}$ without crossing the curves $C_j$ and $C_k$ on the subarcs $P_jP_{j+1}$ and $P_kP_{k+1}$, respectively, of $l$. Since $C_j$ is inside $C_k$ $(j > k)$ and the continuum $K$ is inside $C_j$, it clearly follows that any path $L'$, exiting from the region $h_{jk}$, crosses the subarc $P_jP_{j+1}$ of $l$ with increasing $t$ and the subarc $P_kP_{k+1}$ of $l$ with decreasing $t$. Q.E.D.

Corollary 1. Any path crossing the subarc $P_iP_{i+1}$ of $l$ at $t = t_0$ will cross its subarc $P_{i+1}P_{i+2}$ at some time $t > t_0$.

Corollary 2. There exists a subarc $PA$ of the arc without contact $l$, containing all points $P_i$ for sufficiently large $i$ $(i > I)$, such that any path $L'$ crossing this subarc at $t = t_0$ at a point other than $P$ will cross it again at some $t > t_0$.

Corollary 3. $P$ is an $\omega$-limit point of any path crossing the subarc $P_iP_{i+1}$ of $l$ for sufficiently large $i$ ($i > I$).

Corollary 4. Any path passing through a point of the region $\gamma_i$ will leave it with decreasing $t$ across the subarc $P_iP_{i+1}$ of $l$.

Lemma 10. *If $i$ is sufficiently large, $K$ is the $\omega$-limit continuum of all paths passing through points of the region $\gamma_i$.*

Proof. As before, let $\varepsilon_0 > 0$ be such that the $\varepsilon_0$-neighborhood of the continuum $K$ contains no whole singular path other than the paths making up $K$ itself. Let $i_0$ be so large that the region $\gamma_{i_0}$ is inside $U_{\varepsilon_0}(K)$. Any path through a point of $\gamma_{i_0}$ clearly cannot leave $\gamma_{i_0}$ with increasing $t$ (all paths cross the subarc $P_{i_0}P_{i_0+1}$ of $l$ into this region). Since the closure of $\gamma_{i_0}$ contains no whole singular paths other than the paths making up $K$, it follows that all the $\omega$-limit points of any path through points of $\gamma_{i_0}$ must be points of $K$.

We now show that all the points of $K$ are $\omega$-limit points for any such path.

Let $L_0$ be one of the paths making up $K$ (not an equilibrium state), $P$ a point (the endpoint of the arc $l$) on $L_0$. Let $L_1$, $L_2$, . . ., $L_r$ be all the other paths (not equilibrium states) making up $K$ (by assumption, there are only finitely many singular paths, a fortiori only finitely many of these special singular paths). On each path $L_m$ ($m = 1, 2, . . ., r$) take one point $P^m$ and draw through $P^m$ an arc without contact $l_m$ such that the arcs $l_m$ are all disjoint. On each $l_m$ we have a sequence of points $\{P_j^m\}$ on the semipath $L^+$, tending to the point $P^m$ (and corresponding to indefinitely increasing times $t$). Let $C_j^m$ denote the simple closed curve formed from the turn $P_j^mP_{j+1}^m$ of the semipath $L^+$ and the subarc $P_j^mP_{j+1}^m$ of the arc $l_m$ (Figure 175 shows the case $r = 2$).

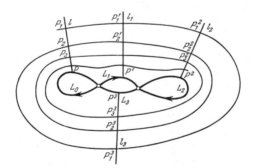

FIGURE 175.

For any fixed $m$ ($m = 1, 2, . . ., r$) the sequence of curves $\{C_j^m\}$ obviously satisfies the same conditions as the previous sequence $\{C_i\}$.

To unify the notation, we put $l_0$ for $l$, $P_i^0$ for $P_i$, and $\{C_i^0\}$ for the sequence of curves previously denoted by $\{C_i\}$. Lemmas 8 and 9 of §15 are valid for $\{C_j^m\}$, $m = 1, 2, . . ., r$. The continuum $K$ lies inside all curves $\{C_j^m\}$, $m = 1, 2, . . ., r$ (from some $j$ on). Indeed, for sufficiently large $t$ the points of $L^+$ are inside the curve $C_i^0$ ($i > i_0$), hence, for any $i > i_0$, there exists $j$ such that the turn $P_j^mP_{j+1}^m$ of the semipath $L^+$, therefore also the curve $C_j^m$, is inside $C_i^0$.

The converse is obviously also true: for any given $m \neq 0$ and fixed $j$ one can always find $i$ so large that the curve $C_i^0$ is inside $C_j^m$. Thus the continuum $K$ lies inside all the curves $C_j^m$ for sufficiently large $j$, for any $m = 0, 1, 2, \ldots, r$.

Now let $i_1 > i$ (where $i$ is fixed as above) be such that $C_{i_1}^1$ lies inside $C_i^0$, $i_2 > i_1$ such that $C_{i_2}^2$ lies inside $C_{i_1}^1$, etc. (it follows from the foregoing discussion that such integers $i_1, i_2, \ldots, i_r$ may always be found). We thus have simple closed curves $C_i^0, C_{i_1}^1, C_{i_2}^2, \ldots, C_{i_r}^r$ each interior to its predecessor in the sequence and interior to the region $\gamma_i$ (Figure 176). Apart from the paths making up $K$, our $\varepsilon_0$-neighborhood of $K$ contains no other whole singular paths. And since the continuum $K$ lies inside (outside) all the curves $C_{i_k}^k$ ($k = 0, 1, \ldots, r$), any path passing through the region enclosed between any two curves $C_{i_k}^k$ and $C_{i_{k+1}}^{k+1}$ must leave this region with both increasing and decreasing $t$; consequently (Lemma 9, Corollary 4), it must cross both subarcs $P_{i_k}^k P_{i_k+1}^k$ and $P_{i_{k+1}}^{k+1} P_{i_{k+1}+1}^{k+1}$ of the arc $l_k$. On the other hand, any path through a point of the region $\gamma_{i_r}$ must cross the subarc $P_{i_r}^r P_{i_r+1}^r$ of $l_r$. Hence it follows that any path passing through some point of $\gamma_i$ crosses some subarc $P_j^k P_{j+1}^k$ of each arc $l_k$.

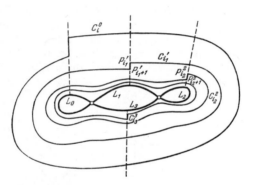

$$C_i^0$$

$$P_{i_1}', \quad C_{i_1}'$$
$$P_{i_1+1}' \qquad P_2^2$$
$$P_{i_2+1}^2$$
$$L_0 \qquad L_1 \qquad L_3 \qquad L_2 \qquad C_{i_2}^2$$
$$C_3^3$$

FIGURE 176.

By Corollary 3 to Lemma 9, each point $P^k$ ($k = 0, 1, \ldots, r$) is a limit point of any such path, and thus each $L_k$ is a limit path for such paths. But every equilibrium state in $K$ is also a limit point for these paths, since any such equilibrium state is a limit point for one of the paths $L_k$. And this means that all points of $K$ are limit points for any path through points of $\gamma_i$. Q.E.D.

*Lemma 11. Let $L^+$ be a nonclosed nonsingular semipath whose limit points are not all equilibrium states. Then each point of $L^+$ has a neighborhood, all paths and semipaths passing through which are nonclosed nonsingular (semi)paths with the same $\omega$-limit points as $L^+$.*

Proof. Let $M$ be any point on $L^+$. By Lemma 7 of §3, there is a neighborhood of $M$ all paths passing through which are nonclosed. Moreover, since the nonsingular paths and semipaths fill out regions, there is a neighborhood of $M$, all paths or semipaths through which are nonsingular.

We now show that the $\omega$-limit points of all paths or semipaths passing through a sufficiently small neighborhood of $M$ are limit points of $L^+$.

Let $K$ be the limit continuum of the semipath $L^+$, $P$ a point in $K$ (not an equilibrium state) and $l$ an arc without contact through $P$. The semipath $L^+$ crosses $l$ at infinitely many points $\{P_i\}$.

Let the closed curves $C_i$ and regions $\gamma_i$ be as before, and suppose $i$ so large that all paths passing through points of $\gamma_i$ have $\omega$-limit continuum $K$ (Lemma 10). Let $M^*$ be any point of $L^+$ inside $\gamma_i$. Then a sufficiently small neighborhood $U_\delta(M^*)$ of $M^*$ is also in $\gamma_i$, and so any path through a point of $U_\delta(M^*)$ has $\omega$-limit continuum $K$. By the continuity of solutions with respect to initial values, the same is true of paths through points of a sufficiently small neighborhood of $M$.

*L e m m a  12.* *Each point of a nonsingular closed path has a neighborhood, all paths through which are nonsingular closed paths.*

P r o o f. Let $L$ be a nonsingular closed path. Since the number of singular elements is finite, all points of $L$ are points of some cell. It is readily seen that there exists $\varepsilon > 0$ such that $U_\varepsilon(L)$ lies in the same cell, i. e., all paths through points of $U_\varepsilon(L)$ are nonsingular.

Let $P$ be any point on $L$. Let $\varepsilon' < \varepsilon$, and choose $\delta > 0$ so small that all paths passing through $U_\delta(P)$ do not leave $U_{\varepsilon'}(L)$. That this is possible follows from the orbital stability of $L$. Suppose that there are nonclosed paths passing through $U_\delta(P)$. Since by assumption they do not leave $U_{\varepsilon'}(L)$, their limit points are also all in or on the boundary of $U_{\varepsilon'}(L)$, hence a fortiori in $U_{\varepsilon'}(L)$. But the set of limit points of a nonclosed path is a union of singular paths. Consequently, there is at least one singular path in $U_\varepsilon(L)$, contradicting the choice of $\varepsilon$.

Thus, all paths through the points of this $\varepsilon$-neighborhood are closed. By Lemma 14 of §3, if $\delta$ is sufficiently small all these paths lie interior to one another. Q.E.D.

R e m a r k. If there are closed paths through any arbitrarily small neighborhood of a closed path $L_0$, then (provided the number of singular elements is finite) $L_0$ must be nonsingular, i. e., all paths passing through a sufficiently small neighborhood are closed.

*T h e o r e m  44.* *All paths passing through some sufficiently small neighborhood of an orbitally stable equilibrium state O are nonsingular closed paths (interior to one another and enclosing O).*

P r o o f. Since $O$ is an orbitally stable equilibrium state, no path can tend to it. Consequently, any neighborhood of $O$ contains closed paths enclosing $O$. Since there are only finitely many singular elements, the equilibrium state always has a neighborhood in which there are no singular paths. But then (see Remark to Lemma 12) all paths in this neighborhood are nonsingular closed paths. Q.E.D.

As we know, in this case the equilibrium state is called a c e n t e r.

The last lemma in this subsection concerns nonsingular whole arcs, i. e., arcs of paths whose endpoints lie on boundary arcs without contact but are not corner points, and all interior points of the arc are points of $\overline{G}^*$ (see §16.1); also treated in this lemma are nonsingular semipaths whose endpoints lie on boundary arcs (or cycles) without contact. The lemma is a direct consequence of Lemma 5 in §3.

L e m m a 13. *a) Each point of a nonsingular whole arc of a path $\Lambda$, not an endpoint of the arc, has a neighborhood through all points of which pass nonsingular whole arcs of paths crossing the same boundary arcs without contact as $\Lambda$. b) Each point of a nonsingular semipath $L^+$ whose endpoint is on a boundary arc (or cycle) without contact has a neighborhood, through all points of which pass nonsingular positive semipaths whose endpoints lie on the same arc (cycle) without contact as that of $L^+$.*

The analog of part *b)* holds for negative semipaths.

R e m a r k. If some of the nonsingular elements of a given cell are nonsingular whole arcs of paths or nonsingular semipaths (with endpoints on a boundary arc without contact), then it is clear that the boundary of the cell necessarily includes a s u b a r c of this boundary arc which contains no points of other singular elements (i. e., boundary arcs of paths or singular semipaths).

The converse is also obvious: If some of the boundary points of the cell are points of boundary arcs without contact belonging to no other singular elements (i. e., boundary arcs of paths or singular semipaths), then the set of nonsingular elements in the cell necessarily contains either nonsingular whole arcs of paths or nonsingular semipaths whose endpoints lie on the aforementioned boundary arc without contact.*

## 7. PROPOSITIONS PERTAINING TO NONCLOSED ORBITALLY STABLE PATHS.

L e m m a 14. *Let $P$ be any point of a nonclosed nonsingular path, $l$ an arc without contact containing $P$ as an interior point. Then one can always find a subarc $l'$ of $l$, also containing $P$ as an interior point, such that all paths crossing $l'$ have only one point in common with $l'$.*

P r o o f. By Lemma 3 of §3, we may assume that $l$ has no point other than $P$ in common with the path $L$. By the Remark to Lemma 3 of §3, there exists $\tau > 0$ such that all paths crossing $l$ at $t = t_0$ (including the path $L$, which crosses $l$ at $P$) have no other points in common with $l$ at any time $t \neq t_0$ such that

$$|t - t_0| \leqslant \tau.$$

Let $P_1$ denote the point of $L$ corresponding to $t_1 = t_0 - \tau$, and $P_2$ the point of $L$ corresponding to $t_2 = t_0 + \tau$ (Figure 177).

Let $\varepsilon_0 > 0$ be so small that the neighborhoods $U_{\varepsilon_0}(L_{P_1}^-)$ and $U_{\varepsilon_0}(L_{P_2}^+)$ are disjoint from $U_{\varepsilon_0}(P)$ (though $U_{\varepsilon_0}(L_{P_1}^-)$ and $U_{\varepsilon_0}(L_{P_2}^+)$ may have points in common with each other). The existence of $\varepsilon_0$ follows from the fact that $L$ is a nonclosed path and therefore $P$ cannot be a limit point of the path. Since $L$ is orbitally stable, there exist $\delta_1 > 0$ such that all paths passing through the $\delta_1$-neighborhood of $P_1$ do not leave the $\varepsilon_0$-neighborhood of $L_{P_1}^-$ with decreasing $t$, and $\delta_2 > 0$ such that all paths passing through the $\delta_2$-neighborhood of $P_2$ do not leave the $\varepsilon_0$-neighborhood of $L_{P_2}^+$ with increasing $t$. Let $l'$ be a subarc of $l$ with midpoint $P$, so small that it is contained in $U_{\varepsilon_0}(P)$ and all paths crossing it at $t = t_0$ pass at $t = t_1$ through points of the $\delta_1$-neighborhood of $P_1$ and at $t = t_2$ through points of the $\delta_2$-neighborhood

---

* It may happen that the endpoints of a boundary arc without contact (i. e., corner points) figure in the boundary of the cell, while none of its interior points are in the boundary of the cell.

of $P_2$. It follows from the choice of $\tau$, $\varepsilon_0$, $\delta_1$ and $\delta_2$ that the subarc $l'$ satisfies our requirements. Q.E.D.

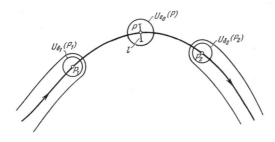

FIGURE 177.

As before, let $L$ be a nonclosed nonsingular (orbitally stable) path and $P$ a point on L.

Lemma 15. *For any* $\varepsilon > 0$, *there exists* $\delta > 0$ *such that if* $L'$ *is an arbitrary path passing through the neighborhood* $U_\delta(P)$ *then there exists a region* $H$, *whose boundary is the union of* $L$, $L'$ *and the* $\alpha$- *and* $\omega$-*limit points of* $L$, *contained entirely in the neighborhood* $U_\varepsilon(L)$.

Proof. Let $\varepsilon_1 > 0$ be a number such that $\varepsilon_1 < \varepsilon$ and the neighborhood $U_{\varepsilon_1}(L)$ and its closure $\overline{U}_{\varepsilon_1}(L)$ do not contain any whole singular path except for the $\alpha$- and $\omega$-limit paths of $L$.

Let $l_0$ be an arc without contact containing $P$ as an interior point, satisfying the following conditions: a) all paths crossing the arc do so at only one point; b) all paths crossing the arc do not leave $U_{\varepsilon_1}(L)$. The arc $l_0$ exists by virtue of Lemma 14 and the fact that $L$ is a nonsingular (i. e., orbitally stable) path.

Let $\delta > 0$ be a number such that all paths passing through $U_\delta(P)$ possess the following properties:

1) Without leaving $U_{\varepsilon_1}(P)$, they cross $l_0$; 2) nonclosed; 3) orbitally stable; 4) have the same $\alpha$- and $\omega$-limit points as the path $L$.

The existence of $\delta$ follows from Lemma 3 of §3 and Lemmas 8 and 11.

Let $L'$ be a path, distinct from $L$, passing through some point of the above $\delta$-neighborhood of $P$, and $P'$ its point of intersection with $l_0$. Consider the set of all points of paths passing through the interior points of the subarc $PP'$ of $l_0$. Denote this set by $H$. By the choice of $l_0$ and $\delta$, $H$ is a subset of the $\varepsilon_1$-neighborhood of $L$.

We claim that $H$ is a region.

Indeed, let $M \in H$, i.e., $M$ is a point on a path crossing the subarc $PP'$ at some point $Q$ distinct from $P$ and $P'$. Then there are paths through a sufficiently small neighborhood of $M$ (see Lemma 5 of §3) which cross the subarc $PP'$ arbitrarily near $Q$, therefore at points distinct from the endpoints $P$ and $P'$. This means that all points of some sufficiently small neighborhood of $M$ are in $H$, so that $H$ is an open set. In addition, it is clear that any two points of $h$ can be joined by a simple arc (for example, a combination of arcs of paths and a subarc of the arc $PP'$) (Figure 178).

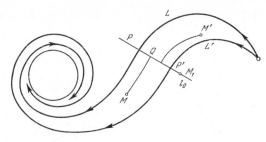

FIGURE 178.

The points of $l_0$ not on the subarc $PP'$ are not in the region $H$. Indeed, suppose that there is a point $M_1$ on $l_0$ outside the subarc $PP'$ but belonging to $H$. Then, by the definition of $H$, the path $L_1$ through $M_1$ passes through some point $Q_1$ of the arc $l_0$ lying between $P$ and $P'$, hence distinct from $M_1$.

Thus the path $L_1$ has at least two points in common with $l_0$, contrary to condition a) above. It follows that $P$ and $P'$ are boundary points of the region $H$, and so (by Lemma 2) all points of the paths $L$ and $L'$ are also boundary points of $H$. But then the $\alpha$- and $\omega$-limit points of $L'$ are also boundary points of $H$, and by the choice of $\varepsilon_1$ and condition b) these coincide with the $\alpha$- and $\omega$- limit points, respectively, of $L$.

We claim that the boundary of $H$ consists precisely of these points — those of $L$ and $L'$ and their $\alpha$- and $\omega$-limit points. Suppose that the boundary of $H$ contains a point $R$ not in one of these categories. Then all points of the path $L_R$ passing through $R$ are also boundary points of $H$ (Lemma 2). The path $L_R$ must be at a positive distance from the subarc $PP'$ of $l_0$. In fact, the distance of $L_R$ from this subarc may vanish only in two cases: when $L_R$ is a path actually crossing the subarc, or when $L_R$ has a limit path crossing it. But neither case is possible because of the assumptions concerning $R$ and the fact that all paths crossing $l_0$ are orbitally stable.

Thus the path $L_R$ is at some positive distance, $d$ say, from the arc $PP'$. But then $L_R$ must be orbitally unstable. Indeed, since the points of $L_R$ are boundary points of the region $H$, it follows that there are paths through an arbitrarily small neighborhood of any point of $L_R$ which cross the arc $PP'$. Consequently, for some suitably chosen number $\sigma > 0$ (e. g., $\sigma < d/3$), they leave the $\sigma$-neighborhood of $L_R$, and this means that $L_R$ is orbitally unstable. But $H$, hence also the path $L_R$, lie entirely within $U_{\varepsilon_1}(L)$, whereas $\varepsilon_1$ was so chosen that $U_{\varepsilon_1}(L)$ contains no orbitally unstable paths other than $\alpha$- and $\omega$-limit paths of $L$. We have derived a contradiction, and so the paths $L$ and $L'$ and their $\alpha$- and $\omega$-limit points exhaust the boundary of $H$, as required. This completes the proof of the lemma.

R e m a r k. In general there may be more than one region whose boundary consists of the paths $L$ and $L'$ and their $\omega$- and $\alpha$-limit points.*

*T h e o r e m  45. All limit points of a singular path $L_0$ (semipath $L_0^+$) which are not equilibrium states are also limit points for the nonsingular paths of any cell whose boundary contains $L_0$.*

---

\* If this region is contained in an $\varepsilon$-neighborhood of $L$, it will be unique provided $\varepsilon$ is sufficiently small.

Proof. Let $P$ be a limit point, not an equilibrium state, of a singular path $L_0$, and let $l$ be an arc without contact through $P$. There exists a sequence of points of $L_0$ on the arc $l$, tending to $P$. Hence there are points on the arc, arbitrarily close to $P$, belonging to any cell whose boundary includes $L_0$. Hence it follows, by Corollary 3 to Lemma 9, that $P$ is a limit point of the nonsingular paths of any such cell. Q.E.D.

## 8. BEHAVIOR OF NONSINGULAR ELEMENTS INSIDE THE SAME CELL.

Theorem 46. *If a cell contains a nonsingular element which is a whole path (or semipath crossing a boundary arc without contact, or arc of path with endpoints on boundary arcs or cycles without contact), then all the nonsingular elements of the cell are whole paths (or, respectively, semipaths crossing a boundary arc without contact, or arcs whose endpoints are on boundary arcs or cycles without contact).*

Proof. Suppose the assertion false: inside some cell containing a whole (nonsingular) path $L$ there is a nonsingular element of another type, say a nonsingular semipath $L'^+$ crossing a boundary arc without contact. Join any point $A$ of $L$ to a point $B$ of $L'^+$ by a simple arc $\lambda$ lying within the cell. On the arc $\lambda$ lie points of two types: through points of the first type pass whole nonsingular paths, and through points of the second there are no whole paths but nonsingular semipaths crossing a boundary arc without contact (or arcs of paths crossing a boundary arc).

By Lemmas 8, 11 and 12, all points sufficiently close to $A$ on $\lambda$ are of the first type, and all points sufficiently close to $B$ on $\lambda$ of the second. Moving along $\lambda$ from $A$ to $B$, we go from points of the first type to points of the second. Consequently, there must be some point $C$ on $\lambda$ which is either the last point of the first type or the first point of the second type. But it follows from Lemmas 8, 11 and 12 that there cannot be a last point of the first type (i. e., a last point through which passes a nonsingular whole path). Consequently, $C$ is the first point of the second type. The nonsingular element through this point is not a whole path, but either a semipath crossing a boundary arc without contact or an arc of a path with endpoints on boundary arcs without contact. But in either case it follows from Lemma 13 that $C$ cannot be the first point on of the second type. Hence all nonsingular elements of the cell are whole paths. Similar arguments apply when the cell contains a semipath or arc of a path crossing a boundary arc without contact. Q.E.D.

Remark. It follows from this theorem and the remark after Lemma 13 that the boundary points of a cell filled by whole paths cannot be points of boundary arcs without contact other than corner points.

## 9. CELLS FILLED BY CLOSED PATHS. We now present a few propositions concerning cells filled by closed paths. The first is proved by the same reasoning as the preceding theorem, and we therefore omit the proof.

Theorem 47. *If a cell contains at least one closed path, then all paths of the cell are closed.*

Theorem 48. *No cell can contain two closed paths exterior to one another.*

Proof. Suppose the assertion false. Let $L$ and $L'$ be two mutually exterior closed paths in the same cell. Let $AB$ be a simple arc joining a point $A$ on $L$ to a point $B$ on $L'$, lying inside the cell and (except for $A$ and $B$) outside $L$ and $L'$. Such an arc always exists.

Indeed, join some point $A'$ on $L$ to a point $B'$ on $L'$ by a simple arc $\lambda$ lying inside the cell. Moving along the arc from $A'$ to $B'$, let $A$ be the last point of $\lambda$ in common with $L$ and $B$ the first point (after $A$) of $\lambda$ in common with $L'$. Then the subarc $AB$ of $\lambda$ satisfies our needs (Figure 179).

Divide the interior points of $AB$ into two types, as follows. The first type includes all points through which pass closed paths containing $L$ but

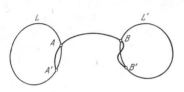

FIGURE 179.

not $L'$ in their interior. For example, all points sufficiently close to $A$ are of this type (Lemma 14, §3). The second type contains all other interior points of the arc $AB$. Such are, for example, all points sufficiently close to $B$ on the arc $AB$. Clearly, there must be a point $C$ on $AB$, any neighborhood of which contains points of both types. Again using Lemma 14 of §3, one readily shows (reasoning exactly as in Theorem 46) that no such point $C$ can exist. Q.E.D.

*Theorem 49. All points lying between two closed paths in the same cell belong to the cell.*

Proof. Suppose on the contrary that there are points between two closed paths $L$ and $L'$ in the same cell which do not belong to the cell, and therefore at least one point belonging to a singular element. Let $M$ be such a point, and suppose, say, that $L$ contains $L'$ in its interior, so that $M$ is inside $L$ and outside $L'$.

Join a point $A$ on $L$ to a point $B$ on $L'$ by a simple arc in the cell, lying (except for its endpoints) outside $L'$ and inside $L$. By Lemma 14 of §3, all paths through points of $AB$ sufficiently close to $B$ do not contain $M$ in their interior, while all paths through points sufficiently close to $B$ will contain $M$.

Continuing as in Theorems 46 and 48, we derive a contradiction. Q.E.D.

*Lemma 16. In any cell filled by closed paths, there exist a sequence of paths $\{L_i\}$ in which each $L_i$ contains its predecessor $L_{i-1}$, and a sequence $\{L'_j\}$ in which each $L'_j$ is inside its predecessor $L'_{j-1}$, such that any path $L^*$ in the cell is interior to some path $L_i$ of the first sequence and exterior to some path $L'_j$ of the second.*

Proof. Let $L$ be some path (closed, of course) in the cell. There must exist points of singular paths both inside and outside this path. Indeed, inside the path there is at least one equilibrium state (Theorem 16, §4), while outside it there are certainly boundary points of the region $G^*$ in which the system is being studied.*

Let $P_1$ and $P_2$ be points of singular paths, $P_1$ outside and $P_2$ inside the closed path $L_1$, and $l$ a simple arc joining $P_1$ and $P_2$. It is clear that $l$ always has a point in common with the path $L$. Let $N$ be one of these common points, and suppose that the first boundary point of the cell encountered when moving along $l$ from $N$ to $P_1$ (from $N$ to $P_2$) is $R_1 (R_2)$ (in particular, $R_1$ may be equal to $P_1$ and $R_2$ to $P_2$). Thus, all interior points of the arc $R_1 R_2$ belong to the cell.

---

* For dynamic systems on the sphere the reasoning is somewhat different: the closed path $L$ divides the sphere into two simply connected regions. By Theorem 16, each of these regions must contain at least one equilibrium state, i.e., point of a singular path.

Let $\{M_n\}$ be some sequence of points on this arc, tending to the point $R_1$. We claim that from any such sequence we can extract a subsequence $\{M_{k_i}\}$ $(k_1 < k_2 < \ldots)$ with the property that for $k_i > k_j$ the path $L_i$ passing through $M_{k_i}$ contains the path $L_j$ through $M_{k_j}$ in its interior.

Indeed, reasoning by induction, suppose that we have already selected $n$ points $M_{k_1}, M_{k_2}, \ldots, M_{k_n}$ with the required property, and let us find an $(n+1)$-th point $M_{k_{n+1}}$. Consider the path $L_n$ through $M_{k_n}$. The point $R_1$, which is a boundary point of the cell and lies outside the path $L$ of this cell, is obviously also outside $L_n$. But then, since the sequence $\{M_n\}$ tends to $R_1$, all of its points, beginning from some sufficiently large $k_{n+1} > k_n$, must also be outside $L_n$.

The path $L_{n+1}$ passing through $M_{k_{n+1}}$ will contain $L_n$ in its interior. We can thus extract the required subsequence from $\{M_n\}$.

Now consider the sequence of paths passing through the points $M_{k_i}$. We claim that this sequence $\{L_i\}$ satisfies the requirements of our lemma. To prove this we must show that for any path $L^*$ in the cell there will always be a path $\tilde{L}_i$ in the sequence $\{L_i\}$ containing $L^*$ in its interior.

Indeed, it is readily seen (via Theorems 47, 48 and 49) that $L^*$ must have a point in common with the arc $R_1R_2$. It then follows by arguments similar to those used to construct the sequence $\{L_i\}$ that all points $M_{k_i}$ with sufficiently large $i$ $(i > I)$ lie outside $L^*$, so that all paths $L_i$ with $i > I$ contain $L^*$ in their interior (see Theorem 48).

Similarly, one proves the existence of the sequence $\{L'_i\}$. This completes the proof of the lemma.

Consider a cell $g$ and one of the two sequences of closed paths whose existence was established in Lemma 16, say the sequence $\{L_i\}$. Every closed path in the cell lies inside some path $L_i$ in the sequence, hence inside all paths of the sequence with sufficiently large $i$.

Let $K$ be the topological limit of the sequence $\{L_i\}$.

*Lemma 17. All points of the continuum $K$ are boundary points of the cell $g$, and apart from these points the paths in the cell enclose no other boundary points of $g$.*

Proof. The points of $K$ are not in $g$, since $K$ lies inside all paths $L_i$ in the sequence $\{L_i\}$ (Theorem 49) and therefore, by the properties of this sequence, inside all paths of $g$. On the other hand, since each point of $K$ is a point of accumulation for points of the paths $L_i$, it follows that all points of $K$ are boundary points of $g$. We claim that any boundary point $N$ of $g$ lying inside all paths of the cell belongs to the continuum $K$.

Consider an arbitrary neighborhood $U_\delta(N)$ of $N$ and some point $Q$ of this neighborhood in the cell. Let $\lambda$ be a simple arc joining $N$ to $Q$ within $U_\delta(N)$. Let $L^*$ be the path passing through $Q$. All paths $L_i$ with sufficiently large $i$ $(i > J)$ lie inside $L^*$. Since $N$ lies inside and $Q$ outside all paths $L_i$ with $i > J$, the arc $\lambda$ must obviously cross each of these paths. Consequently, there are points of all paths $L_i$, $i > J$, on the arc $\lambda$, i. e., in $U_\delta(N)$. Since $U_\delta(N)$ is an arbitrarily small neighborhood, this means that $N$ is a point of the continuum $K$. Q.E.D.

*Theorem 50. Any cell filled by closed paths is doubly connected.*

Proof. Since there are points of singular elements both inside and outside every closed path, a cell filled by closed paths is at least doubly connected. It remains to show that it is at most doubly connected. Let $\{L_i\}$ and $\{L'_j\}$ be sequences of paths with the properties established in Lemma 16. Let $K_1$ be the topological limit of the sequence $\{L_i\}$ and $K_2$

that of the sequence $\{L_i\}$. By Lemma 17, the continuum $K_1$ is the set of all boundary points of $g$ lying inside all paths of the cell, and the continuum $K_2$ the set of all such points lying outside the paths of $g$. By Lemma 17 there are no other boundary points of $g$. Q.E.D.

In view of Lemma 8, we easily see that the boundary of a cell filled by closed paths consists of one or more of the following: whole orbitally unstable paths in $\overline{G}^*$, a closed (orbitally stable) path constituting one of the boundary continua of $\overline{G}^*$, or a closed path consisting of corner and boundary arcs (see, e.g., Figure 180a, b and c).

In this situation one has the following theorem, which is a direct consequence of Lemma 11 and the fact that all paths in the cell are closed.

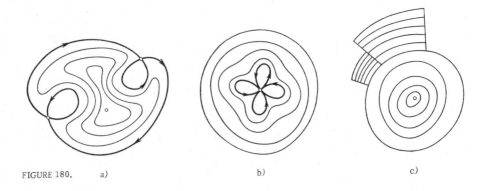

FIGURE 180.        a)                                          b)                                          c)

*Theorem 51. The $\omega$- and $\alpha$- limit points of any singular path in the boundary of a cell filled by closed paths must be equilibrium states.*

**10. CELLS FILLED BY NONCLOSED PATHS.** The proof of the next theorem, using Theorem 46 and Lemma 11, is entirely analogous to that of Theorems 48 and 49 and is therefore omitted.

*Theorem 52. If the paths of a given cell are nonclosed, they have the same $\omega$- limit points and the same $\alpha$- limit points (the limit points of either type may of course be distinct).*

Now let $L$ be a nonclosed path which tends to the same equilibrium state $O$ both as $t \to +\infty$ and as $t \to -\infty$. Then we shall call the simple closed curve formed by $L$ and the point $O$ a l o o p,* saying that the path $L$ f o r m s a l o o p. If two loops with the point $O$ in common bound regions situated interior to one another, we shall say that the loops are n e s t e d (Figure 181); otherwise we shall say that the loops are mutually exterior (Figure 182).

We now devote more attention to the case in which all paths of the cell form loops.

*Theorem 53. If all the paths of a cell form loops, all the loops are nested.*

---

* [The term "oval" is also used /13/.]

FIGURE 181.

FIGURE 182.

Proof. Let $L$ be some path of the cell and $Q$ a point on the path. We shall first show that all paths through points of a sufficiently small neighborhood $U_\delta(Q)$ form nested loops.

Indeed, let $L'$ be the path through a point of $U_\delta(Q)$. By Lemma 15, we can find $\delta > 0$ such that the region $H$ whose boundary is the union of $L$, $L'$ and their $\omega$- and $\alpha$-limit points lie within $U_\varepsilon(L)$, where $\varepsilon$ is any preassigned positive number. In the present case, $H$ is bounded by two simple closed curves — the loops formed by the paths $L$ and $L'$. Were these loops mutually exterior, the region $H$ would be unbounded and could not lie within $U_\varepsilon(L)$, contradicting Lemma 15.

Thus all the paths through a sufficiently small neighborhood of $Q$ form nested loops. But then, using reasoning similar to that used in Theorem 48, one readily sees that all loops formed by paths of the same cell are nested. Q.E.D.

By Lemma 2, the boundary of a cell filled by whole paths is made up of whole paths, either singular or consisting of singular boundary elements.

We now turn to the question of the connectivity of a cell filled by non-closed paths. A first and quite obvious observation is that in this case (finitely many singular paths) a cell must be finitely connected. Indeed, every boundary continuum of the cell is made up of singular elements. Since by assumption there are only finitely many singular elements, it clearly follows that the number of boundary continua is also finite.

Suppose that the cell in question is $n$-connected, its boundary being the union of $n$ disjoint continua $K_1, \ldots, K_n$.

Lemma 18. *If the paths of a cell are nonclosed, each boundary continuum $K_i$ of the cell contains at least one $\omega$- or $\alpha$-limit point of a path of the cell.*

Proof. By assumption, the cell $g$ is $n$-connected and its boundary consists of $n$ continua $K_1, K_2, \ldots, K_n$. Then there exist $n$ simple closed curves $\pi_1, \pi_2, \ldots, \pi_n$ possessing the following properties (see Appendix, §4.4):

1. All the curves $\pi_i$ lie in $g$ and have no points in common.
2. $\pi_1, \ldots, \pi_{n-1}$ are mutually exterior and inside $\pi_n$. 3. Outside $\pi_n$ there are boundary points of the cell $g$. 4. Inside each curve $\pi_i$ $(i = 1, \ldots, n-1)$ there is exactly one (connected) component of the boundary of $g$.

Consider some curve $\pi_i$ $(i = 1, \ldots, n-1)$ and the component $K_i$ of the boundary inside it. If all points of $K_i$ are equilibrium states, then, since $K_i$ is connected and the number of equilibrium states finite, $K_i$ comprises exactly one point, which must be an $\omega$- or $\alpha$-limit point for at least one path of $g$. Indeed, otherwise all paths in some neighborhood of this point would be closed (Theorem 18, §4 and Theorem 47, §16), and clearly it could not be a boundary point of a cell filled by nonclosed paths.

Suppose now that some points of the continuum $K_i$ are not equilibrium states; let $P$ be one of these. Draw through $P$ an arc without contact $l$ inside $\pi_i$.

Since $P$ is a boundary point of $g$ and all paths through a sufficiently small neighborhood of $P$ cross $l$, it follows that one can find a sequence $P_1, P_2, \ldots$ of points of the cell $g$ on $l$, tending to the point $P$.

If $K_i$ contains no $\omega$- or $\alpha$-limit points for paths of $g$, there are no such points within $\pi_i$ either. Then the path $L_j$ passing at $t = \tau$ through $P_j$ must ultimately leave the interior of $\pi_i$, for both $t > \tau$ and $t < \tau$. In so doing, the path $L_j$ cannot cross $l$ twice without leaving the interior of $\pi_i$, since otherwise it would remain there either for $t > \tau$ or for $t < \tau$, and there would be an $\omega$- or $\alpha$-limit point inside $\pi_i$, contrary to assumption.

Let $Q_j$ be the point of intersection of $\pi_i$ and $L_j$ corresponding to the smallest possible value of $t > \tau$. The points $Q_j$ ($j = 1, 2, \ldots$) have at least one point of accumulation $\bar{Q}$ on $\pi_i$.

Let $\bar{L}$ be the path through $\bar{Q}$. Since by assumption $K_i$ contains no $\omega$- or $\alpha$-limit points of paths of $g$, there exists $\varepsilon > 0$ such that the $\varepsilon$-neighborhood of $\bar{L}$ and the $\varepsilon$-neighborhood of $K_i$ are disjoint. On the other hand, through any, arbitrarily small neighborhood of the point $\bar{Q}$ of $\bar{L}$ there are paths (namely, the paths $L_j$ with sufficiently large $j$) which enter the $\varepsilon$-neighborhood of $K_i$ and therefore leave the $\varepsilon$-neighborhood of $\bar{L}$. This means that the path $\bar{L}$ is orbitally unstable; but this is impossible, for $\bar{L}$ lies entirely within the cell $g$ — contradiction.

Similar arguments apply to the boundary continuum $K_n$ outside the curve $\pi_n$, and the proof is complete.

**Theorem 54.** *A cell filled by nonclosed paths is at most doubly connected.*

Proof. Suppose that a cell $g$ filled by nonclosed paths is $n$-tuply connected, $n > 2$. Then, by Lemma 18, there are at least two boundary continua $K_i$ such that either both contain $\omega$-limit points of paths in the cell, or both contain $\alpha$-limit points. To fix ideas, suppose that two boundary continua contain $\omega$-limit points of paths of the cell. Let $K_\omega$ denote the $\omega$-limit continuum of the paths of the cell. Since all points of $K_\omega$ are boundary points of the cell and $K_\omega$ is a connected set, it must be a subset of some boundary continuum, but this contradicts our assumption. Q.E.D.

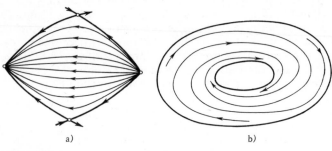

a)             b)

FIGURE 183.

Simple examples show that cells filled by nonclosed paths may be either simply connected (Figure 183a) or doubly connected (Figure 183b).

**11. PROPERTIES OF THE BOUNDARY OF A DOUBLY CONNECTED CELL FILLED BY NONCLOSED PATHS.**

*L e m m a 19.* *Let $g$ be a doubly connected cell filled by nonclosed paths, $K_1$ and $K_2$ the boundary continua of the cell. Then:*

*a) If $L'$ is any path in $g$, the set of points of the cell not on $L'$ is a simply connected region $g'$ whose boundary is the union of the continua $K_1$ and $K_2$ and the path $L'$.*

*b) Any path $L''$ in the region $g'$ separates it into two simply connected regions $g_1''$ and $g_2''$ and its points are boundary points for both.*

*c) Any path $L'''$ in the region $g_1''$ ($g_2''$) separates this region into two simply connected regions and its points are boundary points for both.*

P r o o f. To prove the lemma, we shall first construct a certain auxiliary doubly connected region $g^*$ ($g^* \subset g$) and then show that the region $g'$ of part a) is obtained from $g^*$ by deleting the points of a certain simple arc whose endpoints are on the boundary continua of $g^*$. Hence (see Appendix, §3.4) it will follow that $g'$ is simply connected.

The construction of $g^*$ proceeds as follows. Let $P_1$ and $P_2$ be points on the path $L'$ which correspond (under some fixed motion on $L'$) to times $t_1$ and $t_2$ ($t_2 > t_1$). Let $g^*$ denote the set of points of $g$ which are neither on the negative semipath $L_{P_1}^-$ nor on the positive semipath $L_{P_2}^+$. It is easy to see that $g^*$ is an open set. To show that it is a region, we must prove that any two points of $g^*$ can be joined by a continuum whose points are all in $g^*$. Note first that any two points $A$ and $B$ of $g^*$ can be joined by a simple arc $s$ in $g$. We may assume without loss of generality that $s$ does not pass through the points $P_1$ and $P_2$. If it has no points in common with the semipaths $L_{P_1}^-$ and $L_{P_2}^+$, it lies entirely in $g$ and we are done. If not, suppose, say, that $s$ has points in common with the semipath $L_{P_1}^-$ alone. Suppose that when one moves along $s$ from $A$ to $B$ the first point in common with $L_{P_1}^-$ is $M$ and the last $N$. We may assume that the subarc $MA$ of $s$ is on the negative side of $L_{P_1}^-$ and the subarc $NB$ on its positive side (Figure 184).

Let $R$ be any point on the arc $P_1P_2$ of the path $L'$. Let $l'$ be an arc without contact with endpoint at $M$, lying on the negative side of the path $L'$,

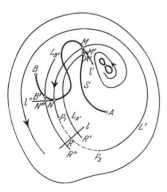

FIGURE 184.

$l''$ an arc without contact with endpoint at $N$, on the positive side of the path $L'$, and $l$ an arc without contact through $R$, containing $R$ as an interior point. Moreover, let the arcs $l'$ and $l''$ be so small that, apart from the points $M$ and $N$, respectively, they have no other points in common with $L'$ or with each other.

All points of $l$ except $R$ are assumed to belong to the region $g^*$; moreover, it is disjoint from $l'$, $l''$ and its only point in common with the path $L'$ is $R$.

By Lemmas 4 and 5 of §3, if $A'$ is a point on the subarc $AM$ sufficiently close to $M$, then the path $L_{A'}$ through this point crosses the arc $l'$ at some point $M'$, subsequently (with increasing $t$) crossing $l$ on the same (i. e., negative) side of $L'$ as $l'$, at some point $R'$. Similarly, if $B'$ is a point on $NB$ sufficiently close to $N$, then the path $L_{B'}$, through $B'$ crosses $l''$ at some point $N'$, subsequently (with increasing $t$) crossing $l$ on the positive side of $L'$ at some point $R''$. Note that the paths $L_{A'}$ and $L_{B'}$ are necessarily distinct from $L'$, since by assumption the arcs $l'$ and $l''$ have no points other than their endpoints $M$ and $N$ in common with $L'$. In addition, by assumption the arc $l$ has no points in common with the semipaths $L_{P_1}^{-}$ and $L_{P_2}^{+}$. Consider the set of points formed by the union of the subarc $AA'$ of $s$, the arc $A'R'$ of the path $L_{A'}$, the subarc $R'R''$ of $l$, the arc $R''B'$ of $L_{B'}$ and the subarc $B'B$ of $s$ (see Figure 184). This set is a continuum, a subset of $g$ having no points in common with the semipaths $L_{P_1}^{-}$ and $L_{P_2}^{+}$, so that it is a subset of the region $g^*$. Proof of the existence of a continuum joining $A$ and $B$ in the region $g^*$ is analogous in case the subarcs $MA$ and $NB$ of $s$ lie on the same side of $L_{P_1}^{-}$, and also when the arc $s$ has points in common with $L_{P_2}^{+}$ alone or with both $L_{P_1}^{-}$ and $L_{P_2}^{+}$. Hence it follows that the set $g^*$ is a region. Using Theorem 54 and the definition of $g^*$, one readily sees that the boundary points of $g^*$ are the points of $K_1$ and $K_2$ and of the semipaths $L_{P_1}^{-}$ and $L_{P_2}^{+}$. The semipath $L_{P_1}^{-}$ and its limit points lying, say, on the continuum $K_1$ form a continuum* which has points in common with $K_1$. Therefore the union of $L_{P_1}^{-}$ and $K_1$ is a continuum, which we denote by $K_1^*$. Similarly, the union of the semipath $L_{P_2}^{+}$ and the continuum $K_2$ is also a continuum, $K_2^*$ say.

The continua $K_1^*$ and $K_2^*$ are disjoint; thus the boundary of $g^*$ is the union of two continua and $g^*$ is a doubly connected region.

The region $g$ of part a) is obtained from $g^*$ by deleting the points on the arc $P_1P_2$ of $L_1$. Now this arc is a simple arc joining two points of the boundary continua $K_1^*$ and $K_2^*$ of the doubly connected region $g^*$. Hence (Appendix, §3.4) the region $g'$ is simply connected, and part a) of the lemma is proved.

We now prove part b). Let $L''$ be any path in $g$ other than $L'$ and $g''$ the set of points of $g'$ not on $L''$. It is readily seen that $g''$ is an open set, whose boundary is the union of the continua $K_1$ and $K_2$ and the paths $L'$ and $L''$.

Let us show that $g''$ is the union of two disjoint regions.

We first prove that $g''$ is not connected. Indeed, otherwise $g''$ is a region. Let $l$ be an arc without contact through a point $Q$ on $L''$, contained in the region $g'$ and having no points other than $Q$ in common with $L''$.

Let $R'$ and $R''$ be two points of $l$, on different sides of the path $L''$ (Figure 185). Since by assumption $g''$ is a region, the points $R'$ and $R''$ can be joined by a simple arc $\lambda$ in $g''$. Moving along $\lambda$ from $R'$ to $R''$, let $P'$ be the last point of $\lambda$ on the subarc $R'Q$ of $l$ and $P''$ the first point of $\lambda$ on the subarc $R''Q$ of $l$. Let $\lambda'$ be the subarc $P'P''$ of $\lambda$. Obviously, $\lambda'$ is a simple arc joining $P'$ and $P''$, lying entirely in the region $g''$ and having no points in common with the subarc $P'P''$ of $l$ other than its endpoints. But then $\lambda'$, combined with the subarc $P'P''$ of $l$, forms a simple closed curve $C$ in the region $g'$. This closed curve has only one point $Q$ in common with the path $L''$, and moreover, since $Q$ is a point of the arc without contact $P'P''$, with increasing $t$ the path $L''$ crosses at this point from one of the regions defined by $C$ into the other, say from the exterior to the interior of $C$. But then the $\omega$-limit points of $L''$, which belong to the continuum $K_2$, must lie inside the curve $C$, while the $\alpha$-limit points, which are points of $K_1$,

* Since the closure of a connected set is connected.

are outside $C$.  Thus, there are boundary points of $g'$ both inside and outside the curve $C$.  Since $C \subset g'$, this implies that $g'$ is a doubly connected region, contrary to part a).  Thus the set $g''$ is not a region.

FIGURE 185.

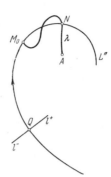

FIGURE 186.

We now show that $g''$ is the union of at most two regions.

Let $l$ be an arc without contact through some point $Q$ on the path $L''$, lying entirely within $g'$ and having no points other than $Q$ in common with $L''$.  Let $l^+$ and $l^-$ denote the subarcs of $l$ on the positive and negative sides of $L''$, respectively (Figure 186).  Let $A$ be an arbitrary point of $g''$.  Join $A$ by a simple arc $\lambda$ in the region $g'$ to some point $M_0$ on $L''$.  Let $N$ be the first point on the arc $AM_0$, moving from $A$ to $M_0$.

Reasoning as in the proof that the region $g^*$ is doubly connected, one readily shows that $A$ may be joined by a simple arc all of whose points are in $g''$ to a point of either $l^+$ or $l^-$.  Hence it is obvious that $g''$ is the union of at most two regions.  This completes the proof of part b).

The proof of part c) is analogous to that of part b) and is therefore omitted.

R e m a r k.  If $g_1''$ and $g_2''$ are the regions into which the path $L''$ separates the region $g'$, then all boundary points of these regions which are not points of $L''$ are boundary points of $g'$.  Similarly, if $g_{11}'''$ and $g_{12}'''$, $g_{21}'''$ and $g_{22}'''$ are the regions into which a path $L'''$ divides the region $g_1''$ or $g_2''$, then all their boundary points which are not on $L'''$ are boundary points of $g_1''$ $(g_2'')$.

T h e o r e m  55.  *All points of one boundary continuum $K_1$ of a doubly connected cell g filled by nonclosed paths are α-limit points of paths of the cell, and all points of the other boundary continuum $K_2$ are ω-limit points of paths of the cell.*

P r o o f.  Suppose the assertion false, so that there are points of the boundary continua $K_1$ and $K_2$ of our doubly connected cell $g$ which are not limit points of paths in the cell.  Let $L'$ be some path in the cell.  By Lemma 19, the complement in $g$ of the path $L'$ is a simply connected region.  As in Lemma 19, denote this region by $g'$.  Through some point $Q$ of $L'$, draw an arc without contact $l$ lying entirely in $g$ and having no points other than $Q$ in common with $L'$.  Let $P'$ and $P''$ be points on $l$, on different sides of $Q$, and join them by a simple arc $s$ in the region $g'$ (Figure 187), in such a way that the subarc $P'P''$ of $l$ and the arc $s$ together form a simple closed curve $C$ (Lemma 19). The curve $C$ has only one point in common with $L'$. With increasing $t$, the path $L'$ crosses at $Q$ from one of the regions defined by $C$ into the other, say from the exterior to the interior of $C$. Consequently, the continuum $K_1$ containing the $\alpha$-limit points of the paths will lie outside $C$, and the continuum $K_2$ containing the $\omega$-limit points of $L'$ inside $C$. But then it is clear that any path in the cell $g$ must have points both outside and inside $C$, and hence must cross the curve $C$. By Lemma 19, any path $L^*$ divides $g'$ into two regions, which we denote by $g_1''(L^*)$ and $g_2''(L^*)$. By definition, the boundary points of $g_1''(L^*)$ and $g_2''(L^*)$ are the points of $K_1$ and $K_2$ and the paths $L'$ and $L^*$. Moreover, the points of the last-named paths appear in the boundaries of both $g_1''(L^*)$ and $g_2''(L^*)$. Suppose that $g_1''(L^*)$ is on the positive side of the path $L'$ and $g_2''(L^*)$ on the negative side of $L'$. Since by assumption there are points of $K_1$ and $K_2$ which are not limit points

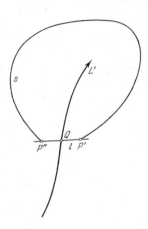

FIGURE 187.

of the paths in $g$, at least one of the two regions $g_1''(L^*)$, $g_2''(L^*)$ must have boundary points which are not points of $L'$, $L^*$ and not $\omega$- or $\alpha$-limit points of these paths. Let the curve $C$ be described in the positive sense, and construct for each point of the curve regions $g_1''(L^*)$ and $g_2''(L^*)$, where $L^*$ is the path through the point of $C$ under consideration. If this procedure is started from the point $Q$, we first go through points on the positive side of the path $L'$. Let us say that a point of $C$ is of the first type if the region $g_1''(L^*)$ has no boundary points other than the points of $L'$, $L^*$ and their $\omega$- and $\alpha$-limit points; then, as shown above, the corresponding region $g_2''(L^*)$ (see Remark after Lemma 19) must have boundary points which are neither points of $L'$, $L^*$ nor their $\omega$- or $\alpha$-limit points. We say that a point of $C$ is of the second type if the region $g_1''(L^*)$ has boundary points which are not on $L^*$, $L'$ and are not their $\omega$- or $\alpha$-limit points. Note that if $g_2''(L^*)$ has no boundary points of this kind, then $g_1''(L^*)$ must have such boundary points, i.e., the corresponding point of $C$ is of the second type.

We claim that all points on $C$ sufficiently close to $Q$, on the positive side of $L'$, are points of the first type. To prove this, consider a path $L^*$ crossing the subarc $QP'$ of the arc without contact $l$ on the positive side of the path $L'$, at some point $Q^*$. It is readily shown that when $Q^*$ is sufficiently close to $Q$, every point of the region $g_1''(L^*)$ lies on a path crossing the arc $QQ^*$ at points other than $Q$ and $Q^*$. Indeed, let $H$ be the region consisting of the points of these paths. Let $Q^*$ be a point so close to $Q$ that (Lemma 15) the boundary of $H$ is the union of the paths $L'$, $L^*$ and their limit points. Clearly, $H \subset g_1''$. Suppose that $g_1''(L^*)$ is distinct from $H$, so that there exist points of $g_1''(L^*)$ not in $H$. Let $B$ be a point such that $B \in g_1''(L^*)$ but $B \notin H$. Join this point to some point $A$ in $H$ (therefore also in $g_1''(L^*)$) by a simple arc $\lambda$ lying entirely in $g_1''(L^*)$. There must be boundary points of $H$ on the arc $\lambda$, since $B \notin H$; these are points of $L'$, $L^*$ or their limit points. But the points of $L'$, $L^*$ and their limit points are not in $g_1''(L^*)$. This contradiction shows that $H = g_1''(L^*)$. Hence it follows that all points sufficiently close to $Q$ on the subarc $P'P''$ of $l$, lying on the positive side of $L'$, are of the first type. On the other hand, by Lemma 15 and the above remark, the points on the negative side are of the second type. As $C$ is described in the negative sense (beginning from $Q$), we must go from points of the first type to points of the second. Consequently, there must be a point $Q_0$ on $C$ which is either the last point of the first type or the first not of the first type, therefore the first point of the second type. Let $L_0$ be the path through $Q_0$ and $g_1''(L_0)$ the corresponding region. If $Q_0$ is of the first type, then $g_1''(L_0)$ has no boundary points other than those of $L'$ and $L^*$ and their limit points. But then, by Lemma 15 and the remark following the preceding lemma, one readily sees that all points sufficiently close to $Q_0$ are also of the first type, contradicting the property of the point $Q_0$. Similarly, one shows that $Q_0$ cannot be of the second type. We have thus arrived at a contradiction, proving the theorem.

The next theorem follows at once from Theorems 48 and 53 and the fact that the number of singular paths is finite.

*Theorem 56. There exist only finitely many mutually exterior closed paths and mutually exterior loops.*

Examples of simply and doubly connected cells filled by whole paths are given in Figure 188.

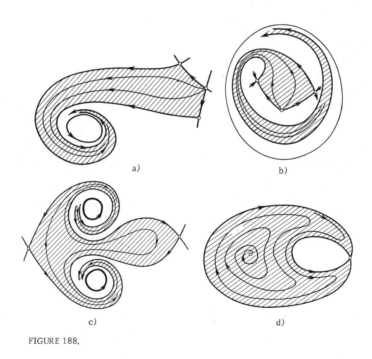

a)      b)

c)      d)

FIGURE 188.

**12. CELLS WHOSE BOUNDARIES INCLUDE BOUNDARY ARCS.** We now consider cells whose boundaries include parts of boundary arcs without contact.

Consider a subarc of a boundary arc without contact, whose endpoints lie on corner arcs or singular semipaths all of whose interior points belong to nonsingular arcs or nonsingular semipaths. We shall call a subarc of this kind a singular $\omega$-arc or singular $\alpha$-arc, depending on whether the semipaths or arcs of paths leave the region $G^*$ across it with increasing or decreasing $t$.

If the boundary of the cell includes a cycle without contact all of whose points are on nonsingular arcs or nonsingular semipaths, we shall call the cycle a singular $\omega$- or $\alpha$-cycle.

Using Lemma 3 one readily verifies that if a singular $\omega$- or $\alpha$-arc (or $\omega$- or $\alpha$-cycle) contains a boundary point of some cell, then all its points are boundary points of this cell. In addition, employing by now familiar arguments, one readily proves the following theorem.

*Theorem 57. a) If a semipath $L^{()}$ (or arc of a path) of a cell crosses a singular $\omega$-($\alpha$-) arc $\lambda$, then all semipaths (arcs of paths) in the cell cross $\lambda$ at its interior points and cannot cross any other $\omega$-($\alpha$-)arc.*

*b) If a semipath (arc of a path) in a cell crosses a singular $\omega$-($\alpha$-) cycle, then all semipaths (arcs of paths) in the cell cross this cycle.*

Another result is the following lemma, whose proof is entirely analogous to that of Lemma 18:

*Lemma 20. Every boundary continuum of a cell, which does not contain points of a singular $\omega$-($\alpha$-)arc or singular cycle, must contain limit points of semipaths of the cell.*

*Theorem 58. A cell whose boundary includes an $\omega$- or $\alpha$-arc is simply connected.*

Proof. If the cell is filled by arcs of paths, the assertion follows directly from Lemma 10 of §3.

Consider now a cell filled by semipaths; to fix ideas, assume that its boundary includes an $\omega$-arc (the case of an $\alpha$-arc is treated similarly).

Suppose that the cell is at least doubly connected, i.e., its boundary is the union of at least two continua $K_1$ and $K_2$. There exists a simple closed curve $C$ in the cell, enclosing one of the continua, say $K_2$. To fix ideas, suppose that the points of the $\omega$-arc $\lambda$ appearing in the boundary of the cell belong to the continuum $K_1$. Let $A$ and $B$ be the endpoints of the arc $\lambda$, which are points of corner arcs or singular semipaths. Let $\{P_i\}$ be a sequence of points on $\lambda$ tending to one of its endpoints, say to $A$. Let $L_i^{()}$ be the semipath with endpoint at $P_i$. It is obvious that any two semipaths and $L_i^{()}$, $L_j^{()}$, are different.

Since the continuum $K_2$ certainly contains limit points of semipaths of the cell, it follows that each semipath $L_i$ ultimately crosses the curve $C$ inward with increasing $t$ and does not leave the interior of the curve again. Let $Q_i$ be the last common point of $C$ and $L_i$ with increasing $t$. We thus obtain a sequence of distinct points $\{Q_i\}$.

Let $Q_0$ be a point of accumulation of this sequence and $L_0$ the path passing through it. Since $Q_0$, like all points of $C$, is a point of the cell, the path $L_0$ necessarily crosses the singular $\omega$-arc $\lambda$ at one of its interior points $P_0$.

Let $\lambda'$ be a subarc of $\lambda$ with midpoint $P_0$, all of whose points are distant some $d > 0$ from the endpoint $A$ of $\lambda$. All semipaths $L_i^{()}$ passing through points $Q_i$ sufficiently close to $Q_0$, i.e., through all points $Q_i$ with sufficiently large $i$, will cross $\lambda$ at points of its subarc $\lambda'$. But apart from the points

$P_i$ the semipath $L_i^{\langle\rangle}$ can have no other points in common with $\lambda$. Consequently, all points $P_i$ with sufficiently large $i$ lie on $\lambda'$ and are therefore distant at least $d$ from $A$. This contradicts the choice of the sequence $P_i$. Q.E.D.

Figure 189a and b illustrates different cases of simply connected cells whose boundaries contain an ω-arc.

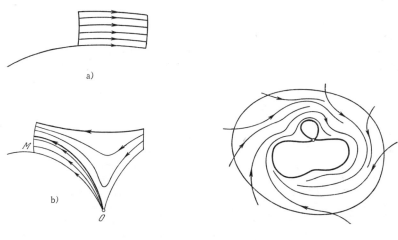

FIGURE 189.                              FIGURE 190.

The proof of the next theorem is analogous and will therefore be omitted.

*T h e o r e m   59. Any cell whose boundary includes a singular* ω (α)*-cycle is doubly connected.*

An example of this situation is shown in Figure 190.

**13. COMPLETE QUALITATIVE INVESTIGATION OF A DYNAMIC SYSTEM. SCHEME OF A DYNAMIC SYSTEM.** The propositions proved in this section highlight the role of the singular paths in the phase portrait as a whole: the singular paths partition the region $\bar{G}^*$ in which the dynamic system is being considered into subregions — cells — in which the paths exhibit the same behavior (in the sense of Theorems 47, 48, 49, 52, 53 and 57).

If one knows the relative position of the cells and the character of the phase portrait in each, it is natural to deem the topological structure of the phase portrait established in full detail.

In this context, clearly, one must be acquainted with the configuration of the singular paths and, in addition, one must know whether the paths in each cell are closed or not. When the paths are not closed, one must know their behavior and, in particular, their limit points.

Nevertheless, the question of exhaustively determining the topological structure of the phase portrait may be attacked from a slightly different vantage point, which does not rely directly on examination of cells.

It is natural to hold that a knowledge of the character (topological structure) of all equilibrium states and of the relative position of the singular paths, accompanied by an indication of those of the singular paths

that make up the limit continua, also constitutes an exhaustive characterization of the topological structure of the phase portrait (hence affording
in addition an unambiguous subdivision into cells).

This approach is adopted in the following chapters for complete
examination of the topological structure of the phase portrait. In this
context the basic elements, which d i r e c t l y condition the qualitative
structure, are the character of the equilibrium states and identification of
the limit continua (not of the position and character of the cells). The
approach is also natural from the standpoint of actual qualitative investigation of specific examples (see Chapter XII).

However, before accomplishing the goal we must specify what we mean
by a knowledge of the character of the equilibrium states and the relative
position of singular paths, etc.

It would seem clear that this phrasing implies a d e s c r i p t i o n of the
character of the equilibrium states, a description of the position of the
singular paths and a description of the limit continua.

A description of this kind will be called a s c h e m e.

In the following chapters we shall first define the (local and global)
s c h e m e   o f   a n   e q u i l i b r i u m   s t a t e, then the (local and global)
s c h e m e   o f   a   l i m i t   c o n t i n u u m, and finally combine these two
partial schemes into the s c h e m e   o f   t h e   d y n a m i c   s y s t e m.

In Chapter XI we shall show that the scheme of a dynamic system
uniquely determines the topological structure of the phase portrait:
if two dynamic systems have identical [or equivalent] schemes, their phase
portraits have the same topological structure.

## Chapter VIII

*SCHEME OF AN EQUILIBRIUM STATE*

INTRODUCTION

In this chapter we consider the neighborhood of an equilibrium state of a system of type (I), on the assumption that the number of singular paths is finite. In Chapter IV we studied s i m p l e equilibrium states of system (I); in other words, we adopted an assumption of a n a l y t i c a l nature, $\Delta \neq 0$. In the present chapter we shall not impose any analytical restrictions on the equilibrium state. Neither shall we offer methods aimed at characterizing an equilibrium state on the basis of its analytical characteristics. Our goal will be, rather, to determine the possible topological structures of an equilibrium state of system (I) when the number of singular paths is finite.

In addition, we shall present a d e s c r i p t i o n of this topological structure, defining the s c h e m e of an equilibrium state.

The chapter comprises four sections. In §17 we examine the neighborhood of an equilibrium state to which at least one semipath tends [or, as we shall sometimes say, which is approached by at least one semipath]. It will be shown that such neighborhoods may be partitioned into regions of three distinct types: r e g u l a r p a r a b o l i c, e l l i p t i c a n d h y p e r b o l i c r e g i o n s. Paraboloc, elliptic and hyperbolic regions, together with elementary rectangles (see §3), will be called e l e m e n t a r y r e g i o n s. §18 is devoted to the proof that any elementary region can be mapped onto another of the same type by a path-preserving mapping (this is geometrically obvious).

In §19 we define the s c h e m e of an equilibrium state approached by at least one semipath, first the l o c a l s c h e m e and then the g l o b a l s c h e m e. The local scheme exploits only such information as can be acquired by direct examination of a sufficiently small neighborhood of the equilibrium state i t s e l f. As we shall see, for the global scheme one needs in addition information about the behavior of the separatrices of other equilibrium states, so that it becomes necessary to devote some attention to the behavior of the singular paths (separatrices) in the large.

§17. EQUILIBRIUM STATE APPROACHED BY
AT LEAST ONE SEMIPATH

**1. AUXILIARY PROPOSITIONS.** Consider an equilibrium state of system (I) to which tend infinitely or finitely many semipaths, but not less than some preassigned number $N^* \geqslant 2$. Consider $N$ $(N \leqslant N^*)$ of these semipaths:

$$L_1^{(\ )}, \; L_2^{(\ )}, \; \ldots, \; L_N^{(\ )} \tag{1}$$

(they may be positive or negative, orbitally stable or orbitally unstable).

Let $C$ be a simple closed curve, not necessarily smooth, containing the equilibrium state $O$ in its interior, such that all the semipaths (1) have points in common with $C$. Let $M_1, M_2, \ldots, M_N$ be their last points in common with $C$ (see Chapter VII, §15).

Recall that, given two points $A$ and $B$ on a simple closed curve $C$, we use the term "arc $AB$" of $C$ for that of the two possible arcs on which motion from $A$ to $B$ induces the positive traversal of $C$ (§10.2). Suppose that the positive traversal of $C$ arranges the points $M_i$ on the curve in the order of their indices, so that $M_k$, $M_{k+1}$ $(k = 1, 2, \ldots, N$, with $M_{N+1}$ understood as $M_1$) are "successive" points, that is to say, no arc $M_k M_{k+1}$ contains a point $M_j$ other than its endpoints. Obviously, if $N \geqslant 3$ the arc $M_{k+1} M_k$ will contain at least one other point $M_j$.

The order in which the points $M_i$ are arranged by the positive traversal of $C$ will be called their c y c l i c   o r d e r   o n   $C$.

Let $s_k$ denote the arc $M_k M_{k+1}$ of $C$. Let $g_k$ be the curvilinear sector (§15.5) bounded by $s_k$ and the arcs $M_k O$ and $M_{k+1} O$ of the semipaths $L_k^{(\ )}$ and $L_{k+1}^{(\ )}$, respectively, and $\sigma_k$ the simple closed curve made up of the arc $s_k$, the semipaths $M_k O$, $M_{k+1} O$ and the point $O$ (i. e., the boundary of $g_k$) (Figure 191).

Let $\tilde{C}$ be a (not necessarily smooth) simple closed curve, distinct from $C$, also containing $O$ in its interior, such that all the semipaths $L_k^{(\ )}$ have points outside it. Let $\tilde{M}_k$ denote the last point in common with $\tilde{C}$ of the semipath $L_k^{(\ )}$ (Figure 192).

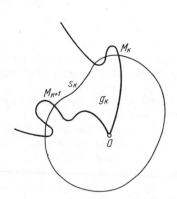

FIGURE 191.

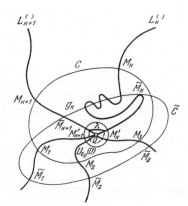

FIGURE 192.

L e m m a   1.* The cyclic order of the points $\tilde{M}_j$ on $\tilde{C}$ is the same as that of the points $M_j$ on $C$ (i. e., if the positive traversal of $C$ induces the order $M_1, M_2, \ldots, M_N$, then that of $\tilde{C}$ induces the order $\tilde{M}_1, \tilde{M}_2, \ldots, \tilde{M}_N$).

* The reader may skip the proof of this geometrically obvious lemma without impairing his understanding of the sequel.

P r o o f. It will suffice to show that no arc $\widetilde{M}_h\widetilde{M}_{h+1}$ contains any point $\widetilde{M}_j$ other than its endpoints.

Consider any two points $M_h$ and $M_{h+1}$ and the corresponding points $\widetilde{M}_h$ and $\widetilde{M}_{h+1}$. Let $\widetilde{s}_h$ denote the arc $\widetilde{M}_h\widetilde{M}_{h+1}$ on $\widetilde{C}$ and $\widetilde{g}_h$ the region (sector) bounded by the simple curve $\widetilde{\sigma}_h$ — the union of $\widetilde{s}_h$, the arcs $\widetilde{M}_hO$ and $\widetilde{M}_{h+1}O$ of the semipaths $L_h^{()}$ and $L_{h+1}^{()}$ and the point $O$.* The sector $\widetilde{g}_h$ is a subregion of the interior of $\widetilde{C}$. By the definition of $\widetilde{g}_h$ and $\widetilde{\sigma}_h$, the direction on $\widetilde{s}_h$ from $\widetilde{M}_h$ to $M_{h+1}$ induces the positive traversal of $\widetilde{\sigma}_h$ (see Appendix, §2.3).

The next part of the construction is based on an elementary property of simple closed curves (see Appendix, §2.2), namely, if $R$ is any point on a simple closed curve $\gamma_0$ and $\varepsilon > 0$ an arbitrary positive number, there exists $\delta(\varepsilon) > 0$ such that any two points $P'$ and $Q'$ on $\gamma_0$ in the neighborhood $U_\delta(R)$ can be joined by a simple arc lying except for its endpoints inside $\gamma_0$ and in the neighborhood $U_\varepsilon(R)$. Now all points sufficiently close to $O$ on the semipaths $L_h^{()}$ and $L_{h+1}^{()}$ are on both curves $\sigma_h$ and $\widetilde{\sigma}_h$ ($\sigma_h$ and $\widetilde{\sigma}_h$ are the boundaries of $g_h$ and $\widetilde{g}_h$, respectively). Hence there exists $\varepsilon_0 > 0$ such that the neighborhood $U_{\varepsilon_0}(O)$ contains no points of $L_h^{()}$ and $L_{h+1}^{()}$ other than those lying on both curves $\sigma_h$ and $\widetilde{\sigma}_h$ and also no points of the curves $C$ and $\widetilde{C}$. By the above-mentioned property, there exists a small neighborhood $U_\delta(O)$ ($\delta(\varepsilon_0) < \varepsilon_0$) (Figure 192) in which any two points $M'_h$ and $M'_{h+1}$, lying on $L_h^{()}$ and $L_{h+1}^{()}$, respectively (hence also on $\sigma_h$), can be joined by a simple arc $\lambda$ lying except for its endpoints in the intersection of $g_h$ and inside $U_{\varepsilon_0}(O)$.

By the choice of $\varepsilon_0$ and $\delta$, the points $M'_h$ and $M'_{h+1}$ are on both $\sigma_h$ and $\widetilde{\sigma}_h$, and the arc $\lambda$ cannot contain any points of $\widetilde{\sigma}_h$ other than its endpoints.

It clearly follows that $\lambda$ is either wholly (except for its endpoints) inside the curve $\widetilde{\sigma}_h$, i. e., in $\widetilde{g}_h$, or wholly outside $\widetilde{\sigma}_h$ (both cases are illustrated in Figure 192).

We claim that the arc $\lambda$, which lies except for its endpoints in $g_h$, is also (again except for its endpoints) in $\widetilde{g}_h$. Prior to proving this, we note that the positive traversal of $\sigma_h$ arranges the points $M_h$, $M_{h+1}$, $M'_h$, $M'_{h+1}$ and $O$ in the order $M_h$, $M_{h+1}$, $M'_{h+1}$, $O$, $M'_h$, $M_h$.

The arc $\lambda$ divides the region $g_h$ into two subregions, whose boundaries are simple closed curves intersecting in the arc $\lambda$. Let $\sigma'_h$ denote that of these curves containing the point $O$ and $g'_h$ the subregion of $g_h$ bounded by $\sigma'_h$ (the arc $s_h$ of $\sigma_h$ cannot be an arc of $\sigma'_h$). In view of the order in which the positive traversal of $\sigma_h$ arranges the points $M_h$, $M_{h+1}$, $M'_h$, $M'_{h+1}$, $O$, we easily conclude (Appendix, §2.3) that the positive traversal of $\sigma'_h$ arranges the points $M'_h$, $M'_{h+1}$ and $O$ in the order $M'_h$, $M'_{h+1}$, $O$, $M'_h$.

Suppose now that, except for its endpoints, the arc $\lambda$ is outside $\widetilde{g}_h$. The curves $\sigma'_h$ and $\widetilde{\sigma}_h$ clearly intersect in an arc which is the union of the arcs $M'_hO$ and $M'_{h+1}O$ of the semipaths $L_h^{()}$ and $L_{h+1}^{()}$. By the Appendix, §2.3, the direction on this arc, opposite to that induced by the positive traversal of $\sigma'_h$, i. e., the direction $M'_hOM'_{h+1}$, must induce the positive traversal of $\widetilde{\sigma}_h$. Then the positive traversal of $\widetilde{\sigma}_h$ must induce the direction on $\widetilde{s}_h$ from $\widetilde{M}_{h+1}$ to $\widetilde{M}_h$. But the positive traversal of $\widetilde{\sigma}_h$ induces on $\widetilde{s}_h$ the same direction as the positive traversal of $\widetilde{C}$, i. e., from $\widetilde{M}_h$ to $\widetilde{M}_{h+1}$. This contradiction implies that the arc $\lambda$ must be, except for its endpoints, wholly in the region $\widetilde{g}_h$.

---

* Of course, we cannot yet assert that the region $\widetilde{g}_h$ does not contain points of any semipath $\widetilde{M}_jO$.

This at once implies the assertion of the lemma. Indeed, apart from its endpoints, the arc $M_h M_{h+1}$ of $C$ contains no other points $M_j$, and the region $g_h$ contains no point of one of the semipaths $M_j O$. Thus, no interior point of $\lambda$ can be a point of one of these semipaths. Now, were the arc $\tilde{M}_h \tilde{M}_{h+1}$ to contain a point $\tilde{M}_j$ other than $\tilde{M}_h$ and $\tilde{M}_{h+1}$, it would follow that $\tilde{g}_h$ contains the semipath $\tilde{M}_j O$, and then there would necessarily be points of this semipath on the arc $\lambda$. Otherwise, as is easily seen, it could not tend to the equilibrium state $O$ without leaving the interior of $\tilde{C}$; by assumption, this is impossible, and the proof is complete.

Remark. The arc $\lambda$, which (except for its endpoints) lies both in $g_h$ and in $\tilde{g}_h$, determines the same subregion $g_h'$ in each of these regions.

Obviously, one can always find $\varepsilon > 0$ such that $U_\varepsilon(O)$ contains no points of $g_h$ and $\tilde{g}_h$ other than those in the region $g_h'$. Thus, for sufficiently small $\varepsilon > 0$ all points of $g_h$ in $U_\varepsilon(O)$ are points of $\tilde{g}_h$, and all points of $\tilde{g}_h$ in $U_\varepsilon(O)$ are also in $g_h$ (Figure 193).

We are now in a position to speak of the cyclic order of the semipaths themselves about the point $O$: we simply take some simple closed curve $C$ enclosing $O$, such that the semipaths have points both inside and outside $C$, and define their cyclic order to be the cyclic order of their last points in common with the curve $C$.

This definition is legitimate, since by Lemma 1 the order is independent of the choice of the curve $C$, i.e., determined by the paths themselves.

We shall say that two semipaths $L_j^{()}$ and $L_h^{()}$ of (1) are successive if the corresponding points $M_j$ and $M_h$ on $C$ are successive, i.e., the arc $M_j M_h$ does not contain other points $M_i$. One of two successive semipaths precedes the other. For example, $L_j^{()}$ precedes $L_h^{()}$ when the arc $M_j M_h$ of $C$ contains no other points $M_i$. If a semipath $L^{*()}$ tends to $O$ and has points outside $C$ ($L^{*()}$ is distinct from $L_j^{()}$ and $L_h^{()}$ but it may — or may not — be one of the semipaths $L_i^{()}$), we shall say that $L^{*()}$ lies between $L_j^{()}$ and $L_h^{()}$ (here the order is clearly important) if the last point of $L^{*()}$ in common with $C$ is on the arc $M_j M_h$. Finally, we shall say that a path or semipath lying entirely inside $C$ is between $L_j^{()}$ and $L_h^{()}$ if it lies inside the sector $h$ bounded by the semipaths $M_j O$, $M_h O$, the point $O$ and the arc $M_j M_h$ of $C$.

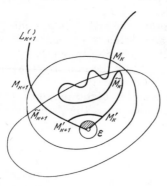

FIGURE 193.

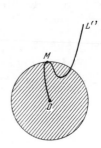

FIGURE 194.

Hitherto we have been considering at least two semipaths tending to the equilibrium state $O$, Later we shall sometimes consider a single semipath tending to an equilibrium state. Let $L^{()}$ be such a semipath, $C$ a simple closed curve, containing the equilibrium state $O$ in its interior, such that its exterior contains points of $L^{()}$. Let $M$ be the last point of $L^{()}$ in common with $C$. Then the region consisting of the points interior to $C$ minus the points of the semipath $MO$ and the equilibrium state $O$ will be called a c u r v i l i n e a r   c y c l i c   s e c t o r. Its boundary is the union of $C$, the semipath $MO$ and the point $O$ (Figure 194).

If $O$ is an equilibrium state approached by at least one semipath, there is a neighborhood of $O$ containing no closed path (Theorem 18). Since moreover we have assumed the number of singular paths to be finite, there always exists $\varepsilon_0 > 0$ such that the closure of the neighborhood $U_{\varepsilon_0}(O)$ contains no closed path and no whole singular path other than the equilibrium state $O$.

L e m m a   2.  *If all points of a path* $L$ *corresponding to times* $t > t_0$ $(t < t_0)$ *lie in a closed neighborhood* $\overline{U_{\varepsilon_0}(O)}$ *containing no closed paths and no whole singular path other than the equilibrium state* $O$, *then* $L$ *tends to* $O$ *as* $t \to +\infty$ $(t \to -\infty)$.

P r o o f.  By assumption, the semipath in question is in $\overline{U_{\varepsilon_0}(O)}$. Therefore, all its $\omega(\alpha)$-limit points must also be in $\overline{U_{\varepsilon_0}(O)}$. By virtue of the choice of $\varepsilon_0$, $L$ cannot be a closed path and hence its $\omega(\alpha)$-limit set is a union of orbitally unstable paths in $\overline{U_{\varepsilon_0}(O)}$, thus coinciding with the equilibrium state $O$. But this means that the semipath tends to $O$. Q.E.D.

Consider the separatrices of the equilibrium state $O$ (see §15). Since system (I) has only finitely many orbitally unstable paths, $O$ has only finitely many separatrices. In this situation we have the following theorem, which is a direct consequence of the finiteness assumption and Lemma 5 of §15.

T h e o r e m   60.  *Any* $\omega(\alpha)$-*separatrix of the equilibrium state* $O$ *is continuable on either the positive or the negative side (or on both).*

P r o o f.  As before, let $\varepsilon_0 > 0$ be such that the closed neighborhood $\overline{U_{\varepsilon_0}(O)}$ contains no closed path and no whole singular path other than $O$. To fix ideas, we consider an $\omega$-separatrix, $L_0^+$ say. This semipath is certainly continuable, say on the positive side, relative to some circle $C$ (Theorem 38, §15). We may assume that this circle $C$ is in $U_{\varepsilon_0}(O)$. Let $C'$ be a circle about $O$ of radius less than that of $C$. Were the continuation of $L_0^+$ relative to $C'$ different from its continuation relative to $C$, it would follow from Lemma 5 of §15 that the interior of $C$, a fortiori the neighborhood $U_{\varepsilon_0}(O)$, contains a singular path which is not an equilibrium state. But this is impossible by assumption. Consequently, the separatrix $L_0^+$ has the same continuation on the positive side relative to any circle about $O$ lying in $\overline{U_{\varepsilon_0}(O)}$. And this means that the semipath $L_0^+$ is continuable relative to the equilibrium state $O$ on the positive side (see Definition XX, Chapter VII). Q.E.D.

Let $C$ be a circle about $O$, of radius less than $\varepsilon_0$ (so that it is contained in the open neighborhood $U_{\varepsilon_0}(O)$). Consider any path passing through a point interior to this circle.

It is clear that there are three possibilities: the path 1) leaves $C$ with both increasing and decreasing $t$, 2) remains inside $C$ with both increasing and decreasing $t$ (though it may have points in common with $C$), or

3) remains inside $C$ with increasing (decreasing) $t$ but ultimately leaves it with decreasing (increasing) $t$.

It follows from Lemma 2 that in case 2) the path, remaining inside $C$ both as $t \to +\infty$ and as $t \to -\infty$, tends to the equilibrium state $O$, while in case 3) it tends to the equilibrium state as $t \to +\infty$ (or as $t \to -\infty$).

Note, moreover, that if the path through a point interior to $C$ is a separatrix of $O$, it necessarily has points outside $C$. This obviously follows from the fact that by assumption the circle $C$ is in the neighborhood $\overline{U}_{\varepsilon_0}(O)$.

## 2. TYPES OF CURVILINEAR SECTOR. HYPERBOLIC (SADDLE), PARABOLIC AND ELLIPTIC SECTOR.

The propositions to be proved in this subsection yield an exhaustive description of the possible behavior of paths in a sector formed by paths tending to an equilibrium state. Let $L_1^{*()}$ and $L_2^{*()}$ be two semipaths tending to an equilibrium state $O$ and having points outside the circle $C$ (the semipaths may be either nonsingular or singular).

Let $M_1$ and $M_2$ be their last points in common with $C$, and $g$ the curvilinear sector whose boundary is the union of the arcs $OM_1$ and $OM_2$ of $L_1^{*()}$ and $L_2^{*()}$, the point $O$ and the arc $M_1M_2$ of $C$.

*Lemma 3. Assume that for any $\varepsilon > 0$ there exists a point of $g$ in $U_\varepsilon(O)$, through which passes a path that leaves $g$ with both increasing and decreasing $t$.*

*Then there exist at least one $\omega$-separatrix and at least one $\alpha$-separatrix of the equilibrium state $O$, each of which is a semipath either lying between $L_1^{*()}$ and $L_2^{*()}$ or coinciding with one of them.*

Proof. By the assumptions of the lemma, there exists a sequence of points $\{N_i\}$ in $g$, tending to $O$, such that the path $L_i$ passing at $t = t_0$ through $N_i$ $(i = 1, 2, \ldots)$ leaves the circle $C$ for both $t < t_0$ and $t > t_0$ (Figure 195).

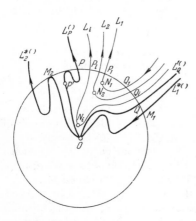

Let $P_i$ denote the point of $L_i$ in common with $C$ corresponding to the smallest possible $t < t_0$, and $Q_i$ the point of $L_i$ in common with $C$ corresponding to the greatest possible $t > t_0$. Let $t_i$ $(t_i < t_0)$ be the parameter value corresponding to $P_i$, and $t_i'$ $(t_i' > t_0)$ that corresponding to $Q_i$. Obviously, the points $P_i$ and $Q_i$ are on the arc $M_1M_2$ of $C$ and are necessarily distinct (otherwise $L_i$ would be a closed path in $\overline{U}_{\varepsilon_0}(O)$, which is impossible).

Except for its endpoints, the arc $P_iN_iQ_i$ of the path $L_i$ (Figure 195) lies wholly within $g$. We may also assume that the arcs $P_iN_iQ_i$ are all distinct, in other words, no arc $P_iN_iQ_i$ contains a point $N_j$ with $j \neq i$. Indeed, each arc $P_iQ_i$ of a path is at a positive distance from $O$, and therefore contains at most finitely many points $N_j$. Thus we can select a subsequence of the points $N_j$ in such a way that each $P_iQ_i$ contains exactly one point $N_j$.

We may assume without loss of generality that each of the sequences $P_i$ and $Q_i$ has one point of accumulation (otherwise we need only consider a suitable subsequence of arcs $P_iQ_i$). Let $P$ be the point of accumulation of $P_i$ and $Q$ that of $Q_i$.

FIGURE 195.

Clearly, the points $P$ and $Q$ are on the arc $M_1 M_2$ of the circle $C$. Let $L$ and $L'$ be the paths through $P$ and $Q$, respectively, $L_P^+$ and $L_Q^-$ the corresponding semipaths.

We claim that these semipaths are the required separatrices of $O$. To verify this, we first prove that $|t_0 - t_i| \to \infty$ and $|t_0 - t_i'| \to \infty$ as $i \to \infty$. Indeed, otherwise there exists $T > 0$ such that for all $i$, say,

$$|t_i - t_0| < T. \tag{2}$$

Suppose that the motion fixed on $L_P^+$ associates the point $P$ with time $t = \tau$, and consider the arc of this semipath defined by $\tau \leqslant t \leqslant \tau + T$. This arc is at a nonzero distance $d_0$ from $O$. Fix a motion on each path $L_i$ such that the point $P_i$ corresponds to the time $t = \tau$, and consider the arcs of these paths defined by $\tau \leqslant t \leqslant \tau + T$. It follows from (2) that the points $N_i$ lie on these arcs. Since $P$ is a point of accumulation of the points $P_i$, we see, taking $\varepsilon < \frac{d_0}{2}$, that there exists $I$ so large that for all $i > I$ the arcs of $L_i$ defined by $\tau \leqslant t \leqslant \tau + T$, hence also the points $N_i$, lie in the $(\frac{d_0}{2})$-neighborhood of the corresponding arc of $L_P^+$. But then, for all $i > I$, the points $N_i$ must be outside the $(\frac{d_0}{2})$-neighborhood of $O$, which is absurd. Consequently, $|t_i - t_0| \to \infty$ as $i \to \infty$. The proof that $|t_i' - t_0| \to \infty$ as $i \to \infty$ is analogous.

We now show that the semipath $L_P^+$ contains no points outside $C$, so that, since it does not leave $C$, it must tend to the equilibrium state $O$. Indeed, suppose that there is a point of $L_P^+$, corresponding to some time $t' > \tau$, outside the circle $C$. It is readily seen that if the motion fixed on each path $L_i$ associates $P_i$ with the time $t = \tau$, then, for sufficiently large $i$, the point on $L_i$ corresponding to $t = t'$ is also outside $C$. But this is impossible, for as we have proved $t' - \tau < t_i - t_0$ for all sufficiently large $i$, and under the motion fixed on $L_i$ all points other than $P_i$, corresponding to times $\tau < t < \tau + (t_i - t_0)$, i. e., the points of the arc $P_i N_i$, lie inside $C$. Thus the semipath $L_P^+$ does not leave $C$; therefore, neither does it leave the closed sector $\bar{g}$ and consequently it tends to the equilibrium state $O$. It remains to show that $L_P^+$ is orbitally unstable.

Let $t' > \tau$ be such that for all $t \geqslant t'$ the semipath $L_P^+$ has no points in common with $C$. (Note that the last point of $L_P^+$ in common with $C$ may be distinct from $P$, since $L_P^+$ has no points outside $C$ but it may have points in common with $C$; see Figure 195.) Let $P'$ be the point on the semipath corresponding to $t = t'$. There clearly exists $\varepsilon > 0$ such that the $\varepsilon$-neighborhood of the arc $P'O$ of $L_P^+$ has no points in common with $C$. But however small a $\sigma$-neighborhood ($\sigma > 0$, $\sigma < \varepsilon$) of $P'$ we choose, for sufficiently large $i$ all paths $L_i$ passing at $t = \tau$ through $P_i$ will pass through the $\sigma$-neighborhood of $P'$. With further increase in $t$, these paths necessarily leave the $\varepsilon$-neighborhood of the arc $P'O$ of $L_P^+$. Indeed, the point $Q_i$ on each path $L_i$ is a point of $C$ lying outside the $\varepsilon$-neighborhood of the arc $P'O$ of $L_P^+$. Under the motion fixed on $L_i$, the point $Q_i$ corresponds to time $T_i = \tau + (t_i' - t_i)$, which is certainly greater than $t'$ for all sufficiently large $i$. This clearly implies that the semipath $L_P^+$ is orbitally unstable. The reasoning for the semipath $L_Q^-$ is analogous. Q.E.D.

Remark 1.  The last points of $L_P^+$ and $L_Q^-$ in common with the circle $C$ are obviously distinct.

Remark 2.  The separatrices $L_Q^-$ and $L_P^+$ are continuations of one another.  This is easily verified by repeating the reasoning used to prove Lemma 3.

If the semipaths $L_1^{*()}$ and $L_2^{*()}$ are separatrices and continuations of one another, and there are paths through points of the sector arbitrarily near $O$ which leave $g$ with both increasing and decreasing $t$, the sector will be called a hyperbolic or saddle sector.

In the next lemma the notation $L_1^{*()}$, $L_2^{*()}$, $M_1$, $M_2$, $g$ has the same meaning as before.

Lemma 4.  *Assume that between the semipaths $L_1^{*()}$ and $L_2^{*()}$ there is no separatrix of $O$, but there is a semipath $L^{()}$ tending to $O$.*

*Then there exists $\varepsilon > 0$ such that there is no path which leaves $g$ through the intersection of $g$ and $U_\varepsilon(O)$ with both increasing and decreasing $t$.*

Proof.  We may assume without loss of generality that the semipath $L^{()}$ in the statement of the lemma contains points outside $C$ (otherwise we need only take a circle $C'$ about $O$, smaller than $C$, outside which there are points of $L^{()}$, and consider the corresponding curvilinear sector $g'$).

Thus, let $M$ be the last point of $L^{()}$ in common with $C$.  Let $g_1$ and $g_2$ denote the curvilinear sectors into which $L_M^{()}$ divides $g$.  Suppose that for any $\varepsilon > 0$ there are points of $g$ in $U_\varepsilon(O)$ through which pass paths that leave $C$ with both increasing and decreasing $t$.  Then, obviously, at least one of the sectors $g_1$ or $g_2$ contains points in any neighborhood of $O$ through which pass paths that leave $C$ with both decreasing and increasing $t$.  Suppose that there are points of this kind in $g_1$.  By Lemma 3, it follows that either $L^{()}$ is a separatrix of the equilibrium state $O$ or there exists a separatrix between $L^{()}$ and $L_1^{()}$.  But this contradicts the assumption.  Q.E.D.

Lemma 5.  *Assume that between the semipaths $L_1^{*()}$ and $L_2^{*()}$ there are no loops and no separatrices of the equilibrium state $O$, but there is a semipath $L^{()}$ between these semipaths which tends to $O$.  Then any path through a sufficiently small neighborhood $U_\varepsilon(O)$ has a semipath between $L_1^{*()}$ and $L_2^{*()}$ (i.e., which tends to $O$ without leaving the sector $g$).  Moreover, the semipaths $L_1^{*()}$ and $L_2^{*()}$, and also $L_1^{*()}$ and $L_2^{*()}$ themselves, are either all positive or all negative, depending on whether $L^{()}$ is positive or negative.*

Proof.  Suppose the assertion false: among the semipaths tending to $O$ and lying between $L_1^{*()}$ and $L_2^{*()}$, and the latter semipaths themselves, there are both positive and negative semipaths.

By the assumptions of our lemma and by Lemma 4, there exists $\varepsilon > 0$ such that no path through the intersection of $g$ and $U_\varepsilon(O)$ leaves the circle $C$ with both increasing and decreasing $t$.  Let $R_1$ and $R_2$ be points on the semipaths $L_1^{*()}$ and $L_2^{*()}$, respectively, so close to $O$ that they can be joined by a simple arc $\lambda$ lying in the intersection of $g$ and the neighborhood $U_\varepsilon(O)$ (see Remark to Lemma 1).  The arc $\lambda$ divides $g$ into two regions, only one of which has $O$ as a boundary point.  It is obvious that any semipath tending to $O$, lying between $L_1^{*()}$ and $L_2^{*()}$ and containing points outside $C$, invariably enters this region, and therefore crosses $\lambda$.  Moreover, by the assumptions of the lemma and the choice of $\varepsilon$ the paths passing through points of $\lambda$ (which by construction lies in $U_\varepsilon(O)$) must lie between $L_1^{*()}$ and $L_2^{*()}$, tend to $O$ and have points outside the circle $C$.

Let us divide the points of $\lambda$ (including its endpoints) into two types, depending on whether the semipaths through them are positive (first type) or negative (second type). By construction, there are points of both types on $\lambda$. Moreover, for any point on $\lambda$ of the first (second) type, all sufficiently close points of $\lambda$ are of the same type. Indeed, let $P$ be a point on $\lambda$ of the first type, say, i. e., there is a positive semipath $L^+$ passing through $P$ at time $t = t_0$. By assumption, there are necessarily points on $L^+$, corresponding to times $t < t_0$, lying outside $C$. By the continuity of solutions with respect to initial values, all semipaths passing at $t = t_0$ through points sufficiently close to $P$ on $\lambda$ also contain points outside $C$, corresponding to times $t < t_0$. But then these semipaths cannot contain points outside $C$, corresponding to times $t > t_0$. For otherwise there would be paths through points of $\lambda$ leaving the circle $C$ for both increasing and decreasing $t$, contradicting the fact that $\lambda$ is in $U_\varepsilon(O)$. Hence it clearly follows that all points sufficiently close to $P$ on $\lambda$ are also of the first type. The same holds for points of the second type. But then, using familiar arguments, we readily show that there cannot be points of both types on $\lambda$. Consequently, the semipaths between $L_1^{*()}$ and $L_2^{*()}$ (and also $L_1^{*()}$ and $L_2^{*()}$ themselves) are either all positive or all negative, depending on whether the semipath $L^{()}$ whose existence is assumed is positive or negative. Q.E.D.

Corollary. If one of the semipaths $L_1^{*()}$, $L_2^{*()}$ is positive and the other negative, then either a) there is at least one separatrix of the equilibrium state $O$ between $L_1^{*()}$ and $L_2^{*()}$, or b) the semipaths $L_1^{*()}$ and $L_2^{*()}$ are separatrices continuing each other, or c) there is a loop in the sector $g$.

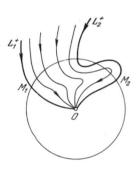

FIGURE 196.

A sector such that all paths through its intersection with a sufficiently small neighborhood of $O$ tend to $O$ as $t \to +\infty$ $(t \to -\infty)$ without leaving the sector and leave the sector with decreasing (increasing) $t$, is known as an $\omega$ ($\alpha$-)-parabolic sector* (Figure 196).

We now turn to the case of a single semipath $L^{*()}$ tending to an equilibrium state $O$. As before, let $\varepsilon_0 > 0$ be such that the closed neighborhood $U_{\varepsilon_0}(O)$ contains no whole singular path other than $O$, and let $C$ be a circle about the equilibrium state, of radius less than $\varepsilon_0$, outside which there are points of $L^{*()}$.

Let $M^*$ be the last point of $L^{*()}$ in common with $C$ and $g$ the corresponding cyclic sector (i. e., the region whose boundary comprises the circle $C$ and the arc $M^*O$ of $L^{*()}$).

Lemmas 6, 7 and 8 below are the analogs of Lemmas 3, 4 and 5 for cyclic sectors; the proofs, which are analogous to those of Lemmas 3, 4 and 5, are omitted.

Lemma 6. *Assume that for any $\varepsilon > 0$ there exists a path through the intersection of $g$ and the $\varepsilon$-neighborhood of $O$ which leaves the circle $C$ with both increasing and decreasing $t$. Then there exist at least one $\omega$-separatrix and at least one $\alpha$-separatrix of $O$, one of which may be $L^{*()}$.*

* [Also known as a fan /13/.]

*Lemma 7. If there is no separatrix of the equilibrium state $O$ through any point of the cyclic sector $g$, but there exists a semipath (not a separatrix) tending to $O$, then there exists $\varepsilon > 0$ such that none of the paths through points of $U_\varepsilon$ $(O)$ leave the circle $C$ with both increasing and decreasing $t$.*

*Lemma 8. Suppose that the cyclic sector $g$ contains no loop and no separatrix of the equilibrium state $O$, but there is a semipath $L^O$ in $g$ (not a separatrix) tending to $O$. Then all paths through a sufficiently small neighborhood $U_\varepsilon$ $(O)$ tend to the equilibrium state and have points outside $C$; moreover, these semipaths are either all positive or all negative, depending on whether $L^O$ is positive or negative.*

If all semipaths through a certain neighborhood of an equilibrium state $O$ are positive (negative) and tend to the equilibrium state, we call the latter a t o p o l o g i c a l   n o d e.* A topological node is said to be s t a b l e if all the semipaths tending to it are positive, u n s t a b l e if they are all negative.

We now discuss regions filled by loops (i. e., paths tending to the equilibrium state $O$ both as $t \to +\infty$ and as $t \to -\infty$). Note that, under our assumption that the number of singular elements is finite, there can be only finitely many cells filled by loops (see Theorem 56, §16).

As before, let $\overline{U_{\varepsilon_0}}$ $(O)$ be a neighborhood containing no whole singular path other than $O$ and suppose that there exists a path $L$, contained entirely in $\overline{U_{\varepsilon_0}}$ $(O)$ (and so tending to $O$ both as $t \to +\infty$ and $t \to -\infty$, thus forming a loop). Let $\sigma$ denote the simple closed curve formed from the path $L$ and the point $O$, and $g_\sigma$ the interior of $\sigma$. By the choice of $U_{\varepsilon_0}$ $(O)$, all paths through points inside $\sigma$ are nonsingular and therefore, by Theorem 53, form loops embedded inside one another (or, as we have called them, n e s t e d loops).

*Lemma 9. For any $\varepsilon > 0$ there is a loop in the intersection of the interior of the loop $\sigma$ and the neighborhood $U_\varepsilon$ $(O)$.*

P r o o f. If this is not true, there exists $\varepsilon > 0$ such that no loop in $g_\sigma$ can lie wholly within the circle $C$ of radius $\varepsilon$ about $O$. We may assume that $\varepsilon < \varepsilon_0$ and that there are points of $L$ outside $C$.

Let $M^+$ and $M^-$ denote the last points of $L$ in common with $C$ for increasing and decreasing $t$, respectively, $g$ and $g'$ the curvilinear sectors into which the semipaths $L^+_{M^+}$ and $L^-_{M^-}$ divide the interior of $C$. The points of $L^+_{M^+}$ and $L^-_{M^-}$ are obviously boundary points for the regions $g$, $g'$ and $g_\sigma$. Hence one readily concludes that all points of $g_\sigma$ sufficiently close to $O$ are points of one (and only one) of the regions $g$ and $g'$, say of $g$. Conversely, all points of $g$ sufficiently close to $O$ are in $g_\sigma$.

According to our assumption, the region $g$ cannot contain any whole loop, for any such loop would lie in the intersection of $g_\sigma$ and the interior of $C$. But the semipaths $L^+_{M^+}$, $L^-_{M^-}$ are components of the boundary of $g$ and are not separatrices (by virtue of the choice of $L$). Moreover, there can be no separatrix between them (otherwise there would be a whole singular path inside $\sigma$, hence in the neighborhood $U_{\varepsilon_0}$ $(O)$, contrary to the choice of $\varepsilon_0$).

It now follows from Lemma 7 that the boundary semipaths $L^+_{M^+}$ and $L^-_{M^-}$ of $g$ must be either both positive or both negative, which is impossible. Q.E.D.

---

* It is clear that simple nodes and simple foci (see Chapter IV) are topological nodes. But multiple foci and other types of multiple equilibrium state are also topological nodes (see Chapter VIII).

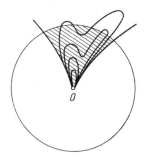

FIGURE 197.

A curvilinear sector containing loops, and moreover only nested loops, is known as an elliptic sector (Figure 197).

The region within the loop formed by a path $L$ will be called a regular elliptic region of the point $O$, or simply an elliptic region of $O$*(Figure 181), if it contains no singular path. Through all points of an elliptic region there are nested loops; moreover, by Lemma 9, one can find such loops within arbitrarily small circles about $O$. The path $L$ which, together with $O$, makes up the boundary $\sigma$ of an elliptic region $g_\sigma$, will sometimes be referred to as the path forming the elliptic region, and we shall speak of the elliptic region formed by the path $L$. The path $L$ may be either singular or nonsingular. If $L$ is a singular path, the corresponding elliptic region $g_\sigma$ is of course a complete cell, while if $L$ is nonsingular $g_\sigma$ is only part of a cell.

Throughout this chapter we confine ourselves to a neighborhood of the equilibrium state $O$ which contains no whole singular path other than $O$ itself. Thus all elliptic regions considered are formed by nonsingular paths $L$.

**3. LEMMAS ON ELLIPTIC REGIONS.** Let $g_\sigma$ and $g_\sigma^*$ be two elliptic regions in $U_{\varepsilon_0}(O)$. There are two possible cases:

1) The regions $g_\sigma$ and $g_\sigma^*$ either coincide or one is a subregion of the other, i.e., the paths $L$ and $L^*$ forming them either coincide or form loops one of which is nested in the other.

2) The regions $g_\sigma$ and $g_\sigma^*$ are disjoint — the paths $L$ and $L^*$ form mutually exterior loops.

It is clear that in case 1) the regions $g_\sigma$ and $g_\sigma^*$ are in the same cell, while in case 2) they are in two distinct cells. In case 1) we shall say that one of the elliptic regions $g_\sigma$ and $g_\sigma^*$ is a part of the other, in case 2) that $g_\sigma$ and $g_\sigma^*$ are distinct. Henceforth, when we say, for example, that an equilibrium state $O$ has $n$ distinct elliptic regions we shall understand the adjective "distinct" in this sense. As in Lemma 9, let $g_\sigma$ be an elliptic region formed by a path $L$.

*Lemma 10. If an equilibrium state $O$ has an elliptic region $g_\sigma$ formed by a path $L$, then it has either another elliptic region distinct from $g_\sigma$, or at least one $\omega$-separatrix and one $\alpha$-separatrix.*

Proof. Let $C$ be a circle about $O$, of radius so small that there are points of the path $L$ outside $C$; let $N$ be one of these points. Let $M_0^+$ and $M_0^-$ be the last points of $L_N^+$ and $L_N^-$ in common with $C$, $g$ and $g'$ the curvilinear sectors into which the semipaths $L^+_{M+}$ and $L^-_{M-}$ divide the interior of $C$. One of these sectors, $g$ say, is such that all its points sufficiently close to $O$ are in $g_\sigma$ (Lemma 9). Consider the other sector $g'$.

One of the semipaths in the boundary of $g'$ is positive, the other negative, and neither is a separatrix. Therefore, by the Corollary to Lemma 5, the sector $g'$ contains either $\alpha$- and $\omega$-separatrices of the equilibrium state $O$,

---

* In the mathematical literature, the term "elliptic region" is often used for a region through all of whose points pass paths that tend to an equilibrium state $O$ both as $t \to +\infty$ and as $t \to -\infty$; some of these paths may be orbitally stable, others orbitally unstable.

or a loop $\sigma'$. In the latter case the elliptic region $g_{\sigma'}$ is distinct from $g_\sigma$, since these are subregions of the disjoint regions $g$ and $g'$, respectively. Q.E.D.

The next lemma deals with paths in elliptic regions. Let $L_1$ and $L_2$ be paths in $U_{\varepsilon_0}(O)$, in the same elliptic region of $O$, so that the loops that they form lie within one another, for example, the loops formed by $L_2$ is nested within that formed by $L_1$. Let $\sigma_1$ denote the simple closed curve formed by the path $L_1$ and the point $O$, and $\sigma_2$ that formed by $L_2$ and $O$. Let $C$ be a circle about $O$, lying in $U_{\varepsilon_0}(O)$, of radius so small that both paths $L_1$ and $L_2$ have points outside it. Let $M_1^+$ and $M_1^-$ denote the last points of $L_1$ in common with $C$ for increasing and decreasing $t$, respectively, and $M_2^+$ and $M_2^-$ the analogous point of $L_2$. The semipaths $L_{1M_1^+}^+$ and $L_{1M_1^-}^-$, $L_{2M_2^+}^+$ and $L_{2M_2^-}^-$ lie entirely within $C$.

Let $g_1$ denote the curvilinear sector bounded by $L_{1M_1^+}^+$ and $L_{1M_1^-}^-$ such that all points sufficiently close to $O$ are inside the curve $\sigma_1$ (see Lemma 9). Assume that the arc of the circle $C$ in the boundary of this sector is $M_1^+M_1^-$. Since the loop formed by $L_2$ is nested in that formed by $L_1$, it is obvious that the points $M_2^+$ and $M_2^-$ are on the arc $M_1^+M_1^-$.

*Lemma 11. The point $M_2^+$ lies on the arc $M_1^+M_1^-$ between $M_1^+$ and $M_2^-$ (or, equivalently, $M_2^-$ lies between $M_2^+$ and $M_1^-$ (Figure 198)).*

P r o o f. Suppose on the contrary that $M_2^+$ is not between $M_1^+$ and $M_2^-$, so that $M_2^-$ is between $M_1^+$ and $M_2^+$.

The semipaths $L_{2M_2^+}^+$ and $L_{2M_2^-}^-$ divide the sector $g$ into three "subsectors." One of these, $g_1'$ say, is bounded by both these semipaths. It is readily seen

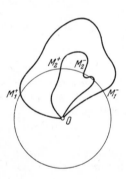

FIGURE 198.

that all points of this sector sufficiently close to $O$ are inside the loop $\sigma_2$ and moreover the sector in question must contain loops nested in $\sigma_2$.

Since we have assumed that $L_{2M_2^-}^-$ is between $L_{1M_1^+}^+$ and $L_{2M_2^+}^+$, it follows that the boundary of one of the subsectors into which $L_{2M_2^+}^+$ and $L_{2M_2^-}^-$ divide $g_1$ contains $L_{1M_1^+}^+$ and $L_{2M_2^-}^-$. Denote this subsector by $g_1''$. All of its points sufficiently close to $O$ are inside the curve $\sigma_1$ and outside $\sigma_2$. Since one of the boundary semipaths of this subsector is positive and the other negative, it follows from the Corollary to Lemma 7 that between these semipaths (i. e., in the region $g_1''$) there is either a separatrix of the equilibrium state $O$ or a loop.

But all points sufficiently close to $O$ in $g$ lie inside $\sigma_1$, that is to say, are in the same cell.

Hence it clearly follows that there are no separatrices between $L_{1M_1^+}^+$ and $L_{2M_2^-}^-$. But neither can there be loops in the region $g_1''$. Indeed, any such loop would lie inside $\sigma_1$ and outside a loop in $g_1'$, which by the foregoing reasoning certainly exists and also lies inside $\sigma_1$. This is impossible, for the interior of $\sigma_1$ is part of a single elliptic region $g_{\sigma_1}$, and all its loops are nested. This contradiction proves the lemma.

C o r o l l a r y. Let $g^*$ be the sector bounded by two positive (negative) semipaths $L_{1M_1^+}^+$ and $L_{2M_2^+}^+$ ( $L_{2M_2^-}^-$ and $L_{1M_1^-}^-$) of paths $L_1$ and $L_2$ in the same elliptic region, such that all points sufficiently close to $O$ are in the same elliptic region.

Then: a) all points of the sector sufficiently close to $O$ lie between the loops formed by $L_1$ and $L_2$; b) the sector is $\omega$ ($\alpha$)-parabolic. The semipaths in this sector are parts of paths forming loops outside the loop formed by $L_2$ and inside the loop formed by $L_1$.

## §18. ELEMENTARY REGIONS. TYPES OF ELEMENTARY REGION

**1. ARC WITHOUT CONTACT IN A PARABOLIC SECTOR.** The three lemmas of this subsection deal with the construction of arcs without contact and certain simple regions defined by these arcs in the neighborhood of an equilibrium state.

*Lemma 1. Let $l$ be a simple arc containing no singular points, and $\varepsilon > 0$, $\Delta > 0$ arbitrary positive numbers. Then there exist a subdivision of the arc $l$ into subarcs $l_1, l_2, \ldots, l_n$ and arcs without contact $\lambda_1, \lambda_2, \ldots, \lambda_l$ such that each arch $\lambda_i$ is in the $\varepsilon$-neighborhood of $l_i$, and every path passing at $t = t_0$ through a point of $l_i$ crosses $\lambda_i$ at some time $t^*$, $|t^* - t_0| < \Delta$, without previously (i.e., at times between $t_0$ and $t^*$) leaving the $\varepsilon$-neighborhood of $l_i$ ($i = 1, 2, \ldots, n$).*

Proof. Suppose the assertion false. Let $d$ be the diameter of the arc $l$. Partition $l$ into finitely many subarcs of diameter $\frac{d}{2}$. Then the assertion must be false for at least one of these subarcs, $l'$ say. Repeating the argument for the arc $l'$, and so on, we get a sequence of nested simple arcs

$$l \supset l' \supset l'' \supset l''' \supset \ldots$$

such that the diameter of $l^j$ is less than $\frac{d}{2^j}$ and the lemma is false for each arc (i. e., for no partition of $l^j$ into subarcs $l_i^{(j)}$ and no arcs without contact $\lambda_i^{(j)}$ in the $\varepsilon$-neighborhoods of $l_i^{(j)}$, respectively, is the assertion of the lemma true). The arcs $l^j$ have a common point — denote it by $M_0$. Let $\lambda$ be some arc without contact containing $M_0$ as an interior point and lying in the neighborhood $U_\varepsilon(M_0)$. By Lemma 1 of §3, any path passing at $t = t_0$ through the points of $l^j$ for sufficiently large $j$ (hence in a sufficiently small neighborhood of $M_0$) will cross the arc $\lambda$ at some time $t^*$, $|t^* - t_0| < \Delta$, without previously leaving $U_\varepsilon(M_0)$. But this obviously implies that the assertion of the lemma is true for the arc $l^j$, contrary to assumption. Q.E.D.

Remark. The analog of this lemma is valid for a simple closed curve $C$ on which there are no equilibrium states. To prove this, simply divide $C$ into two simple arcs $l_1$ and $l_2$ and apply Lemma 1 to each.

We now consider a curvilinear sector $g$ of the interior of a circle $C$. Let $\tilde{L}^{()}$ and $\tilde{\tilde{L}}^{()}$ be the boundary semipaths of the sector, $M_1$ and $M_2$ their last points in common with $C$. Suppose that the sector is $\omega$-parabolic (the treatment of an $\alpha$-parabolic sector is analogous). Then all paths through the intersection of the sector with a sufficiently small neighborhood of $O$ must tend to $O$ as $t \to +\infty$, without leaving the circle $C$, while with decreasing $t$ they leave $C$; the boundary semipaths of the sector are positive semipaths $\tilde{L}_{M_1}^+$ and $\tilde{L}_{M_2}^+$. In this situation:

*L e m m a   2.  If the curvilinear sector g is ω-parabolic, then: a) any two points $P_0$ and $Q_0$ on its boundary semipaths $\tilde{L}_{M_1}$ and $\tilde{\tilde{L}}_{M_2}$, respectively, may be joined by an arc without contact λ lying except for its endpoints in the sector g; b) the arc λ divides g into two regions, and the paths passing through that of these regions whose boundary contains the point O (i.e., the region bounded by the arcs $P_0O$, $Q_0O$ of the semipaths $\tilde{L}^+_{M_1}$ and $\tilde{\tilde{L}}^+_{M_2}$, the point O and the arc λ) tend to O as $t \to +\infty$, while with decreasing t they leave the region across the arc λ, crossing the latter at exactly one point.*

P r o o f.  We first show that there exists at least one arc without contact joining some point of $\tilde{L}^+_{M_1}$ to a point of $\tilde{\tilde{L}}^+_{M_2}$. By the definition of an ω-parabolic sector, there exists $\delta > 0$ such that the paths through all points of the sector in $U_\delta(O)$ tend to O as $t \to +\infty$ without leaving g, while with increasing t they leave the sector.

Let $Q_1$ and $Q_2$ be points on the semipaths $\tilde{L}^+_{M_1}$ and $\tilde{\tilde{L}}^+_{M_2}$, respectively, so close to O that there is a simple arc $l$ (not necessarily without contact) joining them and lying except for its endpoints in the intersection of g and $U_\delta(O)$ (Lemma 1 of §17). It is obvious that all paths that tend to O with increasing t without leaving g have points in common with $l$.

Let $\varepsilon > 0$ be such that the ε-neighborhood of $l$ is in $U_\delta(O)$, $U_\varepsilon(l) \subset U_\delta(O)$. By Lemma 1, given this $\varepsilon > 0$ and any $\Delta > 0$,* we can find a partition of $l$ into subarcs $l_1, l_2, \ldots, l_n$ and arcs without contact $\lambda_1, \lambda_2, \ldots, \lambda_n$, $\lambda_i \subset U_\varepsilon(l_i)$, such that any path passing at $t = t_0$ through a point of $l_i$ crosses $\lambda_i$, without leaving $U_\varepsilon(l_i)$, at some time $t^*$, $|t^* - t_0| < \Delta$.

If some path in the sector g, or one of the semipaths $\tilde{L}^+$ and $\tilde{\tilde{L}}^+$, crosses one of the arcs $\lambda_i$ twice without leaving C, all paths tending to the equilibrium state O must cross this arc. In this case, the subarc of $\lambda_i$ between its points of intersection with $\tilde{L}^+$ and $\tilde{\tilde{L}}^+$ is the required arc without contact. If $n = 1$, i.e., all paths through points of $l$, including $\tilde{L}^+$ and $\tilde{\tilde{L}}^+$, cross one arc without contact λ, then the latter is the required arc without contact. Let $n > 1$. We may assume without loss of generality that the arcs $\lambda_i$ are so chosen that they contain no points outside the sector $\bar{g}$ and no points of the circle C.

Suppose that the endpoint $Q_1$ of the arc $l$ is an endpoint of its subarc $l_1$, so that $l_1$ is on the negative side of the semipath $\tilde{L}^+$.** The endpoint $Q_2$ of $l$ is an endpoint of the subarc $l_n$, and the latter is on the positive side of the semipath $\tilde{\tilde{L}}^+$. Since all paths passing through points of $l_1$ cross the arc without contact $\lambda_1$, the latter must have exactly one point in common with $\tilde{L}^+$ and a path $L_1$ passing through a point distinct from $Q_1$ on $l_1$. Let $A_0$ and $D_1$ be the points of $\tilde{L}^+$ and $L_1$, respectively, in common with $\lambda_1$ (Figure 199).

It is clear that the path $L_1$ tends to O as $t \to +\infty$ without leaving the sector g, while with decreasing t it leaves C.

Let $N'$ denote the last point of $L_1$ in common with C for increasing t. The semipath $L^+_{1N'}$ divides g into two sectors $g_1$ and $g_2$ and is common to the boundary of both these sectors. The semipath $\tilde{L}^+_{M_1}$ figures in the boundary of one sector, $g_1$ say, and the semipath $\tilde{\tilde{L}}^+_{M_2}$ in that of $g_2$. Moreover, all interior points of the subarc $A_0D_1$ of $\lambda_1$ are in the sector $g_1$.

---

*  It will be clear from the sequel that the choice of $\Delta > 0$ is inessential.

**  This follows from the fact that the arc of C in the boundary of g is $M_1M_2$.

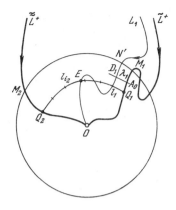

FIGURE 199.

It is clear that all paths passing through the subarc $A_0 D_1$ of $\lambda_1$ tend to $O$ as $t \to +\infty$, without leaving the subregion of $g_1$ bounded by this subarc of $\lambda_1$, the arcs $A_0 O$ and $D_1 O$ of $\tilde{L}^+_{M_1}$ and $L^+_{1N'}$, and the point $O$. On the other hand, all paths which pass through points of $g_1$ and, without leaving this sector, tend to $O$, have points in common with the subarc $A_0 D_1$ of $\lambda_1$.

Moving along the arc $l$ from $Q_1$ to $Q_2$, let the last point of the arc in common with the semipath $L^+_{1N'}$ be $E$. The subarc $EQ_2$ of $l$ thus has no points in common with $L^+_{1N'}$, and all its interior points are in the sector $g_2$. Denote the subarc $EQ_2$ of $l$ by $l'$. The sector $g_2$ and arc $l'$ will now be treated in the same way as $g$ and $l$.

If the point $E$ is on a subarc $l_{i_2}$ ( $i_2 > 1$), it is obvious that the subarc of $l_{i_2}$ with endpoint at $E$ (in the sector $g_2$) and the arcs $l_{i_2+1}, \ldots, l_n$, together with the subarc of $\lambda_{i_2}$ lying in $g_2$ and the arcs $\lambda_{i_2+1}, \ldots, \lambda_n$, constitute a partition of $l'$ and a system of arcs without contact satisfying the conditions of Lemma 1.

If the arc $\lambda_{i_2}$ has no point in common with $\tilde{L}^+$, we can repeat the above argument, with the subarcs of $l_{i_2}$ and $\lambda_{i_2}$ lying in $g_2$ (one of whose endpoints is on $L^+_{1N'}$) and the path $L_2$ through the endpoint of $l_{i_2}$ lying in $g_2$ playing roles analogous to those of $l_1$, $\lambda_1$ and $L_1$ above. Thus we obtain a partition of $g_2$ (by the semipath $L^+_2$) into two sectors, analogous to the previous partition of $g$. When this is done, the subarc of $\lambda_{i_2}$ between its points of intersection with $L_1$ and $L_2$, denoted by $A_1$ and $D_2$, respectively, lies except for its endpoints in the sector which has a boundary semipath $L^+_{1N'}$ in common with the sector $g_1$. Let us denote this sector by $g'_1$.

Since the number of arcs $l_i$ is finite, finitely many repetitions of this procedure yield a partition of the sector $g$ into $k$ ( $k \leqslant n$) subsectors

$$g_1 = g_1^0, \ g'_1, \ \ldots, \ g_1^{(k)}.$$

This partition is produced by semipaths

$$L^+_1, \ L^+_2, \ \ldots, \ L^+_{k-1},$$

on paths $L_1, L_2, \ldots, L_{k-1}$ passing through the endpoints of some of the arcs $l_i$: $l_{i_1}, l_{i_2}, \ldots$ To unify the notation, we denote $\tilde{L}$ by $L^+_0$ and $\tilde{\tilde{L}}$ by $L^+_k$.

$L^+_{\alpha-1}$ and $L^+_\alpha$ are boundary semipaths of the sector $g_1^{(\alpha)}$ (Figure 200). The arc without contact $\lambda_{i_\alpha}$ has points $A_{\alpha-1}$ and $D_\alpha$ in common with the semipaths $L_{\alpha-1}$ and $L_\alpha$, respectively, and the subarc $A_{\alpha-1} D_\alpha$ of $\lambda_{i_\alpha}$ lies except for its endpoints in the sector $g_1^{(\alpha)}$. Each path which tends to the equilibrium state $O$ as $t \to \infty$ without leaving this sector crosses this subarc of $\lambda_{i_\alpha}$ (at exactly one point). We now examine the sector $g_1^{(\alpha)}$ together with its boundary semipaths $L_{\alpha-1}$ and $L_\alpha$.

Suppose that under the motion fixed on the paths crossing the subarc $A_{\alpha-1} D_\alpha$ of $\lambda_{i_\alpha}$ the points of intersection correspond to times $t = t_0^{(\alpha)}$. At no time $t > t_0^{(\alpha)}$ do these paths have points in common with the arc $\lambda_{i_\alpha}$. By Lemma 9 of §3, any two points $P'$ and $P''$ on the semipaths $L^+_{\alpha-1}$ and $L^+_\alpha$, corresponding to times $t' > t_0^{(\alpha)}$ and $t'' > t_0^{(\alpha)}$, respectively, may be joined by an arc without contact, all of whose points lie on paths crossing the arc $A_{\alpha-1}D_\alpha$.

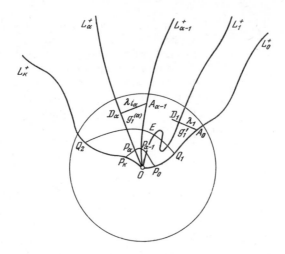

FIGURE 200.

Now let $P_0$ be a point on $L_0^+$ corresponding to some $t > t_0^{(0)}$, $P_k$ a point on $L_k^+$ corresponding to $t > t_0^{(k)}$, and $P_\alpha$ a point on each semipath $L_\alpha^+$ ($\alpha = 1, 2, \ldots, k-1$) corresponding to $t > t_0^{(\alpha)}$.

Note that, depending on whether the semipath is regarded as a boundary semipath of $g_1^{(\alpha-1)}$ or of $g_1^{(\alpha)}$, the motion fixed thereon is either such that $D_{\alpha-1}$ corresponds to $t = t_0^{(\alpha-1)}$ or such that $A_{\alpha-1}$ corresponds to $t = t_0^{(\alpha)}$. These motions need not coincide. With respect to the first, the point $P_\alpha$ corresponds to a time $t_1^{(\alpha-1)} > t_0^{(\alpha-1)}$, with respect to the second — to $t_1^{(\alpha)} > t_0^{(\alpha)}$. By Lemma 8 of §3, any two points $P_\alpha$ and $P_{\alpha+1}$ ($\alpha = 0, 1, 2, \ldots, k-1$) may be joined by an arc without contact lying (except for its endpoints) in the sector $g_1^{(\alpha)}$. In addition, the arcs without contact joining the points $P_\alpha$, $P_{\alpha+1}$ and $P_{\alpha+1}$, $P_{\alpha+2}$ ($\alpha = 1, 2, \ldots, k-1$) may be so chosen (see Remark after Lemma 8 of §3) that they have the same tangent at their common endpoint $P_{\alpha+1}$. But then the union of all the arcs without contact $P_\alpha P_{\alpha+1}$ ($\alpha = 1, \ldots, k-1$) is an arc without contact joining $P_0$ to $P_k$ which is (except for its endpoints) entirely within the sector $g$. We have thus proved the existence of at least one arc without contact joining a point of $\tilde{L}^+$ to a point of $\widetilde{L}^+$ and lying except for its endpoints inside $g$.

Consider the subregion of $g$ whose boundary is the union of the arcs $P_0O$ and $P_kO$ of the semipaths $\tilde{L}^+$ and $\widetilde{L}^+$, the point $O$ and the arc $\lambda$. Denote this region by $g_N$. Clearly, all paths crossing $\lambda$ do so at exactly one point, enter the region $g_N$ with increasing $t$, and tend to $O$ as $t \to +\infty$ without leaving the region. Moreover, by assumption there is no loop inside the sector $g$ or, consequently, inside $g_N$. Therefore, all paths passing through points of $g_N$ leave the region with decreasing $t$ through points of $\lambda$. With increasing $t$, they tend to the equilibrium state, of course without leaving the region $g_N$. Thus the assertion of the lemma is true for the arc $\lambda$.

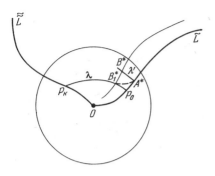

FIGURE 201.

We shall now show that any point $A^*$ on $\widetilde{L}^+_{M_1}$ can be joined to any point $D^*$ on $\widetilde{L}^+_{M_2}$ by an arc without contact lying, except for its endpoints, inside the sector $g$.

It will suffice to show that any point on $\widetilde{L}^+_{M_1}$ and any point on $\widetilde{L}^+_{M_2}$ can be joined to any fixed point of the arc $\lambda$ by arcs without contact, in the sector $g$, with the same tangent at the point in question as the arc $\lambda$.

Suppose that all paths crossing the arc $\lambda$ do so at time $t = t_0$. If the point $A^*$ on $\widetilde{L}^+_{M_1}$ corresponds to time $t > t_0$, the existence of an arc without contact joining $A^*$ to the fixed point on $\lambda$ follows directly from Remark 3 to Lemma 8 of §3. Assume, therefore, that $A^*$ corresponds to $t_1 < t_0$. Draw an auxiliary arc $\lambda'$ with endpoint at $A^*$, lying except for its endpoints in $g$. Let $A^*B^*$ be a subarc of $\lambda'$, so small that all paths crossing it cross $\lambda$ with increasing $t$, without leaving $g$ (see Lemma 5 of §3); let the path through $B^*$ cross $\lambda$ at a point $B^*_1$.

Obviously, one can always join $A^*$ and $B^*_1$ by an arc without contact $\lambda''$ in the elementary quadrangle $A^*B^*B^*_1P_0$ (hence also in $g$). In addition, $\lambda''$ may be so chosen that its tangent at $B^*_1$ coincides with the tangent to $\lambda$ there. Then the union of $\lambda''$ and the subarc $B^*_1P_k$ of $\lambda$ is an arc without contact joining $A^*$ to $P_k$ and meeting our requirements.

Reasoning in analogous fashion for the point $D^*$, one proves the existence of an arc without contact in $g$ joining the points $A^*$ and $D^*$.

The truth of part b) of our lemma for any arc without contact in $g$ joining points of $\widetilde{L}^+$ and $\widetilde{\widetilde{L}}^+$ is established in the same way as for the arc $\lambda$. Q.E.D.

Remark. Since the circle $C$ may be chosen arbitrarily small, it clearly follows that for any $\eta > 0$ there exists an arc without contact, joining points $A^*$ and $D^*$ on the semipaths $\widetilde{L}^+$ and $\widetilde{\widetilde{L}}^+$, respectively, sufficiently close to $0$, which lies inside $U_\eta (0)$.

The region $g_N$ bounded by the arcs $A^*0$ and $D^*0$ of the semipaths $\widetilde{L}^+$ and $\widetilde{\widetilde{L}}^+$, the point $0$ and an arc without contact $\lambda^*$ joining $A^*$ and $D^*$ will be called a regular parabolic sector (sometimes simply parabolic sector, if no confusion can arise). The region will be called an $\omega$-parabolic or $\alpha$-parabolic sector, depending on whether the semipaths through it tend to the equilibrium state $0$ as $t \to +\infty$ or as $t \to -\infty$.

We now consider the case in which the paths through all points of a sufficiently small neighborhood $U_\delta (0)$, other than $0$ itself, tend to $0$ without leaving $C$ as $t \to +\infty$ $(t \to -\infty)$, but leave the interior of $C$ with decreasing (increasing) $t$ (so that $0$ is a topological node). The following analog of Lemma 2 is then valid.

Lemma 3. *There exists a cycle without contact inside the circle $C$, enclosing the point $0$, such that all paths through its interior (except for the equilibrium state $0$) tend to $0$ as $t \to +\infty$ $(t \to -\infty)$, and with decreasing (increasing) $t$ cross the cycle without contact outward.*

Proof. It is sufficient to consider the case in which all paths passing through points of $U_\delta (0)$ tend to the equilibrium state $0$ as $t \to +\infty$ and leave the interior of $t$ with decreasing $C$. Let $L_1^+$ and $L_2^+$ be two of these semipaths, $M_1$ and $M_2$ their last points in common with $C$. The arcs $M_1O$ and $M_2O$ of the semipaths $L_1^+$ and $L_2^+$, together with the point $O$, divide the interior of $C$ into two sectors $g_1$ and $g_2$, for each of which the conclusion of Lemma 2 is valid. Thus there exists an arc without contact $\lambda_1$ joining some point $A$ on $L_1^+$ to a point $B$ on $L_2^+$, lying (except for its endpoints) in the sector $g_1$, and an arc without contact $\lambda_2$ joining $A$ and $B$ in $g_2$. Moreover, these arcs $\lambda_1$ and $\lambda_2$ may always be so chosen that they have the same tangents at $A$ and $B$. Thus the union of $\lambda_1$ and $\lambda_2$ is a cycle without contact $\sigma$ lying interior to the circle $C$. It is obvious that $O$ is inside $\sigma$. Employing arguments analogous to those used to prove part b) of Lemma 2, one readily proves the last assertion of our lemma.

Remark. For any arbitrarily small $\eta > 0$, the above cycle without contact may be so chosen that it lies entirely in $U_\eta(O)$. The region interior to the cycle without contact is called a **full parabolic (nodal) region**, $\omega$-parabolic or $\alpha$-parabolic according as the paths through its points tend to the equilibrium state $O$ as $t \to +\infty$ or as $t \to -\infty$.

**2. ARCS WITHOUT CONTACT IN AN ELLIPTIC REGION.** We now consider arcs without contact in elliptic regions. As before, let the neighborhood $U_{\varepsilon_0}(O)$ contain no whole singular path other than $O$. Suppose that there exist whole paths inside the neighborhood $U_{\varepsilon_0}(O)$, thus forming loops. Let $L_1$ and $L_2$ be two paths in the same cell, therefore forming two nested loops; to fix ideas, let the loop $L_2$ be inside the loop formed by $L_1$. Let $\sigma_1$ and $\sigma_2$ denote the closed curves bounding these loops, and $w$ the region between $\sigma_1$ and $\sigma_2$; all points of $w$ are in the same cell (Figure 202).

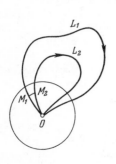

FIGURE 202.

*Lemma 4. There exists an arc without contact joining an arbitrary point on $L_1$ to an arbitrary point on $L_2$, lying except for its endpoints in the region $w$ and cutting all paths passing through $w$ at exactly one point.*

The proof is almost immediate. One first constructs an arc without contact joining any points $M_1$ and $M_2$ (Figure 202) on $L_1$ and $L_2$ in a sufficiently small neighborhood of $O$ (see Lemma 2), and then employs Lemma 5 of §3.

Now let $g_{\sigma_0}$ be a regular elliptic region inside the loop $\sigma_0$ formed by a path $L_0$ and the point $O$.

*Lemma 5. There exists a smooth simple arc joining any point $A_0$ on $L_0$ to $O$, lying entirely (except for its endpoints $A_0$ and $O$) in $g_{\sigma_0}$, which has no contact at any point other than $O$ and cuts all paths through $g_{\sigma_0}$ at exactly one point.*

Proof. Let $\{C_i\}$ be a sequence of circles about $O$ whose radii tend to zero as $i \to \infty$, and $\{L_i\}$ a sequence of paths passing through points of $g_{\sigma_0}$ such that each $L_i$ lies wholly inside the circle $C_i$; the existence of such paths follows from Lemma 9 of §17. Let $\sigma_i$ denote the simple closed curve formed from the path $L_i$ and point $O$, and $w_i$ the region between $\sigma_i$ and $\sigma_{i+1}$. On each path $L_i$, take a point $A_i$. By Lemma 4, we can join the points $A_i$ and $A_{i+1}$ by an arc without contact $\lambda_i$ lying (except for its endpoints $A_i$ and $A_{i+1}$) in the region $w_i$. This arc will cut all paths in $w_i$, each at exactly one point.   Moreover, we may assume that any two arcs $\lambda_i$ and $\lambda_{i+1}$ have a common tangent at their common point $A_{i+1}$.   It is readily seen that the union of all these arcs $\lambda_i$ and the point $O$ is a simple arc possessing the required properties.

**3. REGULAR SADDLE REGION.**   Now let $L_1^+$ and $L_2^-$ be ω - and α-separatrices of an equilibrium state $O$, each a continuation of the other on the positive (negative) side.   As before, let $C$ be a circle about $O$, of radius less than $\varepsilon_0$ (containing no whole singular paths other than $O$), $M_1$ and $M_2$ the last points of $L_1^+$ and $L_2^-$, respectively, in common with $C$ (Figure 203).   Consider arbitrary points $Q$ and $P$ on the arcs $M_1O$ and $M_2O$ of the semipaths $L_1^+$ and $L_2^-$, respectively, and arcs without contact $l_1$ and $l_2$ through these points.   By the definition of separatrices and continuations (see §15.5 and §15.6), there exist subarcs $QA$ and $PB$ of $l_1$ and $l_2$, respectively, such that the following conditions are satisfied:

a) The subarc $QA$ of $l_1$ lies entirely inside the circle $C$, on the positive (negative) side of the semiseparatrix $L_1^+$; b) the subarc $PB$ of $l_2$ lies entirely inside the circle $C$ on the positive (negative) side of $L_2^-$; c) every path passing at $t = t_0$ through a point distinct from $Q$ on $QA$ will cross the arc $PB$ at some time $t > t_0$ (without previously leaving the circle $C$) and with further increase in $t$ will leave $C$; moreover, the path through $A$ crosses $PB$ at $B$; d) every path passing at $t = t_0$ through a point distinct from $P$ on $PB$ will cross the arc $QA$ at some time $t < t_0$ (without previously leaving the circle $C$) and with further decrease in $t$ will leave $C$; moreover, the path through $B$ crosses the arc at $A$.

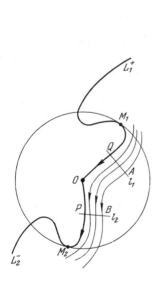

FIGURE 203.

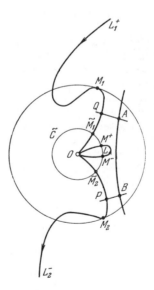

FIGURE 204.

Let $\gamma$ be the simple closed curve inside $C$, consisting of the arc $QO$ of $L_1^+$, the point $O$, the subarc $PO$ of $L_2^-$, the arc without contact $PB$, the arc $AB$ of the path through $A$ and $B$, and the arc without contact $QA$

(Figure 203). Let $g_C$ denote the interior of the curve $\gamma$. This region is obviously a subregion of one of the curvilinear sectors bounded by the semipaths $L_1^+$ and $L_2^-$; denote this sector by $g$.

*Lemma 6.* Any path passing through a point of $g_C$ crosses the arc without contact PB with increasing $t$ and the arc without contact QA with decreasing $t$, each at an interior point.

Proof. Since any path through a point of $g_C$ which crosses one of the arcs $QA$ and $PB$ crosses the other, it will of course suffice to show that there are no paths through a point of $g_C$ which cross neither of the arcs. Suppose that there is such a path, $L$ say. It must lie entirely within $g_C$ (because paths may leave this region only through points of one of the arcs $QA$ and $PB$), and therefore forms a loop in $g_C$ (hence also in the sector $g$). Let $\tilde{C}$ be a circle about $O$, of radius so small that $L$ has points outside it (Figure 204). Let $\tilde{M}_1$ and $\tilde{M}_2$ be the last points in common with $\tilde{C}$ of the semipaths $L_1^+$ and $L_2^-$, respectively, $M^+$ and $M^-$ the last points of $L$ in common with $\tilde{C}$ for increasing and decreasing $t$, respectively. The points $M^+$ and $M^-$ lie on the arc $\tilde{M}_1\tilde{M}_2$ of $\tilde{C}$, and therefore the continuation of the semipath $L_1^+$ on the negative (positive) side is certainly not $L_2^-$. But this contradicts the assumption that $L_2^-$ is the continuation of $L_1^+$ (i. e., the continuation relative to any circle smaller than $C$). Q.E.D.

Remark. For any $\varepsilon > 0$, one can construct a saddle region between the separatrices $L_1^+$ and $L_2^-$, analogous to that considered in the lemma, in the neighborhood $U_\varepsilon(O)$.

We shall call the region $g_C$ a regular hyperbolic (or saddle) region between $L_1^+$ and $L_2^-$, based on the arcs without contact $QA$, $PB$. The arcs without contact $QA$ and $PB$ will be called saddle arcs. In the sequel we shall also have occasion to consider the closure of a saddle region, i. e., a closed hyperbolic region $\overline{g}_C$.

**4. TOPOLOGICAL EQUIVALENCE OF PHASE PORTRAITS IN ELEMENTARY REGIONS OF THE SAME TYPE.** Define an elementary closed region to be any of the following: elementary quadrangle (Figure 205a) (see §3.6), closed regular parabolic sector (Figure 205b), closed regular saddle region (Figure 205c), or closed regular elliptic region (Figure 205d). These elementary closed regions are the "building blocks" from which, as we shall see below, every phase portrait is constructed. In this subsection we consider the (geometrically evident) problem of topological equivalence of the phase portrait in any two elementary closed regions of the same type.

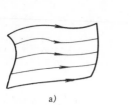   

a)      b)      c)      d)

FIGURE 205.

The elementary proof of the following lemma is omitted:

**L e m m a 7.** *The topological structures of the phase portrait in any two closed elementary regions of different types (elementary quadrangle, regular parabolic sector, regular elliptic region, or regular saddle region) are distinct.*

We now proceed to prove the identity of the topological structure of the phase portrait in any two closed regions of the same type: elementary quadrangles $\bar{\Gamma}$ and $\bar{\Gamma}^*$, regular parabolic sectors $\bar{g}_N$ and $\bar{g}_N^*$, saddle regions $g_C$ and $\bar{g}_C^*$, closed elliptic regions $\bar{g}_\sigma$ and $\bar{g}_\sigma^*$. Each pair of regions is filled by arcs of paths, semipaths and paths, either of two different dynamic systems $D$ and $\tilde{D}^*$ or of the same dynamic system. In the lemmas proved below we establish the topological equivalence of any two regions of the same type, proving the existence of a topological mapping of one region onto the other which preserves a preassigned correspondences between the boundary arcs without contact and boundary arcs of paths, semipaths and paths.

Since throughout we shall be discussing regions of the same type, we shall treat both simultaneously, putting all references to the second region in parentheses.

We begin with elementary quadrangles.

The boundary of the quadrangle $\bar{\Gamma}$ ($\bar{\Gamma}^*$) is made up of two arcs without contact $l_1$ and $l_2$ ($l_1^*$ and $l_2^*$) and two arcs of paths $S_1$ and $S_2$ ($S_1^*$ and $S_2^*$). Let $A_1$ ($A_1^*$) be the common endpoint of $l_1$ and $S_1$, ($l_1^*$ and $S_1^*$), $B_1$ ($B_1^*$) that of $l_1$ and $S_2$ ($l_1^*$ and $S_2^*$), $A_2$ ($A_2^*$) that of $l_2$ and $S_1$ ($l_2^*$ and $S_1^*$), $B_2$ ($B_2^*$) that of $l_2$ and $S_2$ ($l_2^*$ and $S_2^*$).

We assume given a topological correspondence between the points of $l_1$ and $l_1^*$, $S_1$ and $S_1^*$, $S_2$ and $S_2^*$, under which $A_1$ corresponds to $A_1^*$, $B_1$ to $B_1^*$, $A_2$ to $A_2^*$ and $B_2$ to $B_2^*$.

**L e m m a 8.** *There exists a path-preserving topological mapping of $\bar{\Gamma}$ onto $\bar{\Gamma}^*$ which preserves the given topological correspondence between the boundary arcs.*

P r o o f. Let

$$x = f_1(s), \quad y = g_1(s), \quad a_1 \leqslant s \leqslant b_1$$

and

$$x = f_1^*(s^*), \quad y = g_1^*(s^*), \quad a_1^* \leqslant s^* \leqslant b_1^*$$

be parametric equations of the arcs $l_1$ and $l_1^*$ (with the parameter values $a_1$ and $a_1^*$ corresponding to $A_1$ and $A_1^*$, $b_1$ and $b_1^*$ to $B_1$ and $B_1^*$, respectively).

Suppose that under the motion fixed on the paths in $\bar{\Gamma}$ they cross the arc $l_1$ at $t = t_0$, and $l_2$ at $t = \chi(s)$. Similarly, under the motion fixed on the paths in $\bar{\Gamma}^*$ they cross $l_1^*$ at $t^* = t_0^*$, and $l_2^*$ at $t^* = \chi^*(s^*)$. There are regular mappings (see Lemma 10 of §3) of $\bar{\Gamma}$ onto the closed region on the $(t, s)$-plane, defined by inequalities

$$a_1 \leqslant s \leqslant b_1, \quad t_0 \leqslant t \leqslant \chi(s), \tag{1}$$

and of $\bar{\Gamma}^*$ onto the closed region on the $(t^*, s^*)$-plane defined by

$$a_1^* \leqslant s^* \leqslant b_1^*, \quad t_0^* \leqslant t^* \leqslant \chi^*(s^*). \tag{2}$$

305

The prescribed topological mappings of $l_1$ onto $l_1^*$, $S_1$ onto $S_1^*$ and $S_2$ onto $S_2^*$ clearly induce a topological mapping of the corresponding straight-line segments on the $(t,s)$- and $(t^*,s^*)$-planes, bounding the closed regions (1) and (2), respectively. By Lemma 8 of §6.7, Appendix, there exists a topological mapping of the closed regions (1) and (2), preserving the prescribed mappings of the boundary segments and taking segments of the straight lines $s = $ const onto segments of $s^* = $ const. We can now define the required mapping for points of $\bar{\Gamma}$ and $\bar{\Gamma}^*$, via corresponding parameter-value pairs $(t, s)$ and $(t^*, s^*)$. Q.E.D.

Now let $\bar{g}_N$ and $\bar{g}_N^*$ be two parabolic sectors (Figure 206a, b). We may assume without loss of generality that both sectors are $\omega$-parabolic. In case both the sectors are $\alpha$-parabolic or one $\omega$-parabolic and the other $\alpha$-parabolic, we need only replace $t$ by $-t$ to reduce one or both sectors to the case considered.

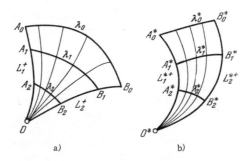

a)                    b)

FIGURE 206.

Let $\lambda$ and $\lambda^*$ be the arcs without contact, $O$ and $O^*$ the equilibrium states, $L_1^+$, $L_2^+$ and $L_1^{*+}$, $L_2^{*+}$ the semipaths making up the boundaries of $\bar{g}_N$ and $\bar{g}_N^*$, respectively. Let $A_0$ and $B_0$ ($A_0^*$ and $B_0^*$) be the common endpoints of the arc $\lambda(\lambda^*)$ and the semipaths $L_1^+$, $L_2^+$ ($L_1^{*+}$, $L_2^{*+}$). We assume given topological mappings of $\lambda$ onto $\lambda^*$, $L_1^+$ onto $L_1^{*+}$ and $L_2^+$ onto $L_2^{*+}$, taking the points $A_0$ and $B_0$ onto $A_0^*$ and $B_0^*$, respectively.

Lemma 9. *There exists a topological mapping of $\bar{g}_N$ onto $\bar{g}_N^*$, which preserves the given topological correspondence between the arcs without contact $\lambda$ and $\lambda^*$ and the semipaths $L_1^+$ and $L_1^{*+}$, $L_2^+$ and $L_2^{*+}$.*

Proof. Let $\varepsilon_1 > \varepsilon_2 > \varepsilon_3 > \ldots$ be a sequence of positive numbers such that $\varepsilon_i \to 0$ as $i \to \infty$. In the sector $\bar{g}_N$, draw an arc without contact $\lambda_1$ with endpoints $A_1$ and $B_1$ on the semipaths $L_1^+$ and $L_2^+$ (distinct from $A_0$ and $B_0$), and in the sector $\bar{g}_N^*$ an arc $\lambda_1^*$ with endpoints (distinct from $A_0^*$ and $B_0^*$) $A_1^*$ and $B_1^*$ on the semipaths $L_1^{*+}$ and $L_2^{*+}$, corresponding to the points $A_1$ and $B_1$ under the given topological mappings of $L_1^+$ onto $L_1^{*+}$ and $L_2^+$ onto $L_2^{*+}$. The arc $\lambda_1$ ($\lambda_1^*$) clearly divides $\bar{g}_N$ ($\bar{g}_N^*$) into two closed regions (of which it is a common boundary arc): a) a regular parabolic sector $\bar{g}_{1N}$ ($\bar{g}_{1N}^*$) whose boundary is made up of the arc $\lambda_1$ ($\lambda_1^*$), the arcs $A_1O$ and $B_1O$ ($A_1^*O^*$ and $B_1^*O^*$) of the semipaths $L_1^+$ and $L_2^+$ ($L_1^{*+}$ and $L_2^{*+}$), and the point $O$ ($O^*$); b) an elementary quadrangle $\bar{\Gamma}_1$ ($\bar{\Gamma}_1^*$) whose boundary is made up of the arcs $\lambda$ and $\lambda_1$ ($\lambda^*$ and $\lambda_1^*$) and the arcs $A_0A_1$ and $B_0B_1$ ($A_0^*A_1^*$ and $B_0^*B_1^*$) of the semipaths

$L_1^+$ and $L_2^+$ ( $L_1^{*+}$ and $L_2^{*+}$ ).  By the Remark to Lemma 2, we may assume that the arcs $\lambda_1$ and $\lambda_1^*$ are so chosen that the sector $\overline{g}_N$ is a subregion of $U_{\varepsilon_1}(O)$ and $\overline{g}_N^*$ a subregion of $U_{\varepsilon_1}(O^*)$.  There exists a path-preserving topological mapping of $\overline{\Gamma}_1$ onto $\overline{\Gamma}_1^*$ which preserves the given topological correspondence between $\lambda$ and $\lambda^*$ and the arcs of $L_1^+$ and $L_1^{*+}$, $L_2^+$ and $L_2^{*+}$ in the boundaries of the quadrangles.  This mapping induces a topological mapping of $\lambda_1$ onto $\lambda_1^*$.

We now treat the parabolic sectors $\overline{g}_{1N}$ and $\overline{g}_{1N}^*$ in the same way as previously $\overline{g}_N$ and $\overline{g}_N^*$.  As before, we divide each of the sectors $\overline{g}_{1N}$ and $\overline{g}_{1N}^*$ by suitable arcs without contact $\lambda_2$ and $\lambda_2^*$ (with endpoints $A_2$, $B_2$ and $A_2^*$, $B_2^*$ on the semipaths $L_1^+$, $L_2^+$, $L_1^{*+}$ and $L_2^{*+}$ corresponding to each other under the given topological mappings) into two closed regions:  a) a closed parabolic sector $\overline{g}_{2N} \subset U_{\varepsilon_2}(O)$ ($\overline{g}_{2N}^* \subset U_{\varepsilon_2}(O^*)$;  b) an elementary quadrangle $\overline{\Gamma}_2$ ($\overline{\Gamma}_2^*$).  As in the case of $\overline{\Gamma}_1$ and $\overline{\Gamma}_1^*$, there exists a path-preserving topological mapping of $\overline{\Gamma}_2$ onto $\overline{\Gamma}_2^*$, preserving the topological correspondence of $\lambda_1$ and $\lambda_1^*$, $L_1^+$ and $L_1^{*+}$, $L_2^+$ and $L_2^{*+}$ already constructed.  Repeating this procedure $k$ times, we obtain parabolic sectors $\overline{g}_{kN}$ and $\overline{g}_{kN}^*$ such that $\overline{g}_{kN} \subset U_{\varepsilon_k}(O)$ and $\overline{g}_{kN}^* \subset U_{\varepsilon_k}(O^*)$, and a topological mapping of $\overline{g}_N \setminus \overline{g}_{kN}$ onto $\overline{g}_N^* \setminus \overline{g}_{kN}^*$.  Finally, we map the point $O$ onto $O^*$.  In this way we have a one-to-one mapping of $\overline{g}_N$ onto $\overline{g}_N^*$, which is easily seen to be bicontinuous at all points of these sectors.  It is bicontinuous at $O$ because the sectors $\overline{g}_{kN}$ and $\overline{g}_{kN}^*$ are contained in $U_{\varepsilon_k}(O)$ and $U_{\varepsilon_k}(O^*)$, respectively, and $\varepsilon_k \to 0$ as $k \to \infty$.  Q.E.D.

Let $\overline{g}_N$ and $\overline{g}_N^*$ be closed parabolic regions whose boundaries are cycles without contact $C$ and $C^*$.  Suppose given a topological mapping of $C$ onto $C^*$.  The following lemma is proved in the same way as Lemma 9.

*Lemma 10.  There exists a topological mapping of $\overline{g}_N$ onto $\overline{g}_N^*$ which preserves the given topological mapping of the cycles without contact $C$ and $C^*$.*

Now let $\overline{g}_C$ and $\overline{g}_C^*$ be two regular closed saddle regions (Figure 207a, b).  The boundary of $\overline{g}_C$ ($\overline{g}_C^*$) is made up of a) arcs $QO$ and $PO$ ($Q^*O^*$ and $P^*O^*$) of separatrices $L_1^+$ and $L_2^-$ ($L_1^{*+}$ and $L_2^{*-}$), each a continuation of the other;  b) arcs without contact $l_1$ and $l_2$ ($l_1^*$ and $l_2^*$) with endpoints $Q$ and $P$ ( $Q^*$ and $P^*$ );  c) an arc $S$ ($S^*$) of a path $L$ ($L^*$), one of whose endpoints $A$ ($A^*$) is an endpoint of $l_1$ ($l_1^*$), the other, $B$ ($B^*$), an endpoint of $l_2$ ($l_2^*$);  d) an equilibrium state $O$ ($O^*$).

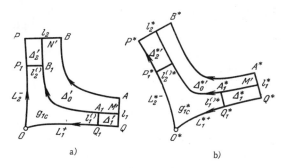

a)                    b)

FIGURE 207.

We suppose given topological mappings of $l_1$ onto $l_1^*$, $S$ onto $S^*$, and also of the arcs of the semipaths $L_1^+$ and $L_2^-$ in the boundary of $\bar{g}_C$ onto the arcs of $L_1^{*+}$ and $L_2^{*-}$ in the boundary of $\bar{g}_C^*$, under which the points $Q$ and $Q^*$, $P$ and $P^*$, $A$ and $A^*$, $B$ and $B^*$ correspond to each other.*

Lemma 11. *There exists a path-preserving topological mapping of $\bar{g}_C$ onto $\bar{g}_C^*$, preserving the given topological correspondence of $l_1$ and $l_1^*$, $S$ and $S^*$, $L_1^+$ and $L_1^{*+}$, $L_2^-$ and $L_2^{*-}$.*

Proof. We again consider a sequence of numbers $\varepsilon_1 > \varepsilon_2 > \dots$ such that $\varepsilon_i \to 0$ as $i \to \infty$. Let $\bar{g}_{1C}$ and $\bar{g}_{1C}^*$ be regular saddle subregions of $\bar{g}_C$ and $\bar{g}_C^*$, respectively, whose boundaries include arcs without contact $l_1^{(1)}$ and $l_2^{(1)}$ ( $l_1^{*(1)}$ and $l_2^{*(2)}$ ) with endpoints $Q_1$ and $P_1$ ( $Q_1^*$ and $P_1^*$ ) distinct from $Q$ and $P$ ( $Q^*$ and $P^*$ ) on the separatrices $L_1^+$ and $L_2^-$ ( $L_1^{*+}$ and $L_2^{*-}$ ). Let the endpoints of these arcs not on the separatrices be $A_1$ and $B_1$ ( $A_1^*$ and $B_1^*$ ); we assume that $A_1$ and $B_1$ are not on the arc $S$ of $L$, and $A_1^*$ and $B_1^*$ not on $S^*$. The path passing through the points $A_1$ and $B_1$ ( $A_1^*$ and $B_1^*$ ) is denoted by $L'$ ($L^*$); $M'$ and $N'$ ( $M^{*\prime}$ and $N^{*\prime}$ ) denote the points at which the path $L'$ ($L^*$) crosses the arcs $l_1$ ($l_1^*$) and $l_2$ ($l_2^*$), respectively (Figure 207).

We may assume that the closed regions $\bar{g}_{1C}$ and $\bar{g}_{1C}^*$ satisfy the following conditions: 1) $\bar{g}_{1C} \subset U_{\varepsilon_1}(O)$ and $\bar{g}_{1C}^* \subset U_{\varepsilon_1}(O^*)$; 2) the points $P_1$ and $P_1^*$, $Q_1$ and $Q_1^*$, $M'$ and $M'^*$ correspond to each other under the given topological mapping of $L_1^+$ onto $L_1^{*+}$, $L_2^+$ onto $L_2^{*+}$ and $l_1$ onto $l_1^*$.

The arc $M'N$ ( $M^{*\prime}N^{*\prime}$ ) of the path $L'$ ($L^*$) and the arcs $l_1^{(q)}$, $l_2^{(q)}$ ( $l_1^{*(q)}$, $l_2^{*(q)}$ ) divide the closed region $\bar{g}_C$ ($\bar{g}_C^*$) into four regions, three elementary quadrangles $\Delta_0'$, $\Delta_1'$, $\Delta_2'$ ( $\Delta_0^{*\prime}$, $\Delta_1^{*\prime}$, $\Delta_2^{*\prime}$ ) and a saddle region $\bar{g}_{1C}$ ($\bar{g}_{1C}^*$). The boundary of $\Delta_0'$ is made up of the arcs $M'A$ and $N'B$ of $l_1$ and $l_2$ and the arcs $AB$ and $M'N'$ of the paths $L$ and $L'$, that of $\Delta_1'$ of the subarc $M'Q_1$ of $l_1$, the arc without contact $l_1^{(1)}$, the arc $M'A_1$ of $L'$ and the arc $QQ_1$ of $L_1^+$, that of $\Delta_2'$ : the subarc $N'P$ of $l_2$, the arc $l_2^{(1)}$, the arc $N'B_1$ of $L'$ and the arc $PP_1$ of $L_2^-$. The boundaries of the quadrangles $\Delta_0^{*\prime}$, $\Delta_1^{*\prime}$ and $\Delta_2^{*\prime}$ are obtained from these by adding asterisks. Construct a path-preserving topological mapping of $\bar{\Delta}_0'$ onto $\bar{\Delta}_0^{*\prime}$, preserving the given correspondence between the arcs $AB$ and $A^*B^*$ of $L$ and $L^*$ and the arcs without contact $M'A$ and $M^{*\prime}A^*$ (Lemma 8), and setting up a correspondence between the arcs $M'N'$ and $M^{*\prime}N^{*\prime}$ under which the points $A_1$ and $A_1^*$, $B_1$ and $B_1^*$ correspond to each other. Similarly, we establish a path-preserving topological mapping of $\bar{\Delta}_1'$ onto $\bar{\Delta}_1^{*\prime}$, preserving the given topological correspondence of the arcs $M'Q$ and $M^{*\prime}Q^*$, $QQ_1$ and $Q^*Q_1^*$, and the correspondence between the arcs $M'A_1$ and $M^{*\prime}A_1^*$ induced by the mapping of $\bar{\Delta}_0'$ onto $\bar{\Delta}_0^{*\prime}$ just constructed. Finally, we construct a topological mapping of $\bar{\Delta}_2'$ onto $\bar{\Delta}_2^{*\prime}$ setting up a correspondence of the arcs $PN'$ and $P^*N^{*\prime}$ such that the paths passing through corresponding points cross the subarcs $M'Q$ and $M^{*\prime}Q^*$ of the arcs $l_1$ and $l_1^*$ at points which correspond to one another under the given topological mapping of the arcs $PP_1$ and $P^*P_1^*$, and moreover the correspondence already established between the arcs $PP_1$ and $P^*P_1^*$, $N'B_1$ and $N^{*\prime}B_1^*$ is also preserved. These mappings of $\Delta_0'$, $\Delta_1'$, $\Delta_2'$ onto $\Delta_0^{*\prime}$, $\Delta_1^{*\prime}$, $\Delta_2^{*\prime}$, respectively, induce a topological correspondence between the arcs and paths figuring in the boundaries of $\bar{g}_{1C}$ and $\bar{g}_{1C}^*$. These two saddle regions may now be treated in the same way as previously $\bar{g}_C$ and $\bar{g}_C^*$. Namely, we determine in $\bar{g}_{1C}$ and $\bar{g}_{1C}^*$ closed saddle regions $g_{2C}$ and $g_{2C}^*$, in the same way as before, such that

$$\bar{g}_{2C} \subset U_{\varepsilon_2}(O), \quad \bar{g}_{2C}^* \subset U_{\varepsilon_2}(O^*).$$

---

* There may also be a mapping of saddle regions taking the positive separatrix $L_1^+$ onto a negative one $L_1^{*-}$ and vice versa. This case may clearly be reduced to that considered here by substituting $-t$ for $t$ in one of the regions.

We obtain in this way elementary quadrangles $\Delta_0''$, $\Delta_1''$, $\Delta_2''$ and $\Delta_0^{*''}$, $\Delta_1^{*''}$, $\Delta_2^{*''}$, analogous to $\Delta_0'$, $\Delta_1'$, $\Delta_2'$ and $\Delta_0^{*'}$, $\Delta_1^{*'}$, $\Delta_2^{*'}$. Proceeding in the same way, we obtain after $k$ repetitions a topological mapping of $\bar{g}_C \setminus \bar{g}_{kC}$ onto $\bar{g}_C^* \setminus \bar{g}_{kC}^*$, where

$$\bar{g}_{kC} \subset U_{\varepsilon_k}(O), \qquad \bar{g}_{kC}^* \subset U_{\varepsilon_k}(O^*).$$

The mapping is completed by mapping $O$ onto $O^*$. In this way we get a one-to-one mapping of $\bar{g}_C$ onto $\bar{g}_C^*$. It is readily seen that this mapping is also bicontinuous (bicontinuity at $O$ follows from the fact that $\bar{g}_{kC} \subset U_{\varepsilon_k}(O)$ and $\bar{g}_{kC}^* \subset U_{\varepsilon_k}(O^*)$ for sufficiently large $k$). Q.E.D.

Now let $\bar{g}_\sigma$ and $\bar{g}_\sigma^*$ be two closed elliptic regions. Let $L$ and $L^*$ be the paths (not equilibrium states), $O$ and $O^*$ the equilibrium states in the boundaries of $\bar{g}_\sigma$ and $\bar{g}_\sigma^*$, respectively. We assume given a topological correspondence between $L$ and $L^*$ (Figure 208a, b).

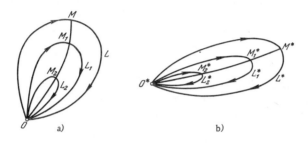

a)                 b)

FIGURE 208.

*Lemma 12. There exists a path-preserving topological mapping of $\bar{g}_\sigma$ onto $\bar{g}_\sigma^*$, preserving the given topological correspondence of the paths $L$ and $L^*$.*

Proof. By Lemma 5, there exists an arc $l$ $(l^*)$ joining any given point $M$ $(M^*)$ on $L$ $(L^*)$ to the equilibrium state $O$ $(O^*)$, which has no contact at points other than $O$ $(O^*)$ and cuts all paths passing through points of $g_\sigma$ $(g_\sigma^*)$. Let $M$ correspond to $M^*$ under the given correspondence between $L$ and $L^*$. Let $\varepsilon_1 > \varepsilon_2 > \ldots$ be a sequence such that $\varepsilon_i \to 0$ as $i \to \infty$.

Let $L_1$ and $L_1^*$ be paths in the elliptic regions $g_\sigma$ and $g_\sigma^*$, respectively (distinct from $L$ and $L^*$), such that $L_1 \subset U_{\varepsilon_1}(O)$, $L_1^* \subset U_{\varepsilon_1}(O^*)$ (see Lemma 9, §17). Let $w_1$ and $w_1^*$ denote the closed regions whose boundaries are the loops formed by the paths $L$ and $L_1$, $L^*$ and $L_1^*$ and the points $O$ and $O^*$, respectively.

Let $M_1$ $(M_1^*)$ be the (unique) point of intersection of the path $L_1$ $(L_1^*)$ and the arc $l$ $(l^*)$. It is easy to construct a path-preserving topological mapping of $w_1$ onto $w_1^*$ which preserves the given correspondence between $L$ and $L^*$. Indeed, the arc without contact $MM_1$ $(M^*M_1^*)$ divides $w$ $(w^*)$ into two regular closed parabolic regions, being a boundary arc for both, and the existence of the required mapping in both follows from Lemma 10.

Now let $L_2$ and $L_2^*$ be paths forming loops nested in the loops formed by the paths $L_1$ and $L_1^*$, respectively, and such that

$$L_2 \subset U_{\varepsilon_2}(O), \qquad L_2^* \subset U_{\varepsilon_2}(O^*),$$

and let $w_2$ and $w_2^*$ be the closed regions bounded by the loops formed by $L_1$ and $L_2$, $L_1^*$ and $L_2^*$, and the points $O$ and $O^*$, respectively.

As before, one constructs a path-preserving topological mapping of $w_2$ onto $w_2^*$, inducing the correspondence just established between the paths $L_1$ and $L_1^*$. Repeating the procedure, and finally mapping $O$ onto $O^*$, we easily obtain the required topological mapping.

## §19. LOCAL AND GLOBAL SCHEME OF AN EQUILIBRIUM STATE

### 1. CYCLIC ORDER OF SEPARATRICES AND ELLIPTIC REGIONS OF AN EQUILIBRIUM STATE WHICH IS NOT A CENTER.

As before, let $\varepsilon_0 > 0$ be such that the neighborhood $U_{\varepsilon_0}(O)$ contains no whole singular path [other than $O$]. On the assumption that the equilibrium state $O$ is not a center, so that at least one semipath tends to $O$, we have the following two possibilities:

1. The equilibrium state has no separatrix and no elliptic region.
2. The equilibrium state has a separatrix or an elliptic region.

In the first case, it follows from Theorems 5 and 8 of §17 that either all paths through a sufficiently small neighborhood of $O$ tend to $O$ as $t \to +\infty$, or they all tend to $O$ as $t \to -\infty$. The equilibrium state is a topological node (§17.2), stable if the paths tend to it as $t \to +\infty$, unstable if they do so as $t \to -\infty$.

We now consider the second case in detail. Let

$$L_1^+, L_2^+, \ldots, L_{n_1}^+ \tag{1}$$

be the $\omega$-separatrices of $O$ and

$$L_1^-, L_2^-, \ldots, L_{n_2}^- \tag{2}$$

its $\alpha$-separatrices (if such exist).

When the point $O$ has elliptic regions (only finitely many of these can exist) (Theorem 56 of §16), we take exactly one path in each region, lying wholly in $U_{\varepsilon_0}(O)$. Let these paths be

$$L_1^*, L_2^*, \ldots, L_{n_3}^*. \tag{3}$$

They form mutually exterior loops. We may assume this system of loops maximal, in the sense that any loop in $U_{\varepsilon_0}(O)$ distinct from the loops formed by the paths (3) is either nested in one of them or conversely. The paths (3) are obviously orbitally stable (since they lie in $U_{\varepsilon_0}(O)$).

We now take positive and negative semipaths on each of the paths (3):

$$L_1^{*+}, L_2^{*+}, \ldots, L_{n_3}^{*+}, \tag{4}$$
$$L_1^{*-}, L_2^{*-}, \ldots, L_{n_3}^{*-} \tag{5}$$

Denote the collection of all the semipaths (1), (2), (4) and (5) by $(L)$.

Let $C$ be a circle about $O$ of radius less than $\varepsilon_0$, outside which there are points of each of the semipaths $(L)$. We denote the last points of these semipaths in common with $C$ by $M_i$ ($i = 1, 2, \ldots, N$, where $N = n_1 + n_2 + 2n_3$) and assume that they are indexed in the order induced by the positive traversal of $C$. In addition, we set $M_{N+1}$ equal to $M_1$. Then no arc $M_iM_{i+1}$ contains any of the points $M_j$ other than its endpoints.

By Lemma 1 of §17, for any simple closed curve interior to $C$ (in particular, a circle about $O$ of radius smaller than that of $C$), the last points of the semipaths $(L)$ in common with the curve are arranged thereon in the same cyclic order as the points $M_i$ ($i = 1, 2, \ldots, N$) on $C$. This order defines the cyclic order of the semipaths $(L)$ (see Lemma 1). Thus we can write all the semipaths in their cyclic order:

$$L_{i_1}^+, \ L_{i_2}^-, \ \ldots, \ L_{i_k}^{*(\ )}. \tag{6}$$

**Lemma 1.** *Any two semipaths $L_i^{*+}$ and $L_i^{*-}$ in the same loop are successive in the cyclic order of the semipaths (L) [briefly: cyclically successive].*

Proof. Let $M_k$ and $M_l$ be the last points of $L_i^{*+}$ and $L_i^{*-}$ in common with the circle. We must show that either $l = k + 1$ or $k = l + 1$. Let $g_\sigma$ be the region interior to the loop formed by $L_i^*$. The subarcs $M_kO$ and $M_lO$ of $L_i^{*+}$ and $L_i^{*-}$ divide the interior of $C$ into two curvilinear sectors $g$ and $g'$.

Suppose that of these two sectors $g$ is the one in which all points sufficiently close to $O$ are in the region $g_\sigma$, and that the arc of $C$ in the boundary of this sector is $M_kM_l$. Were there another point $M_j$ on this arc (apart from its endpoints), there would be another semipath of $(L)$ between $L_i^{*+}$ and $L_i^{*-}$. But then this semipath would have to pass through points of the elliptic region $g_\sigma$; this is clearly absurd, since that region contains no whole singular paths and none of the paths $L_j^*$. Q.E.D.

Consider some path $\overline{L}_i$ inside the loop formed by $L_i^*$, having points outside $C$. Replace the paths (3) by

$$\overline{L}_1, \ \overline{L}_2, \ \ldots, \ \overline{L}_{n_3}. \tag{7}$$

Let

$$\overline{L}_1^+, \ \overline{L}_2^+, \ \ldots, \ \overline{L}_{n_3}^+ \tag{8}$$

and

$$\overline{L}_{\overline{1}}, \ \overline{L}_{\overline{2}}, \ \ldots, \ \overline{L}_{\overline{n}_3} \tag{9}$$

be positive and negative semipaths on these paths. Let us denote the collection of all semipaths (1), (2), (8) and (9) by $(\overline{L})$. By Lemma 11 of §17, if there are no other paths of $(L)$ between $L_i^{*+}$ and $L_i^{*-}$, then the semipaths $\overline{L}_i^+$ and $\overline{L}_{\overline{i}}$ lie between them, and in such a way that $\overline{L}_i^+$ is between $L_i^{*+}$ and $\overline{L}_{\overline{i}}$.

The semipaths $(L)$ and $(\overline{L})$ can be put in correspondence so that the separatrices correspond to themselves and the semipaths $L_i^{*+}, L_i^{*-}$ to the semipaths $\overline{L}_i^+, \overline{L}_{\overline{i}}$ (of paths in the same elliptic region as $L_i^{*+}, L_i^{*-}$). Under this correspondence, of course, any relation of cyclic succession is preserved.

We shall say that the cyclic order of the semipaths $(L)$ and $(\bar{L})$ about the equilibrium state $O$ is the same (or that the semipaths are identically arranged about the equilibrium state $O$).

We shall also say that the cyclic order of the semipaths $(L)$ or $(\bar{L})$ defines a cyclic order of separatrices and regular elliptic regions about $O$.

Let $g_{\sigma_1}, \ldots, g_{\sigma_{n_3}}$ be all distinct elliptic regions of the equilibrium state $O$. We may write the separatrices and elliptic regions of the equilibrium state in their cyclic order:

$$L_{i_1}^+, \; g_{\sigma_{i_2}}, \; L_{i_3}^{()}. \tag{10}$$

It is evident from the foregoing discussion that the sequence (10) differs from the sequence $(L)$ in that each two successive semipaths in the same elliptic region are replaced by the symbol $g_{\sigma i}$ for the elliptic region.

In the sequel we shall use expressions such as "cyclically successive elliptic regions," "a semipath $L^{()}$ tending to $O$ is between elliptic regions $g_{\sigma_k}$ and $g_{\sigma_j}$" (the order is important; see §17.1); the meanings of such expressions should be clear.

For some fixed choice of paths of elliptic regions, consider the collection of semipaths $(L)$ and, as before, let $M_k$ be their last points in common with a circle $C$ (indexed in cyclic order). Let $g_k$ be the curvilinear sectors into which these semipaths divide the interior of $C$. The boundary of $g_k$ consists of the arcs $M_k O$ and $M_{k+1}O$ of two of the semipaths in $(L)$, the point $O$ and the arc $M_k M_{k+1}$ of $C$ (this arc contains none of the points $M_l$ other than its end-points). Thus the interior of $C$ is divided into $N$ curvilinear sectors $g_1, g_2, \ldots, g_N$. The sector $g_1$ will also be denoted by $g_{N+1}$. The sectors $g_k$ are clearly disjoint, $O$ is a boundary point of each one, and any two sectors $g_k$ and $g_{k+1}$ have a common boundary semipath $L_{M_{k+1}}^{()}$. Since the collection $(L)$ includes all separatrices of the equilibrium state $O$, there are no separatrices in any of the sectors $g_k$, i. e., between any two semipaths $L_{M_k}^{()}, L_{M_{k+1}}^{()}$. With regard to $g_k$ there are three possibilities:

1. The boundary semipaths $L_{M_k}^{()}$ and $L_{M_{k+1}}^{()}$ of $g_k$ are semipaths of a path $L_i^*$ of an elliptic region.

Let $g_{\sigma i}$ be the elliptic region bounded by the loop formed by the path $L_i^*$. The intersection of $g_k$ with a sufficiently small neighborhood $U_\varepsilon(O)$ is a subregion of $g_{\sigma i}$, and conversely the intersection of $U_\varepsilon(O)$ with $g_{\sigma i}$ is a subregion of $g_k$. In other words $U_\varepsilon(O) \cap g_k = U_\varepsilon(O) \cap g_{\sigma i}$ (see Remark to Lemma 1, §17). Hence all paths of the elliptic region $g_{\sigma i}$ in the interior of the circle $C$ lie in the sector $g_k$ and, conversely, any path in the sector $g_k$ forms a loop in the region $g_{\sigma i}$.

In the terminology of §17.2, a sector with these properties is an elliptic sector (Figure 197). Since by definition the collection of semipaths $(L)$ includes semipaths on paths of all the different elliptic regions of $O$, it follows that any loop interior to $C$ is in some elliptic region.

2. One of the semipaths $L_{M_k}^{()}, L_{M_{k+1}}^{()}$ is an $\omega$-separatrix, the other an $\alpha$-separatrix of $O$. There is no separatrix in the sector $g_k$, i. e., between $L_{M_k}^{()}$ and $L_{M_{k+1}}^{()}$, neither is there a loop (since any loop interior to $C$ lies in some elliptic sector). Thus the sector in question is certainly not elliptic. But then one readily sees, using Lemma 4 of §17 and the Corollary to Lemma 5 of §17, that all paths passing through points of $g_k$ leave the sector with both increasing and decreasing $t$, and the semipaths $L_{M_k}^{()}$ and $L_{M_{k+1}}^{()}$ are

continuations of each other. Thus we have a h y p e r b o l i c  s e c t o r
(§17.2; Figure 205c).

3. The semipaths $L^{\backslash}_{M_k}$ and $L^{\backslash}_{M_{k+1}}$ in the boundary of $g_k$ either a) are on
distinct paths $L^*_i$ (i. e., in two different loops), or b) are two $\omega$-separatrices
or two $\alpha$-separatrices of the equilibrium state $O$, or c) one of them is on
the path $L^*_i$, the other is a separatrix of $O$.

There are neither separatrices nor loops between $L^{\backslash}_{M_k}$ and $L^{\backslash}_{M_{k+1}}$ (see
case 2). Moreover, by assumption these semipaths cannot be separatrices
which are continuations of each other. By Lemma 5 of §17, it is easy to
see that in this case $g_k$ is an $\omega$- or $\alpha$-parabolic sector (and the semipaths
$L^{\backslash}_{M_k}$ and $L^{\backslash}_{M_{k+1}}$ are either both positive or both negative) (Figure 196).

The above types 1, 2, 3 clearly exhaust all possibilities for the regions
$g_k$. We have the following lemma, which follows at once from the definition
of regions of type 3:

*L e m m a  2.  If $g_k$ is an elliptic sector, then the adjacent sectors $g_{k-1}$
and $g_{k+1}$ are parabolic sectors, one $\omega$-parabolic and the other $\alpha$-parabolic.*

**2. CANONICAL CLOSED CURVE ABOUT AN EQUILIBRIUM STATE.** Using the above-
described partition of the interior of $C$ into curvilinear sectors of different
types, we shall now construct a certain region around the equilibrium state,
which we call a c a n o n i c a l  n e i g h b o r h o o d  of the equilibrium state.
Its boundary is a simple closed curve, the union of finitely many arcs of
paths and arcs without contact, which we call a c a n o n i c a l  c u r v e.

First let the equilibrium state $O$ be a topological node (case 1 in §19.1).
By the Remark following Lemma 3 of §18, there exists $\varepsilon_0 > 0$ such that for
any $\varepsilon < \varepsilon_0$ there is a cycle without contact in $U_\varepsilon(O)$ enclosing the point $O$.
Any such cycle without contact will be a c a n o n i c a l  c l o s e d  c u r v e
for the equilibrium state $O$.

Now suppose that the equilibrium state is neither a center nor a
topological node (case 2). On each semipath $L^{\backslash}_{M_k}$ on a member of the
collection $(L)$ take one point $P_k$ (distinct from $M_k$, hence inside the circle
$C$), and let

$$P_1,\ P_2,\ \ldots,\ P_N,\ P_{N+1} \equiv P_1$$

be all these points. It is clear that when these points are described in this
order the semipaths of $(L)$ are described in their cyclic order. Since there
is one point $P_k$ on each of the semipaths (4) and (5), it follows that each
path $L^*_i$ of an elliptic region contains exactly two points $P_j$ and $P_{j+1}$. Let $S^*_j$
be the arc on $L^*_i$ with endpoints $P_j$ and $P_{j+1}$. We shall call this an e l l i p t i c
a r c. We now construct a regular parabolic region in each of the parabolic
sectors $g_k$, drawing an arc without contact $l_k$ joining the two points $P_k$ and
$P_{k+1}$ and lying (except for its endpoints) inside $g_k$. We shall call $l_k$ a
p a r a b o l i c  a r c  w i t h o u t  c o n t a c t.

Now construct in each hyperbolic sector $g_i$ a regular hyperbolic region,
based on arcs without contact $\lambda^c_i$ and $\lambda^c_{i+1}$ with endpoints $P_i$ and $P_{i+1}$ on the
separatrices bounding the sector $g_i$. Let $S^c_i$ denote the arc of the path in
the boundary of this regular saddle region, whose endpoints are
respectively endpoints of the arcs $\lambda^c_i$ and $\lambda^c_{i+1}$. We shall call $S^c_{i+1}$ a
h y p e r b o l i c  a r c, and the arcs without contact $\lambda^c_i$, $\lambda^c_{i+1}$ will be called
s a d d l e  a r c s  w i t h o u t  c o n t a c t. As before, the arc $\lambda^c_i$ or $\lambda^c_{i+1}$ whose

endpoint is on an $\omega$-separatrix will be called an $\omega$-saddle arc without contact, and the other, whose endpoint is on an $\alpha$-separatrix — an $\alpha$-saddle arc without contact.

In this situation, we have the following propositions:

a) The arcs without contact $l_k$ are outside all elliptic regions $g_{\sigma_j}$ (the boundaries of the latter are the loops formed by the $L_j^*$). Indeed, any path crossing $l_k$ and belonging to some elliptic region $g_{\sigma j}$ must ultimately leave the parabolic sector $g_h$ containing $l_h$ with both increasing and decreasing $t$, subsequently entering an elliptic sector and remaining there. This is clearly impossible, because any path crossing $l_h$ must tend to the equilibrium state $O$ as $t \to +\infty$ (or $t \to -\infty$), never again leaving the parabolic sector $g_k$.

b) The arcs without contact $\lambda_q^c$ and $\lambda_{q+1}^c$ may be so chosen that they have no points in common with any of the elliptic arcs $S_j^*$ on the paths $L_j^*$. Indeed, the boundary semipaths $L_q^+$ and $L_{q+1}^{'-}$ of the hyperbolic sector are certainly not on the paths $L_i^*$, since they are singular whereas all the $L_j^*$ are non-singular. In addition, no points of $L_q^+$ and $L_{q+1}^{'-}$, in particular, the points $P_q$ and $P_{q+1}$, can be limit points of any path $L_j^*$, because the only limit point of $L_j^*$ is $O$. But then $P_q$ and $P_{q+1}$ lie at a positive distance from all paths $L_j^*$, and consequently the arcs $\lambda_q^c$ and $\lambda_{q+1}^c$ may be chosen as indicated.

Now consider the set $H$ defined as the union of all regular parabolic regions (in the parabolic sectors $g_k$), all hyperbolic regions (in the hyperbolic sectors $g_q$) and all elliptic regions $g_{\sigma_j}$ in the loops formed by the paths $L_j^*$ containing the semipaths of $(L)$ (minus the points $P_k$), and $O$. It is readily seen that $H$ is a bounded open set, its boundary being the union of all the arcs $l_k$, $\lambda_q^c$, $S^*$ and $S^c$.

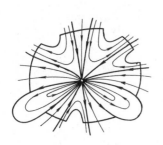

FIGURE 209.

By propositions a) and b), the boundary is a simple closed curve, which we denote by $E$.

*The curve $E$ is called a canonical closed curve of the equilibrium state $O$, the region $H$ interior to this curve a canonical neighborhood of $O$.*

In the sequel we shall mainly consider the closed canonical neighborhood $\bar{H}$ (Figure 209).

There may of course be different canonical neighborhoods, depending on the choice of paths of elliptic regions and the choice of regular parabolic and hyperbolic regions.

The points $P_i$ are the last points of the semipaths of $(L)$ in common with the curve $E$. By the remarks to Lemmas 2 and 6 of §18 and Lemma 9 of §17, for any $\varepsilon > 0$ we can construct a canonical curve in $U_\varepsilon (O)$.

We present without proof an elementary lemma concerning the connection between the positive traversal of the canonical curve $E$ and the loops with the direction of time on the loops. Let $g_\sigma$ be one of the elliptic subregions of the canonical neighborhood $H$ and $S^*$ its elliptic boundary arc.

*Lemma 3. If the positive traversal of $E$ induces on $S^*$ a direction agreeing with that of $t$ (or opposite to that of $t$), then the positive traversal of the loops in $H$ also agrees with the direction of $t$ (is opposite to that of $t$).*

**3. LOCAL SCHEME OF AN EQUILIBRIUM STATE OTHER THAN A CENTER.** We can now define the local scheme of an equilibrium state which is not a center.

*Definition XXI. A local scheme of an equilibrium state $O$ which is not a center is a list containing all its $\omega$-separatrices $(L_1^+, L_2^+, \ldots, L_{n_1}^+)$, all its $\alpha$-separatrices $(L_1^-, L_2^-, \ldots, L_{n_2}^-)$, all its distinct elliptic regions $(g_{\sigma_1}, g_{\sigma_2}, \ldots, g_{\sigma n_3})$, with indication of the cyclic order in which these semipaths and regions are arranged about the equilibrium state $O$.*

The local scheme of an equilibrium state may be specified either in tabular (written) form, or in graphic form — by a s c h e m a t i c  d i a g r a m. The tabular form of a scheme is as follows:

$$O \mid L_{i_1}^-, \; g_{\sigma_{i_2}}, \; L_{i_3}^+, \; \ldots$$

Here the order in which the semipaths and elliptic regions are written out corresponds to the cyclic order in which they are arranged about the equilibrium state. The tabular notation for the scheme may begin with any semipath $L_i^+$ or $L_j^-$ or any elliptic region $g_{\sigma_i}$, so that the scheme is unique to within a cyclic permutation.

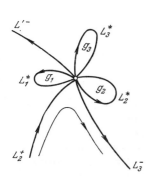

FIGURE 210.

Instead of considering all different elliptic regions of the equilibrium state $O$, we can of course consider all mutually exterior loops or all positive and negative semipaths of these loops ($L_1^{*+}, L_2^{*+}, \ldots, L_{n_3}^{*+}$ and $L_1^{*-}, L_2^{*-}, \ldots, L_{n_3}^{*-}$). In the latter case, the tabular notation for the scheme will be

$$O \mid L_{i_1}^-, \; L_{i_2}^{*-}, \; L_{i_2}^{*+}, \; \ldots$$

The following propositions are obvious:
1) Between any two successive positive (negative) semipaths in the scheme, there is an $\omega$-parabolic ($\alpha$-parabolic) region. 2) Between any two successive semipaths in the scheme, of which one is positive and the other negative, which are not semipaths $L_i^{*+}$ and $L_i^{*-}$ of the same loop, there is a hyperbolic region.*

The scheme is more easily visualized in the form of a schematic diagram. The diagram shows, in their cyclic order, the separatrices of the equilibrium state (with their directions and notations indicated) and one path for each elliptic region (with the direction of $t$ indicated).

---

* The scheme of an equilibrium state, as defined here, involves a specific symbol for each path of an elliptic region. This is necessary for our subsequent description of the scheme of the entire phase portrait "in the large." When one considers the equilibrium state in isolation, it is sufficient to use a plus sign for positive semipaths, a minus sign for negative semipaths, and a zero for elliptic regions. The scheme of the equilibrium state is then a sequence of symbols +, −, 0, a natural term for which would be "abstract local scheme." It is clear, however, that not every sequence of +, − and 0 can be the abstract scheme of an equilibrium state. Indeed, if a + symbol is followed by 0, the latter must be followed either by another 0 or by −. The symbol + corresponds to a separatrix $L_{i_1}^+$ if it is followed by a 0, and this clearly means that in the cyclic order $L_{i_1}^+$ is followed by an elliptic region. If $L_{i_1}^*$ is a path of this elliptic region, then (Lemma 1) $L_{i_1}^+$ in a scheme (10) must be followed by $L_{i_2}^{*+}$ and then $L_{i_2}^{*-}$, while $L_{i_2}^{*-}$ (by the same lemma) is followed either by a negative separatrix or another elliptic region. One can find conditions for a sequence of symbols +, −, 0 to be the abstract scheme of some equilibrium state.

315

Thus, for example, the scheme given in tabular form by

$$O \mid L_1^{'-}, \ g_1, \ L_2^+, \ L_3^{'-}, \ g_2, \ g_3,$$

corresponds to the diagram in Figure 210.

It is clear from 1) and 2) that there is a hyperbolic region (sector) between the semipaths $L_2^+$ and $L_3^{'-}$, while between $L_1^{'-}$ and $L_1^{*-}$, $L_1^{*+}$ and $L_2^+$, etc. there are parabolic sectors ( $L_1^*$, $L_2^*$ and $L_3^*$ are paths of elliptic regions).

We now consider two equilibrium states $O$ and $\tilde{O}$ of two different dynamic systems $D$ and $\tilde{D}$ of type (I) or of the same dynamic system.

Suppose we are given the schemes of these equilibrium states. In other words, we have a list as specified in Definition XXI for $O$, and for $\tilde{O}$, similarly, we have a list of $\omega$-separatrices ( $\tilde{L}_1^+, \tilde{L}_2^+, \ldots, \tilde{L}_{m_1}^+$ ), $\alpha$-separatrices ( $\tilde{L}_1^{'-}, \tilde{L}_2^{'-}, \ldots, \tilde{L}_{m_2}^{'-}$ ), elliptic regions ( $\tilde{g}_{\sigma_1}, \tilde{g}_{\sigma_2}, \ldots, \tilde{g}_{\sigma_{m_3}}$ ), and the cyclic order in which these semipaths and regions are arranged about $\tilde{O}$ is known.

The scheme of the equilibrium state $\tilde{O}$, as before, may be written out in tabular form

$$\tilde{O} \mid \tilde{L}_{j_1}^-, \ \tilde{g}_{\sigma_{j_2}}, \ \ldots$$

or

$$\tilde{O} \mid \tilde{L}_{j_1}^-, \ \tilde{L}_{j_2}^{*+}, \ \tilde{L}_{j_2}^{*-}, \ \ldots,$$

when the regions $\tilde{g}_\sigma$ are replaced by semipaths of their loops.

*Definition XXII. We shall say that the local schemes of equilibrium states $O$ and $\tilde{O}$ are equivalent, with the same orientation and direction of $t$, if there exists a one-to-one correspondence $\theta$ between the $\omega$-separatrices of $O$ ( $L_1^+, L_2^+, \ldots, L_{n_1}^+$ ) and those of $\tilde{O}$ ( $\tilde{L}_1^+, \tilde{L}_2^+, \ldots, \tilde{L}_{m_1}^+$ ), the $\alpha$-separatrices of $O$ ( $L_1^{'-}, L_2^{'-}, \ldots, L_{n_2}^{'-}$ ) and those of $\tilde{O}$ ( $\tilde{L}_1^{'-}, \tilde{L}_2^{'-}, \ldots, \tilde{L}_{m_2}^{'-}$ ), and the distinct elliptic regions of $O$ and $\tilde{O}$ ( $g_{\sigma_1}, g_{\sigma_2}, \ldots, g_{\sigma_{n_3}}$ ) and ( $\tilde{g}_{\sigma_1}, \tilde{g}_{\sigma_2}, \ldots, \tilde{g}_{\sigma_{m_3}}$ ), such that the following condition holds: the scheme of $\tilde{O}$ may be obtained from that of $O$ by replacing the semipaths and elliptic regions of $D$ in the scheme by those of $\tilde{D}$ corresponding to them under $\theta$. Thus, if two semipaths or a semipath and an elliptic region of the system $D$ are consecutive in the cyclic order, then the corresponding semipaths or semipath and elliptic region of the system $\tilde{D}$ are also consecutive in the cyclic order.*

When the schemes of two equilibrium states are equivalent (with the same orientation and direction of $t$), then, of course $n_1 = m_1$, $n_2 = m_2$, $n_3 = m_3$, and, possibly subject to a suitable renumbering of the semipaths and elliptic regions of the equilibrium state $\tilde{O}$, the table describing the scheme of $\tilde{O}$ may be obtained from that describing the scheme of $O$ by simply adding the "tilde."

The converse is also obvious: if two tables describing schemes of equilibrium states differ only in the notation of the semipaths and elliptic regions (e. g., in the addition of a tilde), then the schemes are equivalent. When the schemes are specified in diagrammatic form, the diagrams are identical (except for notation).

If the schemes of two equilibrium states $O$ and $\tilde{O}$ are equivalent, we shall say that the corresponding semipaths and elliptic regions correspond to one another with respect to the scheme [or are in scheme-correspondence]. When the local schemes of $O$ and $\tilde{O}$ are equivalent, there also exists a one-to-one scheme-correspondence between semipaths $L_j^{*+}$, $L_j^{*-}$ and $\tilde{L}_j^{*+}$, $\tilde{L}_j^{*-}$ of scheme-corresponding elliptic regions. Scheme-corresponding semipaths and regions of systems $D$ and $\tilde{D}$ will be assigned the same index.

In similar fashion, mutatis mutandis, one can define equivalence of two schemes with the same orientation and opposite directions of $t$, and with opposite orientation and the same (opposite) directions of $t$.

In the first case (same orientation, opposite directions of $t$), positive semipaths correspond to negative semipaths.

Any of the above cases may be reduced to the original definition by reversing either the orientation or the direction of time (i. e., substituting $-t$ for $t$) for one of the equilibrium states (say $\tilde{O}$). Throughout the sequel, therefore, we shall confine ourselves to equivalence of schemes with the same orientation and direction of $t$.

As before, let the schemes of equilibrium states $O$ and $\tilde{O}$ be equivalent. Consider any canonical neighborhoods of these equilibrium states, $H$ and $\tilde{H}$. Let $E$ and $\tilde{E}$ be the canonical curves bounding them, $l_k$, $\lambda_i^c$, $S_j^*$, $S_q^c$ and $\tilde{l}_k$, $\tilde{\lambda}_i^c$, $\tilde{S}_j^*$, $\tilde{S}_q^c$ their parabolic, elliptic and saddle arcs. The scheme-correspondence between the $\omega$-separatrices and $\alpha$-separatrices, and also between the semipaths of the elliptic regions of $O$ and $\tilde{O}$, clearly induces a natural correspondence between the elementary regions of the canonical neighborhoods $H$ and $\tilde{H}$, the arcs of the canonical curves, and their endpoints. Namely:

1. Elementary regions whose boundary semipaths are scheme-corresponding semipaths correspond to each other. Moreover, corresponding regions are of the same type — both parabolic, both elliptic, or both hyperbolic.

2. Arcs $l_k$, $\lambda_i^c$, $S_j^*$, $S_q^c$ and $\tilde{l}_k$, $\tilde{\lambda}_i^c$, $\tilde{S}_j^*$, $\tilde{S}_q^c$ whose endpoints lie on scheme-corresponding semipaths correspond to each other. Corresponding arcs figure in corresponding regions and are of the same type — both parabolic, both elliptic, both saddle arcs without contact or both saddles arcs of paths.

3. Endpoints of corresponding arcs correspond to each other if they lie on scheme-corresponding semipaths or arcs (elliptic and hyperbolic).

This correspondence between the elementary regions comprising canonical neighborhoods, their boundary arcs, and the endpoints of these arcs, will also be referred to as a scheme-correspondence.

*Theorem 61. If the local schemes of two equilibrium states $O$ and $\tilde{O}$ are equivalent, with the same orientation and direction of $t$, then there exists a path-preserving topological mapping of any closed canonical neighborhood $\bar{H}$ of the first onto any closed canonical neighborhood $\bar{\tilde{H}}$ of the second, preserving orientation and direction of $t$.*

Proof. By the foregoing arguments, we have a one-to-one correspondence between the elementary regions of the canonical neighborhoods $H$ and $\tilde{H}$. Renumber all the elementary regions in such a way that scheme-corresponding regions are assigned the same index.

Let $h_1$, $h_2$, ..., $h_k$ be the elementary regions of $H$ and $\tilde{h}_1$, $\tilde{h}_2$, ..., $\tilde{h}_k$ those of $\tilde{H}$. Scheme-corresponding elementary regions $h_i$ and $\tilde{h}_i$ are of the same type, and their boundaries consist of scheme-corresponding semipaths. Proceeding in accordance with the enumeration of elementary regions, we can construct a topological mapping of $h_i$ onto $\tilde{h}_i$, under which paths go to paths, the direction of $t$ is preserved, and the following additional conditions are satisfied:

1.  Scheme-corresponding boundary semipaths of the regions are mapped onto one another, in such a way that their endpoints $P_i$ and $\widetilde{P}_i$ on the canonical curves $E$ and $\widetilde{E}$ are made to correspond.

2.  If $h_i$ (therefore also $\widetilde{h}_i$) has boundary points in common with previous regions of the sequence, the previously constructed topological mapping and that constructed at the present step coincide at these points. It is clear that conditions 1) and 2) may always be satisfied (see Lemmas 9, 10 and 11 of § 18).

We thus obtain a topological mapping of $H$ onto $\widetilde{H}$ possessing the required properties (the fact that the mapping is one-to-one and bicontinuous on the common boundary of adjacent regions $h_i$ and $h_j$, $\widetilde{h}_i$ and $\widetilde{h}_j$, follows from condition 2). Q.E.D.

R e m a r k  1.  One particular consequence of Theorem 61 is that there exists a path-preserving topological mapping of any canonical neighborhood of an equilibrium state onto any other canonical neighborhood (of the same equilibrium state), preserving orientation and direction of $t$.

R e m a r k  2.  Suppose given a topological mapping of the canonical curves, $E$ onto $\widetilde{E}$, under which scheme-corresponding arcs and endpoints are mapped onto each other.  Then, by Lemmas 9, 10, 11 of § 18, there always exists a topological mapping of $\overline{H}$ onto $\widetilde{\overline{H}}$, possessing the properties enumerated in the theorem, whose restriction to $E$ coincides with the given mapping.  An analogous statement is true when one is given a mapping of individual points of corresponding arcs, provided the cyclic order is preserved by the mapping.

It clearly follows from Remark 1 that if the number of singular paths is finite then any equilibrium state has a well-defined topological structure in the sense of the definition of § 5, and that *the local scheme characterizes the topological structure of the equilibrium state*.  It is obvious in this situation (see Lemma 7 of § 18) that equilibrium states with different local schemes have different topological structures.

**4.  GLOBAL SCHEME OF AN EQUILIBRIUM STATE OTHER THAN A CENTER.**  It may occur that there exist semipaths tending to an equilibrium state which are not separatrices (such as separatrices of other equilibrium states or corner semipaths).  Our consideration of the local scheme made no allowance for such semipaths among the orbitally unstable semipaths tending to an equilibrium state.

We shall now single out these semipaths, thus indicating all singular semipaths tending to an equilibrium state $O$, whether they are $\omega\,(\alpha)$-separatrices or not.  To describe the position of these singular semipaths relative to the separatrices and to each other, we introduce the concept of the global scheme of an equilibrium state, which will play a major role in establishing the topological structure of the phase portrait in the large (and not merely in the neighborhood of the equilibrium state).*

---

* Whereas the local scheme of an equilibrium state can actually be determined by direct examination of its neighborhood (in Chapter VI we presented methods for investigating the type, i.e., scheme, of a simple equilibrium state, and in the next chapter we shall consider methods applicable to certain multiple equilibrium states), the global scheme cannot be constructed on the basis of local properties of separatrices.

We first prove a lemma.

Let $g_{\sigma_1}$ and $g_{\sigma_2}$ be two elliptic regions of the equilibrium state $O$, one succeeding the other in the cyclic order, so that there is no elliptic region between $g_{\sigma_1}$ and $g_{\sigma_2}$.

Lemma 4. *Between any two cyclically successive elliptic regions of an equilibrium state $O$ there must be at least one singular path tending to the point $O$.*

Proof. If there is a separatrix of $O$ between $g_{\sigma_1}$ and $g_{\sigma_2}$, we are done. Suppose, therefore, that there is no separatrix between $g_{\sigma_1}$ and $g_{\sigma_2}$. Let $H$ be some canonical neighborhood of $O$. By Lemma 2, there must be a regular ω- or α-parabolic "sector" between $g_{\sigma_1}$ and $g_{\sigma_2}$ in $H$. To fix ideas, let this be an ω-parabolic sector, $g_\omega$ say. The sector $g_\omega$ is bounded by two positive semipaths $P_k O$ and $P_{k+1} O$, the point $O$ and an arc without contact $l_k$ with endpoints $P_k$ and $P_{k+1}$. The semipaths $P_k O$ and $P_{k+1} O$ are parts of paths $L_1^*$ and $L_2^*$ ("loops") bounding the regions $g_{\sigma_1}$ and $g_{\sigma_2}$. But then the points $P_k$ and $P_{k+1}$, belonging as they do to distinct elliptic regions, also belong to distinct cells. Consequently, the arc $l_k$ joining $P_k$ and $P_{k+1}$ must contain at least one boundary point $M$ of a cell. The path through $M$, $L$ say, is singular by the definition of a cell, and enters the region $g_\omega$ at $M$ with increasing $t$. Consequently, $L_M^+$ is a singular semipath tending to $O$ and situated between $g_{\sigma_1}$ and $g_{\sigma_2}$. Q.E.D.

Corollary. If $L_1^{()}$ and $L_2^{()}$ are two semipaths tending to the equilibrium state $O$, between which there is no singular semipath tending to $O$, then there cannot be two distinct elliptic regions between $L_1^{()}$ and $L_2^{()}$.

Along with the ω- and α-separatrices of the equilibrium state $O$, we now consider all semipaths of orbitally unstable paths in $G$ which tend to this equilibrium state but are not separatrices thereof, and renumber all the semipaths (including separatrices):

$$L_1^{()}, L_2^{()}, \ldots, L_k^{()}, \tag{11}$$

also considering all corner semipaths tending to $O$:

$$\hat{L}_1^{()}, \hat{L}_2^{()}, \ldots, \hat{L}_r^{()}. \tag{12}$$

As before, let $U_{\varepsilon_0}(O)$ be an $\varepsilon_0$-neighborhood of $O$ containing no whole singular paths other than $O$ itself. The curvilinear sectors $g_i$ into which $U_{\varepsilon_0}(O)$ is divided by the separatrices and loop semipaths are now subdivided into smaller curvilinear sectors by the singular semipaths which are not separatrices of $O$. In view of Lemma 5 of §17, one readily sees that between any two cyclically successive singular semipaths there is: a) an ω-parabolic sector, if both the semipaths are positive, and an α-parabolic sector if they are both negative; b) an elliptic or hyperbolic sector if one of the semipaths is positive and the other negative. As before, instead of considering semipaths of the loops we may consider all distinct elliptic regions of the equilibrium state.

Definition XXIII. *The global scheme of an equilibrium state $O$ which is not a center is a list of all singular semipaths tending to $O$ and all its elliptic regions, indicating a) the cyclic order in which they are arranged about the equilibrium state, b) which of the semipaths are corner semipaths. The global scheme may be written in tabular form:*

$$O \mid L_{i_1}^+, \; g_{\sigma_1}, \; L_{\bar{i}_2}^-, \; \hat{L}_{\bar{i}_3}^-, \; \ldots \tag{13}$$

The order of appearance of the semipaths and elliptic regions in this table corresponds to their cyclic order about $O$.  As in the case of the local scheme, the scheme is unique up to a cyclic permutation.  In accordance with previous arguments (see b) above), if no symbol $g_\sigma$ appears in the scheme between two semipaths, of which one is positive and one negative, this means that there is a hyperbolic sector between them.

It is clear that the global scheme determines the local scheme, but the converse is false.  As in the case of the local scheme, one can consider semipaths of loops instead of elliptic regions.

As in the local case, the global scheme may also be given by a schematic diagram, with the notation of the semipaths indicated.  For example, consider the global scheme

$$O \mid L_{\bar{1}}^-, \; g_1, \; \hat{L}_1^+, \; L_5^+, \; L_6^+, \; g_2, \; L_2^-, \; L_3^+, \; g_3, \; L_7^-, \; g_4, \; L_4^+.$$

The corresponding schematic diagram is shown in Figure 211.  As mentioned above, if there is no closed nodal region between two semipaths, of which one is positive and one negative, then there must be a hyperbolic sector between them.  In addition, by Lemma 4, between any two cyclically successive elliptic regions there is at least one singular path.

Let $\bar{H}$ be a closed canonical neighborhood of an equilibrium state $O$ and $E$ its boundary curve.

Clearly, any semipath tending to $O$ which is not a separatrix necessarily crosses one of the parabolic arcs without contact $l_k$.  In particular, the singular paths tending to $O$ (which are not separatrices of $O$) also cross parabolic arcs.

These semipaths divide the parabolic regions into smaller regions, also parabolic, and correspondingly the parabolic arcs $l_k$ in their boundaries are divided into smaller arcs without contact (Figure 212).

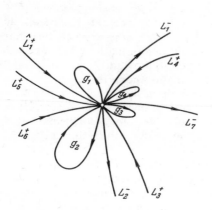

FIGURE 211.

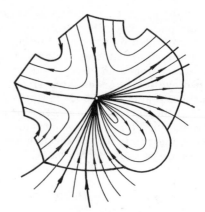

FIGURE 212.

When $l_k$ is $\omega$-parabolic, we shall call these subarcs of $l_k$ $\omega$-arcs and denote them by $a_i$; when $l_k$ is $\alpha$-parabolic, we shall call them $\alpha$-arcs and denote them by $b_j$. Thus, the arcs $a_i$ and $b_j$ have no points in common with any singular semipath. In particular, $a_i$ and $b_j$ may be the entire parabolic arc $l_k$. It is readily seen that at least one of the endpoints of $a_i$ or $b_j$ lies on a singular path. The arcs $a_i$, $b_j$, together with the above-defined elliptic arcs, saddle arcs of paths and saddle arcs without contact, will be called the c a n o n i c a l   a r c s of the canonical curve $E$. Consider a parabolic region whose boundary is the union of an arc $a_i$ (or $b_j$), two semipaths through the endpoints of $a_i$ ( $b_j$) and the point $O$. Any region of this type, or an elliptic or hyperbolic region as defined previously, in the canonical neighborhood $H$ will be called a c a n o n i c a l   r e g i o n of $H$.

Suppose now that $O$ is an equilibrium state of a system $D$, and consider an equilibrium state $\tilde{O}$ of a system $\tilde{D}$ ($D$ and $\tilde{D}$ may of course be the same system). Let the global schemes of these equilibrium states be given.

Thus, for $\tilde{O}$, as for $O$, we have a list of all singular semipaths tending to it, i. e., the $\omega$- and $\alpha$-separatrices of $\tilde{O}$, all semipaths of orbitally unstable paths tending to it which are not separatrices $(\tilde{L}_1^{(\cdot)}, \tilde{L}_2^{(\cdot)}, \ldots, \tilde{L}_{\tilde{k}}^{(\cdot)})$, and all corner semipaths tending to $\tilde{O}$ ( $\hat{\tilde{L}}_1^{(\cdot)}$, $\hat{\tilde{L}}_2^{(\cdot)}$, $\ldots$, $\hat{\tilde{L}}_{\tilde{r}}$). Also listed are all the elliptic regions of $\tilde{O}$ ($\tilde{g}_{\sigma_1}$, $\tilde{g}_{\sigma_2}$, $\ldots$, $\tilde{g}_{\sigma_{\tilde{l}}}$ ). The global scheme of the equilibrium state $\tilde{O}$ is given in tabular form:

$$\tilde{O} | \tilde{L}_1^{(\cdot)}, \ \tilde{g}_{\sigma_1}, \ \ldots \tag{14}$$

The order of the semipaths and elliptic regions corresponds to the cyclic order in which they are arranged about $\tilde{O}$. The following definition of equivalent global schemes is entirely analogous to that of equivalent local schemes.

*D e f i n i t i o n   XXIV. We shall say that the global schemes of the equilibrium states $O$ and $\tilde{O}$ are equivalent, with the same orientation and direction of $t$, if there is a one-to-one correspondence $\theta$ between the $\omega$- and $\alpha$-separatrices of the equilibrium states, the corner and singular semipaths tending to them, and their elliptic regions, satisfying the following conditions: The global scheme of $\tilde{O}$ may be obtained from that of $O$ by replacing semipaths and elliptic regions of $O$ by those of $\tilde{O}$ corresponding to them under the correspondence $\theta$.*

Thus, when the global schemes of equilibrium states $O$ and $\tilde{O}$ are equivalent, it is clear that

$$k = \tilde{k}, \quad r = \tilde{r}, \quad l = \tilde{l},$$

and the tabular notation (14) for the scheme of $\tilde{O}$ may be obtained from the notation of the scheme of $O$ by renumbering (if necessary) and providing each symbol with a tilde. Semipaths and elliptic regions corresponding to one another under $\theta$ will be referred to as scheme-corresponding semipaths (elliptic regions). When the schemes of two equilibrium states are equivalent, we clearly have also a scheme-correspondence between semipaths $L_i^{*+}$ and $\tilde{L}_i^{*+}$, $L_j^{*-}$ and $\tilde{L}_j^{*-}$ of the loops. Now let $H$ ($\tilde{H}$) be a canonical neighborhood of $O$ ( $\tilde{O}$) and $E$ ( $\tilde{E}$) the canonical curve. Exactly as in the case of $O$, the singular semipaths (not separatrices) tending to $\tilde{O}$ divide $\tilde{H}$ into canonical regions and $\tilde{E}$ into canonical arcs $\tilde{a}_i$, $\tilde{b}_i$, $\tilde{S}_c$, $\tilde{S}^*$ and $\tilde{\lambda}_i^c$.

Equivalence of the global schemes of equilibrium states $O$ and $\tilde{O}$ induces a natural one-to-one correspondence between the canonical regions in the canonical neighborhoods $H$ and $\tilde{H}$, the canonical arcs of $E$ and $\tilde{E}$, and their endpoints: 1) canonical regions whose boundaries involve scheme-corresponding singular paths correspond to each other; 2) canonical arcs whose endpoints lie on scheme-corresponding singular semipaths and which figure in the boundaries of scheme-corresponding regions (according to 1)) correspond to each other; 3) endpoints of canonical arcs lying on scheme-corresponding semipaths or arcs correspond to each other.

*Theorem 62. If the global schemes of equilibrium states $O$ and $\tilde{O}$ are equivalent, with the same orientation and direction of $t$, there exists a path-preserving topological mapping of any closed canonical neighborhood $\bar{H}$ of $O$ onto any closed canonical neighborhood $\bar{\tilde{H}}$ of $O$, preserving orientation and direction of $t$, under which scheme-corresponding corner and singular (non-corner) semipaths are mapped onto one another.*

The proof is completely analogous to that of Theorem 61, involving in addition reference to Remark 2 following that theorem.

R e m a r k. If we are given a topological mapping of $E$ onto $\tilde{E}$, under which scheme-corresponding canonical arcs and their endpoints are mapped onto one another, then there exists a topological mapping of $\bar{H}$ onto $\bar{\tilde{H}}$, possessing the properties listed in the theorem, whose restriction to $E$ coincides with the given mapping.

**5. EQUILIBRIUM STATES OF THE CENTER TYPE.** Let us now assume that there are no semipaths tending to the equilibrium state $O$. Then by Theorem 18 of §4, any arbitrarily small neighborhood of $O$ contains a closed path containing $O$ in its interior. By the assumption that the number of orbitally unstable paths is finite, there exists $\varepsilon_0 > 0$ such that the neighborhood contains no whole singular path other than $O$. Let $L$ be a closed path in $U_{\varepsilon_0}(O)$ containing $O$ in its interior (such a path always exists), and $g_L$ the interior of $L$. It is obvious that the region $g_L$ contains no singular path other than $O$. We shall call $g_L$ a c a n o n i c a l   n e i g h b o r h o o d of the center $O$.

The elementary proof of the following lemma is omitted.

*L e m m a  5. All paths passing through points of $g_L$ other than $O$ are closed, inside one another, and contain the point $O$ in their interior.*

Now let $L_1$ and $L_2$ be two closed paths, of which one is interior to the other. Let the region $w$ between them contain no singular point, and suppose that all paths through its points are closed. It is clear that these paths must lie inside one another.

*L e m m a  6. There exists an arc without contact joining some point $M_1$ of $L_1$ to a point $M_2$ on $L_2$, lying (except for its endpoints $M_1$ and $M_2$) inside $w$, and cutting all paths through points of $w$.*

We now return to the region $g_L$ bounded by a closed path $L$.

*L e m m a  7. There exists a simple smooth curve joining $O$ to a point $A$ on $L$, lying except for its endpoints inside $L$, and cutting all paths passing through points distinct from $O$ in $g_L$.*

The proofs of Lemmas 6 and 7 are analogous to those of Lemmas 2 and 5 of §18, and are therefore omitted.

We shall call a closed path $L$ a p a t h  o f  t h e  c e n t e r $O$ if its interior contains no singular element other than $O$ (i. e., no orbitally unstable path and no boundary curve).

*Definition XXV.* *The scheme of a center is specified by indicating whether or not the direction of t on the paths of the center agrees with the positive traversal of these curves.*

Let $O$ and $\tilde{O}$ be centers of two systems $D$ and $\tilde{D}$, respectively.

*Definition XXVI.* *We shall say that the schemes of centers $O$ and $\tilde{O}$ are equivalent, with the same orientation and direction of t, if the direction of t either agrees with the positive traversal on all paths of both, or opposes the positive traversal on all paths of both.*

Let $g_L$ and $\tilde{g}_L$ be arbitrary canonical neighborhoods of these centers, bounded by closed paths $L$ and $\tilde{L}$. The following theorem is easily proved, using Lemma 7.

*Theorem 63.* *If the schemes of centers $O$ and $\tilde{O}$ are equivalent, with the same orientation and direction of t, then there exists a path-preserving topological mapping of $\bar{g}_L$ onto $\tilde{g}_L$, preserving orientation and direction of t.*

Remark. If we are given a topological mapping of $L$ onto $\tilde{L}$, preserving direction of $t$, we can always construct a topological mapping of $\bar{g}_L$ onto $\tilde{g}_L$, possessing the properties listed in the theorem, whose restriction to $L$ coincides with the given mapping.

*Chapter IX*

*INVESTIGATION OF CERTAIN TYPES OF*
*MULTIPLE EQUILIBRIUM STATES**

INTRODUCTION

In this chapter we shall consider a certain class of multiple equilibrium states, which is in a natural sense the most elementary such class: isolated multiple equilibrium states of analytic dynamic systems such that the expansions of the right-hand sides in the neighborhood of the equilibrium state involve at least one first-order term.

The investigation will proceed by systematic examination of the paths tending to the equilibrium state in any one of the possible directions. Accordingly, the first section (§20) considers the directions of approach to multiple equilibrium states (this question was studied for simple equilibrium states in Chapter IV, §9).

In §21 we study equilibrium states (of the type indicated) which have one nonzero characteristic root ($\lambda \neq 0$). In this case the equilibrium state may be either a saddle point, node, or "saddle node" (equilibrium state with one parabolic and two hyperbolic sectors).

In §22 we consider the case in which both characteristic roots vanish ($\sigma = 0$). Here there are seven possibilities: saddle point, node, focus, center, saddle node, degenerate equilibrium state (two hyperbolic sectors) and equilibrium state with elliptic region (one elliptic and one hyperbolic sector).

The method employed in this chapter is due to Bendixson /33/. Though used here for a particular case, it is also applicable to arbitrary equilibrium states of analytic systems.

§20. DIRECTIONS OF APPROACH OF PATHS TO
MULTIPLE EQUILIBRIUM STATES

**1. TRANSFORMATION TO POLAR COORDINATES.** We shall confine ourselves throughout to dynamic systems

$$\frac{dx}{dt} = P(x, y), \qquad \frac{dy}{dt} = Q(x, y) \tag{I}$$

of the analytic class.

* This chapter depends to some extent on the discussion in Chapter VIII of neighborhoods of equilibrium states and makes use of the concepts of parabolic sector, elliptic sector and elliptic region introduced there.

Let $O$ be an isolated multiple equilibrium state ($\Delta = 0$; see §7) at the origin. Following Definition IX of §9, we wish to consider the directions in which paths may tend to this equilibrium state.

Suppose that the Taylor expansions of the functions $P(x, y)$ and $Q(x, y)$ about the equilibrium state $O(0, 0)$ have the form

$$P(x, y) = P_m(x, y) + \varphi(x, y),$$
$$Q(x, y) = Q_m(x, y) + \psi(x, y),$$
(1)

where $m \geqslant 1$, $P_m(x, y)$ and $Q_m(x, y)$ are homogeneous polynomials (the sums of all $m$-th order terms) and the "remainders" $\varphi(x, y)$ and $\psi(x, y)$ involve all higher-order terms. We shall assume that the polynomials $P_m$ and $Q_m$ do not vanish identically. As in §8, we introduce polar coordinates, setting $x = \varrho \cos \theta$, $y = \varrho \sin \theta$. This gives the system

$$\frac{d\varrho}{dt} = P(\varrho \cos \theta, \varrho \sin \theta) \cos \theta + Q(\varrho \cos \theta, \varrho \sin \theta) \sin \theta,$$
$$\frac{d\theta}{dt} = \frac{1}{\varrho} [Q(\varrho \cos \theta, \varrho \sin \theta) \cos \theta - P(\varrho \cos \theta, \varrho \sin \theta) \sin \theta],$$
(2)

which, after simple manipulations, becomes

$$\frac{d\varrho}{dt} = \varrho^m [P_m(\theta) \cos \theta + Q_m(\theta) \sin \theta + \varrho \bar{\varphi}(\varrho, \cos \theta, \sin \theta)],$$
$$\frac{d\theta}{dt} = \varrho^{m-1} [Q_m(\theta) \cos \theta - P_m(\theta) \sin \theta + \varrho \bar{\psi}(\varrho, \cos \theta, \sin \theta)],$$
(3)

where $P_m(\theta)$, $Q_m(\theta)$ are abbreviations for $P_m(\cos \theta, \sin \theta)$, $Q_m(\cos \theta, \sin \theta)$, and $\bar{\varphi}$, $\bar{\psi}$ are analytic functions. The relation between the paths of systems (I) and (3) may be determined exactly as in §8.

Let $\varrho = \varrho(t)$, $\theta = \theta(t)$ be an arbitrary path of system (3) in the strip $\Omega^+$: $[-\infty < \theta < +\infty, 0 < \varrho < \varrho^*]$ on the $(\varrho, \theta)$-plane (where $\varrho^*$ is sufficiently small). We introduce a new parameter $\tau$ on this path by setting

$$\frac{d\tau}{dt} = \varrho^{m-1}(t).$$
(4)

Since $\varrho > 0$, $\tau$ is a monotone increasing function of $t$ and therefore has an inverse $t = t(\tau)$. The equations

$$\varrho = \varrho(t(\tau)), \qquad \theta = \theta(t(\tau))$$
(5)

are also parametric equations of the path.

It follows from (3) and (4) that the functions $\varrho = \varrho(t(\tau))$, $\theta = \theta(t(\tau))$ satisfy the system of equations

$$\frac{d\varrho}{d\tau} = \varrho[P_m(\theta) \cos \theta + Q_m(\theta) \sin \theta + \varrho \bar{\varphi}],$$
$$\frac{d\theta}{d\tau} = Q_m(\theta) \cos \theta - P_m(\theta) \sin \theta + \varrho \bar{\psi}.$$
(6)

Thus, the paths of system (3) in the strip $\Omega^+$ coincide with those of system (6). Assume now that $L$ is a path of system (I) lying in $U_{\varrho^*}(O)$ tending, say, as $t \to +\infty$ to the equilibrium state $O$. Let $L_1$ be a path of system (3) corresponding to the path $L$. We have $\varrho(t) \to 0$ as $t \to +\infty$. We claim that $\tau$ also tends to $+\infty$ as $t \to +\infty$ (with respect to motion on the path $L_1$). In fact, since $\tau$ is a monotone increasing function of $t$, it either tends to $+\infty$ or is bounded above as $t \to +\infty$. Consider the bracketed expression in the first equation of (6). It is readily seen that for sufficiently small $\varrho$ this expression is bounded in absolute value, and so there exists $C > 0$ such that $\frac{d\varrho}{d\tau} > -C\varrho$ for all sufficiently small $\varrho$. Integrating this inequality from $\tau_0$ to $\tau > \tau_0$, we get

$$\ln \varrho(\tau) - \ln \varrho(\tau_0) > -C(\tau - \tau_0),$$

or $\tau > \tau_0 + \frac{\ln \varrho(\tau) - \ln \varrho(\tau_0)}{-C}$. Hence it follows that $\tau \to +\infty$ as $\varrho(\tau) \to 0$.

Note that system (6) may be considered not only in the strip $[0 < \varrho < \varrho^*]$ but also in the strip $[|\varrho| < \varrho^*]$, provided $\varrho^*$ is sufficiently small. It can have equilibrium states in this strip only on the $\theta$-axis ($\varrho = 0$). In fact, if $\varrho \neq 0$ the right-hand sides of system (6) can vanish together only if the right-hand sides of system (2) vanish, i.e., $P(\varrho \cos \theta, \varrho \sin \theta) = 0$, $Q(\varrho \cos \theta, \varrho \sin \theta) = 0$. But since $O$ is an isolated equilibrium state, this cannot hold for small nonzero $\varrho$.

As before, let $L$ be a path of system (I) tending to $O$ as $t \to +\infty$, and $L_1$ the corresponding path of system (6). As we have shown, $\varrho(\tau) \to 0$ as $\tau \to +\infty$ (along the path $L_1$). In order to ascertain whether $L$ tends to $O$ in a definite direction, we must study the function $\theta(\tau)$ corresponding to the path $L_1$. This we now proceed to do.

**2. GENERAL CASE.** Assume first that the polynomial $xQ_m(x, y) - yP_m(x, y)$ does not vanish identically. When this is true, the equation

$$Q_m(\cos \theta, \sin \theta) \cos \theta - P_m(\cos \theta, \sin \theta) \sin \theta = 0 \qquad (7)$$

has either no real roots or at most $m + 1$ roots in the interval $0 \leqslant \theta \leqslant \pi$.[*]

Let these roots be $\theta_1, \theta_2, \ldots, \theta_s$ ($s \leqslant m + 1$). Then all the roots of equation (7) are given by

$$\theta_{kn} = \theta_k + \pi n \qquad (1 \leqslant k \leqslant s, \ -\infty < n < +\infty).$$

The points $\varrho = 0$, $\theta = \theta_k + \pi n$ represent all equilibrium states of system (6) in the strip $|\varrho| < \varrho^*$.

A priori, there are three possibilities: 1) $\theta(\tau) \to +\infty$ or $\theta(\tau) \to -\infty$ as $\tau \to +\infty$; 2) $\theta(\tau)$ remains bounded in absolute value as $\tau \to +\infty$; 3) $\theta(\tau)$ is not bounded as $\tau \to +\infty$, but tends neither to $+\infty$ nor to $-\infty$.

We claim that case 3) is impossible. Suppose that $\theta(\tau)$ is not bounded above and does not tend to $+\infty$. Then there exists $\theta^*$ such that for any $T$ there is a $\tau_1 > T$ for which $\theta(\tau_1) \leqslant \theta^*$. On the other hand, for any $\theta^{**}$ and any $T$ there is a $\tau_2 > T$ such that $\theta(\tau_2) > \theta^{**}$.

---

[*] This is because equation (7) reduces to an $m$-th degree equation in tg $\theta$; we must of course allow for the possibility of infinite roots.

Take $\theta^{**} > \theta^*$, and let $\bar{\theta}$ be a number distinct from any of the $\theta_{kn}$, such that $\theta^* < \bar{\theta} < \theta^{**}$. To fix ideas, let

$$Q_m (\cos \bar{\theta}, \sin \bar{\theta}) \cos \bar{\theta} - P_m (\cos \bar{\theta}, \ \sin \bar{\theta}) \sin \bar{\theta} > 0.$$

Then, by the second equation of (6), $\frac{d\theta}{d\tau} > 0$ for $\theta = \bar{\theta}$ and sufficiently small $\varrho$ $(0 < \varrho < \bar{\varrho})$, i.e., the straight-line segment $\theta = \bar{\theta}$, $0 < \varrho < \bar{\varrho}$ has no contact with the paths of system (6) and the paths of the system may cross it (with increasing $\tau$) only from left to right (Figure 213).

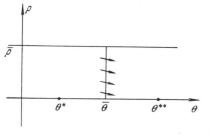

FIGURE 213.

Since the path $L_1$ passes through the strip $0 < \varrho < \bar{\varrho}$ for large $\tau$, it may cross the above segment only in one direction. But then the function $\theta (\tau)$ cannot go from values less than $\theta^*$ to values greater than $\theta^{**}$ or conversely for arbitrarily large $\tau$, which contradicts the choice of $\theta^*$ and $\theta^{**}$. Thus case 3) cannot occur.

In case 1), $\theta (\tau) \to + \infty$ or $\theta (\tau) \to - \infty$ as $\tau \to + \infty$. This implies that as $t \to + \infty$ the representative point moves along the path $L$ toward the equilibrium state $O$ in such a way that $\theta (t) \to + \infty$ or $\theta (t) \to - \infty$, so that $L$ is a spiral.

In case 2), the positive semipath $L_1^+$ of system (6) remains in a bounded region of the plane and therefore has an $\omega$-limit set.

Since $\varrho (\tau) \to 0$ as $\tau \to + \infty$, this limit set must lie on the $\theta$-axis. But then, by Theorem 9 of §4, it is an equilibrium state, i.e., a point with coordinates $\varrho = 0$, $\theta = \theta_{kn}$. Hence a point $M$ moving along $L$ tends to $O$ as $t \to + \infty$, in such a way that $\lim_{t \to +\infty} \theta (t) = \theta^*$ exists and is equal to one of the numbers $\theta_{kn}$, so that $L^+$ tends to $O$ in the direction $\theta_{kn}$. One often says that this direction satisfies the equation

$$xQ_m (x, \ y) - yP_m (x, \ y) = 0 \qquad (8)$$

in the sense that tg $\theta^*$ is one of the quotients $y/x$ determined by this equation.

R e m a r k  1.  If we assume that the direction $\theta^*$ in which the semipath $L^+$ tends to $O$ does not satisfy both equations $P_m (x, y) = 0$, $Q_m (x, y) = 0$ (i.e., the equations $P_m (\cos \theta^*, \sin \theta^*) = 0$, $Q_m (\cos \theta^*, \sin \theta^*) = 0$), then it is readily seen that the (finite or infinite) limit of $\frac{dy}{dx}$ as $t \to + \infty$ exists and this limit is precisely $\lim y/x$, i.e., tg $\theta^*$.

To prove this, consider the equation

$$\frac{dy}{dx} = \frac{Q_m (x, \ y) + \psi (x, \ y)}{P_m (x, \ y) + \varphi (x, \ y)} .$$

Setting $x = \varrho \cos \theta$, $y = \varrho \sin \theta$ and canceling out $\varrho^m$, we have

$$\frac{dy}{dx} = \frac{Q_m (\cos \theta, \ \sin \theta) + \varrho \psi_1 (\varrho, \ \cos \theta, \ \sin \theta)}{P_m (\cos \theta, \ \sin \theta) + \varrho \varphi_1 (\varrho, \ \cos \theta, \ \sin \theta)}$$

($\varphi_1$ and $\psi_1$ are analytic functions).  Since $\varrho \to 0$, $\theta \to \theta^*$ along $L^+$, it follows that

$$\lim \frac{dy}{dx} = \frac{Q_m\,(\cos \theta^*,\ \sin \theta^*)}{P_m\,(\cos \theta^*,\ \sin \theta^*)}\ .$$

Together with the equality

$$\cos \theta^* Q_m\,(\cos \theta^*,\ \sin \theta^*) - \sin \theta^* P_m\,(\cos \theta^*,\ \sin \theta^*) = 0$$

this implies that $\lim\limits_{t \to +\infty} \dfrac{dy}{dx} = \mathrm{tg}\,\theta^* = \lim y/x$, proving our assertion.

The case in which $P_m\,(\cos \theta^*,\ \sin \theta^*)$ and $Q_m\,(\cos \theta^*,\ \sin \theta^*)$ vanish together will not be considered.

Remark 2.  If there is a path spiraling, say as $t \to +\infty$, toward the equilibrium state $O$, then all paths through points of some neighborhood of $O$ are also spirals, so that $O$ is a stable or unstable focus.  Let us prove this.

Take a number $\bar{\theta}$ such that $Q_m\,(\cos \bar{\theta},\ \sin \bar{\theta})\cos \bar{\theta} - P_m\,(\cos \bar{\theta},\ \sin \bar{\theta})\sin \bar{\theta}$ does not vanish.  Then, as is easily seen, for sufficiently small $\varrho$ the segment $\theta = \bar{\theta}, 0 < \varrho < \bar{\varrho}$ has no contact with the paths of system (I).

Indeed, the condition for a path to touch the segment is $\dfrac{dy}{dx} = \mathrm{tg}\,\bar{\theta}$ or

$$Q\,(x,\ y)\cos \bar{\theta} - P\,(x,\ y)\cos \bar{\theta} = 0.$$

Simple manipulations give

$$\varrho^m\,[Q_m\,(\cos \bar{\theta},\ \sin \bar{\theta})\cos \bar{\theta} - P_m\,(\cos \bar{\theta},\ \sin \bar{\theta})\sin \bar{\theta} + \varrho\psi\,(\varrho,\ \bar{\theta})] = 0.$$

By virtue of the choice of $\bar{\theta}$, this expression cannot vanish for small $\varrho$.

Let $L$ be a spiral path of the system, crossing the segment without contact at points $A_1, A_2, A_3, \ldots$ tending monotonically to the point $O$

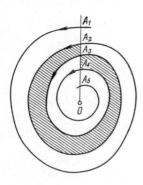

FIGURE 214.

(Figure 214).  Then it is readily seen by Lemma 14 of §3 that any path through the interior of the region bounded by the arc $A_i A_{i+1}$ of the path and the segment without contact $A_{i+1}A_i$ passes successively through all subsequent regions of this type (hatched in the figure) and is therefore a spiral.  This proves the assertion.

**3.  SINGULAR CASE.**  Suppose now that

$$x\,Q_m\,(x,\ y) - y\,P_m\,(x,\ y) \equiv 0.$$

It is obvious that $Q_m\,(x,\ y) = y\,Q_{m-1}\,(x,\ y)$ and $P_m\,(x,\ y) = xQ_{m-1}\,(x,\ y)$, where $Q_{m-1}$ is a homogeneous polynomial of degree $m-1$, not identically zero (recall that $m \geqslant 1$).  Transforming as before to polar coordinates and defining a new parameter $\tau$ by

$$\frac{d\tau}{dt} = \varrho^m, \tag{9}$$

we arrive at the system

$$\frac{d\varrho}{d\tau} = Q_{m-1}(\cos\theta,\ \sin\theta) + \varrho\overline{\varphi}(\varrho,\ \cos\theta,\ \sin\theta),$$

$$\frac{d\theta}{d\tau} = \varrho^{r}[z(\cos\theta,\ \sin\theta) + \varrho\overline{\psi}(\varrho,\ \cos\theta,\ \sin\theta)].$$

(10)

Here $\overline{\varphi}$ and $\overline{\psi}$ are analytic functions, $r$ is some nonnegative integer and $z(\cos\theta,\ \sin\theta)$ is a homogeneous polynomial of degree $m+r+2$.* As before, the paths of system (10) in the strip $0 < \varrho < \varrho^{*}$ correspond to the paths of system (I) in $U_{\rho*}(O)$. In this case, however, in contradistinction to the nonsingular case, we cannot state that $\tau \to +\infty$ as $\varrho \to 0$ along the path $L_1$ (i.e., as $t \to +\infty$). We can only say that, as $\varrho \to 0$, $\tau$ tends either to a finite limit $T$ or to $+\infty$ (since $\frac{d\tau}{dt} > 0$ by (9)).

Consider the system (10) in the strip $|\varrho| < \varrho^{*}$, with $\varrho^{*}$ sufficiently small. The equilibrium states of system (10) in this strip lie on the $\theta$-axis, with coordinates $(0,\ \theta)$ where $\theta$ is a root of the equation

$$Q_{m-1}(\cos\theta,\ \sin\theta) = 0.$$

(11)

Moreover, if $r > 0$ every root $\theta$ of equation (11) may be associated with an equilibrium state $(0,\ \theta)$ of system (10); but if $r = 0$ then $(0,\ \theta)$ is an equilibrium state only if $z(\cos\theta,\ \sin\theta)$ also vanishes. Clearly, system (10) either has no equilibrium states at all in the strip $|\varrho| < \varrho^{*}$, or all its equilibrium states are points $\varrho = 0$, $\theta = \theta_{kn} = \theta_k + n\pi$, where $k$ runs through the values $1, 2, \ldots, s$ ($n$ ranges over all integers and $s$ is an integer, $s \leqslant m-1$). The points $(0,\ \theta_k)$, $k = 1, 2, \ldots, s$, are the equilibrium states of the system on the interval $0 \leqslant \theta < \pi$ of the $\theta$-axis (see the beginning of the preceding subsection).

We shall call directions $\theta$ satisfying equation (11) s i n g u l a r directions.

Suppose that $\theta$ is not a singular direction. Then $\frac{d\varrho}{d\tau} \neq 0$ at the point $(0,\ \theta)$, hence also at all neighboring points. The equation of a path through this

point may be written $\theta = \theta(\varrho)$, and $\frac{d\theta}{d\varrho} \neq \infty$.

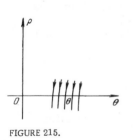

FIGURE 215.

Consequently, the paths of system (10) near the point $(0,\ \theta)$ behave as illustrated in Figure 215. But this means that for each nonsingular direction there is exactly one positive or negative semipath of system (I) which tends to $O$ in the direction $\theta$.

Now let $L^+$ be a semipath of system (I) in $U_{\rho*}(O)$, tending to $O$ but not one of the semipaths just considered (which tend to $O$ in nonsingular directions).

Consider a path of system (10) corresponding on the $(\varrho,\ \theta)$-plane to the semipath $L^+$. Let this path be $\widetilde{L}$ $(\varrho = \widetilde{\varrho}(\tau),\ \theta = \widetilde{\theta}(\tau))$, and suppose that the points of $L^+$ are in one-to-one correspondence with the points of $\widetilde{L}$ for which $\tau_0 < \tau < T$. These points

---

* We are assuming that the derivative $d\theta/d\tau$ does not vanish identically, so that the number $r$ and function $z$ are defined. If $d\theta/d\tau \equiv 0$ the integral curves are the rays $\theta = \text{const}$.

form an arc of the path $\tilde{L}$ which we denote by $\tilde{L}_1$. It is clear that this arc lies in the strip $0 < \varrho < \varrho^*$ on the $(\varrho, \theta)$-plane and as $\tau \to T$ the point $M(\tau)$ on the arc moves so that $\tilde{\varrho}(\tau) \to 0$.

Note that $T$ is either finite or $+\infty$. If $T = +\infty$, then $\tilde{L}_1$ is a positive semipath. If $T$ is finite, there are two possibilities: 1) $\tilde{L}$ is also defined for $\tau \geqslant T$; 2) $\tilde{L}$ is defined only for $\tau < T$.

We first show that the points of the arc $\tilde{L}_1$ of $\tilde{L}$ lie in a finite region of the plane. To this end, we consider an arc $A_0B_0$ of some path of system (10) passing through a point $A_0(0, \theta_0)$, where $\theta_0$ is not a singular direction.

Let us assume that all points of this arc other than $A_0$ correspond to $0 < \varrho \leqslant \varepsilon$, where $\varepsilon$ is sufficiently small $(\varepsilon < \varrho^*)$. Translating the arc $A_0B_0$ along the $\theta$-axis for distances $2\pi k$ $(k = \pm 1, \pm 2, \pm 3, \ldots)$, we get curves $A_kB_k$ which are also arcs of paths of system (10) (Figure 216).

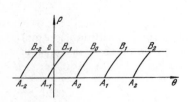

FIGURE 216.

It is obvious that when $\tau$ increases, tending to $T$, so that also $\tilde{\varrho}(\tau) \to 0$, the arc $\tilde{L}_1$ of the path $\tilde{L}$ enters one of the "quadrangles" $A_kB_kB_{k+1}A_{k+1}$ (see Figure 216) and cannot leave it thereafter with increasing $\tau$. In fact, $\tilde{L}_1$ is in the strip $0 < \varrho < \varrho^*$, and so $\varrho(\tau) > 0$. Now, from some time on, $\varrho(\tau)$ falls below $\varepsilon$.

Therefore, with increasing $\tau$ the arc $\tilde{L}_1$ cannot leave the region $A_kB_kB_{k+1}A_{k+1}$ through either of the sides $A_kA_{k+1}$ or $B_kB_{k+1}$. But $\tilde{L}_1$ cannot leave through the sides $A_kB_k$ or $A_{k+1}B_{k+1}$, since they are arcs of other paths of system (10). We have thus shown that as $\varrho \to 0$ the arc $\tilde{L}_1$ (corresponding to the semipath $L^+$ of system (I)) remains in a bounded portion of the strip $0 < \varrho < \varrho^*$.

If $T = +\infty$, $\tilde{L}_1$ is a positive semipath and, as described at the end of §20.2, it tends to one of the equilibrium states $(0, \theta_{kn})$ of system (10) on the $\theta$-axis. This means that the semipath $L^+$ tends to $O$ in the direction $\theta_{kn}$.

If $T$ is finite and there is a point on $\tilde{L}$ for which $\tau = T$, then this point $\varrho = \tilde{\varrho}(T)$, $\theta = \tilde{\theta}(T)$ is on the $\theta$-axis (i. e., $\tilde{\varrho}(T) = 0$) and $\tilde{\theta}(\tau) \to \tilde{\theta}(T)$ as $\tau \to T$ (i. e., as $\varrho \to 0$). It follows that $L^+$ tends to $O$ in the direction $\tilde{\theta}(T)$. Note that $\tilde{\theta}(T)$ is then a singular direction (satisfying equation (11)), but the point $(0, \tilde{\theta}(T))$ is not an equilibrium state of system (10).[*]

The case in which $T$ is finite but $\tilde{L}$ is defined only for $\tau < T$ cannot occur, since this would contradict Theorem 2 of §1.

Thus we see that if

$$x\theta_m(x, y) - yP_m(x, y) \equiv 0,$$

then any path tending to the equilibrium state $O$ does so in a definite direction.

Remark 1. It is easy to see that if $\tilde{\theta}$ is a singular direction but $(0, \tilde{\theta})$ is not an equilibrium state of system (10), i. e., $\varrho = 0$, $z(\cos\tilde{\theta}, \sin\tilde{\theta}) \neq 0$, then the equilibrium state $O$ must be approached in the direction $\tilde{\theta}$ by either one or two semipaths; in the latter case one of the semipaths is a continuation of the other.

---

[*] That is to say, $\varrho = 0$, $z(\cos\tilde{\theta}(T), \sin\tilde{\theta}(T)) \neq 0$.

In fact, in this case $(0, \tilde{\theta})$ is a regular point of the analytic system $(10)$, and the paths passing through it are tangent to the $\theta$-axis (since $\frac{d\varrho}{d\tau} = 0$, $\frac{d\theta}{d\tau} \neq 0$). Hence, the phase portrait in the neighborhood of the point is as illustrated in Figure 217 or Figure 218.

Examining the corresponding paths on the $(x, y)$-plane, we at once verify the assertion.

R e m a r k 2. If $\bar{\theta}$ is a singular direction and $(0, \bar{\theta})$ is an equilibrium state of $(10)$, then (a priori) there are four possibilities: 1) there exists no semipath tending to $O$ in the direction $\bar{\theta}$; 2) there exists one such semipath; 3) there is a finite number (more than one) of such paths; 4) there are infinitely many semipaths tending to O in the direction $\bar{\theta}$. The examples of §20.4 will show that each of these possibilities can actually occur.

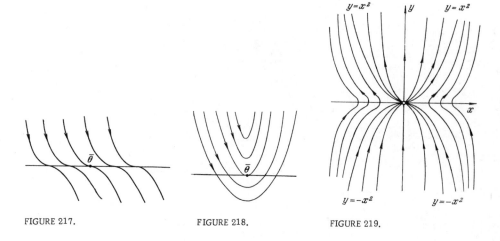

FIGURE 217.                    FIGURE 218.                    FIGURE 219.

The basic results of §20.2 and §20.3 may now be summarized:

*T h e o r e m 64. Any semipath of an analytic system*

$$\frac{dx}{dt} = P_m\,(x,\,y) + \varphi\,(x,\,y),$$

$$\frac{dy}{dt} = Q_m\,(x,\,y) + \psi\,(x,\,y),$$

*which tends to the equilibrium state O $(0, 0)$ is either a spiral or tends to O in a definite direction $\theta^*$.*

*If at least one path of the system is a spiral tending to O as $t \to +\infty$ (or as $t \to -\infty$), then all paths passing through points of some neighborhood of O are also spirals (so that O is a stable or unstable focus).*

*If $xQ_m\,(x,\,y) - yP_m\,(x,\,y) \not\equiv 0$, all directions $\theta^*$ along which the semipaths tend to O satisfy the equation*

$$xQ_m\,(x,\,y) - yP_m\,(x,\,y) = 0$$

*(i. e.,* $\cos\theta^* Q_m\,(\cos\theta^*,\ \sin\theta^*) - \sin\theta^* P_m\,(\cos\theta^*,\ \sin\theta^*) = 0$*).*

If $xQ_m(x, y) - yP_m(x, y) \equiv 0$, *the system assumes the form*

$$\frac{dx}{dt} = xQ_{m-1}(x, y) + \varphi(x, y)$$
$$\frac{dy}{dt} = yQ_{m-1}(x, y) + \psi(x, y). \qquad (Q_{m-1}(x, y) \not\equiv 0),$$

*In this case, for every nonsingular direction* $\theta$ *(i.e., the direction not satisfying the equation* $Q_{m-1}(x, y) = 0$), *there exists exactly one semipath tending to* $O$ *in the direction* $\theta$. *If* $\theta^*$ *is a singular direction, there may be no semipaths tending to* $O$ *in the direction* $\theta^*$, *a finite number, or infinitely many.*

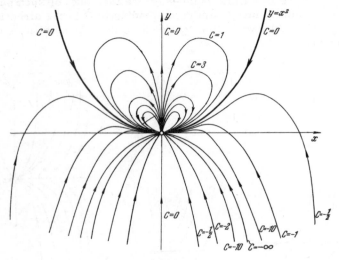

FIGURE 220.

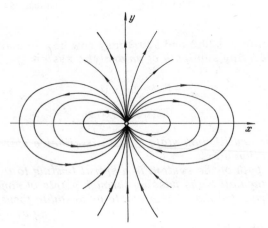

FIGURE 221.

**4. EXAMPLES.** We leave it to the reader to examine the following examples (in all of which the system may be integrated by elementary techniques) and verify that all the cases listed in Theorem 64 are indeed possible.

Example 1.

$$\frac{dx}{dt} = xy; \qquad \frac{dy}{dt} = y^2 + x^4.$$

The system may be integrated with the aid of the substitution $\frac{y}{x} = u$. One can show that the singular directions are $\theta = 0$, $\theta = \pi$, and in each singular direction there are two semipaths approaching $O$ (Figure 219).

Example 2.

$$\frac{dx}{dt} = xy - 3x^3; \qquad \frac{dy}{dt} = y^2 - 6x^2y + x^4.$$

The system is integrated by the substitution $y/x^2 = u$. Examining the resulting integral, one readily verifies that the singular directions are $\theta = 0$ and $\theta = \pi$, and in each of these there are infinitely many paths tending to the equilibrium state (Figure 220).

Example 3.

$$\frac{dx}{dt} = xy, \qquad \frac{dy}{dt} = y^2 - x^4.$$

The paths of this system are the curves $y^2 + x^4 = Cx^2$ $(C > 0)$ and also the half-axes $x = 0$, $y > 0$ and $x = 0$, $y < 0$. The singular directions are $\theta = 0$, $\theta = \pi$. Moreover, there is no path tending to the equilibrium state in either of the singular directions (Figure 221).

## §21. TOPOLOGICAL STRUCTURE OF A MULTIPLE EQUILIBRIUM STATE IN THE CASE $\sigma = P'_x(0, 0) + Q'_y(0, 0) \neq 0$

**1. AUXILIARY TRANSFORMATIONS. LEMMAS.** Consider the transformation defined by

$$x = x, \qquad y = x\eta. \tag{1}$$

This and the analogous transformation

$$x = \xi y, \qquad y = y \tag{2}$$

will be employed systematically in what follows. We shall therefore discuss some of their properties.

a) Consider the $(x, y)$- and $(x, \eta)$-planes. We shall use the term **slit plane** for either of these planes slit along the axis $x = 0$. Transformation (1) defines a topological mapping of the slit $(x, y)$-plane onto the $(x, \eta)$-plane. One sees from formula (1) that points of the first (second, third, fourth) quadrant in the $(x, y)$-plane are mapped respectively onto points of the first (third, second, fourth) quadrant in the $(x, \eta)$-plane.

Formula (1) does not define a mapping of the full (nonslit) $(x, y)$-plane onto the $(x, \eta)$-plane, because the mapping in question is undefined on the axis $x = 0$ of the $(x, y)$-plane $\left(\text{since } \eta = \dfrac{y}{x}\right)$. The "inverse" transformation is defined on the axis $x = 0$ of the $(x, \eta)$-plane, but maps it onto a single point $(0, 0)$ of the $(x, y)$-plane.

b) Let $U_\delta(0)$ be a small [circular] neighborhood of the origin $0$ of the $(x, y)$-plane, of radius $\delta$. On the slit plane, this neighborhood is the union of two regions (semidisks), each of which is mapped by transformation (1) onto a "strip-shaped" region to the right or left of the axis $x = 0$, bounded by the latter and a curve which approaches it asymptotically (Figure 222).

Let $\Gamma$ denote the region on the $(x, \eta)$-plane bounded by these two curves. Transformation (1) maps the slit neighborhood $U_\delta(0)$ topologically onto the slit region $\Gamma$.

c) Consider a dynamic system

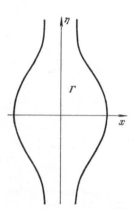

FIGURE 222.

$$\frac{dx}{dt} = P(x, y), \qquad \frac{dy}{dt} = Q(x, y), \qquad (3)$$

defined in $U_\delta(0)$. Transformation (1) may be regarded as a change of variables in system (3). The transformed system is readily seen to be

$$\frac{dx}{dt} = P(x, \eta x), \qquad \frac{d\eta}{dt} = \frac{Q(x, \eta x) - \eta P(x, \eta x)}{x}. \qquad (4)$$

Throughout the sequel we shall assume that system (3) is analytic, with an equilibrium state at $0\,(0, 0)$. Then the numerator in the second equation of (4) must contain a factor $x$, which cancels out. The result is a system identical with system (4) in the slit region $\Gamma$ but defined in the entire (nonslit) region $\Gamma$. However, we cannot say that the change of variables (1) takes the paths of system (4) onto those of system (3) unless both systems are considered only in the slit regions $\Gamma$ and $U_\delta(0)$.

Transformation (2) obviously possesses properties analogous to properties a), b), c). We shall not state them, remarking only that the respective planes must be slit along the axis $y = 0$ in order to make (2) well defined. Transformation (2) maps the points of the first (second, third, fourth) quadrant of the $(\xi, y)$-plane onto points of the first (second, fourth, third) quadrants of the $(x, y)$-plane. The region $\Gamma$ onto which transformation (2) maps the neighborhood $U_\delta(0)$ clearly lies along the $\xi$-axis, and system (3) becomes

$$\frac{d\xi}{dt} = \frac{P(\xi y, y) - \xi Q(\xi y, y)}{y}, \qquad \frac{dy}{dt} = Q(\xi y, y). \qquad (5)$$

We now prove two lemmas which are basic for what follows. Assume that the right-hand sides of system (3) do not vanish identically. Then system (3) may be written

$$\frac{dx}{dt} = P_m (x, y) + \varphi (x, y), \qquad \frac{dy}{dt} = Q_m (x, y) + \psi (x, y), \qquad (6)$$

where $m \geqslant 1$, $P_m (x, y)$ and $Q_m (x, y)$ are homogeneous $m$-th degree polynomials (one of which may vanish identically), $\varphi (x, y)$ and $\psi (x, y)$ are functions whose series expansions in $U_\delta (O)$ begin with terms of order at least $m + 1$.

Applying transformation (1) to system (6), we get system (4); it is readily seen that the right-hand sides of the latter involve a common factor $x^{m-1}$. Substituting $x^{m-1}dt = d\tau$, we obtain the system

$$\frac{dx}{d\tau} = xP_m (1, \eta) + x^2 P^* (x, \eta), \qquad \frac{d\eta}{d\tau} = Q_m (1, \eta) - \eta P_m (1, \eta) + xQ^* (x, \eta), \quad (7)$$

where $P^* (x, \eta)$, $Q^* (x, \eta)$ are analytic in the region $\Gamma$ (the image of $U_\delta (O)$ under (1)). When $m = 1$ systems (4) and (7) are of course the same. When $m > 1$ the paths of system (7) may be either whole paths of system (4) or unions of several paths of system (4), so that a whole path of system (4) may be only an arc of a path of system (7) (see § 1).

Lemma 1. *Let $O (0, 0)$ be an isolated equilibrium state of system (6) and*

$$x = x (t), \qquad y = y (t) \qquad (8)$$

*a semipath of the system tending to the equilibrium state in the direction $\theta = $ arctg $k$ or $\theta = \pi + $ arctg $k$, where $k \neq \infty$. Then:*
*1) The point $\tilde{O} (0, k)$ on the $(x, \eta)$-plane is an equilibrium state of system (7). 2) The semipath (8) corresponds to a semipath of system (7) in the slit $(x, \eta)$-plane, which tends to the point $\tilde{O} (0, k)$. 3) Conversely, to any semipath of system (7) on the slit $(x, \eta)$-plane which tends to the equilibrium state $\tilde{O} (0, k)$ corresponds a semipath of system (6) which tends to the equilibrium state $O (0, 0)$ in the direction $\theta = $ arctg $k$ or $\theta = \pi + $ arctg $k$.*

Proof. To fix ideas, suppose that (8) is a positive semipath tending to $O (0, 0)$ in the direction $\theta = $ arctg $k$ or $\theta = \pi + $ arctg $k$. By Theorem 64 (§20), we have $Q_m (1, k) - kP_m (1, k) = 0$,[*] whence it follows at once that $(0, k)$ is an equilibrium state of system (1), proving part 1). The semipath (8) tends to $O (0, 0)$ in a direction $\theta \neq \frac{\pi}{2} + n\pi$, and hence, for sufficiently large $t$, it lies in the slit $(x, y)$-plane. Corresponding to (8), we have a semipath or arc of a path $\tilde{L}$ of system (7) in the slit $(x, \eta)$-plane.

Let $M (x (t), y (t))$ be a point on the semipath (8), and $\tilde{M} (x (\tau), \eta (\tau))$ its image on $\tilde{L}$ under transformation (1). We have $x \to 0$ and $\eta = \frac{y}{x} \to k$ as $t \to +\infty$, and consequently the point $\tilde{M}$ tends to the equilibrium state $(0, k)$ as $t \to \infty$ (so that also $\tau \to \infty$). This also implies that the semipath (8)

---

[*] We are assuming here that $xQ_m - yP_m \not\equiv 0$. But if $xQ_m - yP_m \equiv 0$, it is also true that $Q_m (1, k) - kP_m (1, k) = 0$.

corresponds to a s e m i p a t h $\tilde{L}$ of system (7), and not merely an arc of a path (see §1.7). Note that $\tilde{L}$ may be a negative semipath. This proves part 2).

The third part of the lemma is obvious. Q.E.D.

Suppose now that system (6) has a semipath tending to the equilibrium state $O$ in the direction $\theta = \frac{\pi}{2}$ or $\frac{3}{2}\pi$. In this case we apply transformation (2). The role of system (7) is taken over by the system

$$\tfrac{d\xi}{dt} = P_m(\xi, 1) - \xi Q_m(\xi, 1) + y\bar{P}(\xi, y), \qquad \tfrac{dy}{dt} = yQ_m(\xi, 1) + y^2\bar{Q}(\xi, y). \qquad (9)$$

L e m m a 2. Let $O(0, 0)$ be an isolated equilibrium state of system (6), and $x = x(t)$, $y = y(t)$ a semipath of this system tending to $O$ in the direction $\theta = \frac{\pi}{2}$ or $\theta = \frac{3}{2}\pi$. Then:

1) The point $\tilde{O}(0, 0)$ on the $(\xi, y)$-plane is an equilibrium state of system (9). The semipath $x = x(t)$, $y = y(t)$ corresponds to a semipath of system (9) in the slit $(\xi, y)$-plane which tends to $\tilde{O}(0, 0)$. 2) Conversely, to any semipath of system (9) in the slit $(\xi, y)$-plane which tends to $\tilde{O}(0, 0)$ corresponds a semipath of system (6) which tends to the equilibrium state $O(0, 0)$ in the direction $\theta = \frac{\pi}{2}$ or $\theta = \frac{3}{2}\pi$.

Recall that in this context the plane is slit along the axis $y = 0$.

The proof of Lemma 2 is analogous to that of Lemma 1.

The next lemma, though unrelated to the transformations studied above, is also necessary for the sequel.

L e m m a 3. If $f_2(x, y)$ is analytic in the neighborhood of $O$ and its expansion in powers of $x$ and $y$ begins with terms of degree at least 2, $m$ is an odd number, $m \geqslant 2$ and $\Delta_m < 0$, then the system

$$\tfrac{dx}{dt} = \Delta_m x^m, \qquad \tfrac{dy}{dt} = y - ax + f_2(x, y) \qquad (A)$$

1) cannot have more than one semipath tending to $O$ in the direction arctg $a$, or more than one semipath tending to $O$ in the direction $\pi +$ arctg $a$;

2) has exactly two semipaths tending to $O$, one in the direction $\frac{\pi}{2}$ and the other in the direction $\frac{3}{2}\pi$, and these paths are the positive and negative $y$-axis or segments thereof bounded on one side by $O$; 3) has no paths tending to the point $O$ in other directions (except for $0, \pi, \frac{\pi}{2}, \frac{3}{2}\pi$).

P r o o f. The directions in which the paths of system (A) approach the equilibrium state $O$ are defined by the equation $x(y - ax) = 0$ (or $\cos\theta(\sin\theta - a\cos\theta) = 0$) (see Theorem 64, §20). The third assertion of the theorem now follows immediately.

We now prove the first part of the lemma. To fix ideas, let us consider paths tending to $O$ in the direction $\theta =$ arctg $a$. Let $L_1$ be such a path. For sufficiently large $t$, all points of this path will lie to the right of the $y$-axis. The first equation of (A) shows that then $\frac{dx}{dt}$ does not vanish, $x(t)$ is a monotone function and so the equation of $L_1$ may be written near $O$ in closed form $y = y_1(x)$, where $y_1(x)$ satisfies the differential equation

$$\frac{dy_1}{dx} = \frac{y_1 - ax + f_2(x, y_1)}{\Delta_m x^m}. \tag{B}$$

Suppose now that there exist two semipaths, $y = y_1(x)$ and $y = y_2(x)$, tending to the point $O$ in the direction $\theta = \mathrm{arctg}\, a$. Denote $y_1(x) - y_2(x) = z(x)$. Since distinct paths never intersect, it follows that for all sufficiently small $x$ the sign of the difference $z(x)$ cannot change, since $z(x) > 0$. Since both semipaths satisfy equation (B),

$$\frac{dz}{dx} = \frac{z + f_2(x, y_1(x)) - f_2(x, y_2(x))}{\Delta_m x^m}.$$

We can write this equality as

$$\frac{dz}{dx} = \frac{z}{x} \cdot \frac{1 + o(1)}{\Delta_m x^{m-1}}$$
$$(x > 0, \ o(1) \to 0 \ \text{as} \ x \to 0).$$

Now $\Delta_m < 0$, and the rest of the proof coincides word for word with that of the uniqueness of the separatrix of a simple saddle point tending to it in the direction $\theta$ (§7.3). This completes the proof of the first part.

In order to ascertain what semipaths can tend to $O$ in the directions $\frac{\pi}{2}$ and $\frac{3}{2}\pi$, apply the transformation $x = \xi y$, $y = y$ to system (A). We get a system of type (5):

$$\frac{d\xi}{dt} = -\xi + \tilde{P}_2(\xi y, y), \qquad \frac{dy}{dt} = y - a\xi y + f_2(\xi y, y), \tag{$\tilde{A}$}$$

where $\tilde{P}_2$ involves only terms of at least second order. It is self-evident that the origin $\tilde{O}(0, 0)$ of the $(\xi, y)$-plane is a simple saddle point of system $(\tilde{A})$, its separatrices being the semi-axes $y = 0$, $\xi > 0$ and $y = 0$, $\xi < 0$ (or segments thereof bounded on one side by $\tilde{O}$). Consequently, there exist only two more semipaths tending to the point $\tilde{O}$: the remaining separatrices of the saddle point $\tilde{O}$. These two semipaths clearly lie in the slit $(\xi, y)$-plane on different sides of the axis $y = 0$. But then it follows from the properties of transformation (2) and from Lemma 2 that there exists exactly one semipath of system (A), which tends to $O$ in the direction $\frac{\pi}{2}$, and exactly one which tends to $O$ in the direction $\frac{3}{2}\pi$. Since the half-axes $x = 0$, $y > 0$ and $x = 0$, $y < 0$ (or segments thereof) are semipaths answering to this description, this completes the proof of the second part and thus of the whole lemma.

**2. TOPOLOGICAL STRUCTURE OF A MULTIPLE EQUILIBRIUM STATE, $\sigma \neq 0$.** We assume till further notice that $O(0, 0)$ is an isolated equilibrium state with at least one nonzero characteristic root. Then our system becomes

$$\frac{dx}{dt} = ax + by + P_2(x, y), \qquad \frac{dy}{dt} = cx + dy + Q_2(x, y), \tag{10}$$

where $P_2(x, y)$, $Q_2(x, y)$ are analytic in the neighborhood of the origin and their series expansions involve only terms of at least second order;

$$\Delta = ad - bc = 0, \qquad \sigma = a + d \neq 0.$$

It is readily shown that under these assumptions there exists a nonsingular linear transformation reducing the system to the form

$$\frac{d\bar{x}}{dt} = \bar{P}_2(\bar{x}, \bar{y}), \qquad \frac{d\bar{y}}{dt} = \bar{y} + \bar{Q}_2(\bar{x}, \bar{y}),$$

where $\bar{t} = \varkappa t$ ($\varkappa$ a constant), and the functions $\bar{P}_2$ and $\bar{Q}_2$ satisfy the same conditions as the functions $P_2(x, y)$ and $Q_2(x, y)$.* Hence there is no loss of generality in investigating only a special case of system (10):

$$\frac{dx}{dt} = P_2(x, y) = P(x, y), \qquad \frac{dy}{dt} = y + Q_2(x, y) = Q(x, y). \qquad (11)$$

All our deliberations relate to some sufficiently small neighborhood $U_\delta(O)$, containing no equilibrium states other than $O$. Let

$$\sigma(x, y) = \frac{\partial P(x, y)}{\partial x} + \frac{\partial Q(x, y)}{\partial y} \qquad (12)$$

($P$ and $Q$ are the functions on the right of system (11)).

The function $\sigma(x, y)$ is continuous, $\sigma(0, 0) = \sigma = 1$. We may therefore assume that $\sigma(x, y) > 0$ at all points of $U_\delta(O)$, and hence, by Bendixson's test (see §12), the neighborhood $U_\delta(O)$ contains neither closed paths nor loops. Thus the point $O(0, 0)$ is not a center and it has no elliptic sectors. Hence there must exist semipaths that tend to the equilibrium state. We wish to investigate these semipaths.

Applying transformation (2) and reasoning as in the proof of part 2) of Lemma 3, we first ascertain that there exists exactly one semipath of system (11) which tends to $O$ in the direction $\frac{\pi}{2}$, and exactly one which does so in the direction $\frac{3}{2}\pi$. Denote these semipaths by $L_1$ and $L_2$, respectively.

It follows from the existence of these semipaths and from Remark 2 in §20 that the point $O(0, 0)$ cannot be a focus for system (11). Finally, by Theorem 64, any semipaths other than $L_1$ and $L_2$ that tend to $O$ (if such exist) must do so in the direction $0$ or $\pi$ (since for system (11) we have $xQ_m(x, y) - yP_m(x, y) = xy$). Our task is now to determine the number and nature of these paths.

We first express system (11) in another form. Consider the equation

$$y + Q_2(x, y) = 0. \qquad (13)$$

---

* See §6.1. If $b \neq 0$, a suitable transformation is $\bar{x} = -dx + by$, $\bar{y} = ax + by$; if $b = 0$, and $a = 0$, one can take $\bar{x} = x$, $\bar{y} = \frac{c}{d}x + y$. Finally, if $b = d = 0$ the transformation is $\bar{x} = -\frac{c}{a}x + y$, $\bar{y} = x$. Since $\sigma \neq 0$, $a$ and $d$ cannot both vanish.

By the implicit function theorem, this equation has a solution $y = \varphi(x)$ in a small neighborhood of $O$, where $\varphi(x)$ is an analytic function such that

$$\varphi(0) = 0, \qquad \varphi'(0) = 0. \tag{14}$$

Define a function $\psi(x)$ by

$$\psi(x) = P_2(x, \varphi(x)). \tag{15}$$

This function cannot vanish identically. Indeed, were $\psi(x) \equiv 0$, it would follow from (15) and the definition of $\varphi(x)$ that all points of the curve $y = \varphi(x)$ are equilibrium states of system (11), whereas we have assumed that $O$ is an isolated equilibrium state. Therefore the series expansion of the function $\psi(x)$ has the form

$$\psi(x) = \Delta_m x^m + \ldots, \tag{16}$$

where $m \geqslant 2$, $\Delta_m \neq 0$. The obvious equalities

$$P_2(x, y) = \psi(x) + P_2(x, y) - P_2(x, \varphi(x)),$$
$$y + Q_2(x, y) = y - \varphi(x) + Q_2(x, y) - Q_2(x, \varphi(x))$$

imply that system (11) can be written

$$\frac{dx}{dt} = \psi(x) + [y - \varphi(x)] \cdot \bar{P}(x, y),$$
$$\frac{dy}{dt} = [y - \varphi(x)][1 + \bar{Q}(x, y)], \tag{17}$$

where $\bar{P}$ and $\bar{Q}$ are analytic in a sufficiently small neighborhood $U_\delta(O)$ and $\bar{P}(0, 0) = \bar{Q}(0, 0) = 0$.

The curve $y = \varphi(x)$ is an isocline of horizontal directions for system (11). Let $C$ denote the circle bounding the neighborhood $U_\delta(O)$,* $P_1$ and $P_2$ the points of intersection of the curve $y = \varphi(x)$ and $C$. The arc $P_1 P_2$ of the curve $y = \varphi(x)$ divides $U_\delta(O)$ into two regions. It follows from equations (17) that $\frac{dy}{dt} > 0$ at all points of the upper region, i.e., the field vectors at these points point upward; similarly, the field vectors at points of the lower region point downward. Hence it follows at once that the semipaths $L_1$ and $L_2$, lying respectively above and below the curve $y = \varphi(x)$, are negative semipaths. Denote the first points of these semipaths (with increasing $t$) in common with $C$ by $M_1$ and $M_2$, respectively (Figure 223).

Since $\varphi(x)$ is analytic, it follows that either $y = \varphi(x)$ coincides with the $x$-axis $((\varphi(x) \equiv 0)$, or we may assume that it has no points in common with the $x$-axis other than $O$ and its tangent is horizontal only at $O$ (this is accomplished by taking $U_\delta(O)$ sufficiently small). In the first case, the segments $OP_1$ and $OP_2$ of the curve (i.e., of the $x$-axis) are clearly semipaths of system (11). In the second case, any arc of $y = \varphi(x)$ between

---

* In the subsequent arguments it is often necessary to replace $U_\delta(O)$ by a smaller neighborhood in order to meet various requirements. Throughout the sequel we shall always assume that this has been done wherever necessary, without changing the notation $U_\delta(O)$.

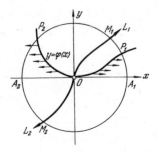

FIGURE 223.

the points $P_1$ and $P_2$ which does not contain $O$ is an arc without contact (since $y = \varphi(x)$ is an isocline of horizontal directions). The direction of the field vector at points of $y = \varphi(x)$ is determined by the first of equations (17), which has the form $\frac{dx}{dt} = \psi(x)$. In view of (16), we see that this direction depends on the sign of $\Delta_m$ and the parity of $m$. Therefore, there are exactly four possible cases. Figure 223 shows the directions of the field vectors at points of the curve $y = \varphi(x)$ in case $m$ is odd and $\Delta_m > 0$.*

We now establish the main theorem of this section, which describes the possible topological structure of an equilibrium state $O(0, 0)$ of system (11).

*Theorem 65. Let $O(0, 0)$ be an isolated equilibrium state of system (11). Let $y = \varphi(x)$ be a solution of the equation $y + Q_2(x, y) = 0$ in the neighborhood of $O(0, 0)$, and assume that the series expansion of the function $\psi(x) = P_2(x, \varphi(x))$ has the form $\psi(x) = \Delta_m x^m + \ldots$, where $m \geqslant 2$, $\Delta_m \neq 0$. Then: 1) If $m$ is odd and $\Delta_m > 0$, $O(0, 0)$ is a topological node. 2) If $m$ is odd and $\Delta_m < 0$, $O(0, 0)$, is a topological saddle point, two of whose separatrices tend to $O$ in the directions $0$ and $\pi$, the other two in the directions $\frac{\pi}{2}$ and $\frac{3}{2}\pi$. 3) If $m$ is even, $O(0, 0)$ is a saddle node, i.e.,*

*an equilibrium state whose canonical neighborhood is the union of one parabolic and two hyperbolic sectors. If $\Delta_m < 0$, the hyperbolic sectors contain a segment of the positive x-axis bordering on the origin $O$ (Figure 224), and if $\Delta_m > 0$ they contain a segment of the negative x-axis (Figure 225).*

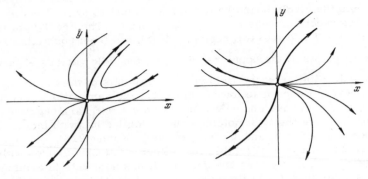

FIGURE 224.    FIGURE 225.

Proof. We first consider case 1). The directions of the field vectors at points of the curve $y = \varphi(x)$ are shown in Figure 223. Let $A_1$, $A_2$ denote the points at which the $x$-axis is cut by the circle $C$. The curves $OM_1$, $OP_2$, $OA_2$,

* In Figure 223, the arcs $OP_1$ and $OP_2$ of $y = \varphi(x)$ lie above the $x$-axis. Either arc (or both) may lie below the $x$-axis, but the proofs as we give them remain valid in essentials and there is no need for a separate discussion.

etc. divide $U_\delta(0)$ into curvilinear sectors, which we denote by $OP_1M_1$, $OM_1P_2$, etc. (stipulating throughout that the arcs $P_1M_1$, $M_1P_2$, etc. of $C$ figuring in the boundaries of the sectors are those on which the directions from $P_1$ to $M_1$, $M_1$ to $P_2$, etc. are induced by the positive traversal of $C$).

Suppose first that $\varphi(x) \equiv 0$. Then the segments $OA_1$ and $OA_2$ of the $x$-axis are negative semipaths (Figure 226). We claim that no positive semipath can tend to $O$. Indeed, otherwise let $L^+$ be a positive semipath tending to the origin, passing within the sector $OA_1M_1$ or $OM_1A_2$. Then the ordinate of a point $M$ on this semipath must tend to zero as $t \to +\infty$. But this contradicts the fact that $\frac{dy}{dt} > 0$ along any path situated above the curve $y = \varphi(x)$ (as follows from equation (17); see above). Similarly, one shows that no positive semipath tending to $O$ can pass through the sectors $OA_2M_2$ and $OM_2A_1$. Now suppose that $\varphi(x)$ is not identically zero, so that the curve $y = \varphi(x)$ has only the one point $O$ in common with the $x$-axis (Figure 227). Let $L^+$ be a positive semipath tending to the equilibrium state $O$. The same reasoning as used above (for the case $\varphi(x) \equiv 0$) shows that $L^+$ cannot lie entirely within the sectors $OP_1M_1$, $OM_2P_2$, $OA_2M_2$, $OM_2A_1$. Suppose that $L^+$ is inside the sector $OA_1P_1$. Then, by the Corollary to Lemma 5 of § 17, since there are no elliptic sectors in the neighborhood of $O$ there must be at least one hyperbolic sector $H$ between semipaths $L^+$ and $L_1^-$. Denote the separatrices in the boundary of this sector by $L_H^+$ and $L_H^-$. Noting the sign of the derivative $\frac{dy}{dt}$ along the path, we see (as before) that the separatrix (negative semipath) $L_H^-$ is either identical with $L_1^-$ or inside the sector $OP_1M_1$, and the separatrix $L_H^+$ is inside the sector $OA_1P_1$ (Figure 227). But then the paths through points of the hyperbolic sector $H$ will obviously cross the arc $OP_1$ of the curve $y = \varphi(x)$ in the direction opposite to that of the field at the points of the curve, which is impossible. One shows in similar fashion that $L^+$ cannot lie inside the sector $OA_2P_2$. Thus there are no positive semipaths tending to the point $O$. Hence any path through the neighborhood $U_\delta(0)$, other than $L_1$ and $L_2$, tends to $O$ as $t \to -\infty$ (in the direction $\theta = 0$ or $\theta = \pi$). But this means that the canonical neighborhood of the equilibrium state $O$ contains no elliptic or hyperbolic sectors; in other words, every path through a point of $U_\delta(0)$ tends to $O$ as $t \to -\infty$. This proves the first part of the theorem. (Note that all semipaths except $L_1^-$ and $L_2^-$ tend to $O$ in the direction 0 to $\pi$).

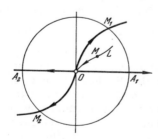

FIGURE 226.

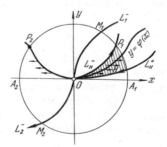

FIGURE 227.

We now proceed to case 2), in which the field vectors on the curve $y = \varphi(x)$ are directed opposite to those indicated in Figure 223. We have already shown that system (17) has exactly one (negative) semipath $L_1^-$ tending to $O$ in the direction $\frac{\pi}{2}$, and exactly one (again negative) semipath $L_2^-$ tending to $O$ in the direction $\frac{3}{2}\pi$. All other semipaths tending to $O$ (if such exist) must do so in the directions $0$ and $\pi$.

Since the semipaths $L_1^-$ and $L_2^-$ are both negative, it follows at once that the sector $OM_2M_1$ contains at least one semipath tending to $O$, therefore tending to $O$ in the direction $0$, and the sector $OM_1M_2$ at least one semipath, tending to $O$ in the direction $\pi$. We shall prove that there cannot be two semipaths tending to $O$ in the direction $0$, and the same holds for the direction $\pi$. This will clearly complete the proof of this part of the theorem.

Our system may be written either in the form (11):

$$\frac{dx}{dt} = P_2(x, y) = P(x, y),$$
$$\frac{dy}{dt} = y + Q_2(x, y) = Q(x, y), \qquad \text{(C)}$$

or in the form (17):

$$\frac{dx}{dt} = \psi(x) + [y - \varphi(x)]\,\overline{P}(x, y) = P(x, y),$$
$$\frac{dy}{dt} = [y - \varphi(x)]\,[1 + \overline{Q}(x, y)] = Q(x, y) \qquad \text{(D)}$$
$$(\overline{P}(0, 0) = \overline{Q}(0, 0) = 0).$$

Applying the transformation $x = x$, $y = \eta_1 x$ to system (11), we obtain

$$\frac{dx}{dt} = P_2(x, \eta_1 x) = P^{(1)}(x, \eta_1),$$
$$\frac{d\eta_1}{dt} = \eta_1 + \frac{Q_2(x, \eta_1 x) - \eta_1 P_2(x, \eta_1 x)}{x} = \eta_1 + Q^*(x, \eta_1) = Q^{(1)}(x, \eta_1). \qquad \text{(C}_1\text{)}$$

If the series expansion of the function $Q^*(x, \eta_1)$ contains no linear terms, system $(C_1)$ is similar in form to system (C) and therefore possesses the same properties. Apply the transformation $x = x$, $\eta_1 = \eta_2 x$, and call the result $(C_2)$. Repeating the operation, we get systems $(C_3)$, $(C_4)$, ..., $(C_r)$, ... In any case, this can be done as long as the new system $(C_r)$ has the same form as the original system (C). All the systems $(C_1)$, $(C_2)$, ... resemble system (C) in that they have exactly one path tending to the origin in each of the directions $\frac{\pi}{2}$ and $\frac{3}{2}\pi$; moreover, in regard to system $(C_r)$ these paths are clearly the two halves of the axis $x = 0$ or segments thereof bounded on one side by the point $O_r(0, 0)$. Furthermore, all other semipaths of system $(C_r)$ that tend to the origin do so in the direction $0$ or $\pi$. It follows from these properties of systems $(C_r)$ and from Lemma 1 that there is a one-to-one correspondence between the semipaths of system (C) that tend to the equilibrium state $O$ in the directions $0$ and $\pi$, on the one hand, and the semipaths of each of systems $(C_r)$ that tend to the points $O_r$ in the directions $0$ and $\pi$, on the other.

Let us examine the form of systems $(C_r)$ more closely. To this end, we shall use the form $(D)$ of the original system. Simple manipulations show that application of the transformation $x = x$, $y = \eta_1 x$; $x = x$, $\eta_1 = \eta_2 x$; ... $x = x$, $\eta_{r-1} = \eta_r x$ to system $(D)$ yields

$$\frac{dx}{dt} = \psi(x) + x^r \left[ \overline{\eta_r} - \frac{\varphi(x)}{x^r} \right] \cdot \overline{P}(x, \eta_r x^r),$$

$$\frac{d\eta_r}{dt} = \left[ \eta_r - \frac{\varphi(x)}{x^r} \right] [1 + \overline{Q}(x, \eta_r x^r)] - r \cdot \eta_r \frac{P_2(x, \eta_r x^r)}{x}. \qquad (D_r)$$

System $(D_r)$ differs from $(C_r)$ only in notation. By assumption, $\psi(x) = \Delta_m x^m + \ldots$, where $m > 2$ and $\Delta_m \leqslant 0$.

As for the function $\varphi(x)$, it follows from the relations $\varphi(0) = \varphi'(0) = 0$ that either $\varphi(x) \equiv 0$ or the series expansion is

$$\varphi(x) = a_0 x^l + a_1 x^{l+1} + \ldots, \qquad (18)$$

where $l > 2$, $a_0 \neq 0$.

We first consider the case in which either $\varphi(x) \equiv 0$ or $l > m$. In this case it is readily verified that systems $(D_1)$, $(D_2)$, ..., $(D_{m-2})$ have the same form as system $(D)$. System $(D_{m-1})$ may be written

$$\frac{dx}{dt} = \Delta_m x^m [1 + f_1(x, \eta_{m-1})],$$

$$\frac{d\eta_{m-1}}{dt} = \eta_{m-1} - ax + g_2(x, \eta_{m-1}), \qquad (19)$$

where $a$ may be equal to 0, $f_1(0, 0) = 0$, and the expansion of the function $g_2(x, \eta_{m-1})$ does not involve linear terms.*

In the neighborhood of the point $(0, 0)$,

$$\frac{1}{1 + f_1(x, \eta_{m-1})} = 1 + \overline{f_1}(x, \eta_{m-1}),$$

where $\overline{f_1}(0, 0) = 0$. Consider the system

$$\frac{dx}{dt} = \Delta_m x^m,$$

$$\frac{d\eta_{m-1}}{dt} = [\eta_{m-1} - ax + g_2(x, \eta_{m-1})] \frac{1}{1 + f_1(x, \eta_{m-1})}$$

or, equivalently, the system

$$\frac{dx}{dt} = \Delta_m x^m,$$

$$\frac{d\eta_{m-1}}{dt} = [\eta_{m-1} - ax + g_2(x, \eta_{m-1})] [1 + \overline{f_1}(x, \eta_{m-1})]. \qquad (20)$$

In the neighborhood of $O(0, 0)$, systems (19) and (20) have the same paths (since they correspond to the same differential equation; see §1.7). But system (19) has the form $(A)$ of Lemma 3. Hence system (20), and therefore also (19), cannot have more than one semipath on the slit (along the axis $x = 0$) plane which tends to the point $O_{m-1}(0, 0)$ and lies in the

---

* System $(D_{m-1})$ always has the form (19). It may happen, however, that even the original system $(D)$ or some $(D_j)$, $1 \leqslant j < m - 1$, has the form (19). It may then be used in the sequel instead of (19).

right half-plane ($x > 0$), or more than one semipath tending to $Q_{m-1}$ and lying in the left half-plane ($x < 0$). But then systems $(D_{m-2})$, $(D_{m-3})$, . . ., $(D_1)$ and finally (D) cannot have two semipaths tending to $O(0, 0)$ in the direction 0 or two semipaths tending to $O(0, 0)$ in the direction $\pi$.  This proves part 2) of our theorem for the case in question.

Now consider the case $l < m$.  Write the function $\varphi(x)$ as

$$\varphi(x) = a_0 x^l + a_1 x^{l+1} + \ldots, \tag{21}$$

where $a_0 \neq 0$.  Again, one easily sees that systems $(D_1)$, $(D_2)$, . . ., $(D_{l-2})$ have the same form as system (C).  Let us write out system $(D_{l-1})$ in detail:

$$\frac{dx}{dt} = \psi(x) + x^{l-1}[\eta_{l-1} - a_0 x - a_1 x^2 - \ldots]\overline{P}(x, \eta_{l-1} x^{l-1}),$$

$$\frac{d\eta_{l-1}}{dt} = [\eta_{l-1} - a_0 x - a_1 x^2 - \ldots][1 + \overline{Q}(x, \eta_{l-1} x^{l-1})] - \tag{$D_{l-1}$}$$

$$- (l-1)\eta_{l-1}\frac{P_2(x, \eta_{l-1} x^{l-1})}{x}.$$

It may happen that the original system (D) or one of the systems $(D_1)$, $(D_2)$, . . ., $(D_{l-1})$ already has the form (19), in which case we proceed as before.

Now suppose that none of these systems is of the form (19).  Consider the system $(D_{l-1})$.  We must determine how many paths of this system can tend to the equilibrium state $O_{l-1}$ in the slit plane.  The directions in which they tend to $O_{l-1}$ are determined from the equation

$$x(\eta_{l-1} - a_0 x) = 0.$$

Only two paths can tend to $O_{l-1}$ in the directions $\frac{\pi}{2}$ and $\frac{3}{2}\pi$, both situated on the axis $x = 0$ (this is established as in the proof of Lemma 3).  We need therefore consider only paths tending to $O_{l-1}$ in directions arctg $a_0$ and $\pi + $ arctg $a_0$.  We first apply to system $(D_{l-1})$ the transformation

$$\eta_{l-1} = \overline{\eta}_l x. \tag{22}$$

The result is a certain system $(\overline{D}_l)$.  Using Lemma 1, we must determine how many paths of system $(\overline{D}_l)$ can tend to the equilibrium state $\overline{O}_l(0, a_0)$ in the slit $(x, \overline{\eta}_l)$-plane.  Apply to system $(D_l)$ the transformation

$$x = x, \quad \overline{\eta}_l = \eta_l + a_0, \tag{23}$$

whose properties are obvious, obtaining from $(\overline{D}_l)$ the system $(D_l)$, and determine how many paths of system $(D_l)$ can tend to the equilibrium state $O_l(0, 0)$ in the slit $(x, \eta_l)$-plane.  Transformations (22) and (23) can be replaced by a single transformation

$$x = x, \quad \eta_{l-1} = (\eta_l + a_0)x, \tag{24}$$

which converts system $(D_{l-1})$ into the system

$$\frac{dx}{dt} = \psi(x) + x^l[\eta_l - a_1 x - a_2 x^2 \ldots]\,\overline{P}(x, (\eta_l + a_0)\,x^l),$$

$$\frac{d\eta_l}{dt} = [\eta_l - a_1 x - a_2 x^2 - \ldots][1 + \overline{Q}(x, (\eta_l + a_0)\,x^l)] -$$

$$-l(\eta_l + a_0)\,\frac{P_2(x, (\eta_l + a_0)\,x^l)}{x}. \qquad (D_l)$$

If system $(D_l)$ has the form (19), the continuation is the same as in the case $l \geqslant m$. Otherwise, we successively apply transformations $x = x, \ \eta_l = (\eta_{l+1} + a_1)\,x$; $x = x, \ \eta_{l+1} = (\eta_{l+2} + a_2)\,x$, etc., obtaining systems $(D_{l+1}), (D_{l+2}), \ldots$ System $(D_{l+k})$ is easily seen to have the form

$$\frac{dx}{dt} = \psi(x) + x^{l+k}[\eta_{l+k} - a_{k+1}x \ldots]\,xP^*(x, \eta_{l+k}),$$

$$\frac{d\eta_{l+k}}{dt} = [\eta_{l+k} - a_{k+1}x - \ldots][1 + xQ^*(x, \eta)] - R^*(x, \eta_{l+k}). \qquad (D_{l+k})$$

Here $P^*$ and $Q^*$ are analytic in a neighborhood of the origin and

$$R^*(x, \eta_{l+k}) = h(x, \eta_{l+k})\,\frac{P_2(x, (a_0 + a_1 x + \ldots + a_k x^k + \eta_{l+k} x^k)\,x^l)}{x^{k+1}}, \qquad (25)$$

where $h(x, \eta_{l+k})$ is a polynomial, whose exact form is unimportant. Recall that $P_2(x, y)$ is the right-hand side of the first equation in the original system (C):

$$\varphi(x) = a_0 x^l + a_1 x^{l+1} + \ldots, \ \psi(x) = \Delta_m x^m + \ldots = P_2(x, \varphi(x)).$$

Expressing the numerator of the fraction in (25) as

$$P_2(x, \varphi(x)) + [P_2(x, a_0 x^l + \ldots + a_k x^{k+l} + \eta_{l+k} x^{k+l}) -$$
$$- P_2(x, a_0 x^l + a_1 x^{l+1} + \ldots + a_k x^{k+l} + a_{k+1} x^{k+l+1} + \ldots)],$$

and applying Taylor's formula to the differences in square brackets, we see that all terms of the numerator involve $x$ to degree at least $\min\{m, k + l + 1\}$.

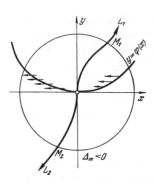

FIGURE 228.

Therefore, if $1 \leqslant k \leqslant m - l - 1$,[*] all terms in $R^*(x, \eta_{l+k})$ involve the factor $x$ to degree at least 2 (since $l \geqslant 2$). Hence it follows, in view of the form of the right-hand side of system $(D_{l+k})$, that at least one of the systems $(D_{l+1}), (D_{l+2}), \ldots, (D_{m-1})$ has the form (19). One can then continue as in the case $l \geqslant m$. This completes the proof of part 2).

3) If $m$ is even and $\Delta_m < 0$, the vector field on the curve $y = \varphi(x)$ has the direction shown in Figure 228, while if $\Delta_m > 0$ it has the opposite direction. Thus, if $\Delta_m < 0$ ($\Delta_m > 0$), there are no positive semipaths inside the sector $OM_1M_2$ ($OM_2M_1$) that tend to the point $O$ — this is proved exactly as in the proof of

---

[*] If $m = l + 1$, we have $k = 0$, since then the system $(D_l)$ itself is readily seen to have the form (19).

part 1) — and this is therefore a parabolic sector. Inside the sector $OM_2M_1$ $(OM_1M_2)$ there cannot be two semipaths that tend to $O$— this is proved as in the proof of part 2) — and so this sector is the union of two hyperbolic sectors. Q.E.D.

## §22. TOPOLOGICAL STRUCTURE OF A MULTIPLE EQUILIBRIUM STATE IN THE CASE $\sigma = 0$

**1. AUXILIARY LEMMAS.** In this section we shall consider an isolated equilibrium state $O(0, 0)$ of an analytic system

$$\frac{dx}{dt} = ax + by + P_2(x, y), \qquad \frac{dy}{dt} = cx + dy + Q_2(x, y) \qquad (1)$$

on the assumption that the following conditions hold

$$|a| + |b| + |c| + |d| \neq 0, \qquad (2)$$
$$\sigma = a + d = 0, \qquad (3)$$
$$\Delta = ad - bc = 0, \qquad (4)$$

in other words, we have an isolated multiple equilibrium state with zero characteristic roots and nontrivial linear terms on the right. As in the preceding section, we may confine ourselves without loss of generality to the special case*

$$\frac{dx}{dt} = y + P_2(x, y), \qquad \frac{dy}{dt} = Q_2(x, y). \qquad (5)$$

As before, $P_2(x, y)$ and $Q_2(x, y)$ are analytic in the neighborhood of $O(0, 0)$, and their series expansions involve only terms of at least second order. System (5) can be simplified. Consider the transformation

$$\xi = x, \qquad \eta = y + P_2(x, y). \qquad (6)$$

Since the Jacobian of this transformation at $O(0, 0)$ is unity, it maps a certain neighborhood of $O(0, 0)$ on the $(x, y)$-plane in one-to-one fashion onto a neighborhood of $\bar{O}(0, 0)$ on the $(\xi, \eta)$-plane, in such a way that $O$ is mapped onto $\bar{O}$. The inverse transformation has the form

$$x = \xi, \qquad y = f(\xi, \eta), \qquad (7)$$

where $f$ is analytic, $f(0,0) = 0.$** System (5) is reduced by transformation (6) to the system

---

* If $a \neq 0$, it follows from (3) and (4) that $b \neq 0$, $c \neq 0$, and system (1) can be brought to the form (5) by the transformation $\bar{x} = -y$, $\bar{y} = -cx + ay$. If $a = 0$ but $b \neq 0$, system (5) is obtained by a simple time transformation $\bar{t} = bt$. Finally, if $a = 0$, $b = 0$, then $c \neq 0$, and we need the transformation $\bar{x} = y$, $\bar{y} = cx$.

** This follows directly from the implicit function theorem (see Appendix, §4.3).

$$\frac{d\xi}{dt} = \eta,$$

$$\frac{d\eta}{dt} = Q_2(\xi, f(\xi, \eta)) + P'_{2x}(\xi, f(\xi, \eta))\eta + P'_{2y}(\xi, f(\xi, \eta)) Q_2(\xi, f(\xi, \eta)).$$

The function on the right of the second equation is analytic and its expansion in powers of $\xi$, $\eta$ involves only terms of at least second order. Denote the function by $\widetilde{Q}_2(\xi, \eta)$. Thus system (5) has been reduced to the form

$$\frac{d\xi}{dt} = \eta, \qquad \frac{d\eta}{dt} = \widetilde{Q}_2(\xi, \eta). \tag{8}$$

The paths of system (6) passing through a sufficiently small neighborhood of $O$ correspond in one-to-one fashion to the paths of system (8) near $\widetilde{O}$. Moreover, $\widetilde{O}$ is clearly also an isolated equilibrium state of system (8), corresponding to the point $O$.

Let $L^+$ be a semipath of system (5), positive say, tending to the equilibrium state $O$ in a definite direction $\theta$. By Theorem 64, $\theta$ is either 0 or $\pi$, i.e., $\mathrm{tg}\ \theta = 0$. Then the corresponding semipath $\widetilde{L}^+$ of system (8) tends to $\widetilde{O}$, also in a definite direction (0 or $\pi$). Indeed, if $x = x(t)$, $y = y(t)$ are the equations of $L^+$ and $\xi = \xi(t)$, $\eta = \eta(t)$ those of $\widetilde{L}^+$, then by assumption, as $t \to +\infty$,

$$\lim x(t) = \lim y(t) = 0, \qquad \lim \frac{y(t)}{x(t)} = \mathrm{tg}\ \theta = 0.$$

It follows from (6) that as $t \to +\infty$

$$\lim \xi(t) = \lim \eta(t) = 0,$$

and

$$\lim = \frac{\eta(t)}{\xi(t)} = \lim \frac{y + P_2(x, y)}{x} = 0.$$

This proves our assertion. Hence it follows, in particular, that if $O$ is a node or focus of system (5), then $\widetilde{O}$ is a node or focus, respectively, of system (8). Thus, we may study system (8) instead of (5). Returning to the original notation, we write the system

$$\frac{dx}{dt} = y, \qquad \frac{dy}{dt} = Q_2(x, y). \tag{9}$$

Let us apply to system (9) the transformation

$$x = x, \qquad y = \eta x,$$

considered in §21. The result is

$$\frac{dx}{dt} = \eta x, \qquad \frac{d\eta}{dt} = \frac{Q_2(x, \eta x)}{x} - \eta^2. \tag{10}$$

It is immediately clear that this system has a unique equilibrium state $\widetilde{O}(0, 0)$ on the axis $x = 0$, and that the semi-axes $x = 0$, $\eta > 0$ and $x = 0$, $\eta < 0$ are positive and negative semipaths, respectively, of system (10).

Let $u$ be a [circular] neighborhood of the equilibrium state $O(0, 0)$ and $\Gamma$ the corresponding region on the $(x, \eta)$-plane (see §21.1b). Suppose there exist in the $(x, \eta)$-plane semipaths $\tilde{L}_1^{()}$ and $\tilde{L}_2^{()}$ of system (10) which tend to the equilibrium state $\tilde{O}$ and lie respectively to the right and left of the axis $x = 0$. We shall assume that the semipaths cross the boundary of $\Gamma$ at points $\tilde{N}_1$ and $\tilde{N}_2$, in such a way that all their points between $\tilde{O}$ and $\tilde{N}_1$, ($\tilde{O}$ and $\tilde{N}_2$) are interior points of $\Gamma$.

It follows from §21.1 that the semipaths $\tilde{L}_1^{()}$ and $\tilde{L}_2^{()}$ correspond to semipaths $L_1^{()}$ and $L_2^{()}$ of system (9) on the $(x, y)$-plane which lie to the right and left, respectively, of $x = 0$ and tend to the equilibrium state $O$ in directions $\theta = 0$ and $\theta = \pi$ (Figure 229a, b).

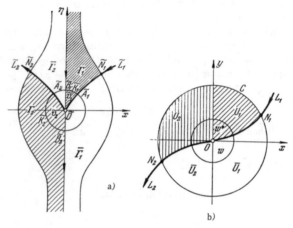

a)

b)

FIGURE 229.

Let $N_1$ and $N_2$ be the points corresponding to $\tilde{N}_1$ and $\tilde{N}_2$ on the bounding circle $C$ of the neighborhood $u$. Let $\Gamma_1$ denote the subregion of $\Gamma$ bounded by the semipath $\eta > 0$, $x = 0$, the arc $\tilde{N}_1\tilde{O}$ of the semipath $\tilde{L}_1^{()}$ and the corresponding portion of the boundary of $\Gamma$ (Figure 229a). Similarly, let $\Gamma_2$ denote the subregion of $\Gamma$ bounded by the semipath $\eta < 0$, $x = 0$, the semipath $\tilde{L}_2^{()}$ and the boundary of $\Gamma$. The regions $\Gamma_1$ and $\Gamma_2$ are the images of two curvilinear sectors, $u_1$ and $u_2$ say, on the $(x, y)$-plane.

Let $\tilde{v}$ be a small neighborhood of $\tilde{O}$, lying entirely within $\Gamma$, and $\tilde{v}_1$, $\tilde{v}_2$ the intersections of $\tilde{v}$ with $\Gamma_1$ and $\Gamma_2$, respectively: $\tilde{v}_1 = \tilde{v} \cap \Gamma_1$; $\tilde{v}_2 = \tilde{v} \cap \Gamma_2$. Let $w$ be a small neighborhood of $O$ and $w^*$ its intersection with the union of $u_1$, $u_2$ and the positive $y$-axis.

Lemma 1. A) If $\tilde{v}_1$ and $\tilde{v}_2$ are parabolic sectors of the equilibrium state $\tilde{O}$, then $w^*$ is the union of one elliptic region and two parabolic sectors (bordering on the elliptic region and the semipaths $L_1^{()}$ and $L_2^{()}$).

B) If $\tilde{v}_1$ and $\tilde{v}_2$ are hyperbolic regions, then $w^*$ is a hyperbolic region.

C) If one of the regions $\tilde{v}_1$, $\tilde{v}_2$ is hyperbolic and the other parabolic, then $w^*$ is a parabolic region.

Proof. A) Note first that all paths of system (9) that cross the positive $y$-axis near $O$ do so from left to right (with increasing $t$), almost horizontally. This follows at once from examination of system (9) $\left(\dfrac{dx}{dt} > 0,\right.$ $\dfrac{dy}{dx}$ small$\left.\right)$.

Let $\tilde{v}_1$ and $\tilde{v}_2$ be parabolic regions. Then $\tilde{L}_1^{()}$, hence also $L_1^{()}$, are positive semipaths, while $\tilde{L}_2^{()}$ and $L_2^{()}$ are negative semipaths. The mapping (10) maps the region $\tilde{v}_1$, bounded by the curvilinear triangle $\tilde{O}\tilde{A}_1\tilde{K}_1\tilde{B}_1$, onto the region $v_1$ bounded by the "loop" $OA_1K_1O$, since the segment $\tilde{O}\tilde{B}_1$ on the $\eta$-axis is mapped onto the point $O$ (Figure 230). Since $\tilde{A}_1\tilde{K}_1\tilde{B}_1$ is an arc without contact

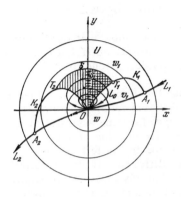

FIGURE 230.

and the mapping (10) has Jacobian $\begin{vmatrix} 1 & 0 \\ \eta & x \end{vmatrix} = x,$ which does not vanish at any point of this arc other than $\tilde{B}_1$, it follows that the corresponding portion $A_1K_1O$ of the "loop" is also an arc without contact (excluding the point $O$; see Appendix, §6.4).

Note that the region $v_1$, as the image of $\tilde{v}_1$, has the following properties:

1) Every path through an interior point of $v_1$ remains in $v_1$ with increasing $t$ and tends to the equilibrium state $O$ as $t \to +\infty$.

2) Conversely, if $L$ is any semipath of system (9) in the right half-plane of the $(x, y)$-plane, lying above $L_1$ and tending to $O$, then its preimage must be in $\tilde{v}_1$. But then $L^{()}$ is a positive semipath and therefore remains inside $v_1$ from some time on.

Let $w_1$ be a sufficiently small neighborhood of $O$. Let $ST_1$ be a simple arc satisfying the following conditions: I) its endpoint $S$ is on the positive $y$-axis, its endpoint $T_1$ on the arc without contact $A_1K_1O$; II) $ST_1$ has no points in common with the curve $A_1K_1O$ and the $y$-axis other than its endpoints; III) $ST_1$ is inside the neighborhood $w_1$, sufficiently near $O$, and (except for $S$) entirely in the right half-plane. The existence of such an arc $ST_1$ was proved in §18, Lemma 2.

Consider the curvilinear triangle $OT_1S$, cross-hatched in Figure 230. All paths through interior points of the segment $OS$ of the $y$-axis enter this triangle with increasing $t$. Let $L$ be one of these paths. $L$ cannot remain within the triangle with increasing $t$, for otherwise it would have to tend to the equilibrium state $O$; then, by properties 1) and 2) of the region $v_1$, $L$ would enter this region, contrary to assumption.

Thus, all paths crossing the segment $OS$ of the $y$-axis at its interior points enter the triangle $OT_1S$ and leave it with increasing $t$.

We claim that at least one of these paths leaves the triangle through a point of the arc $T_1O$ of the curve $A_1K_1T_1O$. Indeed, if this is not true, all paths $L$ leave the triangle $OT_1S$ through the arc $T_1S$. Consider the first points of these paths in common with the arc $T_1S$, and let $P$ be one of their points of accumulation ($P$ may coincide with $T_1$), and $L_P$ the path passing through $P$, at time $t = t_0$ say. At times $t < t_0$ the path $L_P$ cannot cross the segment $OS$ of the $y$-axis, since otherwise it would be one of the paths $L$; neither can $L_P$ cross the arc $OA_1K_1T_1$ with decreasing $t$, as follows from properties 1) and 2). But then this path must either a) lie outside the triangle $OT_1S$ for all $t < t_0$, or b) lie inside the triangle for all $t < t_0$. The first case is impossible by virtue of the behavior of $L$ and the continuity of solutions with respect to initial values. The second is also impossible, because were the path $L_P$ to remain inside $OT_1S$ it would tend to $O$, which is impossible by properties 1) and 2).

Consequently, there exists a path $L_0$ entering the triangle $OT_1S$ through a point $S_1$ on $OS$ and leaving it through a point on the arc $T_1O$ of the curve $A_1K_1T_1O$. With further increase in $t$ this path, remaining inside $v_1$, tends to $O$ as $t \to +\infty$.

It is clear that all paths crossing the $y$-axis at points of $OS$ behave in this way.

Treating the region $v_2$ in analogous fashion, one can show that under our conditions all paths crossing the segment $OS$ of the $y$-axis sufficiently near $O$ enter $v_2$ with decreasing $t$ and then, remaining inside $v_2$, tend to $O$ as $t \to -\infty$.

We may assume without loss of generality that the above-mentioned path $L_0$ through the point $S_1$ also has this property, so that it forms a loop. All paths interior to this loop must be of the same type, forming a family of nested loops. Indeed, by 1) and 2) there cannot be loops lying entirely to the right of the $y$-axis. Similarly, there are no loops to the left of the $y$-axis. Hence it follows that there cannot be two different elliptic regions between the paths $L_1$ and $L_2$. But then it is clear that any sufficiently small neighborhood of $w$ of the point $O$ satisfies the first assertion of our lemma.

B) Suppose now that $\tilde{v}_1$ and $\tilde{v}_2$ are regular hyperbolic regions. Then $\tilde{L}_1^{()}$ and $L_2^{()}$ are negative semipaths, $\tilde{L}_2^{()}$ and $L_2^{()}$ positive semipaths (Figure 231). We claim that if $w$ is a sufficiently small neighborhood of $O$, then its subregion $w^*$ cut out by the semipaths $L_1^{()}$ and $L_2^{()}$ is a hyperbolic region. Indeed, if there is a semipath through a point of $w^*$ which tends to $O$, then, for sufficiently large $t$, it will lie entirely within one of the curvilinear sectors $u_1$ or $u_2$. The corresponding semipath on the $(x, \eta)$-plane tends to $\tilde{O}$ and, for sufficiently large $t$, will lie entirely in the sector $\tilde{v}_1$ or $\tilde{v}_2$. But this contradicts the assumption that $\tilde{v}_1$ and $\tilde{v}_2$ are hyperbolic sectors. Consequently, all paths through points of $w^*$ leave this region, and $w^*$ is a hyperbolic region.

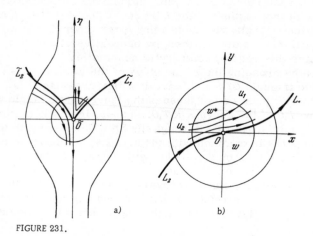

a)  b)

FIGURE 231.

C) In the third case, one of the regions, say $\tilde{v}_1$, is parabolic, and the other, $\tilde{v}_2$, is hyperbolic. Then $\tilde{L}_1^{()}$ and $L_1^{()}$, as well as $\tilde{L}_2^{()}$ and $L_2^{()}$, are positive semipaths. The previous arguments imply that any path passing through $w^*$ tends to $O$ in the direction $\theta = 0$ with increasing $t$ (as in case A)), and leaves $w^*$ with decreasing $t$ (as in case B)). Consequently, $w^*$ is a parabolic region (Figure 232). This completes the proof of the lemma.

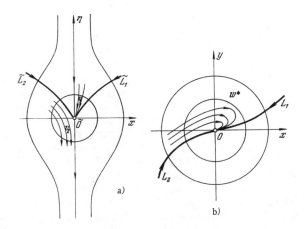

a)

b)

FIGURE 232.

Remark 1. Instead of the regions $\Gamma_1$ and $\Gamma_2$, we could have considered the regions $\bar{\Gamma}_1$ and $\bar{\Gamma}_2$ on the $(x, \eta)$-plane and the corresponding regions $\bar{u}_1$ and $\bar{u}_2$ on the $(x, y)$-plane (Figure 229). Lemma 1 clearly remains valid (with obvious modifications) for these regions.

Now consider the system

$$\frac{dx}{dt} = x \cdot X (x, y), \qquad \frac{dy}{dt} = Y (x, y), \tag{11}$$

where $X$ and $Y$ are analytic in the neighborhood of $O(0, 0)$, $X(0, 0) = Y(0, 0) = 0$. The point $O(0, 0)$ is an isolated equilibrium state, the series expansion of $Y$ involves only terms of at least second order and at least one term of exactly second order. It follows from the form of system (11) and the fact that the equilibrium state is isolated that the semi-axes $y > 0$, $x = 0$ and $y < 0$, $x = 0$ of the $y$-axis (or segments thereof bordering on the point $O$) are semipaths of system (11) tending to $O$ in directions $\theta = \frac{\pi}{2}$ and $\theta = \frac{3}{2}\pi$, respectively. We shall assume that these are the only such semipaths of system (11). In addition, we shall assume that there exist exactly four directions, $\theta = \text{arctg } k_1$, $\theta = \pi + \text{arctg } k_1$, $\theta = \text{arctg } k_2$, $\theta = \pi + \text{arctg } k_2$, in each of which there is at least one semipath of system (11) tending to $O$.

Let us apply the transformation $x = x$, $y = \eta x$ to system (11). The result is

351

$$\frac{dx}{dt} = x \cdot X(x, \eta x),$$

$$\frac{d\eta}{dt} = \frac{Y(x, \eta x)}{x} - \eta X(x, \eta x). \tag{12}$$

By the properties of $X$ and $Y$, the right-hand sides of system (12) have a common factor $x$. Suppose that the function on the right of the second equation in (12) is not divisible by $x^2$. Transforming variables by $x \cdot dt = d\tau$, we obtain the system

$$\frac{dx}{d\tau} = X(x, \eta x),$$

$$\frac{d\eta}{d\tau} = \frac{Y(x, \eta x)}{x^2} - \eta \frac{X(x, \eta x)}{x}. \tag{13}$$

Since the function on the right of the first equation in (13) has the factor $x$, while that on the second does not (by assumption), it follows that the axis $x = 0$ is not a singular curve but a union of paths of the system, and the equilibrium states on the axis are isolated (by the assumption that the functions on the right are analytic; see Introduction to Chapter IV).

It follows from Lemma 1 of §21 and our assumptions that the points $\tilde{O}_1 (0, k_1)$ and $\tilde{O}_2 (0, k_2)$ are equilibrium states of system (13).

Remark 2. System (12) is obtained from system (1) by the transformation $x = x$, $y = \eta x$. Let $L$ be a path of system (11) in the slit plane, and $\tilde{L}$ the corresponding path of system (12), which is at the same time a path of system (13) ($\tilde{L}$ is also on the slit plane). As a point $M$ moves along the path $L$ in the positive direction (increasing $t$), the corresponding point $\tilde{M}$ on $\tilde{L}$ also moves in the positive direction, provided $\tilde{L}$ is treated as a path of system (12). But if $\tilde{L}$ is treated as a path of system (13), $\tilde{M}$ moves in the positive direction only provided $L$ (therefore also $\tilde{L}$) lies in the right half-plane ($x > 0$), and in the negative direction if $L$ is in the left half-plane ($x < 0$). This follows from the fact that the functions on the right of system (12) involve a factor $x$ not present in system (13).

Lemma 2. Let $\tilde{O}_1 (0, k_1)$ be a simple saddle point of system (13). Then: 1) If the equilibrium state $\tilde{O}_2 (0, k_2)$ is a node, the canonical neighborhood of the equilibrium state $O (0, 0)$ is the union of two hyperbolic sectors and two parabolic sectors. 2) If $\tilde{O}_2 (0, k_2)$ is a saddle point, two of whose separatrices lie on different sides of the $\eta$-axis, then the canonical neighborhood is the union of six hyperbolic sectors. 3) If $\tilde{O}_2 (0, k_2)$ is a saddle node, both of whose saddle regions lie on one side of the $\eta$-axis, the canonical neighborhood of $O$ is the union of four hyperbolic sectors and one parabolic sector.

Proof. Let $u$ be a small neighborhood of $O$ and $\Gamma$ the corresponding region (see §21.1 and Figure 222). We shall assume that $u$ contains no equilibrium states other than $O$. Then all equilibrium states of system (13) in $\Gamma$ lie on the $\eta$-axis. By Lemma 1 of §21 and our assumptions on system (11), any semipath $L^{()}$ of system (11) in the slit $(x, y)$-plane which tends to the point $O$ corresponds to a semipath $\tilde{L}^{()}$ in the slit $(x, \eta)$-plane tending to $\tilde{O}_1$ or $\tilde{O}_2$; conversely, to every semipath $\tilde{L}^{()}$ tending to $\tilde{O}_1$ or $\tilde{O}_2$ corresponds a semipath $L^{()}$ tending to $O$. Since $\tilde{O}_1 (0, k_1)$ is a saddle point, it has four separatrices. Two of them are segments of the $\eta$-axis. The other two lie on different sides of the $\eta$-axis; this follows from the fact that by assumption system (11) has semipaths tending to $O$ in both directions $\theta = \text{arctg } k_1$ and $\theta = \pi + \text{arctg } k_1$. Denote these separatrices by

393

$\tilde{L}_1^{()}$ and $\tilde{L}_2^{()}$, and the corresponding semipaths on the $(x, y)$-plane by $L_1^{()}$ and $L_2^{()}$. To complete the proof of the lemma, it now suffices to consider, in each of the above three cases, the relative positions of the paths in the region $\Gamma$ and their images under the transformation $x = x$, $y = \eta x$. We leave the details to the reader. The arguments involve the properties of the transformation and Remark 2 above.

Figures 233, 234 and 235 illustrate cases 1), 2) and 3), respectively, on the assumption that $k_1 > k_2$. We now proceed to the next lemmas.

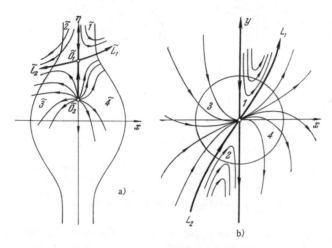

FIGURE 233.

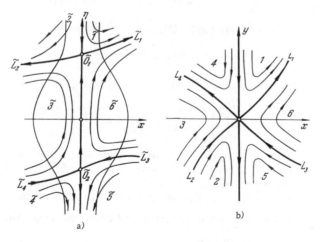

FIGURE 234.

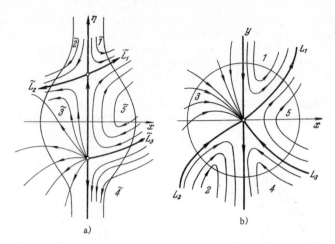

FIGURE 235.

*Lemma 3. The system*

$$\frac{dx}{dt} = xy,$$

$$\frac{dy}{dt} = -ky^2 + x^2 f(x, y) + xy f_1(x, y),$$

(14)

*where $f(x, y)$ and $f_1(x, y)$ are analytic in the neighborhood of the origin, $k > 0$, and $f(x, 0) \not\equiv 0$, has exactly two semipaths tending to the point $O(0, 0)$ in the directions $\theta = \frac{\pi}{2}$ and $\theta = \frac{3}{2}\pi$, namely, the positive and negative $y$-axes.*

Proof. Applying the transformation $x = \xi y$, $y = y$ to system (14) and transforming parameters by $y \, dt = d\tau$, we obtain the system

$$\frac{d\xi}{d\tau} = (1 + k)\xi - \xi^3 f(\xi y, y) - \xi^2 f_1(\xi y, y),$$

$$\frac{dy}{d\tau} = -ky + \xi^2 y f(\xi y, y) + \xi y f_1(\xi y, y).$$

(15)

For the equilibrium state $\tilde{O}(0, 0)$ of this system, we have

$$\Delta = \begin{vmatrix} 1+k & 0 \\ 0 & -k \end{vmatrix} = -k(1+k) < 0,$$

i. e., $\tilde{O}(0, 0)$ is a simple saddle point of system (15). The separatrices are $\xi = 0$ and $y = 0$ or segments of these semi-axes bordering on the point $\tilde{O}$. System (15) has no other semipaths that tend to $\tilde{O}$. By Lemma 2 of §21, there exist exactly two semipaths of system (14) that tend to $O$ in the directions $\frac{\pi}{2}$ and $\frac{3}{2}\pi$. Since the positive and negative $y$-axes have this property, the proof is complete.

*Lemma 4. Let $f(x, y)$ and $\varphi(x)$ be analytic in the neighborhood of $O(0, 0)$, $a \neq 0$; $b < 0$ and $\varphi(0) = 0$. Then the system*

354

$$\frac{dx}{dt} = xy,$$

$$\frac{dy}{dt} = ax\,[1 + \varphi\,(x)] + by^2 + xyf\,(x, y) \qquad (16)$$

has a saddle point at $O\,(0, 0)$, whose separatrices are the positive and negative $y$-axes and two other semipaths which also tend to $O\,(0, 0)$ in the directions $\frac{\pi}{2}$ and $\frac{3}{2}\pi$; the latter lie respectively in the first and fourth quadrants if $a > 0$ (Figure 236), in the second and third if $a < 0$.

Proof. By Theorem 64, there exist exactly two directions in which paths of system (16) can tend to the equilibrium state $O\,(0, 0)$ — the directions $\theta = \frac{\pi}{2}$ and $\theta = \frac{3}{2}\pi$. The semi-axes $x = 0$, $y > 0$ and $x = 0$, $y < 0$ have this property. To determine all such semipaths, we first apply the transformation $x = \xi y$, $y = y$ to system (16), obtaining

$$\frac{d\xi}{dt} = (1 - b)\,\xi y - a\xi^2 - a\xi^2\varphi\,(\xi y) - \xi^2 yf\,(\xi y, y),$$

$$\frac{dy}{dt} = a\xi y + by^2 + a\xi y\varphi\,(\xi y) + \xi y^2 f\,(\xi y, y). \qquad (17)$$

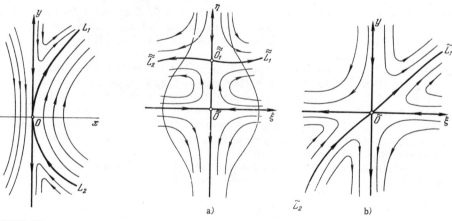

FIGURE 236.          FIGURE 237.

a)          b)

The point $\tilde{O}\,(0, 0)$ is an isolated equilibrium state for this system. By Lemma 2, in order to determine all semipaths of system (16) tending to $O$ we must find all paths of system (17) that tend to $\tilde{O}$ and lie in the $(\xi, y)$-plane slit along the axis $y = 0$. We shall try to find all paths of system (17) that tend to $\tilde{O}$. It is clear that the semi-axes $\xi = 0$, $y > 0$ and $\xi = 0$, $y < 0$ are semipaths of system (17) and tend to $\tilde{O}$ in the directions $\theta = \frac{\pi}{2}$ and $\theta = \frac{3}{2}\pi$. Applying the transformation $x = \xi y$, $y = y$ to system (17), one shows exactly as in the proof of Lemma 3 that these are the only semipaths of system (17) tending to $\tilde{O}$ in these directions. We now determine the

355

semipaths of system (17) that tend to $\tilde{O}$ in other directions. To this end, we apply to system (17) the transformation $\xi = \xi$, $y = \eta\xi$ and transform the parameter by $\xi \, dt = d\tau$, obtaining

$$\frac{d\xi}{d\tau} = -a\xi + (1-b)\,\xi\eta - a\xi\varphi\,(\xi^2\eta) - \xi^2\eta f\,(\xi^2\eta,\ \xi\eta),$$

$$\frac{d\eta}{d\tau} = 2a\eta - (1-2b)\,\eta^2 + 2a\eta\varphi\,(\xi^2\eta) + 2\xi\eta^2 f\,(\xi^2\eta,\ \xi\eta). \tag{18}$$

It is easy to see that system (18) has exactly two equilibrium states on the axis $\xi = 0$: $\tilde{O}\,(0,\ 0)$ and $\tilde{O}_1\left(0,\frac{2a}{1-2b}\right)$. For the first, $\Delta$ is equal to $-2a^2 < 0$, for the second $\Delta = -\frac{2a^2}{1-2b} < 0$. Consequently, both points are simple saddle points. All four separatrices of the saddle point $\tilde{O}$ are segments of the coordinate axes $\xi = 0$ and $\eta = 0$ (since these axes are unions of paths of system (18)). Two separatrices of the saddle point $\tilde{O}_1$ are also on the axis $\xi = 0$, while the other two, $\tilde{L}_1$ and $\tilde{L}_2$ say, lie on either side of this axis. Thus the position of the paths of system (18) near the $\eta$-axis is as illustrated in Figure 237a for the case $a > 0$.

We can now apply Lemma 2. System (18) satisfies the assumptions of the second part of Lemma 2, and therefore the canonical neighborhood of the equilibrium state $\tilde{O}$ of system (17) is the union of six hyperbolic regions. Thus the point $\tilde{O}$ is approached by exactly six semipaths, four of which are the halves of the coordinate axes $\xi = 0$ and $y = 0$. The other two correspond to the semipaths $\tilde{L}_1$ and $\tilde{L}_2$; denote them by $\tilde{L}_1$ and $\tilde{L}_2$. When $a > 0$ the configuration of paths in the neighborhood of $\tilde{O}$ is as illustrated in Figure 237b.

System (17) was obtained from system (16) by the transformation $x = \xi y$, $y = y$. It follows from the properties of this transformation and from Lemma 2 of §21 that there are exactly four semipaths of system (16) tending to the equilibrium state $O$. Two of these are the halves of the $y$-axis, the other two, $L_1^{()}$ and $L_2^{()}$, correspond to $\tilde{L}_1^{()}$ and $\tilde{L}_2^{()}$; their position is as in Figure 236 if $a > 0$, symmetrically opposite if $a < 0$. This completes the proof of our lemma.

## 2. TOPOLOGICAL STRUCTURE OF A MULTIPLE EQUILIBRIUM STATE IN THE CASE $\sigma = 0$.

We now return to the original system (9):

$$\frac{dx}{dt} = y, \qquad \frac{dy}{dt} = Q_2\,(x,\ y).$$

Since the point $O\,(0,\ 0)$ is by assumption an isolated equilibrium state, we can write the system as

$$\frac{dx}{dt} = y,$$

$$\frac{dy}{dt} = a_k x^k\,[1 + h\,(x)] + b_n x^n y\,[1 + g\,(x)] + y^2 f\,(x,\ y), \tag{A}$$

where $h\,(x)$, $g\,(x)$, $f\,(x,\ y)$ are analytic in the neighborhood of the origin, $h\,(0) = g\,(0) = 0$, $k > 2$, $a_k \neq 0$. The coefficient $b_n$ may vanish; if $b_n \neq 0$, then $n \geqslant 1$.

The possible topological structure of the equilibrium state $O(0, 0)$ of system (A) is established in Theorems 66 and 67 below. An equilibrium state whose canonical neighborhood is the union of two hyperbolic sectors will be called a d e g e n e r a t e   e q u i l i b r i u m   s t a t e. If the canonical neighborhood of $O$ consists of one hyperbolic and one elliptic sector, we shall call $O$ an e q u i l i b r i u m   s t a t e   w i t h   e l l i p t i c   r e g i o n.

*T h e o r e m   66. Let the number $k$ in system (A) be odd, $k = 2m + 1$ $(m \geqslant 1)$, and $\lambda = b_n^2 + 4(m + 1) a_{2m+1}$.*

*Then if $a_{2m+1} = a_k > 0$, the equilibrium state $O$ of system (9) is a topological saddle point (Figure 238). But if $a_k < 0$, the point $O$ is: 1) a focus or center if $b_n = 0$, or if $b_n \neq 0$ and $n > m$, or if $b_n \neq 0$, $n = m$ and $\lambda < 0$; 2) a topological node if $b_n \neq 0$, $n$ is even and $n < m$, or if $b_n \neq 0$, $n$ is even, $n = m$ and $\lambda > 0$; 3) an equilibrium state with elliptic region if $b_n \neq 0$, $n$ is odd and $n < m$, or if $b_n \neq 0$, $n$ is odd, $n = m$ and $\lambda > 0$ (Figure 239).*

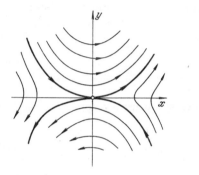

FIGURE 238.

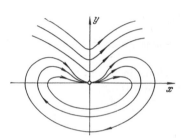

FIGURE 239.

R e m a r k.   Figure 239 illustrates the case $b_n > 0$; if $b_n < 0$ the phase portrait is obtained by reflection in the $x$-axis.

P r o o f.   There are exactly two directions in which paths of system (A) can tend to the equilibrium state $O(0, 0)$; these are the directions 0 and $\pi$. We apply the successive transformations $x = x$, $y = \eta_1 x$; $x = x$, $\eta_1 = \eta_2 x$; ...; $x = x$, $\eta_{r-1} = \eta_r x$, eliminating in every intermediate system from the second on (i. e., after the transformation $x = x$, $\eta_1 = \eta_2 x$) the common factor $x$ appearing on the right of the system.* We obtain a series of systems $(A_1)$, $(A_2)$, ..., $(A_r)$; easy computations show that $(A_r)$ is the system

$$\frac{dx}{dt} = \eta_r x,$$
$$\frac{d\eta_r}{dt} = -r\eta_r^2 + a_{2m+1} x^{2m-2r+2}[1 + h(x)] + b_n x^{n-r+1}\eta_r[1 + g(x)] + \eta_r^2 x \cdot f(x, \eta_r x^r). \qquad (A_r)$$

We consider several cases.

---

* This means that a new parameter is introduced at each stage, but we shall retain the same notation for the parameters.

1) Assume that either $b_n = 0$, or $b_n \neq 0$ and $n > m$. For all $r$ such that $1 \leqslant r \leqslant m - 1$ we have

$$2m - 2r + 2 > 4, \qquad n - r + 1 \geqslant n - m + 2 \geqslant 3.$$

It follows from Theorem 64 that for each such $r$ there are exactly four directions in which paths of system $(A_r)$ can tend to the equilibrium state $O_r (0, 0)$: $0, \ \pi, \frac{\pi}{2}, \frac{3}{2}\pi$. By Lemma 3, for $1 \leqslant r \leqslant m$ there are only two semi-

paths tending to $O_r$ in the directions $\frac{\pi}{2}$ and $\frac{3}{2}\pi$: $x = 0$, $\eta_r > 0$ and $x = 0$, $\eta_r < 0$. Now, by the properties of the transformations and Lemma 1 of §21, to each semipath $L$ of system $(A)$ in the slit plane which tends to $O$ (in the direction $0$ or $\pi$) corresponds a semipath $L_r$ of each of the systems $(A_r)$ $(r = 1, 2, \ldots, m - 1, m)$ in the slit plane which tends to the equilibrium state $O_r$, and moreover this correspondence is one-to-one. Hence it follows that investigation of the equilibrium state $O$ of the original system $(A)$ reduces to that of the equilibrium state $O_m (0, 0)$ of system $(A_m)$:

$$\frac{dx}{dt} = \eta_m x,$$
$$\frac{d\eta_m}{dt} = -m\eta_m^2 + a_{2m+1} x^2 [1 + h(x)] + b_n x^{n-m+1} \eta_m [1 + g(x)] + \eta_m^2 x f(x, \eta_m x^m). \qquad (A_m)$$

The directions in which the paths of this system may approach $O_m (0, 0)$ are determined from the equation

$$x [(m + 1) \eta_m^2 - a_{2m+1} x^2] = 0. \qquad (19)$$

If $a_h = a_{2m+1} < 0$, the semipaths of system $(A_m)$ can tend to $O_m$ only in the directions $\frac{\pi}{2}$ and $\frac{3}{2}\pi$. As we have indicated, there are exactly two such

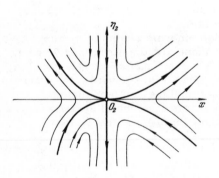

FIGURE 240.

semipaths — the halves of the axis $x = 0$. Consequently, there are no paths on the slit $(x, \eta_m)$-plane that tend to $O_m (0, 0)$. But then none of the systems $(A_{m-1})$, $(A_{m-2})$, ..., $(A_1)$ has such semipaths, and the original system $(A)$ has no semipaths tending to the point $O (0, 0)$ in a definite direction. This means that $O$ is either a focus or a center for system $(A)$.

Now let $a_h = a_{2m+1} > 0$. It follows from equation $(19)$ that in this case there are four additional directions, other than $\frac{\pi}{2}$ and $\frac{3}{2}\pi$, in which the semipaths of system $(A_m)$ can tend to the equilibrium state $O_m$:

$$\text{arctg } \sqrt{\frac{a_{2m+1}}{m+1}}, \quad \pi + \text{arctg } \sqrt{\frac{a_{2m+1}}{m+1}}, \quad \text{arctg}\left(-\sqrt{\frac{a_{2m+1}}{m+1}}\right), \quad \pi + \text{arctg}\left(-\sqrt{\frac{a_{2m+1}}{m+1}}\right).$$

It is immediate that the points $\left(0, \sqrt{\frac{a_{2m+1}}{m+1}}\right)$ and $\left(0, -\sqrt{\frac{a_{2m+1}}{m+1}}\right)$ are simple saddle points for system $(A_{m+1})$. Therefore, by Lemma 2, the canonical neighborhood of the equilibrium state $O_m$ of system $(A_m)$ is the union of six hyperbolic sectors and the point $O_m$ has six separatrices, two of which are the semi-axes $x = 0$, $\eta_m > 0$ and $x = 0$, $\eta_m < 0$. The phase portrait in the neighborhood of $O_m$ and the directions on the separatrices are the same as in Figure 234. The phase portrait of system $(A_r)$ in the neighborhood of $O_r$ $(r = m-1, m-2, \ldots, 2, 1)$ and the directions on the separatrices are shown in Figure 240 (the difference between this and the case $r = m$ is that the separatrices not on the axis $x = 0$ tend to $O_r$ in the directions 0 and $\pi$). Finally, going from system $(A_1)$ to the original system $(A)$ and using Lemma 1 and Remark 2 of §22.1, we see that $O(0, 0)$ is a saddle point of system $(A)$, the phase portrait being as in Figure 238.

2) Now let $b_n \neq 0$, $n < m$. When $r = n$, system $(A_r)$ is

$$\frac{dx}{dt} = \eta_n x,$$

$$\frac{d\eta_n}{dt} = -n\eta_n^2 + b_n x \eta_n [1 + g(x)] + \qquad (A_n)$$

$$+ a_{2m+1} x^{2m-2n+2} [1 + h(x)] + \eta_n^2 x f(x, \eta_n x^n).$$

Here $2m - 2n + 2 > 4$, and so the directions in which semipaths of system $(A_n)$ can tend to the equilibrium state $O_n(0, 0)$ are determined by the equation

$$x \eta_n [\eta_n (1+n) - b_n x] = 0. \qquad (20)$$

This equation yields six directions $\frac{\pi}{2}$, $\frac{3}{2}\pi$, 0, $\pi$, $\operatorname{arctg} \frac{b_n}{1+n}$, $\pi + \operatorname{arctg} \frac{b_n}{1+n}$. As before (by Lemma 3), there are two, and only two, semipaths tending to $O_n$ in the directions $\frac{\pi}{2}$ and $\frac{3}{2}\pi$: $x = 0$, $\eta_n > 0$, and $x = 0$, $\eta_n < 0$. To study the remaining semipaths tending to $O_n$, we consider the system

$$\frac{dx}{dt} = \eta_{n+1} x,$$

$$\frac{d\eta_{n+1}}{dt} = b_n \eta_{n+1} [1 + g(x)] - (n+1) \eta_{n+1}^2 + \qquad (A_{n+1})$$

$$+ a_{2m+1} x^{2m-2n} [1 + h(x)] + \eta_{n+1}^2 x f(x, \eta_{n+1} x^{n+1}).$$

It is readily seen that the equilibrium states of this system are the points $O_{n+1}(0, 0)$ and $\bar{O}(0, \frac{b_n}{1+n})$. Direct computation shows that for the point $\bar{O}$ we have $\Delta = -\frac{b_n}{n+1} < 0$, so that $\bar{O}$ is a simple saddle point of system $(A_{n+1})$. It has two separatrices along the axis $x = 0$, and another two on either side of this axis. For the equilibrium state $O_{n+1}$, we have $\Delta = 0$ and $\sigma = b_n \neq 0$. We can therefore apply Theorem 65. We must first bring system $(A_{n+1})$ to the form (11) of §21. This can be done by introducing a new time parameter, $\tau = b_n t$. The first term in the expansion of the function $\varphi(x)$ in this case (see Theorem 65) is easily found (using undetermined coefficients) to be $-\frac{a_{2n+1}}{b_n} x^{2m-2n}$. Hence the first term in the

expansion of $\psi(x)$ is $-\frac{a_{2m+1}}{b_n^2} x^{2m-2n+1}$. It now follows from Theorem 65 that
the equilibrium state $O_{n+1}$ of system $(A_{n+1})$ is a node if $a_{2m+1} < 0$ and a
saddle point if $a_{2m+1} > 0$. In the latter case, there are two separatrices
along the axis $x = 0$ and the other two lie on different sides of the axis.
Applying Lemma 2, we can now determine the nature of the equilibrium
state for system $(A_n)$. By the second part of Lemma 2, if $a_{2m+1} > 0$ the
canonical neighborhood of $O_n$ is the union of six hyperbolic sectors
(Figure 240). By the first part of the lemma, if $a_{2m+1} < 0$ the canonical
neighborhood of $O_n$ is the union of two hyperbolic and two parabolic
sectors (Figure 241, for the case $b_n > 0$; the situation for $b_n < 0$ is
obtained by reflection in the $\eta_n$-axis). We can now characterize the
equilibrium state $O(0, 0)$ of the original system $(A)$, proceeding
successively through the sequence of systems $(A_{n-1}), (A_{n-2}), \ldots, (A_r), \ldots, (A_1)$,
$(A)$. Note that in the case $n < m$ under consideration the equality
$2m - 2r + 2 > 4$ holds for all $r$, $1 \leqslant r \leqslant n - 1$. Thus the transition from
system $(A_n)$ to $(A_{n-1}), (A_{n-2}), \ldots, (A_1), (A)$ is carried out just as in the
case $n > m$, and we at once conclude that if $a_{2m+1} > 0$ the point $O(0, 0)$ is a
saddle point (the phase portrait in its neighborhood is illustrated in
Figure 238). If $a_{2m+1} < 0$ the parity of the number $n$ comes into play.
Indeed, the transformation $x = x$, $\eta_{r-1} = \eta_r \cdot x$ maps points of the second and
third quadrants in the $(x, \eta_r)$-plane onto points of the third and second
quadrants, respectively, in the $(x, \eta_{r-1})$-plane. Thus the phase portrait of
system $(A_{n-1})$ in the neighborhood of $O_{n-1}$ is that shown in Figure 242
if $b_n < 0$; the phase portrait for $b_n > 0$ is obtained by reflection in the
$x$-axis.

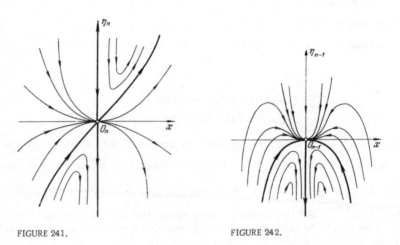

FIGURE 241.                    FIGURE 242.

It is clear that the phase portraits of systems $(A_{n-2}), (A_{n-3})$, etc. in
the neighborhood of the origin are described alternately by Figures 241
and 242. It follows that the phase portrait of system $(A_1)$ near its

equilibrium state $O_1(0,\ 0)$ is like that of $(A_n)$ for odd $n$, like that of $(A_{n-1})$ for even $n$.

It remains to go from system $(A_1)$ to system $(A)$ by means of Lemma 1 and Remark 1. The result is that in this case $(n < m,\ b_n \neq 0$ and $a_{2m+1} < 0)$ the equilibrium state $O(0,\ 0)$ of system $(A)$ is a topological node if $n$ is even and an equilibrium state with elliptic region if $n$ is odd. If $b_n > 0$ $(b_n < 0)$ the elliptic region lies below (above) the hyperbolic region (see Figure 239 for the case $b_n > 0$).

3) It remains to examine the case $m = n$. For $r = m = n$ system $(A_r)$ is

$$\frac{dx}{dt} = \eta_m\, x,$$

$$\frac{d\eta_m}{dt} = -\,m\eta_m^2 + a_{2m+1}\, x^2\, (1 + h\,(x)) + b_n x \eta_m\, [1 + g\,(x)] + \eta_m^2\, xf\,(x,\ \eta_m x^m).$$

The directions in which semipaths of system $(A_m)$ may tend to the equilibrium state $O_m(0,\ 0)$ are determined from the equation

$$x\, [\eta_m^2\, (1 + m) - b_n \eta_m x - a_{2m+1}\, x^2] = 0. \tag{21}$$

By Lemma 3, the only semipaths that tend to $O_m\,(0,\ 0)$ in the directions $\frac{\pi}{2}$ and $\frac{3}{2}\pi$ are $x = 0,\ \eta_m > 0$ and $x = 0,\ \eta_m < 0$. Let $\lambda = b_n^2 + 4\,(m+1)\,a_{2m+1}$. We distinguish three cases.

a) $\lambda < 0$. By equation (21), $\frac{\pi}{2}$ and $\frac{3}{2}\pi$ are the only directions in which semipaths of $(A_m)$ can tend to $O_m$. But then, as in case 1 $(n > m)$, none of the systems $(A_m)$, $(A_{m-1})$, $(A_{m-2})$, . . ., $(A_1)$, $(A)$ have semipaths on the slit plane tending to the origin, so that the equilibrium state $O(0,\ 0)$ of system $(A)$ is either a focus or a center.

b) $\lambda > 0$. Consider the system

$$\frac{dx}{dt} = \eta_{m+1}\, x,$$

$$\frac{d\eta_{m+1}}{dt} = a_{2m+1}\, [1 + h\,(x)] - (m+1)\,\eta_{m+1}^2 + \qquad\qquad (A_{m+1})$$

$$+\, b_n \eta_{m+1}\, [1 + g\,(x)] + \eta_{m+1}^2\, xf\,(x,\ \eta_{m+1} x^{m+1}).$$

It has two equilibrium states $\tilde{O}_1\,(0,\ k_1)$ and $\tilde{O}_2\,(0,\ k_2)$, where

$$k_1 = \frac{b_n + \sqrt{\lambda}}{2\,(m+1)}\ , \qquad k_2 = \frac{b_n - \sqrt{\lambda}}{2\,(m+1)}\ .$$

Direct computation shows that

$$\Delta\,(0,\ k_1) = -k_1\sqrt{\lambda}, \qquad \Delta\,(0,\ k_2) = k_2\sqrt{\lambda}. \tag{22}$$

If $a_{2m+1} > 0$, then $\sqrt{\lambda} > |b_n|$, $k_1 > 0$, $k_2 < 0$, and both points $\tilde{O}_1$ and $\tilde{O}_2$ are simple saddle points of system $(A_{m+1})$. Hence, as in the preceding cases (for $a_{2m+1} > 0$), we conclude that $O\,(0,\ 0)$ is a saddle point of system $(A)$.

Now let $a_{2m+1} < 0$. Then $\sqrt{\lambda} < |b_n|$, and the numbers $k_1$ and $k_2$ are of the same sign. It follows from (22) that in this case one of the equilibrium states $\tilde{O}_1$ and $\tilde{O}_2$ is a simple saddle point of system $(A_{m+1})$ and the other a simple node; moreover, if $b_n > 0$ the saddle point lies on the $\eta_{m+1}$-axis above the node, and conversely if $b_n < 0$. The reasoning now proceeds exactly as in case 2 $(n < m,\ a_{2m+1} < 0)$. The final conclusion is that if $b_n \neq 0$, $\lambda > 0$, $a_{2m+1} < 0$ and $m = n$, the point $O(0,\ 0)$ is a topological node for even $n$ and an equilibrium state with elliptic region for odd $n$ (Figure 239).

c) $\lambda = 0$. In this case it is readily seen that in a certain region $\Gamma$ containing the $\eta_{m+1}$-axis the system $(A_{m+1})$ has a unique equilibrium state $\tilde{O}\left(0,\ \dfrac{b_m}{2(m+1)}\right)$. Direct computations show that for this point $\Delta = 0$, $\sigma(0,0) \neq 0$. Thus the phase portrait in the vicinity of the point $\tilde{O}$ may be investigated using Theorem 65, provided system $(A_{m+1})$ is first brought to the form (11) of §21. The investigation (which is left to the reader) shows that the equilibrium state $\tilde{O}$ of system $(A_{m+1})$ is a saddle node; the phase portrait near $\tilde{O}$ is as in Figure 243 for $b_n > 0$, in Figure 244 for $b_n < 0$. We may now proceed as in the proof of Lemma 2 to characterize the equilibrium state $O_m$ of system $(A_m)$. It is not hard to see that if $b_n > 0$ the phase portrait near $O_m$ is that shown in Figure 241, and the phase portrait for $b_n < 0$ is obtained by reflection in the $x$-axis. The rest of the proof is the same as in case 2 $(n < m,\ a_{2m+1} < 0)$, and the final conclusion is that if $b_n \neq 0$, $m = n$, $\lambda = 0$, the equilibrium state $O(0,\ 0)$ of system $(A)$ is

1) a topological node for even $n$;
2) an equilibrium state with elliptic region for odd $n$ (Figure 239).

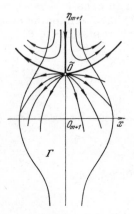

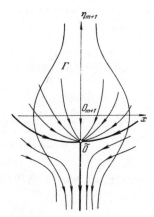

FIGURE 243.    FIGURE 244.

Combining all the above results, we see that the theorem has been proved in its entirety.

We now consider the case of even $k$.

**Theorem 67.** *Let the number $k$ in system $(A)$ be even, $k = 2m$, $(m \geqslant 1)$. Then the equilibrium state $O(0,\ 0)$ is:*

1) *a degenerate equilibrium state if $b_n = 0$, or $b_n \neq 0$ and $n \geqslant m$ (Figure 245); 2) a saddle node if $b_n \neq 0$ and $n < m$ (Figure 246).*

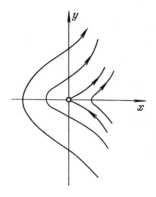

FIGURE 245.

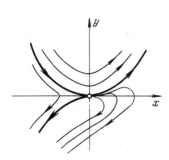

FIGURE 246.

Figure 245 illustrates the case $a_{2m} > 0$, and Figure 246 the case $b_n > 0$, $a_{2m} < 0$.

The proof is analogous to that of Theorem 66, and will therefore be omitted.* We remark only that the proof also uses Lemma 4.

**3. SIMPLIFIED VERSION. EXAMPLES.** We shall first show how investigation of the topological structure of the equilibrium state $O(0, 0)$ can be simplified, for system (5):

$$\frac{dx}{dt} = y + P_2(x, y), \qquad \frac{dy}{dt} = Q_2(x, y).$$

As set forth in this section, the investigation involves the following steps.
a) Apply transformation (6)

$$\xi = x, \qquad \eta = y + P_2(x, y)$$

and find the inverse transformation (7)

$$x = \xi, \qquad y = f(\xi, \eta).$$

The function $f(\xi, \eta)$ satisfies the identity

$$\eta \equiv f(\xi, \eta) + P_2(\xi, f(\xi, \eta)). \qquad (23)$$

As a result we obtain system (8):

$$\frac{d\xi}{dt} = \eta, \qquad \frac{d\eta}{dt} = \widetilde{Q}_2(\xi, \eta).$$

Return to the old notation, replacing $\xi$ and $\eta$ respectively by $x$ and $y$. The identity (23) is written

$$y \equiv f(x, y) + P_2(x, f(x, y)). \qquad (24)$$

* A full account of the proof of Theorem 67 may be found in /84/.

363

Denote

$$\varphi(x) = f(x, 0). \tag{25}$$

It follows from (24) that

$$\varphi(x) + P_2(x, \varphi(x)) = 0, \tag{26}$$

and $\varphi(0) = 0$. In the changed notation, system (8) is

$$\frac{dx}{dt} = y, \quad \frac{dy}{dt} = \tilde{Q}_2(x, y).$$

The expression for $\tilde{Q}_2$,

$$\tilde{Q}_2(x, y) = Q_2(x, f(x, y)) + P'_{2x}(x, f(x, y)) y +$$
$$+ P'_{2y}(x, f(x, y)) \cdot Q_2(x, f(x, y)), \tag{27}$$

was encountered at the beginning of §22.1.

b) The next step is to express the function $\tilde{Q}_2(x, y)$ in the form

$$\tilde{Q}_2(x, y) = a_k x^k + a_{k+1} x^{k+1} + \ldots + y(b_n x^n + b_{n+1} x^{n+1} + \ldots) + y^2 f(x, y). \tag{28}$$

c) According as $k$ is odd or even, we now apply Theorem 66 or 67 to determine the type of equilibrium state $O(0, 0)$ on the basis of the numbers $k$, $n$, $a_k$ and $b_n$.*

It is evident from (28) that $a_k x^k$ is the first (lowest-order) term in the expansion of the function $\tilde{Q}_2(x, 0)$ in powers of $x$, and $b_n x^n$ the first term in the expansion of $\frac{\partial \tilde{Q}_2(x, 0)}{\partial y}$. Consequently, by (25) and (27) $a_k x^k$ is the first term in the expansion of

$$Q_2(x, \varphi(x))[1 + P'_{2y}(x, \varphi(x))], \tag{29}$$

and $b_n x^n$ the first term in the expansion of

$$P'_{2x}(x, \varphi(x)) + Q'_{2y}(x, \varphi(x))[1 + P'_{2y}(x, \varphi(x))]\frac{\partial f(x, 0)}{\partial y} +$$
$$+ P''_{2yy}(x, \varphi(x)) \cdot Q_2(x, \varphi(x))\frac{\partial f(x, 0)}{\partial y}. \tag{30}$$

It follows from the identity (24) that

$$\frac{\partial f(x, 0)}{\partial y} = \frac{1}{1 + P'_{2y}(x, \varphi(x))}.$$

We may therefore write (30) in the form

$$P'_{2x}(x, \varphi(x)) + Q'_{2y}(x, \varphi(x)) + \frac{P''_{2yy}(x, \varphi(x)) \cdot Q_2(x, \varphi(x))}{1 + P'_{2y}(x, \varphi(x))}. \tag{31}$$

---

* If $\tilde{Q}_2(x, y)$ involves no terms containing $y$ to the first power, $b_n$ is taken equal to zero.

It is obvious from (29) that the first term in the expansion of (29), viz., $a_k x^k$, also serves as the first term in the expansion of $Q_2(x, \varphi(x))$.

Let $B_N x^N$ denote the first term in the expansion of

$$\sigma(x) = P'_{2x}(x, \varphi(x)) + Q'_{2y}(x, \varphi(x)). \qquad (32)$$

If the function (32) is identically zero, we take $B_N = 0$.

We claim that when using Theorems 66 and 67 to determine the topological structure of an equilibrium state $O(0, 0)$ of system (5) *one can replace the numbers $n$ and $b_n$ by $N$ and $B_N$, respectively*.

To prove this, we note first that if $B_N \neq 0$ and $N \leq m$, then the lowest-order terms in the expansions of the functions (31) and (32) simply coincide, i.e., $b_n = B_N$ and $n = N$. If $B_N = 0$, then either $b_n = 0$ or $n > m$. In either case it is immaterial whether $n$ and $b_n$ or (respectively) $N$ and $B_N$ are used in Theorems 66 and 67. Finally, if $B_N \neq 0$ and $N > m$, then there are three possibilities: a) the lowest-order terms of the expansions of the functions (31) and (32) coincide, b) $b_n = 0$, or c) $n > m$ (though the lowest-order terms of the two expansions are distinct). In all these cases the result of applying Theorems 66 and 67 is unchanged if $n$ and $b_n$ are replaced by $N$ and $B_N$.

The assertion is thus proved. It yields the following simple rule.

The topological structure of the equilibrium state $O(0, 0)$ of the system

$$\frac{dx}{dt} = y + P_2(x, y), \qquad \frac{dy}{dt} = Q_2(x, y) \qquad (33)$$

can be investigated as follows.

a) Determine the expansion coefficients of the function $y = \varphi(x)$, which is the solution of the equation

$$y + P_2(x, y) = 0.$$

To do this, one substitutes the expression

$$\varphi(x) = a_1 x + a_2 x^2 + \ldots + a_k x^k + \ldots$$

into the equation for $\varphi(x)$, collects the coefficients of like powers of $x$ and equates the results to zero. Thus the coefficients $a_1, a_2, \ldots, a_n, \ldots$ are found successively.

b) As the appropriate $a_i$ are determined, substitute into the functions

$$\psi(x) = Q_2(x, \varphi(x)) \quad \text{and} \quad \sigma(x) = P'_{2x}(x, \varphi(x)) + Q'_{2y}(x, \varphi(x))$$

the truncated series for $\varphi(x)$, $a_1 x + a_2 x^2 + \ldots + a_n x^n$, in order to determine the first nonvanishing coefficient in the expansion of $\psi(x)$, $a_k x^k$ ($a_k \neq 0$, $a_i = 0$, $i < k$), and the first nonvanishing coefficient in the expansion of $\sigma(x)$, $b_n x^n$ ($b_n \neq 0$, $b_j = 0$, $j < n$) (if $\sigma(x) \equiv 0$, then of course $b_n = 0$).

c) If $k$ is odd (even), apply Theorem 66 (67), thus characterizing the equilibrium state $O(0, 0)$ according to the values of $k$, $n$, $a_k$ and $b_n$.

Figures 238, 239, 245 and 246 illustrate the phase portrait in the neighborhood of the point $O(0, 0)$ in case the directions of approach of the paths to the equilibrium state are the directions of the coordinate axes. In the general case the situation will of course be slightly different, but the topological structure remains the same.

Remark. In a sufficiently small neighborhood of the origin, the transformation

$$\xi = x, \quad \eta = y - \varphi(x) \tag{34}$$

has the same effect as transformation (6), converting system (5) to the form (8). Therefore, if the local phase portrait of the transformed system (8)

$$\frac{d\xi}{dt} = \eta, \quad \frac{d\eta}{dt} = \tilde{Q}_2(\xi, \eta)$$

is known, we can return to the original system (5)

$$\frac{dx}{dt} = y + P_2(x, y), \quad \frac{dy}{dt} = Q_2(x, y)$$

by means of the transformation

$$x = \xi, \quad y = \eta + \varphi(\xi), \tag{35}$$

which is simply the inverse of (34).

In all the examples below, the systems are not necessarily reducible to canonical form and it is required to determine the type of the equilibrium state $O(0, 0)$ for each system, using one of the three theorems of this chapter.

1. $$\frac{dx}{dt} = x(\beta x - y), \quad \frac{dy}{dt} = -\frac{1}{a} y - y^2 + ax^2, \tag{36}$$

where $\beta > 0$, $a > 0$. Here $a = b = c = 0$, $d = -\frac{1}{a}$. Consequently, $\Delta = 0$,

$\delta = -\frac{1}{a} \neq 0$, and we shall use Theorem 65, after first reducing the system

to the form (11) of §21. To this end we need only introduce a new

parameter $\bar{t} = -\frac{1}{a} t$, and system (36) becomes

$$\frac{dx}{d\bar{t}} = -a\beta x^2 + axy, \quad \frac{dy}{d\bar{t}} = y + ay^2 - a^2 x^2. \tag{37}$$

We now find the function $y = \varphi(x)$, the solution of the equation $y + ay^2 - a^2 x^2 = 0$. Let

$$\varphi(x) = c_1 x + c_2 x^2 + c_3 x^3 + c_4 x^4 + \cdots$$

It is readily seen that $c_1 = c_3 = 0$, $c_2 = a^2$, $c_4 = -a^5$, ... Then

$$\varphi(x) = a^2 x^2 - a^5 x^4 + \cdots, \quad \psi(x) = P_2(x, \varphi(x)) = -a\beta x^2 + a^3 x^3 + \cdots$$

We get $m = 2$ and $\Delta_m = -\alpha\beta < 0$. By Theorem 65, the equilibrium state of system (37), hence also of system (36), is a saddle node, whose hyperbolic sectors enclose a segment of the positive $x$-axis.

Thus, the topological structure of the equilibrium state has been determined. The phase portrait in the neighborhood of the equilibrium state will be clarified by some additional information. The $y$-axis is the union of paths of system (36), and all other paths tending to $O$ approach it in the directions $\theta = 0$ and $\theta = \pi$ (see §21.2). Setting $x = 0$ in the second equation of (36), we see that both halves of the $y$-axis are positive semipaths. Setting $y = 0$ in both equations (36), we see that the separatrices of the hyperbolic sectors lie above the $x$-axis. On the basis of this information, we get the phase portrait illustrated in Figure 247

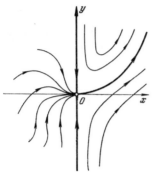

(remember that we have substituted $\bar{t} = -\frac{1}{a}t$).

FIGURE 247.

2.
$$\frac{dx}{dt} = x(-y + y^2 + 3xy), \qquad \frac{dy}{dt} = 3x + y - x^2 + y^3 + 3xy^2. \qquad (38)$$

Here $a = b = 0$, $c = 3$, $d = 1$. Hence $\Delta = 0$, $\delta = 1 \neq 0$ and the type of the equilibrium state is determined by Theorem 65. To reduce system (38) to the form (11) of §21, we follow §21.2 and apply the transformation

$$\bar{x} = x, \qquad \bar{y} = 3x + y.$$

We obtain the system

$$\frac{d\bar{x}}{dt} = -\bar{x}\bar{y} + 3\bar{x}^2 - 3\bar{x}^2\bar{y} + \bar{x}\bar{y}^2,$$

$$\frac{d\bar{y}}{dt} = \bar{y} + 8\bar{x}^2 - 3\bar{x}\bar{y} - 3\bar{x}\bar{y}^2 + \bar{y}^3. \qquad (39)$$

Solving the equation $\bar{y} + 8\bar{x}^2 - 3\bar{x}\bar{y} - 3\bar{x}\bar{y}^2 + \bar{y}^3 = 0$, we find

$$\varphi(\bar{x}) = -8\bar{x}^2 - 24\bar{x}^3 - \ldots,$$
$$\psi(\bar{x}) = \overline{P}_2(\bar{x}, \varphi(\bar{x})) = 3\bar{x}^2 + 8\bar{x}^3 + \ldots$$

Here $m = 2$, $\Delta_m = 3 > 0$ and by Theorem 65 the point $\overline{O}(0, 0)$ is a saddle node for system (39); hence the same is true of $O(0, 0)$ for system (38).

In addition, the $\bar{y}$-axis consists of paths of system (39); by Theorem 65 the hyperbolic sectors lie to the left of the $\bar{y}$-axis and the separatrices bounding them lie below the $\bar{x}$-axis (this is readily seen by examining the direction of the field on the $\bar{x}$-axis). Now, as in Example 1, we conclude that all paths tending to the point $\overline{O}(0, 0)$, except for the halves of the $\bar{y}$-axis, approach it in the directions $\theta = 0$ and $\theta = \pi$, i. e., touch the $\bar{x}$-axis at the point $\overline{O}(0, 0)$.

To determine the phase portrait of system (38) in the neighborhood of $O(0, 0)$, recall that the linear transformation applied was $\bar{x} = x$, $\bar{y} = 3x + y$.

Thus all paths (except those lying on the $y$-axis) tending to the equilibrium state $O$ of system (38) do so touching the straight line $y = -3x$, i. e., in the directions arctg $(-3)$ and $\pi + $ arctg $(-3)$ (Figure 248).

3.
$$\frac{dx}{dt} = y - \frac{1}{2}xy - 3x^2, \quad \frac{dy}{dt} = -y\left(x + \frac{3}{2}y\right). \tag{40}$$

In this case $a = c = d = 0$, $b = 1$. Consequently, $\Delta = 0$, $\delta = 0$, and the type of the equilibrium state is to be found using Theorem 66 or 67.

System (40) is already in the form (5), with

$$P_2(x, y) = -\frac{1}{2}xy - 3x^2, \quad Q_2(x, y) = -xy - \frac{3}{2}y^2.$$

The solution of the equation $y + P_2(x, y) = 0$ is

$$\varphi(x) = 3x^2 + \frac{3}{2}x^3 + \dots$$

Then

$$\psi(x) = Q_2(x, \varphi(x)) = -3x^3 - 15x^4 - \dots,$$
$$\delta(x) = P'_{2x}(x, \varphi(x)) + Q'_{2y}(x, \varphi(x)) = -7x - \frac{21}{2}x^2 - \dots$$

Consequently, $k = 2m + 1 = 3$, $m = 1$, $a_k = -3 < 0$, $n = 1$, $b_n = -7 < 0$ and $\lambda = b_n^2 + 4(m+1)a_{2m+1} = 25 > 0$. Since $k$ is odd, $a_k < 0$, $m = n$, $\lambda > 0$ and $n$ is odd, it follows from Theorem 66 that the equilibrium state $O$ of system (40) is an equilibrium state with elliptic region. The topological structure of the equilibrium state is thus established. It is easy to see that the $x$-axis is an integral curve of system (40). Examining the situation further with the aid of the transformation $x = x$, $y = \eta x$, one can show that the halves of the $x$-axis are separatrices. The phase portrait is shown in Figure 249.

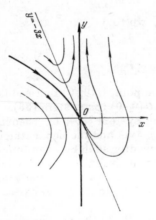

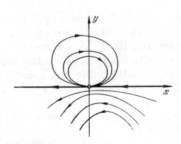

FIGURE 248.                    FIGURE 249.

4.
$$\frac{dx}{dt} = x\,(2y+x), \qquad \frac{dy}{dt} = x+(2+\beta)\,xy+y^2. \tag{41}$$

Here $a = b = 0$, $c = 1$, $d = 0$. Consequently, $\Delta = 0$, $\delta = 0$ and the type of the equilibrium state is determined by Theorem 66 or 67. To reduce the system to the form (5), according to §22.1, we must apply the transformation $\bar{x} = y$, $\bar{y} = x$. System (41) then becomes

$$\frac{d\bar{x}}{dt} = \bar{y}+(2+\beta)\,\bar{x}\bar{y}+\bar{x}^2, \qquad \frac{d\bar{y}}{dt} = 2\bar{x}\bar{y}+\bar{y}^2. \tag{42}$$

It is easy to see that

$$\varphi(\bar{x}) = -\bar{x}^2+\dots, \qquad \psi(\bar{x}) = -2\bar{x}^3+\dots, \qquad \delta(\bar{x}) = 4\bar{x}-(4+\beta)\,\bar{x}^2+\dots$$

Then $k = 2m+1 = 3$, $m = 1$, $a_k = -2 < 0$, $n = 1$, $b_n = 4 > 0$, $n$ is odd and $\lambda = 0$. Under these conditions it follows from Theorem 66 that $\bar{O}\,(0, 0)$ is an equilibrium state with elliptic region for system (42).

5.
$$\frac{dx}{dt} = y+y^2-x^3, \qquad \frac{dy}{dt} = 3x^2y+y^3-3x^5. \tag{43}$$

This system is already in the form (5). It is readily seen that

$$\varphi(x) = x^3-x^6+2x^9+\dots, \qquad \psi(x) = -3x^8+x^9+\dots, \qquad \sigma(x) = 3x^6-18x^9+\dots$$

Here $k = 2m$ is an even number and so the type of equilibrium state is determined by Theorem 67. Since $m = 4$, $n = 6$, we have $n > m$ and $O\,(0, 0)$ is a degenerate equilibrium state. Since $a_{2m} = -3$, we readily conclude from the Remark in §22.3 that the phase portrait is as illustrated in Figure 250.

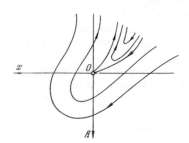

This example shows that a knowledge of the lowest-order term of the series expansion of the function $\varphi\,(x)$ is not sufficient to determine the lowest-order terms of the expansions of $\psi\,(x)$ and $\sigma\,(x)$.

The detailed discussion of Examples 6 and 7 is left to the reader.

FIGURE 250.

6.
$$\frac{dx}{dt} = \left(y+\frac{\lambda}{n-1}\right)x, \qquad \frac{dy}{dt} = x+\frac{1}{1-n}\,y^2,$$

where $n > 1$, $\lambda > 0$. The equilibrium state $O\,(0, 0)$ is a saddle node.

7.
$$\frac{dx}{dt} = y+\frac{n}{1-n}\,x^2+\frac{\lambda}{1-n}\,xy, \qquad \frac{dy}{dt} = y\left(\frac{\lambda}{n-1}\,y-x\right),$$

where $n > 1$, $\lambda > 0$. The equilibrium state $O\,(0, 0)$ is an equilibrium state with elliptic region.

*Chapter X*

SCHEME OF LIMIT CONTINUUM AND
BOUNDARY SCHEME OF REGION *G**

INTRODUCTION

This chapter, which is a direct sequel to Chapter VIII, investigates properties of ω- and α-limit continua and boundary continua of cells filled by closed paths, culminating in a description of the schemes of these continua. Also considered in this chapter is the scheme of the boundary of a region *G** with normal boundary (or boundary scheme). Together with the global schemes of equilibrium states, global schemes of limit continua and boundary schemes constitute the basic components of our ultimate goal, the description of the configuration of singular paths (indicating which of them are limit paths) embodied in the scheme of a dynamic system, which, as we shall see in the next chapter, fully characterizes the topological structure of the phase portrait.

Chapter X comprises four sections. In §23 we consider ω- and α-limit continua bounding cells filled by closed paths. When these continua are not equilibrium states (the case in which they are may of course be studied directly as in Chapter VIII), we shall call them zero-limit continua.

In §24 we define the local scheme of a limit continuum, which singles out those of the singular paths of the system that make up the continuum and describes their relative positions.

§25 is devoted to the global scheme of a limit continuum. The global scheme characterizes the disposition of the limit continuum on the plane by delineating the relative position of the simple closed curves, whose common points are equilibrium states, formed by the constituent paths of the continuum, indicating moreover all singular paths that tend to the continuum.

In §26 we consider the scheme of the boundary of a region *G** ⊂ *G*.

In Chapter VIII we mentioned that the global scheme of an equilibrium state is considerably more difficult to determine than its local scheme. The latter, as its name implies, may be constructed by examining only a small neighborhood of the equilibrium state (investigation "in the small"). Fairly general methods are available to this end (such as the methods of Chapters IV and IX). On the other hand, determination of the global scheme requires additional information, concerning the behavior of the singular paths as a whole ("in the large").

In the case of limit continua it is clear that even the local scheme is conditioned by the behavior in the large of singular paths, and, in particular, of limit cycles. As we have seen, only a few special methods to this end are at present known and general methods are lacking. In practice, therefore, determination of the local scheme of a limit continuum is a problem of a vastly different order of difficulty as compared with the analogous problem for equilibrium states.

## §23. PROPERTIES OF LIMIT CONTINUA AND BOUNDARY CONTINUA OF CELLS FILLED BY CLOSED PATHS

### 1. PROPERTIES OF ω- AND α-LIMIT CONTINUA WHICH ARE NOT EQUILIBRIUM STATES.

Let us consider the ω- or α-limit continua of a nonclosed path, on the assumption that they are not equilibrium states. Since the boundary $\Gamma$ of the region $\bar{G}^*$ (in which the dynamic system is being studied) is assumed normal, no limit continuum can have points in common with $\Gamma$ (see property 2) of normal boundaries, §16.2).

By §4.6, the limit continuum of a nonclosed semipath, if not an equilibrium state, is either a closed path or the union of equilibrium states and orbitally unstable whole paths tending to the equilibrium states both as $t \to +\infty$ and as $t \to -\infty$ (i.e., separatrices; see §15). The limit continuum of a positive semipath $L^+$ (or the ω-limit continuum) will be denoted by $K_\omega$, and the limit continuum of a negative semipath $L^-$ (α-limit continuum) by $K_\alpha$.

To fix ideas, we shall confine ourselves henceforth to the limit continuum $K_\omega$ of a positive semipath. All results obtained for $K_\omega$ are of course valid, mutatis mutandis, for the limit continuum $K_\alpha$ of a negative semipath. We shall therefore omit both the derivation and formulation of the results for continua $K_\alpha$.

We first recall a few properties of limit paths, established in §4.5. Let $L_0$ be a limit path (not an equilibrium state) of a semipath $L^+$, i.e., a component of some continuum $K_\omega$. Let $M_0$ be a point on this path and $l_0$ an arc without contact containing $M_0$ as an interior point. By Corollary 1 to Lemma 2 of §3.4, $l_0$ cannot contain any points of $L_0$ other than $M_0$. By Corollary 2 to the same lemma, the points at which $L^+$ crosses $l_0$ are all on a subarc lying on one side (positive or negative) of $L_0$.

To fix ideas, suppose that all points of $L^+$ in common with $l_0$ lie on the subarc of $l_0$ on the positive side of $L_0$. Let $M$ be any point other than $M_0$ on $L_0$ and $l$ an arc without contact through $M$, containing $M$ as an interior point and having no points other than $M$ in common with $L_0$. Then all points of intersection of $L^+$ and $l$ also lie on the subarc of the latter on the positive side of $L_0$.

According to the definition of §4.5, $L_0$ is then called a positively situated limit path of $L^+$ (or ω-limit path of $L$).

Similarly, one speaks of a negatively situated limit path of $L^+$ ($L^-$) (or ω (α)-limit path of $L$).

Proceeding now to investigation of the properties of limit continua, we recall that if a limit continuum contains no equilibrium states it consists of a single closed path (Theorem 14, Chapter II).

The situation is more interesting when the continuum $K_\omega$ contains at least one equilibrium state.

*Lemma 1. Let $K_\omega$ be an ω-limit continuum, not an equilibrium state
or closed path, $L_0$ be a constituent path (not an equilibrium state) of $K_\omega$,
which is a positively (negatively) situated limit path of a semipath $L^+$.*

*Then a) $L_0$ is ω- and α-continuable on the positive (negative) side; b) the
ω- and α-continuations of $L_0$ on the positive (negative) side are also
positively (negatively) situated limit paths of $L^+$.*

Proof. To fix ideas, let us assume that $L_0$ is a positively situated limit
path of $L^+$. The path $L_0$ tends to an equilibrium state both as $t \to +\infty$ and as
$t \to -\infty$. Let the equilibrium state to which it tends as $t \to +\infty$ be $O_1$, that to
which it tends as $t \to -\infty$ being $O_2$ ($O_1$ and $O_2$ may coincide). We first
consider $O_1$.

Let $C_1$ be a circle about $O_1$, so small that its interior contains no whole
orbitally unstable path other than $O_1$. Let $Q$ be a point on $L_0$, beyond the
last point of $L_0$ (with increasing $t$) in common with $C_1$. Let $QA$ be an arc
without contact through $Q$, entirely within $C_1$ and on the positive side of $L_0$
(Figure 251). Since $L_0$ is a positively situated limit path of $L^+$, the arc $QA$
contains an infinite sequence of points of $L^+$, $\{Q_i\}$ say, tending to $Q$,
corresponding to an unbounded increasing sequence of $t$-values and lying
on $QA$ in order of increasing $t$.

We first prove part a).

Suppose that $L_0$ has no ω-continuation on the positive side. Then all paths
crossing some subarc of the arc $QA$ tend to $O_1$ as $t \to +\infty$ without leaving $C_1$,
never again crossing $QA$. But then the sequence $\{Q_i\}$ cannot lie on $QA$,
contrary to assumption. Similarly one proves, considering the equilibrium
state $O_2$, that $L_0$ is α-continuable.

We now prove b). Let $L_1$ be the ω-continuation of $L_0$ on the positive side.
Let $P$ be some point on $L_1$ beyond its last point (with decreasing $t$) in
common with the circle $C_1$. Let $PB$ be an arc without contact through $P$,
lying inside $C$ on the positive side of $L_1$ (Figure 252). Since the arc $QA$
contains a sequence of points of $L^+$ corresponding to indefinitely increasing
times $t$ and tending to $Q$, it follows (Definition XIX) that the arc $PB$ also
contains a sequence of points of $L^+$ corresponding to indefinitely increasing
times $t$ and tending to $P$. But this means that $P$ is a limit point of the
semipath $L^+$. Hence part b) follows easily. Q.E.D.

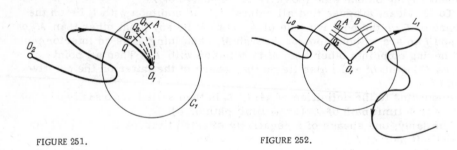

FIGURE 251.                              FIGURE 252.

Assuming as before that the continuum $K_\omega$ is neither an equilibrium
state nor a closed path, we prove the following theorem.

*Theorem 68.\** *Let one of the components (not an equilibrium state) of $K_\omega$ be a positively (negatively) situated limit path of $L^+$. Then all the constituent paths of $K_\omega$ (other than equilibrium states) are continuations of one another on the positive (negative) side, are positively situated limit paths of $L^+$, and may be so ordered that any two consecutive paths, including the last and first, are $\omega$-continuations of each other on the positive (negative) side.*

Proof. Let $N$ be the number of constituent paths of $K_\omega$ which are not equilibrium states. Let $L_1$ be one of these paths, say a positively situated limit path of $L^+$.

Let $O_1$ be the equilibrium state which is the $\omega$-limit point of $L_1$. By Lemma 1, $L_1$ is continuable on the positive side, and its $\omega$-continuation $L_2$ is also a positively situated limit path of $L^+$, so that $L_2$ is also a component of $K_\omega$. Let $O_2$ denote the $\omega$-limit point (an equilibrium state, of course) of $L_2$ (the points $O_1$ and $O_2$ may, in particular, coincide). Proceeding in this way, we obtain a sequence of constituent paths of $K_\omega$: $L_1, L_2, \ldots$, in which each path is the $\omega$-continuation of its predecessor in the sequence (on the positive side). Each path $L_i$ tends to an equilibrium state $O_i$ $(i = 1, 2, 3, \ldots)$ as $t \to +\infty$. Since the number of paths in $K_\omega$ is finite (by assumption, the number of singular paths is finite), there exists a smallest natural number $R$ such that the paths $L_1, L_2, \ldots, L_R$ are distinct and $L_{R+1}$ is one of the paths $L_i$, $1 \leqslant i \leqslant R$. But then necessarily $i = 1$, for if $i > 1$ one easily sees that $L_i$ has two distinct $\alpha$-continuations on the positive side ($L_{i-1}$ and $L_R$), which is absurd. Hence $L_1$ and $L_{R+1}$ coincide, and we get a chain of pairwise distinct paths

$$L_1, L_2, \ldots, L_R, \tag{1}$$

which are positively situated limit paths of $L^+$ and continuations of one another in the cyclic order (also on the positive side).

The limit points $O_1, O_2, \ldots, O_R$ of these paths may be distinct or coincide. Each $O_i$ is an $\omega$-limit point of $L_i$ and an $\alpha$-limit point of $L_{i+1}$ ($i = 1, 2, \ldots, R$; $L_{R+1} \equiv L_1$). Let $K'_\omega$ denote the closed set consisting of all points of the paths (1) and the points $O_i$ ($i = 1, 2, \ldots, R$).

To prove the theorem, it will suffice to show that $K'_\omega$ contains all the limit points of the semipath $L^+$, so that $K'_\omega \equiv K_\omega$ and hence $R = N$.

We first show that for any $\varepsilon > 0$ and sufficiently large $T$ all points of $L^+$ corresponding to times $t > T$ are in the $\varepsilon$-neighborhood of $K'_\omega$ (compare with the theorem on closed limit paths, §4). Consider the point $O_i$ and the paths that tend to it, $L_i$ and $L_{i+1}$ ($i = 1, 2, \ldots, R$; $L_{R+1} \equiv L_1$). Since $L_{i+1}$ is an $\omega$-continuation of $L_i$ on the positive side, it follows that for any $\varepsilon > 0$ there exist arcs without contact $\lambda_i^+$ and $\lambda_{i+1}^-$ with the following properties (see §15): a) one of the endpoints of $\lambda_i^+$, $M_i^+$ say, is on $L_i$, and one of the endpoints of $\lambda_{i+1}^-$, $M_{i+1}^-$ say, on $L_{i+1}$; the arcs have no points in common and lie on the positive side of $L_i$ and $L_{i+1}$, respectively (excluding the points $M_i^+$ and $M_i^-$, neither arc has points in common with the continuum $K'_\omega$; see Corollary 1 to Lemma 11 in §3); b) the arcs $\lambda_i^+$ and $\lambda_{i+1}^-$ lie entirely in $U_\varepsilon(O_i)$, and any path crossing $\lambda_i^+$ at $t = t_0$ will cross $\lambda_{i+1}^-$ at some time $T > t_0$, remaining at all times $t_0 \leqslant t \leqslant T$ in $U_\varepsilon(O_i)$ (Figure 253).\*\*

---

\* This theorem together with Lemma 1 constitute a sharpened version of a theorem of Bendixson.

\*\* It should be clear that the theorem may be proved using only one arc without contact through a point of each $L_i$, rather than two as in the present proof. We are considering two arcs with endpoints on each $L_i$ in order to permit direct reference to a previously established proposition (Lemma 6 of §18).

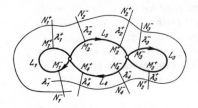

FIGURE 253.

Consider the arc $\lambda_1^-$. By property b), for any given $\varepsilon > 0$ there exists a subarc $M_R^+ N_R^+$ of $\lambda_R^+$ such that any path crossing it at some time $t = t_0$, at a point other than $M_R^+$, crosses $\lambda_R^-$ at some larger $t$, remaining at intermediate times within $U_\varepsilon\,(O_R)$. Then there exists a subarc $M_R^- N_R^-$ of $\lambda_R^-$ such that paths crossing it with increasing $t$ cross the arc $M_R^+ N_R^+$ without previously leaving the $\varepsilon$-neighborhood of the subarc $M_R^- M_R^+$ of $L_R$ (and without crossing the arcs $\lambda_R^-$ and $\lambda_R^+$). Continuing the construction in this way, we get a sequence of arcs without contact $M_{R-1}^+ N_{R-1}^+,\ M_{R-1}^- N_{R-1}^-, \ldots$ $\ldots, M_2^- N_2^-,\ M_2^+ N_2^+$, which are subarcs of $\lambda_{R-1}^+,\ \lambda_{R-1}^-, \ldots$, ending with a subarc $M_1^- N_1^-$ of $\lambda_1^-$. By construction, the arc $M_1^- N_1^-$ possesses the following properties: any path crossing this arc at $t = t_1$ at a point other than $M_1^-$ will cross $\lambda_1^-$ again at some $t_2 > t_1$, without crossing $\lambda_1^-$ at times $t$ between $t_1$ and $t_2$ and without leaving the $\varepsilon$-neighborhood of $K_\omega'$.

Now consider the semipath $L^+$. Since $M_1^-$ is a limit point of $L^+$ (positively situated), the arc $M_1^- N_1^-$ has points in common with $L^+$. Let $P_1$ be one of these points, corresponding to time $t_1$, say. By the preceding argument, $L^+$ crosses the arc $\lambda_1^-$ at some point $P_2$ at a time $t = t_2 > t_1$, in such a way that the entire arc $P_1 P_2$ of $L^+$ is in the $\varepsilon$-neighborhood of $K_\omega'$. Clearly, $P_2$ lies on $\lambda_1^-$ between $P_1$ and $M_1^-$. But then $P_2$ itself is on the arc $M_1^- N_1^-$ and the preceding argument may again be applied. Thus we obtain an infinite sequence of points $P_1,\ P_2,\ P_3, \ldots$ on $L^+$, corresponding to times $t_1, t_2, \ldots,$ such that each arc $P_n P_{n+1}$ of $L^+$ is in $U_\varepsilon\,(K_\omega')$. Since $t_n \to \infty$ as $n \to \infty$, this means that the entire part of $L^+$ corresponding to times $t > t_1$ is in the $\varepsilon$-neighborhood of $K_\omega'$.

Hence it follows that $L^+$ has no limit points not in $K_\omega'$. Indeed, any point $Q$ not in $K_\omega'$ lies at a nonzero distance $\varrho$ from $K_\omega'$. Take $\varepsilon < \varrho/2$. From some time $T\,(T = T\,(\varepsilon))$ and thereafter, all points of $L^+$ will lie in the $\varepsilon$-neighborhood of $K_\omega'$ and hence, by the choice of $\varepsilon$, at a distance greater than $\frac{\varrho}{2}$ from $Q$, so that no point $Q$ not in $K_\omega'$ can be a limit point of $L^+$.

Thus $K_\omega' \equiv K_\omega$.   Q.E.D.

Remark. To fix ideas, suppose that all constituent paths of the continuum $K_\omega$ are positively situated limit paths of $L^+$, and let $l_0$ be an arc without contact through a point $M_0$ of one of these paths $L_0$ (not an equilibrium state). Then, by Lemma 8 of §15 and Lemma 9 of §16, $K_\omega$ is the $\omega$-limit continuum not only of $L^+$ but also of all paths crossing a sufficiently small subarc $M_0 A$ of $l_0$ on the positive side of $L_0$.

In case all paths in $K_\omega$ are positively situated limit paths of $L^+$, we shall say that $K_\omega$ is a p o s i t i v e  s i t u a t e d  l i m i t  c o n t i n u u m  of $L^+$.

We shall also say that $K_\omega$ is a p o s i t i v e l y  s i t u a t e d  $\omega$-l i m i t  c o n t i n u u m, without mentioning the path, understanding thereby that (by Theorem 68, Remark) $K_\omega$ is the $\omega$-limit continuum for all paths crossing an arc without contact $l$ passing through a point $M_j$ of any constituent path $L_j$ (not an equilibrium state) of $K_\omega$, on a sufficiently small subarc $M_j A_j$ of $l$ on the positive side of $L_j$.

Throughout the sequel we shall also have to consider not simply continua $K_\omega$ but positively (or negatively) situated limit continua $K_\omega$. In this connection, we introduce the following notation.

$K_\omega^+$ ($K_\omega^-$) will denote a positively (negatively) situated $\omega$-limit continuum, treated as "$\omega$-limiting" on the positive (negative) side alone. The symbols $K_a^+$ and $K_a^-$ have an analogous meaning.

If we do not wish specifically to indicate that an $\omega(\alpha)$-limit continuum $K$ is positively (negatively) situated but only to emphasize that one of these two alternatives is meant, we shall use the symbol $K^{(\cdot)}$.

These conventions are necessary because there exist "two-sided" limit continua, i.e., limit continua which are both positively and negatively situated. For such continua both notations $K^+$ and $K^-$ are of course valid.

**2. ZERO-LIMIT CONTINUA AND THEIR PROPERTIES.** Let $K_0$ be a continuum which is part of the boundary of a cell filled by closed paths. We shall call such continua 0-limit (zero-limit) continua. By Theorem 50, any cell $w$ filled by closed paths is doubly connected, so that its boundary is the union of two 0-limit continua.

Suppose that $K_0$ is not a single equilibrium state (center), and let $L_0$ be some constituent path (not an equilibrium state) of $K_0$.

*Lemma 2. a) Let P be a point on $L_0$ and $l$ an arc without contact containing P as an interior point, $A_0$ and $B_0$ the endpoints of $l$ lying respectively on the positive and negative sides of $L_0$. Then all points of $l$ in the cell w are on one of the subarcs $PA_0$ or $PB_0$ of $l$, and there exists a subarc PA of $PA_0$ (a subarc PB of $PB_0$) all of whose points (except P) are in w.*

*b) If all points of the cell w on an arc without contact through a point of $L_0$ lie on the positive (negative) side of $L_0$, then the same is true of any arc without contact through any point of $L_0$.*

Proof. We first prove part a).

Let $A$ be some point of $w$ lying, say, on the subarc $PA_0$ of $l$. We claim that all points of the subarc $PA$ of $PA_0$ are in $w$. In fact, otherwise, let $Q$ be an interior point of $PA$ not in $w$. Since $P$ is a boundary point of $w$, there must be points of $w$ on the subarc $PQ$ of $l$. Let $R$ be one of these. The paths $L_A$ and $L_R$ passing through $A$ and $R$, respectively, are in the cell $w$ and are therefore closed paths, one inside the other, with the point $Q$ lying between them. But this point is not in $w$, and this contradicts Theorem 49.

Part b) follows directly from Remark 1 to Lemma 9 in §3 (Figure 45). Q.E.D.

If all points of the cell $w$ on every arc without contact passing through a point of a boundary path $L_0$ of the cell lie on the positive (negative) side of $L_0$, we shall say that $L_0$ is a positively (negatively) situated zero-boundary path for $w$.

Let us assume first that one of the components of $K_0$ is a closed path. In particular, this closed path may consist of boundary and corner arcs. The following theorem is valid:

*Theorem 69. If one of the components of $K_0$ is a closed path $L_0$, this closed path exhausts the entire continuum $K_0$.*

Proof. Let $l$ be an arc without contact through some point $P$ of $L_0$. To fix ideas, suppose that the points of the cell $w$ on this arc lie on the positive side of $L_0$. Then, by Lemma 2, we may assume that all points of $l$ on the positive side of $L_0$ (excluding $P$) are points of $w$.

By Lemma 5 of §3, given any $\varepsilon > 0$ we can find a subarc $PA$ of $l$ such that all paths crossing it at points other than $P$ are closed paths of $w$ lying entirely within the $\varepsilon$-neighborhood of $L_0$. To fix ideas, let the continuum $K_0$, hence also the path $L_0$, be interior to the closed paths of the cell. Suppose that there is a point $Q$ of $K_0$ which is not on $L_0$, and is therefore at a positive distance $\varrho$ from $L_0$. Take $\varepsilon < \frac{\varrho}{2}$. On all paths crossing the arc $PA$ at points sufficiently near (but distinct from) $P$ there must be points in $U_\varepsilon(Q)$. But, by the preceding argument, all paths through points on a sufficiently small subarc $PA$ of $l$ are in the $\varepsilon$-neighborhood $L_0$, and by the choice of $\varepsilon$ cannot lie in $U_\varepsilon(Q)$. This contradiction proves the theorem.

Suppose now that $K_0$ comprises both paths proper and equilibrium states, so that it is not a closed path.

*L e m m a   3.   Let $K_0$ be a 0-limit continuum which is not a closed path, and $L_0$ a nonclosed constituent path (not an equilibrium state) of $K_0$, which is a positively (negatively) situated boundary path for $w$. Then: a) the path $L_0$ is $\omega$- and $\alpha$-continuable on the positive (negative) side; b) the $\omega$- and $\alpha$-continuations of $L_0$ on the positive (negative) side are also positively (negatively) situated boundary paths of the cell and therefore are components of $K_0$.*

The proof is entirely analogous (mutatis mutandis) to that of Lemma 1, and is therefore omitted.

*T h e o r e m   70.   Let one of the constituent paths (not an equilibrium state) of $K_0$ be a positively (negatively) situated boundary path of $w$. Then all constituent paths of $K_0$ which are not equilibrium states are continuations of one another on the positive (negative) side, positively (negatively) situated boundary paths of $w$, and may be ordered in such a way that of any two successive paths (including the last and the first) one is the $\omega$-continuation of the other on the positive (negative) side.*

P r o o f.   Suppose that $L_1$, $L_2$, . . ., $L_N$ are all the constituent paths of $K_0$ (excluding equilibrium states). Reasoning as in Theorem 68, we find that there exists a chain of constituent paths of $K_0$, $L_1$, $L_2$, . . ., $L_R$ $(R \leqslant N)$, in which each path $L_{i+1}$ is the $\omega$-continuation of $L_1$ on the positive side, $L_1$ is that of $L_R$, and all these paths are positively situated boundary paths of $w$.

Let $O_i$ $(i = 1, 2, . . ., R)$ denote the equilibrium state to which the path $L_i$ tends as $t \to +\infty$ and the path $L_{i+1}$ as $t \to -\infty$ (some or all of the equilibrium states $O_i$ may coincide). Let $K_0'$ denote the continuum formed by the paths $L_i$ $(i = 1, 2, . . ., R)$ and equilibrium states $O_i$.

As in Theorem 68, it will suffice to show that $K_0' \equiv K_0$ and that therefore $R = N$. To this end, fix $\varepsilon > 0$ and let $\lambda_i^+$ and $\lambda_i^-$ be arcs without contact with endpoints on the paths $L_i$ $(i = 1, 2, . . ., R)$, possessing the same properties a) and b) as in Theorem 68.

Consider $\lambda_1^-$. We may assume without loss of generality that all its points except the endpoint $M_1^-$ lie in $w$. Suitably modifying the arguments of Theorem 68, we readily show that there exists a subarc $M_1^- N_1^-$ of $\lambda_1^-$ such that all

FIGURE 254.

paths crossing it (which are of course closed paths in $w$) lie entirely in the $\varepsilon$-neighborhood of $K'_0$.

To fix ideas, suppose moreover that the continuum $K'_0$ lies inside the closed paths of $w$. If $K'_0 \not\equiv K_0$, there must be a boundary point $Q$ of $w$ lying inside all the closed paths of the cell at a positive distance from $K'_0$. Reasoning now as in the proof of Theorem 68, one readily shows that $K_0 \equiv K'_0$, Q.E.D.

When the points of the paths in a cell $w$ filled by closed paths lie on an arc without contact through a point on a constituent path $L_i$ (not an equilibrium state) of $K_0$ on the positive side of the path, we shall say that $K_0$ is a positively situated boundary (or 0-limit) continuum of $w$, denoting it by $K_0^+$.

We shall also sometimes say that $K_0^+$ is a 0-limit continuum for a path $L$, meaning thereby that $L$ is a path in a cell for which $K_0^+$ is a boundary continuum. Similarly, we shall speak of a negatively situated 0-limit continuum $K_0$, denoting it by $K_0^-$.

As in the case of $\omega$- and $\alpha$-limit continua, we shall use the notation $K_0^{(\,)}$ if there is no need to specify whether it is a positively or negatively situated 0-limit continuum. In particular, $K_0$ may be both positively and negatively situated (though of course for different regions), in which case it will be treated as both $K_0^+$ and as $K_0^-$ (Figure 254).

## 3. CONTINUUM CONSISTING OF SINGULAR PATHS WHICH ARE CONTINUATIONS OF ONE ANOTHER.

Theorems 68 and 70 show that a continuum $K_\omega$, $K_\alpha$ or $K_0$ which is neither an equilibrium state nor a closed path has the following structure:
a) it is the union of finitely many nonclosed paths tending to equilibrium states both as $t \to +\infty$ and as $t \to -\infty$, and of these equilibrium states;
b) the nonclosed paths may be numbered in such a way that each path is the $\omega$-continuation of its predecessor (and the first the $\omega$-continuation of the last) on the same (positive or negative) side.

The next theorem is in a sense the converse of Theorems 68 and 70.

*Theorem 71. Any continuum $K$ satisfying conditions a) and b) is a positively (negatively) situated $\omega$-, $\alpha$- or 0-limit continuum.*

Proof. To fix ideas, suppose that each of the constituent paths $L_1, L_2, \ldots, L_N$ (not equilibrium states) of $K$ is the $\omega$-continuation of another path of the sequence on the positive side. We may assume them so numbered that the $\omega$-continuation of each path $L_i$ tending to an equilibrium state $O_i$ as $t \to +\infty$ is the path $L_{i+1}$, which of course tends to the same equilibrium state as $t \to -\infty$ ($i = 1, 2, \ldots, N; L_{N+1} \equiv L_1$). Some or all of the equilibrium states $O_i$ may coincide.

It is clear that all the paths $L_i$ are singular (orbitally unstable), and the distance $\varrho$ between $K$ and the boundary of the region $G^*$ in which the dynamic system is being considered is positive (because the boundary is normal). Let $\varepsilon < \frac{\varrho}{2}$ be a positive number such that $U_\varepsilon(K)$ is disjoint from the boundary $\Gamma$ and contains no singular paths other than those making up $K$. For the given $\varepsilon > 0$, let $\lambda_1^-$ and $\lambda_1^+$ be arcs with the same properties as in the proof of Theorem 68. By the choice of $\varepsilon$, the arc $\lambda_1^-$ contains no point of a singular path lying in $U_\varepsilon(K)$ other than its endpoint $M_1^-$, which lies on $L_1$.

As in Theorem 68, we can find a subarc $M_1^- N_1^-$ of $\lambda_1^-$ such that any path crossing it at time $t_0$ will cross $\lambda_1^-$ again at some time $t_1 > t_0$, without

leaving the $\varepsilon$-neighborhood of $K$ at intermediate times $t_0 \leqslant t \leqslant t_1$. In particular, let $N_1^*$ be the point at which the path $L^*$ passing through $N_1^-$ crosses
$\lambda_1^-$ again.

Conversely, any path crossing the subarc $M_1^- N_1^*$ of $\lambda_1^-$ at $t = t_1'$ must cross
this arc at some previous time $t_2'$ $(t_2' < t_1')$, without leaving the $\varepsilon$-neighborhood
of $K$ at times $t_1' \geqslant t \geqslant t_2'$. Let us consider all possible cases.

1) $N_1^*$ is distinct from $N_1^-$ and lies on $\lambda_1^-$ between $N_1^-$ and $M_1^-$. The path
$L^*$ is nonclosed and at some time $t_2 > t_1$ crosses the arc $\lambda_1^-$ at a point $N_2^*$,
which is of course between $N_1^*$ and $M_1^-$. Continuing in this way, we establish
the existence of an infinite sequence of points $\{N_i^*\}$ of $L^*$ on $\lambda_1^-$, corresponding
to a monotone increasing sequence $t_1, t_2, \ldots, t_n, \ldots$ It is readily seen that
$t_n \to \infty$ as $n \to \infty$, and $N_k^*$ lies on the arc $M_1^- N_1^-$ between $M_1^-$ and $N_{k-1}^*$. Since
each arc $N_k^* N_{k+1}^*$ of the path $L^*$ is in $U_\varepsilon(K)$, this at once implies that the
positive semipath $L^+$ of $L^*$ beginning at the point $N_1^-$ lies entirely within
$U_\varepsilon(K)$. In addition, by the Corollary to Lemma 2 of §4 the sequence $\{N_i^*\}$
has a single point of accumulation, and this point, which is a limit point
of the semipath $L^+$, must lie on a singular path in $U_\varepsilon(K)$. But by the choice
of $\varepsilon$ the only point on $\lambda_1^-$ which is a point of a singular path in $U_\varepsilon(K)$ is $M_1^-$.
Consequently, the point of accumulation of the sequence $\{N_i^*\}$ must be $M_1^-$.
This clearly implies that $L_1$ is an $\omega$-limit path for $L^*$, and also for all paths
crossing the subarc of $\lambda_1^-$ on the positive side of $L_1$. Hence (Theorem 68)
the same is true of the continuum $K$. In other words, $K$ is a positively
situated $\omega$-limit continuum.

2) $N_1^*$ is distinct from $N_1^-$ but $N_1^-$ lies between $N_1^*$ and $M_1^-$. One then
proves as in case 1) that $K$ is a positively situated $\alpha$-limit continuum.

3) $N_1^*$ coincides with $N_1^-$. Then $L^*$ is a closed path. It is easy to see
that then all paths crossing the subarc $M_1^- N_1^-$ of $\lambda_1^-$ are also closed. Indeed,
suppose that there are points other than $M_1^-$ on $M_1^- N_1^-$ through which pass
closed paths and also points through which pass nonclosed paths. Since
closed and nonclosed paths necessarily belong to different cells, it follows
that the arc $M_1^- N_1^-$ must contain a point of a singular path other than $M_1^-$.
And this contradicts the choice of $\lambda_1^-$.

Thus all paths through the points of the subarc $M_1^- N_1^-$ of $\lambda_1^-$ are closed
and belong to the same cell $w$. The point $M_1^-$, on the path $L_1$, is a boundary
point of the cell. Consequently, $L_1$ is a boundary path of the cell $w$, and
moreover positively situated. Therefore (see Theorem 70) the same holds
for the entire continuum $K$, and the proof is complete.

§24. LOCAL SCHEME OF A LIMIT CONTINUUM
AND CANONICAL NEIGHBORHOOD

1. $\omega$ ($\alpha$)-ENUMERATION OF $\omega$-, $\alpha$- AND 0-LIMIT CONTINUA. Let $K^0$ be a
positively situated $\omega$-, $\alpha$- or 0-limit continuum which is not an equilibrium
state. Let $K$ contain at least one equilibrium state, and suppose its
constituent paths written out in a sequence.

$$L_1, \ O_{i_1}, \ L_2, \ O_{i_2}, \ \ldots, \ L_N, \ O_{i_N},$$

with the following properties: a) each path $L_{k+1}$ is the $\omega$-continuation of
$L_k$ $(k = 1, 2, \ldots, N, \ L_{N+1} \equiv L_1)$ on the positive (negative) side; b) each

equilibrium state $O_i$ is an $\omega$-limit point of $L_k$ and an $\alpha$-limit point of $L_{k+1}$ (some or even all of the equilibrium states may coincide, i.e., we may have $O_k \equiv O_m$ for $k \neq m$).

The above sequence will be called an $\omega$-e n u m e r a t i o n of the continuum $K$. It is unique up to a cyclic permutation of its symbols.

Similarly, one can define an $\alpha$-enumeration of the continuum $K$. Clearly, an $\alpha$-enumeration is obtained from an $\omega$-enumeration by simply inverting the order of the symbols. If $K$ is a closed path $L_0$, it has exactly one $\omega$-enumeration, which is also its (unique) $\alpha$-enumeration, consisting of the path $L_0$ alone. Henceforth we shall confine ourselves to $\omega$-enumerations of a continuum $K$.

As shown previously, it follows from property 3) of a normal boundary that an $\omega$- or $\alpha$-limit continuum $K$ cannot have points in common with the boundary, but a $0$-limit continuum may have such points.

It is readily seen that if this is so the $0$-limit continuum in question is either a closed (orbitally stable) path, constituting one of the boundary continua of the region $\bar{G}^*$, or a closed (orbitally stable) path (lying entirely in $\bar{G}^*$) consisting of boundary and corner arcs. In the first case, the $\omega$-enumeration of the continuum contains a single closed path $L_0$; in the second, an $\omega$-enumeration of the continuum is obtained by enumerating all its corner and boundary arcs in the $\omega$-direction (i.e., the direction of increasing $t$):

$$l_1, \ \hat{l}_2, \ l_3, \ \hat{l}_4, \ \ldots \tag{1}$$

We cite a few simple examples.

E x a m p l e 1. The $\omega$-enumeration of the $\omega$-limit continuum $K$ shown in Figure 255 is

$$\omega | O_1, \ L_1, \ O_2, \ L_2, \ O_2, \ L_3, \ O_3, \ L_4, \ O_4, \ L_5, \ O_4, \ L_6, \ O_3, \ L_7, \ O_5, \ L_8.$$

Note that the $\omega$-limit continuum of Figure 256 has the same $\omega$-enumeration. The constituent paths of the continuum in this example are numbered in order of their $\omega$-enumeration.

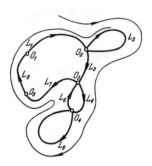

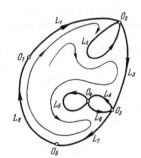

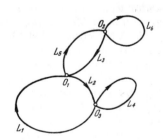

FIGURE 255.        FIGURE 256.        FIGURE 257.

Example 2. The $\omega$-limit continuum illustrated in Figure 257 has the $\omega$-enumeration

$$\omega \,|\, L_1,\ O_1,\ L_5,\ O_2,\ L_6,\ O_2,\ L_3,\ O_1,\ L_2,\ O_3,\ L_4,\ O_3.$$

(Here the paths are not numbered in order of their $\omega$-enumeration.)

**2. EQUIVALENCE OF ENUMERATIONS OF TWO LIMIT CONTINUA.** Let $K^{()}$ and $K'^{()}$ be two $\omega$- (or two $\alpha$-, or two 0-) limit continua.

Let

$$\begin{aligned} O_1,\ O_2,\ \ldots,\ O_n, \qquad\qquad &\text{(O)}\\ L_1,\ L_2,\ \ldots,\ L_N \qquad\qquad &\text{(L)} \end{aligned}$$

be the equilibrium states and paths proper figuring in the continuum $K^{()}$,

$$\begin{aligned} O'_1,\ O'_2,\ \ldots,\ O'_n, \qquad\qquad &\text{(O')}\\ L'_1,\ L'_2,\ \ldots,\ L'_N \qquad\qquad &\text{(L')} \end{aligned}$$

those figuring in $K'^{()}$.

Suppose that $\omega$-enumerations of $K^{()}$ and $K'^{()}$ are given:

$$\begin{aligned} L_{j_1},\ O_{i_1},\ L_{j_2},\ O_{i_2},\ \ldots,\\ L'_{j'_1},\ O'_{i'_1},\ L'_{j'_2},\ O'_{i'_2},\ \ldots \end{aligned}$$

*Definition.* $\omega$-enumerations of continua $K^{()}$ and $K'^{()}$ are said to be equivalent if there is a one-to-one correspondence between the equilibrium states (O) and (O'), and between the paths (L) and (L'), figuring in the continua, such that: a) any two successive paths in the $\omega$-enumeration of $K^{()}$ correspond to two successive paths in the $\omega$-enumeration of $K'^{()}$; b) any two identical equilibrium states (O) in the $\omega$-enumeration of $K^{()}$ correspond to two identical equilibrium states (O') in the $\omega$-enumeration of $K'^{()}$.*

Thus, if the $\omega$-enumerations of $K^{()}$ and $K'^{()}$ are equivalent, then $n = n'$, $N = N'$, and, after suitable renumbering (if necessary) of the paths of $K'^{()}$, the $\omega$-enumeration of $K'^{()}$ may be obtained from that of $K^{()}$ by simply adding primes. The elementary proof of the following lemma is omitted.

*Lemma 1. If two limit continua $K^{()}$ and $K'^{()}$ have equivalent $\omega$-enumerations, there exists a topological mapping of one onto the other under which corresponding paths are mapped onto each other and the direction of increasing $t$ is preserved.*

**3. "ONE-SIDED" CANONICAL NEIGHBORHOOD OF A LIMIT CONTINUUM.** Let $K$ be an $\omega$-limit continuum which is not an equilibrium state. To fix ideas, we assume that it is positively situated, denoting it by $K^+$. Let $L_1$ be some constituent path of $K^+$, not an equilibrium state, and $P$ some point on $L_1$; let $l$ be an arc without contact containing $P$ as an interior point (apart from $P$, the arc $l$ can have no points in common with $K$; see Corollary 1 to Lemma 2, §3).

*Lemma 2. Through any point sufficiently close to $P$ on $l$, on the positive side of the path $L_1$, there is a cycle without contact $C$ satisfying the following conditions: 1) there exists exactly one region whose boundary is the union of $C$ and $K^+$; 2) through all points of this region pass paths which tend to $K^+$ as $t \to +\infty$ and leave the region with decreasing $t$, hence crossing $C$; 3) any semipath for which $K^+$ is an $\omega$-limit continuum crosses $C$.*

Proof. $K$ is a positively situated $\omega$-limit continuum for any path $L$ crossing the arc $l$ at a point sufficiently close to $P$ on the positive side of $L_0$ (the path through $P$). Hence there is an infinite sequence of points $\{P_i\}$ of $L$ on $l$, on the positive side of $L_0$, corresponding to an indefinitely increasing sequence of times $t$ and tending to the point $P$. As in §3, let $C_i$ denote the simple closed curve formed by the turn $P_iP_{i+1}$ of $L$ and the subarc $P_iP_{i+1}$ of $l$. By Lemma 8 of §15, for sufficiently large $i$ $(i > J)$ the curves $C_i$ lie inside (outside) one another (not counting their common points) and the continuum $K^+$ lies interior (exterior) to all these curves. To fix ideas, we suppose that $K^+$ is interior to the curves $C_i$.

By Lemma 17 of §3, for every $i > J$ we can draw through $P_{i+1}$ a cycle without contact $C^{(i)}$ lying between $C_i$ and $C_{i+1}$ (i. e., all points of $C^{(i)}$ other than $P_{i+1}$ are inside $C_i$ and all points of $C_{i+1}$ other than $P_{i+1}$ are inside $C^{(i)}$). It is obvious that instead of $P_{i+1}$ we may take any point $P$ on $l$, on the positive side of $L_0$.* Let $\gamma_i$ be the region bounded by the union of $C_i$ and $K^+$, not containing any whole singular path. The cycle without contact $C^{(i)}$ through $P_{i+1}$ divides this region into two subregions, one of which is bounded by the union of $C^{(i)}$ and $K^+$. Denote the latter region by $\gamma_i'$. By Lemma 10 of §16, for sufficiently large $i$, $K^+$ is a positively situated $\omega$-limit continuum for all paths through $\gamma_i$, hence also for all paths through the region $\gamma_i'$.

Now, by definition, all paths passing through points of $\gamma_i$ leave it with decreasing $t$; hence it is clear that all paths through $\gamma_i'$ a fortiori leave this region, crossing the cycle without contact $C^{(i)}$ (since they cannot cross $K^+$). Further, any path for which $K^+$ is an $\omega$-limit continuum will at any rate cross the subarc $PP_{i+1}$ of $l$ at points arbitrarily close to $P$; in other words, any such path will certainly have points lying in $\gamma_i'$. Hence follows part 3) of the theorem, and the proof is complete.

Remark 1. For any sufficiently large $i$, the cycle without contact $C^{(i)}$ lies between the curves $C_i$ and $C_{i+1}$; hence it obviously follows that each curve $C_{i+1}$ lies between two (nested) cycles without contact $C^{(i)}$ and $C^{(i+1)}$.

Remark 2. For any $\varepsilon > 0$, there is a region $\gamma_i'$ contained in the $\varepsilon$-neighborhood of $K^+$ (see Lemma 8, §15).

Any region whose boundary is the union of a cycle without contact $C$ and a continuum $K^+$, such that $K^+$ is the (positively situated) $\omega$-limit continuum

FIGURE 258.

of all paths passing through the region, will be called a canonical neighborhood of $K^+$, and denoted by $\gamma_c$ or simply $\gamma$. The cycle without contact $C$ figuring in the boundary of the canonical neighborhood will be called a cycle without contact of the continuum $K^+$ (Figure 258).

It is clear that a canonical neighborhood of a continuum $K^+$ contains no singular paths, and all paths passing through it leave it with decreasing $t$ across the cycle without contact $C$. In the sequel we shall generally consider the closure of $\gamma_c$, i. e., the closed canonical neighborhood $\overline{\gamma}_c$.

We now return to the curves $C_i$ (formed by turns $P_iP_{i+1}$ of the path $L$ and subarcs $P_iP_{i+1}$ of the arc without contact $l$). The traversal of $C_i$

---

* Note that there may be more than one cycle without contact through $P_{i+1}$.

induced by the direction of $t$ on the arc $P_iP_{i+1}$ of $L$ will be called the $t$-t r a v e r s a l of $C_i$, and the opposite traversal will be called the a n t i - $t$-t r a v e r s a l.

The direction induced by the $t$-traversal of $C_i$ on the subarc $P_iP_{i+1}$ of $l$ is the direction from $P_{i+1}$ to $P_i$. The $t$-traversal of $C_i$ also induces a certain traversal (either positive or negative) on all closed curves, hence also on every cycle without contact $C$ of the continuum $K^+$. We shall call this the $t$-c o m p a t i b l e  t r a v e r s a l of (any) cycle without contact of $K$.

All assertions in this section pertaining to an $\omega$-limit continuum $K^+$ carry over with obvious modifications to an $\omega$-limit continuum $K^-$, and also to $\alpha$-limit continua $K^+$ or $K^-$.

Now let $K_0^+$ be a 0-limit continuum (to fix ideas, assume that it is positively situated), and $L$ one of the closed paths in the cell $w$ for which $K_0^+$ is a positively situated boundary continuum.

Any region bounded by the union of the closed path $L$ and the continuum $K_0^+$ will be called a c a n o n i c a l  n e i g h b o r h o o d  o f  t h e  c o n t i n u u m $K_0^+$ and denoted by $\gamma_L$. Clearly, $\gamma_L$ is a sub-region of a cell, so that through all points of $\gamma_L$ pass closed paths of the cell $w$ (Figure 259).

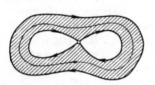

It is an obvious consequence of Lemma 2 that for any $\varepsilon > 0$ the 0-limit continuum has a canonical neighborhood $\gamma_L$ lying entirely in the $\varepsilon$-neighborhood of $K_0^+$.

Henceforth we shall consider mainly closed canonical neighborhoods $\bar{\gamma}_L$. In the sequel we shall use the term c a n o n i c a l  c u r v e of an $\omega$-, $\alpha$- or 0-limit continuum $K^{()}$ for any simple closed curve $C$ figuring in the boundary of

FIGURE 259.

a canonical neighborhood — a cycle without contact if $K^{()}$ is an $\omega$- or $\alpha$-limit continuum and a closed path if it is a 0-limit continuum.*

## 4. LOCAL SCHEMES OF $\omega$-, $\alpha$-AND 0-LIMIT CONTINUA. EQUIVALENCE OF PHASE PORTRAITS IN CANONICAL NEIGHBORHOODS WITH EQUIVALENT LOCAL SCHEMES.

*D e f i n i t i o n  XXVII.  A local scheme of an $\omega$-, $\alpha$- or 0-limit continuum $K^{()}$ is a specification of a) whether $K^{()}$ is an $\omega$-, $\alpha$- or 0-limit continuum, and b) an $\omega$-enumeration of the continuum $K^{()}$.*

*We shall say that the local schemes of two limit continua $K^{()}$ and $K'^{()}$ are equivalent with the same direction of $t$ if: a) both $K^{()}$ and $K'^{()}$ are $\omega(\alpha, 0)$-limit continua; b) their $\omega$-enumerations are equivalent.*

Paths of $K^{()}$ and $K'^{()}$ (or corner and boundary arcs, in the case of schemes of type (1)) which correspond to each other in equivalent schemes will be called s c h e m e - c o r r e s p o n d i n g paths (arcs).

In similar fashion, one defines equivalence of two schemes with reversed direction of $t$.

It is readily seen that specification of a local scheme yields no information concerning the disposition of the limit continuum on the plane. For example, the continua illustrated in Figures 255 and 256 have the same local scheme.

*T h e o r e m  72.  If the local schemes of two $\omega(\alpha, 0)$-limit continua $K^{()}$ and $K'^{()}$ of two dynamic systems (which may coincide) are equivalent, then*

---

* [In the sequel we shall often deal with a specific canonical neighborhood of each limit continuum, referring to it (with a certain abuse of rigor) as "the" canonical neighborhood.]

*the topological structures of the phase portraits in any two closed canonical neighborhoods of the continua are equivalent.*

Proof. Let the two dynamic systems be $D$ and $D^*$. Let $K^+$ be an $\omega$-limit continuum for paths of $D$ and $K^{*+}$ an $\omega$-limit continuum for paths of $D^*$; assume that the local schemes of $K^+$ and $K^{*+}$ are equivalent with the same direction of $t$.

Let $L_1$, $L_2$, ... $L_R$ be the constituent paths of $K^+$ and $L_1^*$, $L_2^*$, ... $L_{R^*}^*$ those of $K^{*+}$. Since $K^+$ and $K^{*+}$ have equivalent local schemes, $R = R^*$.

Let $\gamma$ be a canonical neighborhood of $K^+$, $C$ the cycle without contact figuring in its boundary, $\gamma^*$ a canonical neighborhood of $K^{*+}$ and $C^*$ the cycle without contact figuring in its boundary.

We shall assume that the scheme-corresponding paths are identically indexed, $L_i$ and $L_i^*$. Their $\omega$- and $\alpha$-limit equilibrium states are also assumed to be scheme-corresponding. As in Theorem 68, let $\lambda_i^-$ and $\lambda_i^+$ ($i = 1, 2, 3, \ldots$) be arcs without contact for paths of system $D$, with endpoints $M_i^-$ and $M_i^+$ on the path $L_i$. Let $N_i^-$ and $N_i^+$ be points on these arcs (again as in Theorem 68) with the following properties: all paths crossing the subarc $M_i^- N_i^-$ of $\lambda_i^-$ a) have $\omega$-limit continuum $K$; b) with increasing $t$, cross the subarcs $M_1^+ N_1^+$, $M_2^- N_2^-$, ... of the arcs $\lambda_1^+$, $\lambda_2^-$, $\lambda_2^+$, ... in succession, and ultimately cross $\lambda_1^-$ again at a point of its subarc $M_1^- P_1$ (where $P_1$ is between $M_1^-$ and $N_1^-$) (Figure 260).

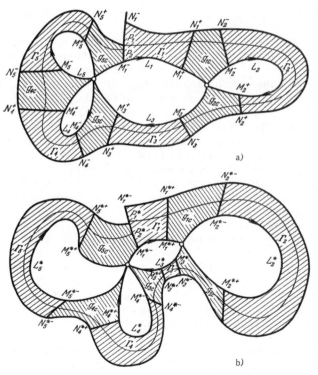

FIGURE 260.

In particular, it follows from a) that with increasing $t$ the path $L$ through $N_1^-$ crosses the arcs $\lambda_1^-$ and $\lambda_1^+$ successively at the points $N_1^-$ and $N_1^+$ and crosses $\lambda_1^-$ again at $P_1$; subsequently, with further increase in $t$, it crosses $\lambda_1^-$ at an infinite sequence of points $\{P_i\}$ corresponding to an indefinitely increasing sequence of times $t$ and tending to the point $M_1^-$. Let $C_1$ be the simple closed curve formed by the turn $N_1^-P_1$ of the path $L$ and the subarc $N_1^-P_1$ of the arc $\lambda_1^-$. The point $N_1^-$ may always be so chosen that $C_1$ lies inside the region $\gamma$, so that the region $\gamma_1$ bounded by the union of $C_1$ and $K$ is a subregion of the canonical neighborhood $\gamma$. Let $\Delta$ be the region bounded by the union of $C_1$ and the arc without contact $C$ ($\gamma_1$ and $\Delta$ are the regions into which $C_1$ divides the canonical neighborhood $\gamma$).

Since the schemes of the continua $K$ and $K^*$ are equivalent, each path $L_i$ corresponds (in one-to-one fashion) to the path $L_i^*$ of system $D^*$. Hence we may treat the situation in exactly the same way for the system $D^*$, obtaining arcs without contact $\lambda_i^{*-}$, $\lambda_i^{*+}$, points $M_i^{*-}$, $M_i^{*+}$, $N_i^{*-}$, $N_i^{*+}$, $P_i^*$, a path $L^*$ whose points in common with the arcs $\lambda_i^{*-}$ are $\{P_i^*\}$, and finally a simple closed curve $C_1^*$ and regions $\gamma_1^*$ and $\gamma^*$ (Figure 260b).

In order to define a topological mapping of $\bar\gamma$ onto $\bar\gamma^*$, we shall first map $\gamma_1$ onto $\gamma_1^*$ and then, preserving the correspondence induced by this mapping between the curves $C_1$ and $C_1^*$, define a mapping of the closed region $\bar\Delta$ onto $\bar\Delta$ *.

The subarcs $M_i^+N_i^+$ and $M_i^-N_i^-$ of $\lambda_i^+$ and $\lambda_i^-$ (subarcs $M_i^{*+}N_i^{*+}$ and $M_i^{*-}N_i^{*-}$ of $\lambda_i^{*+}$ and $\lambda_i^{*-}$) clearly divide the closed region $\bar\gamma_1$ ($\bar\gamma_1^*$) into closed elementary quadrangles $\Gamma_i$ ($\Gamma_i^*$) and closed saddle regions $\bar g_{ic}$ ($\bar g_{ic}^*$), which have common boundary arcs $M_i^+N_i^+$, $M_i^-N_i^-$ ($i = 1, 2, \ldots, R$) ($M_i^{*+}N_i^{*+}$, $M_i^{*-}N_i^{*-}$), or $P_1M_1^-$ ($P_1^*M_1^{*-}$) (Figure 260). Let us say that the elementary quadrangles and saddle regions of systems $D$ and $D^*$, respectively, whose boundaries involve identically indexed arcs (i.e., arcs differing only by the addition of an asterisk), are "scheme-corresponding" regions. We can now define our mapping of $\bar\gamma_1$ onto $\bar\gamma_1^*$. First define a topological mapping of the subarcs $M_i^-N_i^-$ and $M_i^{*-}N_i^{*-}$ of $\lambda_i^-$ and $\lambda_i^{*-}$, under which the points $M_i^-$ and $M_i^{*-}$, $N_i^-$ and $N_i^{*-}$ correspond to each other, and points on the same path of system $D$ correspond to points on the same path of system $D^*$. This is done as follows. First let $\Phi_0$ be an arbitrary topological mapping of the subarcs $N_1^-P_1$ and $N_1^{*-}P_1^*$ of $\lambda_1^-$ and $\lambda_1^{*-}$, under which $N_1^-$ corresponds to $N_1^{*-}$ and $P_1$ to $P_1^*$. Since all paths crossing the subarc $N_1^-P_1$ of $\lambda_1^-$ (the subarc $N_1^{*-}P_1$ of $\lambda_1^{*-}$) cross each subarc $P_kP_{k+1}$ ($P_k^*P_{k+1}^*$) exactly once (see Lemma 8 of §3), we can define a topological mapping of the subarcs $P_kP_{k+1}$ and $P_k^*P_{k+1}^*$ of $\lambda_1^-$ and $\lambda_1^{*-}$ by stipulating that corresponding points lie on paths that cross the subarcs $N_1^-P_1$ and $N_1^{*-}P_1^*$ at points corresponding to each other under the mapping $\Phi_0$.

In other words, corresponding points on the subarcs $P_1P_2$ and $P_1^*P_2^*$ are the "successors" of corresponding points on the subarcs $N_1^-P_1$ and $N_1^{*-}P_1^*$, corresponding points on $P_2P_3$ and $P_2^*P_3^*$ are successors of corresponding points on $P_1P_2$ and $P_1^*P_2^*$, and so on (see Figure 261, where $Q_1$ and $Q_1^*$, $Q_2$ and $Q_2^*$, $\ldots$ are pairs of corresponding points). Finally, map the point $M_1^-$ onto $M_1^{*-}$.

As a result, we obtain a mapping of the subarcs $M_1^-N_1^-$ and $M_1^{*-}N_1^{*-}$ of $\lambda_1^-$ and $\lambda_1^{*-}$, which is obviously one-to-one and bicontinuous, i.e., a topological mapping (bicontinuity at points on these arcs outside the subarcs $N_1^-P_1$ and $N_1^{*-}P_1^*$ follows directly from the continuity of solutions with respect to initial values).

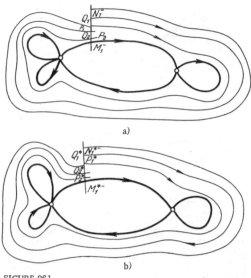

a)

b)

FIGURE 261.

The correspondence we have constructed between the subarcs $M_{\bar{1}}N_{\bar{1}}$ and $M_1^* N_1^{*-}$ of $\lambda_{\bar{1}}$ and $\lambda_1^{*-}$ induces a topological correspondence between the arcs $M_{\bar{i}}N_{\bar{i}}$ and $M_i^{*-}N_i^{*-}$, $M_{\bar{i}}^{\dagger}N_{\bar{i}}^{\dagger}$ and $M_i^{*+}N_i^{*+}$. It is clear that this correspondence pairs off points on these arcs at which paths crossing $M_{\bar{1}}N_{\bar{1}}$ and $M_1^{*-}N_1^{*-}$ at corresponding points cross the arcs $\lambda_{\bar{i}}$, $\lambda_i^{*-}$ and $\lambda_{\bar{i}}^{\dagger}$, $\lambda_i^{*+}$ (without previously crossing     and $\lambda_i^{*-}$ again). Now define a path-preserving topological mapping between scheme-corresponding elementary quadrangles $\Gamma_i$ and $\bar{\Gamma}_i^*$ and saddle regions $\bar{g}_{ic}$ and $\bar{g}_{ic}^*$, preserving the correspondence just defined between the arcs $M_{\bar{i}}^{\dagger}N_{\bar{i}}^{\dagger}$ and $M_i^{*+}N_i^{*+}$, $M_{\bar{i}}N_{\bar{i}}$ and $M_i^{*-}N_i^{*-}$. In combination, this yields a topological mapping of $\bar{\gamma}_1$ onto $\bar{\gamma}_1^*$ under which paths of system $D$ go onto paths of system $D^*$. That this mapping is one-to-one and bicontinuous on the continua $K^+$ and $K^{*+}$ clearly follows from Lemma 1.

We now proceed to the closed regions $\bar{\Delta}$ and $\bar{\Delta}^*$. First consider the region $\Delta$ (the treatment of $\Delta^*$ is entirely analogous). Clearly, all paths crossing the subarc $N_{\bar{1}}P_1$ of $\lambda_{\bar{1}}$ will cross the cycle without contact $C$, and, conversely, all paths crossing $C$ will cross $N_{\bar{1}}P_1$. In addition, all paths passing through points of $\Delta$ cross the subarc $N_{\bar{1}}P_1$ of $\lambda_{\bar{1}}$ with decreasing $t$. Let $Q$ be the point at which $L$ crosses $C$.

Let $R$ be a point on the arc $N_{\bar{1}}P_1$ (distinct from $N_{\bar{1}}$ and $P_1$), $L_1$ the path through this point and $S$ its point of intersection with $C$. The points $Q$ and $S$ divide $C$ into two arcs without contact, while the arcs $QN_{\bar{1}}$ and $RS$ of the paths $L$ and $L_1$, respectively, divide the region $\Delta$ into two elementary quadrangles $\Pi_1$ and $\Pi_2$ (Figure 262a).

Let $R^*$ be the point on the subarc $N_1^{*-}P_1^*$ of $\lambda_1^{*-}$, corresponding to $R$ under the mapping established above of $\bar{\gamma}_1$ onto $\bar{\gamma}_1^*$, and $L_1^*$ the path of system $D^*$ through this point. Let $Q^*$ and $S^*$ denote the points of intersection of $L^*$ and $L_1^*$, respectively, with the cycle $C^*$, and $\Pi_1^*$, $\Pi_2^*$ the elementary quadrangles (similar to $\Pi_1$, $\Pi_2$) into which the arcs $R^*N_1^*$ and $R^*S^*$ of $L^*$ and $L_1^*$ divide $\Delta^*$) (Figure 262b).

385

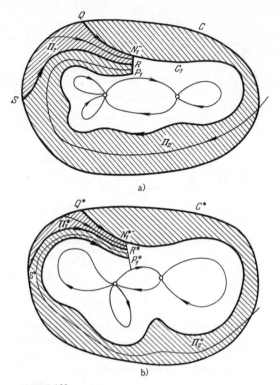

FIGURE 262.

We first map the elementary quadrangles $\overline{\Pi}_1$ and $\overline{\Pi}_1^*$ onto each other, in such a way that paths of systems $D$ and $D^*$ are made to correspond and the correspondence between the arcs $N_1^-P$ and $N_1^*$-$P^*$ induced by our mapping of $\overline{\gamma}_1$ and $\overline{\gamma}_1^*$ is also preserved.

We then define a topological mapping of the elementary quadrangles $\overline{\Pi}_2$ and $\overline{\Pi}_2^*$, preserving the already existing correspondence between their boundary arcs (the arcs without contact $RP_1$ and $R^*P_1^*$, the arcs $P_1N_1^-$ and $P_1^*N_1^{*-}$ of paths $L$ and $L^*$) and between the arcs $RS$ and $R^*S^*$ of paths $L$ and $L^*$. We thus obtain a topological mapping of $\overline{\Delta}$ onto $\overline{\Delta}^*$ which agrees on the curves $C_1$ and $C_1^*$ with the already constructed mapping of $\overline{\gamma}_1$ onto $\overline{\gamma}^*$.

It is now clear that we have constructed a topological mapping of the closed canonical neighborhoods $\overline{\gamma}$ and $\overline{\gamma}^*$, which maps paths of systems $D$ and $D^*$ onto one another.

Thus the theorem is proved for $\omega$-limit continua $K_\omega^+$ and $K_\omega^{*+}$. The proof for $\alpha$-limit continua $K_\alpha^{()}$ and $K_\alpha^{*()}$ is analogous.

Now let $K_0^+$ and $K_0^{*+}$ be $0$-limit continua, containing no corner and boundary arcs, $\gamma$ and $\gamma^*$ canonical neighborhoods of these continua whose boundaries are closed paths $L$ and $L^*$, respectively.

Let $\lambda_1^+$, $\lambda_1^-$ and $\lambda_1^{*+}$, $\lambda_1^{*-}$ be arcs without contact with the same properties as before. Let $L_1$ $(L_1^*)$ be a closed path crossing all the arcs $\lambda_1^+$, $\lambda_1^-$ $(\lambda_1^{*+}$, $\lambda_1^{*-})$. The canonical neighborhood $\overline{\gamma}_1$ $(\overline{\gamma}_1^*)$ bounded by the path $L_1$ and the continuum

$K_0^+$ (the path $L_1^*$ and continuum $K_0^{*+}$) is a subregion of $\gamma\,(\gamma^*)$ and is divided by the arcs $\lambda_i^{()}\,(\lambda_i^{*()})$ into elementary quadrangles and saddle regions. As in the case of $\omega$-limit continua (with a few simplifications), one now defines a topological mapping for each pair of scheme-corresponding elementary quadrangles and saddle regions (all the mappings agreeing on the arcs $\lambda_i^+$, $\lambda_i^{*+}$). One then obtains a topological mapping of the annular regions bounded by the paths $L$ and $L_1$, on the one hand, and $L^*$ and $L_1^*$, on the other, preserving the mapping just defined for $L_1$ and $L_1^*$.

Finally, we obtain a path-preserving topological mapping of $\bar{\gamma}$ onto $\bar{\gamma}^*$.

If $K_0^+$ and $K_0^{*+}$ are unions of boundary and corner arcs, we simply consider disjoint arcs without contact through the endpoints of the arcs and proceed as before.

This completes the proof of Theorem 72.

R e m a r k  I.  Let $K_\omega^+$ and $K_\omega^{*+}$ be continua with equivalent local schemes, and suppose we are given a topological mapping of one onto the other (under which scheme-corresponding paths are mapped onto each other), which moreover preserves directions on paths. Assume moreover that a topological correspondence of the cycles without contact $C$ and $C^*$ is given, preserving the $t$-compatible traversals of both cycles (i. e., if a point describes $C$ in the direction compatible with that of $t$, the same holds for the corresponding point describing $C^*$). Then the topological mapping of $\bar{\gamma}$ onto $\bar{\gamma}^*$ may always be constructed in such a way as to preserve the given correspondence of the continua $K_\omega^+$ and $K_\omega^{*+}$ and the cycles without contact $C$ and $C^*$. To do this, one must clearly choose the endpoints $M_i^{*+}$ and $M_i^{*-}$ of the arcs $\lambda_i^{*+}$ and $\lambda_i^{*-}$ at points corresponding under the given mapping to $M_i^+$ and $M_i^-$, and the correspondence between the arcs without contact $\lambda_i^{()}$ and $\lambda_i^{*()}$ must be that induced by the given correspondence of $C$ and $C^*$. Finally, the correspondence between scheme-corresponding elementary quadrangles and saddle regions must preserve the given correspondence at points of the continua $K_\omega^+$ and $K_\omega^{*+}$ (see Remark to the lemmas of Chapter VIII concerning the equivalence of elementary regions). A similar remark is valid for $\alpha$-limit continua $K_a^+$ and $K_a^{*+}$, or $0$-limit continua $K_0^+$ and $K_0^{*+}$.

R e m a r k  II.  It follows at once from Theorem 72 that the phase portraits in any two canonical neighborhoods of a continuum $K_\omega^{()}\,(K_a^{()},\,K_0^{()})$ are equivalent.

## §25. GLOBAL SCHEME OF A LIMIT CONTINUUM

**1. SIMPLE CLOSED CURVES FORMED BY CONSTITUENT PATHS OF A LIMIT CONTINUUM.** The following lemmas will add to our knowledge of the structure of limit continua.

L e m m a  1.  *Any limit continuum $K$ is the union of finitely many simple closed curves*

$$S_1,\ S_2,\ \ldots,\ S_p, \tag{S}$$

*formed by the paths of $K$, which possess the following properties: a) any path of $K$ which is not an equilibrium state is a component of exactly one curve of $(S)$; b) every curve of $(S)$ has a point (equilibrium state) in* *common with at least one of the other curves of* (S) *(if* $p > 1$*), and no two of these curves can have more than one point in common; c) the curves* (S) *are either mutually exterior\* or one of them contains all the others, which are mutually exterior; d) any simple closed curve whose points are all in* K *appears in the sequence* (S).

Proof. First let $K$ be an $\omega$-limit continuum. Let us consider simple closed curves all of whose points are in $K_\omega$. The continuum $K_\omega$ is the union of finitely many nonclosed paths $L_i$ and equilibrium states $O_j$ to which they tend. Since paths cannot intersect, it is obvious that if $S$ is a simple closed curve all of whose points are in $K_\omega$ then any nonclosed path $L$ is either a component of $S$ or has no points in common with it. This means that any simple closed curve of this kind is a union of whole paths. Since $K$ is the union of finitely many paths, it follows immediately that the number of simple closed curves all of whose points are in $K_\omega$ is also finite (if such curves exist at all).

We now prove that any nonclosed path $L_i$ is a component of some simple closed curve $S$ whose points are all in $K_\omega$. Let $L_1, O_1, L_2, O_2, \ldots, L_R, O_R$ be an $\omega$-enumeration of $K_\omega$. (Some or all of the points $O_i$ may coincide.) If $O_1 \equiv O_R$, then the path $L_1$ and the point $O_1$ form the required curve $S$. If $O_1$ and $O_R$ are distinct points, let $p$, $1 \leqslant p < R$, be the maximum number such that $O_p$ coincides with $O_1$. If $O_{p+1} \equiv O_R$, the paths $L_1, O_1, L_{p+1}, O_R$ form the required curve $S$. If $O_{p+1}$ and $O_R$ are distinct, let $q$, $p + 1 \leqslant q < R$, be the maximum number such that $O_q$ coincides with $O_{p+1}$. If $O_{q+1}$ and $O_R$ coincide, the paths $L_1, O_1, L_{p+1}, O_{p+1}, L_q, O_R$ form the required curve $S$, and if not we can continue in this way. The procedure must terminate after finitely many steps in a simple closed curve containing the path $L_1$. Thus all the nonclosed paths $L_i$, hence also all the equilibrium states $O_j$ (as limit points of the paths $L_i$), are components of some simple closed curve all of whose points are in $K_\omega$. Let $S_1, S_2, \ldots, S_l$ be all such curves (we have already established that their number is finite). As we have seen, these curves exhaust all the constituent paths of $K_\omega$. We claim that each nonclosed path $L_i$ appears in exactly one of the curves $(S)$. Suppose that $L_i$ is a component of two curves, say $S_1$ and $S_2$. Let $L^*$ be a path for which $K$ is an $\omega$ ($\alpha$ or 0)-limit continuum. Clearly, $L^*$ cannot cross any of the closed curves $S_i$, so that it is either wholly inside $S_i$ or wholly outside. it. On the other hand, all the points of the curves $S_i$ are limit points of $L^*$. Hence it follows at once that the curves $S_1$ and $S_2$ are either mutually exterior (excluding their common points) or one of them lies inside the other (again excluding common points), say $S_2$ inside $S_1$. In the first case $L^*$ passes outside both curves $S_1$ and $S_2$. But since $L_i$ borders on either side on the region interior to $S_1$ and $S_2$, it cannot be a limit path of $L^*$ — contradiction. In the second case, $L^*$ passes inside $S_1$ and outside $S_2$. But then\*\* $L_i$ again cannot be a limit path for $L^*$. We have thus proved part a) of our lemma. It follows from part a) that the common points of the curves $(S)$ are necessarily equilibrium states. If $p > 1$, each of the curves $(S)$ has a point in common with at least one of the others, in view of the fact that $K_\omega$, as a continuum, is a connected set.

---

\* Except for their common points.

\*\* $L_i$, together with the equilibrium states to which it tends, is clearly a simple arc, and we may therefore apply the propositions of §2.2 in the Appendix.

Suppose now that two of the curves, say $S_1$ and $S_2$, have more than one point in common. Then there obviously exists an arc $\lambda$ consisting of paths of $S_1$ such that 1) the endpoints $O_1$ and $O_2$ of $\lambda$ are equilibrium states which are also points of $S_2$, 2) the interior points of $\lambda$ are not points of the curve $S_2$.

The points $O_1$ and $O_2$ divide $S_2$ into two arcs $S_2'$ and $S_2''$. Suppose that $\lambda$ (excluding its endpoints) lies entirely inside $S_2$. Then it divides the interior of the curve into two regions $g_2'$, $g_2''$, bounded respectively by the arcs $\lambda$ and $S_2'$, $\lambda$ and $S_2''$. Since $\lambda$ is a limit arc for the path $L^*$, the latter must pass through the interior of $S_2$, hence inside either $g_2'$ or $g_2''$. In the first case $S_2''$, and in the second $S_2'$, cannot be a limit arc of $L^*$. This contradicts the fact that $K_\omega$ is a limit continuum of $L^*$.

The case in which $\lambda$ lies outside $S_2$ is treated in analogous fashion. This proves part b).

To prove part c), we denote the region bounded by $S_1$ by $g_1$ and assume that $g_2 \subset g_1$. If some curve, say $S_3$, lies inside $g_2$ or outside $g_1$, then the three curves $S_1, S_2, S_3$ cannot all be unions of limit paths of $L^*$. This contradicts the fact that they are components of the limit continuum $K_\omega$ of $L^*$, and completes the proof of part c). Part d) is a direct consequence of the fact that $(S)$ is by definition the set of all simple closed curves figuring in the continuum $K_\omega$.

The case of a 0-limit continuum $K$ is treated in similar fashion. Q.E.D.

R e m a r k. It follows from the proof of part a) of the lemma that each simple closed curve $S_i$ is the union of certain paths

$$L_{q_1}, O_{j_1}, L_{q_2}, O_{j_2}, \ldots, L_{q_k}, O_{j_k}, \tag{1}$$

which form a subsequence of the $\omega$-enumeration of the continuum $K$. Moreover, a point moving along the paths in the above order (their order in the $\omega$-enumeration) describes the curve $S_i$ in either the positive or the negative direction. But then the following assertion is obviously valid. When the simple closed curve $S_i$ $(i = 1, 2, \ldots, l)$ is described in the positive direction (positive traversal), its constituent nonclosed paths are described either all in the direction of increasing $t$ or all in the direction of decreasing $t$.

E x a m p l e. Let the $\omega$-enumeration of $K$ be

$$\omega \,|\, O_1, L_1, O_2, L_2, O_2, L_3, O_3, L_4, O_4, L_5, O_4, L_6, O_3, L_7, O_5, L_8.$$

The curve $S_1$ containing $L_1$ is

$$S_1\colon O_1, L_1, O_2, L_3, O_3, L_7, O_5, L_8.$$

The path $L_2$ does not appear here. The curve containing it is $S_2\colon O_2, L_2$. Similarly, $S_3\colon O_3, L_4, O_4, L_6$ and $S_4\colon O_4, L_5$. All paths of $K$ are now exhausted by the curves $S_1, S_2, S_3, S_4$. As we have stated, specification of only one sequence of paths does not fully determine the relative positions of the curves $S_i$; neither does it indicate whether the positive traversal of a curve $S_i$ agrees with the direction of increasing or decreasing $t$ (see the preceding remark). Thus, the limit continua illustrated in Figures 255 and 256 have the same $\omega$-enumeration, viz. that considered in the above example. But in the first continuum the curves $S_i$ $(i = 1, 2, 3, 4)$ are mutually exterior, while in the second there is one curve $S_i$ containing all the rest. In the continuum of Figure 255, the positive traversal on all the curves $S_i$ agrees with the direction of decreasing $t$, in that of Figure 256 — increasing $t$.

**2. ONE-SIDED AND TWO-SIDED LIMIT CONTINUA.** Let $K$ be an $\omega$-, $\alpha$- or $0$-limit continuum. We shall call $K$ a o n e - s i d e d limit continuum if it is only positively (negatively) situated, t w o - s i d e d if it is both positively and negatively situated (naturally, for different paths). In the first case only one of the notations $K^+$ or $K^-$ applies, and in the second — both.

Even if the continuum $K^+$ is one-sided, say a positively situated $\omega$-limit continuum, some or even all of its constituent paths may also be negatively situated limit paths. Consequently, each such path may be a component of some negatively situated $\omega$-, $\alpha$- or $0$-limit continuum. The latter, however, is necessarily distinct from $K^+$ (even though the two continua have paths in common). The elementary proof of the following lemma is omitted.

L e m m a 2. *A limit continuum* $K$ *consisting of more than one simple closed curve* $S$ *is one-sided. Moreover, if the curves* $S_i$ *are mutually exterior, the paths for which* $K$ *is an* $\omega$-, $\alpha$- *or* $0$-*limit continuum lie outside all the curves* $S_i$. *But if one of the curves,* $S_1$ *say, encloses all the others, the paths for which* $K$ *is an* $\omega$-, $\alpha$- *or* $0$-*limit continuum lie inside* $S_1$ *and outside all the other* $S_i$ *(Figures 255 and 256).*

L e m m a 3. *Let* $K^+$ *be an* $\omega$ ($\alpha$ or $0$)-*limit continuum, containing at least one equilibrium state, which consists of a single simple closed curve* $S$. *Let* $L$ *be a path, not a component of* $K^+$, *which tends to an equilibrium state* $O$ *in* $K^+$. *Then* $K^+$ *is one-sided, and moreover if* $L$ *is outside (inside) the curve* $S$ *the paths for which* $K^+$ *is an* $\omega$ ($\alpha$ or $0$)-*limit continuum lie inside (outside)* $S$.

P r o o f. Since $S$ is a simple closed curve, there exist only two semi-paths $L_1^+$ and $L_2^-$ in $K^+$ which tend to $O$ and are continuations of each other on the positive side. The semipaths $L_1^+$ and $L_2^-$ cannot be continuations of each other from the negative side as well, for if they were there could be no path $L$ not a component of $K^+$ but tending to $O$. Therefore, either the path $L_1^+$ has no continuation on the negative side or its continuation on the negative side is not $L_2^-$ (Figure 263a, b). In the first case (Figure 263a) $L_1^+$ cannot be a negatively situated limit semipath, and in the second (Figure 263b) it may appear in some $\omega$-, $\alpha$- or $0$-limit continuum $K'^-$ other than $K^+$. This means that $K$ is a one-sided continuum.

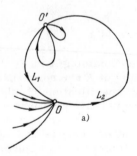

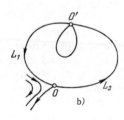

FIGURE 263.

Moreover, it is clear that if the points on the positive side of the semipaths $L_1^+$ and $L_2^-$ lie inside (outside) the curve $S$, then $L$ lies outside (inside) $S$, for otherwise $L_1^+$ and $L_2^-$ could not be continuations of each other on the positive side. Q.E.D.

*Corollary. A continuum $K$ is a two-sided $\omega$-, $\alpha$- or $0$-limit continuum if and only if it is either a closed path, or a simple closed curve made up of nonclosed paths tending to equilibrium states and having the same continuations on both the positive and the negative side. In the latter case, no paths other than those appearing in $K$ can tend to the equilibrium states of $K$.*

Let the continuum $K^0$ be the union of more than one simple closed curve $S_i$, and let $0$ be one of its equilibrium states.

Suppose that the point $0$ is common to more than two of the curves $S_i$, say $S_1, S_2, \ldots, S_m$ (after suitable renumbering). It is clear that each of these curves $S_j$ must include two semipaths tending to $0$, a positive semi-path $L_i^+$ and a negative one $L_i^-$ (there are only two such semipaths, in the sense that any semipath tending to $0$ and figuring in the curve $S_j$ is either part of one of $L_i^+$, $L_i^-$ or contains it).

Reasoning as in Lemma 11 of §17 and recalling that any constituent semipath of $K$ has a continuation, we readily prove the following assertions. 1) The region between any two semipaths $L_j^+$ and $L_j^-$ figuring in the same curve $S_p$ contains either none or all of the semipaths $L_i^{\prime}$ $(i \neq j)$. 2) Of any two cyclically successive semipaths $L_i^{\prime}$, one is always positive and one is negative.

Now consider any two of the curves $S_j$ $(j = 1, 2, \ldots, l)$, say $S_\mu$ and $S_\lambda$. By 1) and 2), it is readily seen that the semipaths

$$L_\mu^+, \ L_\mu^{\prime -} \ \text{and} \ L_\lambda^+, \ L_\lambda^{\prime -} \tag{2}$$

may be arranged about the point $0$ in either of the cyclic orders

$$L_\mu^+, \ L_\mu^{\prime -}, \ L_\lambda^+, \ L_\lambda^{\prime -}, \tag{3}$$

or

$$L_\mu^+, \ L_\lambda^{\prime -}, \ L_\lambda^+, \ L_\mu^{\prime -}. \tag{4}$$

In the following lemmas we shall establish the relation between which of the orders (3) or (4) applies, on the one hand, and the question of whether the positive traversals of $S_\mu$ and $S_\lambda$ are $t$-traversals or anti-$t$-traversals, and whether $S_\mu$ and $S_\lambda$ lie inside or outside one another.

*Lemma 4. Let the curves $S_\mu$ and $S_\lambda$ be mutually exterior. Then: a) if the semipaths (2) are arranged about $0$ in the cyclic order (3)*

$$L_\mu^+, \ L_\mu^{\prime -}, \ L_\lambda^+, \ L_\lambda^{\prime -},$$

*the positive traversal on both curves is the anti-$t$-traversal; b) if the semipaths (2) are arranged about $0$ in cyclic order (4)*

$$L_\mu^+, \ L_\lambda^{\prime -}, \ L_\lambda^+, \ L_\mu^{\prime -},$$

*the positive traversal of both curves is the $t$-traversal.*

Proof. a) Let σ be a simple closed curve containing the equilibrium state $O$ in its interior, cutting each of the semipaths (3) in exactly one point $M_\mu^+$, $M_\mu'^-$, $M_\lambda^+$, $M_\lambda'$ , respectively, and having no other points in common with $S_\mu$ and $S_\lambda$.

We may assume that σ is a canonical curve of the equilibrium state $O$. Indeed, the semipaths (2) are separatrices of $O$ and it is readily seen that any canonical curve of $O$ (which of course has points in common with the semipaths (2)) whose saddle arcs are sufficiently small has exactly one point in common with each of the semipaths (2). In addition, all constituent semipaths of $S_\mu$ and $S_\lambda$ tend to only two equilibrium states, semipaths distinct from (3) tending to equilibrium states other than $O$ (since $S_\mu$ and $S_\lambda$ are simple closed curves). Hence it follows that the points of $S_\mu$ and $S_\lambda$ not on the semipaths (2) lie at a positive distance from the point $O$. Hence any canonical curve in a sufficiently small neighborhood of $O$ with sufficiently small saddle arcs will have no points in common with $S_\mu$ and $S_\lambda$ other than those in which it cuts the semipaths (2). We can now prove part a). By assumption, $S_\mu$ and $S_\lambda$ are mutually exterior (not counting their common point $O$) and their points of intersection with the curve σ lie in cyclic order $M_\mu^+$, $M_\mu'^-$, $M_\lambda^+$, $M_\lambda'^-$ (Figure 264).

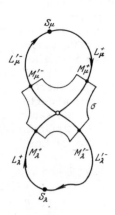

FIGURE 264.

Since $M_\lambda^+$ and $M_\lambda'^-$ are points of $S_\lambda$, they lie outside $S_\mu$, and so all interior points of the arc $M_\mu^+ M_\mu'^-$ (the subarc of σ containing $M_\mu^+$ and $M_\mu'^-$) lie inside $S_\mu$. Therefore, if $\sigma_1$ is the simple closed curve formed from the arc $M_\mu^+ M_\mu'^-$ of σ and the arc of $S_\mu$ consisting of the arc $M_\mu^+O$ of $L_\mu^+$, the arc $M_\mu'-O$ of $L_\mu'$ and the point $O$, then the regions interior to $\sigma_1$ and $S_\mu$ lie on one side of their common arc $M_\mu'-OM_\mu^+$ (see Appendix, §2.2).

Since the positive traversal of the curve $\sigma_1$ induces the direction $M_\mu'-OM_\mu^+$ on the common curve, it follows that the positive traversal of $S_\mu$ induces the direction $M_\mu'-OM_\mu^+$; this is the direction of decreasing $t$ on the paths $L_\mu$ and $L_\mu'$, and consequently on all paths included in $S_\mu$ which are not equilibrium states (see Remark to Lemma 1).

Now let $\sigma_2$ be the simple closed curve made up of the arc $M_\lambda^+M_\lambda'^-$ of σ and the arc $M_\lambda^+OM_\lambda'^-$ of $S_\lambda$; treating this curve and $S_\mu$ as before, we conclude that the positive traversal of $S_\lambda$ agrees with the direction of decreasing $t$ on all paths in $S_\lambda$ which are not equilibrium states.

The proof of part b) is analogous.

The proof of the next lemma is similar, and is left to the reader.

Lemma 5. Suppose that one of the curves $S_\mu$ and $S_\lambda$ lies inside the other. a) If the semipaths (2) are arranged about $O$ in cyclic order (4) $L_\mu^+$, $L_\lambda'^-$, $L_\lambda^+$, $L_\mu'^-$ , then the positive traversal of the inner curve is the $t$-traversal, that of the outer curve the anti-$t$-traversal (Figure 265). b) If the semipaths (2) are arranged about $O$ in cyclic order (3) $L_\mu^+$, $L_\mu'^-$, $L_\lambda^+$, $L_\mu'^-$ , then the positive traversal of the outer curve is the $t$-traversal, that of the inner curve the anti-$t$-traversal.

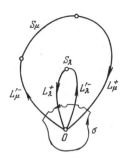

FIGURE 265.

The following converse is also valid:

**Lemma 6.** *a) If the positive traversal of both $S_\mu$ and $S_\lambda$ is the t-traversal, they are mutually exterior and the semipaths (2) are arranged about $O$ in the cyclic order $L_\mu^{\prime-}, L_\mu^+, L_\lambda^-, L_\lambda^{\prime+}$. b) If the positive traversal of both curves $S_\mu$ and $S_\lambda$ is the anti-t-traversal, they are mutually exterior and the semipaths (2) are arranged about $O$ in cyclic order $L_\lambda^-, L_\mu^+, L_\mu^{\prime-}, L_\lambda^{\prime+}$.*

Proof. The lemma follows from Lemmas 4 and 5 by an obvious indirect argument.

The proof of the following lemma is analogous.

**Lemma 7.** *Suppose that the positive traversal of one of the curves $S_\lambda$, $S_\mu$ is the t-traversal, that* of the other the anti-t-traversal. *Then one of them lies inside the other and a) if the semipaths (2) are arranged about $O$ in cyclic order $L_\mu^{\prime-}$, $L_\mu^+$, $L_\lambda^-$, $L_\lambda^+$ then the positive traversal of the outer (inner) curve is the (anti-) t-traversal; b) if the semipaths (2) are arranged about $O$ in cyclic order $L_\mu^+$, $L_\lambda^-$, $L_\lambda^+$, $L_\mu^{\prime-}$, the positive traversal of the inner (outer) curve is the (anti-)t-traversal.*

**Theorem 73.** *Let $K^0$ be an $\omega$-, $\alpha$- or $0$-limit continuum consisting of more than one simple closed curve $S_i$. a) If all the curves $S_i$ are mutually exterior, their positive traversals are either all t-traversals or all anti-t-traversals; b) if one of the curves $S_i$, say $S_j$, encloses all the others, then either the positive traversal of $S_j$ is the t-traversal and the positive traversals of all the other $S_i (i \neq j)$ are anti-t-traversals, or vice versa.*

Proof. If $K^0$ is the union of two simple closed curves $S_i$, the assertion follows directly from Lemmas 4 through 7. If $K^0$ comprises $n$ simple closed curves

$$S_1, S_2, S_3, \ldots, S_n,$$

where $n > 2$, we need only consider the pairs of curves $S_1$ and $S_2$, $S_2$ and $S_3, \ldots, S_{n-1}$ and $S_n$ successively, applying Lemmas 4 through 7. Q.E.D.

Remark. Let $K^0$ be a positively (negatively) situated $\omega$-, $\alpha$- or $0$-limit continuum and $L^*$ a path for which $K^0$ is an $\omega$- or $\alpha$-limit continuum on, if $K^0$ is a $0$-limit continuum, a closed path in the cell for which it is a boundary continuum. Then: a) if the positive traversal of a constituent curve $S_i$ of $K^0$ is the t-traversal, the path $L^*$ lies inside (outside) $S_i$; b) if the positive traversal of $S_i$ is the anti-t-traversal, the path $L^*$ lies outside (inside) $S_i$.

**3. RELATIVE POSITIONS OF CONTINUA AND THEIR CANONICAL CURVES.** Let $C$ be the canonical curve of a limit continuum $K^0$ (see §24.3), i.e., a cycle without contact or a closed path, according as $K^0$ is an $\omega$-, $\alpha$ or $0$-limit continuum.

**Lemma 8.** *a) If $K^0$ lies inside the canonical curve $C$, it is either a simple closed curve or the union of several mutually exterior simple closed curves. b) If $K^0$ lies outside $C$, there exists a constituent simple closed curve of $K^0$ containing $C$ in its interior, and all the other simple closed curves $S_i$ of $K^0$ (if such exist) lie inside it and do not contain $C$ in their interior.*

Proof. a) Suppose there exist two simple closed curves of $K^0$, $S_\mu$ and $S_\lambda$, such that $S_\mu$ lies inside $S_\lambda$. Since by assumption $K^0$ is inside the canonical curve $C$, $S_\lambda$ is also inside $C$ and the points of any canonical neighborhood of $K^0$ are clearly outside $S_\lambda$. But then the points of $S_\mu$ (which is inside $S_\lambda$) can be neither $\omega(\alpha)$-limit points for paths outside $S_\lambda$ nor boundary points of a cell filled by closed paths lying outside $S_\lambda$. This contradiction proves part a).

b) Now let $K^0$ be outside $C$. Suppose that $C$ is exterior to all constituent curves $S_i$ of $K^0$. By definition, the region bounded by $C$ and $K^0$ is a canonical neighborhood of $K^0$. Under our assumptions, the only possible region whose boundary includes both $C$ and $K^0$ is outside $C$ and outside all the curves $S_i$. Moreover, its boundary must also include the boundary of the region $G^*$ (in which the dynamic system is being considered). But by the definition of a canonical neighborhood its boundary cannot contain any point not in $C$ or $K^0$. This contradiction implies that $C$ must be inside one of the curves $S_i$, say $S_1$. But then none of the curves $S_i$ $(i \neq 1)$ can lie outside $S_i$, for otherwise there could be no region bounded only by $C$ and $K^0$, contrary to the assumption that $C$ is a canonical curve of $K^0$. Consequently, all the curves $S_i$ $(i \neq 1)$ are inside $S_1$, and it is clear that they are mutually exterior (excluding their common points) and exterior to $C$. Q.E.D.

Remark. Suppose that the continuum $K^0$ is, say, a continuum of type $K_\omega^+$. Let $l$ be an arc without contact with endpoint $P$ on $K_\omega^+$, lying on the positive side of $K_\omega^+$, $L$ a path tending to $K_\omega^+, P_1, P_2, \ldots$ the points at which it crosses the arc $l$ and $C_i$ the closed curves familiar from previous discussions (each the union of the turn $P_i P_{i+1}$ of $L$ and the subarc $P_i P_{i+1}$ of $l$). By Remark 2 following Lemma 2 of §24, it is clear that if $K_\omega^+$ is inside (outside) its cycle without contact, then it is also inside (outside) all curves $C_i$ for sufficiently large $i$. Conversely, if $K_\omega^+$ lies inside (outside) all the $C_i$ for sufficiently large $i$, it is also inside (outside) any of its cycles without contact.

When the continuum $K_\omega^+$ consists of more than one simple closed curve $S_i$ $(i = 1, 2, \ldots)$, we have the following converse, which follows at once from Lemma 8:

Lemma 9. *Let the limit continuum $K^0$ be the union of more than one simple closed curve $S_i$. a) If all the $S_i$ are mutually exterior, the canonical curve $C$ contains $K^0$ in its interior. b) If one of the $S_i$, say $S_1$, contains all the others in its interior, then any canonical curve $C$ is inside $S_1$ and outside all the other $S_i$ $(i \neq 1)$.*

As before, let $C$ be the cycle without contact of an $\omega$- or $\alpha$-limit continuum $K^0$. The next lemma clarifies the relation between the $t$-traversal of the constituent curves $S_i$ of $K^0$ and the $t$-compatible traversal of its cycles without contact (see §25.3).

Lemma 10. *a) Let $K^0$ consist of a single simple closed curve or of several mutually exterior (except for common points) closed curves $S_i$. If the positive traversals of the curves $S_i$ are the $t$-traversals (anti-$t$-traversals), then the positive traversal of any cycle without contact of $K^0$ is $t$-compatible ($t$-incompatible). b) Let one of the simple closed curves $S_i$ of $K^0$, say $S_1$, contain all the others in its interior. Then if the positive traversal of $S_1$ is the $t$-traversal (anti-$t$-traversal), the positive traversal of any cycle without contact of $K^0$ is $t$-compatible ($t$-incompatible).*

Proof. To fix ideas, we consider a continuum $K_\omega^+$.

a) Let $C$ be some cycle without contact of $K_\omega^+$ and $\gamma$ the corresponding canonical neighborhood. Let $L_1$ be a constituent path of $K_\omega^+$ and $S_1$ the simple closed curve of $K_\omega^+$ containing $L_1$. Fixing a motion on $L_1$, choose four points $A$, $A_1$, $A_2$, $A_3$ on $L_1$ corresponding to times $t_0$, $t_1$, $t_2$, $t_3$ such that $t_0 < t_1 < t_2 < t_3$. Let $l_0$, $l_1$, $l_2$, $l_3$ be arcs without contact, each with one endpoint at $A$, $A_1$, $A_2$, $A_3$, respectively, on the positive side of $L_1$ and disjoint from each other (Figure 266). Any path $L$ passing at $t = t_0$ through a point

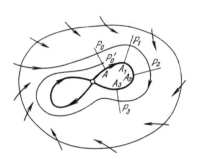

FIGURE 266.

$P_0$ on $l_0$ sufficiently close to $A$ will cross $l_1$ at some time $t_1' > t_0$ at a point $P_1$, without previously crossing either $l_0$ or $l_1$; then, at some time $t_2' > t_1'$, it will cross $l_2$ at a point $P_2$ and $l_3$ at a point $P_3$ at time $t_3' > t_2'$. Finally, it will cross $l_0$ again at some point $P_0'$. Let $C'$ be the simple closed curve formed by the turn $P_0 P_0'$ of $L$ and the subarc $P_0 P_0'$ of $l_0$. Clearly, the subarcs $P_1 A_1$, $P_2 A_2$ and $P_3 A_3$ of $l_1$, $l_2$ and $l_3$, respectively, have no interior points in common with the curves $S_1$ and $C'$ and join the three points $P_1$, $P_2$, $P_3$ of $C'$ to the points $A_1$, $A_2$ $A_3$ of $S_1$ (Figure 266). (In the figure, the positive traversal is the $t$-traversal.)

By the preceding lemma and the remark following it, the continuum $K_\omega^+$, hence also the curve $S_1$, lie inside the curve $C'$, and this clearly implies the assertion of part a) (see Appendix, §2.7). The proof of part b) is analogous.

If $K^0$ is a 0-limit continuum, the canonical curve $C$ is any closed path of the cell for which $K^0$ is a boundary continuum, and the modifications needed in the proof are obvious.

*Lemma 11. a) Let $K^0$ be a 0-limit continuum consisting of a single simple closed curve or several mutually exterior curves $S_i$. Then, if the positive traversal of all the $S_i$ is the (anti-)$t$-traversal, the positive traversal of any closed path of the cell $w$ for which $K^0$ is a boundary continuum is also the (anti-)$t$-traversal. b) Suppose that one of the constituent curves $S_i$ of $K^0$, $S_1$ say, contains all the others in its interior. Then, if the positive traversal of $S_1$ is the (anti-)$t$-traversal, the positive traversal of any closed path in the cell $t$ is also the (anti-)$t$-traversal.*

**4. FREE AND NONFREE CONTINUA.** Let $K^0$ be an $\omega$- or $\alpha$-limit continuum. The set of paths for which it is a limit continuum may contain singular semipaths (i. e., orbitally unstable semipaths, separatrices or corner semipaths.

$K^0$ is said to be *free* if it is not an $\omega$- or $\alpha$-limit continuum for any singular semipath, *nonfree* if there is at least one singular semipath tending to it.

Let $K^0$ be a nonfree $\omega$-limit continuum, and $L_1^{*0}$, $L_2^{*0}$, ..., $L_R^{*0}$ all the singular semipaths tending to $K^0$. It is evident that each of these semipaths is either a semipath of a separatrix of some equilibrium state not appearing in $K^0$, or a corner semipath. Any cycle without contact of the continuum $K^0$ is crossed by all the paths $L_i^{*0}$. In this situation, we have the following

*Lemma 12.  The cyclic order of the crossing points of the paths $L_i^{*()}$ on any cycle without contact of the continuum $K^()$ is independent of the specific cycle chosen.*

P r o o f.  Let $C_1$ and $C_2$ be any two cycles without contact of $K^()$. Suppose first that they are disjoint. Since each path $L_i^{*()}$ crosses both $C_1$ and $C_2$, and does so only once for each cycle, the assertion is self-evident in this case.

Now let the cycles $C_1$ and $C_2$ intersect. We can always find a cycle without contact of $K^()$ which is disjoint from both $C_1$ and $C_2$ (e. g., any cycle without contact $C'$ in a sufficiently small neighborhood of $K^()$). We can then apply the previous argument to the pairs $C_1$ and $C'$, $C_2$ and $C'$, proving the lemma.

R e m a r k.  Let $S$ be a simple closed curve, the union of a turn of some path distinct from any of the paths $L_i^*$ (i. e., a nonsingular path) and an arc without contact contained in some canonical neighborhood of $K^()$. It is clear that all the semipaths $L_i^{*()}$ cross this arc without contact. Moreover, the cyclic order of their crossing points on $S$ is the same as their cyclic order on any cycle without contact of $K^()$.

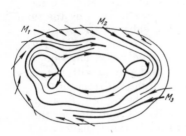

FIGURE 267.

Lemma 12 yields certain information on the cyclic order of the singular semipaths $L_i^{*()}$ $(i = 1, 2, 3, \ldots, n)$ that tend to the continuum $K^()$: it is the same as the cyclic order of the points $M_i$ at which these paths cross any cycle without contact of the continuum (Figure 267).

**5.  GLOBAL SCHEME OF A LIMIT CONTINUUM.**  We first recall the definition of the local scheme of a limit continuum. According to §24.3, the local scheme of a limit continuum $K^+$ or $K^-$ is a list giving an enumeration of its paths and indicating whether it is an ω-, α- or 0-limit continuum. By Theorem 72, the local scheme uniquely determines the topological structure of the phase portrait in any closed canonical neighborhood of $K^()$. By Lemma 1, the local scheme enables us to determine the "path-composition" of the simple closed curves $S_i$ that make up the continuum.

However, as we have mentioned, the local scheme does not determine the relative positions of the curves $S_i$ on the plane, neither does it indicate which of the possible traversals of the curves is the $t$-traversal or what singular semipaths tend to $K^()$. The global scheme, which we are not going to define, fills these gaps.

*Definition XXVIII.  A global scheme of a limit continuum $K^()$ is a list specifying 1) whether the continuum is positively or negatively situated (i. e., which of the signs + or − is to replace the parentheses in the notation $K^()$); 2) a local scheme of the continuum, i. e., an indication as to whether it is an ω-, α- or 0-limit continuum and an ω-enumeration of its constituent paths; 3) whether the positive traversal on each of the simple closed curves $S_i$ in $K^()$ is the $t$-traversal or anti-$t$-traversal (the curves $S_i$ are uniquely determined by the local scheme; see Remark to Lemma 1 above); 4) if $K^()$ is an ω- or α-limit continuum, a list of all the singular semipaths that tend to it and their cyclic order, indicating which of them are corner semipaths and which semipaths of orbitally unstable paths.*

Let $O_1$, $O_2$, ..., $O_m$ $(O)$ be all the equilibrium states, $L_1$, $L_2$, . . ., $L_R$ $(L)$ all the nonclosed paths, and $S_1$, $S_2$, ..., $S_l$ $(S)$ all the simple closed curves of the continuum $K^{()}$.

Let $\tilde{L}_1$, $\tilde{L}_2^{()}$, ..., $\tilde{L}_p^{()}$ $(\tilde{L})$ be all the semipaths of orbitally unstable paths tending to $K^{()}$, and $\hat{L}_1^{()}$ $\hat{L}_2^{()}$, ..., $\hat{L}_q^{()}$ $(\hat{L})$ all the corner semipaths tending to $K^{()}$.

We write $S_i^+$ for $S_i$ if the positive traversal of the curve is the $t$-traversal, $S_i^-$ otherwise; we write $S_i^{()}$ if either of the symbols + or − is implied but it is immaterial (or not known) which. With these conventions, the global scheme of a continuum $K^{()}$ is specified as follows.

I. Indicate which of the symbols + or − is to replace the parentheses in the notation for $K$.

II. Indicate whether $K^{()}$ is an ω-, α- or 0-limit continuum.

III. Specify an ω-enumeration of $K^{()}$.

IV. Construct a table specifying the $t$-traversals of the curves $S_i$ (which are fully defined by the ω-enumeration), i.e., listing the constituent paths of $S_i^{()}$ and indicating the symbol (+ or −) to replace the parentheses in its notation:

$$S_i^{()} \,|\, L_{a_i'}, \; O_{\beta_i'}, \; ..., \; L_{a_i(n)}, \; O_{\beta_i(k)}, \qquad i = 1,\, 2,\, ...,\, r.$$

In the special case of a continuum consisting of a single closed path $L_0$, there is only one closed curve $S^{()}$, coinciding with the path, and the table is simply $S^{()}|L_0$.

V. If $K^{()}$ is not free, list (in cyclic order) all singular semipaths tending to the continuum (indicating which of them are corner semipaths).

Remark 1. The global scheme uniquely determines the relative positions of the constituent curves $S_i$ of $K^{()}$, and also the positions relative to the $S_i$ of the paths for which $K^{()}$ is an ω-, α- or 0-limit continuum. If $K^{()}$ consists of a single closed path $L_0$, this follows at once from the definition of the "positive side" of a path and from the information concerning the relation between the positive traversal of $L_0$ and its $t$-traversal. If $K^{()}$ is the union of several $S_i$, it follows directly from the Remark to Lemma 8.

Remark 2. If $K^{()}$ is the union of more than one curve $S_i$, the global scheme makes it possible to determine the cyclic order of the semipaths of $K^{()}$ tending to any equilibrium state $O_j$ of the continuum.

Indeed, the ω-enumeration of $K^{()}$ specifies which semipaths tend to the equilibrium state $O_j$, and their cyclic order is determined by using Lemmas 4 through 7. Conversely, if it is not known a priori whether $K^{()}$ is positively or negatively situated, but the cyclic order of the semipaths about each equilibrium state $O_j$ is given (e.g., by global schemes of the equilibrium states), Lemmas 4 through 7 will supply the missing information.

The global scheme of a limit continuum may be given the form of a schematic diagram* with suitable notation for the paths. This version of the global scheme is considerably more intuitive and graphic than the tabular form.

---

* The local scheme may also be given in diagrammatic form. We shall not dwell on this, however, since the information given by the local scheme as regards the configuration of the limit continuum is extremely incomplete.

For example, let the scheme of a continuum $K_\omega^{(-)}$ be given by the table

$$S_1^{(+)} | L_1, \, O_1;$$
$$S_2^{(-)} | O_1, \, L_2, \, O_2, \, L_3;$$
$$S_3^{(-)} | O_2, \, L_4.$$

This table corresponds to Figure 268.  As another example, consider the scheme of a continuum $K_0^{(-)}$:

$$S_1^{(+)} | L_1, \, O_1, \, L_2, \, O_2;$$
$$S_2^{(+)} | L_3, \, O_1;$$
$$S_3^{+} | L_4, \, O_2.$$

The corresponding schematic diagram is shown in Figure 269.

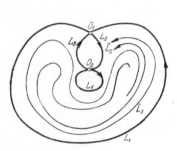

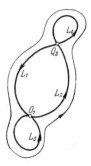

FIGURE 268.                         FIGURE 269.

Now let $K^{()}$ and $K^{*()}$ be two ω-, α- or 0-limit continua (consisting of paths of one dynamic system or two different ones). We retain the notation introduced above for the constituent paths of $K^{()}$ and the semipaths tending to it.

The constituent paths of the continuum $K^{*()}$, the semipaths tending to it, and its simple closed curves will be denoted by the same symbols as those of $K^{()}$, plus stars. Similarly, the numbers $m^*$, $R^*$, $l^*$, $p^*$, $q^*$ will have the same meanings for $K^{*()}$ as do $m, R, l, p, q$ for $K^{()}$.

Suppose that the local schemes of the continua are equivalent (with the same direction of $t$). We can then establish a scheme-correspondence (possibly in more than one way) between the equilibrium states $(O_j)$ and $(O_j^*)$ and paths $(L)$ and $(L^*)$ other than equilibrium states.  Any such scheme-correspondence induces a natural correspondence between the constituent curves $S_i$ of $K^{()}$ and $S_i^*$ of $K^{*()}$, which will also be termed a scheme-correspondence between the curves $S_i$ and $S_i^*$.

Note that even if the local schemes of the two continua are equivalent the curves $S_i$ may be mutually exterior but one of the curves $S_i^*$ encloses all the others.  In other words, the continua $K^{()}$ and $K^{*()}$ with equivalent local schemes may have constituent curves $S_i$ and $S_i^*$ with different configurations (see the examples in Figures 255 and 256).

*Definition XXIX.* We shall say that the global schemes of two ω-, α- or 0-limit continua $K^{()}$ and $K^{*()}$ are equivalent, with the same orientation and direction of t, if: 1) their local schemes are equivalent; 2) the continua are both ω (or α, or 0)-limit continua, and both positively (or negatively) situated; 3) there exists a (local) scheme-correspondence between the paths of $K^{()}$ and $K^{*()}$ under which the positive traversals of corresponding curves $S_i$ and $S_i^*$ are both (anti-) t-traversals (i. e., $S_i^+$ corresponds to $S_i^{*+}$, $S_j^-$ to $S_j^{*-}$), so that the curves $S_i$ and $S_i^*$ occupy the same relative positions; 4) either both continua $K^{()}$ and $K^{*()}$ are free, or both are nonfree and there exists a correspondence between the singular semipaths tending to $K^{()}$ and $K^{*()}$, respectively, corner semipaths corresponding to corner semipaths and cyclic order being preserved.

As usual, equivalence of the schemes of two continua $K^{()}$ and $K^{*()}$ induces a natural s c h e m e - c o r r e s p o n d e n c e between the paths $(L)$ and $(L^*)$, $(O)$ and $(O^*)$, semipaths $(\widetilde{L}^{()})$ and $(\widetilde{L}^{*()})$, $(\hat{L}^{()})$ and $(\hat{L}^{*()})$, and simple closed curves $S_i$ and $S_i^*$.

When the schemes of the continua are equivalent, it is clear that $m = m^*$, $R = R^*$, $l = l^*$ (if $K^{()}$ and $K^{*()}$ are nonfree we also have $p = p^*$, $q = q^*$), and the global scheme of $K^{*()}$ may be obtained from that of $K^{()}$ by simply starring all symbols.

The converse is also true: if the global scheme of a continuum $K^{*()}$ may be obtained from that of $K^{()}$ by starring the symbols for the paths $(L)$ and $(O)$, semipaths $(\widetilde{L}^{()})$ and $(\hat{L}^{()})$ and curves $S_i^{()}$, then the schemes are equivalent. The question of whether two global schemes are equivalent becomes a purely combinatorial problem and may be solved in a finite number of trials. If the schemes of continua $K^{()}$ and $K^{*()}$ are equivalent, they may be represented by the same diagram.

Let $C$ and $C^*$ be canonical curves of continua $K^{()}$ and $K^{*()}$, i. e., either cycles without contact or closed paths, according as $K^{()}$ and $K^{*()}$ are ω-, α- or 0-limit continua. Let γ and γ* be the canonical neighborhoods bounded by $C$ and $C^*$, respectively.

We shall say that the continua $K^{()}$ and $K^{*()}$ are i d e n t i c a l l y  s i t u a t e d relative to the canonical curves $C$ and $C^*$ if both $K^{()}$ and $K^{*()}$ lie inside (outside) $C$ and $C^*$.

*Theorem 74.* Let $K^{()}$ and $K^{*()}$ be two limit continua, $C$ and $C^*$ their canonical curves and γ and γ* the canonical neighborhoods bounded by $C$ and $C^*$. If the global schemes of $K^{()}$ and $K^{*()}$ are equivalent, then: 1) the continua $K^{()}$ and $K^{*()}$ are identically situated relative to their canonical curves $C$ and $C^*$; 2) there exists a path-preserving topological mapping of $\bar{γ}$ onto $\bar{γ}^*$ which also preserves scheme-correspondence of singular paths and semipaths (in the case of nonfree continua).

P r o o f. The first assertion follows directly from part 3) of Definition XXIX, Lemma 8 and Theorem 73.

When the continua $K^{()}$ and $K^{*()}$ are free, the second part of the theorem follows directly from Theorem 72. If $K^{()}$ and $K^{*()}$ are nonfree ω- or α-limit continua, it follows from part 4) of Definition XXIX that there is a topological correspondence between the cycles without contact $C$ and $C^*$ which preserves the correspondence of their points of intersection with scheme-corresponding semipaths $(\widetilde{L}^{()})$, $(\hat{L}^{()})$ and $(\widetilde{L}^{*()})$. $(\hat{L}^{*()})$. By the Remark following Theorem 72, there exists a topological mapping of $\bar{γ}$ onto $\bar{γ}^*$ which preserves this correspondence between $C$ and $C^*$. Q.E.D.

Remark. If the continua $K^{()}$ and $K^{*()}$ are free, the topological mapping of $\bar{\gamma}$ onto $\bar{\gamma}^*$ may always be constructed in such a way that any given correspondence between the cycles without contact $C$ and $C^*$ is preserved (the relation between the traversals of $C$ and $C^*$ and the $t$-traversals is also preserved).

If $K^{()}$ and $K^{*()}$ are nonfree, the required topological mapping may be constructed in such a way as to induce any topological mapping of $C$ onto $C^*$ which preserves the correspondence of their points of intersection with scheme-corresponding singular semipaths (see Remark to Theorem 72).


## §26.  SCHEME OF THE BOUNDARY OF A REGION

**1. CORNER POINTS OF BOUNDARY CURVES.**  We have been assuming throughout that, although our dynamic system $D$ is defined in a plane region $G$, we are confining attention to a closed subregion $\bar{G}^*$ of $G$, with normal boundary (see §16.2).

Recall that a point common to a boundary arc without contact and a boundary arc of a path is called a c o r n e r   p o i n t, and an arc (of a path or semipath) lying inside $\bar{G}^*$, at least one of whose endpoints is a corner point, is called a c o r n e r   a r c  or  c o r n e r   s e m i p a t h.  Clearly, every corner arc $\hat{l}$ of a path is the continuation of a boundary arc of a path (or of two such arcs) (Figure 270a, b).  Every corner semipath $\hat{L}$ is also the continuation of a boundary arc of a path (Figure 271).

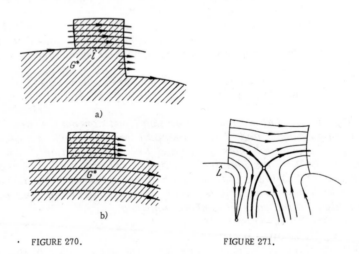

a)

b)

· FIGURE 270.

FIGURE 271.

Retaining our previous notation for corner semipaths and non-corner singular semipaths ($\hat{L}_r$ and $\tilde{L}_s$, respectively), we denote the boundary arcs of paths by $l_q$, boundary arcs without contact by $\lambda_q$, and corner arcs by $\hat{l}_p$; a whole path figuring in the boundary will be denoted by $L_0$.  A closed boundary curve $\Gamma$ of $\bar{G}^*$ will be denoted by $\Gamma^+$ or $\Gamma^-$, depending on whether the region $\bar{G}^*$ is in its interior or its exterior.  A boundary arc of a path $l_i$ figuring in a boundary curve $\Gamma$ will be denoted by $l^+$ if its direction of increasing $t$ is that induced by the positive traversal of $\Gamma$, by $l^-$ otherwise. The endpoint of $l_i$ corresponding to the larger value of $t$ will be called its $\omega$-endpoint and denoted by $M_i^\omega$, the endpoint corresponding to the smaller $t$-value is its $\alpha$-endpoint $M_i$. The corner point which is the endpoint of a positive (negative) corner semipath will be called its $\alpha$-endpoint ($\omega$-endpoint).

Let $M$ be a corner point which is an endpoint of an arc $l$ of a path $L$. Suppose that $M$ corresponds to time $t_0$ (for some fixed motion on $L$). To fix ideas, let us assume that $M$ is the $\omega$-endpoint of $l$. It is easily seen that there are only two possible cases:

1) All points of $L$ corresponding to times $t$ sufficiently near and greater than $t_0$ lie outside $\bar{G}*$.

2) All such points lie in $\bar{G}*$.

In the first case we shall call $M$ an $\omega$-o u t e r  c o r n e r  p o i n t (Figure 272a), in the second an $\omega$-i n n e r  c o r n e r  p o i n t (Figure 272b). The concepts of $\alpha$-outer ($\alpha$-inner) corner point for the $\alpha$-endpoint $M$ of $l$ are defined similarly.

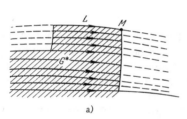
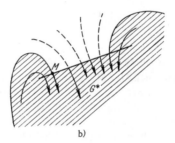

a)         b)

FIGURE 272.

It is clear that an  i n n e r  corner point $M$ which is the $\omega$ ($\alpha$)-endpoint of a boundary arc $l$ is at the same time the $\alpha$ ($\omega$)-endpoint of a corner arc or corner semipath (also on the path $L$). We shall call the latter corner arc or semipath the $\omega$ ($\alpha$)-c o n t i n u a t i o n of the boundary arc $l$. The latter in turn will be referred to as the $\alpha$ ($\omega$)-continuation of the corner arc (semipath). Thus every corner point $M$ is either $\omega$- or $\alpha$-outer, $\omega$- or $\alpha$-inner.

Let $\lambda$ be an arc without contact. The paths through interior points of $\lambda$ either all enter or all leave $\bar{G}*$ with increasing $t$. In the first case we shall call $\lambda$ a  p o s i t i v e  b o u n d a r y  a r c  w i t h o u t  c o n t a c t, in the second a  n e g a t i v e  b o u n d a r y  a r c  w i t h o u t  c o n t a c t. The definitions of positive and negative boundary cycles without contact are analogous.

L e m m a  1.  *Let $\lambda$ be a boundary arc without contact, $M$ one of its corner points. If $\lambda$ is a positive arc without contact, then $M$ is either $\omega$-outer or $\alpha$-inner. If $\lambda$ is negative, $M$ is either $\alpha$-outer or $\omega$-inner.*

P r o o f.  Let $\lambda$ be a positive arc without contact, $L$ the path through $M$, $t_0$ the parameter value corresponding to $M$, $AM$ an arc of $L$ corresponding to times $t < t_0$ close to $t_0$, and $BM$ a similar arc for $t > t_0$. Each of the arcs $AM$ and $BM$ is either on the boundary of $\bar{G}*$, inside the region, or outside it, but only one of them can belong to the boundary of $\overline{G}^*$.  Since $\lambda$ is a positive arc without contact, one readily sees that $AM$ cannot be outside $\overline{G}^*$ or $BM$ inside it.  Thus, either $AM$ is on the boundary of $\overline{G}^*$ and $BM$ outside, or $AM$ is inside $\overline{G}^*$ and $BM$ on the boundary.  In the first case (Figure 273a) $M$ is an $\omega$-outer corner point, and in the second (Figure 273b) $\alpha$-inner.

The case of a negative arc without contact $\lambda$ is treated similarly.  Q.E.D.

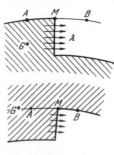

FIGURE 273.

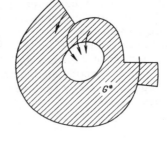

FIGURE 274.

## 2. SCHEME OF A BOUNDARY CURVE, SCHEME OF THE BOUNDARY, EQUIVALENCE OF TWO BOUNDARY SCHEMES.

Let $\Gamma_j$ be some closed boundary curve.  Let us examine all the singular semipaths and corner arcs having points in common with $\Gamma_j$.  If $\Gamma_j$ is not a cycle without contact, these semipaths and arcs either cross $\Gamma_j$ at interior points of boundary arcs without contact, or have an inner corner point in common with $\Gamma_j$.  It is evident that the other endpoint of any corner arc crossing $\Gamma_j$ (which is also a corner point) may lie either on the same boundary curve $\Gamma_j$ or on some other boundary curve (see, e. g., Figure 274).  Corner semipaths with an endpoint at a corner point are also singular semipaths having points in common [with the boundary curve].

*Definition XXX. A scheme (or global scheme) of a boundary curve $\Gamma_j$ is a list specifying 1) whether the curve is an outer or inner boundary curve of $\overline{G}^*$ (i. e., which of the symbols + or − is to be assigned it as a superscript); 2) whether $\Gamma_j$ is a cycle without contact, closed path, or union of arcs of paths and arcs without contact; 3) if $\Gamma_j$ is a cycle without contact, whether it is positive or negative; if $\Gamma_j$ is a union of arcs of paths and arcs without contact, these arcs are enumerated in cyclic order, indicating for each whether the direction of increasing $t$ does or does not coincide with that induced by the positive traversal of $\Gamma_j$, and finally indicating all inner corner points of $\Gamma_j$; 4) if $\Gamma_j$ is a cycle without contact, all singular paths and corner arcs intersecting $\Gamma_j$ are enumerated in cyclic order; if $\Gamma_j$ is a union of arcs of paths and arcs without contact, all the singular semipaths and corner arcs intersecting each arc without contact $\lambda_i$ in $\Gamma_j$ are enumerated, in their order along the arc in the direction induced by the positive traversal of $\Gamma_j$; 5) if $\Gamma_j$ is a closed path, the list specifies whether its positive traversal is the $t$- traversal or not.*

If $\Gamma_j$ is a cycle without contact, its scheme may be specified in tabular form

$$\Gamma_{\bar{j}} \mid \tilde{L}_i^+, \ \hat{L}_j^+, \ \hat{\imath}_k, \ \ldots$$

(Of course, when $\Gamma_j$ is a negative cycle all the semipaths crossing it are positive, and when $\Gamma_j$ is positive, they are all negative.)

If $\Gamma_j$ is a union of arcs without contact and arcs of paths, its scheme may be specified by the following tables.

First, a table

$$\Gamma_p^+ \,|\, l_i^{()}, \ \lambda_j, \ M_{k_1}^\omega, \ l_{i_2}^{()}, \ \lambda_{i_2}, \ \ldots \tag{1}$$

where the constituent arcs of $\Gamma_j$ are numbered and written out in their cyclic order and each inner corner point (and only inner points!) is inserted between the arcs $\lambda_i$, $l_{i+1}^{()}$ or $l_k^{()}$, $\lambda_{k+1}$ whose common endpoint it is. All the corner points not figuring in this table are outer.

Second, a table listing the singular semipaths and corner arcs intersecting each constituent arc $\lambda_j$ of $\Gamma_j$, enumerated in their order along $\lambda_j$ in the direction induced by the positive traversal of $\Gamma_j$:

$$\lambda_j \,|\, \hat{l}_{\beta_j}, \ \widetilde{L}_{\alpha_j}^{()}, \ \ldots, \ \hat{L}_{\delta_j'}, \ \ldots \tag{2}$$

Remark 1. The first table in the scheme of the boundary curve (table (1)) also tells us which of the arcs without contact $\lambda_i$ are positive and which negative. Indeed, let $\lambda_k$ be one of these arcs, and $M_k^{()}$ the corner point common to $\lambda_k$ and $l_k$. If $M_k^{()}$ appears in table (1), it is an inner corner point, and if it does not it is outer. Moreover, the table indicates whether $M_k^{()}$ is an $\omega$- or $\alpha$-endpoint of the arc $l_h$. Thus, given the local scheme, one can state for each corner point whether it is $\omega$- or $\alpha$-inner, $\omega$- or $\alpha$-outer. Lemma 1 then states whether $\lambda_k$ is a positive or negative arc without contact.

Every corner arc or corner semipath passes through an inner corner point.

For example, suppose that the scheme of a boundary curve is given by the following tables:

$$\Gamma^+ \,|\, l_1^+, \ \lambda_1, \ M_2^\omega, \ l_2^-, \ M_2^\alpha, \ \lambda_2, \ M_3^\alpha, \ l_3^+, \lambda_3, l_4^-, \lambda_4, \ \ldots, \ \lambda_j, M_j^\omega, \tag{3}$$

$$\lambda_1 \,|\, \hat{L}_{i_{11}}, \ \widetilde{L}_{i_{12}}, \ \hat{l}_{i_{13}}, \ \ldots, \ \hat{l}_{i_{1s}},$$
$$\lambda_2 \,|\, \hat{L}_{i_{21}}, \ L_{i_{22}}, \ \ldots \tag{4}$$

Consider the point $M_2^\omega$. We see from table (3) that $M_2^\omega$ is an inner corner point on $\lambda_1$; moreover, it is the last point of $\lambda_1$ (when this arc is described in the direction induced by the positive traversal of $\Gamma^+$). Consequently, examination of table (4) shows that the corner arc $\hat{l}_{1s}$ passes through $M_2^\omega$. All other singular semipaths and corner arcs crossing $\lambda_1$ pass through its interior points. The other inner corner points of $\Gamma^+$ may be considered in similar fashion.

Remark 2. The data presented in tables (3) and (4) are not entirely independent. For example, by Lemma 1 the notation $\ldots l_3^+, \lambda_3, l_4^{()}, \ldots$ alone implies that the parentheses in $l_4$ should be replaced by a minus sign. Thus the minus sign at this point in table (3) could have been omitted. In the second table it is not essential to indicate whether the singular semipaths appearing therein are positive or negative. For example, consider the first row of table (4). We know that $\lambda_1$ is a positive arc without contact. But then it is obvious that the semipaths $\hat{L}_{i_{11}}, \tilde{L}_{i_{12}}$ crossing this arc are negative.

*Definition XXXI. The scheme of the boundary of $G^*$ is an enumeration of all its constituent simple closed curves, each provided with its scheme.*

Thus the scheme of the boundary of $G^*$ (or boundary scheme) consists of finitely many tables of types (1) and (2). Of course, the boundary scheme may also be specified in schematic (diagram) form rather than tabular form. The details are left to the reader.

Now suppose that we have two (not necessarily different) dynamic systems $D$ and $D'$, defined in regions $G$ and $G'$, respectively. These systems are to be considered in closed regions $\bar{G}^*$ and $\bar{G}'^*$, respectively ($\bar{G}^* \subset G$, $\bar{G}'^* \subset G'$), with normal boundaries. One can now define the equivalence of the two corresponding boundary schemes, as in the case of schemes of limit continua (Definition XXIX). For brevity's sake we shall confine ourselves to a more formal definition.

Let $\Gamma_1$, $\Gamma_2$, ..., $\Gamma_r (\Gamma)$ be all the closed boundary curves of $\bar{G}^*$, $\hat{L}_1^{( )}$, $\hat{L}_2^{( )}$, ..., $\hat{L}_q^{( )}$ $(\hat{L}^{( )})$ the corner semipaths of system $D$, $\tilde{L}_1^{( )}$, $\tilde{L}_2^{( )}$, ..., $\tilde{L}_p^{( )}$ $(\tilde{L}^{( )})$ the orbitally unstable semipaths of system $D$ whose endpoints lie on the boundary of $G^*$,

$$l_1, l_2, \ldots, l_n \quad (l)$$

the boundary arcs of paths of system $D$,

$$\hat{l}_1, \hat{l}_2, \ldots, \hat{l}_m \quad (\hat{l})$$

the corner arcs, $\lambda_1$, $\lambda_2$, ..., $\lambda_s (\lambda)$ the boundary arcs without contact of $\bar{G}^*$ and $M_{i_1}^\omega$, $M_{i_2}^\omega$, ..., $M_{i_\mu}^\omega$, $M_{j_1}^\alpha$, $M_{j_2}^\alpha$, ..., $M_{j_\nu}^\alpha (M)$ the inner corner points of the curves of $\Gamma$.*

Let $(\Gamma')$, $(\hat{L}'^{( )})$, $(\tilde{L}'^{( )})$, $(l')$, $(\hat{l}')$ and $(\lambda')$ be the corresponding sets for system $D'$ in the region $\bar{G}'^*$, and $r'$, $q'$, $p'$, $n'$, $m'$, $s'$, $\mu'$, $\nu'$ the numbers corresponding to $r$, $q$, $p$, $n$, $m$, $s$, $\mu$, $\nu$.

*Definition XXXII. The boundary schemes of the regions $G^*$ and $G'^*$ are said to be equivalent if there exists a one-to-one correspondence $\theta$ between all singular elements in the sets*

$$(\Gamma), \ (\hat{L}^{( )}), \ (\tilde{L}^{( )}), \ (l), \hat{(l)}, \ (\lambda), (M)$$

*on the one hand, and all singular elements in the sets*

$$(\Gamma'), \ (\hat{L}'^{( )}), (\tilde{L}'^{( )}), \ (l'), \ (\hat{l}'), \ (\lambda'), \ (M')$$

*on the other, under which the cycles without contact in $(\Gamma)$ correspond to the cycles without contact in $(\Gamma')$, the semipaths $(\hat{L}^{( )})$ to the semipaths $(\hat{L}'^{( )})$, the arcs without contact $(\lambda)$ to the arcs without contact $(\lambda')$ and so on, and the following condition is satisfied: if each element of system $D'$ is denoted by the same symbol as the element of system $D$ corresponding to it under $\theta$ with a prime, then the boundary scheme of $G'^*$ (i.e., the table describing*

---

* The notation for these points is coordinated with that for the boundary arcs $l_i$.

*the scheme) is obtained from that of G\* by adding a prime to the symbol for each element.*

It is clear that the boundary schemes of systems $D$ and $D'$ with equivalent boundary schemes are described by the same schematic diagram.

*Chapter XI*

## SCHEME OF A DYNAMIC SYSTEM AND FUNDAMENTAL THEOREM

### INTRODUCTION

In the preceding chapters we considered the local and global schemes of equilibrium states and limit continua, and the scheme of the boundary.

We are now ready to introduce the global scheme of a dynamic system with finitely many singular paths. The global scheme of a dynamic system includes the global schemes of the equilibrium states and limit continua as component parts. It provides an exhaustive description of the relative positions of the singular elements and fully determines the topological structure of the phase portrait. The fundamental theorem of this chapter is as follows. If the schemes of dynamic systems $D$ and $D'$, in closed regions $\bar{G}*$ and $\bar{G}*'$ are equivalent, with the same orientation and direction of $t$, then the topological structures of the phase portraits of $D$ and $D'$ in $\bar{G}*$ and $\bar{G}*'$, respectively, are identical. The proof proceeds by construction of an identifying mapping, i. e., a path-preserving topological mapping of $\bar{G}*$ onto $\bar{G}*'$.

In other words, the global scheme is a topological invariant of a dynamic system. The numerous concrete examples cited throughout previous chapters have demonstrated that the topological structure of the phase portrait is inextricably bound up with the "nature of the equilibrium states, the number and relative position of closed paths, and the behavior of separatrices." The concept of the scheme of a dynamic system in effect provides this intuitive but very vague turn of speech with a rigorous meaning.

The global scheme is a quite natural summary of that data concerning a dynamic system which is indispensable for its complete qualitative investigation.

We shall consider only dynamic systems with finitely many singular paths (finiteness assumption), restricting ourselves further to systems in a bounded plane region with normal boundary. Any dynamic system on the sphere which has finitely many singular paths may be reduced to a system of this type. Indeed, every dynamic system on the sphere has at least one equilibrium state. By the finiteness assumption, all equilibrium states of our systems are isolated. Let $O$ be one of them and consider a canonical neighborhood of $O$. The canonical curve $\sigma$ bounding this neighborhood is a normal boundary which divides the sphere into two regions, each of which can be mapped (e. g., by stereographic projection) onto a plane region with normal boundary. Thus the study of a dynamic system on the sphere reduces to study of two dynamic systems in plane regions with normal boundary.

The chapter comprises three sections. In §27, which is auxiliary in nature, we define a certain system of canonical neighborhoods of equilibrium states and limit continua satisfying certain natural requirements, a r e g u l a r system of canonical neighborhoods.

We then consider the subarcs into which arcs and cycles without contact are divided by their points of intersection with singular paths. These subarcs will be called elementary ω- and α-arcs. Cycles without contact having no points in common with singular paths will be called free ω- and α-cycles. Elementary ω-, α-arcs and ω-, α-cycles play a major role in constructing the topological mapping required by the fundamental theorem.

§28 is devoted to a discussion of the relative position of free continua. Also studied here are the properties of certain natural subregions of the region between canonical neighborhoods.

In §29 we define the global scheme of a dynamic system; apart from the schemes of all equilibrium states and limit continua, the scheme also describes the relative position of the free limit continua. Finally, the fundamental theorem is proved.

## §27. REGULAR SYSTEM OF CANONICAL NEIGHBORHOODS. ω (α)-ARCS AND ω (α)-CYCLES

**1. NOTATION FOR SINGULAR ELEMENTS OF A DYNAMIC SYSTEM.** We first recall some notation. Let $D$ be a dynamic system, defined in a region $G$ and considered in a closed subregion $\bar{G}^*$ with normal boundary. Let $O_1, O_2, \ldots,$ $O_m$ $(O)$ be all the equilibrium states of system $D$ in $\bar{G}^*$, $g_1, g_2, \ldots, g_m$ $(g)$ their canonical neighborhoods (see §19.2), and $\sigma_1, \sigma_2, \ldots, \sigma_m$ $(\sigma)$ the corresponding canonical curves of the equilibrium states (i. e., the boundaries of the neighborhoods $g_i$).

Let $K_1^{()}, K_2^{()}, \ldots, K_N^{()}$ $(K)$ be all the (one-sided) limit continua of system $D$ (not equilibrium states) in $\bar{G}^*$, $\gamma_1, \gamma_2, \ldots, \gamma_N$ $(\gamma)$ their canonical neighbor-hoods, $C_1, C_2, \ldots, C_N$ $(C)$ the corresponding canonical curves of $(K)$; each curve $C_i$ is either a cycle without contact or a closed path, and in combination with $K_i$ itself forms the boundary of $\gamma_i$.

**2. REGULAR SYSTEMS OF CANONICAL NEIGHBORHOODS.** In the sequel we shall generally consider c l o s e d canonical neighborhoods $\bar{g}_i$ and $\bar{\gamma}_i$. If the canonical neighborhoods are chosen arbitrarily, canonical neighborhoods of different equilibrium states (or of equilibrium states and limit continua) may of course intersect. However, this can be avoided by suitable choice of the canonical neighborhoods (not counting saddle regions based on separatrices figuring in the limit continua, which always have points in common with the canonical neighborhoods). This is shown in the following

L e m m a  1. *The canonical neighborhoods of equilibrium states and of limit continua other than equilibrium states may always be chosen in such a way that the following conditions are all satisfied: a) none of the canonical curves (σ) and (C) intersect the boundary of $\bar{G}^*$ or corner arcs (but they may intersect corner semipaths); b) the canonical curves (σ) of the equilibrium states are disjoint, mutually exterior, and disjoint from the canonical curves (C) of limit continua other than equilibrium states; c) the canonical curves (C) of different limit continua (not equilibrium states) and their canonical neighborhoods (γ) are pairwise disjoint; d) every* *elementary region (elliptic, parabolic or saddle region) figuring in the canonical neighborhood of an equilibrium state, with the exception of a saddle region bordering on a limit path, is disjoint from all canonical neighborhoods of limit continua which are not equilibrium states.*

P r o o f. We first prove that condition a) can always be satisfied. By the definition of a normal boundary, its constituent arcs of paths, hence also their continuations (corner arcs), cannot be parts of orbitally unstable paths or semipaths lying wholly in $\bar{G}*$. In particular, there are no equilibrium states on the boundary. The union E of all boundary and corner arcs is clearly a closed set. Thus any equilibrium state $O_i$ lies at a positive distance from this set, and every canonical neighborhood in a sufficiently small neighborhood $U_\varepsilon (O_i)$ is obviously disjoint from E.

Now let $K_i^{\langle\rangle}$ be a limit continuum which is not an equilibrium state. No point of the boundary or of a corner arc can lie on a limit continuum, the only exceptions occurring when the boundary curve is an orbitally stable closed path and when a closed path made up of boundary and corner arcs is the boundary continuum of some cell $w$ filled by closed paths (see §24.1). But in this case any closed path in $w$ is a canonical curve for $K_i^{\langle\rangle}$, and neither this curve nor the corresponding canonical neighborhood (whose points are all in $w$) can have points in common with E. In all other cases the limit continuum $K_i^{\langle\rangle}$ is made up of orbitally unstable paths and its distance from E is positive. It clearly follows that any canonical neighborhood of $K_i^{\langle\rangle}$ lying together with its boundary (canonical curve) in $U_\varepsilon (K_i^{\langle\rangle})$ for sufficiently small $\varepsilon > 0$ has no points in common with E.

The truth of condition b) clearly follows from the finiteness of the number of equilibrium states and continua other than equilibrium states in $\bar{G}*$ and from the fact that by definition there are no equilibrium states on the canonical curves $(C)$.

Now for condition c). Let $K_i^{\langle\rangle}$ and $K_j^{\langle\rangle}$ be two distinct limit continua which are not equilibrium states. We are considering o n e - s i d e d limit continua (see Chapter IX), and so d i s t i n c t limit continua may either 1) have no common points, 2) have some but not all of their points in common, or 3) coincide as point sets. In the latter case, one of the continua must be a $K^+$ and the other a $K^-$, so that all their paths must be positively situated limit paths in one and negatively situated in the other. When the continua $K_i^{\langle\rangle}$ and $K_j^{\langle\rangle}$ are disjoint (case 1)) condition c) can clearly be satisfied.

Suppose that $K_i^{\langle\rangle}$ and $K_j^{\langle\rangle}$ have common points. Since by definition all paths through the points of a canonical neighborhood of an $\omega$ ($\alpha$)-limit continuum are nonclosed while all those through points of a canonical neighborhood of a 0-limit continuum are closed, it clearly follows that the canonical neighborhoods of an $\omega$ ($\alpha$)-limit continuum and a 0-limit continuum cannot intersect. It thus remains to consider the following cases: 1) both $K_i^{\langle\rangle}$ and $K_j^{\langle\rangle}$ are $\omega$ ($\alpha$)-limit continua; 2) both $K_i^{\langle\rangle}$ and $K_j^{\langle\rangle}$ are 0-limit continua; 3) $K_i^{\langle\rangle}$ (say) is an $\omega$-limit continuum and $K_j^{\langle\rangle}$ an $\alpha$-limit continuum.

C a s e 1. To fix ideas, suppose both $K_i^{\langle\rangle}$ and $K_j^{\langle\rangle}$ are $\omega$-limit continua. Assume first that the continua are distinct as point sets. Since all paths through points of any canonical neighborhood of $K_i^{\langle\rangle}$ and its bounding cycle $C_i$ tend to $K_i^{\langle\rangle}$ as an $\omega$-limit continuum, while all paths through a canonical neighborhood of $K_j^{\langle\rangle}$ and the cycle with contact bounding it tend to $K_j^{\langle\rangle}$, it is clear that in this case the canonical neighborhoods and cycles without contact cannot intersect. Now suppose that $K_i^l$ and $K_j^l$ coincide as point sets, so that one of them is $K^+$ and the other $K^-$. Let $L_0$ be one of their constituent paths (not an equilibrium state). If the canonical neighborhoods of $K^+$ and $K^-$ intersect, then $L_0$ is a both positively and negatively situated limit path for any path $L$ through any of their common points. But this is impossible (Corollary 2 to Lemma 2 of §4). Thus the canonical neighborhoods of two distinct $\omega (\alpha)$-limit continua are disjoint.

Case 2. All points of a canonical neighborhood of a 0-limit continuum and the closed path bounding it lie in the same cell. Hence if the canonical neighborhoods of $K_i^l$ and $K_j^l$ have common points, these continua must be boundary continua of one cell $w$ filled by closed paths. If $K_j^l$ and $K_i^l$ have common points but do not coincide as point sets, this is impossible, because the boundary of a cell filled by closed paths is the union of d i s j o i n t continua. Now suppose that $K_i^l$ and $K_j^l$ coincide as point sets, so that one of them, $K_i^l$ say, is a continuum $K^+$, and the other, $K_j^l$ is a $K^-$. Again, both continua are boundary continua of the same cell $w$ filled by closed paths. Then any constituent path (not an equilibrium state) of the continua is both a positively and negatively situated boundary path for $w$. But this is impossible by Lemma 2 of §23. Thus, if two distinct limit continua have common points they figure in the boundaries of distinct cells and their canonical neighborhoods must be disjoint.

Case 3. Suppose that the canonical neighborhoods of the continua $K_j^\alpha$ and $K_i^\omega$ have a common point $M$. Let $L_M$ be the path through this point. Then $K_i^\omega$ is a limit continuum for the semipath $L_M^+$ and $K_j^\alpha$ for $L_M^-$. The continuum $K_i^\omega$ is either inside or outside the canonical curve $C_i$. To fix ideas, suppose that $K_i^\omega$ is o u t s i d e $C_i$ (the other case is treated similarly). By the properties of canonical neighborhoods, all points of the semipath $L_M^-$ ultimately lie i n s i d e $C_i$, and so its limit continuum $K_j^\alpha$ must lie inside $C_i$. But then $K_i^\omega$ and $K_j^\alpha$ must be disjoint — contradiction.

We now consider condition d). For elliptic and parabolic regions the proof follows the same lines as for condition c).

Now consider a saddle region $g_c$. Assume that the separatrices in the boundary of $g_c$ are not limit paths on the side corresponding to this saddle region. Assume moreover that condition b) is satisfied. Let $\gamma$ be some canonical neighborhood. If $g_c$ contains both points of $\gamma$ and points not in $\gamma$, it must also contain points of the cycle without contact $C$ bounding $\gamma$. But this is impossible, because of condition b). Hence, $g_c$ is either disjoint from $\gamma$ or a subregion of $\gamma$. In the latter case, all points of any arc without contact in the boundary of $g_c$, except its endpoint lying on the boundary separatrices, must also be in $\gamma$. But this obviously means that the boundary separatrices of $g_c$ are also limit paths on the side corresponding to $g_c$, contradicting our assumption. This completes the proof of Lemma 1.

In particular, there may even be infinitely many points of a saddle arc without contact (see §18.3, §19.2) on singular semipaths, namely, when the saddle region is bounded by separatrices which are limit paths of some singular semipath (or of several singular semipaths).

If $g_c$ is an arbitrary saddle region, one can always find a saddle region $g_c'$ contained in $g_c$ whose boundary arc is not an arc of a singular path or semipath.

*Definition. A regular system of canonical neighborhoods (or regions) is a system of canonical neighborhoods satisfying conditions a), b), c) and d) of Lemma 1, and the additional condition that no arc of a path in the boundary of a saddle region be an arc of a nonsingular path.*

Henceforth we shall consider only regular systems of canonical neighborhoods. In any regular system, the canonical neighborhoods of ω-limit continua and stable nodes, as well as ω-parabolic sectors, are sometimes known as regions of attraction or sinks. Canonical neighborhoods of α-limit continua and unstable nodes, and also α-parabolic sectors, are regions of repulsion or sources.

As implied above, all saddle regions considered henceforth will be such that the arcs of paths figuring in their boundaries are arcs of nonsingular paths. Each saddle arc without contact in the boundary of such a saddle region has only one endpoint on a singular path or semipath — the endpoint of one of the boundary semipaths (separatrices) of the region. Interior points of the saddle arcs without contact may lie on singular semipaths.

**3. ELEMENTARY ARCS AND FREE CYCLES WITHOUT CONTACT.** Fix some regular system of canonical neighborhoods. Throughout the sequel we shall denote the canonical neighborhoods by $(\gamma)$ and $(g)$, the canonical curves of the system by $(C)$ and $(\sigma)$, and the parabolic arcs of the curves $(\sigma)$ by $(l)$. In accordance with our notational conventions, we shall also denote simple closed boundary curves by $(\Gamma)$, boundary arcs without contact by $(\lambda)$, and saddle arcs (i. e., arcs without contact in the boundaries of hyperbolic sectors; see §18.3) by $(\lambda_c)$. As before (§19.2), a saddle arc is an ω(α)-saddle arc if the paths through its interior points enter (leave) the saddle region. Each saddle region $g_c$ has one ω-saddle arc and one α-saddle arc in its boundary. Since our system of canonical neighborhoods is regular, only one endpoint of any saddle arc can lie on a singular semipath. The endpoint of the α-saddle arc in the boundary of $g_c$ is also the "endpoint" of the α-separatrix in the boundary, and the endpoint of the ω-saddle arc is the "endpoint" of the ω-separatrix.

Consider the cycles without contact $(C)$ and $(\sigma)$ in the boundaries of the canonical neighborhoods of the limit continua, both those which are not equilibrium states and those which are (nodes). Any of these cycles without contact having no point in common with any singular semipath will be called a free cycle without contact. It is clear that every free cycle without contact $(C)$ or $(\sigma)$ is in the boundary of a canonical neighborhood $\gamma_i$ of a limit continuum which is not an equilibrium state or of a canonical neighborhood $g_i$ of a free node. A free cycle without contact will be called an ω- or α-cycle according as it appears in the boundary of the canonical neighborhood of an ω- or α-limit continuum (in particular, of a stable or unstable node). If a boundary curve $(\Gamma)$ is a cycle without contact and none of its points is an endpoint of a singular semipath or endpoint of a singular (nodal) arc, we shall call it a free boundary cycle without contact, or simply a free cycle without contact together with those defined above.

A free boundary cycle without contact will be called an ω- or α-cycle according as it is an ω- or α-boundary cycle.

Cycles without contact $(C)$ and $(\sigma)$ and boundary cycles which are not free will be called nonfree cycles without contact. A nonfree cycle has points in common with singular semipaths and (if it is a boundary cycle) corner arcs.

We have three groups: a) all simple closed curves $(C)$, $(\sigma)$, $(\Gamma)$ which are nonfree cycles without contact, b) all parabolic arcs without contact $(l)$ figuring in the canonical curves $(\sigma)$ of equilibrium states which are not nodes, c) all boundary arcs without contact $(\lambda)$. Let us consider each of these groups in turn.

a) Consider some nonfree cycle without contact which has more than one point in common with singular semipaths or (if it is a boundary cycle) with singular semipaths and corner arcs. These points of singular elements divide the cycle into finitely many simple arcs without contact, each of which has no interior points in common with singular semipaths or corner arcs. We shall call these elementary arcs.

If the nonfree cycle without contact has only one point in common with a singular semipath or corner arc, we shall call the entire cycle a cyclic elementary arc, and its point of intersection with a singular element will be called its endpoint.

b) and c): Parabolic arcs without contact $(l)$ and boundary arcs without contact $(\lambda)$. They may have points in common with singular semipaths, and the arcs $(\lambda)$ may also intersect corner arcs. Again, these common points divide each arc into subarcs without contact which have no interior points in common with singular elements. These subarcs will also be called elementary arcs (in particular, the entire arc $l$ or $\lambda$ may be an elementary arc). At least one endpoint of each elementary arc lies on a singular semipath or singular arc of a path (i.e., boundary or corner arc). This is evident when the elementary arc is part of a cycle without contact $(C)$ or arc $(\lambda)$. In the case of a parabolic arc $(l)$, the assertion follows from Lemma 4 of §19 and the definition of a parabolic arc. An elementary arc will be called an elementary $\omega\,(\alpha)$-arc if it is part of a cycle without contact $(C)$ or $(\sigma)$ in an $\omega\,(\alpha)$-limit continuum (in particular, a node), or part of a parabolic arc $(l)$ in an $\omega\,(\alpha)$-parabolic sector, or a positive (negative) boundary arc (or cycle) without contact. Henceforth we shall call elementary $\omega\,(\alpha)$-arcs simply $\omega\,(\alpha)$-arcs, provided no confusion can arise.

Since elementary $\omega$- and $\alpha$-arcs are subarcs of parabolic arcs and cycles without contact, it is clear that no path or semipath can have more than one point in common with an $\omega\,(\alpha)$-arc or $\omega\,(\alpha)$-cycle.

Lemma 2. *Any nonsingular nonclosed path (entirely contained in $\bar{G}*\acute{)}$, which is not a loop contained in a closed canonical neighborhood of an equilibrium state, crosses exactly one $\omega$-arc or one free $\omega$-cycle and exactly one $\alpha$-arc or free $\alpha$-cycle.*

Proof. Let $L$ be a path satisfying the assumptions of the lemma. Note first that any such path must have points outside all canonical neighborhoods $\gamma$ of $\omega$- and $\alpha$-limit continua and outside all parabolic sectors and regions $g$ of equilibrium states. Indeed, any path passing at $t = t_0$ through a point of $\gamma$ or $g$ or through a point on a boundary cycle or arc without contact of one of these regions necessarily leaves the region either at $t > t_0$ or at $t < t_0$ (this follows from the definition of $\gamma$ and $g$). Since our system of canonical neighborhoods is regular, it is readily seen that in either case all points of $L$ corresponding to times $t < t_0$ (or $t > t_0$) sufficiently near $t_0$ will lie outside all regions $\gamma_i$ and $g_i$. Let $M$ be one of these points and $\tau$ the corresponding time (relative to the relevant motion). As $t$ increases (i.e., at some $t > \tau$) $L$ either crosses the boundary of $\bar{G}*$ or tends to an equilibrium state or continuum $K_\omega^0$ which is not an equilibrium state. In this situation, any nonsingular path tending to an equilibrium state as $t \to +\infty$ will invariably enter a parabolic sector of the equilibrium state (or a parabolic region if the equilibrium state is a node).

Thus the path $L$ must cross either a free $\omega$-cycle or an elementary $\omega$-arc at an interior point. (The relevant elementary arc may be a subarc either of a canonical curve $\sigma_i$ or $C_i$ or of a boundary curve $\Gamma_i$.) It remains to show that $L$ can cross only one free $\omega$-cycle or $\omega$-arc. Let $\tau_1$ be the first time $t > \tau$ at which $L$ crosses an $\omega$-cycle or $\omega$-arc (it exists by virtue of the fact that the number of elementary arcs and free cycles is finite) and $M_1$ the point on $L$ corresponding to $t = \tau_1$. If $M_1$ is on the boundary of $\overline{G}^*$, it is the last point of $L$ in the region. If $M_1$ is on some $\sigma_i$ or $C_i$, all points of $L$ corresponding to $t > \tau_1$ lie inside the corresponding $\gamma_i$ or $g_i$, and therefore cannot be a point of any elementary $\omega$-arc or $\omega$-cycle. Thus $\tau_1$ is the only time at which $L$ crosses an elementary $\omega$-arc or free $\omega$-cycle.

One proves in similar fashion that a path crosses exactly one elementary $\alpha$-arc or free $\alpha$-cycle. Q.E.D.

Remark 1. Since any path crossing a free $\omega(\alpha)$-cycle without contact or elementary $\omega(\alpha)$-arc at an interior point satisfies the assumption of Lemma 2, it follows that any such path crosses exactly one free $\omega(\alpha)$-cycle or $\omega(\alpha)$-arc.

Remark 2. A path through the endpoint of an elementary $\omega(\alpha)$-arc can cross a free $\alpha(\omega)$-cycle or $\alpha(\omega)$-arc only at one of the latter's endpoints, for the endpoint of an elementary arc lies either on a singular element or on an elliptic arc of the canonical curve of an equilibrium state.

## 4. CONJUGATE ELEMENTARY $\omega$- AND $\alpha$-ARCS AND CONJUGATE FREE $\omega$- AND $\alpha$-CYCLES.

We first prove the following

*Lemma 3. All paths through points of the same free $\omega(\alpha)$-cycle or through interior points of the same element $\omega(\alpha)$-arc cross either the same free $\alpha(\omega)$-cycle or the same $\alpha(\omega)$-arc at interior points.*

Proof. We first consider a free $\omega$-cycle $C_\omega$. Suppose the assertion false, i.e., paths crossing this cycle cross either at least two different $\alpha$-cycles, a free $\alpha$-cycle and at least one $\alpha$-arc, or at least two different $\alpha$-arcs. To fix ideas, let the path in question cross two $\alpha$-cycles $C_{1\alpha}$ and $C_{2\alpha}$. Let the path through some point $A$ of $C_\omega$ cross $C_{1\alpha}$ at a point $A'$. Then all paths crossing $C_\omega$ sufficiently near $A$ will cross $C_{1\alpha}$ at points arbitrarily near $A'$. Let $B$ be a point of $C_\omega$ through which passes a path crossing $C_{2\alpha}$. The points $A$ and $B$ divide $C_\omega$ into two arcs; consider one of these. Clearly, there must be a point $P$ on this arc which is either the last point (moving along the arc $AB$ from $A$ to $B$) through which there is a path crossing $C_{1\alpha}$, or the first point through which there is a path not crossing $C_{1\alpha}$, therefore crossing $C_{2\alpha}$. In either case, let $Q$ denote the point at which the path $L$ through $P$ crosses $C_{1\alpha}$ or $C_{2\alpha}$, respectively. Now, since all the paths crossing $C_\omega$ at points sufficiently near $P$ are orbitally stable, they must cross the cycle $C_{1\alpha}$ (or $C_{2\alpha}$) at points arbitrarily close to $Q$. And this means that, contrary to assumption, $Q$ can be neither the last point through which there is a path crossing $C_{1\alpha}$ nor the first point through which there is a path crossing $C_{2\alpha}$ — contradiction. The proof for the other cases is analogous. Q.E.D.

*Lemma 4. A path crossing an $\omega(\alpha)$-arc cannot cross a free cycle without contact.*

Proof. Suppose on the contrary that paths through points of an $\omega$-arc without contact $\lambda_\omega$ cross a free $\alpha$-cycle without contact $C_\alpha$. Then the paths through all interior points of $\lambda_\omega$ will cross $C_\alpha$ and paths through all points of $C_\alpha$ will cross $\lambda_\omega$ at interior points (see Remark 2 to Lemma 2).

Let $Q$ be an endpoint of $\lambda_\omega$, $\{Q_i\}$ any sequence of points on $\lambda_\omega$ tending to $Q$, and $\{P_i\}$ the sequence of points on $C_\alpha$ at which the paths through $Q_i$ cross the cycle. The points $P_i$ are of course pairwise distinct. In addition, we may assume without loss of generality that $\{P_i\}$ has only one point of accumulation (otherwise, we need only consider a suitable subsequence of the points $Q_i$). Let $P$ be the point of accumulation and $L_P$ the path through $P$. $L_P$ is clearly a nonsingular path (since $C_\alpha$ is a free cycle) and by Lemma 3 it must cross $\lambda_\omega$ at some interior point $Q'$ (distinct, of course, from the endpoint $Q$). But all paths through points sufficiently near $P$ must cross $\lambda_\omega$ at points arbitrarily near $Q'$ (by the continuity of solutions with respect to initial values). In particular, all paths through points $P_i$ sufficiently near $P$ will cross $\lambda_\omega$ at points $Q'_i$ for sufficiently large $i$, which lie arbitrarily close to $Q'$. Since $Q'$ is distinct from $Q$, it follows that the points $Q'_i$ are also distinct from $Q_i$ for sufficiently large $i$. On the other hand, the points $Q_i$ and $Q'_i$ must lie on the same path — that passing through $P_i$, and by Lemma 2 no path can have more than one point in common with an $\omega$-arc. This contradiction proves the lemma.

Corollary 1. All paths crossing a free $\omega$ ($\alpha$)-cycle without contact cross exactly one free $\alpha$ ($\omega$)-cycle without contact.

Corollary 2. All paths crossing an $\omega$ ($\alpha$)-arc without contact cross exactly one (simple or cyclic) $\alpha$ ($\omega$)-arc without contact.

We shall say that an $\omega$-arc ($\omega$-cycle) $a$ and $\alpha$-arc ($\alpha$-cycle) $b$ are conjugate if all paths crossing one of them also cross the other (Figures 275, 276, 277, 278).

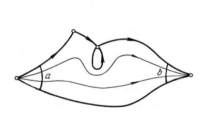

FIGURE 275.

FIGURE 276.

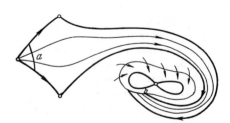

FIGURE 277.

FIGURE 278.

FIGURE 279.

413

Obviously, paths crossing conjugate ω- and α-arcs at interior points, or ω- and α-cycles, belong to the same cell. Thus all elementary arcs and all free cycles are partitioned into pairs of conjugates. Note that a cyclic elementary arc and a noncyclic one may be conjugate; a simple example is shown in Figure 279. Of any two conjugate free cycles without contact, one or even both may be components of the boundary.

Lemma 5. *Two paths, semipaths or arcs of paths in the same cell cross the same ω-arc and the same α-arc.*

Proof. Otherwise, there exist two paths $L_1$ and $L_2$ in the same cell which cross two distinct α (or ω)-arcs. Join any point $A$ on $L_1$ to a point $B$ on $L_2$ by a simple arc within the cell. Reasoning as in the proofs of Lemmas 3 and 4, one readily derives a contradiction. Q.E.D.

## §28. CONJUGATE FREE ω-, α- AND 0-LIMIT CONTINUA AND THE REGIONS BETWEEN THEIR CANONICAL NEIGHBORHOODS

**1. RELATIVE POSITION OF TWO CONJUGATE FREE ω- AND α-CYCLES.** Our first lemma characterizes the relative position of two conjugate free ω- and α-cycles.

Lemma 1. *Of any two conjugate free ω- and α-cycles, one must be inside the other.*

Proof. Suppose the contrary, i.e., let $C$ and $C'$ be two mutually exterior conjugate cycles. Each of these cycles either is or is not a boundary curve Γ. If one of the cycles $C$ and $C'$ is not a Γ, then, since it is free, all its points belong to the same cell $w$. Within any such cycle there must be a boundary continuum of $w$. On the other hand, if one of $C$ and $C'$, say $C$, is a Γ, then the points of $\bar{G}^*$ are either all inside $C$ or all outside $C$. But $C$ is conjugate to $C'$, which is exterior to it, i.e., arcs of paths in $G^*$ connect points of $C$ to points of $C'$, which are outside it. Hence it is clear that the points of $G^*$ lie outside $C$, and since $C$ is free it follows that all points exterior to it in a sufficiently small neighborhood belong to the same cell $w$. Thus the cycle $C$ itself is a boundary continuum of $w$. In either case, all points sufficiently near $C$ and $C'$ and outside them belong to the same cell $w$. But it is obvious that outside the cycles $C$ and $C'$ there are both points of $w$ and points of other cells. Hence there must be boundary points, therefore also a boundary continuum, of the cell $w$ outside both $C$ and $C'$. This continuum is distinct from both $C$ and $C'$. But then the boundary of the cell is the union of at least three disjoint continua: one inside each of $C$ and $C'$ (or identical to the respective cycle), and a third outside both which is a boundary continuum of $\bar{G}^*$. This contradicts the theorems of §16, and completes the proof of the lemma.

Remark. Through all points of the annular region between two conjugate cycles $C_\alpha$ and $C_\omega$ there are paths which cross $C_\alpha$ with decreasing $t$ and cross $C_\omega$ with increasing $t$ (see Lemma 16, §3). The converse is also obvious: if the annular region between two cycles $C_\omega$ and $C_\alpha$ in a regular system of canonical neighborhoods contains no singular paths, so that any path through the region crosses $C_\alpha$ with decreasing $t$ and $C_\omega$ with increasing $t$, then the cycles $C_\alpha$ and $C_\omega$ are conjugate.

By Lemma 1, for any two conjugate ω (α)-cycles without contact, we can define the o u t e r   c y c l e and the (conjugate) i n n e r   c y c l e.

As indicated previously, one of any two conjugate cycles (or even both) may be a boundary curve. Let us consider the case in which at least one of two conjugate free cycles is not a boundary curve Γ. Then there exists an ω - or α-limit continuum (which may in particular be a node) to which the cycle in question belongs (i. e., a continuum $K_\omega$ or $K_\alpha$ in the boundary of whose canonical neighborhood this cycle appears as a component).

*L e m m a  2.* *If the outer (inner) of two conjugate cycles without contact is not a boundary curve* Γ, *then the* ω (α)-*limit continuum to which it belongs lies outside (inside) it.*

P r o o f. Let $C$ and $C'$ be two conjugate cycles without contact, and suppose that the outer cycle $C$ is not a boundary curve Γ. If our assertion is false, the continuum $K^{()}$ to which $C$ belongs lies outside this cycle. But the cycle $C'$ is inside $C$. The annular region between $C$ and $C'$ cannot contain any singular path (see Remark to Lemma 1). Hence it is clear that the continuum $K^{()}$, to which $C$ belongs, is also inside $C$. But this means that $C$ is in the canonical neighborhood $\gamma$ of $K^{()}$ bounded by $C'$. If $C'$ is a boundary curve Γ, this is impossible, since by definition the canonical neighborhood of a limit continuum cannot contain a Γ. It is also impossible if $C'$ is not a boundary curve, since our system of canonical neighborhoods is regular. This contradiction proves the assertion of the lemma concerning the outer cycle. The proof for the conjugate inner cycle is analogous. Q.E.D.

**2. CONJUGATE ω- AND α-LIMIT CONTINUA.** A limit continuum $K^{()}$ lying outside (inside) the cycle without contact that belongs to it will be called an o u t e r (i n n e r)  l i m i t   c o n t i n u u m.

If $C$ and $C'$ are conjugate ω - and α-cycles which are not boundary curves Γ, the ω-limit and α-limit continua $K^{()}$ and $K'^{()}$ to which they belong will be called c o n j u g a t e limit continua. We shall also say that $K^{()}$ ($K'^{()}$) is conjugate to $K'^{()}$ ($K^{()}$). If one of the conjugate cycles $C'$ is a boundary curve Γ and the other $C$ is not, we shall say that the continuum $K^{()}$ to which $C$ belongs is c o n j u g a t e to the boundary cycle $C'$. The boundary cycle $C'$ itself will be called o u t e r or i n n e r, according as the points of $G^*$ lie outside or inside it. It is clear that any outer ω (α)-limit continuum is conjugate either to an inner α (ω)-limit continuum or to an inner boundary cycle without contact. Any inner ω (α)-limit continuum is conjugate either to an outer α (ω)-limit continuum or to an outer boundary cycle. Finally, if both conjugates $C$ and $C'$ are boundary cycles, one of them is inner and the other outer.

Bearing in mind that the canonical curve of an equilibrium state may be a cycle without contact only in the case of a node, and using Lemma 1 of §25, one readily sees that 1) an inner continuum $K_\omega^{()}$ is either a node, a simple closed curve (in particular, a closed path) or the union of several simple closed curves $S_i$ which are mutually exterior except for their common points; 2) an outer continuum $K_\omega^{()}$ ($K_\alpha^{()}$) is either a simple closed curve (in particular, a closed path) or the union of several simple closed curves, in which case one of them, $S_0$ say, contains all the others in its interior, and the latter are mutually exterior. If $K'^{()}$ and $K^{()}$ are two conjugate limit continua, then clearly the inner continuum $K^{()}$ lies inside the curve $S_0$ of the outer continuum $K'^{()}$.

The elementary proof of the following lemma is omitted.

*L e m m a 3. a) Any two conjugate ω- and α-limit continua are boundary continua of the same doubly connected cell filled by whole paths. b) An ω (α)-limit continuum and a conjugate boundary cycle without contact are boundary continua of the same doubly connected cell filled by semipaths. c) Any two conjugate boundary cycles without contact are boundary continua of the same doubly connected cell filled by arcs of paths.*

Figure 278 illustrates the case of conjugate ω- and α-limit continua, Figure 280 that of a limit continuum conjugate to a boundary cycle without contact. The last possibility is two conjugate boundary cycles without contact, in which case the region between the two conjugate cycles exhausts all of $G^*$.

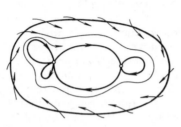

FIGURE 280.

### 3. CONJUGATE 0-LIMIT CONTINUA.

We now consider nonsingular closed paths. Any cell filled by such paths is doubly connected and its boundary is the union of two continua which, as before, we shall call 0-limit continua $K_0^{()}$. In particular, one of them may be an equilibrium state (center). The canonical curve bounding a canonical neighborhood of $K_0^{()}$ is a closed path of the cell for which the latter is a boundary continuum. Since we are dealing with a regular system of canonical neighborhoods, the canonical neighborhoods of distinct continua are disjoint and the canonical curves bounding them (closed paths) distinct. Canonical curves $C$ and $C'$ of two continua $K_0^{()}$ are said to be c o n j u g a t e if they are paths in the same cell.

By Theorem 49 (§16), one of any two conjugate canonical curves is always inside the other and the annular region between them does not contain points of any singular path. The following lemma is the analog of Lemma 2 for 0-limit continua; the proof is omitted.

*L e m m a 4. Given two conjugate canonical curves, the 0-limit continuum to which the outer one belongs lies outside it, and the 0-limit continuum to which the inner one belongs lies inside it.*

As in the case of ω- and α-limit continua, a continuum $K_0^{()}$ is said to be outer or inner according as it lies outside or inside the canonical curve that belongs to it. Two zero-limit continua are said to be conjugate if the canonical curves which belong to them are conjugate. Of any two conjugate 0-limit continua, one is always outer, the other inner. For every outer continuum $K_0^{()}$ there is a conjugate inner continuum, and conversely. Moreover, by Lemma 9 of §25, 1) an inner continuum $K_0^{()}$ is either a center, a simple closed curve, or the union of several simple closed curves which are mutually exterior (not counting their common points); 2) an outer continuum $K_0''^{()}$ is either a simple closed curve or the union of several simple closed curves, in which case one of them, $S_0$ say, contains all the others and the latter are mutually exterior.

If $K_0^{()}$ and $K_0''^{()}$ are conjugate continua and $K_0^{()}$ is inner, the latter lies inside the curve $S_0$ of the outer continuum $K_0''^{()}$. Under these assumptions, we have the following simple lemma (proof omitted):

*L e m m a 5. Conjugate continua $K_0^{()}$ and $K_0''^{()}$ are boundary continua of the same cell.*

The next lemma concerns conjugate free ω-, α-limit continua, boundary cycles without contact, and 0-limit continua. Let $K''^{()}$ be an outer ω-, α- or 0-limit continuum or an outer boundary cycle without contact. Of course, $K''^{()}$ may be a simple closed curve, e. g., when it is a boundary cycle without contact or a limit cycle. In that case we shall denote this simple closed curve (identical with $K''^{()}$ as a point set) by $S'_0$. But if $K''^{()}$ is not a simple closed curve, we know that it is a limit continuum one of whose simple closed curves contains all the others in its interior, and then the relevant (outer) curve will be denoted by $S'_0$.

L e m m a  6. *If $K^{()}$ is the α-, ω-or 0-limit continuum or boundary cycle without contact conjugate to $K''^{()}$, then no constituent simple closed curve $S$ of any limit continuum (and no boundary curve Γ) can lie inside $S'_0$ and at the same time contain $K^{()}$ in its interior.*

P r o o f. This is obvious for a boundary curve Γ, for if Γ is inside $S'_0$ and its interior contains $K^{()}$, then there are points of $G^*$ both inside and outside Γ, contrary to the definition of a boundary curve. Now suppose that there exists a simple closed curve $S^*$, figuring in some limit continuum $K^{*()}$, which is inside $S'_0$ and contains $K^{()}$ in its interior. To fix ideas, let $K''^{()}$ be an ω- or α-limit continuum, the union of $S'_0$ and certain mutually exterior simple closed curves $S'_1, S'_2, \ldots, S'_p$ inside $S'_0$, and let $K^{()}$ be an α- or ω-limit continuum, the union of mutually exterior simple closed curves $S_1, S_2, \ldots, S_q$. Let $C$ and γ ($C'$ and γ′) be the canonical curve and neighborhood of $K^{()}(K''^{()})$. The curve $S^*$ cannot lie inside any of the curves $S'_1, \ldots, S'_p$ or $S_1, \ldots, S_q$, since then the continuum $K^{()}$ would also lie inside that curve, and this is impossible. Neither can $S^*$ have points in common with the neighborhoods γ′ and γ, since there are no points of singular paths in these neighborhoods. Consequently, $S^*$ must lie wholly in the region $\bar{R}$ bounded by $C'$ and $C$. But this is impossible (see Lemma 16 of §3). The proof is similar with $K^{()}$ and $K''^{()}$ are conjugate 0-limit continua and when one or both of them is a boundary cycle without contact. Q.E.D.

**4. PATHS PASSING THROUGH ENDPOINTS OF CONJUGATE ω- AND α-ARCS.** Before investigating conjugate ω- and α-arcs, we devote some attention to ω- and α-saddle arcs which are subarcs of canonical curves σ of equilibrium states in the regular system of canonical neighborhoods. Recall that an ω (α)-saddle arc is a saddle arc across which paths enter (leave) the corresponding saddle region. Clearly, whereas elementary ω- and α-arcs bound regions of attraction (sinks) or regions of repulsion (sources) (ω- and α-parabolic regions and canonical neighborhoods of ω- and α-limit continua), which a path, having once entered, never leaves, saddle arcs do not delimit such regions. Nevertheless, they play a role somewhat analogous to that of elementary arcs in relation to singular paths which are not equilibrium states. The following lemma is stated for ω-arcs and ω-saddle arcs; the analog for α-arcs and α-saddle arcs is also valid.

L e m m a  7. *Any nonclosed singular orbitally unstable path $L$ lying wholly within the closed region $\bar{G}^*$ passes through the endpoints of at most two ω-arcs or ω-saddle arcs. If $L$ passes through the endpoint of only one such arc, this arc must be cyclic; if $L$ passes through the endpoints of two arcs, it passes through the common endpoint of two arcs, one on its positive side and one on its negative side, and each of these arcs may be either an ω-arc or an ω-saddle arc.*

(As mentioned, an analogous statement holds for orbitally unstable paths passing through endpoints of α-arcs or α-saddle arcs.)

Proof. As $t \to +\infty$, the path $L$ tends either to an equilibrium state or to a limit continuum including at least one path which is not an equilibrium state. If $L$ tends to an equilibrium state $O$ as $t \to +\infty$, it will ultimately enter a canonical neighborhood of the point and remain there. Since $L$ is a singular path, it must clearly pass through an endpoint of at least one ω-arc or ω-saddle arc figuring in the canonical curve σ of the equilibrium state $O$. Moreover, since we are assuming the system of canonical neighborhoods to be regular, only one endpoint of any ω-saddle arc can lie on a singular path, and this endpoint is also the endpoint of a semipath (ω-separatrix) figuring in the boundary of the saddle region. It is thus clear from the structure of canonical curves of an equilibrium state that $L$, having once passed through the endpoint of an ω-arc or ω-saddle arc, can no longer pass through the endpoint of any other ω-arc or ω-saddle arc. But if $L$ does not pass through the endpoint of a cyclic arc, it must pass through the common endpoint of two arcs, one on its positive and one on its negative side, and each of these arcs is either an ω-arc or an ω-saddle arc. In this case, therefore, the proof is complete.

If $L$ tends to a limit continuum (not an equilibrium state) $K^{()}$ as $t \to +\infty$, it necessarily reaches some canonical neighborhood γ of $K^{()}$ and must cross the boundary cycle $C$ of this neighborhood at a point which is the endpoint of some ω-arc. Moreover, $L$ cannot pass through the endpoint of any ω-saddle arc, since otherwise it would follow from our assumption concerning saddle arcs that $L$ is an ω-separatrix for some equilibrium state, which is impossible. Hence, the properties of canonical neighborhoods of limit continua which are not equilibrium states imply the assertion of the lemma in this case too.

Remark 1. At least one of the elementary ω- and α-arcs or ω- and α-saddle arcs through whose endpoints a singular orbitally unstable path passes must be a saddle arc. Indeed, otherwise $L$ could not be orbitally unstable. In particular, if an orbitally unstable path $L$ passes through an endpoint of an ω (α)-cyclic arc, then it cannot go through the endpoint of an α (ω) -cyclic arc.

Remark 2. The parameter value corresponding to the endpoint of an α-arc or α-saddle arc on a path $L$ is smaller than that corresponding to the endpoint of an ω-arc or ω-saddle arc.

The following lemma is also valid; the elementary proof, based on the fact that the boundary of the region is normal and the system of canonical neighborhoods regular, is omitted.

Lemma 8. *An orbitally unstable positive semipath* $L^+$ *in* $\overline{G}^*$, *whose endpoint lies on the boundary of* $\overline{G}^*$, *passes 1)* *through the endpoint of one cyclic boundary α-arc or through the common endpoint of two noncyclic boundary arcs lying on different sides of* $L^+$, *but not through the endpoint of an α-saddle arc; 2)* *through the common endpoint of two ω-arcs or ω-saddle arcs on different sides of* $L^+$, *at least one of which is a saddle arc.*

(An analogous assertion is true for an orbitally unstable negative semipath whose endpoint is on the boundary of $\overline{G}^*$; see Figure 281a, b.)

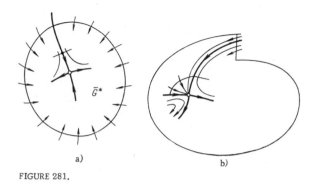

FIGURE 281.

## 5. LEMMA ON BOUNDARY SINGULAR ELEMENTS AND ω- AND α-ARCS WHICH ARE SUBARCS OF BOUNDARY ARCS WITHOUT CONTACT.

*L e m m a 9. Each endpoint of a boundary arc $l$ of a path is an endpoint of exactly one elementary arc, to be precise: the endpoint of a noncyclic boundary ω- or α-arc. The interior points of $l$ are not endpoints of ω- or α-arcs.*

In Figures 282 and 283 the elementary arcs are denoted by $\lambda_i$ and $\lambda_i'$. A perusal of these figures shows that two elementary arcs without contact whose endpoints are endpoints of a boundary arc $l$ may lie either on different sides of $l$ or on the same side.

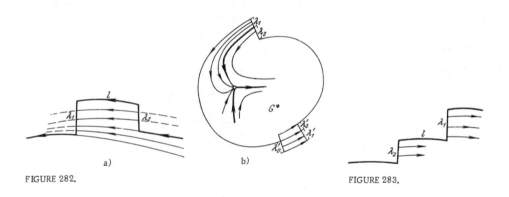

FIGURE 282.

FIGURE 283.

Remark 1. Let $B$ be a corner point which is the ω-endpoint of a boundary arc of a path $l$ and an endpoint of an elementary arc without contact $\lambda$. It is easy to see that if $\lambda$ is an ω-arc, then $\lambda$ and $G^*$ are both on the same side of $l$, whereas if $\lambda$ is an α-arc they are on different sides (Figure 284a, b). Similarly, if $A$ is the α-endpoint of an arc $l$ and an endpoint of an elementary arc $\lambda$, then $\lambda$ and $G^*$ are on the same side of $l$, whereas if $\lambda$ is an α-arc they are on different sides.

Remark 2. If the ω-endpoint $B$ of a boundary arc $l$ is an endpoint of an elementary α-arc, it follows from Lemma 1 of §26 that there exists a corner arc or corner semipath with endpoint $B$ which is an ω-continuation of $l$. A similar assertion holds for an α-endpoint $A$ of $l$ (Figure 285).

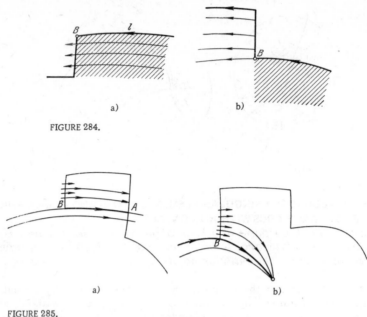

FIGURE 284.

a)   b)

FIGURE 285.

a)   b)

Now consider a corner arc $\hat{l}$. Let $A$ and $B$ be its $\alpha$- and $\omega$-endpoint, respectively. By definition, both $A$ and $B$ are on the boundary of $G^*$, and either one of them is a corner point of the boundary (Figure 286a) or both are (Figure 286b).

a)   b)

FIGURE 286.

*Lemma 10. If the $\alpha$-endpoint $A$ of a corner arc $\hat{l}$ is a corner point of the boundary, then $A$ is an endpoint of exactly one $\alpha$-arc, namely, the endpoint of an elementary noncyclic boundary $\alpha$-arc (Figure 286a). If $A$ is not a corner point, it is the endpoint of either two or one $\alpha$-arcs. In the first case both $\alpha$-arcs are elementary noncyclic boundary arcs on different sides of $\hat{l}$ (Figure 287a), in the second the $\alpha$-arc is an elementary cyclic boundary arc (Figure 287b, c).*

An analogous statement holds for the $\omega$-endpoint of $\hat{l}$. No interior point of $\hat{l}$ can be an endpoint of an $\omega$- or $\alpha$-arc without contact.

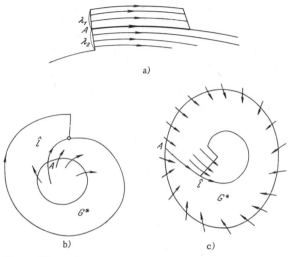

FIGURE 287.

We now consider a corner semipath. Note first that no corner semipath can pass through the endpoint of a saddle arc, since then it would clearly have to be orbitally unstable, contradicting the normality of the boundary of $G^*$.

**Lemma 11.** *Let $\hat{L}^+$ be a positive corner semipath with endpoint $A$ ( $A$ is a corner point of the boundary and there is only one α-arc with an endpoint on $\hat{L}^+$ — an elementary boundary arc on one side of $\hat{L}^+$). Then the $\hat{L}^+$ passes either 1) through the endpoint of exactly one (non-boundary) ω-arc, and then this arc is cyclic, or 2) through the common endpoint of two noncyclic ω-arcs on different sides of $\hat{L}^+$.*

(An analogous statement holds for a negative corner semipath $\hat{L}^-$.)

**6. CHAINS OF SINGULAR ELEMENTS, PATHS AND BOUNDARY ARCS JOINING THE ENDPOINTS OF CONJUGATE ω- AND α-ARCS.** We are now going to consider pairs of conjugate α- and ω-arcs and the singular elements passing through their endpoints. It follows directly from the definition of α- and ω-arcs that an endpoint of an α (or ω)-arc may belong to either 1) an orbitally unstable path lying wholly within $G^*$, or an orbitally unstable semipath whose endpoint is on the boundary of $\overline{G}^*$ (in the latter case the arc $a$ may be an elementary boundary arc); 2) a boundary or corner arc of a path (in this case $a$ is a boundary arc without contact); 3) a corner semipath (in this case $a$ may be an arc without contact, either on the boundary or not); or, finally 4) a nonsingular semipath in an elliptic region of some equilibrium state $O$ (in which case the endpoint of $a$ is an endpoint of an elliptic arc).

We first consider a simple α-arc $a$, and let $b$ be the conjugate ω-arc. All the lemmas below are stated for the case in which $a$ is on the positive side of the singular element* (singular path or semipath, corner semipath, boundary or corner arc of path) containing one of its endpoints. Analogous assertions are valid when the simple α-arc is on the negative side of the element, and for simple ω-arcs.

---

* Which is distinct from $a$ when $a$ is a boundary α-arc.

*Lemma 12. Let $a$ be a simple $\alpha$-arc and $b$ the conjugate $\omega$-arc, simple or cyclic, and suppose that one endpoint of $a$ is either on an orbitally unstable path $\tilde{L}_0$ in $G^*$ or an endpoint (on the boundary of $G^*$) of an orbitally unstable semipath $\tilde{L}_0^+$,* the arc $a$ lying on the positive side of $\tilde{L}_0$ $(\tilde{L}_0^+)$. Then: either 1) $\tilde{L}_0$ $(\tilde{L}_0^+)$ passes through an endpoint of the conjugate $\omega$-arc $b$, which is either on the positive side of $\tilde{L}_0$ $(\tilde{L}_0^+)$ or cyclic, or 2) there exists a finite chain of pairwise distinct orbitally unstable paths, beginning with $\tilde{L}_0$ $(\tilde{L}_0^+)$ and possibly ending with a semipath,*

$$\tilde{L}_0\,(\,or\ \tilde{L}_0^+),\ \tilde{L}_1,\ \ldots,\ \tilde{L}_{R-1},\ \tilde{L}_R\,(\,or\ \tilde{L}_R^-),$$

*such that $\tilde{L}_{i+1}$ is the $\omega$-continuation of $\tilde{L}_i$ on the positive side $(i = 1, 2, \ldots, R-1)$, and the last path $\tilde{L}_R$ (or semipath $\tilde{L}_R^-$) passes through an endpoint of the conjugate $\omega$-arc $b$, which either lies on the positive side of $\tilde{L}_R(\tilde{L}_R^-)$ or is a cyclic $\omega$-arc (Figure 288).*

Proof. Let $A$ be the endpoint of the $\alpha$-arc $a$ on the orbitally unstable path $\tilde{L}_0$, corresponding under some fixed motion on $\tilde{L}_0$ to time $t = t_0$.

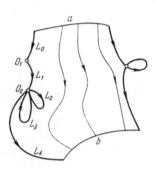

FIGURE 288.

Assume first that $\tilde{L}_0$ is not $\omega$-continuable on the positive side. By Lemma 7, $\tilde{L}_0$ must pass at some time $t > t_0$ either through an endpoint $B$ of a cyclic $\omega$-arc, or through the common endpoint $B$ of two arcs on different sides of $\tilde{L}_0$, each of which may be either an elementary $\omega$-arc or an $\omega$-saddle arc.

In the first case, it is clear from Lemma 5 of §3 that all paths crossing the arc $a$ at time $t = t_0$, sufficiently near $A$, will cross the cyclic $\omega$-arc with endpoint $B$ at $t > t_0$. This means that the cyclic $\omega$-arc in question is conjugate to $a$. Since the $\omega$-arc conjugate to a given $\alpha$-arc is unique, this cyclic arc is precisely $b$. In the second case, consider the arc with endpoint $B$ which lies on the positive side of $\tilde{L}_0$. Since by assumption $\tilde{L}_0$ is not $\omega$-continuable on the positive side, this arc cannot be an $\omega$-saddle arc and is therefore an elementary $\omega$-arc. But then, by Lemma 5 of §3, it is an $\omega$-arc conjugate to $a$, therefore coinciding with $b$.

Assume now that $\tilde{L}_0$ is continuable on the positive side. Let $O_1$ be the equilibrium state to which it tends as $t \to +\infty$, and $\tilde{L}_1$ its $\omega$-continuation on the positive side. $\tilde{L}_1$ may or may not be $\omega$-continuable on the positive side; if it is, we consider its $\omega$-limit point $O_2$ and its $\omega$-continuation on the positive side. Continuing in this way, we obtain a sequence of orbitally unstable paths and equilibrium states:

$$\tilde{L}_0,\ O_1,\ \tilde{L}_1,\ O_2,\ \ldots, \tag{1}$$

in which a) each point $O_{i+1}$ is an $\omega$-limit point of $\tilde{L}_i$ and an $\alpha$-limit point of $\tilde{L}_{i+1}$; b) each path $\tilde{L}_{i+1}$ is the $\omega$-continuation of $\tilde{L}_i$ on the positive side. We claim that any such sequence will ultimately reach

---

* In this case the points of the $\alpha$-arc are obviously points of a boundary arc without contact, while the semipath $\tilde{L}^{(\ )}$(which is orbitally unstable) is certainly not a corner semipath.

a path $\tilde{L}_R$ which is not ω-continuable on the positive side or a semipath $\tilde{L}_R^-$ with endpoint on the boundary of $\bar{G}^*$, at which the sequence breaks off. Indeed, suppose that the sequence does not break off, in other words, it never reaches a path which is not ω-continuable on the positive side. Since the number of orbitally unstable paths is finite, the paths $\tilde{L}_0$, $\tilde{L}_1$, . . . cannot all be different. Let $R$ be the smallest positive integer such that $\tilde{L}_0$, $\tilde{L}_1$, . . ., $\tilde{L}_R$ are different and $\tilde{L}_{R+1}$ coincides with one of them, say $\tilde{L}_p$, $0 \leqslant p \leqslant R$. If $p \neq 0$, the paths $\tilde{L}_R$ and $\tilde{L}_{p-1}$ have the same ω-continuation $\tilde{L}_{R+1}$ on the positive side, i.e., $\tilde{L}_{R+1}$ has two distinct α-continuations on the positive side, which is impossible. Suppose that $p = 0$. Then the ω-continuation of $\tilde{L}_R$ on the positive side is $\tilde{L}_0$, so that $\tilde{L}_0$ is continuable on the positive side. But this contradicts the fact that $\tilde{L}_0$ passes through the endpoint of the elementary α-arc $a$, which is on the positive side of $\tilde{L}_0$. Thus any sequence of type (1) invariably breaks off at some path $\tilde{L}_R$ which is not ω-continuable or at a semipath $\tilde{L}_R^-$ with endpoint on the boundary of $\bar{G}^*$. Using similar reasoning, one readily shows that the paths $\tilde{L}_0$, $\tilde{L}_1$, . . ., $\tilde{L}_R$ are all distinct.

Some or all of the equilibrium states $O_1$, $O_2$, . . ., $O_R$ may coincide. Since $\tilde{L}_R$ is not continuable on the positive side, one can prove, repeating the reasoning used for $\tilde{L}_0$, that $\tilde{L}_R$ passes through an endpoint of the ω-arc $b$ conjugate to $a$, which is either on the positive side of $\tilde{L}_R$ or cyclic. If the last term of (1) is a semipath $\tilde{L}_R^-$ with endpoint on the boundary of $\bar{G}^*$, then (Lemma 8) it passes either through the endpoint of a simple ω-arc on its positive side or through the endpoint of a cyclic boundary ω-arc.

In either case one readily verifies the assertion of the lemma. Q.E.D.

R e m a r k 1. It follows from the above proof that the converse of Lemma 12 is also valid, in the following sense. Let the orbitally unstable path $\tilde{L}_0$ pass through the endpoint of an α-arc $a$ on its positive side. a) If $\tilde{L}_0$ is not ω-continuable on the positive side, there exists an ω-arc $b$ with endpoint on $\tilde{L}_0$, either on the positive side of $\tilde{L}_0$ or cyclic, and the arcs $a$ and $b$ are conjugate. b) If $\tilde{L}_0$ is ω-continuable on the positive side, there exist a chain of paths (1) and an ω-arc $b$ with endpoint on $\tilde{L}_R$ (or $\tilde{L}_R^-$), either on the positive side of $\tilde{L}_R$ ($\tilde{L}_R^-$) or cyclic, and the arcs $a$ and $b$ are conjugate.

R e m a r k 2. All paths in the sequence are boundary paths of the same cell — the cell whose paths cross the relevant ω- and α-arcs at their interior points. An analogous statement holds true if a semipath $\tilde{L}_0^+$ passes through the endpoint of $a$.

The endpoints $A$ and $B$ of conjugate simple α- and ω-arcs, lying either on the same singular path $\tilde{L}_0$ or semipath $\tilde{L}_0^+$, or on the first and last (semi)path in the sequence (1), will be called c o n j u g a t e e n d p o i n t s of the conjugate arcs.

We now consider the case in which the endpoint of one of two conjugate arcs $a$ and $b$, say the endpoint of a simple α-arc $a$, is the endpoint of a corner arc of a path or of a corner semipath.

L e m m a 13. *Let the endpoint of a simple α-arc $a$ lie on a boundary or corner arc $l_0$ or on a corner semipath $\hat{L}^+$, in such a way that $a$ is on the positive side of $l_0$ or $\hat{L}^+$. Then the following alternatives are possible. a) The ω-endpoint of $l_0$ or a point of the semipath $\hat{L}^+$, respectively, is an endpoint of the arc $b$ conjugate to $a$, which is either on the positive side of $l_0$ ($\hat{L}_0^+$) or cyclic. b) There exists a finite chain of alternating corner and boundary arcs of paths $l_0$, $l_1$, $l_2$, . . ., $l_{R-1}$, satisfying the* *following conditions: 1) $l_1$ is an ω-continuation of $l_0$ (see §26.1), each $l_i$ is an ω-continuation of $l_{i-1}$ ($i = 1, 2, \ldots, R-1$), and there exists either another boundary or corner arc of a path $l_R$ which is an ω-continuation of $l_{R-1}$ and itself has no continuation, or a corner semipath $\hat{L}_R^{\pm}$ which is a continuation of $l_{R-1}$; 2) all elementary arcs other than $a$ and $b$ whose endpoints are endpoints of $l_0, l_1, \ldots, l_{R-1}$ or the ω-endpoint of $l_R$, are not cyclic and lie on the negative side of these arcs; 3) if the ω-continuation of $l_{R-1}$ is a boundary or corner arc $l_R$, the ω-endpoint of the latter is an endpoint of the ω-arc $b$ conjugate to $a$, which is a boundary ω-arc, either on the positive side of $l_R$ or cyclic. If the ω-continuation of $l_{R-1}$ is a semipath $\hat{L}_R^{\pm}$, the latter passes through an endpoint of the ω-arc $b$ conjugate to $a$ (which is not a boundary arc), which is either on the positive side of $\hat{L}_R$ or cyclic.*

An analogous assertion is true for a simple ω-arc $b$.

The proof follows directly from the basic definitions (of corner arcs, boundary arcs, etc.) and the normality of the boundary. Figures 289, 290 and 291 illustrate the various possibilities.

Remark 1. By the definition of boundary and corner arcs, and the definition of corner semipaths which are continuations of one another,

FIGURE 289.

all the arcs $l_1, l_2, \ldots, l_{R-1}$, as well as $l_0$ and the arc $l_R$ (semipath $\hat{L}_R^{\pm}$), are arcs (semipaths) of the same path $L$ in the domain of definition $G$ of system (I).

Remark 2. The converse of Lemma 13 is also valid: Suppose that the endpoint of a simple boundary $a$-arc $a$ is the endpoint of a boundary or corner arc $l_0$ or a corner semipath $\hat{L}_0^+$, $a$ lying on the positive side of $l_0$ or $\hat{L}_0^+$, as the case may be.

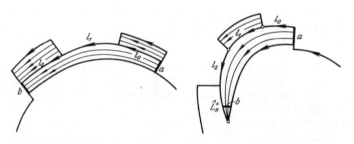

FIGURE 290.                    FIGURE 291.

Then the alternatives are as follows: 1) The ω-endpoint of $l_0$ is an endpoint of an ω-arc $b$ which is either on the positive side of $l_0$ or cyclic (or, in the case of $\hat{L}_0^+$, the latter passes through an endpoint of an ω-arc $b$ on its positive side), and the arcs $a$ and $b$ are conjugate. 2) There exists a chain of alternating boundary and corner arcs as described above, whose last term is a boundary corner arc or corner semipath passing through an endpoint of an ω-arc $b$ which is either on the positive side of this last term, and the arcs $a$ and $b$ are conjugate.

The next lemma is concerned with the case of a corner semipath $\hat{L}_0^-$ passing through an endpoint of a simple $a$-arc $a$ which is not a boundary arc. Part of the lemma duplicates the statement of Lemma 13 word for word, and the proof is omitted.

*Lemma 14.* Let *a be a simple a-arc, not on the boundary, through whose endpoint passes a corner semipath* $\hat{L}_0^-$. *Suppose that a lies on the positive side of* $\hat{L}_0^-$. *Then the following alternatives are possible: a) The endpoint of* $\hat{L}_0^-$ *is an endpoint of the arc b conjugate to a, which is a boundary arc on the positive side of* $\hat{L}_0^-$. *b) There exists a chain of alternating boundary and corner arcs of paths* $l_0, l_1, l_2, \ldots, l_{R-1}$, *such that 1)* $l_0$ *is the ω-continuation of the semipath* $\hat{L}_0^-$, *each* $l_i$ *is the ω-continuation of* $l_{i-1}$ *(* $i = 1, 2, \ldots, R$ *), and there is either another boundary or corner arc* $l_R$ *which is an ω-continuation of* $l_{R-1}$ *and itself has no continuation, or a corner semipath* $\hat{L}_R^+$ *which is a continuation of* $l_{R-1}$; *2) all elementary arcs other than a and b, whose endpoints are endpoints of the arcs* $l_0, l_1, \ldots, l_{R-1}$, *are not cyclic and lie on the negative side of these arcs; 3) if the ω-continuation of* $l_{R-1}$ *is a boundary or corner arc* $l_R$, *its ω-endpoint is an endpoint of the ω-arc b conjugate to a, which is a simple boundary ω-arc, either on the positive side of* $l_R$ *or cyclic. If the ω-continuation of* $l_{R-1}$ *is a corner semipath* $\hat{L}_1^+$, *the latter passes through an endpoint of the arc b conjugate to a, which is a simple nonboundary arc on the positive (negative) side of* $\hat{L}_R^+$.

An analogous assertion is true for an *ω- arc b.*

The proof of Lemma 14 is omitted. Figures 292 and 293 illustrate the various possibilities.

R e m a r k  1. As in Lemma 13, the semipath $\hat{L}_0^-$, arcs $l_i$ and semipath $\hat{L}_R^+$ all lie on the same path *L* in the domain of definition *G.*

R e m a r k  2. The converse is also valid; the precise formulation is entirely analogous to that of the converse of Lemma 13 (see Remark 2 above).

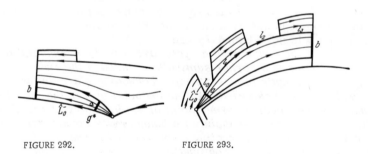

FIGURE 292.                    FIGURE 293.

It was assumed in the preceding lemmas that, of the two relevant conjugate arcs, the arc with an endpoint on the singular element is simple. We now formulate two lemmas for the case in which the arc in question is cyclic.

*Lemma 15.* Let *a be a cyclic a-arc, one of whose endpoints is either a point of an orbitally unstable path* $\tilde{L}_0$ *or a boundary point of G\* which is the endpoint of an orbitally unstable semipath* $\tilde{L}_0^+$.* *Then there exist two finite chains of paths (the last term of each may be a semipath)*

$$\tilde{L}_0, \ (\tilde{L}_0^+), \ \tilde{L}_1, \ \ldots, \ \tilde{L}_R \ (\tilde{L}_R^-), \tag{2}$$
$$\tilde{L}_0, \ (\tilde{L}_0^+), \ \tilde{L}_1', \ \ldots, \ \tilde{L}_S' \ (\tilde{L}_S'^-), \tag{3}$$

---

* In this case *a is obviously a boundary cyclic a-arc.*

*in the first (second) of which each path is the ω-continuation of its predeces-*
*sor on the positive (negative) side, and the following conditions are*
*satisfied: 1) if the arc b conjugate to a is simple, then the last (semi)paths*
$\tilde{L}_R$, $\tilde{L}_S$ ($\tilde{L}_R^-$, $\tilde{L}_S^-$) *in sequences (2) and (3) are distinct and pass through*
*distinct endpoints of b (Figure 279); 2) if b is a cyclic arc, the last*
*terms in sequences (2) and (3) coincide and pass through the endpoint of b*
*(Figure 294).*

*Of the two chains (2) and (3), only one can consist of a single path $\tilde{L}_0$.*

The proof is analogous to that of Lemma 12.

R e m a r k. The converse is also valid: Let an orbitally unstable path
pass through the endpoint of a cyclic α-arc a. Then there exist two chains

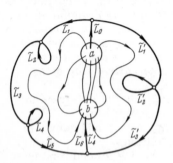

(2) and (3) of paths beginning with $\tilde{L}_0$, as in
Lemma 15; if the last paths $\tilde{L}_R$ and $\tilde{L}_S'$ (or
semipaths) are distinct they pass through
distinct endpoints of the same simple ω-arc
b, and if they coincide they pass through
the endpoint of a cyclic arc b. In either
case, the arcs a and b are conjugate.

The same assertion holds true for an
orbitally unstable semipath whose endpoint
on the boundary of G* is the endpoint of a
boundary cyclic arc.

If the arc a is cyclic, it is clear that
there is another alternative: there is a
corner arc through its endpoint (and then a
is a boundary arc). The endpoint of a may
also lie on a corner semipath; in this case,

FIGURE 294.

however, the endpoint in question cannot be the endpoint of the corner
semipath, so that a is not a boundary arc.

L e m m a 16. *If the endpoint of a cyclic arc a is the α-endpoint of a*
*corner arc $l_0$ or there is a corner semipath $\hat{L}_0^-$ through the endpoint of a,*
*then the ω-arc b conjugate to a is a simple boundary ω-arc and the*
*following conditions are satisfied: 1) One endpoint of b is the ω-endpoint of*
*$l_0$ or $\hat{L}_0$, and it is moreover a corner point of the boundary of G*. 2) The*
*other endpoint of b is the ω-endpoint of a boundary or corner arc $l_R$ which is*
*the last term in a chain of alternating boundary and corner arcs $l_1$, $l_2$, ..., $l_R$,*
*where the first term $l_1$ is an ω-continuation of $l_0$ or of $\hat{L}_0^-$, respectively,*
*and each $l_i$ is a continuation of $l_{i-1}$ (Figures 295 and 287).*

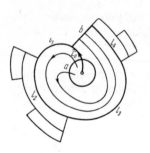

To end this subsection, we consider the case
in which the endpoint of the elementary arc is on
a nonsingular path.

L e m m a 17. *If an endpoint A of an*
*elementary α-arc a lies on a nonsingular path L,*
*then L is a loop figuring in the boundary of an*
*elliptic sector of some equilibrium state. The*
*endpoint of the conjugate elementary arc b is on*
*the same loop L, and both arcs are noncyclic*
*and situated on the same side of L.*

The assertion is obvious and the proof is
therefore omitted (Figure 276).

FIGURE 295.

The endpoints $A$ and $B$ of arcs $a$ and $b$ which are endpoints of the same elliptic arc will also be called conjugate endpoints of $a$ and $b$.

**7. REGIONS BETWEEN CONJUGATE CANONICAL CURVES OR ELEMENTARY ARCS.** Retaining our assumption that the system of canonical neighborhoods is regular, we now examine the points of $G^*$ which are neither in canonical neighborhoods nor on their boundaries.

We first consider points not lying on singular elements. Any such point may lie 1) on an arc of an orbitally stable path or semipath, or on a nonsingular whole arc of a path between its points of intersection with two conjugate arcs; 2) on an arc of an orbitally stable path or semipath, or on a nonsingular whole arc of a path between its points of intersection with two conjugate cycles without contact; 3) on a closed (orbitally stable) path between conjugate canonical curves (closed paths) of two conjugate 0-limit continua.

Let $a$ and $b$ be conjugate arcs and $l$ an arc of an orbitally stable path or nonsingular whole arc in $G^*$ with endpoints $P$ and $Q$ on $a$ and $b$, respectively.

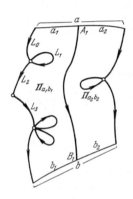

FIGURE 296.

The points $P$ and $Q$ are clearly interior points of $a$ and $b$. Let $M$ be an interior point of $l$. Consider the set of all such points $M$ when $l$ ranges over all possible arcs whose endpoints are interior points of $a$ and $b$. Denote this set by $\Pi_{ab}$ (Figures 288 and 296). $\Pi_{ab}$ is clearly a subset of the cell that contains the interior points of $a$ and $b$.

The following lemma is geometrically evident.

*Lemma 18. The set $\Pi_{ab}$ is a region, whose boundary is the union of the conjugate arcs $a$ and $b$ and the chains joining their endpoints. Each chain joining endpoints of $a$ and $b$ may consist of points of orbitally unstable paths or semipaths (in particular, of one orbitally unstable path) (see Lemma 12), of boundary and corner arcs of paths and corner semipaths (see Lemmas 13, 14 and 15), or of an arc of a path forming a loop.*

Proof. By the definition of $\Pi_{ab}$, the endpoints of $l$ are interior points of the arcs $a$ and $b$. Hence it is readily seen, by Lemma 10 of §3, that $\Pi_{ab}$ is a region. Now consider its boundary. Let $A$ be an endpoint of $a$ and $(L)$ the chain joining it to the conjugate endpoint $B$ of $b$. Using Lemma 10 of §3 and properties of the chain $(L)$ as described in Lemmas 12 through 15, one readily shows that any arc $l$ with endpoint sufficiently close to $A$ lies in an arbitrarily small neighborhood of the chain $(L)$. Hence it clearly follows that all points of $(L)$ are boundary points of $\Pi_{ab}$. Thus, all points of chains joining conjugate endpoints of $a$ and $b$, and the points of $a$ and $b$ themselves (denote the set of all these points by $\Gamma_{ab}$), are boundary points of $\Pi_{ab}$.

We claim that $\Pi_{ab}$ has no boundary points other than those of $\Gamma_{ab}$. If this is false, let $R$ be a boundary point of $\Pi_{ab}$ which is not in $\Gamma_{ab}$ and is therefore at a positive distance $d$ $(d > 0)$ from the set $\Gamma_{ab}$ (which is of course closed). There exists a sequence $\{Q_n\}$ of points of $\Pi_{ab}$ that tend to $R$. By the definition of $\Pi_{ab}$, each point $Q_i$ lies on an arc of a path $l_i$ joining points of $a$ and $b$. Let $M_i$ denote the endpoint of $l_i$ on $a$. We may assume without loss of generality that the points $M_i$ have a unique point of accumulation $M^*$. $M^*$ cannot be an endpoint of $a$. Indeed, were $M^*$ an endpoint of $a$, all arcs $l_i$ with sufficiently large $i$ would lie in an arbitrarily small neighborhood of the chain joining $M^*$ to the conjugate endpoint of $b$, and the points $Q_i$ of these arcs $l_i$ could not tend to the point $R$, which is at a positive distance $d$ from $\Gamma_{ab}$. Thus $M^*$ is not an endpoint of $a$. Let $l^*$ be the arc, whose interior points lie in $\Pi_{ab}$, with endpoint at $M^*$. It is readily seen that the points $Q_i$ can tend only to points of the arc $l^*$, so that $R$ must lie on $l^*$. But this clearly contradicts our assumptions concerning $R$. Q.E.D.

We shall also call $\Pi_{ab}$ the r e g i o n  b e t w e e n  t h e  c o n j u g a t e  a r c s. When these regions are indexed, we shall use the notation $\Pi_{ab}^i$. In the sequel we shall usually consider the closure of $\Pi_{ab}$, i. e., the closed region $\bar{\Pi}_{ab}$.

The region between two conjugate cycles without contact will be denoted by $\Xi_{a\omega}$, and the region between two conjugate canonical curves which are closed paths by $Z_{00}$.

In view of Lemmas 7 and 18, we easily obtain the following proposition:

*Any point of $G^*$ which is not a point of a singular element and does not lie in any canonical neighborhood or on a canonical curve is in some region $\Pi_{ab}^i$, $\Xi_{a\omega}^j$, or $Z_{00}^k$.*

We now consider the points of $G^*$ on singular elements (other than the singular elements of the boundary), in other words, points of orbitally unstable semipaths, corner semipaths or corner arcs.

*L e m m a  19. Any point of a singular element of $G^*$ which is not in any closed canonical neighborhood of an equilibrium state or on any $\omega$-, $\alpha$- or $0$-limit continuum is a boundary point of one of the regions $\Pi_{ab}^i$.*

P r o o f. Let $P$ be a point of the relevant singular element, lying neither on a canonical curve nor in a canonical neighborhood. If this singular element (orbitally unstable path or semipath, corner semipath or corner arc) contains an endpoint of an $\alpha$- or $\omega$-arc (possibly on the boundary), the truth of our assertion follows directly from the continuity of solutions with respect to initial conditions. In particular, the assertion is always true when $P$ is on a corner arc, corner semipath or orbitally unstable semipath with endpoint on the boundary of $G^*$. (In all these cases, the singular element in question has a point on the boundary of $G^*$, which is an endpoint of an elementary boundary arc.)

To prove the lemma, therefore, it remains to consider the case of a point $P$ on an orbitally unstable path $\tilde{L}_0$ which does not pass through an endpoint of any $\alpha$- or $\omega$-arc. The path $\tilde{L}_0$ must obviously be both $\omega$- and $\alpha$-continuable, and on both positive and negative sides. We first consider the $\omega$-continuation of the path $\tilde{L}_0$ on the positive side, $\tilde{L}_1$ say. Let $\tilde{L}_2$ be the $\omega$-continuation of $\tilde{L}_1$, and so on. We obtain a chain of paths

$$\tilde{L}_0, \tilde{L}_1, \ldots, \tilde{L}_R, \tag{4}$$

in which each term is the $\omega$-continuation of its predecessor on the positive side. The chain will either reach a path $\tilde{L}_R$ which has no $\omega$-continuation on the positive side, or extend indefinitely (Theorem 71, §23). In the first case, it follows from Lemma 7 that $\tilde{L}_R$ must pass through an endpoint of an $\omega$-arc, in which case $\tilde{L}_0$ is a term in a chain joining endpoints of conjugate arcs. In the second case, repeating the now familiar argument, we see that there exists a path $\tilde{L}_R$ such that all paths $\tilde{L}_0, \ldots, \tilde{L}_R$ are distinct and $\tilde{L}_0$ is the $\omega$-continuation of $\tilde{L}_R$ on the positive side. But then it follows from Theorem 71 that $\tilde{L}_0$ is a constituent path of some $\omega$-, $\alpha$- or 0-limit continuum and all its points lie on the boundary of some canonical neighborhood of the limit continuum, contrary to assumption. Q.E.D.

R e m a r k. The same orbitally unstable path $\tilde{L}_0$ may of course be a limit path on one side and a boundary path for a region of type $\Pi_{ab}$ on the other.

§29. SCHEME OF A DYNAMIC SYSTEM AND
FUNDAMENTAL THEOREM

1. **SCHEME OF A DYNAMIC SYSTEM.** In this subsection we introduce the s c h e m e  o f  a  d y n a m i c  s y s t e m and define equivalence of schemes. Our definition will of course lean heavily on previously defined concepts — the global schemes of equilibrium states, limit continua and boundary of the region.

Let $D$ be a dynamic system defined, as always, in a bounded region $G$ and considered in a closed subregion $\overline{G}*$ with normal boundary.

*D e f i n i t i o n  XXXIII.  The scheme of the dynamic system $D$ consists of the following components:*

I. *Lists of all the singular elements of system $D$ in $\overline{G}*$:*

1) *all equilibrium states* $O_1, O_2, \ldots, O_m$ $(O)$;

2) *all orbitally unstable paths\** $L_1, L_2, \ldots, L_k$ $(L)$;

3) *all orbitally unstable semipaths* $\tilde{L}_1^{()}, \tilde{L}_2^{()}, \ldots, \tilde{L}_p^{()}$ $(\tilde{L}^{()})$ *with endpoints on the boundary of the region;*

4) *all corner semipaths* $\hat{L}_1^{()}, \hat{L}_2^{()}, \ldots, \hat{L}_q^{()}$ $(\hat{L}^{()})$;

5) *all corner arcs of paths* $\hat{l}_1, \hat{l}_2, \ldots, \hat{l}_s$ $(\hat{l})$;

6) *all boundary arcs of paths\*\** $l_1, l_2, \ldots, l_n$ $(l)$;

7) *all boundary arcs without contact* $\lambda_1, \lambda_2, \ldots, \lambda_h$ $(\lambda)$;

8) *all boundary cycles without contact (if such exist)* $C_1, C_2, \ldots, C_r$ $(C)$.

*With the notation introduced above for the singular elements:*

II. *Global schemes of all equilibrium states* $(O_i)$ *(§19.4).*

III. *Lists of all (one-sided) $\omega$-, $\alpha$- and 0-limit continua* $K_1^{()}, K_2^{()}, \ldots, K_N^{()}$ $(K^{()})$ *together with their global schemes. In particular, each of these schemes (§25.5) specifies for each continuum $K_i^{()}$ the simple closed curves* $S_{a_1}^i, S_{a_2}^i, \ldots, S_{aj}^i$ $(i = 1, 2, \ldots, N)$ *making up the continuum.*

IV. *A boundary scheme for the region $G*$, i.e., a list of all simple closed boundary curves* $\Gamma_1, \Gamma_2, \ldots, \Gamma_\beta$ $(\Gamma)$, *together with their schemes (§26).*

V. *A list of all pairs of conjugate free $\omega$-, $\alpha$- and 0-limit continua and boundary cycles without contact, indicating for each pair which of its elements is inner and which outer.*

The schemes listed above clearly determine which of the paths $(L)$ and semipaths $(\tilde{L}^{()})$ and $(\hat{L}^{()})$ tend to which equilibrium states, which of the paths $(L)$ are components of which limit continua, and so on. It follows

_____

\* The corresponding semipaths will be denoted by $L_i^+$ and $L_i^-$.

\*\* For corner points of the boundary, we retain the notation introduced in §26.1: a corner point which is the $\omega$ $(\alpha)$-endpoint of a boundary arc of a path $l$ will be denoted by $M_i^\omega$ $(M_i^\alpha)$, and the set of all corner points by $(M)$.

from the definition that the scheme of a dynamic system may be given the form of a collection of tables:

I.  Tables listing all singular elements $(O)$, $(L)$, $(\tilde{L}^{()})$, $(\hat{L}^{()})$, $(l)$, $(\lambda)$, $(\hat{l})$, $(\tilde{\Gamma})$.

II.  Tables specifying the boundary scheme of the region.

III.  Tables specifying the global schemes of all equilibrium states.

IV.  Tables specifying the global schemes of all one-sided limit continua $(K^{()})$ (§25.4). The continua should be suitably numbered.

V.  Tables listing all pairs of conjugate free $\omega$-, $\alpha$- and 0-limit continua, boundary cycles without contact and free nodes, also indicating the relative position of each pair.

Only the tables of type V require specification. We introduce the following notation: If $K^{()}$ is an outer continuum or boundary cycle without contact $\tilde{\Gamma}$, and $K'^{()}$ or $\tilde{\Gamma}'$ the conjugate inner continuum or cycle, we shall write

$$K'^{()} \supset K^{()}, \quad \tilde{\Gamma}' \supset \tilde{\Gamma}^{()}.$$

With this notation, tables of type V may be written as $K_3^{(+)} \supset K_2^{(-)}, K_4^{(-)} \supset K_3^{(+)}$, etc.

We shall use the term s c h e m e   e l e m e n t for the singular elements $(L)$, $(\tilde{L}^{()})$, $(\hat{L}^{()})$, $(l)$, $(\lambda)$, $(\hat{l})$, $(\Gamma)$, also for the limit continua, boundary curves, corner points, curves $S_i$, etc. Like schemes of equilibrium states and limit continua, the global scheme of a dynamic system may also be given the form of a schematic diagram, which is extremely convenient and intuitive. This form of the scheme sometimes makes it possible to do away with individual notation for the paths figuring therein; indeed, the purpose of such notation (as in the tables) is only to determine the relative positions of the singular paths. For example, if some singular path has a negative semipath in the scheme of an equilibrium state and a positive semipath in the scheme of a limit continuum, this means that the singular path goes from the equilibrium state to the limit continuum as $t \to +\infty$. Now this behavior of the path is quite easy to display in a schematic diagram, so that the notation for the path may be omitted. We present a few simple examples of schemes in both tabular and schematic form.

E x a m p l e  1. Boundary of region: $\alpha$-cycle without contact. Equilibrium states: $O_1, O_2, O_3$. Orbitally unstable paths: $L_0, L_1, L_2, L_3, L_4$;

$$O_1 \mid L_1^+ \quad \text{(topological node)};$$
$$O_2 \mid L_1^-, \ L_2^+, \ L_3^-, \ L_4^+ \quad \text{(saddle point)};$$
$$O_3 \mid L_3^+ \quad \text{(topological node)};$$

$K^-$: $\alpha$-limit continuum consisting of a single path $L_0$, positive traversal is anti-$t$-traversal; $K^+$: $\alpha$-limit continuum consisting of a single path $L_0$, positive traversal is anti-$t$-traversal. Singular paths tending to $K^{(+)}$: $L_2^-, L_4^-$. The corresponding schematic diagram is Figure 297 (see Example 15 of Chapter XII).

E x a m p l e  2. Boundary: $\alpha$-cycle without contact. Equilibrium states: $O_1, O_2, O_3$. Singular paths: $L_1, L_2$.

$O_2 | L_1^{(+)}, L_1^-, L_2^+, L_2^-$ (saddle point);

$K_1^{(+)} | L_1, L_2$ (α-limit continuum);

$K_2^{(+)} | L_2, O_2$ (α-limit continuum);

$K_3^- | L_1, O_2, L_2, O_2$ (α-limit continuum);

$O_1$ and $O_3$: stable free topological nodes; $K_1^+ \supset O_1$, $K_2^+ \supset O_3$. The schematic diagram is given in Figure 298.

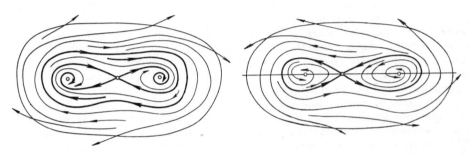

FIGURE 297.                    FIGURE 298.

Now let $D$ and $D'$ be two (not necessarily different) dynamic systems defined in regions $G$ and $G'$, respectively. Let the closed regions of interest be $\overline{G}^*$ and $\overline{G'}^*$, $\overline{G}^* \subset G$, $\overline{G'}^* \subset G'$, with normal boundaries. We now define equivalence of the schemes of systems $D$ and $D'$ in regions $\overline{G}^*$ and $\overline{G'}^*$.

The notation for the singular elements of system $D$ will be as before. The notation for those of system $D'$ is obtained by adding primes at the appropriate places:

$$(O'), (L'), (\tilde{L}'^{(\,)}), (\hat{L}'^{(\,)}), (\hat{l'}), (l'), (\lambda'), (C').$$

This gives all singular elements of system $D'$ and also a corresponding system of numbers $m', k', p', q', s', n', h', r'$, analogous to the numbers $m, k, p, q, s, n, h, r$, for system $D$.

As in the case of schemes of equilibrium states, the definition given below pertains to equivalence of schemes with the same orientation and direction of $t$. Obvious modifications yield the definitions of schemes of two dynamic systems with reversed orientation and the same direction of $t$, with the same orientation and reversed direction of $t$, and finally with both orientation and direction of $t$ reversed.

*Definition XXXIV. The schemes of two dynamic systems $D$ and $D'$ are said to be equivalent, with the same orientation and direction of $t$, if there exists a one-to-one correspondence θ between all the singular elements of $D$ and those of $D'$, under which equilibrium states $O$ correspond to equilibrium states $O'$, paths $L$ to paths $L'$, positive (negative) semipaths $(\tilde{L}^{(\,)})$ to positive (negative) semipaths $(\tilde{L}'^{(\,)})$, etc., and the following condition is satisfied: the scheme of system $D'$ is obtained from that of system $D$ by replacing each singular element of $D$ by that of $D'$ corresponding to it under θ.*

431

If systems $D$ and $D'$ have equivalent schemes, then $m' = m$, $k' = k$, $p' = p$, $q' = q$, $s' = s$, $n' = n$, $h' = h$, $r' = r$, etc. It is clear that a finite number of trials is sufficient to determine whether two given schemes are equivalent or not. The correspondence $\theta$ of the definition will be called a s c h e m e  c o r r e s p o n d e n c e, and we shall speak as usual of s c h e m e - c o r r e s p o n d i n g  e l e m e n t s. Note that there may be more than one scheme correspondence between the elements of two systems with equivalent schemes (at least one must exist, of course). For this reason, when discussing scheme-corresponding singular elements of two dynamic systems (with equivalent schemes), we shall always assume given a certain well-defined scheme correspondence $\theta$. Any singular element of $D'$ will be denoted by the same symbol as the scheme-corresponding element of $D$, with a prime added. We shall also use such obvious notation as $\theta(O_i) = O_i'$, $\theta(L_i) = L_i'$, and so on. It follows from Definition XXXIV that the scheme correspondence between the singular elements of systems $D$ and $D'$ induces a one-to-one correspondence 1) between the $\omega$-, $\alpha$- and 0-limit continua of the systems which are unions of scheme-corresponding singular elements, and also between the constituent simple closed curves $S_i$ of the continua; 2) between the boundary curves which are unions of scheme-corresponding boundary arcs of paths and arcs without contact, and also between scheme-corresponding boundary cycles without contact.

This induced correspondence between the limit continua and boundary curves will also be refered to as a s c h e m e  c o r r e s p o n d e n c e and denoted by the same symbol $\theta$. The continua and boundary curves of system $D'$ will be denoted by the same symbols as the corresponding continua and boundary curves of $D$, with primes. We shall also write $\theta(K_j^{()}) = K_j^{()'}$, $\theta(\Gamma_i) = \Gamma_i'$, and so on.

In case the schemes of systems $D$ and $D'$ are described in tabular form as above and the schemes are equivalent, the scheme of system $D'$ is obtained from that of system $D$ by simply adding primes in the symbol for each element of the scheme.

The following elementary propositions are direct consequences of the definition.

I. The boundary schemes of the regions $\bar{G}^*$ and $\bar{G'}^*$ in which systems $D$ and $D'$, respectively, are being considered are equivalent, and the schemes of scheme-corresponding boundary curves $\Gamma_i$ and $\Gamma_i'$ are also equivalent. The correspondence $\theta$ between the singular elements of systems $D$ and $D'$ induces a scheme correspondence between the singular elements figuring in $\Gamma_i$ and $\Gamma_i'$.

II. Any two scheme-corresponding equilibrium states $O_i$ and $O_i'$ ($\omega$-, $\alpha$- or 0-limit continua $K_i^{()}$ and $K_i^{()'}$) have equivalent global schemes, and the correspondence between the singular elements of their global schemes is precisely that induced by the equivalence of their global schemes. In particular, free $\omega$- and $\alpha$-limit continua of system $D$ correspond to free $\omega$- and $\alpha$-limit continua of system $D'$.

III. If free $\omega$- and $\alpha$-limit continua or 0-limit continua $K_i^{()}$ and $K_j^{()}$ of system $D$ are conjugate, so are the scheme-corresponding continua $K_i^{()'}$ and $K_j^{()'}$ of system $D'$.

IV. If $L_i$ and $L_i'$ are scheme-corresponding singular paths of systems $D$ and $D'$, i. e., $L_i' = \theta(L_i)$, then: a) their $\omega$- and $\alpha$-limit continua are scheme-corresponding (in particular, scheme-corresponding equilibrium states);

b) the two paths are either both ω (α)-continuable or both noncontinuable, on the same (positive or negative) side; if they are ω (α)-continuable, their ω (α)-continuations on the appropriate side are scheme-corresponding paths.

V. If a corner arc $\hat{l}$ or corner semipath $\hat{L}^{()}$ of system $D$ is an ω-continuation of a boundary arc $l$, then the corner arc $\hat{l}'$ or corner semipath $\hat{L}'^{()}$ of system $D'$ is an ω-continuation of the boundary arc $l'$ scheme-corresponding to $l$ (in particular, outer corner points correspond to outer ones, inner corner points to inner ones).

VI. Any scheme correspondences between singular elements of dynamic systems induces a natural correspondence between their cells and between the boundaries of the cells.

The foregoing discussion pertains to the scheme of the dynamic system as a whole. One can also speak of the scheme of a certain combination of cells or the scheme in some subregion $\overline{H}^* \subset \overline{G}^*$ (with normal boundary). Obvious examples are the local schemes of equilibrium states and limit continua.

If the schemes of two dynamic systems are given in schematic diagram form and are equivalent with the same orientation and direction of $t$, they are represented by the same schematic diagram with identically directed arrows on the paths.

**2. SCHEME CORRESPONDENCE BETWEEN CANONICAL CURVES AND ARCS OF CANONICAL CURVES.** As before, let $D$ and $D'$ be dynamic systems, having equivalent schemes with the same orientation and direction of $t$, and $\theta$ a scheme correspondence between their singular elements. The notation for the singular elements and scheme elements is as before. Assume that a regular system of canonical neighborhoods is provided for each system. The notation for the canonical neighborhoods and canonical curves is again retained from the preceding sections ($\gamma_i$, $C_i$, $g_i$, $\sigma_i$), those of system $D'$ being denoted by the same symbols with primes. The scheme correspondence between equilibrium states $O_i$ and $O'_i$ and limit continua $K_i^{()}$ and $K'^{()}_i$ of systems $D$ and $D'$ induces a natural one-to-one correspondence between the canonical neighborhoods, canonical curves, and canonical arcs (parabolic, elliptic and saddle arcs without contact and saddle arcs of paths) and elementary ω- and α-arcs. In other words:

*Lemma 1. If the schemes of dynamic systems $D$ and $D'$ are equivalent and $\Sigma$ and $\Sigma'$ are their regular systems of canonical neighborhoods, the scheme correspondence induces the following one-to-one correspondence between the canonical regions, sectors, curves and arcs:*

*1) Canonical neighborhoods $g_i$ and $g'_i$ and canonical curves $\sigma_i$ and $\sigma'_i$ of scheme-corresponding equilibrium states $O_i$ and $O'_i$ correspond to each other.*

*2) In canonical neighborhoods $g_i$ and $g'_i$ that correspond according to 1), canonical regions of the same type (elliptic, parabolic and hyperbolic) correspond to each other, as do the arcs of the canonical curves $\sigma_i$ and $\sigma'_i$ in the boundaries of these sectors (i.e., elliptic and parabolic arcs, saddle arcs of paths and saddle arcs without contact) and their endpoints. Under this correspondence: a) corresponding endpoints of corresponding parabolic arcs belong either to corresponding singular elements (paths or semipaths) or to corresponding elliptic arcs; b) endpoints of corresponding saddle arcs belong to corresponding singular elements, and ω (α)-saddle arcs correspond to ω (α)-saddle arcs.*

*3) Canonical neighborhoods $\gamma_i$ and $\gamma'_i$ and canonical curves $C_i$ and $C'_i$ of scheme-corresponding limit continua $K_i^{()}$ and $K'^{()}_i$ correspond to each other.*

433

*Under this correspondence: a) if $K_i^{()}$ is outside (inside) $C_i$, then $K_i'^{()}$ is also outside (inside) $C_i'$; b) if the canonical curves $C_j$ and $C_i$ of free $\alpha$- and $\omega$-limit or 0-limit continua are conjugate, then the corresponding canonical curves $C_j'$ and $C_i'$ are also conjugate.*

*4) $\omega(\alpha)$-cycles and elementary $\omega(\alpha)$-arcs whose points are on scheme-corresponding curves $C$ and $C'$ or $\sigma$ and $\sigma'$ correspond to each other, in such a way that: a) $\omega(\alpha)$-cycles correspond to $\omega(\alpha)$-cycles and are cycles without contact of corresponding free $\omega(\alpha)$-limit continua (i.e., of corresponding curves $C$ and $C'$ or $\sigma$ and $\sigma'$); b) cyclic $\omega(\alpha)$-arcs correspond to cyclic $\omega(\alpha)$-arcs and points of corresponding cyclic arcs belong to canonical curves $C$ and $C'$ or $\sigma$ and $\sigma'$ which correspond to each other according to 1) and 3); c) simple $\omega(\alpha)$-arcs correspond to simple $\omega(\alpha)$-arcs, corresponding arcs being arcs of canonical curves $C$ and $C'$, $\sigma$ and $\sigma'$ that correspond to each other according to 1) and 2), with endpoints on corresponding singular elements or on elliptic arcs that correspond to each other according to 2).*

Proof. Part 1) is obvious. Part 2) follows from the equivalence of local schemes of scheme-corresponding equilibrium states (see 1)). Part 3) follows from the equivalence of local schemes of scheme-corresponding limit continua. Part 4) follows from the equivalence of global schemes of scheme-corresponding equilibrium states and limit continua.

Remark 1. In particular, it follows from Lemma 1 that a one-to-one correspondence as described exists for the neighborhoods, regions, curves and arcs of any two different regular systems of canonical neighborhoods of one system $D$.

Remark 2. There is a one-to-one correspondence between the boundary $\omega$- and $\alpha$-arcs of systems $D$ and $D'$ and between their endpoints, induced by the equivalence of the schemes of the boundary curves.

Remark 3. Let $l$ and $l'$ be corresponding arcs of canonical curves, which are either $\omega$- and $\alpha$-arcs, parabolic arcs, or elliptic arcs, etc. Let $M$ and $M'$, $N$ and $N'$ be the corresponding endpoints of the arcs (see part 2) of Lemma 1). Then the direction on $l$ from $M$ to $N$ and the direction on $l'$ from $M'$ to $N'$ either both agree or both disagree with the positive traversal of the canonical curves of which $l$ and $l'$ are subarcs. In addition, if $l$ and $l'$ are arcs of paths, the directions indicated above either both agree or both disagree with the direction of increasing $t$. This follows immediately from the fact that the schemes of systems $D$ and $D'$ are equivalent with the same orientation and direction of $t$.

As usual, we shall continue to use the term "scheme correspondence" for the correspondences induced according to Lemma 1 by the scheme correspondence proper $\theta$, also employing the obvious notation $\theta(g_i) = g_i'$, $\theta(\gamma_i) = \gamma_i'$, $\theta(\sigma_i) = \sigma_i'$, $\theta(C_i) = C_i'$; also $\theta(a) = a'$, $\theta(b) = b'$, $\theta(l) = l'$, $\theta(A) = A'$, where $a$ and $a'$, $b$ and $b'$, $l$ and $l'$ are scheme-corresponding arcs, $A$ and $A'$ scheme-corresponding endpoints of corresponding arcs. We state one more lemma, which follows directly from Lemma 1 and the above remarks.

Lemma 2. *Scheme-corresponding singular paths $L$ and $L'$ ($L' = \theta(L)$) contain endpoints of corresponding $\omega$- and $\alpha$-arcs or saddle arcs. Moreover, if an elementary or saddle arc $a$ with endpoint on $L$ is simple and on the positive (negative) side of $L$, then the simple arc $a'$ corresponding to $a$ ($a' = \theta(a)$), whose endpoint is on the path $L' = \theta(L)$, lies on the positive (negative) side of $L'$.*

An analogous assertion is true for other pairs of scheme-corresponding singular elements which contain endpoints of elementary and saddle arcs: orbitally unstable semipaths $\tilde{L}^{()}$ and $\tilde{L}'^{()} = \theta\,(\tilde{L}^{()})$ with endpoints on the boundaries of $G^*$ and $G'^*$, corner semipaths $\hat{L}^{()}$ and $\hat{L}'^{()} = \theta\,(\hat{L}^{()})$, boundary and corner arcs of paths $l$ and $l' = \theta\,(l)$, $\hat{l}$ and $\hat{l}' = \theta\,(\hat{l})$, and finally scheme-corresponding elliptic arcs. By Lemma 17 of § 28, one endpoint of any elliptic arc is always an endpoint of an $\alpha$-arc and the other of an $\omega$-arc.

### 3. CONJUGATE $\omega$- AND $\alpha$-ARCS OF SYSTEMS $D$ AND $D'$ WITH EQUIVALENT SCHEMES.

By part 3) of Lemma 1, any two conjugate canonical curves of system $D$ correspond under $\theta$ to conjugate canonical curves of system $D'$. In particular, conjugate $\omega$- and $\alpha$-cycles without contact of $D$ correspond to conjugate $\omega$- and $\alpha$-cycles of $D'$. What can we say about conjugate $\omega$- and $\alpha$-arcs of system $D$ and the corresponding $\omega$- and $\alpha$-arcs of system $D'$?

Lemma 3. *If $a$ and $b$ are two conjugate elementary arcs of system $D$, then the scheme-corresponding elementary arcs $a'$ and $b'$ are also conjugate. Moreover, conjugate endpoints of $a$ and $b$ correspond to conjugate endpoints of $a'$ and $b'$.*

Proof. An endpoint $A$ of an elementary $\alpha$-arc $a$ of system $D$ may lie on one of the following: 1) an orbitally unstable path $L_0$ or orbitally unstable semipath $\tilde{L}^{()}$ (with endpoint on the boundary of $G^*$); 2) a boundary or corner arc or corner semipath; 3) an elliptic arc of the canonical curve of an equilibrium state.

We consider case 1), assuming to fix ideas that $a$ is a simple $\alpha$-arc on the positive side of $L_0$. The arc $b$ may be either simple or cyclic. By Lemma 12 of § 28, there is a finite chain of pairwise distinct orbitally unstable paths

$$L_0, L_1, \ldots, L_R\,(L_R^{\pm}), \tag{1}$$

in which $L_{i+1}$ is the $\omega$-continuation of $L_i$ on the positive side ($i = 0, 1, 2, \ldots, R-1$), and the last path $L_R$ (or semipath) passes through an endpoint $B$ of the conjugate elementary $\omega$-arc $b$. In particular, the chain (1) may comprise only one term $L_0$. To fix ideas, suppose that the last term is a path $L_R$ (not a semipath). The arc $b$ is either on the positive side of $L_R$ or cyclic. $A$ and $B$ are conjugate endpoints of $a$ and $b$. Consider the scheme-corresponding arcs $a'$ and $b'$, $a' = \theta\,(a)$, $b' = \theta\,(b)$, the path $L'_0$ of system $D'$, $L'_0 = \theta\,(L_0)$, passing through the endpoint $A'$ of $a'$ corresponding to the endpoint $A$ of $a$, and finally the paths $L'_i$ of system $D'$ corresponding to the paths $L_i$, $L'_i = \theta\,(L_i)$. By proposition IVb) in § 28.1, each path in the chain

$$L'_0, L'_1, \ldots, L'_R, \tag{2}$$

is the continuation of its predecessor on the positive side; by parts 4b) and 4c) of Lemma 1, the path $L'_R$ passes through the endpoint $B'$ of $b'$. If $b$ is a simple arc on the positive side of $L_R$, then $b'$ is also a simple arc, on the positive side of $L'_R$. But the chain (2) clearly possesses the same properties as the chain (1). Hence, by the Remark to Lemma 12 of § 28, the arcs $a'$ and $b'$ are conjugate and $A'$ and $B'$ are conjugate endpoints. This proves the lemma in case 1) for a simple arc. If $a$ is cyclic the arguments are similar, except that Lemma 15 of § 28 is used.

The proof in cases 2) and 3) is analogous and will therefore be omitted. In case 2) the proof employs Remark 2 to Lemmas 13 and Lemma 14 of § 28, and in case 3) Lemma 17 of § 28.

We now consider the regions between conjugate canonical curves and between conjugate canonical arcs of systems $D$ and $D'$. As before ($\S 28.8$), these regions for system $D$ will be denoted by $\Pi_{ab}^i$, $\Xi_{a\omega}^j$, $Z_{00}^k$, and for system $D'$ by $\Pi_{a'b'}^{'i}$, $\Xi_{a\omega}^{'j}$, $Z_{00}^{'k}$. The correspondence between conjugate canonical curves and elementary arcs clearly induces a natural correspondence between these regions (and their closures $\overline{\Pi}_{ab}^i$ and $\overline{\Pi}_{a'b'}^{'i}$, $\overline{\Xi}_{a\omega}^j$ and $\overline{\Xi}_{a\omega}^{'j}$, $\overline{Z}_{00}^k$ and $\overline{Z}_{00}^{'k}$). Thus:

Lemma 4. *If the schemes of systems $D$ and $D'$ are equivalent, there is a one-to-one correspondence between the regions $\Pi_{ab}^i$ and $\Pi_{a'b'}^{'i}$, $\Xi_{a\omega}^j$, and $\Xi_{a\omega}^{'j}$, $Z_{00}^k$ and $Z_{00}^{'k}$ between scheme-corresponding conjugate canonical curves and conjugate elementary arcs. Moreover, the singular elements figuring in the boundaries of corresponding regions $\Pi_{ab}^i$ and $\Pi_{a'b'}^{'i}$ are scheme-corresponding.*

This correspondence between the regions will again be termed a scheme correspondence and denoted by the same symbol $\theta$. Throughout the sequel, we shall assume that scheme-corresponding regions of the above types are identically numbered for both systems $D$ and $D'$.

**4. FUNDAMENTAL THEOREM.** The goal of this subsection is to show that *the topological structure of a dynamic system is completely determined by its scheme.* In other words, if two dynamic systems have equivalent schemes, their phase portraits possess the same topological structure. This result is stated in Theorem 76. Assuming that the schemes of two systems $D$ and $D'$, considered in closed regions $\overline{G}^* \subset G$ and $\overline{G}'^* \subset G'$ with normal boundary, are equivalent, we shall construct a path-preserving mapping of $\overline{G}^*$ and $\overline{G}'^*$, preserving both orientation and direction of $t$ on the paths of systems $D$ and $D'$. A few preliminary remarks will explain how the mapping is constructed. Using regular systems of canonical neighborhoods of systems $D$ and $D'$, we partition the closed regions $\overline{G}^*$ and $\overline{G}'^*$ into closed subregions with common boundaries — canonical neighborhoods and regions between conjugate canonical curves and conjugate elementary arcs (regions of types $\Pi_{ab}$, $\Xi_{a\omega}$ and $Z_{00}$). There is a natural scheme correspondence between any two such subregions of systems $D$ and $D'$. By previous theorems and Lemma 5 below, there exists an orientation-preserving topological mapping of each such region onto the scheme-corresponding region, also preserving paths and the direction of increasing $t$ on paths. In addition, the mappings of the various regions may be so constructed that they agree on the common boundaries, so that the final result is a path-preserving, orientation-preserving topological mapping of $\overline{G}^*$ onto $\overline{G}'^*$, which also preserves the direction of $t$ on paths. In other words, if the systems $D$ and $D'$ have equivalent schemes, with the same orientation and direction of $t$, their phase portraits in the regions $G^*$ and $G'^*$ are topologically equivalent.

After these general remarks, we take up the detailed proof of the fundamental theorem. The phase portraits of scheme-corresponding canonical neighborhoods are topologically equivalent by Theorem 72, and the same is readily proved for regions of type $\Xi_{a\omega}$ and $\Xi_{a\omega}'$, $Z_{00}$ and $Z_{00}'$, by first drawing certain auxiliary arcs (for regions $\Xi_{a\omega}$ these are arcs of paths joining cycles without contact, and for $Z_{00}$ arcs without contact joining closed boundary curves; see Lemma 7 of $\S 19$) and then using Lemma 8 of $\S 18$ (topological equivalence of phase portraits in elementary quadrangles).

*Lemma 5. If the schemes of systems D and D′ are equivalent, with the same orientation and direction of t, and $\Pi_{ab}$ and $\Pi'_{a'b'}$ are scheme-corresponding regions, then the phase portraits in these regions are topologically equivalent, with the same orientation and direction of t.*

Proof. We divide each of $\Pi_{ab}$ and $\Pi'_{a'b'}$ into two subregions, as follows. First consider $\Pi_{ab}$. Let $A_1$ be an arbitrary interior point of the arc $a$. Whether $a$ is a simple or cyclic arc, it is divided by $A_1$ into two simple arcs, $a_1$ and $a_2$, say. Let $L_{A_1}$ be the path through $A_1$ and $B_1$ the point at which it crosses the arc $b$. $B_1$ divides $b$ into two simple arcs $b_1$ and $b_2$. It is readily seen that all paths passing through interior points of $a_1$ cross one of these arcs, say $b_1$, also at interior points; all paths passing through interior points of $a_2$ cross the other $(b_2)$ at interior points. Thus the arc $A_1B_1$ of $L_{A_1}$ divides $\Pi_{ab}$ into two disjoint regions, which we denote by $\Pi_{a_1b_1}$ and $\Pi_{a_2b_2}$ (Figure 296). The region $\Pi_{a_1b_1}$ is the set of all points on paths of system $D$ that pass through interior points of the arc $a_1$ and lie between the points at which these paths cross $a_1$ and $b_1$. The region $\Pi_{a_2b_2}$ is similar. Each point of $\Pi_{ab}$ is of course either in $\Pi_{a_1b_1}$, in $\Pi_{a_2b_2}$ or an interior point of the arc $A_1B_1$ of $L_{A_1}$. Let $A$ denote the common endpoint of $a_1$ and $a$. To fix ideas, let us assume that $a_1$ is on the positive side of the singular element containing $A$ (or of the relevant elliptic arc, if this be the case). Then $a_2$ is clearly on the negative side of the singular element (elliptic arc) through its endpoint other than $A_1$. Repeating the argument used in Lemma 12 of §28, one readily sees that the boundary of $\Pi_{a_1b_1}$ is the union of the arcs $a_1$ and $b_1$, the arc $A_1B_1$ of $L_{A_1}$ and the chain joining the endpoint $A$ of $a$ (which is also an endpoint of $a_1$) to the conjugate endpoint $B$ of $b$ (the endpoint of $b_1$ other than $B_1$) (Figure 299). A similar statement holds for the boundary of $\Pi_{a_2b_2}$. The points of one of the chains joining conjugate endpoints of $a$ and $b$ are boundary points of $\Pi_{a_1b_1}$, those of the other chain boundary points of $\Pi_{a_2b_2}$, and each point in the closed region $\overline{\Pi}_{ab}$ lies in one of the closed regions $\overline{\Pi}_{a_1b_1}$ and $\overline{\Pi}_{a_2b_2}$. Any point of $\overline{\Pi}_{ab}$ not on the arc $A_1B_1$ of $L_{A_1}$ may be in both $\overline{\Pi}_{a_1b_1}$ and $\overline{\Pi}_{a_2b_2}$ only if the chains figuring in the boundary of $\Pi_{ab}$ have common points. Similarly, taking an arbitrary interior point $A'_1$ of the arc $a'$ and considering the path $L_{A'_1}$ of system $D′$ through this point, we divide the region $\Pi'_{a'b'}$ into two subregions.

Denote the simple arcs into which $A'_1$ divides the arc $a'$ by $a'_1$ and $a'_2$. If $a'$ (hence also $a$) is a simple arc, we let $a'_1$ denote the arc whose endpoint $A'$ other than $A'_1$ (which is an endpoint of $a'$) corresponds (in the scheme) to the endpoint $A$ of $a$. Then, by Lemma 2, the arc $a'_1$ (hence also $a_1$) is on the positive side of the singular element (or elliptic arc) passing through the point $A'$. If $a'$ (hence also $a$) is a cyclic arc, then each of the arcs into which $A'_1$ divides $a'$ has an endpoint coinciding with the endpoint $A'$ of the (cyclic) arc $a'$. In this case $a'_1$ will denote the subarc on the positive side of

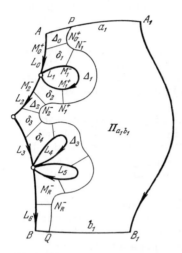

FIGURE 299.

437

the singular element through $A'$. Let $B_1'$ denote the point at which $L_{A_1'}'$ crosses the arc $b'$, $b_1'$ and $b_2'$ the arcs into which this point divides $b'$ (analogous to the arcs $b_1$ and $b_2$), and $\Pi_{a_1'b_1'}'$ and $\Pi_{a_2'b_2'}'$ the regions (analogous to $\Pi_{a_1b_1}$ and $\Pi_{a_2b_2}$) into which the arc $A_1'B_1'$ of $L_{A_1'}'$ divides the region $\Pi_{a'b'}'$. By the corollaries of Definition XXXIV (§29.1), the chains of paths joining the endpoints of $a'$ and $b'$ and figuring in the boundaries of $\Pi_{a'b'}'$ and $\Pi_{a_2'b_2'}'$ correspond to the analogous chains in the boundaries of $\Pi_{a_1b_1}$ and $\Pi_{a_2b_2}$, respectively.

We first prove that there are topological mappings of $\Pi_{a_1b_1}$ and $\Pi_{a_2'b_2'}$ and of $\Pi_{a_2b_2}$ onto $\Pi_{a_2'b_2'}'$, mapping paths onto paths and preserving the direction of $t$. Since one endpoint of $a_1$ is an endpoint of $a$, the following cases may occur (see Lemma 1): 1) the endpoint $A$ of $a_1$ lies on an orbitally unstable path $L$ or semipath $\overline{L}^0$; 2) the endpoint $A$ lies on a boundary or corner arc or corner semipath; 3) $A$ is the endpoint of an elliptic arc figuring in the canonical neighborhood of an equilibrium state. The scheme-corresponding endpoint $A'$ of $a'$ is also an endpoint of $a_1'$, by the definition of the latter. Consequently, the endpoint $A'$ of $a_1'$ lies on the singular element (or elliptic arc) of system $D'$ corresponding (in the scheme) to the singular element (elliptic arc) containing the endpoint $A$ of $a_1$.

Case 1. To fix ideas, let us assume that the singular element through $A$ is a path $L$ (not a semipath). If the chain joining the endpoints $A$ and $B$ of $a_1$ and $b_1$ (hence also that joining the endpoints $A'$ and $B'$ of $a_1'$ and $b_1'$) consists of one singular path $L_0(L_0')$, then the regions $\Pi_{a_1b_1}$ and $\Pi_{a_1'b_1'}'$ are obviously elementary quadrangles, and the existence of the required topological mapping follows from Lemma 8 of §18.

Suppose now that the chain joining the endpoints $A$ and $B$ of $a_1$ and $b_1$ (joining the endpoints $A'$ and $B'$ of $a_1'$ and $b_1'$) consists of more than one path. Let these paths be $L_0, L_1, \ldots, L_R$ $(L_0', L_1', \ldots, L_R')$, arranged in order of $\omega$-continuation. The paths $L_i$ and $L_i'$ in these sequences are scheme-corresponding and their $\omega$- and $\alpha$-limit points are scheme-corresponding equilibrium states. The path $L_0$ $(L_0')$ passes through an endpoint of $a_1$ $(a_1')$, the path $L_R(L_R')$ through an endpoint of $b_1$ $(b_1')$. Moreover, by assumption the arc $a_1$ $(a_1')$ is on the positive side of $L_0$ $(L_0')$, and $b_1$ $(b_1')$ on the positive side of $L_R(L_R')$. By Lemma 12 of §28, the path $L_0$ $(L_0')$ passes through the endpoint of an $\omega$-saddle arc $\lambda_0^+$ $(\lambda_0'^+)$ on its positive side, and the path $L_R$ $(L_R')$ through an endpoint of an $\alpha$-saddle arc $\lambda_R^-$ $(\lambda_R'^-)$ on its positive side. In addition, again by Lemma 12, each of the paths $L_i$ $(L_i')$, $i = 1, 2, \ldots, R-1$, passes through the endpoints of two saddle arcs, both on its positive side, an $\alpha$-saddle arc $\lambda_i^-$ $(\lambda_i'^-)$ and an $\omega$-saddle arc $\lambda_i^+$ $(\lambda_i'^+)$.

Let $M_i^-(M_i'^-)$ and $M_i^+(M_i'^+)$ denote the endpoints of $\lambda_i^-(\lambda_i'^-)$ and $\lambda_i^+(\lambda_i'^+)$, respectively, on the path $L_i$ $(L_i')$. By Remark 2 to Lemma 7 of §28, the point $M_0^+$ $(M_0'^+)$ corresponds on $L_0$ $(L_0')$ to a time $t$ greater than the point $A$ $(A')$, the point $B$ $(B')$ to a time $t$ greater than $M_R$ $(M_R')$, and each point $M_i^+$ $(M_i'^+)$ to a time $t$ greater than the point $M_i^-$ $(M_i'^-)$. One can always choose a point $P$ $(P')$ on the arc $a_1$ $(a_1')$ so close to its endpoint $A$ $(A')$ that: a) the path $L_P$ $(L_{P'})$ through $P$ $(P')$ crosses all the arcs $\lambda_i^+(\lambda_i'^+)$ and $\lambda_i^-$ $(\lambda_i'^-)$ in succession, ultimately crossing the arc $b_1$ $(b_1')$. Let $N_i^+$ $(N_i'^+)$ and $N_i^-$ $(N_i'^-)$ denote the points at which $L_P$ $(L_{P'})$ crosses $\lambda_i^+$ $(\lambda_i'^+)$ and $\lambda_i^-$ $(\lambda_i'^-)$, and $Q$ $(Q')$ the point at which it crosses $b_1$ $(b_1')$; b) with increasing $t$, all paths crossing the subarc $AP$ $(A'P')$ of $a_1$ $(a_1')$ cross the subarc $M_0^+N_0^+$ $(M_0'^+N_0'^+)$ of $\lambda_0^+$ $(\lambda_0'^+)$, then the subarc $M_1^-N_1^-$ $(M_1'^-N_1'^-)$ of $\lambda_1^-$ $(\lambda_1'^-)$, and so on, finally crossing the subarc $BQ$ $(B'Q')$ of $b_1$ $(b_1')$ (see the proof of Theorem 72).

The arc $PQ$ of $L_P$ clearly divides the region $\Pi_{a_1 b_1}$ into two subregions; one of these, $\Pi_1$ say, is of exactly the same type as $\Pi_{a_1 b_1}$, while the other, $\Pi_2$ say, is an elementary quadrangle. The arc $P'Q'$ of $L_{P'}$ divides $\Pi'_{a_1' b_1'}$ into analogous regions $\Pi_1'$ and $\Pi_2'$. Consider the closures of $\Pi_1$ $(\Pi_1')$ and $\Pi_2$ $(\Pi_2')$: $\overline{\Pi}_1$ $(\overline{\Pi}_1')$ and $\overline{\Pi}_2$ $(\overline{\Pi}_2')$. It is clear that the region $\overline{\Pi}_1$ $(\overline{\Pi}_1')$ is divided by the subarcs $M_i^+ N_i^+$ $(M_i'^+ N_i'^+)$ of $\lambda_i^+$ $(\lambda_i'^+)$ and the subarcs $M_i^- N_i^-$ $(M_i'^- N_i'^-)$ of $\lambda_i^-$ $(\lambda_i'^-)$ into closed elementary quadrangles and saddle regions, each pair of which have a common boundary arc — either $M_i^- N_i^- (M_i'^- N_i'^-)$ or $M_i^+ N_i^+ (M_i'^+ N_i'^+)$. (A detailed description of the boundaries of these regions is not needed since it is quite obvious; see Figure 299.) Denote the elementary quadrangle whose boundary includes the arc $M_i^+ N_i^+$ $(M_i'^+ N_i'^+)$ by $\Delta_i$ $(\Delta_i')$, and the saddle region adjacent to it, whose boundary includes the same arc, by $\delta_{i+1}$ $(\delta_{i+1}')$, $i = 0, 1, 2, \ldots, R-1$; $\Delta_R$ $(\Delta_R')$ will denote the elementary quadrangle whose boundary includes the arc $M_R^- N_R^-$ $(M_R'^- N_R'^-)$. It is clear that the equilibrium states, arcs of semipaths and subarcs of saddle arcs figuring in the boundaries of identically numbered elementary quadrangles $\Delta_i$ and $\Delta_i'$ and saddle regions $\delta_j$ and $\delta_j'$ are scheme-corresponding.

We first establish an identifying topological mapping for the elementary quadrangles $\Delta_0$ and $\Delta_0'$, under which the relevant arcs of $L_0$ and $L_0'$ are mapped onto each other (with the direction of $t$ preserved). This clearly induces a topological correspondence of the arcs $M_0^- N_0^-$ and $M_0'^- N_0'^-$. Preserving this correspondence, we construct an identifying mapping for the saddle regions $\delta_1$ and $\delta_1'$, preserving direction of $t$. The next step is to construct an identifying mapping for the elementary quadrangles $\Delta_1$ and $\Delta_1'$, and so on, finally constructing a mapping for the elementary quadrangles $\Delta_R$ and $\Delta_R'$. At each step, the correspondence between the arcs $M_i^- N_i^-$ and $M_i'^- N_i'^-$ or $M_i^+ N_i^+$ and $M_i'^+ N_i'^+$ carried over from the preceding step is preserved. We thus obtain a topological mapping of $\overline{\Pi}_1$ onto $\overline{\Pi}_1'$, possessing the required properties, which moreover induces a certain mapping of the arc $PQ$ of $L_P$ onto the arc $P'Q'$ of $L_{P'}$. We now define a mapping of $\overline{\Pi}_2$ onto $\overline{\Pi}_2'$, preserving this correspondence. We thus obtain the required topological mapping of $\overline{\Pi}_{a_1 b_1}$ onto $\overline{\Pi}'_{a_1' b_1'}$.

Now consider cases 2) and 3): the endpoint $A$ of $a_1$ ($A'$ of $a_1'$) lies on a boundary or corner arc of a path, a corner semipath or an elliptic arc. Now, boundary and corner arcs of paths and corner semipaths which are continuations of one another are parts of the same path in the region $G$. Consequently, in cases 2) and 3) the endpoints $A$ and $B$ of $a_1$ and $b_1$, respectively, lie on the same path $L_0$ of system $D$; similarly, the endpoints $A'$ and $B'$ of $a_1'$ and $b_1'$, respectively, lie on the same path $L_0'$ of system $D'$. In case 2) these arcs consist of scheme-corresponding corner and boundary arcs, and in case 3) they are scheme-corresponding elliptic arcs. In either case the regions $\Pi_{a_1 b_1}$ and $\Pi'_{a_1' b_1'}$ are elementary quadrangles. By Lemma 10 of §3, in case 2) there exists a path-preserving topological mapping of $\overline{\Pi}_{a_1 b_1}$ onto $\overline{\Pi}'_{a_1' b_1'}$, preserving direction of $t$ and mapping the points of $AB$, which are points of boundary and corner arcs of paths, onto the scheme-corresponding points of $A'B'$. The existence of the required mapping in case 3) follows directly from Lemma 17 of §28.

In all cases, therefore, there exists a topological mapping of $\overline{\Pi}_{a_1 b_1}$ onto $\overline{\Pi}'_{a_1' b_1'}$ possessing the required properties.

All that has been said concerning the closed regions $\overline{\Pi}_{a_1 b_1}$ and $\overline{\Pi}'_{a_1' b_1'}$ is valid for $\overline{\Pi}_{a_2 b_2}$ and $\overline{\Pi}'_{a_2' b_2'}$ as well. Construct an identifying mapping of $\overline{\Pi}_{a_2 b_2}$ onto $\overline{\Pi}'_{a_2' b_2'}$, preserving the direction of $t$ and satisfying the following conditions: a) it preserves the correspondence between the arcs $A_1B_1$ and $A_1'B_1'$ induced by the mapping just constructed of $\overline{\Pi}_{a_1b_1}$ onto $\overline{\Pi}'_{a_1'b_1'}$; b) if the boundaries of the closed regions $\overline{\Pi}_{a_2b_2}, \overline{\Pi}'_{a_2'b_2'}$ include the same paths $L_i$ $(L_i')$ or semipaths as the boundaries of $\overline{\Pi}_{a_1b_1}, \overline{\Pi}'_{a_1'b_1'}$,* then the correspondence between these paths induced by the mapping of $\overline{\Pi}_{a_1b_1}$ onto $\overline{\Pi}'_{a_1'b_1'}$ is also preserved. This clearly defines a mapping of the original closed region $\overline{\Pi}_{ab}$ onto $\overline{\Pi}'_{a'b'}$ satisfying all the conditions of the lemma. Q.E.D.

Remark 1. Let $T$ be a preassigned topological mapping of the arc $a$ onto $a'$ (or $b$ onto $b'$), taking $A$ onto $A'$, and also of some (or all) of the boundary curves of $\Pi_{ab}$ and $\Pi'_{a'b'}$ and of scheme-corresponding paths $L_i$ and $L_i'$ (or between semipaths $L_i'$ $'$ and $L_i''$ or arcs thereof), preserving the direction of $t$. It is readily seen that the required topological mapping of $\overline{\Pi}_{ab}$ onto $\overline{\Pi}'_{a'b'}$ may be so constructed that it preserves these preassigned correspondences. To prove this, we need only introduce obvious modifications in the proof of Lemma 5; to be precise, the points $A_1'$ and $P'$ can no longer be chosen at will on the arc $a'$, but must be the points corresponding to $A_1$ and $P$ under the given correspondence; similarly, construction of the mappings for the closed regions $\overline{\Delta}_i$ and $\overline{\Delta}_i'$, $\delta_j$ and $\delta_j'$ must preserve the correspondence $T$ between the arcs $a_1$ and $a_1'$, and also between the singular paths $L_i$ and $L_i'$.

Remark 2. Let $T$ be a preassigned topological correspondence between the arcs $a$ and $a'$, under which the scheme-corresponding endpoints of the arcs correspond to one another. Suppose that this correspondence $T$ is preserved by the mapping of $\overline{\Pi}_{ab}$ onto $\overline{\Pi}'_{a'b'}$. Then, in particular, the latter mapping induces a topological correspondence of the arcs $b$ and $b'$ (under which corresponding points belong to paths crossing the arcs $a$ and $a'$ at points that correspond under $T$, and corresponding endpoints of $b$ and $b'$ are conjugate to corresponding endpoints of $a$ and $a'$).

This topological mapping of one of the arcs $b$ and $b'$ conjugate to $a$ and $a'$ onto the other will be called the topological mapping induced by the given mapping of $a$ onto $a'$.

Now let topological mappings of $a$ onto $a'$ and of $b$ onto $b'$ be given. Then it is clear that an identifying mapping for the scheme-corresponding closed regions $\overline{\Pi}_{ab}$ and $\overline{\Pi}'_{a'b'}$, preserving the given mappings of $a$ onto $a'$, $b$ onto $b'$, can be constructed if and only if the mapping of $b$ onto $b'$ is that induced according to the above definition.

We can now prove the fundamental theorem. Under the previous assumptions concerning systems $D$ and $D'$:

Theorem 75. *If the schemes of two dynamic systems $D$ and $D'$, considered in closed regions $\overline{G}*$ and $\overline{G}'*$, respectively, are equivalent with the same orientation and direction of $t$, then their phase portraits in $\overline{G}*$ and $\overline{G}'*$ are topologically equivalent, with the same orientation and direction of $t$.*

Proof. It will suffice to show that there exists a path-preserving (i.e., identifying) topological mapping of $\overline{G}*$ onto $\overline{G}'*$, preserving also orientation and direction of $t$. Fix a regular system of canonical neighborhoods for each of systems $D$ and $D'$. For both systems $D$ and $D'$, we consider the following families of closed regions (all either canonical neighborhoods or subregions thereof):

---

* This is clearly possible; see Figure 294.

1) all closed elliptic regions of canonical neighborhoods of equilibrium states;

2) all regular parabolic sectors of these neighborhoods;

3) all closed canonical neighborhoods of free nodes;

4) all canonical neighborhoods of centers;

5) all canonical neighborhoods of $\omega$-, $\alpha$- and 0-limit continua;

6) all closed annular regions bounded by pairs of conjugate $\omega$- and $\alpha$-cycles;

7) all closed annular regions bounded by conjugate canonical curves of 0-limit continua;

8) all closed regions between conjugate $\omega$- and $\alpha$-arcs (regions of type $\Pi_{ab}$).

All the saddle sectors of canonical neighborhoods of equilibrium states appear in one of the regions of families 4), 5) or 8).

No two of the closed regions listed above have points in common other than boundary points, and any point of $\bar{G}^*$ $(\bar{G}'^*)$ lies in one of them. Since the schemes of systems $D$ and $D'$ are equivalent, there is a one-to-one scheme correspondence between all the above regions for system $D$, on the one hand, and those for system $D'$, on the other, such that singular boundary elements of corresponding regions are scheme-corresponding singular elements of systems $D$ and $D'$. Moreover, we know that there is a path-preserving topological mapping for any two scheme-corresponding regions of types 1) through 8), also preserving orientation and direction of $t$ (see § 18.4). In order to complete the proof we must obviously show that these various "partial" mappings can be suitably coordinated on the common boundaries of the regions. To do this, we first consider all $\alpha$-cycles and $\alpha$-arcs of systems $D$ and $D'$ and define a topological mapping for each pair of scheme-corresponding $\alpha$-cycles and $\alpha$-arcs.

In other words, we define 1) an arbitrary orientation-preserving topological mapping of any two scheme-corresponding $\alpha$-cycles $C_\alpha$ and $C'_\alpha$; 2) an arbitrary orientation-preserving topological mapping of any two scheme-corresponding cyclic arcs $a$ and $a'$, mapping their endpoints onto one another; 3) an arbitrary topological mapping of any two scheme-corresponding simple $\alpha$-arcs $a$ and $a'$, preserving scheme correspondence of their endpoints.

Once this has been done for all $\alpha$-cycles and $\alpha$-arcs, we define a (no longer arbitrary) mapping for each pair of scheme-corresponding $\omega$-cycles $C_\omega$ and $C'_\omega$ or $\omega$-arcs $b$ and $b'$, namely, the mapping induced by the mapping of the conjugate $\alpha$-cycles or $\alpha$-arcs. In other words, points of $\omega$-cycles $C_\omega$ and $C'_\omega$ (interior points of $\omega$-arcs $b$ and $b'$) will correspond to each other if they lie on paths crossing the conjugate $\alpha$-cycles ($\alpha$-arcs) at corresponding points. Endpoints of $\omega$-arcs $b$ and $b'$ will correspond if they are conjugate to corresponding endpoints of the conjugate $\alpha$-arcs $a$ and $a'$. These mappings of $\omega$- and $\alpha$-cycles, $\omega$-arcs and $\alpha$-arcs, generate a mapping of all scheme-corresponding cycles without contact among the curves $C$ and $\gamma$, both free and nonfree (figuring in scheme-corresponding limit continua and free nodes). In addition, we obtain a mapping of all scheme-corresponding parabolic arcs and boundary arcs without contact. Denote this mapping of scheme-corresponding $\omega$- and $\alpha$-cycles, $\omega$- and $\alpha$-arcs, nonfree cycles without contact and parabolic arcs by $T$. It is clear that $T$ defines a mapping for all points of singular elements of systems $D$ and $D'$, i.e., for the endpoints of scheme-corresponding elementary arcs. It also defines a mapping for the endpoints of scheme-corresponding elliptic arcs.

Assume now that the closed regions 1) through 8) are numbered in both systems $D$ and $D'$, in such a way that corresponding regions are identically numbered.

Let $h_1$, $h_2$, ..., $h_N$ and $h_1' = \theta(h_1)$, $h_2' = \theta(h_2)$, ..., $h_N' = \theta(h_N)$ be these closed regions, arranged in order of the above-mentioned enumeration (each of $h_i$ and $h_i'$ is a closed region of one of our eight types). In order of the enumeration, we now define a path-preserving topological mapping of each $h_i$ onto the scheme-corresponding $h_i'$, preserving orientation and direction of $t$ and satisfying the following additional conditions: a) the restriction of the mapping to the $\alpha$- and $\omega$-arcs, $\alpha$- and $\omega$-cycles figuring in the boundaries of $h_i$ and $h_i'$ coincides with the mapping $T$ (in particular, the mapping reduces to $T$ at points of singular elements which are endpoints of elementary arcs, and at endpoints of elliptic arcs); b) if regions $h_{i_0}$ and $h_{i_0}'$ have boundary points in common with regions $h_i$ and $h_i'$ already disposed of, the restriction of the mapping to such points must coincide with the previously defined mapping of $h_i$ onto $h_i'$, $i = 1, 2, ..., i_0 - 1$.

Conditions a) and b) can always be satisfied. Thus, we obtain an identifying mapping of $\bar{G}^*$ onto $\bar{G}'^*$, preserving orientation and direction of $t$, and the theorem is proved.

The fundamental theorem, stated below, now follows immediately from Theorem 75 via elementary arguments, which are omitted.

*Fundamental Theorem 76. The phase portraits of systems $D$ and $D'$ in closed regions $\bar{G}^*$ and $\bar{G}^{*\prime}$ are equivalent if and only if their schemes are equivalent.*

## 5. SCHEME OF A DYNAMIC SYSTEM ON THE SPHERE. SCHEME OF A DYNAMIC SYSTEM DEFINED ON THE PLANE AND MAPPED ONTO THE POINCARÉ SPHERE.

Chapters VIII, X and XI have been devoted to dynamic systems in a bounded plane region. All concepts defined therein carry over with only a few obvious modifications to dynamic systems on the sphere (in the sense of §2).

In particular, the scheme of a dynamic system on the sphere is defined in analogous fashion, in either tabular or diagrammatic form. Clearly, the scheme of a dynamic system on the sphere does not involve a boundary scheme.

When a dynamic system defined on the place is studied on the Poincaré sphere, we can also proceed in the same way, considering the "scheme of a dynamic system on the Poincaré sphere." One necessary modification in this case is the need to define the "scheme of the equator." We shall not dwell upon the fairly obvious details of this procedure, mentioning only the following points: 1) the scheme indicates whether the equator is a limit cycle or not and, if it is, lists all singular paths that tend to it; 2) the scheme lists all equilibrium states on the equator (when the latter is not a limit cycle), and provides the global schemes of these equilibrium states.

In the next and last chapter, we shall consider some concrete examples of dynamic systems, considered on the Poincaré sphere. We shall use the schematic-diagram form of the scheme rather than the tabular form, in view of its greater appeal to intuition.

*Chapter XII*

## QUALITATIVE INVESTIGATION OF CONCRETE DYNAMIC SYSTEMS "IN THE LARGE"

This chapter comprises a single section, in which, utilizing the tools for investigation of dynamic systems developed in Chapters IV and VI, we shall determine the schemes of several concrete dynamic systems. Almost all these examples stem from applications in various fields of physics.

### §30. EXAMPLES

Example 1 /67/.

$$\dot{x} = 2xy \equiv P(x, y), \quad \dot{y} = 1 + y - x^2 + y^2 \equiv Q(x, y).$$

Equilibrium states: $A(1, 0)$ and $B(-1, 0)$ — unstable foci. It is easy to see that the straight line $x = 0$ is an integral curve. The system is symmetric about the $y$-axis. There are no limit cycles, by Dulac's test: setting $F(x, y) = x^{-2}$, we see that the expression $\frac{\partial (PF)}{\partial x} + \frac{\partial (QF)}{\partial y} = \frac{1}{x^2}$ does not change sign.

Consideration of the system on the Poincaré sphere reveals two saddle points on the equator: the ends of the $y$-axis (Figure 300).*

Example 2 /67/.

$$\dot{x} = 2x(1 + x^2 - 2y^2) \equiv P(x, y),$$
$$\dot{y} = -y(1 - 4x^2 + 3y^2) \equiv Q(x, y).$$

Equilibrium states $O(0, 0)$ — saddle point, $A_1(1, 1)$, $A_2(1, -1)$, $A_3(-1, 1)$, $A_4(-1, -1)$ — stable foci. It is readily seen that the straight lines $x = 0$ and $y = 0$ are integral curves, and the vector field of the system is symmetric about the coordinate axes. By Dulac's test, there are no limit cycles: if $F(x, y) = x^{-3/2}y^{-2}$, then the expression

FIGURE 300.

---

* Here and below we consider the projection of the hemisphere onto a disk (see §13).

443

$$\frac{\partial (PF)}{\partial x} + \frac{\partial (QF)}{\partial y} = - \frac{x^2 + y^2}{x^3/2 y^2}$$

does not change sign in any quadrant of the $(x, y)$-plane.

It can be shown that there are two pairs of saddle points on the equator of the Poincaré sphere — the ends of the $x$- and $y$-axes (Figure 301).

Example 3 /68/.

$$\dot{x} = x \left[ (x^2 + y^2 + 1)(x^2 + y^2 - 1) - 4y^2 \right] \equiv P(x, y),$$
$$\dot{y} = y \left[ (x^2 + y^2 + 1)(x^2 + y^2 - 1) + 4x^2 \right] \equiv Q(x, y) .$$

Equilibrium states: $O (0, 0)$ — saddle node, $A (0, 1)$ and $B (0, -1)$ — saddle point, $C (1, 0)$ and $D (-1, 0)$ — unstable nodes. The axes $x = 0$ and $y = 0$ are integral curves. The phase portrait is symmetric about the coordinate axes. There are no limit cycles, since there are integral curves through all equilibrium states.

FIGURE 301.

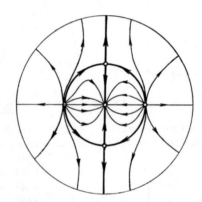

FIGURE 302.

It remains to determine the behavior of the separatrices. Consider the family of circles $x^2 + y^2 = C$. Differentiating along paths of the system (§3.13), we get

$$\frac{d (x^2 + y^2)}{dt} = 2 (x^2 + y^2)(x^2 + y^2 + 1)(x^2 + y^2 - 1).$$

Hence $x^2 + y^2 = 1$ is an integral curve. Outside this circle $\frac{d (x^2 + y^2)}{dt} > 0$, and inside it $\frac{d (x^2 + y^2)}{dt} < 0$, so that all the circles $x^2 + y^2 = C \ (C \neq 1)$ are circles without contact. Consequently, infinity is absolutely stable (Figure 302).

Example 4 /69/.

$$\dot{x} = (x - y)^2 - 1, \qquad \dot{y} = (x + y)^2 - 1. \qquad (1)$$

It is readily seen that system (1) possesses central symmetry. Equilibrium states: $A (0, 1)$ and $B (0, -1)$ — saddle points, $C (1, 0)$ — unstable focus,

$D$ $(-1, 0)$ — stable focus. Rotating the vector field of system (1) through 45°, we get

$$\dot{x} = -2xy, \qquad \dot{y} = x^2 + y^2 - 1. \tag{2}$$

System (2) is integrable. Its general solution is $x \left(\frac{x^2}{3} + y^2 - 1\right) = C$

(Figure 303a). Treating the closed curves of system (2) as a topographic system, we show that system (1) has no limit cycles. In fact, since system (1) is centrally symmetric, it may have only an even number of cycles about the foci $C$ and $D$. But each such cycle would have to cross the closed curves of the topographic system both inward and outward with increasing $t$, which is impossible since they are curves without contact. To compute the slopes of the separatrices at the saddle $A$ $(0, 1)$, we get the quadratic equation $k^2 - 2k - 1 = 0$, whence $k_{1,2} = 1 \pm \sqrt{2}$. By symmetry, the separatrices of the saddle point $B$ $(0, -1)$ have the same slope.

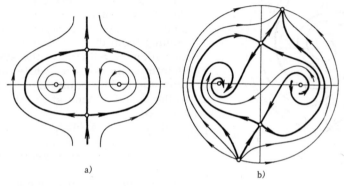

a)                    b)

FIGURE 303.

Note that for $x > 0$ the paths of system (1) cross the curves of the topographic system outward with increasing $t$, while for $x < 0$ they cross the curves inward with increasing $t$. Hence it follows that one $\omega$-separatrix of the saddle point $A$ and one $\omega$-separatrix of the saddle point $B$ tend to the focus $C$ $(1, 0)$ as $t \to -\infty$. Similarly, one $\alpha$-separatrix of $A$ and one $\alpha$-separatrix of $B$ tend to the focus $D(-1, 0)$ as $t \to +\infty$. The remaining four separatrices cannot tend to the foci. We shall show that they tend to infinity.

Consider the behavior of the paths of the system at infinity. Applying the Poincaré transformation $y = \frac{1}{z}$, $x = \frac{\tau}{z}$ and multiplying the resulting equations by $z$, we get

$$\frac{d\tau}{dt} = 1 - 3\tau - \tau^2 - z^2 - \tau^3 + \tau z^2 \equiv P(\tau, z),$$

$$\frac{dz}{dt} = -z - 2\tau z - \tau^2 z + z^3 \equiv Q(\tau, z). \tag{3}$$

445

502

Setting $z = 0$, we obtain an equation for the equilibrium states,
$f(\tau) \equiv \tau^3 + \tau^2 - 3\tau - 1 = 0$. We have $f'(\tau) = 3\tau^2 + 2\tau + 3$. This quadratic
trinomial does not change sign, so that the equation $f(\tau) = 0$ has only one
real root; moreover, the root $\tau_0$ must be positive since $f(0) = -1$,
$f'(\tau) > 0$. Computing the partial derivatives of the right-hand sides of
system (3) at $z = 0$, $\tau = \tau_0$, we get

$$\Delta \equiv (P'_\tau Q'_z - P'_z Q'_\tau)_0 = (1 + 2\tau_0 + \tau_0^2)(3 + 2\tau_0 + 3\tau_0^2) > 0,$$
$$\sigma \equiv (P'_\tau + Q'_z)_0 = -4(\tau_0^2 + \tau_0 + 1) < 0.$$

Thus $z = 0$, $\tau = \tau_0$ is a stable node. It can be shown that the system

obtained by applying the Poincaré transformation $x = \frac{1}{z}$, $y = \frac{\tau}{z}$ has no

equilibrium states at the ends of the $x$-axis. Thus system (1) has two nodes
on the equator of the Poincaré sphere,
one stable and one unstable (Figure 303b).

The behavior of the separatrices of
the saddle points $A$ and $B$ which do not
tend to the foci is now clear. In view of
the sign of the expression for $x$ on the
$y$-axis, we see that the two separatrices
issuing from the saddle points into the
half-plane $x > 0$ tend to the stable node
at infinity as $t \to +\infty$, while those going
from the saddle points into $x < 0$ tend
to the unstable node at infinity as
$t \to -\infty$.

Example 5 /62/.

FIGURE 304.

$$\dot{x} = y, \quad \dot{y} = -x - ay - \mu x^2 - y^2,$$

where $\mu$ and $\alpha$ are positive parameters.
The system has two equilibrium states: $O(0, 0)$ — stable focus or node,

$A(-\frac{1}{\mu}, 0)$ — saddle point. There are no limit cycles or closed contours

consisting of paths (see §12.6, Example 6).

An examination of the behavior of paths at infinity shows that the system
has a pair of nodes on the equator of the Poincaré sphere: the positive
end of the $y$-axis is an unstable node, the negative end of the $y$-axis a stable
node. We can now determine the behavior of the separatrices of the saddle

point $A(-\frac{1}{\mu}, 0)$. Both $\omega$-separatrices come from a node at infinity, since

there are no other $\alpha$-limit sets. One $\alpha$-separatrix goes to the node at
infinity, since it cannot pass through the point $O$ without crossing an
$\omega$-separatrix, and the other $\alpha$-separatrix cannot reach the node at infinity
and tends to the focus or node $O(0, 0)$ (Figure 304).

Example 6 /70/.

$$\dot{x} = x(3 - x - ny) \equiv P(x, y), \quad \dot{y} = y(-1 + x + y) \equiv Q(x, y).$$

We consider the qualitative appearance of this system for $n > 3$.

Equilibrium states: $O(0, 0)$, $A(0, 1)$, $B(3, 0)$ — saddle points; $C\left(\frac{n-3}{n-1}, \frac{2}{n-1}\right)$,

three cases: 1) $n > 5$, stable focus; 2) $\frac{11+\sqrt{128}}{7} < n < 5$, unstable focus;

3) $3 < n < \frac{11+\sqrt{128}}{7}$, unstable node. Since nodes and foci are topologically equivalent, we distinguish only two cases: $n > 5$ and $3 < n < 5$. It is clear that the axes $x = 0$ and $y = 0$ are integral curves. The system has no closed contours consisting of paths (see §12.6, Example 7).

A direct check shows that when $n = 5$ the system has a solution

$$xy^3 \left(\frac{x}{3} + y - 1\right)^2 = C.$$ Investigation of the curves of this family reveals the existence of a continuum of closed paths, so that the equilibrium state $C$ is a center when $n = 5$. In this case the separatrix goes from the saddle point $A$ to the saddle point $B$ (Figure 305a).

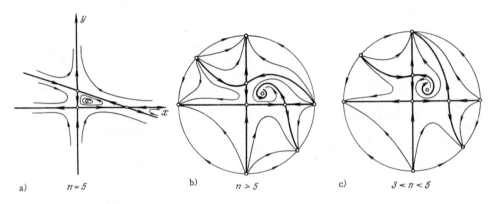

a)   $n = 5$     b)   $n > 5$     c)   $3 < n < 5$

FIGURE 305.

It was shown previously (see §13.3, Example) that the equilibrium states on the equator of the Poincaré sphere are: 1) the ends of the $y$-axis — stable node and unstable node; 2) the ends of the $x$-axis — stable node and unstable node; 3) a pair of points in the second and fourth quadrants — stable node (fourth quadrant) and unstable node (second quadrant).

It is now easy to determine the behavior of the separatrices. 1) $n > 5$. Since the system has no closed contours consisting of whole paths, the $\omega$-separatrix of the saddle point $A$ cannot issue from the saddle point $B$. Thus it issues from the unstable node at the end of the $x$-axis, since the focus $C$ and the end of the $y$-axis are stable; the $\omega$-separatrix of $B$ tends to the focus $C$ as $t \to \infty$, since it can neither go to infinity nor cross a separatrix of $A$ (Figure 305b). 2) $3 < n < 5$. The equilibrium state $C$ is unstable; the $\omega$-separatrix of the saddle point $B$ tends to the end of the $y$-axis, for there are no other possibilities. The separatrix of the saddle point $A$ tends to $C$ as $t \to -\infty$, since it cannot go to infinity without crossing the separatrix of $B$. The behavior of the remaining separatrices is now uniquely determined (Figure 305c).

Example 7 /70/.

$$\dot{x} = x(3-x-y), \quad \dot{y} = y(x-1).$$

The system has three equilibrium states: $O(0, 0)$ and $A(3, 0)$ — saddle points, $C(1, 2)$ — stable focus. The axes $x = 0$ and $y = 0$ are clearly integral curves. The system has no closed contours consisting of paths (see § 12.6, Example 7).

Now consider the behavior of the paths at infinity. Applying the Poincaré transformation $x = \frac{\tau}{z}$, $y = \frac{1}{z}$ and multiplying both sides of the resulting equations by $z$, we get

$$\frac{d\tau}{dt} = -\tau - 2\tau^2 + 4\tau z, \quad \frac{dz}{dt} = z(z-\tau).$$

This system has two equilibrium states for $z = 0$: 1) $\tau_1 = 0$, $z = 0$, 2) $\tau_2 = -\frac{1}{2}$, $z = 0$. The characteristic roots for the first equilibrium state are $\lambda_1 = -1$, $\lambda_2 = 0$. The coordinate axes are clearly integral curves.

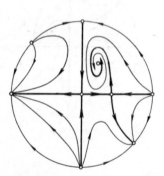

FIGURE 306.

Bearing in mind the direction of motion on the axes and the possible types of equilibrium state thereon (Theorem 65, §21), we conclude that the equilibrium state in question is a saddle node. It has a nodal region in the half-plane $z < 0$. Since there are infinitely many paths tending to this equilibrium state in the direction $\tau = 0$ but only two in the direction $z = 0$, the halves of the axis $z = 0$ are separatrices. Thus there are saddle regions in the half-plane $z > 0$. Two points on the equator of the Poincaré sphere correspond to this equilibrium state — the ends of the $y$-axis. Consequently, the positive end of the $y$-axis is a saddle point, and the negative end an unstable node. It is readily shown that the

equilibrium state $z = 0$, $\tau = -\frac{1}{2}$ is an unstable node, and the corresponding pair of points on the equator of the Poincaré sphere comprises an unstable node in the second quadrant and a stable node in the fourth. Using the other Poincaré transformation $x = \frac{1}{z}$, $y = \frac{\tau}{z}$, one can show that the positive end of the $x$-axis is an unstable node, the negative end a stable node. The behavior of all separatrices is now determined (Figure 306).

Example 8 /71/.

$$\dot{x} = x(y-\beta), \quad \dot{y} = \beta(a-y) - kxy.$$

All the parameters are positive. Applying the transformation $x_1 = \alpha\beta kx$ and then returning to the original notation for the variables, we get

$$\dot{x} = x\,(y-\beta), \qquad \dot{y} = a - \beta y - \frac{1}{a}\,xy,$$

where $a = \alpha\beta$. This system has two equilibrium states; $A\,(0, \frac{a}{\beta})$ and
$B\,[\frac{a}{\beta}\,(a - \beta^2),\ \beta]$. If $a < \beta^2$, $A$ is a stable node, $B$ a saddle point; if $a > \beta^2$,
$A$ is a saddle point and $B$ a stable node.* Obviously, $x = 0$ is an integral
curve. It is easy to see that the straight line $\beta x + a\beta y - a^2 = 0$ through the
points $A$ and $B$ is also an integral curve. The existence of two integral
curves which are straight lines implies that there are no limit cycles.

Let us consider the behavior of the paths at infinity. Applying the
Poincaré transformation $x = \frac{\tau}{z}$, $y = \frac{1}{z}$ and multiplying the resulting
equations by $z$, we get

$$\frac{d\tau}{dt} = \tau + \frac{1}{a}\,\tau^2 - a\tau z^2, \qquad \frac{dz}{dt} = \frac{1}{a}\,\tau z + \beta z^2 - az^3.$$

When $z = 0$ this system has two equilibrium states: 1) $\tau_1 = 0$, $z = 0$,
2) $\tau_2 = -a$, $z = 0$. Arguments similar to those used for Example 7 show that
the point $\tau = z = 0$ is a saddle node, with the nodal region in the half-plane
$z > 0$ and saddle regions in $z < 0$; the halves of the axis $z = 0$ are
separatrices. The equilibrium state $z = 0$, $\tau = -a$ is a stable node. Using
the Poincaré transformation $x = \frac{1}{z}$, $y = \frac{\tau}{z}$ and multiplying by $z$, we get

$$\frac{d\tau}{dt} = -\frac{1}{a}\,\tau - \tau^2 + az^2, \qquad \frac{dz}{dt} = z\,(\beta z - \tau).$$

As shown in Example 1 of §22, the equilibrium state $z = \tau = 0$ of this system
is a saddle node with nodal region in $z < 0$ and saddle regions in $z > 0$.
Equilibrium states on the equator of the Poincaré sphere: 1) the ends of
the $x$-axis — saddle point and unstable node, 2) the ends of the $y$-axis —
unstable node and saddle point, 3) the ends of the straight line $x + ay = 0$ —
stable and unstable nodes (Figure 307a, b).

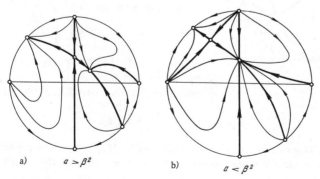

a)     $a > \beta^2$        b)      $a < \beta^2$

FIGURE 307.

* When $a = \beta^2$ the equilibrium states $A$ and $B$ combine to form a multiple equilibrium state.

Example 9 /72/.

$$\dot{x}=x\left(y+\tfrac{3}{2}\right), \quad \dot{y}=x+y-2y^2.$$

Equilibrium states: $O(0, 0)$ — unstable node, $A(0,\tfrac{1}{2})$ and $B(6, -\tfrac{3}{2})$ —

saddle points. There are no limit cycles, since any limit cycles would have to encircle the node $O$, which is impossible because $x = 0$ is an integral curve.

Now for the behavior of the paths at infinity. The Poincaré transformation $x = \tfrac{1}{z}$, $y = \tfrac{\tau}{z}$ reduces the system to the form

$$\frac{d\tau}{dt} = z - 3\tau^2 - \tfrac{1}{2}\tau z, \quad \frac{dz}{dt} = -z\left(\tau + \tfrac{3}{2}z\right).$$

When $z = 0$ this system has one equilibrium state, $z = \tau = 0$, which, as shown in Example 3 of §22, is a multiple equilibrium state with a closed nodal region. Its separatrices are the halves of the axis $z = 0$; in the half-plane $z < 0$ no path tends to the equilibrium state as $t \to \infty$ or as

$t \to -\infty$. Using the other Poincaré transformation $x = \tfrac{\tau}{z}$, $y = \tfrac{1}{z}$, one can

show that the positive end of the $y$-axis is an unstable node and the negative end a stable node.

The behavior of the separatrices can now be determined unambiguously. It is easy to see that the $\omega$-separatrix of the saddle point $A$ is the $y$-axis.

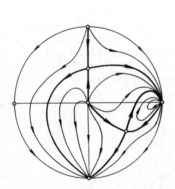

As for its $\alpha$-separatrices, the separatrix in the half-plane $x < 0$ tends to the negative end of the $y$-axis, since all other equilibrium states with $x \leqslant 0$ are unstable. Since $y > 0$ on the positive $x$-axis, the separatrix in the half-plane $x > 0$ tends to the positive end of the $x$-axis. The behavior of the four separatrices of the saddle point $B$ is also uniquely determined (Figure 308).

Example 10 /73/.

$$\dot{x} = -x(2+y), \quad \dot{y} = \alpha x + \beta y,$$

where $\alpha \neq 0$, $\beta \neq 0$. Transforming variables by $\bar{x} = \alpha x$ and then returning to the previous notation for the variables, we get

FIGURE 308.

$$\dot{x} = -x(y+2), \dot{y} = x+\beta y.$$

Equilibrium states: $O(0, 0)$ and $A(2\beta, -2)$. If $\beta > 0$, $O$ is a saddle point, if $\beta < 0$ it is a stable node. The slopes of the paths at the point $O$ are

$k_1 = -\dfrac{1}{2+\beta}$ and $k_2 = \infty$ (see §9.7, Example 3). The equilibrium state $A$ is

1) a saddle point if $\beta < 0$, 2) an unstable focus if $0 < \beta < 8$, 3) an unstable node if $\beta > 8$.

The axis $x = 0$ is an integral curve. The system has no limit cycles (see §12.6, Example 8).

To determine the behavior of the paths at infinity, we apply the Poincaré transformation $x = \frac{1}{z}$, $y = \frac{\tau}{z}$, to get

$$\frac{d\tau}{dt} = z + (2+\beta)\,\tau z + \tau^2, \qquad \frac{dz}{dt} = z\,(2\tau + z).$$

When $z = 0$ this system has one equilibrium state, $z = \tau = 0$, which, as shown previously (Example 4 of §22), is a multiple equilibrium state with a closed nodal region. It can be shown that the half-plane $z > 0$ contains no paths tending to this point. Using the other Poincaré transformation $x = \frac{\tau}{z}$, $y = \frac{1}{z}$, we get

$$\frac{dz}{dt} = -z^2\,(\beta + \tau),$$

$$\frac{d\tau}{dt} = -\tau\,[1 + (2+\beta)\,z + \tau z].$$

Again, there is one equilibrium state on $z = 0$: $z = \tau = 0$. The characteristic roots are $\lambda_1 = -1$, $\lambda_2 = 0$. Recalling the possible types of such equilibrium states (Theorem 65, §21) and the direction of motion along the $z$- and $\tau$-axes, which are integral curves, we conclude that this equilibrium state is a saddle node, with the halves of the axis $z = 0$ as its separatrices. It is readily seen that when $\beta > 0$ the nodal region is in $z > 0$, the saddle regions in $z < 0$; when $\beta < 0$ the situation is reversed.

The behavior of the separatrices is now determined (Figure 309a, b).

Example 11 /72/.

$$\dot{x} = x\left(y + \frac{\lambda}{n-1}\right), \qquad \dot{y} = x + \frac{1}{1-n}\,y^2.$$

Let $n > 1$, $\lambda > 0$. The system has two equilibrium states: $O\,(0,\,0)$ and $A\left(\frac{\lambda^2}{(n-1)^3}, \frac{\lambda}{1-n}\right)$. It can be shown that $O$ is a saddle node (Example 6 in §22). The slopes of the separatrices at $O$ are $k_1 = \frac{n-1}{\lambda}$, $k_2 = \infty$. It is easy to see that $A$ is a saddle point. There are no limit cycles, as evidenced by the nature of the equilibrium states.

To determine the behavior at infinity, we apply the Poincaré transformation $x = \frac{1}{z}$, $y = \frac{\tau}{z}$ and multiply the resulting equations by $z$:

$$\frac{d\tau}{dt} = z + \frac{n}{1-n}\,\tau^2 + \frac{\lambda}{1-n}\,\tau z, \qquad \frac{dz}{dt} = z\left(\frac{\lambda}{n-1}\,z - \tau\right).$$

When $z = 0$ this system has one equilibrium state, $\tau = z = 0$, which can be shown to be a multiple equilibrium state with closed nodal region (Example 7 in §22). As in Example 9, one can show that in the half-plane $z < 0$ all the paths remain at a finite distance from the equilibrium state.

Using the other Poincaré transformation $y = \frac{1}{z}$, $x = \frac{\tau}{z}$, one can show that the ends of the $y$-axis are nodes — one stable and one unstable. In view of the direction of motion on the paths on the isoclines of vertical and horizontal directions and on the axis $y = 0$, the behavior of the separatrices of the saddle node $O$ and saddle point $A$ is now uniquely determined (Figure 310).

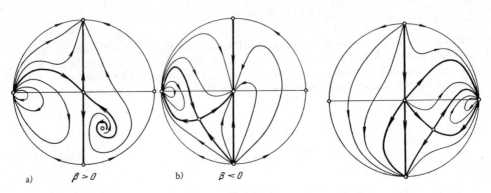

a)      $\beta > 0$      b)      $\beta < 0$

FIGURE 309.                     FIGURE 310.

Example 12 /74/.

$$\dot{x} = y, \quad \dot{y} = 7y + x\,(x^{1/2} - 12),$$

where $x \geqslant 0$. In §9.8 we extended the definition of this system to the half-plane $x < 0$ and found its equilibrium states in the closed half-plane $x \geqslant 0$: $O\,(0, 0)$ — unstable node, $A\,(144, 0)$ — saddle point. The system has no closed contours consisting of paths (Bendixson's test).

The slopes of the separatrices at the saddle point are determined from the equation $k^2 - 7k - 6 = 0$, whence $k_{1,2} = \frac{7}{2} \pm \sqrt{\frac{49}{4} + 6}$. The coordinate axes are isoclines — the $x$-axis an isocline of vertical directions, the $y$-axis an isocline of directions with slope 7.

Let us consider the logical possibilities for the separatrices $L_1$, $L_2$, $L_3$, $L_4$ (Figure 311a). It is clear that $L_1$ goes to infinity in the half-plane $y > 0$, since were $L_1$ to cross the $x$-axis for $x < 144$ the separatrix $L_4$ would have no $\alpha$-limit points. $L_2$ comes from infinity to the half-plane $x < 0$, since were $L_2$ to cross the $x$-axis for $x < 144$ the separatrix $L_3$ would have no $\omega$-limit points.

The separatrix $L_3$ either goes to infinity or leaves the region through the $y$-axis. We claim that the second alternative holds: $\dot{x} < 0$ for $y < 0$, and so $x$ decreases along $L_3$. Suppose that $y \to -\infty$, $x \to x_0$ along $L_3$, i.e., that $L_3$ is asymptotic to the straight line $x = x_0$. Then $\sup \left| \frac{dy}{dx} \right| = \infty$. But

$$\frac{dy}{dx} = 7 + \frac{x}{y}\,(x^{1/2} - 12) \to 7 \text{ as } x \to x_0,\ y \to -\infty, \text{ and so } L_3 \text{ crosses the } y\text{-axis.}$$

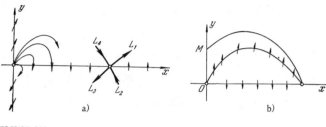

FIGURE 311.

For the separatrix $L_4$ we have a priori three alternatives: 1) $L_4$ issues from the node $0$, 2) $L_4$ crosses the $y$-axis with decreasing $t$, 3) $L_4$ goes to infinity. The third alternative is impossible. Indeed, $\dot{x} > 0$ for $y > 0$, so that $x$ decreases with decreasing $t$ on $L_4$ and $x \to x_0 < 144$. But if $x$ is bounded, then $\dot{y} > 0$ for sufficiently large $y$, so that $y$ decreases with decreasing $t$. Let us find the critical directions at the node $0$. They are the roots of the equation $k^2 - 7k + 12 = 0$: $k_1 = 3$, $k_2 = 4$. Consider the isocline corresponding to slope 4, whose equation is $3y = x(12 - \sqrt{x})$. On this isocline we have $y' = 4 - \frac{1}{2}\sqrt{x}$. Thus the slope of the paths is greater than that of the isocline at all the latter's points, except $x = 0$. Thus no path can come into the region from without (Figure 311b). Suppose that $L_4$ issues not from the node $0$ but from some other point $M$ on the $y$-axis. Then the path issuing from some point of the segment $OM$ would enter the closed region bounded by the isocline, the separatrix and the segment $OM$ with increasing $t$, never leaving this region again. This contradiction shows that the second alternative is also impossible. We have thus determined the behavior of all the separatrices (Figure 312).

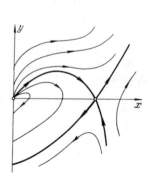

FIGURE 312.

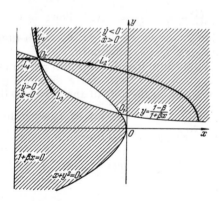

FIGURE 313.

Example 13 /75/.

$$\dot{x} = y - \frac{1-\beta}{1+\beta x} \equiv P(x, y), \quad \dot{y} = -\varepsilon(x+y^2) \equiv Q(x, y).$$

453

This system is physically meaningful in the plane region defined by the inequality $1 + \beta x > 0$, with $\varepsilon > 0$, $0 < \beta < \frac{1}{3}$.

The system has two equilibrium states in the above region: $O_1(-1, 1)$ and $O_2(x_2, y_2)$, where $x_2 = \frac{1}{2} - \frac{1}{\beta} + \sqrt{\frac{1}{\beta} - \frac{3}{4}}$, $y_2 = -\frac{1}{2} + \sqrt{\frac{1}{\beta} - \frac{3}{4}}$. The equilibrium state $O_1(-1, 1)$ is a node or focus, stable if $\varepsilon > \frac{\beta}{2(1-\beta)}$ and unstable if $\varepsilon < \frac{\beta}{2(1-\beta)}$. The equilibrium state $O_2(x_2, y_2)$ is a saddle point. What can we say about its separatrices (Figure 313)? The isoclines of vertical and horizontal directions $y = \frac{1-\beta}{1+\beta x}$ (A) and $x + y^2 = 0$ (B) divide the region of interest into subregions in which $\dot{x}$ and $\dot{y}$ do not change sign. The slopes of the separatrices at the saddle point are determined from the equation

$$k^2 + \frac{\beta y_2 + 2\varepsilon (1-\beta)}{1 + \beta x_2} k + \varepsilon = 0.$$

Hence $k_1$ and $k_2$ are both negative, since $\varepsilon > 0$, $1 + \beta x > 0$, $y_2 > 0$, $0 < \beta < \frac{1}{3}$. The separatrix $L_1$ goes to infinity as $t \to -\infty$, while $L_4$ crosses the straight line $x = -\frac{1}{\beta}$ with increasing $t$. $L_2$ issues from the saddle point into the region $\dot{x} > 0$, $\dot{y} < 0$. There are five alternatives: 1) $L_2$ goes to infinity, tending asymptotically to the $x$-axis; 2) $L_2$ leaves the region across the straight line $1 + \beta x = 0$ or at infinity (in the half-plane $y < 0$); 3) $L_2 \to O_1$; 4) $L_2$ tends to a limit cycle about $O_1$; 5) $L_2 \to O_2$.

The first alternative is impossible, since if $\dot{x} > 0$, $\dot{y} < 0$, then for sufficiently large $x$ and $0 < y < y_2$ there exists $N > 0$ such that $\frac{dy}{dx} = \frac{-\varepsilon (x + y^2)(1 + \beta x)}{-(1-\beta) + y(1 + \beta x)} < -N$. Thus $L_2$ crosses the isocline (B) and reaches a region in which $\dot{x} < 0$, $\dot{y} < 0$. With further increase in $t$, one of alternatives 2) — 5) may hold.

We now show that for certain parameter values the system has no limit cycles. Any limit cycle must lie either to the right of the straight line $x = x_2$, or to the left of the straight line $x = x^0$, where $x^0$ is the abscissa of the point of intersection of $L_2$ with the isocline $y = \frac{1-\beta}{1+\beta x}$. Using Dulac's test with $F(x, y) = e^{2\varepsilon x}$, we get

$$\frac{\partial (PF)}{\partial x} + \frac{\partial (QF)}{\partial y} = \frac{1-\beta}{1+\beta x} e^{2\varepsilon x} \left( \frac{\beta}{1+\beta x} - 2\varepsilon \right).$$

This expression vanishes on the straight line $x = \frac{\beta - 2\varepsilon}{2\varepsilon\beta}$, which is on the left of the straight line $x = x_2$ if $\frac{1}{2\varepsilon} - \frac{1}{\beta} < \frac{1}{2} - \frac{1}{\beta} + \sqrt{\frac{1}{\beta} - \frac{3}{4}}$, whence $4\varepsilon (1-\beta) > -\beta + \sqrt{\beta (4 - 3\beta)}$. Thus there can be no limit cycles.

Now consider the equation $\frac{dy}{dx} = -\varepsilon\frac{x+y^2}{y}$, obtained from the original

equation by setting $\beta = 1$. Its general integral is $e^{2\varepsilon x}\left(\frac{1}{2\varepsilon} - x - y^2\right) = h$. Let
$\lambda$ be an arc of the path of this equation passing through the saddle point $O_2$
and extending up to the intersection with the isocline (B). In the region
$x > 0$, $\dot{y} < 0$, this is an arc without contact for system (1). Indeed,

$$\left(\frac{dy}{dx}\right)_{\beta < \frac{1}{3}} - \left(\frac{dy}{dx}\right)_{\beta = 1} = \frac{\varepsilon(1-\beta)(x+y^2)}{y(1+\beta x)[(1-\beta)/(1+\beta x)-y]} < 0.$$

The slope of $\lambda$ at $O_2$ is zero, while that of $L_2$ is negative. Thus $L_2$ crosses
the isocline (B) to the left of $\lambda$. Let $x^*$ be the abscissa of the point at which
$\lambda$ intersects the isocline (A):

$$e^{2\varepsilon x^*}\left(\frac{1}{2\varepsilon} - x^* - \frac{(1-\beta)^2}{(1+\beta x^*)^2}\right) = \frac{1}{2\varepsilon}e^{2\varepsilon x_2}.$$

Hence

$$1 - 2\varepsilon\left(x^* - \frac{(1-\beta)^2}{(1+\beta x^*)^2}\right) = \exp\left\{-2\varepsilon\left(x^* - \frac{1}{2} + \frac{1}{\beta} - \sqrt{\frac{1}{\beta} - \frac{3}{4}}\right)\right\}. \quad (C)$$

Thus the system has no limit cycles for parameter values $0 < \beta < \frac{1}{3}$,
$2\varepsilon(1 + \beta x^*) \leqslant \beta$, where $x^*$ is the positive root of equation (C). It follows
that for parameter $\varepsilon$ and $\beta$ such that the system has no limit cycles the
behavior of the separatrix $L_2$ depends on the stability properties of the
equilibrium state $O_1(-1, 1)$.

The behavior of the separatrix $L_3$ depends on that of $L_2$ (Figure 314a,b).

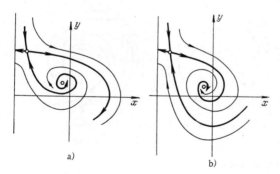

a)  b)

FIGURE 314.

Example 14 /76/.

$$\dot{x} = y(x+2) + x^2 + y^2 - 1, \quad \dot{y} = -x(x+2).$$

The system has two equilibrium states:   $A\ (0, -1+\sqrt{2})$, $B\ (0, -1-\sqrt{2})$.
$A$ is an unstable focus, $B$ a saddle point. It is readily shown that the circle
$x^2 + y^2 = 1$ is a path. The point $A$ is clearly inside this circle, $B$ outside it.
The circle $x^2 + y^2 = 1$ is a limit cycle, since the system is analytic.
    We claim that this is the only limit cycle. Introduce a new parameter:

$$\dot{x} = y\,(ax+2)+x^2+y^2-1, \quad \dot{y} = -x\,(ax+2).$$

When $a = 1$ this gives the original system. When $a = 0$ the system is
integrable. Its general solution is $(x^2 + y^2 - 1)\,e^y = C$ (Figure 315a). Let us
treat the closed paths of this system as a topographic system for the
original one. The contact curves for our system and the topographic
system are

$$x^2\,(x^2 + y^2 - 1) = 0,$$

$x = 0$ is a "false" contact, $x = 0$ being an isocline of horizontal directions
for both systems. The left-hand side of this expression changes sign on
the circle $x^2 + y^2 = 1$. A limit cycle of the original system other than
$x^2 + y^2 = 1$ would have to enclose one equilibrium state $A$. But then it would
cross the closed curves of the topographic system, both inward and outward.
Now all these closed curves, except $x^2 + y^2 = 1$, are closed single-crossing
curves for the paths of our system, and so this is impossible. Consequently,
the limit cycle is unique. Since the focus is unstable, the paths of our
system wind onto the limit cycle from both sides. Consequently,
$x^2 + y^2 = 1$ is a stable limit cycle. It can be shown that there are two nodes
on the equator of the Poincaré sphere, one stable and one unstable.

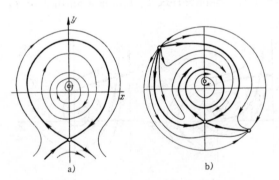

a)                              b)

FIGURE 315.

    We can now determine the behavior of the separatrices of the saddle
point. Its ω-separatrices issue from the unstable node at infinity, since
there are no other α-limit sets. It is readily seen that the limit cycle
lies between the ω-separatrices. But then one of the α-separatrices of
the saddle point must tend to the stable limit cycle and the other to the
stable node at infinity (Figure 315b).

Example 15.

$$\dot{x} = y - \mu \left[ y^2 - x^2 \left( a^2 - \frac{1}{2} x^2 \right) - a \right] x (a^2 - x^2),$$

$$\dot{y} = x (a^2 - x^2) + \mu y \left[ y^2 - x^2 \left( a^2 - \frac{1}{2} x^2 \right) - a \right],$$

(A)

where $\mu$ and $a$ are sufficiently small, $\mu < 0$. System (A) is obtained by rotating the vector field of the system

$$\dot{x} = y, \quad \dot{y} = x (a^2 - x^2)$$

(B)

through an angle $\varphi (x, y) = \operatorname{arctg} \mu \left[ y^2 - x^2 \left( a^2 - \frac{1}{2} x^2 \right) - a \right]$. System (B) is

integrable: its general solution is $y^2 - x^2 \left( a^2 - \frac{x^2}{2} \right) = C$ (Figure 316a). It

is not hard to see that system (B) is a topographic system for system (A), with contact on the curve

$$y^2 - x^2 \left( a^2 - \frac{1}{2} x^2 \right) = a,$$

(C)

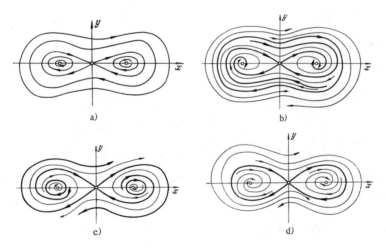

a)                    b)

c)                    d)

FIGURE 316.

which is a solution of both systems (A) and (B). All other closed curves of system (B) are curves without contact for system (A). We consider the following cases: 1) $a > 0$. Equation (C) defines one closed curve. Inside this curve the angle between the vector fields of systems (A) and (B) is negative, while outside it is positive. Thus the curve (C) is an unstable limit cycle of system (A), clearly the only limit cycle. It encloses three equilibrium states — two foci and a saddle point (Figure 316b). 2) $a < 0$. Equation (C) defines two closed curves, lying within the closed contour formed by the separatrices of system (B) and symmetric with respect to the origin. These two curves are unstable limit cycles for system (A), and moreover the only limit cycles of (A) (Figure 316c). 3) $\alpha = 0$. Equation (C) defines the closed contour formed by the separatrices of system (B). This closed curve is a closed limit contour for the paths of system (A), which tend to it as $t \to -\infty$. There are no other limit cycles (Figure 316d).

Example 16 /77/.

$$\dot{x} = xy - \mu \left[ -\frac{1}{3} (x-1)(x+2) + \frac{1}{2} y^2 + \frac{1}{3} xy + \frac{1}{3} y \right],$$
$$\dot{y} = -\frac{1}{3} (x-1)(x+2) + \frac{1}{2} y^2 + \frac{1}{3} xy + \frac{1}{3} y + \mu xy,$$

$$\text{(A)}$$

where $\mu < 0$ is sufficiently small [in absolute value]. This system is obtained by rotating the vector field of the system

$$\dot{x} = xy, \quad \dot{y} = -\frac{1}{3}(x-1)(x+2) + \frac{1}{2} y^2 + \frac{1}{3} xy + \frac{1}{3} y \qquad \text{(B)}$$

through the angle arc tg $\mu$. The equilibrium states of systems (A) and (B) coincide in position: $A (1, 0)$ and $B (-2, 0)$. It is readily shown that $A (1,0)$ is an unstable focus and $B (-2, 0)$ a stable focus. Note that on the axis $x = 0$ system (A) satisfies the inequality

$$\dot{x} = -(3y^2 + 2y + 4)\frac{\mu}{6} > 0.$$

Consider the behavior of system (A) at infinity. Applying the Poincaré transformation $x = \frac{\tau}{z}$, $y = \frac{1}{z}$ and multiplying the result by $z$, we get

$$\frac{d\tau}{dt} = -\frac{\mu}{2} + \left( \frac{1}{2} - \frac{\mu}{3} \right) \tau + \frac{\mu}{3} z - \left( \frac{1}{3} + \frac{2}{3} \mu \right) \tau^2 - \frac{2}{3} \mu z^2 -$$
$$- \left( \frac{1}{3} - \frac{\mu}{3} \right) \tau z + \frac{1}{3} \tau^3 + \frac{1}{3} \tau^2 z - \frac{2}{3} \tau z^2 \equiv P(\tau, z),$$
$$\frac{dz}{dt} = \frac{1}{2} z - \left( \mu + \frac{1}{3} \right) \tau z - \frac{1}{3} z^2 + \frac{1}{3} \tau^2 z + \frac{1}{3} \tau z^2 - \frac{2}{3} \tau^3 \equiv Q(\tau, z).$$

Clearly, $z = 0$ is an integral curve. When $z = 0$ the $\tau$-coordinate of the equilibrium state is a solution of the cubic equation

$$f(\tau) = -\frac{\mu}{2} + \left( \frac{1}{2} - \frac{\mu}{3} \right) \tau - \left( \frac{1}{3} + \frac{2}{3} \mu \right) \tau^2 + \frac{1}{3} \tau^3 = 0.$$

We have $f'(\tau) \equiv \tau^2 - \frac{2}{3} (1 + 2\mu) \tau + \frac{1}{2} - \frac{\mu}{3}$. For sufficiently small $\mu$ this quadratic trinomial does not change sign. Thus the equation $f(\tau) = 0$ has one real root $\tau = \tau_0$, whose sign coincides with that of $\mu$ since $f(\tau)$ is monotone increasing and $f(0) = -\frac{\mu}{2}$. We have $\Delta \equiv P'_\tau Q'_z - P'_z Q'_\tau < 0$ for $\tau = \tau_0$, $z = 0$. Thus the point $\tau = \tau_0$, $z = 0$ is a saddle point for sufficiently small $\mu$. The direction of motion on the paths on the equator of the Poincaré sphere is easily established, e. g., by setting $\tau = z = 0$ in the equation $\dot{\tau} = P(\tau, z)$. The result is $\dot{\tau} = -\frac{\mu}{2} > 0$.

It can be shown that the system obtained by applying the other Poincaré transformation $x = \frac{1}{z}$, $y = \frac{\tau}{z}$ has no other equilibrium states.

We have already mentioned that $\dot{x} > 0$ on the $y$-axis. The focus $A(1, 0)$ is unstable. Consequently, there must be an $\omega$-limit set in the half-plane

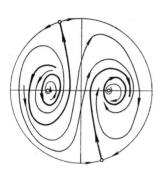

$x > 0$. Since this half-plane contains one equilibrium state, a focus, it follows that the only possible $\omega$-limit set is a stable limit cycle. Thus there is at least one limit cycle about the focus $A$ (in general, the number of limit cycles must be odd). A similar argument establishes the existence of an unstable limit cycle (in general, an odd number thereof) in the half-plane $x < 0$.

The separatrices of the saddle points on the equator of the Poincaré sphere cannot go from saddle point to saddle point, since this would contradict the direction of motion on the paths crossing the $y$-axis. The $\omega$-separatrix of the saddle point at infinity must tend to an unstable limit cycle as $t \to -\infty$, and its $\alpha$-separatrix tends to a stable limit cycle as $t \to +\infty$ (Figure 317).

FIGURE 317.

Example 17 /78/.

$$\dot{x} = x\left(ke^{-\frac{\mu}{y}} - e^{-\frac{1}{y}}\right) \equiv P(x, y), \qquad \dot{y} = xe^{-\frac{1}{y}} - \beta(y - y_0) \equiv Q(x, y), \qquad (A)$$

where $\beta$, $k$, $\mu$, $y_0$ are positive parameters, $\mu < 1$. The system is physically meaningful only in the region $G$ of the phase plane in which $x \geqslant 0$, $y > \varepsilon$, where $\varepsilon > 0$ is sufficiently small.

There are two equilibrium states:

$$A(0, y_0) \text{ and } B\left[\frac{\mu - 1 - y_0 \ln k}{\ln k} \beta k^{\frac{1}{\mu - 1}}, \frac{\mu - 1}{\ln k}\right].$$

It is easy to see that when $k > 1$ or $k < e^{\frac{\mu - 1}{y_0}}$ the region $G$ contains the one equilibrium state $A(0, y_0)$. If $e^{\frac{\mu - 1}{y_0}} < k < 1$, both equilibrium states $A$ and $B$ lie in $G$. These equilibrium states may be of various types, depending on the values of the parameters (see below, a)–g)).

Let us study the behavior of the paths at infinity. Applying the Poincaré transformation $x = \frac{\tau}{z}$, $y = \frac{1}{z}$, we get

$$\frac{d\tau}{dt} = \beta\tau + k\tau e^{-\mu z} - \tau e^{-z} - \tau^2 e^{-z} - \beta y_0 \tau z,$$

$$\frac{dz}{dt} = \beta z - \tau z e^{-z} - \beta y_0 z^2. \qquad (B)$$

When $z = 0$ system (B) has two equilibrium states in the region of interest: 1) $\tau = z = 0$, 2) $z = 0$, $\tau = k + \beta - 1$ if $k + \beta > 1$, but only one equilibrium state,

$\tau = z = 0$, if $k + \beta < 1$. Investigation of the equilibrium states in accordance with the values of the parameters yields to the following cases:

a) $k > 1$. One equilibrium state in the finite part of the plane: $A(0, y_0)$ – saddle point. Two nodes at infinity: stable and unstable.

b) $k < e^{\frac{\mu-1}{y_0}}$, $k + \beta < 1$. One equilibrium state in the finite part of the plane: $A(0, y_0)$ – stable node. One saddle point at infinity.

c) $k < e^{\frac{\mu-1}{y_0}}$, $k + \beta > 1$. One equilibrium state in the finite part of the plane: $A(0, y_0)$ – stable node. At infinity: unstable node and saddle point.

d) $e^{\frac{\mu-1}{y_0}} < k < 1$, $\sigma < 0$, $k + \beta > 1$, where $\sigma \equiv P'_x + Q'_y = -\beta \left[ y_0 \left( \frac{\ln k}{\mu-1} \right)^2 - \frac{\ln k}{\mu-1} + 1 \right]$.

$A$ is a saddle point, $B$ a stable node or focus. At infinity: unstable node and saddle point.

e) $e^{\frac{\mu-1}{y_0}} < k < 1$, $\sigma > 0$, $k + \beta > 1$. $A$ is a saddle point, $B$ an unstable node or focus. At infinity: unstable node and saddle point.

f) $e^{\frac{\mu-1}{y_0}} < k < 1$, $\sigma < 0$, $k + \beta < 0$. $A$ is a saddle point, $B$ a stable node or focus. One saddle point at infinity.

g) $e^{\frac{\mu-1}{y_0}} < k < 1$, $\sigma > 0$, $k + \beta < 0$. $A$ is a saddle point, $B$ an unstable node or focus. One saddle point at infinity. It is not difficult to see that the straight line $y = \varepsilon$ is a line without contact for the system; with increasing $t$ the paths crossing this straight line enter the region of interest.

To summarize: when there is only one equilibrium state in the finite part of the region $G$, the qualitative appearance of the phase portrait is uniquely determined by the values of the parameters. But if the finite part of $G$ contains two equilibrium states, one degree of freedom remains: the (even) number of limit cycles. Taking note of the nature of the equilibrium state(s) at infinity and the direction of the vector field on the straight line $y = \varepsilon$, we see that when $B$ is a stable equilibrium state the system has either no limit cycles or an even number thereof. If $B$ is unstable, there exists at least one stable limit cycle (and in general there must be an odd number) (Figure 318). There is one saddle point at infinity.

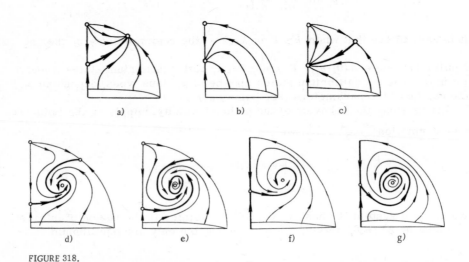

a)     b)     c)

d)     e)     f)     g)

FIGURE 318.

460

Example 18 /79/.

$$\dot{x}=ax+by-x\,(x^2+y^2), \quad \dot{y}=cx+dy-y\,(x^2+y^2).$$

It is readily seen that infinity is absolutely unstable. In fact, differentiating the expression $x^2+y^2=R^2$ along the paths of the system, we get

$$\frac{d\,(x^2+y^2)}{dt}=ax^2+bxy-x^2\,(x^2+y^2)+cxy+dy^2-y^2\,(x^2+y^2)<0$$

for sufficiently large $R$. We can therefore confine the investigation to the interior of some cycle without contact.

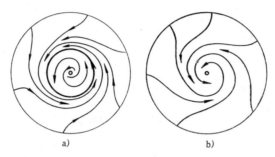

a)                              b)

FIGURE 319.

The system is symmetric about the origin.

Let us assume that $\Delta\equiv(a-d)^2+4bc<0$. In this case the system has a single equilibrium state $O\,(0,\ 0)$. If $a+d>0$ this is an unstable focus. Since infinity is absolutely unstable, there must be at least one stable limit cycle. Using Dulac's test, we show that the limit cycle is unique.

Let $F\,(x,\ y)=\frac{1}{by^2-cx^2+(a-d)\,xy}$. Since $\Delta\equiv(a-d)^2+4bc<0$, it follows that $bc<0$ and $4bc>(a-d)^2$, so that the denominator of $F\,(x,\ y)$ does not change sign in any annular region about the origin. $F\,(x,\ y)$ is single-valued, continuous and differentiable in this region. We have

$$\frac{d\,(PF)}{dx}+\frac{d\,(QF)}{dy}=\frac{-2\,(x^2+y^2)}{by^2-cx^2+(a-d)\,xy}\ .$$

It is easy to see that this expression does not change sign in any annular region about the origin, and so the limit cycle is unique (Figure 319a). It also follows that if $a+d<0$, when $O$ is a stable focus, there are no limit cycles, since then the number of limit cycles must be even (Figure 319b).

## APPENDIX*

### §1. ELEMENTARY THEORY OF SETS IN EUCLIDEAN SPACE

**1. NOTATION.** Sets of points in $n$-dimensional euclidean space $E_n$ (in particular, on the plane $E_2$) will be denoted by capital letters $M, N, \ldots$, points by lower case letters $a, b, \ldots$.** The standard set-theoretic symbols will be used:

$a \in M$ : the point $a$ is an element of the set $M$.

$M \subset N$ : all points of $M$ are also points of $N$, i.e., $M$ is a subset of $N$.

$M \cap N$ : the i n t e r s e c t i o n of sets $M$ and $N$, i.e., the set of all points common to $M$ and $N$.

$M \cup N$ : the u n i o n of the sets, i.e., the set whose elements are all points of $M$ and all points of $N$.

$N \setminus M$ : the set of all points of $N$ not in $M$, called the d i f f e r e n c e of $N$ and $M$. $N \setminus M$ is also known as the c o m p l e m e n t of $M$ in $N$.

**2. CLOSED AND OPEN INTERVALS.** Let $x_1, x_2, \ldots, x_n$ be the coordinates of a point in a euclidean space $E_n$. A set of points in $E_1$ (i.e., on the straight line) with coordinates $a \leqslant x \leqslant b$ is called a c l o s e d  i n t e r v a l (sometimes: a s e g m e n t), denoted by $[a, b]$.

The set of all points with coordinates $a < x < b$ is called an o p e n i n t e r v a l, denoted by $(a, b)$.

A set of points $a < x \leqslant b$ or $a \leqslant x < b$ is called a half-open interval and denoted by $(a, b]$ or $[a, b)$, respectively.

A c l o s e d  p a r a l l e l e p i p e d in $E_n$ is a set of points $M$ $(x_1, x_2, \ldots, x_n)$ with coordinates $a_i \leqslant x_i \leqslant b_i$ (where $a_i, b_i, i = 1, 2, \ldots, n$ are fixed numbers, $a_i < b_i$ for all $i$).

**3. POINT OF ACCUMULATION, BOUNDARY AND INTERIOR POINTS.** A set $M$ in euclidean space $E_n$ is said to be b o u n d e d if the set of distances $\varrho(x, y)$ between any two of its points $x$ and $y$ is bounded above. The supremum of the numbers $\varrho(x, y)$ is called the d i a m e t e r of the set. The $\varepsilon$-neighborhood of a point $a$ in $E_n$ is the set of all points distant less than $\varepsilon$ from $a$. The $\varepsilon$-neighborhood of $a$ is denoted by $U_\varepsilon(a)$.

A point $a$ in $E_n$ is:

a p o i n t  o f  a c c u m u l a t i o n † of a set $K$ if every neighborhood of $a$ contains infinitely many points of $K$,

---

\* [All cross-references by section number are to sections of the Appendix, except in §§9, 10.]

\*\* [This convention is not followed consistently.]

† The term l i m i t  p o i n t is also widely used. In this book, however, we need the term l i m i t  p o i n t of a path in a connotation quite different from the set-theoretic one. We shall therefore use the term "point of accumulation" invariably.

a b o u n d a r y  p o i n t of $K$ if every neighborhood of $a$ contains both points of $K$ and points not of $K$,

an i n t e r i o r  p o i n t of $K$ if it has a neighborhood containing only points of $K$.

An interior point of $K$ necessarily belongs to $K$, but a point of accumulation or boundary point may or may not belong to $K$. A boundary point of $K$ which is not in $K$ must be a point of accumulation of $K$.

Every bounded set $K$ in euclidean space has at least one point of accumulation, which need not belong to $K$ (Bolzano-Weierstrass theorem).

**4. OPEN AND CLOSED SETS.** A set $K$ is said to be o p e n if it does not contain any of its boundary points (so that all its points are interior points), and c l o s e d if it contains all its boundary points (of course, there are sets which are neither open nor closed). The entire space $E_n$ is both open and closed.

The c l o s u r e of a set $K$ is the set $K$ consisting of all points of $K$ and all boundary points of $K$ (including those not in $K$). Since all boundary points of $K$ are points of accumulation of $K$, the closure of $K$ is the set of all points of $K$ and all its points of accumulation.

The set of all boundary points of $K$ is called its b o u n d a r y. The boundary of a set is always a closed set.

If $K_1$ is an open set (in particular, the space $E_n$) and $K_2$ a closed subset of $K_1$, then the complement of $K_2$ in $K_1$ is an open set. If $K_1$ is a closed set and $K_2$ an open subset of $K_1$, then the complement of $K_2$ in $K_1$ is a closed set.

The union of any number and intersection of a finite number of open sets is an open set. The intersection of any number and union of a finite number of closed sets is a closed set.

**5. DISTANCE BETWEEN SETS.** If $K_1$ and $K_2$ are two disjoint (i. e., non-intersecting) closed sets, at least one of which is bounded, then the infimum of the distances between points $a_1$ of $K_1$ and $a_2$ of $K_2$ is positive. This infimum, denoted by $d(K_1, K_2)$, is known as the d i s t a n c e  b e t w e e n  t h e  s e t s $K_1$ and $K_2$.

A closed set $F \subseteq E_n$ is said to be c o m p a c t if any infinite sequence of points of $F$ has a point of accumulation.

Any bounded [closed] set in $E_n$ is compact.

**6. CONNECTED SETS. CONTINUUM AND REGION.** A set $K$ is said to be c o n n e c t e d if it cannot be expressed as the union of two nonempty disjoint sets $K_1$ and $K_2$, each of which contains all of its boundary points that belong to $K$. In particular, a closed set is connected if it cannot be expressed as the union of two nonempty disjoint closed sets, and an open set is connected if it cannot be expressed as the union of two disjoint nonempty open sets.

A closed connected set in $E_n$ is called a c o n t i n u u m. An open connected set is called a r e g i o n. Any open set can be expressed as the union of finitely or infinitely many disjoint regions.

The union of finitely or infinitely many connected sets, with the property that any set is accessible from any other through a finite sequence of sets such that any two consecutive sets have nonempty inter-section, is a connected set. In particular, the union of finitely many continua (finitely or infinitely many regions) with the above property is a continuum (region).

The intersection of a descending sequence of continua $K_1 \supset K_2 \supset K_3 \cdots$ is a continuum. If the diameters of the sets $K_i$ tend to zero, the intersec-tion of the sequence $K_i$ contains exactly one point.

If $a$ and $b$ are points of a continuum $K$, we shall also say that the continuum $K$ c o n n e c t s or j o i n s the points $a$ and $b$.

If $g$ is a region, the c l o s e d   r e g i o n $\overline{g}$ is the closure of $g$, i. e., the set of all points of $g$ and all boundary points of $g$ ($\overline{g}$ is a closed connected set, i. e., a continuum).

**7. REGIONS WITH COMMON BOUNDARY.** Let $g$ and $g'$ be regions having the same boundary points. Then, if they have at least one point in common, they must coincide.

If $K$ is a continuum containing a point of a region $g$ and a point not in $g$, then $K$ must contain a point of the boundary of $g$. In other words, a continuum joining an interior point of an open set to an exterior point must cut the boundary of the set.

**8. DENSE AND NOWHERE DENSE SETS.** A set $K_1$ is said to be d e n s e relative to a set $K_2$ if the closure of $K_1$ contains $K_2$, $\overline{K}_1 \supset K_2$. If moreover $K_1 \subset K_2$, we say that $K_1$ is (e v e r y w h e r e) d e n s e   i n $K_2$. If a set $K$ is dense in some region $g$ of $E_2$ (in particular, $g$ may be the entire space $E_n$), then every point $R \in g$ is a point of accumulation of $K$ (not necessarily an element of $K$).

A set $K$ is said to be n o w h e r e   d e n s e in a region $g$ if the complement of its closure is dense in $g$. A set $K$ is nowhere dense in $g$ if and only if it has no interior points.

**9. NEIGHBORHOODS, COVERS.** The $\varepsilon$-neighborhood of a set $K$ is defined as the union of the $\varepsilon$-neighborhoods of all its points, denoted by $U_\varepsilon(K)$. The $\varepsilon$-neighborhood of any set is an open set.

If $\Gamma$ is the set of boundary points of a set $K$ (irrespective of whether they belong to $K$ or not), then the $\varepsilon$-neighborhood of $K$ always contains the $\varepsilon$-neighborhood of $\Gamma$.

Apart from the $\varepsilon$-neighborhood of a point $a$, we shall sometimes use the term n e i g h b o r h o o d for any region containing $a$.

A collection of open sets (in particular, regions) $g$ is called a c o v e r of a set $K$ in $E_n$ if every point $a \in K$ lies in at least one of these open sets. If the cover consists of countably (finitely) many sets, it is called a countable (finite) cover.

H e i n e - B o r e l   t h e o r e m. Any cover of a compact set $F \subseteq E_n$ contains a finite subcover.

**10. TOPOLOGICAL LIMIT.** Let

$$M_1, M_2, \ldots, M_n, \ldots \qquad (1)$$

be a sequence of sets in $E_n$.* The u p p e r   t o p o l o g i c a l   l i m i t $\overline{lt}M_k$ of the sequence (1) is defined as the set of all points each of whose neighborhoods contains infinitely many points of the sets $M_i$. The l o w e r t o p o l o g i c a l   l i m i t $\underline{lt}\, M_k$ is the set of points each of whose neighborhoods contains points of all but a finite number of sets $M_i$. If

$$A = \overline{lt}M_i = \underline{lt}M_i,$$

the sequence (1) is said to be t o p o l o g i c a l l y   c o n v e r g e n t and the set $A$ is known as its t o p o l o g i c a l   l i m i t, denoted $A = lt M_k$. The topological limit of connected sets is connected.

* The sequence will be denoted by $\{M_i\}$.

**11. MAPPINGS OF SETS.**   Let $K_1$ and $K_2$ be sets in $E_n$ (possibly coinciding with the entire space $E_n$), and suppose that to each point $M$ of $K_1$ corresponds exactly one point $M'$ of $K_2$. This correspondence is called a (single-valued) mapping $T$ of $K_1$ onto $K_2$.   The point $M' \in K_2$ corresponding to a point $M$ is called the i m a g e of $M$, and we write $M' = T(M)$. If every point of $K_2$ is the image of some point of $K_1$, we say that $T$ is a mapping of $K_1$ o n t o $K_2$ and $K_2$ is called the image of $K_1$, $K_2$ the pre-image of $K_1$.

Given a mapping $T$ of $K_1$ into $K_2$ and a mapping $U$ of $K_2$ into $K_3$, we can define a mapping $S$ of $K_1$ into $K_3$ in a natural way. The mapping $S$ is called the product (or composition) of the mappings $T$ and $U$ and denoted by $UT$. If any two distinct points of $K_1$ correspond under a mapping $T$ of $K_1$ onto $K_2$ to distinct points of $K_2$, $T$ is said to be a o n e - t o - o n e   m a p p i n g. In this case there exists an inverse mapping $T^{-1}$ of $K_2$ onto $K_1$, taking each point $m'$ of $K_2$ onto a point $m$ of $K_1 : m = T^{-1}(m')$. (We shall also say that the points of the sets $K_1$ and $K_2$ are in one-to-one correspondence.)

Let $m_1 \in K_1$, $m_1' = T(m_1)$.  The mapping $T$ is said to be c o n t i n u o u s at $m_1$ if for any $\varepsilon > 0$ there exists $\delta > 0$ such that

$$T(U_\delta(m_1) \cap K_1) \subset U_\varepsilon(m_1'). \qquad (2)$$

A mapping of $K_1$ into $K_2$ is said to be c o n t i n u o u s if it is continuous at each point of $K_1$.  A continuous mapping of $K_1$ onto $K_2$ is said to be u n i f o r m l y   c o n t i n u o u s if for any $\varepsilon > 0$ there exists $\delta > 0$, i n d e p e n d e n t   o f   t h e   p o i n t $m_1 \in K_1$, such that (2) holds.  The product of two continuous mappings is a continuous mapping.

The following propositions, presented without proof, are fundamental:

A.   The image of a bounded closed set in $E_n$ under a one-to-one contin-uous mapping is a bounded closed set.

B.   The image of a connected set under a one-to-one continuous mapping is a connected set.

**12. TOPOLOGICAL MAPPINGS.**   Let $T$ be a one-to-one continuous mapping of $K_1$ onto $K_2$. If the inverse mapping $T^{-1}$ is also continuous, then $T$ is called a one-to-one bicontinuous mapping of $K_1$ onto $K_2$, or a t o p o l o g i c a l m a p p i n g (or h o m e o m o r p h i s m) of $K_1$ onto $K_2$. Clearly, under these conditions $T^{-1}$ is a topological mapping of $K_2$ onto $K_1$. The set $K_2(K_1)$ is called the topological image of $K_1(K_2)$. We shall also say that $T$ defines a topological correspondence between the points of $K_1$ and $K_2$.  The product of two topological mappings is a topological mapping.

If there exists a topological mapping of $K_1$ onto $K_2$, the sets $K_1$ and $K_2$ are said to be h o m e o m o r p h i c (or topologically equivalent, or of the same topological structure).

The following two propositions are fundamental:

*T h e o r e m   I. A one-to-one continuous (not a priori bicontinuous) mapping of a bounded closed set $K_1$ in $E_n$ onto a set $K_2$ is always bicontinuous, hence a topological mapping.*

**13. BROUWER'S THEOREM ON INVARIANCE OF REGIONS.**   Let $T$ be a topological mapping of a set $K_1$ in a space $E_n$ onto a set $K_1'$ in a space $E_n'$. If $m$ is an interior point of $K'$, then $m' = T(m)$ is an interior point of $K_1'$, and if $m$ is a boundary point of $K_1$, then $m' = T(m)$ is a boundary point of $K_1'$. Consequently, if $K_1$ is a region, then so is $K_1'$.

**14. SYSTEMS OF FUNCTIONS DEFINING MAPPINGS.** Let $T$ be a mapping of a subset $K$ of an $n$-dimensional euclidean space $E_n$ with cartesian coordinates $x_1, \ldots, x_n$ onto a subset $K'$ of a space $E'_n$ with cartesian coordinates $y_1, \ldots, y_n$ (in particular, $E'_n$ may coincide with $E_n$). In terms of coordinates, this mapping is described by equations

$$y_i = f_i(x_1, \ldots, x_n), \qquad i = 1, 2, \ldots, n,$$

where the $f_i$ are functions defined on $K$. The mapping $T$ is continuous if and only if all the functions $f_i(x_1, \ldots, x_n)$ are continuous. If $K$ is a bounded closed set, the functions $f_i$ are uniformly continuous and $T$ is a uniformly continuous mapping (by Theorem A of §1.11 the image of $K$ is a bounded closed set).

**15. SIMPLE ARCS.** Let $x_i = f_i(t)$ $(i = 1, 2, \ldots, n)$ be functions defined for all $t$, $t_0 \leqslant t \leqslant T$, with the properties: a) $f_i(t)$ is continuous for all $t$, $t_0 \leqslant t \leqslant T$; b) there is no pair $t_1$, $t_2$, $t_1 \neq t_2$, such that

$$\sum [f_i(t_1) - f_i(t_2)]^2 = 0.$$

Then the set of points $M(t)$ in $E_n$ with coordinates $x_i = f_i(t)$ is called a p a r a m e t r i z e d  s i m p l e  a r c. The points $M_0 [f_1(t_0), \ldots, f_n(t_0)]$ and $M_1 [f_1(T), \ldots, f_n(T)]$ are known as the e n d p o i n t s of the arc; [all its other points will be called i n t e r i o r  p o i n t s].

It is clear from b) that two distinct values of $t$ define two distinct points of a parametrized simple arc. Moreover, again by b), for any positive $\delta_0 < T - t_0$ there exists $\varepsilon_0 > 0$ such that if $|t_1 - t_2| \geqslant \delta_0$ then $\varrho(M(t_1), M(t_2)) > \varepsilon_0$.

A  s i m p l e  a r c is a set of points which, for a suitable choice of functions $x_i = f_i(t)$ satisfying conditions a) and b), is a parametrized simple arc. A simple arc is obviously homeomorphic to a closed interval. It may possess different parametrizations, but for any parametrization the endpoints of a simple arc remain the same — they are independent of the parametrization.

Let $l$ be a parametrized simple arc and $t$ $[t_0 \leqslant t \leqslant T]$ the parameter. One can define two directions on $l$, one corresponding to increasing $t$ and the other to decreasing $t$.

Let $\tau_1$ and $\tau_2$, $\tau_1 < \tau_2$, be any two numbers in the interval $[t_0, T)$, such that at least one of the endpoints of the interval $[\tau_1, \tau_2]$ is not an endpoint of $[t_0, T]$. Then the set of points $M [f_1(t), f_2(t) \ldots]$, where $\tau_1 \leqslant t \leqslant \tau_2$, is also a simple arc — a subarc of $l$. If $\lambda$ is a subarc of $l$, then the direction in which $\lambda$ is described when $l$ is described in a given direction is called the direction i n d u c e d on $\lambda$ by the direction on $l$. We shall also use the expression "the direction on $l$ is induced by the direction on $\lambda$," which needs no explanation.

**16. SIMPLE CLOSED CURVES.** Let $x_i = \varphi_i(t)$ be functions defined for $t_0 \leqslant t \leqslant T$, single-valued and continuous on that interval, such that $\varphi_i(t_0) = \varphi_i(T)$, and moreover the equality $\varphi_i(t_1) = \varphi_i(t_2)$ $(t_0 \leqslant t_1 < t_2 \leqslant T)$ may occur only for $t_1 = t_0$, $t_2 = T$. Then the set of points $M(x_1, \ldots, x_n)$, where $x_1 = \varphi_1(t), \ldots, x_n = \varphi_n(t)$, $t$, $t_0 \leqslant t \leqslant T$, is a p a r a m e t r i z e d  s i m p l e  c l o s e d  c u r v e. A  s i m p l e  c l o s e d  c u r v e is a set of points which, for a suitable choice of functions $\varphi_i(t)$ satisfying the above conditions, is a parametrized simple closed curve. A simple closed curve is obviously homeomorphic to a circle. It may possess different parametrizations. On any parametrized simple closed curve one can define a positive t r a v e r s a l* by stipulating that the curve $C$ is traversed (or described) in the sense of either increasing or decreasing $t$. Any two points $M_1$ and $M_2$ on a simple closed curve divide it into two simple arcs with common endpoints.

Let $l_1$, $l_2$, ..., $l_n$ be simple arcs, and let $M_1^i$ and $M_2^i$ be the endpoints of $l_i$. Suppose that each arc $l_i$ has one endpoint in common with $l_{i+1}$, $M_2^{i-1} = M_1^i$, $M_2^i = M_1^{i+1}$, and $M_1^1 = M_2^n$, and that apart from these common endpoints the arcs $l_i$ have no other points in common. Then the union of the arcs is a s i m p l e   c l o s e d   c u r v e.** In particular, if the $l_i$ are straight-line segments, their union is a polygon.

Simple arcs and simple closed curves are c o n t i n u a.

The following proposition is obvious:

*T h e o r e m   II.  A simple arc and a simple closed curve in $E_n$ are nowhere dense in $E_n$ $(n \geqslant 2)$.*

§2. SIMPLE CLOSED CURVES AND SIMPLE ARCS ON THE PLANE. ORIENTATION OF THE PLANE (TRAVERSAL OF SIMPLE CLOSED CURVES). TYPES OF TOPOLOGICAL MAPPING

In this section we present some fundamental material and state without proof a number of theorems pertaining to the relative position of simple arcs and simple closed curves on the plane.

Sets on the plane, in particular, curves and regions, will be referred to as p l a n e   s e t s (p l a n e   c u r v e s and p l a n e   r e g i o n s). Only plane sets will be considered in this section.

### 1. TWO FUNDAMENTAL THEOREMS.

*T h e o r e m   III.  A simple arc does not separate the plane (i.e., the set consisting of the points of the plane minus the points of the arc is a region).*

*T h e o r e m   IV (Jordan curve theorem).  A simple closed curve c divides the plane into two regions and is the boundary of each of these regions. One of these regions (the interior of c) is bounded and the other (the exterior of c) unbounded.*†

### 2. LEMMAS ON SIMPLE CLOSED CURVES.

We cite without proof a number of elementary propositions (see /57/).

A boundary point $M$ of a region $G$ is said to be a c c e s s i b l e in $G$ if there exists a simple arc, one of whose endpoints is $M$ and all its other points lie in $G$.

A.  All points of a simple closed curve $C$ are accessible both in the interior of $C$ and in the exterior of $C$.

---

\*  [Literally, "sense of circulation"; our term is borrowed from Whyburn, G. T., Topological Analysis, Princeton, N. J., Princeton University Press. 1958.]

\*\*  An elementary proof of this statement is obtained by mapping the simple arcs $l_i$ onto arcs of a circle.

†  For a proof of these propositions, which seem intuitively quite obvious, see, e. g., /58/ [also Whyburn, G. T., op. cit.].

B. Let $C$ be a simple closed curve, $M$ an arbitrary point on $C$. Then:
1) For any $\varepsilon > 0$ there exists $\delta > 0$ such that any two points $P$ and $Q$ in $U_\delta (M)$, both inside (outside) $C$, can be joined within $U_\varepsilon (M)$ by a simple arc lying entirely inside (outside) $C$ (Figure 320). 2) For any $\varepsilon > 0$ there exists $\delta > 0$ such that any two points $R$ and $N$ on $C$, lying in $U_\delta (M)$, can be joined inside $U_\varepsilon (M)$ both by a simple arc all of whose points except $R$ and $N$ are inside $C$ and by a simple arc all of whose points except $R$ and $N$ are outside $C$.

C. Let $l$ be a simple arc with endpoints $M_1$ and $M_2$ on the curve $C$ and no other points in common with $C$. Let $s_1$ and $s_2$ be the simple arcs into which $M_1$ and $M_2$ divide $C$. Then: 1) If the interior points of $l$ lie inside $C$, $l$ divides the interior $G$ of $C$ into two regions $G_1$ and $G_2$; the boundary of $G_1$ is a simple closed curve $C_1$, the union of the arcs $s_1$ and $l$, and the boundary of $G_2$ is a simple closed curve $C_2$, the union of $s_2$ and $l$ (Figure 321). 2) If the interior points of $l$ lie outside $C$, the union of $l$ with $s_1$ $(s_2)$ is a simple closed curve $C_1$ $(C_2)$; one of these curves, $C_2$ say, contains all points of one of the arcs $s_1$, $s_2$, say $s_1$, so that the interior of $C_1$ is a subregion of the interior of $C_2$ (Figure 322).

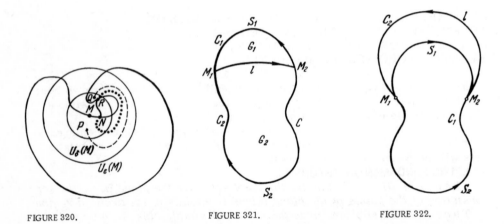

FIGURE 320.

FIGURE 321.

FIGURE 322.

D. Let $M_1$ and $M_2$ be two points on a simple closed curve $C$, joined by two arcs $l$ and $l_1$ such that all points of $l$ except its endpoints $(M_1$ and $M_2)$ lie inside $C$; all points of $l_1$ except its endpoints $(M_1$ and $M_2)$ lie outside $C$. Then the union of $l$ and $l_1$ is a simple closed arc $C_1$; all interior points of one of the arcs $s_1$ and $s_2$ into which $C$ is divided by $M_1$ and $M_2$ are inside $C_1$, and all interior points of the other outside $C_1$ (Figure 323).

E. Let $C_1$ be a simple closed curve which has one point $M$ in common with a closed curve $C$, all points of $C_1$ except $M$ lying inside $C$. Then $C_1$ divides the interior $G$ of $C$ into two regions: one is the interior of $C_2$ and the boundary of the other is the union of $C$ and $C_1$ (Figure 324).

*Theorem V.* Let $T$ be a topological mapping of a simple closed curve $C_1$ onto a simple closed curve $C_2$. Then there is a topological mapping $\tilde{T}$ of the plane onto itself which coincides with $T$ on $C$ and maps the interior (exterior) of $C_1$ onto the interior (exterior) of $C_2$.

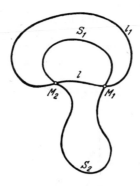

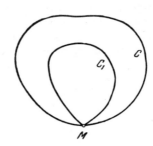

FIGURE 323.                    FIGURE 324.

### 3. TRAVERSAL OF SIMPLE CLOSED CURVES. CYCLIC ORDER ON A SIMPLE CLOSED CURVE.

On any simple closed curve one can define two different traversals. Let $C$ be a simple closed curve, with a prescribed traversal, and $M_1$, $M_2$, $M_3$ three (distinct) points on $C$. When the curve is described in accordance with the prescribed traversal, these points, starting say from $M_1$, are met in a certain order, so that the traversal of $C$ induces a well-defined cyclic order of the points, namely, either

$$(M_1,\ M_2,\ M_3) \equiv (M_2,\ M_3,\ M_1) \equiv (M_3,\ M_1,\ M_2), \qquad (1)$$

or

$$(M_1,\ M_3,\ M_2) \equiv (M_3,\ M_2,\ M_1) \equiv (M_2,\ M_1,\ M_3). \qquad (2)$$

Conversely, if $M_1$, $M_2$, $M_3$ are three given points on a simple closed curve, specification of some cyclic order (i. e., the order (1) or (2)) induces a definite traversal of the curve (the traversal for which the points $M_1$, $M_2$, $M_3$ have the specified cyclic order). Now let $M_1$, $M_2$, . . ., $M_n$, $n > 3$, be points on a simple closed curve $C$, indexed in agreement with a certain traversal of $C$. We shall then speak of the cyclic order of these points induced by the prescribed traversal of $C_1$ (or: the points $M_1$ are indexed in the cyclic order induced by the traversal on $C$).

### 4. INDUCED DIRECTION ON A SIMPLE SUBARC OF A SIMPLE CLOSED CURVE.

Let $s$ be a simple arc lying on a curve $C$, with endpoints $M_1$ and $M_2$ (besides $s$ there is one other subarc of $C$ with the same endpoints). Let a traversal be prescribed on $C$ and suppose that under this traversal the arc $s$ is described in the direction from $M_1$ to $M_2$. We shall say that this direction on the arc $s$ is induced by the prescribed traversal of $C$.

Conversely, let a positive direction be prescribed on a simple arc $s$ with endpoints $M_1$ and $M_2$, which is a subarc of a simple closed curve $C$, say from $M_1$ to $M_2$. Then the traversal of $C$ which induces on $s$ the direction from $M_1$ to $M_2$ on $s$ is called the traversal induced by the prescribed positive direction on $s$.

Let $\lambda$ and $\lambda'$ be simple arcs with common endpoints into which points $M_1$ and $M_2$ divide a simple closed arc $C$. If the direction on $\lambda$ $(\lambda')$ induced by a prescribed traversal of $C$ is the direction from $M_1$ to $M_2$ (from $M_2$ to $M_1$), we shall also denote the arc $\lambda$ by $M_1M_2$ (and the arc $\lambda'$ by $M_2M_1$; the order of the letters is essential). We shall then say that a point on $\lambda$ lies between $M_1$ and $M_2$, a point on $\lambda'$ between $M_2$ and $M_1$ (Figure 325).

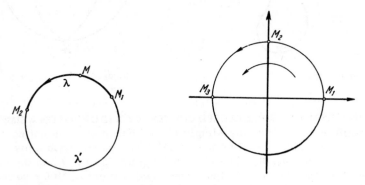

FIGURE 325.                    FIGURE 326.

## 5. ORIENTATION OF THE PLANE. In many of the questions studied in this book, the orientation of the plane plays an essential role.

An orientation is said to be defined on the plane (or the plane is oriented) if one of the two possible traversals of simple closed arcs (or, equivalently, one of the two possible "senses of rotation") is singled out and declared positive. The opposite traversal or sense of rotation is then negative.

To define an orientation, one first prescribes the positive traversal on an arbitrary but fixed simple closed curve (for example, as shown below, a certain circle), and then "transfers" this traversal to all other simple closed curves.* When this is done, the positive traversal may be chosen

---

* The details of this transfer procedure will be omitted. It may be based, for example, on suitable construction of auxiliary arcs joining points of nonintersecting closed curves (the dashed arcs in Figure 327), followed by application of the lemmas in the next subsection (relation between traversals of two simple curves having a common arc).

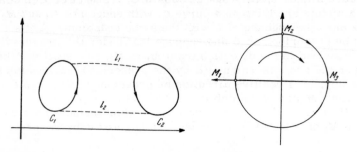

FIGURE 327.                    FIGURE 328.

with reference to a cartesian coordinate system in the plane. One way of doing this is as follows.

Let the plane be referred to a cartesian coordinate system $x$, $y$ and stipulate that the positive direction on the axes corresponds to increase in the corresponding coordinate. Let $C$ be a circle about the origin $(0, 0)$. Let $M_1$, $M_2$ and $M_3$ be the points at which it cuts the positive $x$-axis, positive $y$-axis and negative $x$-axis, respectively (Figure 326).

To fix ideas, we stipulate that the positive traversal on $C$ is that inducing the cyclic order $M_1, M_2, M_3$. Thus fixed, the positive traversal is then trasferred to all simple closed curves.*

Once the positive traversal — and therefore also positive sense of rotation — has been fixed, we can state that the angle between the positive $x$-axis and any other straight line is p o s i t i v e when measured f r o m t h e p o s i t i v e $x$ - a x i s i n t h e p o s i t i v e s e n s e, and n e g a t i v e otherwise.

With regard to two planes $E$ and $E'$, one can always stipulate that the positive traversal has been chosen in the above manner on each of them (i. e., the positive traversal on a circle about the origin is that inducing the cyclic order $M_1 M_2$, $M_3$ of the above three points). We shall then say that the choice of positive traversal on the planes is c o m p a t i b l e.

**6. PROPOSITIONS ON TRAVERSALS OF SIMPLE CLOSED CURVES WITH A COMMON ARC OR COMMON POINT.** Let $C_1$ and $C_2$ be two simple closed curves on the oriented plane, having a common arc $l$ with endpoints $M_1$ and $M_2$. Apart from this common arc $l$, the curves have no other points in common. Suppose that the positive traversal of $C_1$ induces the direction from $M_1$ to $M_2$ on $l$; this will be the positive direction on $l$.

Then: 1) If the region bounded by $C_2$ is a subregion of the interior of $C_1$ (Figure 322), the traversal of $C_2$ induced by the positive direction on $l$ is positive; 2) if the regions bounded by $C_1$ and $C_2$ are mutually exterior, the traversal of $C_2$ induced by the positive direction on $l$ is negative.

The situation is entirely analogous for the negative traversal of a curve $C$.

The converse is also true, in the following sense. 1) If the positive direction on $l$ induces on $C_1$ and $C_2$ a positive (negative) traversal, then one of the regions bounded by $C_1$ and $C_2$ lies inside the other; 2) if the positive direction on $l$ induces the positive traversal on one of the curves and the negative traversal on the other (or conversely), the regions bounded by $C_1$ and $C_2$ are mutually exterior.

Assume now that $C_1$ and $C_2$ are simple closed curves with exactly one point $O$ in common. The regions bounded by $C_1$ and $C_2$ may be either mutually exterior or one interior to the other.

Let $\sigma$ be a simple closed curve containing $O$ in its interior, not lying entirely inside either of the curves $C_1$ and $C_2$ and having exactly two points

---

\*　If the positive directions on the coordinate axes are chosen as in Figure 328, the positive traversal will coincide with clockwise rotation about the circle.

However, we shall make no use of the intuitive conception of clockwise or anticlockwise "rotation," which has no mathematical meaning in the present context. Note that the positive traversal in Figure 326 is anticlockwise. There is, however, no mathematical distinction between the two coordinate systems of Figures 326 and 328 on the plane (they may be treated as identical coordinate systems viewed from "opposite sides" of the plane).

If the two coordinate systems are introduced o n t h e s a m e p l a n e, however, we can always distinguish between the traversals they induce.

in common with each. Denote the common points of $\sigma$ and $C_1$ by $M_1$ and $M_1'$, those of $\sigma$ and $C_2$ by $M_2$ and $M_2'$. The points $M_1$ and $M_1'$ divide $\sigma$ into two arcs $\lambda_1$ and $\lambda_1'$ and are the common endpoints of these arcs (Figure 329a and b).

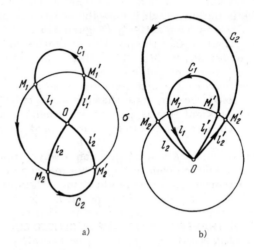

a)                                                         b)

FIGURE 329.

Clearly, all interior points of one of these curves, $\lambda_1'$ say, are in the interior of $C_1$, while all interior points of $\lambda_1$ are in the exterior of $C_1$. Similarly, $M_2$ and $M_2'$ divide $\sigma$ into two arcs $\lambda_2$ and $\lambda_2'$; all interior points of one of these arcs are in the interior of $C_2$, all interior points of the other arc in the exterior of $C_2$. One of the arcs $\lambda_1$ and $\lambda_1'$ contains neither of the points $M_2$, $M_2'$, the other contains both. An analogous assertion holds for the arcs $\lambda_2$ and $\lambda_2'$ and the points $M_1$ and $M_1'$. Suppose that the positive traversal of the curve $\sigma$ induces the cyclic order, $M_1$, $M_2$, $M_2'$, $M_1'$. Let $l_1$ and $l_1'$ denote the subarcs $OM_1$ and $OM_1'$ of $C_1$, all of whose points other than $M_1$ and $M_1'$ lie inside $\sigma$. Similarly, $l_2$ and $l_2'$ will be the subarcs $OM_2$ and $OM_2'$ of $C_2$ all of whose points except $M_2$ and $M_2'$ lie inside $\sigma$.

With the above notation, we have the following propositions, cited without proof.

I. If the interiors of $C_1$ and $C_2$ are mutually exterior, the positive traversal of $C_1$ induces on $l_1$ the direction from $M_1$ to $O$, and the positive traversal of $C_2$ induces on $l_2$ the direction from $O$ to $M_2$, so that both these directions on $l_1$ and $l_2$ induce the same direction (from $M_1$ to $M_2$) on the union of $l_1$ and $l_2$, which is an arc with endpoints $M_1$ and $M_2$ (Figure 329a).

II. If one of the regions interior to $C_1$ and $C_2$ lies inside the other and the positive traversal of $C_1$ induces on $l_1$ the direction from $M_1$ to $O$ (from $O$ to $M_1$), then the positive traversal of $C_2$ induces on $l_2$ the direction from $M_2$ to $O$ (from $O$ to $M_2$) (so that these directions on $l_1$ and $l_2$ induce opposite directions on the union of $l_1$ and $l_2$) (Figure 329b).

The following lemmas are in a sense the converse of the above.

III. If the positive traversal of $C_1$ induces on $l_1$ the direction from $M_1$ to $O$ and the positive traversal on $C_2$ induces on $l_2$ the direction from

$O$ to $M_2$, then the interiors of the closed curves $C_1$ and $C_2$ are mutually exterior (Figure 329a).

IV. If the positive traversal of $C_1$ induces on $l_1$ the direction from $M_1$ to $O$ and the positive traversal of $C_2$ induces on $l_2$ the direction from $M_2$ to $O$, then one of the regions interior to the closed curves $C_1$ and $C_2$ lies inside the other (Figure 329b).

**7. TWO PROPOSITIONS ON THE RELATION BETWEEN ORDER OF POINTS ON DISJOINT SIMPLE CLOSED CURVES.** Let $C_1$ and $C_2$ be two simple closed curves with no points in common. Let $P_1, P_2, \ldots, P_N, N \geqslant 3$, be points on $C_1$, indexed in cyclic order with respect to the positive traversal of $C_1$. Then:

I. Irrespective of whether $C_1$ and $C_2$ are mutually exterior or one inside the other, there exist simple arcs $l_i$, each with one endpoint $P_i$ on $C_1$ and the other a point $Q_i$ on $C_2$, such that apart from the endpoints $P_i$, $Q_i$ the arcs $l_i$ have no points in common with $C_1$ and $C_2$ or with each other (Figure 330a, b).

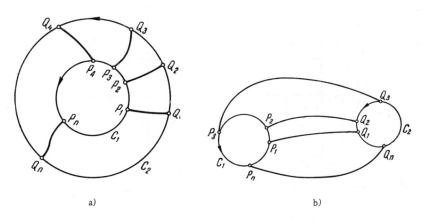

a)                                                      b)

FIGURE 330.

The next two propositions follow from the propositions of the preceding subsection.

II. If one of the simple closed curves $C_1$ and $C_2$ lies inside the other, the positive traversal of $C_2$ induces the order $Q_1, Q_2, \ldots, Q_N$. If the curves are mutually exterior, the positive traversal of $C_2$ arranges the points $Q_i$ in cyclic order $Q_N, Q_{N-1}, \ldots, Q_1$.

The converse is also true:

III. If the positive traversals of the curves $C_1$ and $C_2$ arrange the points $P_i$ and $Q_i$, respectively, in cyclic order $P_1, P_2, \ldots, P_N$ and $Q_1, Q_2, \ldots, Q_N$, then one of the curves $C_1$ and $C_2$ lies inside the other. If the positive traversals of $C_1$ and $C_2$ arrange $P_i$ and $Q_i$ in cyclic order $P_1, P_2, \ldots, P_N$ and $Q_N, Q_{N-1}, \ldots, Q_1$, respectively, then the curves $C_1$ and $C_2$ are mutually exterior.

**8. TWO TYPES OF TOPOLOGICAL MAPPING OF A PLANE (ORIENTATION-PRESERVING AND ORIENTATION-REVERSING).** We now consider topological mappings of a plane $E_2$ onto itself or onto another plane $E_2'$. The orientations of $E_2$ and $E_2'$ are assumed to be compatible (see §2.5).

It is obvious that any topological mapping maps a simple closed curve $C$ onto a simple closed curve $C'$. We can then divide the topological mappings of a plane onto itself into two types.

First type: orientation-preserving topological mappings.

Under a mapping of this type, a point describing any simple closed curve $C$ in the positive direction is mapped onto a point describing the image $C'$ of $C$ in the positive direction.

Second type: orientation-reversing topological mappings.

Under a mapping of this type, a point describing any simple closed curve $C$ in the positive direction is mapped onto a point describing the image $C'$ of $C$ in the negative direction. Mirror reflection is an orientation-reversing topological mapping of the plane onto itself (Figure 331a, b).

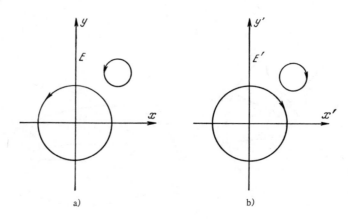

a)                                    b)

FIGURE 331.

## §3. POSITIVE AND NEGATIVE "SIDES" OF A SIMPLE ARC

**1. REGIONS DEFINING DIFFERENT "SIDES" OF A SIMPLE ARC.** Let $l$ be a simple arc and $l'$ another simple arc with exactly one point $M$ in common with $l$. Assume moreover that $M$ is not an endpoint of $l$, though it may be an endpoint of $l'$.

In the cases illustrated in Figures 332a and b, it is natural to say that the arc $l'$ "lies on one side of the arc $l$," while Figures 332c and d show $l'$ crossing from one side of $l$ to the other.

In this section we shall define these geometrically intuitive but rather vague concepts, which play an essential role in the main text.

We shall assume throughout that the plane is oriented (i.e., a positive traversal is prescribed for simple closed curves). Let $M_1$ and $M_2$ be the endpoints of a simple arc $l$ and suppose that the positive direction on $l$ is from $M_1$ to $M_2$.

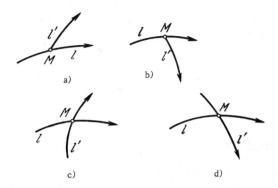

a)   b)   c)   d)

FIGURE 332.

Join the endpoints $M_1$ and $M_2$ of $l$ by simple arcs $l_1$ and $l_2$ such that a) $l_1$ and $l_2$ have no points other than $M_1$ and $M_2$ in common with $l$; b) $l_1$ and $l_2$ have no common points other than $M_1$ and $M_2$, so that their union is a simple closed curve $C$; c) all points of $l$ except $M_1$ and $M_2$ (which lie on the curve $C$) are in the interior of $C$ (Figure 333). Let $G$ be the interior of $C$. The arc $l$ (see §2.6) divides $G$ into two regions $G_1$ and $G_2$; the boundary of one of them, $G_1$ say, is a simple closed curve $C_1$, the union of $l$ and $l_1$, while the boundary of $G_2$ is a simple closed curve $C_2$, the union of $l$ and $l_2$. Suppose that the traversal of $C_1$ induced by the positive direction on $l$ is positive. Then the traversal of $C_2$ induced by the positive direction on $l$ is negative. Thus the regions $G_1$ and $G_2$ in a certain sense characterize different "sides" of the arc.*

In this situation, we shall say that the interior $G_1$ of the curve $C_1$ on which the traversal induced by the positive direction on $l$ is positive defines the positive side of the arc $l$; the interior $G_2$ of $C_2$ defines the negative side of $l$.

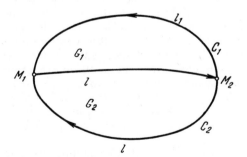

FIGURE 333.

* It will follow easily from the sequel that the different "sides" of a simple arc may be characterized using only one of the simple closed curves $C_1$ or $C_2$, hence only one of the arcs $l_1$ or $l_2$. Consideration of two closed curves, however, makes the discussion completely symmetric.

If we take different arcs satisfying conditions a), b) and c), we obviously obtain different closed curves $C_1$ and $C_2$ and different regions $G_1$ and $G_2$ defining the sides of the arc $l$.

Now let $\lambda$ be an arc whose only point in common with $l$ is one of its endpoints $M$ (Figure 332a, b).

Let

$$x = f(u), \qquad y = g(u), \qquad a \leqslant u \leqslant b, \tag{1}$$

be parametric equations of the arc $\lambda$, the endpoint $M$ corresponding to $u = a$. The following proposition is obvious:

*Theorem V'. Suppose that for some choice of arcs $l_1$ and $l_2$ satisfying conditions a), b) and c) all points of $\lambda$ sufficiently close to $M$ (i.e., points such that $|u - a| < \varepsilon$, where $\varepsilon$ is some positive number) lie in the region $G_1$ ($G_2$) defining the positive (negative) side of $l$. Then for any other choice of arcs satisfying conditions a), b) and c) there exists $\varepsilon^* > 0$ ($\varepsilon^* \leqslant \varepsilon$) such that all points on the arc $\lambda$ with $|u - a| < \varepsilon^*$ also lie respectively in the region defining the positive (negative) side of $l$.\**

By virtue of this theorem, the fact that all points of $\lambda$ sufficiently close to $M$ lie in the region defining the positive (negative) side of $l$ d e p e n d s  o n l y  o n  t h e  r e l a t i v e  p o s i t i o n  o f  t h e  a r c s $l$ and $\lambda$ and not on the choice of the arcs $l_1$ and $l_2$, i.e., on the choice of the regions $G_1$ and $G_2$.

In the case considered here, where the arc $\lambda$ has no points in common with $l$ other than one of its endpoints, which is an interior point of $l$, this justifies the following definition.

*Definition I. We shall say that all points of $\lambda$ other than the endpoint $M$ (which is on $l$) lie o n  t h e  p o s i t i v e  ( n e g a t i v e )  s i d e  o f $l$, or simply that the arc $\lambda$ lies on the positive (negative) side of $l$, if for some choice of arcs $l_1$ and $l_2$ satisfying conditions a), b) and c) all points of $\lambda$ sufficiently close to $M$ lie in the region $G_1$ ($G_2$) defining the positive (negative) side of $l$.*

Up to now we have assumed that the only common point $M$ of $l$ and $\lambda$ (which is not an endpoint of $l$) is an endpoint of $\lambda$. We now assume that this point is an endpoint of neither $l$ nor $\lambda$. On the basis of the above definition, we shall now describe all possible relative positions of $l$ and $\lambda$.

The point $M$ divides $\lambda$ into two subarcs $AM$ and $A_2M$, where $A$ and $A_2$ are the endpoints of $\lambda$. Suppose that the positive direction on $\lambda$, i.e., the direction of increasing parameter in equations (1), is from $A$ to $A_2$.

If the subarc $AM$ of $\lambda$ lies on the negative side of $l$ and the subarc $A_2M$ on the positive side of $l$ (or vice versa), we shall say that $\lambda$ c r o s s e s  a t $M$  f r o m  t h e  n e g a t i v e  s i d e  o f $l$  t o  i t s  p o s i t i v e  s i d e  (or vice versa) (Figure 334a).

If both subarcs $AM$ and $A_2M$ of $\lambda$ lie on the same (positive or negative) side of $l$, we shall say that $\lambda$ lies on one side of $l$ (Figure 334b).

**2. DEFINITION OF REGIONS DEFINING DIFFERENT SIDES OF A SIMPLE ARC BY USING CURVILINEAR COORDINATES.** In many cases (including those considered in this book) the most natural definition of "regions" characterizing different sides of a simple arc is not by auxiliary arcs $l_1$ and $l_2$ as above, but by the following different method.

---

\* An easy proof of this proposition may be derived from the propositions of §3.6.

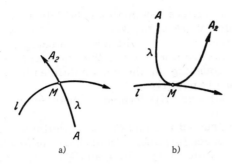

FIGURE 334.

Let $x = \varphi(s, t)$, $y = \psi(s, t)$ be functions defined in a rectangle $\sigma$ on the $(s, t)$-plane (Figure 335a, b),

$$a \leqslant s \leqslant b, \qquad |t| \leqslant a \qquad (a > 0), \tag{2}$$

and suppose that these functions define a topological mapping $T$ of the rectangle into the $(x, y)$-plane, taking the interval $a \leqslant s \leqslant b$ on the axis $t = 0$ (the interval $AB$) onto the given simple arc $l$.* The entire rectangle $\sigma$ is mapped onto some closed region $G$ in the $(x, y)$-plane whose interior contains all interior points of $l$.

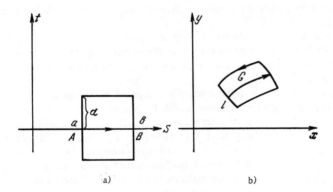

FIGURE 335.

The parameters $s$ and $t$ satisfying (2) may be treated as c u r v i l i n e a r c o o r d i n a t e s of points in $G$.

Under these conditions, we shall say that the subregion of $G$ corresponding to $t > 0$ ($G_1$, say) defines one (e. g., the positive) side of the arc $l$, and the subregion $G_2$ corresponding to $t < 0$ defines the other, negative side of $l$. $G_1$ and $G_2$ are clearly the subregions into which the arc $l$ divides the region $G$.

* The functions $\varphi(s, t)$, $\psi(s, t)$ are thus single-valued and continuous at all points of the rectangle (2), and are such that to any two distinct pairs $(s, t)$ correspond distinct pairs $(\varphi(s, t), \psi(s, t))$ (the last condition assures that the mapping $T$ is one-to-one).

477

In addition, we may always assume that the following conditions are satisfied: a) the segment $AB$ of the axis $t = 0$, $a \leqslant s \leqslant b$, is mapped onto the simple arc $l$ in such a way that the arc is described in a prescribed direction when $s$ varies from $a$ to $b$; b) the traversal induced on the simple closed curve bounding $G_1$ (the region defined by $t > 0$) by the positive direction on $l$ is positive.

If one of these conditions fails to hold, we need only replace $s$ by $-s$ or $t$ by $-t$, respectively.

We have thus constructed regions $G_1$ and $G_2$ defining the sides of a simple arc as in the preceding subsection; here, however, these regions were determined not by auxiliary arcs $l_1$ and $l_2$ but by a certain mapping or, equivalently, by introducing a certain system of curvilinear coordinates in the neighborhood of the arc $l$. As before, by assuming that the only common point $M$ of $l$ and $\lambda$ other than the endpoints of $l$ is an endpoint of $\lambda$, we arrive at the following definition:

We shall say that the arc $\lambda$ lies on the positive (negative) side of the arc $l$ if, for any choice of functions $\varphi\,(s,\,t)$, $\psi\,(s,\,t)$ satisfying the above conditions, all points of $\lambda$ sufficiently close to $M$ correspond to values $t > 0$ $(t < 0)$.

One can obviously describe the relative position of arcs $l$ and $\lambda$ as in Figure 334 using Definition I in this new formulation.

### 3. PROPOSITIONS ON THE RELATIVE POSITION OF ARCS AND SIMPLE CLOSED CURVES.
The following elementary propositions are cited without proof.

I. Let $l'$ be a subarc of a simple arc $l$ and $P$ a point on $l'$ (therefore also on $l$) which is not an endpoint of either $l$ or $l'$. Let $\lambda$ be a simple arc with endpoint $P$ and no other points in common with $l$. If $\lambda$ lies on the positive (negative) side of $l'$ (or $l$), then it lies on the positive (negative) side of $l$ $(l')$.

II. If $l$ and $\lambda$ are simple arcs with exactly one common point $P$, not an endpoint of either, and $\lambda$ crosses at $P$ from the negative side of $l$ to its positive side, then $l$ crosses at $P$ from the positive side of $\lambda$ to its negative side.

Let $\lambda$ be a simple arc with endpoints $A_1$ and $A_2$, with positive direction from $A_1$ to $A_2$, having exactly one point $P$, not one of its endpoints, in common with a simple closed curve $C$. We shall say that $\lambda$ crosses at $P$ from the interior (exterior) of $C$ to the exterior (interior) of $C$ if its subarc $A_1P$ minus the endpoint $P$ lies inside (outside) $C$ and the subarc $PA_2$ minus $P$ lies outside (inside) $C$.

Fix some traversal, either positive or negative, on a simple closed curve $C$. Let $l$ be a simple subarc of $C$; the positive direction on $l$ will be that induced by the positive traversal of $C$.

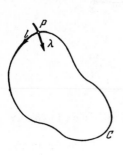

Let $\lambda$ be a simple arc whose only point $P$ in common with $l$ is not an endpoint of either $l$ or $\lambda$ (Figure 336).

III. If $\lambda$ crosses at $P$ from the positive side of $l$ to its negative side, then if the fixed traversal of $C$ is positive (negative), the arc $\lambda$ crosses at $P$ from the interior of $C$ to the exterior of $C$ (from the exterior of $C$ to the interior of $C$). This follows at once from Definition I above.

FIGURE 336.

478

IV. If the arc $\lambda$ crosses at $P$ from the positive side of $l$ to its negative side, and in so doing crosses from the interior of $C$ to the exterior of $C$ (from the exterior of $C$ to the interior of $C$), then the fixed traversal of $C$ is positive (negative).

**4. ROUNDED REGIONS IN THE PLANE.** First note the following elementary properties of regions:

A) Any two points of a region $G$ can be joined by a simple arc all of whose whose points are in $G$, and any points that can be joined to some point of $G$ by a simple arc not containing boundary points of $G$ are also points of $G$.

B) Any continuum (in particular, a simple arc) joining a point of $G$ to a point not in $G$ must contain at least one boundary point of $G$.

C) No simple arc $l$, one of whose endpoints lies on the boundary of a region $G$, which contains no other points not in $G$, can separate the region $G$ (i.e., the set obtained by deleting the points of $l$ from $G$ is a region).

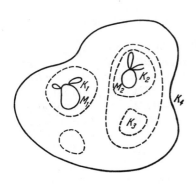

FIGURE 337.

D) Any simple closed curve $l$ whose points are all in a region $G$ divides $G$ into two regions; the boundary of one of these regions is $l$, that of the other the union of $l$ and the boundary of $G$.

A region $G$ is said to be **simply connected** if, for any simple closed curve $C$ lying entirely within $G$, all points interior to $C$ are points of $G$.

A region $G$ is said to be $n$-**tuply connected** $(n > 1)$ if: a) for any $n$ mutually exterior simple closed curves in $G$, there is at least one such that all points interior to it belong to $G$; b) there exist $n - 1$ mutually exterior simple closed curves in $G$ whose interiors contain points not belonging to $G$ (Figure 337).

The boundary of an $n$-tuply connected region is the union of $n$ disjoint continua $K_1, K_2, \ldots, K_n$, called the boundary continua of the region. One of these boundary continua is the **outer** boundary continuum, the others are **inner**. Any closed curve lying entirely within $G$ lies in the interior of the outer boundary continuum, but there always exists a simple closed curve in $G$ enclosing any one of the inner boundary continua, and one containing any 2, $\ldots$, $n - 1$ inner continua.

It is obvious that there exist regions which are not finitely connected. Let $G$ be an $n$-tuply connected region and $K_1, K_2, \ldots, K_n$ its boundary continua.

I. Let $l$ be a simple arc whose endpoints $M_1$ and $M_2$ lie on one boundary continuum $K_i$ and all of whose other points lie in $G$. Then $l$ divides $G$ into two regions and its points are boundary points of each of these regions.

II. A simple closed arc containing exactly one boundary point $M_0$ of $G$, all of whose other points lie in $G$, divides $G$ into two regions and its points are boundary points of each of these regions.

III. Let $n \geqslant 2$ and let $l$ be a simple arc with one endpoint on a boundary continuum $K_i$, the other endpoint on another boundary continuum $K_j$ $(i \neq j)$, and all other points in $G$. Let $G'$ be the subregion of $G$ whose boundary is the union of the continua $K_i$ and the arc $l$. Then $G'$ is an $(n-1)$-tuply

479

connected region. One of the boundary continua of this subregion is $K_i \cup l \cup K_j$. We shall then say that $l$ converts the $n$-tuply connected region into an $(n-1)$-tuply connected region.

## §4. HADAMARD'S LEMMA AND THE IMPLICIT FUNCTION THEOREM

**1. CLASSES OF FUNCTIONS.** Let $z = F(x_1, \ldots, x_n)$ be a function defined in some region $G$ of $E_n$. In particular, $G$ may be the entire space $E$ and $n$ any given natural number; when $n = 1$ we get a function of one variable.

We shall say that $F(x_1, \ldots, x_n)$ is of class $C_k$ (analytic) if it has continuous partial derivatives up to order $k$ (of all orders) at all points of $G$.

**2. HADAMARD'S LEMMA. \***   In Chapter IV repeated use is made of the following lemma:

*Hadamard's Lemma. Let $G$ be a region in the space $E_{n+m}$ $(x_1, \ldots, x_n \ z_1, \ldots, z_m)$ which is convex with respect to $x_1, x_2, \ldots, x_n$. Let $F(x_1, \ldots, x_n, z_1, \ldots, z_m)$ be a function having continuous derivatives with respect to $x_1, \ldots, x_n$ up to some order $k > 0$ in $G$. Then there exist $n$ functions $\varphi_i(x_1, \ldots, x_n, y_1, \ldots, y_n, z_1, \ldots, z_m)$, $i = 1, 2, \ldots, n$, continuously differentiable with respect to $x_1, \ldots, x_n, y_1, \ldots, y_n$ up to order $k - 1$, such that*

$$F(y_1, \ldots, y_n, z_1, \ldots, z_m) - F(x_1, \ldots, x_n, z_1, \ldots, z_m) =$$

$$= \sum_{i=1}^{n} \varphi_i(x_1, \ldots, x_n, y_1, \ldots, y_n, z_1, \ldots, z_m)(y_i - x_i).$$

**3. THE IMPLICIT FUNCTION THEOREM.** We first state the implicit function theorem, which is used repeatedly in the main text, for one function of three variables. The formulation for any number of variables is entirely analogous.

Let $F(x, y, z)$ be a function of class $C_k$ (or analytic) in a region $G$, such that at some point $M(x_0, y_0, z_0)$ of $G$

$$F(x_0, y_0, z_0) = 0, \quad F'_z(x_0, y_0, z_0) \neq 0. ^{**}$$

Then the equation $F(x_0, y_0, z_0) = 0$ can be  s o l v e d  for $z$ in the neighborhood of $x_0, y_0, z_0$. In rigorous terms:

*T h e o r e m  VI. There exists exactly one function $z = \varphi(x, y)$, defined for all $x$ and $y$ such that $|x - x_0| \leqslant \alpha_0$, $|y - y_0| \leqslant \beta_0$, where $\alpha_0 > 0$, $\beta_0 > 0$ are suitably chosen constants, satisfying the equation*

$$F(x, y, z) = 0, \tag{1}$$

*taking the value $z_0$ for $x = x_0, y = y_0$, and belonging to class $C_k$ (or analytic).*

---

\*  For a proof, see /61/. Hadamard's Lemma is a sharpened version of Lagrange's formula.

\*\*  In special cases the function $F(x, y, z)$  may be independent of $x$ or $y$ or both. However, since $F'_z \neq 0$, it must depend on $z$.

Remark I. There always exist positive numbers $a \leqslant a_0$, $\beta \leqslant \beta_0$ and $\gamma$ such that for all $x$, $y$ ($|x - x_0| \leqslant a$, $|y - y_0| \leqslant \beta$)

$$|\varphi(x, y) - z_0| < \gamma, \tag{2}$$

and moreover for all $x, y$ ($|x - x_0| \leqslant a$, $|y - y_0| \leqslant \beta$) there are no values of $z$ other than $z = \varphi(x, y)$ which satisfy equation (1) and inequality (2). Thus, in the special case that $F(x, y, z)$ does not depend on $x$ and $y$ (i.e., is a function of one variable $z$, $F(x, y, z) \equiv \Phi(z)$), we at once infer that the root $z_0$ is isolated (of course, this can also be proved directly).

Remark II. The partial derivatives of the function $z = \varphi(x, y)$ can be determined successively from the relations

$$F'_x + \varphi'_x F'_z = 0, \quad F'_y + \varphi'_y F'_z = 0,$$
$$F''_{xx} + \varphi'_x F''_{xz} + \varphi''_x, \quad F''_{2x} + \varphi''_{xx} F'_z = 0,$$
$$\cdots \cdots \cdots \cdots \cdots \cdots$$

We now state the implicit function theorem for two functions (the formulation for more variables is entirely analogous).

Let $F_1(x, y, u, v)$, $F_2(x, y, u, v)$ be functions* of class $C_k$ (or analytic) in some region $G$ of the space $E_n(x, y, u, v)$, such that at some point $M_0(x_0, y_0, u_0, v_0)$ of $G$

$$D = \begin{vmatrix} \dfrac{\partial F_1}{\partial u} & \dfrac{\partial F_1}{\partial v} \\ \dfrac{\partial F_2}{\partial u} & \dfrac{\partial F_2}{\partial v} \end{vmatrix} \neq 0. \tag{3}$$

Under these assumptions, we have

Theorem VII. *There exists exactly one pair of continuous functions*

$$u = \varphi_1(x, y), \qquad v = \varphi_2(x, y),$$

*defined for all $x$, $y$ such that*

$$|x - x_0| \leqslant a_0, \qquad |y - y_0| \leqslant \beta_0,$$

*where $a_0 > 0$, $\beta_0 > 0$ are suitably chosen constants, which satisfies the equations*

$$F_1(x, y, u, v) = 0, \qquad F_2 = (x, y, u, v) = 0 \tag{4}$$

*and the condition*

$$u_0 = \varphi_1(x_0, y_0), \qquad v_0 = \varphi_2(x_0, y_0).$$

*The functions $\varphi_1(x, y)$ and $\varphi_2(x, y)$ are of class $C_k$ (or analytic).*

Remark I. As in the case of one implicit function (see Remark I to Theorem VI), one can always find positive numbers $a \leqslant a_0$, $\beta \leqslant \beta_0$, $\gamma$ and $\delta$ such that for all $x$, $y$ with

---

* In special cases the functions $F_1$ and $F_2$ may be independent of $x$, $y$ or both, but by virtue of (3) they must depend on $u$ and $v$.

$$|x-x_0| \leqslant \alpha, \qquad |y-y_0| \leqslant \beta,$$

we have

$$|\varphi_1(x, y)-u_0| < \gamma, \qquad |\varphi_2(x, y)-v_0| < \delta, \tag{5}$$

and moreover for all $x$, $y$, $|x-x_0| \leqslant \alpha$, $|y-y_0| \leqslant \beta$, there are no values of $u$ and $v$ other than $u=\varphi_1(x, y)$, $v=\psi_2(x, y)$ which satisfy equations (4) and conditions (5).

Remark II. In the special case that the functions $F_1(x, y, u, v)$ and $F_2(x, y, u, v)$ do not depend on $x$, $y$, so that $F_1(x, y, u, v) \equiv \Phi_1(u, v)$, $F_2(x, y, u, v) \equiv \Phi_2(u, v)$, the system (4) is

$$\Phi_1(u, v)=0, \qquad \Phi_2(u, v)=0, \tag{6}$$

with

$$\Phi_1(u_0, v_0)=\Phi_2(u_0, v_0)=0$$

and

$$D = \begin{vmatrix} \Phi'_{1u}(u_0, v_0) & \Phi'_{1v}(u_0, v_0) \\ \Phi'_{2u}(u_0, v_0) & \Phi'_{2v}(u_0, v_0) \end{vmatrix} \neq 0. \tag{7}$$

Then it follows from Remark I that the solution $u_0$, $v_0$ of system (6) is the only solution such that

$$|u-u_0| < \gamma, \qquad |v-v_0| < \delta,$$

where $\gamma > 0$ and $\delta > 0$ are suitably chosen constants. Regarding $\Phi_1(u, v)=0$ and $\Phi_2(u, v)=0$ as curves on the $(u, v)$-plane, we can give our assertion a geometric interpretation: under condition (7), $(u_0, v_0)$ is an isolated common point of the curves (6).

Remark III. The Jacobians

$$\frac{D(F_1, F_2)}{D(x, y)} = \begin{vmatrix} F'_{1x} & F'_{1y} \\ F'_{2x} & F'_{2y} \end{vmatrix}, \qquad \frac{D(F_1, F_2)}{D(u, v)} = \begin{vmatrix} F'_{1u} & F'_{1v} \\ F'_{2u} & F'_{2v} \end{vmatrix}$$

and

$$\frac{D(\varphi_1, \varphi_2)}{D(x, y)} = \begin{vmatrix} \varphi'_{1x} & \varphi'_{2x} \\ \varphi'_{1y} & \varphi'_{2y} \end{vmatrix}$$

satisfy the relation

$$\frac{D(F_1, F_2)}{D(x, y)} = \frac{D(F_1, F_2)}{D(u, v)} \frac{D(\varphi_1, \varphi_2)}{D(x, y)}. \tag{8}$$

## §5. ANGLE BETWEEN VECTORS. SMOOTH SIMPLE ARC AND SMOOTH SIMPLE CLOSED CURVE. ANGLE BETWEEN TWO SMOOTH ARCS

**1. ANGLE BETWEEN VECTORS.** Let $E_2$ be the euclidean plane, referred to cartesian coordinates $x$, $y$. Let $M$ be a vector at a point $M_0$, with components $M_x$ and $M_y$.

We call $M$ a z e r o  v e c t o r if $M_x = M_y = 0$, i. e., its length $\sqrt{M_x^2 + M_y^2}$ is equal to zero. Let $M$ and $N$ be two nonzero vectors, with components $M_x$, $M_y$ and $N_x$, $N_y$, respectively. The angle between $M$ and $N$ is defined to be the number $a$, $-\pi < a \leqslant \pi$ such that

$$\cos a = \frac{M_x N_x + M_y N_y}{\sqrt{M_x^2 + M_y^2}\,\sqrt{N_x^2 + N_y^2}}, \qquad \sin a = \frac{M_x N_y - M_y N_x}{\sqrt{M_x^2 + M_y^2}\,\sqrt{N_x^2 + N_y^2}}.$$

Thus, the angle between $M$ and $N$ is the smallest angle (in absolute value) through which $M$ must be rotated to make it coincide in direction with $N$. When this angle is positive the rotation is in the positive sense; otherwise the rotation is in the negative sense (see §2.5). We shall denote the angle between $M$ and $N$ by $MN$ (the order is important!).

It is obvious that $MN = -NM$.

Let $M$, $N$ and $K$ be three vectors at a point $O$, the angle between $M$ and $N$ being positive. We shall then say that the vector $K$ lies b e t w e e n $M$ and $N$ if the angle between $K$ and $M$ is negative and the angle between $K$ and $N$ positive. The angle between vectors is obviously defined only for nonzero vectors.

If $L_1$ is the straight line defined by the vector $M$, $L_2$ the line defined by $N$, and the positive directions on these lines coincide with the directions of $M$ and $N$ (or: "are induced" by the vectors $M$ and $N$), the angle between $L_1$ and $L_2$ is defined to be the angle between $M$ and $N$.

**2. SMOOTH SIMPLE ARC.** A simple arc $l$ is said to be s m o o t h if it has a parametric representation $x = \varphi(t)$, $y = \psi(t)$ in which the functions $\varphi(t)$, $\psi(t)$ satisfy the following conditions: 1) they are single-valued and continuous for all $t$, $a \leqslant t \leqslant b$ (where $a$ and $b$, $b > a$, are given numbers such that $M_0(\varphi(a), \psi(a))$ and $M_1(\varphi(b), \psi(b))$ are the endpoints of $l$), and for any two numbers $t_1$, $t_2$, $t_1 \neq t_2$ we have $[\varphi(t_1) - \psi(t_2)]^2 + [\psi(t_1) - \psi(t_2)]^2 \neq 0$, i. e., to distinct values of $t$ correspond distinct points of the arc $l$; 2) the functions have continuous derivatives $\varphi'(t)$, $\psi'(t)$ which do not vanish simultaneously for any $t$, $a \leqslant t \leqslant b$.

Condition 1) assures that the set of points $M(\varphi(t), \psi(t))$ is a simple arc, while 2) is the smoothness condition. It follows readily from condition 2) (e. g., by reductio ad absurdum) that there exists $A_0 > 0$ such that $\varphi'^2(t) + \psi'^2(t) > A_0$ for all $t$, $a \leqslant t \leqslant b$.

A simple arc is said to be smooth of class $C_k$ (analytic) if the functions $\varphi(t)$, $\psi(t)$ are of class $C_k$ (analytic).

Suppose that for given functions $\varphi(t)$, $\psi(t)$ the positive direction on the smooth arc $l$ is that of increasing $t$. At each point of $l$ we can define a t a n g e n t  v e c t o r $k$, the vector with components $k_x' = \varphi'(t)$; $k_y' = \psi'(t)$. The straight line defined by this vector is tangent to the arc. The tangent of the angle between the $x$-axis and this straight line, the positive direction on the latter being that induced by $k$, is called the s l o p e of the line.

### 3. SMOOTH SIMPLE CLOSED CURVE, PIECEWISE-SMOOTH SIMPLE CLOSED
CURVE. A simple closed curve is said to be s m o o t h if it has a parametric representation $x = \varphi(t)$, $y = \psi(t)$ in which the functions $\varphi(t)$ and $\psi(t)$ satisfy the following conditions: a) they are single-valued and continuous for all $t$, $t_0 \leqslant t \leqslant T$ (where $t_0$ and $T$ are certain fixed numbers), $\varphi(t_0) = \varphi(T)$, $\psi(t_0) = \psi(T)$, and moreover if $t_1$ and $t_2$ are such that $t_0 \leqslant$ $\leqslant t_1 < t_2 \leqslant T$ then the equalities $\varphi(t_1) = \varphi(t_2)$, $\psi(t_1) = \psi(t_2)$ can hold only when $t_1 = t_0$, $t_2 = T$; b) the functions have continuous derivatives $\varphi'(t)$, $\psi'(t)$ which do not vanish simultaneously for any $t$ ($t_0 \leqslant t \leqslant T$), and moreover $\varphi'(t_0) = \varphi'(T)$, $\psi'(t_0) = \psi'(T)$.

Condition a) means that the set of points $M(\varphi(t), \psi(t))$ is a simple closed curve, while b) is the smoothness condition. These conditions are satisfied, in particular, if $\varphi(t)$, $\psi(t)$ are periodic functions with $\tau = T - t_0$, with continuous derivatives which cannot vanish simultaneously. As in the case of a smooth simple arc, it follows from condition b) that $\varphi'^2(t) + \psi'^2(t) > A_0$ for all $t$, $t_0 \leqslant t \leqslant T$, for some positive constant $A_0$. Any two points on a smooth simple closed curve $C$ clearly divide it into two smooth simple arcs. As in the case of arcs, a tangent vector is defined at each point of $C$.

A smooth simple closed curve is said to be of class $C_k$, $k \geqslant 1$ (analytic) if the functions $\varphi(t)$, $\psi(t)$ are of class $C_k$ (analytic) and moreover their derivatives satisfy the relations

$$\varphi^{(i)}(t_0) = \psi^{(i)}(T),$$
$$\psi^{(i)}(t_0) = \psi^{(i)}(T),$$

where $i$ takes on all values from 1 to $k$ ($k = \infty$ in the analytic case). A simple closed curve which is not smooth is said to be p i e c e w i s e - s m o o t h if it is the union of finitely many smooth arcs.

### 4. SMOOTH CURVES. Let $x = \varphi(t)$, $y = \psi(t)$ be functions defined on an
o p e n interval $(a, b)$ (in particular, we may have $a = -\infty$, $b = +\infty$), which have continuous derivatives $\varphi'(t), \psi'(t)$, $\varphi'(t)^2 + \psi'(t)^2 \geqslant 0$ [sic], for all $t$ in this interval. Then the set of points $M(\varphi(t), \psi(t))$, $t \in (a, b)$, will be called a s m o o t h c u r v e or simply a c u r v e.*

### 5. SMOOTH SIMPLE ARCS WITH A COMMON POINT. Let $l_1$ and $l_2$ be two
smooth simple arcs, with parametric equations $x = \varphi_1(t)$, $y = \psi_1(t)$, $t \in [a_1, b_1]$ ($a_1 < b_1$), and $x = \varphi_2(u)$, $y = \psi_2(u)$, $u \in [a_2, b_2]$ ($a_2 < b_2$) respectively, where $\varphi_1$, $\psi_1$ and $\varphi_2$, $\psi_2$ are functions of class $C_1$ satisfying conditions 1) and 2) of §5.2. Suppose that the positive direction on each of these arcs is that of increasing parameters $t$ and $u$.

Let $M_0(x_0, y_0)$ be a common point of $l_1$ and $l_2$, corresponding on $l_1$ to a parameter value $t_0 \in [a_1, b_1]$ and on $l_2$ to $u_0 \in [a_2, b_2]$, so that $x_0 = \varphi_1(t_0) = \varphi_2(u_0)$, $y_0 = \psi_1(t_0) = \psi_2(u_0)$ . (The point $M_0$ may or may not be an endpoint of either or both of the arcs $l_1$ and $l_2$.)

The angle $\alpha$ between the arcs $l_1$ and $l_2$ (in this order!) at their common point $M_0$ is defined to be the angle between the tangent vector to $l_1$ and the tangent vector to $l_2$ at the point $M_0$. Thus the angle $\alpha$ is determined from the formulas

---

* [At this point the authors use the Russian word "liniya" instead of "krivaya" as previously for "curve." The distinction seems to be that here the functions $\varphi(t)$, $\psi(t)$ are defined on an open interval, rather than a closed interval as in the case of an arc or simple closed curve, and moreover the "simplicity" condition (see §1.15—16) is not required. Our translation will continue to use "curve" for both concepts; the distinction will always be clear from the context.]

$$\sin \alpha = \frac{\varphi_1'(t_0)\,\psi_2'(u_0) - \varphi_2'(u_0)\,\psi_1'(t_0)}{\sqrt{\varphi_1'^2(t_0) + \psi_1'^2(t_0)}\,\sqrt{\varphi_2'^2(u_0) + \psi_2'^2(u_0)}},$$

$$\cos \alpha = \frac{\varphi_1'(t_0)\,\varphi_2'(u_0) + \psi_1'(t_0)\,\psi_2'(u_0)}{\sqrt{\varphi_1'^2(t_0) + \psi_1'^2(t_0)}\,\sqrt{\varphi_2'^2(u_0) + \psi_2'^2(u_0)}}.$$

The sign of the angle between arcs $l_1$ and $l_2$ is therefore that of the determinant

$$D_0 = \begin{vmatrix} \varphi_1'(t_0) & \psi_1'(t_0) \\ \varphi_2'(u_0) & \psi_2'(u_0) \end{vmatrix}.$$

It is clear that the angle between $l_2$ and $l_1$ is equal in absolute value and opposite in sign to that between $l_1$ and $l_2$.

Two smooth simple arcs i n t e r s e c t (or one intersects or cuts the other) if the angle between them at their common point is not zero, $\sin \alpha \neq 0$. Two smooth simple arcs t o u c h (or are t a n g e n t) at their common point if the angle between them is 0 or $\pi$, $\sin \alpha = 0$.

L e m m a 1. *Let $l_1$ and $l_2$ be smooth simple arcs intersecting at a point* $M_0$ $(x_0, y_0)$ $(x_0 = \varphi_1(t_0) = \varphi_2(u_0),\ y_0 = \psi_1(t_0) = \psi_2(u_0))$. *Then there exists* $\Delta > 0$ *such that the subarc of $l_2$ defined by $|u - u_0| \leqslant \Delta$ has no points in common with $l_1$ other than $M_0$, and similarly the subarc of $l_1$ defined by $|t - t_0| \leqslant \Delta$ has no points in common with $l_2$ other than $M_0$.*

P r o o f. We first take some positive number $\delta < \dfrac{b_1 - a_1}{2}$ and consider the set of points on $l_1$ such that $|t - t_0| \geqslant \delta$. This set is the union of two simple arcs $\lambda_1$ and $\lambda_2$ (the points of $\lambda_1$ are defined, say, by $a_1 \leqslant t \leqslant t_0 - \delta$, those of $\lambda_2$ by $t_0 + \delta \leqslant t \leqslant b_1$).

Let $\eta$ denote the infimum of the distances of the point $M_0$ from points of $\lambda_1$ and $\lambda_2$. It is clear that $\eta > 0$ (for the union of $\lambda_1$ and $\lambda_2$ is a closed set not containing $M_0$). Now let $\delta' > 0$ be so small that the subarc of $l_2$ defined by $|u - u_0| \leqslant \delta'$ lies entirely in $u_{\eta/2}(M_0)$. This subarc of $l_2$ obviously has no points in common with the arcs $\lambda_1$ and $\lambda_2$. Thus it remains to show that for sufficiently small $\delta > 0$ and $\delta' > 0$ the subarc of $l_2$ defined by $|u - u_0| < \delta'$ has no points other than $M_0$ in common with the subarc of $l_1$ defined by $|t - t_0| < \delta$.

To prove this, observe first that by assumption

$$\varphi_1(t_0) = \varphi_2(u_0), \qquad \psi_1(t_0) = \psi_2(u_0), \tag{1}$$

$$D = \varphi_1'(t_0)\,\psi_2'(u_0) - \psi_1'(t_0)\,\psi_2'(u_0) \neq 0. \tag{2}$$

Suppose that for every $\delta > 0$ there exist $t_1$ and $u_1$, $t_1 \neq t_0$, $u_1 \neq u_0$, such that $|t_1 - t_0| < \delta$, $|u_1 - u_0| < \delta$ and at the same time

$$\varphi_1(t_1) = \varphi_2(u_1), \qquad \psi_1(t_1) = \psi_2(u_1). \tag{3}$$

Subtracting (1) from (3), we get

$$\varphi_1(t_1) - \varphi_1(t_0) = \varphi_2(u_1) - \varphi_2(u_0), \qquad \psi_1(t_1) - \psi_1(t_0) = \psi_2(u_1) - \psi_2(u_0)$$

and, by the mean-value theorem,

$$\varphi_1'(\xi_1)(t_1-t_0)=\varphi_2'(\xi_2)(u_1-u_0), \qquad \psi_1'(\eta_1)(t-t_0)=\psi_2'(\eta_2)(u_1-u_0). \quad (4)$$

Since by assumption $D_0 \neq 0$ and the functions $\varphi_1'$, $\varphi_2'$, $\psi_1'$, $\psi_2'$ are continuous, it follows that for all $\xi_1$, $\eta_1$ sufficiently close to $t_0$ and $\xi_2$, $\eta_2$ sufficiently close to $u_0$, i.e., for sufficiently small $\delta > 0$,

$$D=\varphi_1'(\xi_1)\,\psi_2'(\eta_2)-\varphi_2'(\xi_2)\,\psi_1'(\eta_1) \neq 0.$$

Now, we can treat (4) as a system of linear equations in $u_1 - u_0$ and $t_1 - t_0$. Since $D \neq 0$, this system cannot have a nontrivial solution $u_1 - u_0$, $t_1 - t_0$. And this clearly contradicts our assumption. Q.E.D.

R e m a r k. If the functions $\varphi_1$ and $\psi_1$ are defined only for $t > t_0$, or the functions $\varphi_2$ and $\psi_2$ only for $u > u_0$ (i.e., when the common point of the arcs $l_1$ and $l_2$ is an endpoint of one or both of them), certain obvious modifications must be made in the above proof.


## §6. REGULAR MAPPINGS. CURVILINEAR COORDINATES. SOME PROPOSITIONS ON SMOOTH ARCS AND SMOOTH CLOSED CURVES

**1. REGULAR MAPPINGS.** In this section we define and discuss r e g u l a r  m a p p i n g s, a special type of topological mapping. We shall restrict ourselves to the case $n = 2$ (mappings of sets on a euclidean plane into sets of the same or another euclidean plane). Subject to obvious changes, the entire discussion carries over to the case $n > 2$.

Let the functions

$$x=f(u, v), \qquad y=g(u, v) \tag{1}$$

be defined in some region $G$ of the $(u, v)$-plane, single-valued, continuous and continuously differentiable up to some order $p \geqslant 1$ at all points of the region. We shall regard $u$, $v$ as cartesian coordinates on the $(u, v)$-plane and $x$, $y$ as cartesian coordinates on the $(x, y)$-plane.

Let $M$ be some subset of $G$ (for example, an open or closed region $\Gamma$ contained together with its boundary in $G$). The functions (1) clearly define a mapping $T$ of the set $M$ on the $(u, v)$-plane onto some set $M^*$ on the $(x, y)$-plane.

The mapping $T$ is said to be r e g u l a r on $M$ if the following conditions are satisfied: 1) $T$ is a topological mapping of $M$, i.e., the functions (1) have single-valued inverses $u = \tilde{f}(x, y)$, $v = \tilde{g}(x, y)$ at the points of $M$, defining there the inverse mapping $T^{-1}$; 2) the Jacobian

$$\frac{D(f, g)}{D(u, v)} = \begin{vmatrix} f_u' & f_v' \\ g_u' & g_v' \end{vmatrix}$$

does not vanish at any point of $M$.*

---

\* If functions (1) are defined on a set $M$ and $\dfrac{D(x, y)}{D(u, v)} \neq 0$ at all points of the set, this does not necessarily imply that the mapping of $M$ onto its image $M^*$ is one-to-one.

A regular mapping is said to be of class $C_k$ (analytic) if the functions $f(u, v)$, $g(u, v)$ are of class $C_k$ (analytic).

*Theorem VIII.* *Let T be a regular mapping on a set M. Then:*
*1) The inverse mapping $T^{-1}$ is regular on the image $M^*$ of $M$, and if $T$ is of class $C_k$ (analytic) then so is $T^{-1}$. 2) $T$ maps interior points of $M$ onto interior points of $M^*$. 3) $T$ maps boundary points of $M$ onto boundary points of $M^*$.*

The proof of this theorem, which is omitted here* (see /59/, /60/), is based on the implicit function theorem. In particular, the fact that the Jacobian of the inverse mapping $u = \tilde{f}(x, y)$, $v = \tilde{g}(x, y)$ is also nonvanishing follows from the fact that by formula (7) of §4

$$\frac{D(\tilde{f}, \tilde{g})}{D(x, y)} = \frac{1}{\dfrac{D(f, g)}{D(u, v)}} .$$

R e m a r k I. Let $T$ be a mapping defined by functions of type (1), defined at all points of a closed region $\bar{H}$ but regular only at interior points of $H$ (so that the mapping may fail to be one-to-one or the Jacobian may vanish on the boundary of $H$). Then part 2) of the theorem obviously remains valid at the interior points of $H$.

R e m a r k II. A regular mapping of class $C_k$ (or analytic) maps smooth curves of class $C_{k'}$, $k' \leqslant k$ (or analytic curves) onto smooth curves of the same class.

**2. CURVILINEAR COORDINATES.** The interpretation of the system of functions (1) as a mapping $T$ of a set $M$ on the $(u, v)$-plane onto a set $M^*$ on the $(x, y)$-plane is closely bound up with another interpretation of these functions, as a t r a n s f o r m a t i o n   t o   c u r v i l i n e a r   c o o r d i n a t e s. Let

$$u = \tilde{f}(x, y), \qquad v = \tilde{g}(x, y) \tag{2}$$

be the functions inverse to (1).

With each point $m^*(x, y)$ in the set $M^*$ on the $(x, y)$-plane we can associate the pair of numbers $u$, $v$ — the cartesian coordinates of the point $m$ on the $(u, v)$-plane which is mapped onto $m^*$ (the pre-image of $m^*$).

The numbers $u$, $v$ are c u r v i l i n e a r   c o o r d i n a t e s of the point $m^*(x, y)$. Treating $u$ and $v$ as curvilinear coordinates of the points $m^*(x, y)$ of the set $M^*$, we shall say that equations (1) define on $M^*$ a regular t r a n s f o r m a t i o n   o f   c o o r d i n a t e s (of class $k$ or analytic, as the case may be). Equations (1) will be called the transformation formulas, the determinant $D$ the determinant (or Jacobian) of the transformation. Straight lines $u = C$, $v = C$ on the $(u, v)$-plane are mapped onto curves on the $(x, y)$-plane whose parametric equations are, respectively,

$$x = f(C, v), \qquad y = g(C, v), \tag{3}$$
$$x = f(u, C), \qquad y = g(u, C) \tag{4}$$

(the parameters are $u$, $v$). These are the coordinate curves of the curvilinear coordinate system.

---

* Since a regular mapping is a topological mapping [parts 2) and 3)] follow at once from Brouwer's theorem.

Since a regular mapping is one-to-one, there is exactly one curve of each family (3) and (4) through each point of the set $M^*$.

The system of coordinates $u$, $v$ obtained by a regular transformation (1) will also be referred to as a r e g u l a r system of curvilinear coordinates.*

**3. TRANSFORMATION OF COMPONENTS OF A VECTOR UNDER REGULAR MAPPINGS. CONTRAVARIANT VECTOR. TRANSFORMATION OF THE TANGENT VECTOR.** As before, let the functions $x = f(u, v)$, $y = g(u, v)$ define a regular mapping $T$ of some region $H$ on the $(u, v)$-plane onto a region $G$ on the $(x, y)$-plane (where $x$, $y$ and $u$, $v$ are cartesian coordinates).

Let $\widetilde{v}$ be a vector with components $U$, $V$ at some point $\widetilde{P}_0(u_0, v_0)$. Let $P_0$ be the image of $\widetilde{P}_0$ on the $(x, y)$-plane, $P_0 = T(\widetilde{P}_0)$, and let $v$ be the vector at $P_0$ with components

$$X = f'_u(u_0, v_0) U + f'_u(u_0, v_0) U, \qquad Y = g'_u(u_0, v_0) U + g'_v(u_0, v_0) U. \quad (5)$$

We shall say that $v$ corresponds to $\widetilde{v}$ under the mapping $T$, or, alternatively, that t h e  c o m p o n e n t s  of  $\widetilde{v}$  t r a n s f o r m  u n d e r  $T$  a c c o r d i n g  to  f o r m u l a s  (5).

If we treat $u$ and $v$ as curvilinear coordinates of points on the $(x, y)$-plane and $u = \widetilde{f}(x, y)$, $v = \widetilde{g}(x, y)$ are the inverses of the functions (1), then $U$ and $V$ will be the components of $v$ in the curvilinear system, and

$$U = \widetilde{f}'_x(x_0, y_0) X + \widetilde{f}'_y(x_0, y_0) Y, \qquad V = \widetilde{g}'_x(x_0, y_0) X + \widetilde{g}'_y(x_0, y_0) Y. \quad (6)$$

(Formulas (6) are completely symmetric to formulas (5).)

Now let $\widetilde{l}(u = \widetilde{\varphi}(t), v = \widetilde{\psi}(t))$ be a smooth simple arc in the $(u, v)$-plane. We claim that the tangent vector to this arc transforms under $T$ according to formulas (5) (i. e., the tangent vector is a contravariant vector). Indeed, $T$ clearly maps $\widetilde{l}$ onto a smooth simple arc $l$ in the $(x, y)$-plane, defined by

$$x = f(\widetilde{\varphi}(t), \widetilde{\psi}(t)) = \varphi(t), \qquad y = g(\widetilde{\varphi}(t), \widetilde{\psi}(t)) = \psi(t).$$

Let $\widetilde{P}_0(u_0, v_0)$ be a point on $\widetilde{l}$, corresponding to some $t = t_0$, so that $u_0 = \varphi(t_0)$, $v_0 = \psi(t_0)$. The components of the tangent vector at this point are $\widetilde{\varphi}'(t_0)$, $\widetilde{\psi}'(t_0)$. If $P_0$ is the image of $\widetilde{P}_0$ on the $(x, y)$-plane ($P_0 = T(\widetilde{P}_0)$), the components of the tangent vector to $l$ at $P_0$ are obviously

$$\varphi'(t_0) = f'_u(u_0, v_0) \widetilde{\varphi}'(t_0) + f'_v(u_0, v_0) \widetilde{\psi}'(t_0),$$
$$\psi'(t_0) = g'_u(u_0, v_0) \widetilde{\varphi}'(t_0) + g'_v(u_0, v_0) \widetilde{\psi}'(t_0),$$

so that the components of the tangent vector indeed transform according to to (6).

**4. CHANGE OF ANGLE BETWEEN VECTORS UNDER A REGULAR MAPPING. THE  ROLE OF THE JACOBIAN.** Consider two vectors at a point $P_0$ on the $(u, v)$-plane, $\widetilde{v}_1$ with components $U_1$, $V_1$ and $\widetilde{v}_2$ with components $U_2$, $V_2$.

---

* The term "curvilinear coordinates" is employed not only in the r e g u l a r case but also in a more general setting; for example, we speak of "polar coordinates" although the polar coordinate system is not regular. Nevertheless, in the case of polar coordinates the corresponding mapping of a region $0 \leqslant \varphi < 2\pi$, $\varrho > \varrho_0$, $\varrho_0 > 0$, of the $(\varphi, \varrho)$-plane into the $(x, y)$-plane is regular.

By §5.5, the sine of the angle $\widetilde{a}$ between $\widetilde{v}_1$ and $\widetilde{v}_2$ (in this order!) is given by

$$\sin \widetilde{a} = \frac{U_1 V_2 - V_1 U_2}{\sqrt{U_1^2 + V_1^2} \, \sqrt{U_2^2 + V_2^2}} .$$

Now let the images of the vectors $\widetilde{v}_1$ and $\widetilde{v}_2$ on the $(x, y)$-plane under the transformation $T$ be vectors $v_1$ (components $X_1$, $Y_1$) and $v_2$ (components $X_2$, $Y_2$).

The sine of the angle $a$ between $v_1$ and $v_2$ is

$$\sin a = \frac{X_1 Y_2 - X_2 Y_1}{\sqrt{X_1^2 + Y_1^2} \, \sqrt{X_2^2 + Y_2^2}} .$$

Replacing $X_1$, $Y_1$ and $X_2$, $Y_2$ by their expressions in terms of $U_1, V_1$ and $U_2, V_2$ (formula (5)), we get

$$\sin a = \frac{D_0 \sin \widetilde{a}}{\sqrt{X_1^2 + Y_1^2} \sqrt{X_2^2 + Y_2^2}} ,$$

where $D_0$ is the Jacobian of $T$ at the point $P_0$.

Since $D_0 \neq 0$, this clearly implies the following proposition (the $(u, v)$-plane and $(x, y)$-plane are assumed to be compatibly oriented; see §2.5):

Lemma 1. *If the angle between the vectors $\widetilde{v}_1$ and $\widetilde{v}_2$ is not zero, then the angle between the corresponding vectors $\widetilde{v}_1$ and $\widetilde{v}_2$ is also not zero and moreover a) if $D_0 > 0$, the angles $a$ and $\widetilde{a}$ have the same sign (since $\sin a$ and $\sin \widetilde{a}$ have the same sign); b) if $D_0 < 0$, the angles $a$ and $\widetilde{a}$ have opposite signs (since then $\sin a$ and $\sin \widetilde{a}$ have opposite signs).*

The following proposition, which is related to Lemma 1, connects it with the concept of orientation-preserving and orientation-reversing topological mappings.

Lemma 2. *Let $T$ be a regular mapping of a simply connected region $H$ of the $(u, v)$-plane onto a simply connected region $G$ of the $(x, y)$-plane. If $D_0 > 0$, $T$ is an orientation-preserving mapping, and if $D_0 < 0$ it is an orientation-reversing mapping.* *

5. REGULAR MAPPINGS AND REGIONS DEFINING SIDES OF A SMOOTH SIMPLE ARC. As shown in §3.2, the regions defining the different sides of a simple arc may be introduced by means of functions $x = \varphi (s, t)$, $y = \psi (s, t)$ defining a topological mapping of a rectangle in the $(s, t)$-plane with sides parallel to the $t$- and $s$-axes onto a closed region $\overline{G}$ of the $(x, y)$-plane containing the arc $l$.

Suppose that the simple arc $l$, with parametric equations $x = f (s), y = g (s)$, $s \in [a, b]$, is smooth, so that the functions $f (s)$ and $g (s)$ are of class $C_1$. It is then frequently quite natural to require that the functions defining the mapping of the rectangle on the $(s, t)$-plane define a regular mapping of the rectangle $\sigma$.

---

* It is essential that the regions be simply connected. For example, the Jacobian of an inversion $\varrho = 1/\varrho_1$ is negative ($D = -1$), but any positively described circle $C'$ about the origin is mapped into a circle about the origin which is also positively described. In this case, however, the regular transformation is not defined at the origin. But if we consider a simple closed curve not containing the origin in its interior, the lemma implies that the mapping $\varrho = 1/\varrho_1$ inverts the traversal.

Thus, we consider functions $x = \varphi(s, t)$, $y = \psi(s, t)$ satisfying the following conditions:

1) They are defined in a rectangle $\sigma : a \leqslant s \leqslant b$, $|t| \leqslant a$, and are of class $C_1$ in $\sigma$.

2) They define a topological mapping $T$ of $\sigma$ onto a closed region $\bar{G}$ on the $(x, y)$-plane, with $\varphi(s, 0) = f(s)$, $\psi(s, 0) = g(s)$, i.e., the segment $t = 0$, $s \in [a, b]$ is mapped onto the arc $l$.

3) At no point of the rectangle $\sigma$ does the Jacobian $D = \dfrac{D(\varphi, \psi)}{D(s, t)}$ vanish. To fix ideas, we shall assume that $D > 0$ (the case $D < 0$ is treated in analogous fashion).

As mentioned above, we may treat $s$ and $t$ as curvilinear coordinates in the region $\bar{G}$ on the $(x, y)$-plane. The curvilinear coordinate curves $t = \text{const}$ are smooth arcs

$$x = \varphi(s, C), \quad y = \psi(s, C), \tag{7}$$

and these formulas give the arc $l$ when $C = 0$; the curvilinear coordinate curves $s = \text{const}$ are the arcs

$$x = \varphi(C, t), \quad y = \psi(C, t). \tag{8}$$

Since $D > 0$ by assumption, the angle between the coordinate curves (7) and (8) at their common points is always positive, so that $T$ is an orientation-preserving mapping. Hence, by §3.2, we may stipulate that the points of $\bar{G}$ for which $t > 0$ define the positive side of $l$, and the points of $\bar{G}$ for which $t < 0$ define its negative side.

**6. CONSTRUCTION OF FUNCTIONS** $x = \varphi(s, t)$, $y = \psi(s, t)$. In many problems functions $x = \varphi(s, t)$, $y = \psi(s, t)$ satisfying conditions 1) − 3) of the preceding subsection may be constructed in the following natural way. As before, let $x = f(s)$, $y = g(s)$, $s \in [a, b]$, be the parametric equations of the arc $l$. Let $\varphi(s, t)$, $\psi(s, t)$ be functions defined for all $s$ and $t$ such that

$$s \in [a, b], \quad |t| \leqslant \tau \quad (\tau > 0), \tag{9}$$

satisfying the following conditions:

a) They are single-valued, continuous and continuously differentiable for all $s$ and $t$ in the above region;

b)
$$\varphi(s, 0) \equiv f(s), \quad \psi(s, 0) \equiv g(s);$$

c)
$$D_0 = \begin{vmatrix} \varphi_s'(s, 0) & \varphi_t'(s, 0) \\ \psi_s'(s, 0) & \psi_t'(s, 0) \end{vmatrix} \neq 0.$$

To fix ideas, we shall assume that $D_0 > 0$ (the case $D_0 < 0$ is analogous). Evidently, one cannot assume from the start that any functions satisfying conditions a), b), c) will automatically satisfy conditions 1), 2), 3) of §6.5 for those values of $s$ and $t$ for which they are defined. Indeed, under conditions a), b), c) the functions $x = \varphi(s, t)$, $y = \psi(s, t)$ define a family of smooth arcs

$$x = \varphi(C, t) \quad y = \psi(C, t), \tag{10}$$

which are not tangent to the arc $l$ at their common points with $l$ (corresponding to $t = 0$). But when $t \neq 0$, $|t| < \tau$, curves of the family (10) defined by distinct $C$'s may intersect. Condition 2) will then clearly fail to hold. It can be shown, however, that under conditions a), b), c) there exists a number $a \in [0, \tau]$ such that the curves (10) do not intersect for any $|t| \leqslant a$. Namely:

L e m m a  3.  *If the functions*

$$x = \varphi(s, t), \qquad y = \psi(s, t) \tag{11}$$

*satisfy conditions a), b), c), there exists a positive number $a < \tau$ such that for all $s$ and $t$ with*

$$s \in [a, b], \quad |t| \leqslant a, \tag{12}$$

*the functions satisfy conditions 1), 2), 3), thus defining a regular mapping of the rectangle (12) onto a closed region $\bar{g}$ bounded by a simple closed arc whose interior contains all points of $l$ except its endpoints (which lie on the boundary of $\bar{g}$). The segment $t = 0$, $s \in [a, b]$ is mapped onto the arc $l$.*

Proof.  By condition c), there exists $a > 0$ such that for all $s \in [a, b]$, $|t| \leqslant a$, the Jacobian of the functions (11) does not vanish. It remains to show that for all sufficiently small $t$ the mapping defined by (11) is topological, i. e., one-to-one. Since the functions $\varphi(s, t)$ and $\psi(s, t)$ are single-valued, it will suffice to show that for a suitably chosen $a$ any two distinct points $M'(s', t')$ and $M''(s'', t'')$ with $s' \in [a, b]$, $s'' \in [a, b]$, $|t'| < a$, $|t''| < a$ are mapped onto two distinct points $M'(x', y')$ and $M''(x'', y'')$, where

$$x' = \varphi(s', t'), \qquad y' = \psi(t', s'),$$
$$x'' = \varphi(s'', t''), \qquad y'' = \psi(t'', s'').$$

To prove this, we first observe that by condition c) ($D_0 > 0$) there exist $\Delta > 0$ and $a_0 > 0$ so small that

$$D = \begin{vmatrix} \varphi'_s(s_1, t_1) & \psi'_s(s_2, t_2) \\ \varphi'_t(s_3, t_3) & \psi'_t(s_4, t_4) \end{vmatrix} > 0 \tag{13}$$

for arbitrary $s_i$ and $t_j$ such that

$$|s_i - s_k| < \Delta \qquad (i = 1, 2, 3, 4; \; k = 1, 2, 3, 4),$$
$$|t_j| < a_0 \qquad (j = 1, 2, 3, 4). \tag{14}$$

Reasoning indirectly, suppose that for any $a \in (0, a_0)$ there exist points $M'(s', t')$ and $M''(s'', t'')$ with $|t'| < a$, $|t''| < a$ and either $s' \neq s''$ or $t' \neq t''$ which are mapped onto the same point of the $(x, y)$-plane. This means that

$$\varphi(s', t') = \varphi(s'', t''), \qquad \psi(s', t') = \psi(s'', t''), \tag{15}$$

where either $s' \neq s''$ or $t' \neq t''$, and $|t'| < a$, $|t''| < a$. We distinguish between two cases: $|s' - s''| < \Delta$ and $|s' - s''| > \Delta$, where $\Delta$ is the number figuring in (14). In the first case, using (15) and the mean-value theorem we obtain

$$\varphi'_s(\xi_1, \eta_1)(s'-s'')+\varphi'_t(\xi_2, \eta_2)(t'-t'')=0,$$
$$\psi'_s(\xi_3, \eta_3)(s'-s'')+\psi'_t(\xi_3, \eta_3)(t'-t'')=0, \qquad (16)$$

where $s'<\xi_i<s''$, $t'<\eta_i<t''$ $(i=1, 2, 3, 4)$, $|t'|<a$, $|t''|<a$. It follows from (14) and (13) that $|\xi_i-\xi_k|<\Delta$, $|u_j|<a$, and so

$$D=\begin{vmatrix} \varphi'_s(\xi_1, \eta_1) & \psi'_s(\xi_2, \eta_2) \\ \varphi'_t(\xi_3, \eta_3) & \psi'_t(\xi_4, \eta_4) \end{vmatrix} \neq 0$$

by the choice of $\Delta$ and $a$.

But then (16) can be true only when $s'=s''$ and $t'=t''$, contrary to assumption. Thus (15) cannot hold for $|s'-s''|<\Delta$. Now suppose that for arbitrarily small $a$ there exist $s'$, $t'$, $s''$, $t''$ such that $|s'-s''|>\Delta$, $|t'|<a$, $|t''|<a$, for which (15) holds. Then one readily sees that there must exist a sequence of pairs $s'_k$, $t'_k$ $(s''_k, t''_k)$ such that $|s'_k-s''_k|>\Delta$ and

$$\lim_{k\to-\infty} t'_k=0, \qquad \lim_{k\to\infty} t''_k=0,$$

and for all $k$

$$\varphi(s'_k,t'_k)=\varphi(s''_k, t''_k); \qquad \psi(s'_k,t'_k)=\psi(s''_k, t''_k).$$

We may assume without loss of generality that $s'_k$, $s''_k$ converge as $k\to\infty$ to limits $s'_0$ and $s''_0$ (otherwise we need only extract convergent subsequences from the sequences $s'_k$ and $s''_k$). Since $|s'_k-s''_k|>\Delta$ for any $k$, it follows that also $|s'_0-s''_0|>\Delta$. Moreover, by the continuity of the functions $\varphi(t, s)$, $\psi(t, s)$ and condition b), we must have

$$\varphi(s'_0, 0)=f(s'_0)=\varphi(s''_0, 0)=f(s''_0), \qquad \psi(s'_0, 0)=g(s'_0)=\psi(s''_0, 0)=g(s''_0).$$

But $l$ is a simple arc, so that for no $s'_0$, $s''_0$ with $s'_0 \neq s''_0$ can it be true that

$$f(s'_0)=f(s''_0), \qquad g(s_0)=g(s''_0).$$

Thus we again arrive at a contradiction, and so for some choice of $a>0$ the functions (11) define a regular mapping of the rectangle $s \in [a, b]$, $|t| \leqslant a$ onto some closed region $\bar{g}$ on the $(x, y)$-plane. Moreover, the boundary of the rectangle is mapped onto the boundary of $\bar{g}$, which is a simple closed curve, and the straight-line segment $t=0$, $s \in [a, b]$, is mapped onto $l$. Q.E.D.

In particular, it follows immediately from the lemma that for a simple arc of class $C_2$* and sufficiently small $t$ $(|t|<a, a>0)$ the functions

$$x=f(s)+tg'(s), \qquad y=g(s)-tf'(s)$$

---

* This is generally the case for the arcs considered in the text. If the functions $f(s)$ and $g(s)$ are differentiable to first but not to second order, the functions (17) will not satisfy conditions a), b), c). One can then replace $f'(s)$ and $g'(s)$ by trigonometric polynomials $P_1(s)$ and $P_2(s)$ approximating them to within arbitrary accuracy, and then the functions

$$x=f(s)+tP_2(s), \; y=g(s)-tP_1(s)$$

will satisfy conditions 1), 2), 3).

(in this case the arcs $s =$ const are segments on the normals to $l$) satisfy conditions 1), 2), 3).

**7. INTERSECTION OF TWO SMOOTH ARCS AND INTERSECTION OF A SMOOTH ARC WITH A SMOOTH OR PIECEWISE-SMOOTH SIMPLE CLOSED CURVE.** Let $l$ be a smooth simple arc,

$$x = f(s), \qquad y = g(s), \qquad s \in [a, b],$$

having a point $M_0$ in common with a simple smooth arc $\lambda$ (not an endpoint of either $l$ or $\lambda$). Let

$$x = F(u), \qquad y = G(u), \qquad u \in [a_1, b_1]$$

be the parametric equations of $\lambda$. Let the point $M_0$ correspond to parameter values $s_0 \in (a, b)$ and $u_0 \in (a_1, b_1)$. By Lemma 1 of §5, the arc $\lambda$ has no points other than $M_0$ in common with $l$, provided $u$ is sufficiently near $u_0$. Let $\varphi(s, t)$, $\psi(s, t)$ be two functions defining the different sides of $l$ and satisfying conditions 1), 2), 3) of §6.5.

*Lemma 4. Under the above assumptions, if the angle between $l$ and $\lambda$ at $M_0$ is positive (negative), then $\lambda$ crosses from the negative (positive) side of $l$ to its positive (negative) side.*

Proof. Let $\alpha$ be the angle between $l$ and $\lambda$ at $M_0$. Assume that $\alpha > 0$ (the case $\alpha < 0$ is treated similarly). Then

$$\sin \alpha = \frac{f'(s_0) G'(u_0) - g'(s_0) F'(u_0)}{\sqrt{f'(s_0)^2 + g'(s_0)^2} \, \sqrt{F'(u_0)^2 + G'(u_0)^2}} > 0.$$

For all $u$ sufficiently close to $u_0$, each point of $\lambda$ determines a pair $s, t$ such that

$$\varphi(s, t) = F(u), \qquad \psi(s, t) = G(u). \tag{17}$$

Since $M_0$ is common to $l$ and $\lambda$, these equations are satisfied for $t = 0$, $s = s_0, u = u_0$. Now $t$ and $s$ may be determined from (17) as functions of $u$. To prove the lemma, therefore, it is enough to show that $dt/du > 0$ at $M_0$, i. e., for $u = u_0$. But

$$\varphi_s'(s, t) \frac{ds}{du} + \varphi_t'(s, t) \frac{dt}{du} = F'(u),$$

$$\psi_s'(s, t) \frac{ds}{du} + \psi_t'(s, t) \frac{dt}{du} = G'(u). \tag{18}$$

Setting $t = 0$, $s = s_0$, $u = u_0$ in these equations and recalling that by the conditions imposed on the functions $\varphi(s, t)$, $\psi(s, t)$ we have

$$\varphi_s'(s, 0) = f'(s), \qquad \psi_s'(s, 0) = g'(s),$$

we obtain from equations (18)

$$\left( \frac{dt}{du} \right)_{u = u_0} = \frac{f'(s_0) G'(u_0) - g'(s_0) F'(u_0)}{D(s_0, t_0)} > 0,$$

which proves the lemma.

Let $C$ be a simple closed curve, on which we have a smooth simple subarc $l$. This is obviously possible when $C$ is smooth or piecewise-smooth. But even when $C$ is neither smooth nor piecewise-smooth it may contain a smooth arc (for example, if it is the union of one smooth arc and one nonsmooth arc which is not even piecewise-smooth).

Let $\lambda$ be a simple smooth arc and

$$x = f(u), \qquad y = g(u), \qquad u \in [a, b]$$

its parametric equations. Let $\lambda$ intersect $C$ at an interior point $M_0$ of $l$, for $u = u_0$. We stipulate that the positive traversal on the curve is that defined as in §2.5, and the positive direction on the arc $l$ that induced by the positive traversal of $C$.

The following lemma follows from Lemma 3 and Propositions III and IV in §3.3:

*Lemma 5. If the angle between $l$ and $\lambda$ at $M_0$ is positive, then, with increasing $u$ (from $u < u_0$ to $u > u_0$) the arc $\lambda$ enters the curve $C$; if the angle between $l$ and $\lambda$ is negative the arc $\lambda$ leaves $C$ with increasing $u$.*

*Converse: If an arc $\lambda$ intersecting the arc $l$ at a point $M_0$ enters $C$ with increasing $u$, then the angle between $\lambda$ and $l$ is positive; if $\lambda$ leaves $C$ with increasing $u$, then the angle is negative.*

Now assume that the curve $C$ is piecewise-smooth. Let $l_1$ and $l_2$ be two simple smooth subarcs of $C$, with a common endpoint $O$. Assuming the positive direction on $l_1$ and $l_2$ to be that induced by the positive traversal of $C$, suppose that the angle between $l_1$ and $l_2$ at $O$ is neither zero nor $\pi$. Let $\lambda$ be a simple smooth arc whose parametric equations are

$$x = f(u), \qquad y = g(u),$$

having the point $O$ in common with $C$. Let the angles between $l_1$ and $\lambda$ and between $l_2$ and $\lambda$ at $O$ be positive. Then:

*Lemma 6. Under the above assumptions, the arc $\lambda$ enters $C$ at $O$ with increasing $u$.*

The proof proceeds by construction of suitable functions satisfying conditions 1), 2), 3) of §6.5 for each of the arcs.

**8. CONSTRUCTION OF FUNCTIONS WITH PRESCRIBED PROPERTIES.** In this subsection we cite without proof two propositions on the construction of functions. Let $M_1 (x_1, y_1)$ and $M_2 (x_2, y_2)$ be two points on the $(x, y)$-plane such that $x_1 < x_2$. Let $k_1$ and $k_2$ be two arbitrary numbers.

*Lemma 7. For any $h > 0$, there exists a function $y = f(x)$, defined for $x \in [x_1, x_2]$, of class $C_1$ and such that: a) $f(x_1) = y_1$, $f(x_2) = y_2$, $f'(x_1) = k_1$, $f'(x_2) = k_2$; b) if $y_1 \leqslant y_2$, then $y_1 - h \leqslant f(x) \leqslant y_2 + h$, and if $y_1 > y_2$, then $y_1 + h \geqslant f(x) \geqslant y_2 - h$ (Figure 338).*

Let $\tau(x)$ be a function defined and of class $C_1$ on the interval $[x_1, x_2]$. Let $y_2$ be such that $y_2 < \tau(x)$ for all $x \in [x_1, x_2]$ and $k_2$ an arbitrary negative number.

*Lemma 8. There exists a single-valued function $f(x)$ defined for $x \in [x_1, x_2]$, of class $C_1$, such that:*

*a) $f(x_1) = \tau(x_1)$, $f(x_2) = y_2$;*

*b) $f'(x_1) = k_2 + \tau'(x_1)$, $f'(x_2) = k_2$;*

*c) $y_2 < f(x) < \tau(x)$ for all $x \in (x_1, x_2)$ (Figure 339).*

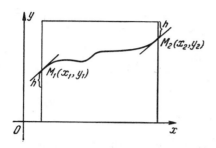

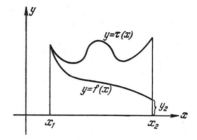

FIGURE 338.

FIGURE 339.

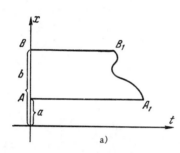

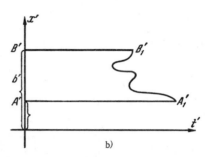

a)

b)

FIGURE 340.

*Lemma 9. Let $\bar{g}$ be a closed region on the $(t, x)$-plane, defined by $a \leqslant x \leqslant b$, $0 \leqslant t \leqslant \varphi(x)$ (Figure 340a), and $\bar{g}'$ a closed region on the $(t',x')$-plane, defined by $a' \leqslant x' \leqslant b'$ $0 \leqslant t' \leqslant \psi(x')$ (Figure 340b). The functions $\varphi(x)$ and $\psi(x')$ are assumed to be single-valued and continuous. Assume that the following conditions hold:*

*1) There exists a topological mapping of $[a, b]$ onto $[a', b']$ (i.e., of the segment $AB$ onto the segment $A'B'$; see Figure 340), defined by a function $x' = \gamma(x)$ (so that $x$ is uniquely determined by $x'$, $x = \gamma^{-1}(x')$ ($\gamma(x)$ and $\gamma^{-1}(x')$, are single-valued and continuous, and moreover $\gamma(a) = a'$, $\gamma(b) = b'$).*

*2) There exists a topological mapping, defined by a function $t' = f_1(t)$, of the segment $0 \leqslant t \leqslant \varphi(a)$ on the straight line $x = a$ onto the segment $0 \leqslant t' \leqslant \psi(a')$ on the straight line $x' = a'$ (the segments $AA_1$ and $A'A_1'$ in Figure 340) (so that $t$ is uniquely determined by $t'$, $t = f_1^{-1}(t)$; $f_1(t)$ and $f_1^{-1}(t')$ are continuous and single-valued, and moreover $f_1(0)=0$, $f_1(\varphi(a)) = \psi(a')$).*

*3) There exists a topological mapping, defined by a function $t' = f_2(t)$, of the segment $0 \leqslant t \leqslant \varphi(b)$ on the straight line $x = b$ onto the segment $0 \leqslant t' \leqslant \psi(b')$ on the straight line $x' = b'$ (the segments $BB_1$ and $B'B_1'$ in Figure 340) (so that $t$ is uniquely determined by $t'$, $t = f_2^{-1}(t')$; $f_2(t)$ and $f_2^{-1}(t')$ are single-valued and continuous and moreover*

$$f_2(0)=0, \qquad f_2(\varphi(b))=\psi(b')).$$

*Then there exists a topological mapping of $\bar{g}$ onto $\bar{g}'$ whose restrictions to the segments specified in conditions 1), 2), 3) are the given mappings.*

495

Proof. We shall simply indicate functions defining the required mapping. For example, consider the functions

$$t' = \psi\left(\gamma(x)\right) \left\{ \frac{f_1\left(\frac{t}{\varphi(x)}\varphi(a)\right)}{\psi(a')} + \frac{x-a}{b-a}\left[ \frac{f_2\left(\frac{t}{\varphi(x)}\varphi(b)\right)}{\psi(b')} - \frac{f_1\left(\frac{t}{\varphi(x)}\varphi(a)\right)}{\psi(a')} \right]\right\} = F(t, x) \tag{19}$$

$$(t \in [0, \varphi(x)], \qquad x \in [a, b] .$$

These functions are single-valued and continuous and define a mapping of $\bar{g}$ onto $\bar{g}'$. We claim that this mapping is one-to-one. Suppose that this is false, so that two distinct points $(x_1, t_1)$ and $(x_2, t_2)$ are mapped onto the same point $(x_0', t_0')$. But by the properties of $\gamma(x)$ (see condition 1)) no two values $x_1$ and $x_2$ can correspond to the same $x_0$, so that necessarily $x_1 = x_2$. Since by assumption $(x_1, t_1)$ and $(x_2, t_2)$ are distinct points, $t_1 \neq t_2$. To fix ideas, let $t_2 > t_1$. Thus, for some $x \in [a, b]$

$$F(t_1, x) = F(t_2, x). \tag{20}$$

Using (19) and rearranging, we get

$$\frac{b-x}{x-a} = -\frac{\psi(a')\left[f_2\left(\frac{t_2}{\varphi(x)}\varphi(b)\right) - f_2\left(\frac{t_1}{\varphi(x)}\varphi(b)\right)\right]}{\psi(b')\left[f_1\left(\frac{t_2}{\varphi(x)}\varphi(a)\right) - f_1\left(\frac{t_1}{\varphi(x)}\varphi(b)\right)\right]}. \tag{21}$$

The left-hand side of this equality is clearly positive (since $x \in [a, b]$). In addition, the functions $f_1\left(\frac{t}{\varphi(x)}\varphi(a)\right)$ and $f_2\left(\frac{t}{\varphi(x)}\varphi(b)\right)$ are increasing (for otherwise the mappings defined in 2) and 3) would not be one-to-one). It follows that the right-hand side of (21) is negative, which is absurd. Thus the function (19) defines a topological mapping of $\bar{g}$ onto $\bar{g}'$. Q.E.D.


§7. THE SPHERE IN EUCLIDEAN SPACE

1. **NEIGHBORHOODS ON THE SPHERE.** Let $S$ be a sphere of radius $r$ in a euclidean space $E_3$ referred to cartesian coordinates $x, y, z$.

A n e i g h b o r h o o d of a point $P$ on the sphere is defined to be the intersection of a neighborhood of $P$ in $E_3$ with sphere $S$. Once this definition has been introduced, we can obviously define on the sphere points of accumulation, interior and boundary points, closed and open sets, regions, and so on. The sphere is compact, i.e., any infinite sequence of points on the sphere has at least one point of accumulation. The sphere as a whole is simultaneously a region and a closed set.*

Use is made in the text of the "stereographic projection of the sphere," which we now proceed to describe. Without loss of generality, we may assume that the center of $S$ is at the origin, so that its equation is

---

* [The reader conversant with topology will recognize that the author is simply defining the relative topology on the sphere.]

$$x^2 + y^2 + z^2 = r^2.$$

Join the point $N$ $(0, 0, r)$ (the n o r t h  p o l e) by a straight line to any other point $P$ on the sphere. The point $P'$ at which this straight line cuts the plane $z = 0$ (the e q u a t o r i a l  p l a n e) is called the stereographic projection of $P$ o n  t h e  e q u a t o r i a l  p l a n e. It is obvious that every point of the sphere except $N$ (which is known as the c e n t e r  o f  t h e  p r o j e c t i o n) has a stereographic projection.

Conversely, if $P'$ is any point on the $(x, y)$-plane, the straight line through $P'$ and the point $N$ on the sphere cuts the sphere at exactly one point $P$. This point is known as the stereographic projection of $P'$ o n the s p h e r e. Letting $\xi$ and $\eta$ denote the cartesian coordinates of a point $P'$ on the equatorial plane, the stereographic projection of a point $P$ $(x, y, z)$ of the sphere, we have

$$\xi = \frac{xr}{r-z}, \qquad \eta = \frac{yr}{r-z} \tag{1}$$

and conversely

$$x = \frac{2r^2\xi}{\xi^2+\eta^2+r^2}, \qquad y = \frac{2r^2\eta}{\xi^2+\eta^2+r^2}, \qquad z = \frac{(\xi^2+\eta^2-r^2)\,r}{\xi^2+\eta^2+r^2}. \tag{2}$$

The functions $(1)$ are single-valued and continuous for $z \neq r$ and therefore define a topological mapping of the sphere punctured at $N$ onto the plane. One can also consider the stereographic projection of points of the sphere not onto the equatorial plane but onto any other plane parallel to the equator. A frequently used variant is stereographic projection onto a plane parallel to the equator and tangent to the sphere at the point antipodal to the north pole (the center of projection).

Equations $(2)$ may be treated as parametric equations of the sphere.

If $M$ is any set on the sphere and $M'$ the set on the plane $z = 0$ consisting of the stereographic projections of the points of $M$, then $M'$ is called the stereographic projection of the set $M$ on the $(x, y)$-plane. Conversely, $M$ is the stereographic projection of the plane set $M'$ on the $(x, y)$-plane onto the sphere $S$. Clearly, the stereographic projection of an interior (boundary) point of $M$ is an interior (boundary) point of $M'$, and conversely.

T h e o r e m  IX. *Any proper subset $M$ of the sphere is homeomorphic to a plane set.*

P r o o f. Let $M_0$ be a point not in the set $M$. The coordinates $x, y, z$ may always be so chosen that $M_0$ has coordinates $(0, 0, z)$ (the north pole). The theorem is then proved by stereographic projection of the sphere from $M_0$ onto the plane.

**2. SIMPLE ARCS AND SIMPLE CLOSED CURVES ON THE SPHERE.** A simple arc (simple closed curve) on the sphere $S$ is a simple arc (simple closed curve) in space all of whose points lie on $s$. The stereographic projection of a simple arc (simple closed curve) on the sphere is a simple arc (simple closed curve) on the plane, and conversely. We have the following propositions:

T h e o r e m  X. *A simple arc and simple closed curve on the sphere are nowhere dense on the sphere.*

*Theorem XI. A simple arc does not separate the sphere (i.e., the complement of a simple arc in the sphere is a region).*

*Theorem XII. A simple closed curve defines two regions on the sphere and is the common boundary of these two regions.*

The properties of the sphere formulated in the last theorem are topologically invariant, i.e., are preserved under any topological mapping of the sphere.

A surface homeomorphic to the sphere is called a surface of genus zero (or simply connected surface). Only on such surfaces does every simple closed curve define two regions of which it is the common boundary.

**3. ATLAS ON THE SPHERE. COORDINATES ON THE SPHERE.** We now examine the introduction of coordinates on the sphere. First we need the concept of a cover of the sphere.

By a finite open cover $\Sigma$ of the sphere we mean a finite collection of regions $G_1, G_2, \ldots, G_N$ on the sphere, possessing the following properties: 1) no region $G_i$ is the entire sphere; 2) each point of the sphere lies in at least one of these regions. In the sequel we shall usually omit the adjectives and speak simply of a cover of the sphere.*

Since by definition each member $G_i$ of a cover of the sphere is a proper subset of the sphere, $G_i$ is homeomorphic to a plane region. Every cover contains at least two regions (for otherwise the single element of the cover would have to coincide with the entire sphere, contrary to the definition).

In this book we frequently employ a special cover consisting of two regions: the sphere punctured at a point $M$, and any region containing $M$ (such as the sphere punctured at one point other than $M$).

Let $G$ be a region in a given cover of the sphere; it may be any proper subregion of the sphere. We shall say that a regular system of local coordinates $u$, $v$ of class $C_k$ (or analytic) is defined in $G$ if we have a mapping $T$:

$$x = \varphi(u, v), \quad y = \psi(u, v), \quad z = \chi(u, v) \tag{3}$$

of some region $H$ on the $(u, v)$-plane onto $G$, with the following properties: a) $T$ is a topological mapping of $H$ onto $G$; b) the functions $\varphi(u, v), \psi(u, v), \chi(u, v)$ are of class $C_k$ (or analytic); c) at no point of $H$ do the Jacobians

$$D_1 = \frac{D(\psi, \chi)}{D(u, v)}, \quad D_2 = \frac{D(\chi, \varphi)}{D(u, v)}, \quad D_3 = \frac{D(\varphi, \psi)}{D(u, v)}$$

vanish simultaneously.

Equations (3) may be treated as parametric equations of the region $G$ on the sphere.

If regular coordinates of class $C_k$ (analytic) are defined in all regions of a given cover, we shall say that an atlas of class $C_k$ (analytic) is defined on the sphere, and the regions in the cover will be called **charts**

---

* One can of course consider not only finite but also infinite covers of the sphere (i.e., consisting of an infinite collection of regions). But by the Heine-Borel theorem any such cover must contain a finite subcover.

(or coordinate neighborhoods).*,** The sphere may be provided with different atlases, depending on the specific cover and mappings (3) used.

Now let $G$ and $\tilde{G}$ be two charts with local coordinates $u$, $v$ and $\tilde{u}$, $\tilde{v}$, respectively, and assume that their intersection $W = G \cap \tilde{G}$ is not empty ($G$ and $\tilde{G}$ may be charts of the same or different atlases).

The coordinates $u$, $v$ are defined in $G$ by (3) and in $\tilde{G}$ by formulas

$$x = \tilde{\varphi}(\tilde{u}, \tilde{v}), \qquad y = \tilde{\psi}(\tilde{u}, \tilde{v}), \qquad y = \tilde{\chi}(\tilde{u}, \tilde{v}), \tag{4}$$

defining a mapping $\tilde{T}$ of a region $\tilde{H}$ on the $(\tilde{u}, \tilde{v})$-plane onto the region $\tilde{G}$ of the sphere. The functions $\tilde{\varphi}, \tilde{\psi}, \tilde{\chi}$ satisfy the same conditions as $\varphi, \psi, \chi$. In particular, at least one of the Jacobians $\tilde{D}_1, \tilde{D}_2, \tilde{D}_3$ does not vanish.

Let $M$ be an arbitrary point in the region $W$. Then:

Lemma 1. *If* $D_1 \neq 0$ ($D_2 \neq 0$, $D_3 \neq 0$, respectively) *at the point* $M$, *then* $\tilde{D}_1 \neq 0$ ($\tilde{D}_2 \neq 0$, $\tilde{D}_3 \neq 0$, respectively) *at* $M$.

Proof. Observe first that if

$$F(x, y, z) = (x-a)^2 + (y-b)^2 + (z-c)^2 - r^2 = 0$$

is the equation of the sphere in cartesian coordinates, then

$$F(\varphi, \psi, \chi) \equiv 0, \qquad F(\tilde{\varphi}, \tilde{\psi}, \tilde{\chi}) \equiv 0. \tag{5}$$

Hence

$$F'_x \varphi'_u + F'_y \psi'_u + F'_z \chi'_u = 0, \qquad F'_x \varphi'_v + F'_y \psi'_v + F'_z \chi'_v = 0 \tag{6}$$

and

$$F'_x \tilde{\varphi}'_{\tilde{u}} + F'_y \tilde{\psi}'_{\tilde{u}} + F'_z \tilde{\chi}'_{\tilde{u}} = 0, \qquad F'_x \tilde{\varphi}'_{\tilde{v}} + F'_y \tilde{\psi}'_{\tilde{v}} + F'_z \tilde{\chi}'_{\tilde{v}} = 0. \tag{7}$$

It follows from (6) and (7) that

$$F'_x : F'_y : F'_z = D_1 : -D_2 : D_3 \tag{8}$$

and

$$F'_x : F'_y : F'_z = \tilde{D}_1 : -\tilde{D}_2 : \tilde{D}_3, \tag{9}$$

whence, in view of the fact that

$$(F'_x)^2 + (F'_y)^2 + (F'_z)^2 \neq 0$$

the assertion follows. Q.E.D.

---

* [Rather than the author's original "coordinate cover," "region of a coordinate cover," we prefer to use the more compact general terminology of the theory of manifolds, albeit rather loosely. Strictly speaking, a chart is a pair $(G, T^{-1})$ (in the above notation) and an atlas is a collection of charts satisfying certain additional conditions (see, e. g., Lang, S., Introduction to Differentiable Manifolds. New York, Interscience Publishers. 1962, p.16).]

** With an atlas available on the sphere, one can define "neighborhoods" of points as the images of neighborhoods of points on the $(u, v)$-plane. This definition is essentially the same as the previous one.

Again, let $W$ be the intersection of two charts $G$ and $\tilde{G}$. With each point $M$ of $W$ we can associate two coordinate pairs, $u$, $v$ and $\tilde{u}$, $\tilde{v}$. Thus, associating with the coordinates $u$, $v$ of a point $M \in W$ its coordinates $\tilde{u}$, $\tilde{v}$, we clearly obtain a one-to-one correspondence between the pairs $u$, $v$ and $\tilde{u}$, $\tilde{v}$ for points of $W$. This correspondence can be written either

$$\tilde{u} = f(u, v), \qquad \tilde{v} = g(u, v),$$

or

$$u = f^{-1}(\tilde{u}, \tilde{v}), \qquad v = g^{-1}(\tilde{u}, \tilde{v}).$$

The functions $f$, $g$, $f^{-1}$, $g^{-1}$ may be determined from the equations

$$\varphi(u, v) = \tilde{\varphi}(\tilde{u}, \tilde{v}), \qquad \psi(u, v) = \tilde{\psi}(\tilde{u}, \tilde{v}), \qquad \chi(u, v) = \tilde{\chi}(\tilde{u}, \tilde{v}),$$

so that we have identities

$$\varphi(u, v) \equiv \tilde{\varphi}(f, g), \qquad \psi(u, v) \equiv \tilde{\psi}(f, g), \qquad \chi(u, v) \equiv \tilde{\chi}(f, g)$$

and analogous identities for $f^{-1}$, $g^{-1}$.

With this notation, we have

Lemma 2. *If the functions $\varphi$, $\psi$, $\chi$ and $\tilde{\varphi}$, $\tilde{\psi}$, $\tilde{\chi}$ satisfy conditions a), b), c), then the functions*

$$\tilde{u} = f(u, v), \qquad \tilde{v} = g(u, v) \tag{10}$$

$$(or \ u = f^{-1}(\tilde{u}, \tilde{v}), \quad v = g^{-1}(\tilde{u}, \tilde{v})) \tag{11}$$

*define a regular transformation of coordinates in the region $W$, of class $C_k$ or analytic, as the case may be, so that*

$$\frac{D(f, g)}{D(u, v)} \neq 0. \tag{12}$$

Proof. By the implicit function theorem, it will clearly suffice to prove (12). By condition c), at least one of the Jacobians $D_1$, $D_2$, $D_3$ does not vanish, say $D_1 \neq 0$. Then by Lemma 1 also $\tilde{D}_1 \neq 0$. But then the first two equations of (3) imply (see Remark III to Theorem VII)

$$\frac{D(\varphi, \psi)}{D(u, v)} = \frac{D(\tilde{\varphi}, \tilde{\psi})}{D(f, g)} \frac{D(f, g)}{D(u, v)}.$$

It follows from this equality that $\dfrac{D(f, g)}{D(u, v)} = \dfrac{D_1}{\tilde{D}_1} \neq 0$. Q.E.D.

Remark. If $h$ and $\tilde{h}$ are the subregions of the regions $H$ and $\tilde{H}$ on the $(u, v)$- and $(\tilde{u}, \tilde{v})$-planes onto which $W$ is mapped by the functions (3) and (4), respectively, then the functions (10) clearly define a regular mapping of $h$ onto $\tilde{h}$.

**4. SIMPLE ATLASES ON THE SPHERE.** When considering examples, we shall use a certain special system of local coordinates on the sphere, which we now describe. Let $N$ and $\tilde{N}$ be two antipodal points on the sphere, $\sigma$ and $\tilde{\sigma}$

the planes tangent to the sphere at these points. Let $G\,(\widetilde{G})$ be the sphere punctured at $N$ (at $\widetilde{N}$). $G$ and $\widetilde{G}$ form a cover of the sphere, all points other than the poles $N$ and $\widetilde{N}$ lying in both regions. We refer the planes $\sigma$ and $\widetilde{\sigma}$ to cartesian coordinate systems $(u,\ v)$ and $\widetilde{u},\ \widetilde{v}$), which are compatible with each other (the $u$-axis parallel to the $\widetilde{u}$-axis, the $v$-axis parallel to the $\widetilde{v}$-axis) (see Figure 135 in Chapter VI, where $u$, $v$ are denoted by $x$, $y$ and $\widetilde{u},\widetilde{v}$ by $u$, $v$). The coordinates $u$, $v$ of a point $M_0$ in the region $G$ are defined as the coordinates $u$, $v$ of its stereographic projection $M$ from $\widetilde{N}$ as center onto the plane $\sigma$. Similarly, the coordinates $\widetilde{u},\ \widetilde{v}$ of a point $M_0$ in $\widetilde{G}$ are the coordinates $\widetilde{u},\ \widetilde{v}$ of its stereographic projection $\widetilde{M}$ from $N$ as center. We thus get an atlas of the sphere consisting of exactly two charts $G$ and $\widetilde{G}$ with local coordinates $u$, $v$ and $\widetilde{u},\ \widetilde{v}$, respectively. We shall use the [ad hoc] term simple atlas for this atlas. It is easy to see that a simple atlas is analytic. Elementary geometric arguments (similarity of triangles $NM\widetilde{N}$ and $N\widetilde{N}\widetilde{M}$) yield

$$\frac{\varrho}{2r} = \frac{2r}{\widetilde{\varrho}}, \quad \widetilde{\varrho}^2 = \widetilde{u}^2 + \widetilde{v}^2, \quad \varrho^2 = u^2 + v^2,$$

and

$$\cos\varphi = \frac{u}{\varrho} = \frac{\widetilde{u}}{\widetilde{\varrho}}, \quad \sin\varphi = \frac{v}{\varrho} = \frac{\widetilde{v}}{\widetilde{\varrho}}.$$

At any point $M$ lying in both charts $G$ and $\widetilde{G}$, the coordinates $u$, $v$ and $\widetilde{u},\widetilde{v}$ are related by

$$\widetilde{u} = \frac{4ur^2}{u^2+v^2}, \quad \widetilde{v} = \frac{4vr^2}{u^2+v^2}$$

or, equivalently,

$$u = \frac{4\widetilde{u}\,r^2}{\widetilde{u}^2 + \widetilde{v}^2}, \quad v = \frac{4\widetilde{v}\,r^2}{\widetilde{u}^2 + \widetilde{v}^2}.$$

For any proper subset of the sphere (e.g., a simple arc or simple closed curve), one can always choose an atlas, even a simple atlas, one of whose charts contains the entire subset. It is readily shown that the local coordinates $u$ and $v$ ($\widetilde{u}$ and $\widetilde{v}$) are defined on the sphere by means of the parametric equations of the sphere given in §7.1.

**5. ORIENTATION OF THE SPHERE AND TYPES OF TOPOLOGICAL MAPPING ON THE SPHERE.** As on the plane, the sphere may be oriented, in other words, one of the two possible traversals of simple closed curves on the sphere is declared positive. Again, topological mappings of the sphere into itself fall into two categories:

I. Orientation-preserving topological mappings.

II. Orientation-reversing topological mappings.

**6. FUNCTIONS DEFINED ON THE SPHERE.** A function

$$\omega = F\,(Q)$$

is defined on the sphere if a number $\omega$ is associated with each point $Q$ on the sphere. Given any atlas of the sphere, the function $F(Q)$ is of course a function of the local coordinates $u_i$, $v_i$ in each chart $G_i$:

$$\omega = F(Q) = f_i(u_i, v_i).$$

If $G_i$ and $G_k$ are charts with nonempty intersection, with the local coordinates in the intersection related by

$$u_i = \varphi(u_k, v_k), \qquad v_i = \psi(u_k, v_k),$$

then in the region $G_k$

$$\omega = F(Q) = f_k(u_k, v_k) \equiv f_i(\varphi(u_k, v_k), \psi(u_k, v_k)).$$

We shall say that a function $F(Q)$ on the sphere is of class $C_k$ (analytic) if for some atlas of class $C_k$ (analytic atlas) on the sphere, and hence for any atlas of the same class, the functions $f_i(u_i, v_i) \equiv F(Q)$ are of class $C_k$ (analytic) as functions of $u_i$, $v_i$.

## §8. FUNDAMENTAL THEOREMS OF THE THEORY OF DIFFERENTIAL EQUATIONS

In this section we state without proof the fundamental theorems of the theory of differential equations (existence-uniqueness theorem, continuity of solution with respect to initial conditions, etc.), which are utilized in the main text. The proofs may be found, e.g., in /11/, /12/, /61/.

The variables are regarded throughout as cartesian or curvilinear coordinates of points in euclidean space $E_{N+1}$. We consider a system of differential equations

$$\left.\begin{aligned}
\frac{dx_1}{dt} &= X_1(t, x_1, \ldots, x_n), \\
&\cdot \cdot \cdot \cdot \cdot \cdot \cdot \cdot \cdot \cdot \cdot \cdot \\
\frac{dx_n}{dt} &= X_n(t, x_1, \ldots, x_n),
\end{aligned}\right\} \tag{1}$$

where $X_k(t, x_1, \ldots, x_n)$ are functions defined in some (open) region $R$ of the space $(t, x_1, \ldots, x_n)$, continuous and continuously differentiable there with respect to $x_1, \ldots, x_n$. Instead of saying that the "right-hand sides of system (1) are defined in the open region $R$," we shall simply say that "system (1) is defined in the open region $R$."

### 1. EXISTENCE-UNIQUENESS THEOREM.

*Theorem A. For any point $M_0(t_0, x_1^0, \ldots, x_n^0)$ of the (open) region $R$, there exist an interval $[t_1, t_2]$ containing $t_0$ and a system of differentiable functions*

$$x_k = \varphi_k(t) \quad (k = 1, 2, \ldots, n), \tag{2}$$

*defined on $t_1$, $t_2$ such that: a) $\varphi_k(t_0) = x_k^0$; b) for all $t \in [t_1, t_2]$ the points $M(t, \varphi_1(t), \ldots, \varphi_n(t))$ lie in R; c) $\varphi_k'(t) \equiv X_k(t, \varphi_1(t), \ldots, \varphi_n(t))$ for all $t$ in*

$[t_1, t_2]$. *This system of functions is unique; namely, if $[t'_1, t'_2]$ is any closed interval containing $t_0$ and contained in $[t_1, t_2]$ ($t_1 \leqslant t'_1 < t_0$, $t_0 < t'_2 \leqslant t_2$), then any system of differentiable functions defined on $[t'_1, t'_2]$ and satisfying conditions* a), b), c) *must coincide on $[t'_1, t'_2]$ with $\varphi_k (t)$ ($k = 1, 2, \ldots, n$).*

The numbers $t_0, x_1^0, \ldots, x_n^0$ are called i n i t i a l  v a l u e s, the point $M (t_0, x_1^0, \ldots, x_n^0)$ the i n i t i a l  p o i n t, and the solution $x_k = \varphi_k (t)$ is called the  s o l u t i o n  of  system  (1)  for  initial  values  $(t_0, x_1^0, \ldots, x_n^0)$ (or the solution satisfying the initial conditions: $x_2 = x_2^0, \ldots, x_n = x_n^0$ for $t = t_0$), defined on the interval $[t_1, t_2]$.

Since by assumption $R$ is open and by Theorem A the system of functions $\varphi_k (t)$ is defined on a closed interval $[t_1, t_2]$, it follows that the solution (2) can be  c o n t i n u e d  for both $t < t_1$ and $t > t_2$. In other words, taking $M_1 (t_1, \varphi_1 (t_1), \ldots, \varphi_n (t_1))$  or  $M_2 (t_2, \varphi_1 (t_2), \ldots, \varphi_n (t_2))$ as initial point, we use Theorem A to construct a solution of system (1) identical to the solution on $[t_1, t_2]$ but defined on some larger interval $[t_1^*, t_2^*]$ ($t_1^* < t_1$, $t_2^* > t_2$). Further, taking $M_1^* (t_1^*, \varphi_1 (t_1^*), \ldots, \varphi_n (t_1^*))$ or $M_2^* (t_2^*, \varphi_1 (t_2^*), \ldots, \varphi_n (t_2^*))$ as new initial points, we again continue the solution, and so on.

We shall say that the solution $x_k = \varphi_k (t)$ of system (1) has been c o n t i n u e d  t o  t h e  m a x i m u m  p o s s i b l e  i n t e r v a l  $(\tau, T)$ if there is no solution identical with $x_k = \varphi_k (t)$ on the interval $(\tau, T)$ and defined on a larger interval $(\tau', T')$ (i. e., such that either $\tau' < \tau$ or $T' > T$).

T h e o r e m  A'. *If system (1) is defined in an open region $R$, the maximum possible interval $(\tau, T)$ to which the solution can be continued is necessarily open; moreover, for any closed bounded region $\bar{R}_1$ contained (together with its boundary) in $R$, there exist $t$, $t'$ and $t''$, $t' > \tau$ and $t'' < T$, such that the points $M_1 (t', \varphi_1, (t'), \ldots, \varphi_n (t'))$ and $M_2 (t'', \varphi_1 (t''), \ldots, \varphi_n t'')$ are not in $\bar{R}_1$.*

The last property is often phrased as follows:  t h e  s o l u t i o n  c a n  b e  c o n t i n u e d  u p  t o  t h e  b o u n d a r y  o f  i t s  d o m a i n  o f  d e f i n i t i o n.

If the solution is defined for all $t$, $-\infty < t < +\infty$, we shall write $\tau = -\infty$, $T = +\infty$.

*Throughout this book, the term "solution" for a system of type (1) will always mean the solution defined on the maximum possible $t$-interval.*

The set of points $M (t, \varphi_1 (t), \ldots, \varphi_n (t))$, $t \in (\tau, T)$, is known as an integral curve of system (1). By Theorem A, exactly one integral curve passes through each point $M (t_0, x_1^0, \ldots, x_n^0)$ of the region $R$.

The solution of system (1) depends on the initial values $t_0, x_1^0, \ldots, x_n^0$. A natural notation for the solution is therefore

$$x_k = \varphi_k (t, t_0, x_1^0, \ldots, x_n^0), \qquad (3)$$

where the functions $\varphi_k$ are defined at all points $M_0 (t_0, x_1^0, \ldots, x_n^0)$ of $R$ and for all $t$ in some interval $(\tau, T)$, which generally depends on the point $M_0$. Thus the functions (3) are defined throughout some region of the $(n+2)$-dimensional space $(t, t_0, x_1^0, \ldots, x_n^0)$. For any fixed $t_0, x_1^0, \ldots, x_n^0$, they are solutions of system (1) satisfying the initial conditions: $x_k = x_k^0$ ($k = 1, 2, \ldots$) for $t = t_0$. By definition, we have

$$\varphi_k (t_0, t_0, x_1^0, \ldots, x_n^0) = x_k^0.$$

If $t_0$, $x_1^0$, ..., $x_n^0$ are treated as arbitrary parameters (subject, of course, to the condition that the point $M_0$ $(t_0, x_1^0, ..., x_n^0)$ lie in the region $R$), the system of functions is sometimes referred to as the general solution of system (1). When $t_0$, $x_1^0$, ..., $x_n^0$ are fixed, we shall sometimes call the functions (2) a particular solution (so that "solution" and "particular solution" are synonymous terms). If

$$x_k = \varphi_k (t, t_0, x_1^0, ..., x_n^0) \qquad (4)$$

is a solution such that for some $t = t_1$ in its interval of definition

$$\varphi_k (t_1, t_0, x_1^0, ..., x_n^0) = x_k', \qquad (5)$$

then the solution (4) may also be written as follows (in view of the uniqueness of the solution for given initial values):

$$x_k = \varphi_k (t, t_1, x_1', ..., x_n'). \qquad (6)$$

The next theorem is of fundamental importance.

## 2. CONTINUITY OF THE SOLUTION WITH RESPECT TO INITIAL VALUES.

*Theorem B. Let* $x_k = \varphi_k (t, t^*, x_1^*, x_2^*, ..., x_n^*)$ *be any solution of system (1), defined for all* $t \in (t_1, t_2)$, *and let* $\tau_1$ *and* $\tau_2$ *be arbitrary numbers in this interval,* $\tau_1 < \tau_2$. *Then, for any* $\varepsilon > 0$, *there exists* $\delta > 0$ $(\delta = \delta (\varepsilon, \tau_1, \tau_2))$ *such that, for all* $t_0, x_1^0, ..., x_n^0$ *satisfying the inequalities* $|t_0 - t^*| < \delta, |x_i^* - x_i^0| < \delta (i = 1, 2, ..., n)$, *the solution* $x_k = \varphi_k (t, t_0, x_1^0, ..., x_n^0)$ *is defined for all* $t \in [\tau_1, \tau_2]$ *and for all these values of*

$$| \varphi_k (t, t_0, x_1^0, ..., x_n^0) - \varphi_k (t, t^*, x_1^*, ..., x_n^*) | < \varepsilon.$$

Corollary. The functions $x = \varphi_k (t, t_0, x_1^0, ..., x_n^0)$ are jointly continuous in $t$, $t_0$, $x_1^0$, ..., $x_n^0$ throughout their domain of definition.

## 3. DERIVATIVES WITH RESPECT TO INDEPENDENT VARIABLE AND INITIAL VALUES.

We have assumed hitherto that the functions $X_k (t, x_1, ..., x_n)$ on the right of the equations of system (1) are continuously differentiable. We now assume that the functions $X_k (t, x_1, ..., x_n)$ have continuous derivatives with respect to $t$ and $x_k$ of order up to $p$ for some $p \geqslant 1$.

*Theorem B'. Under the above assumption, the functions* $x_k = \varphi_k (t)$ $(k = 1, 2, ..., n)$ *have continuous derivatives with respect to* $t$ *of order up to* $p + 1$.

Considering the solution as a function of the initial values, we have

*Theorem B''. If the functions* $X_k (t, x_1, ..., x_n)$ *have partial derivatives with respect to* $x_1, x_2, ..., x_n, t$ *of order up to* $p$, *then the functions* $x_k = \varphi_k(t, t_0, x_1^0, ..., x_n^0)$ *have jointly continuous derivatives throughout their domains of definition: a) of order up to* $p + 1$ *with respect to* $t$ *and* $t_0$, *b) of order up to* $p$ *with respect to the variables* $x_i^0$, *c) of order up to* $p + 1$ *with respect to* $t$, $t_0$ *and the variables* $x_i^0$ *provided at least one differentiation is with respect to* $t$ *or* $t_0$.

These partial derivatives satisfy a certain system of differential equations:

$$\frac{dx_k}{dt} = X_k(t, x_1, \ldots, x_n),$$

$$\frac{d}{dt}\left(\frac{\partial x_k}{\partial t_0}\right) = \sum_{j=1}^{n} \frac{\partial X_k}{\partial x_j}\frac{\partial x_j}{\partial t_0},$$

$$\frac{d}{dt}\left(\frac{\partial x_k}{\partial x_i^0}\right) = \sum_{j=1}^{n} \frac{\partial X_k}{\partial x_j}\frac{\partial x_j}{\partial x_i^0},$$

$$\frac{d}{dt}\left(\frac{\partial^q x_k}{(\partial x_1^0)^{j_1}(\partial x_2^0)^{j_2}\ldots(\partial x_n^0)^{j_n}}\right) = \sum_{j=1}^{n} \frac{\partial X_k}{\partial x_j}\frac{\partial^q x_j}{(\partial x_1^0)^{j_1}(\partial x_2^0)^{j_2}\ldots(\partial x_n^0)^{j_n}} +$$

$$+ G^q_{j_1,\ldots,j_n}\left(x_i, \frac{\partial x_i}{\partial x_j^0}, \ldots\right)$$

$$(j_1 + j_2 + \ldots + j_n = q \leqslant p;\quad k = 1, 2, \ldots, n;\quad i = 1, \ldots, n);$$

$$\frac{d}{dt}\left(\frac{\partial^{q+1} x_k}{\partial t^l\,(\partial x_1^0)^{j_1}\ldots(\partial x_n^0)^{j_n}}\right) = \sum_{j=1}^{n} \frac{\partial X_k}{\partial x_j}\frac{\partial^{q+1} x_j}{\partial t^l\,(\partial x_1^0)^{j_1}\ldots(\partial x_n^0)^{j_n}} +$$

$$+ G^{q+1}_{l,\,j_1,\ldots,j_n}\left(x_i, \frac{\partial x_i}{\partial t_0^l}, \frac{\partial x_i}{\partial x_j^0},\ldots\right)$$

$$(q = 1, 2, \ldots, p;\quad l + j_1 + j_2 + \ldots + j_n = q + 1),$$

where the functions $G^q_{j_1,\ldots,j_n}$ depend on $x_i$ and the derivatives of $x_j$ of order less than $q$ with respect to $x_i^0$, and $G_{l,\,j_1,\ldots,j_n}$ depend on $x_i$ and its derivatives with respect to $t_0$ and $x_j^0$ of order less than $q+1$.

*Theorem C.* If $X_k(t, x_1, \ldots, x_n)$ *are analytic functions in the neighborhood of any point of R, then the functions* $x_k = \varphi_k(t, t_0, x_1^0, \ldots, x_n^0)$ $(k = 1, 2, \ldots, n)$ *are analytic in the neighborhood of any point* $(t, t_0, x_1^0, \ldots, x_n^0)$ *of their domain of definition.*

# §9. ON THE CONCEPT OF "QUALITATIVE STRUCTURE" OF THE PHASE PORTRAIT AND THE CONCEPT OF SINGULAR AND NONSINGULAR PATHS

## 1. COMPARISON OF INVARIANTS OF TOPOLOGICAL AND REGULAR MAPPINGS.

The fundamental problem of the qualitative theory of differential equations on the plane is to determine the properties of the phase portrait of a system

$$\dot{x} = P(x, y), \quad \dot{y} = Q(x, y), \tag{1}$$

which remain invariant under all topological mappings of the plane onto itself (or of some region of the plane onto itself, of the sphere onto itself in the case of dynamic systems on the sphere, etc.).

Now there is a classical field of mathematics — differential geometry — which studies the invariants of regular mappings. It is thus natural to investigate the relevance of invariants of regular mappings for dynamic systems. Though we do not wish to go into details (especially as the very concept of "relevance" is hardly meaningful in this context), we shall nevertheless outline the classification of dynamic systems according to invariants of regular mappings. The reader will readily convince himself that this classification is of little interest for problems arising from applications, such as the classical problem of stability or instability of equilibrium states.

For example, consider an equilibrium state. It was shown in §6, Chapter VI, that the characteristic roots are invariants of a regular mapping, but only as to absolute value and not as to sign. Thus, the classification of equilibrium states by invariants of arbitrary regular mappings yields continuum-many classes, each class consisting of equilibrium states with the same characteristic roots. It is clear that from the standpoint of applications this is of little interest. It is true that determination of the absolute values of the characteristic roots or (what is the same) the directions along which the paths tend to equilibrium states is sometimes valuable (e. g., in approximation of the separatrices, see §15). But in applications one frequently wants to know whether the paths tend to an equilibrium state along definite directions or whether they are spirals. One is then interested in the m e r e f a c t of whether the roots are real or complex, not in their a b s o l u t e v a l u e s. For applications, therefore, classification of equilibrium states by invariance under regular trans-formations is of no interest.

The same is true of limit cycles. Classification of limit cycles by invariance under regular mappings again yields a continuum of classes. Indeed, let $l$ be some arc without contact passing through a point on a closed path, and let $s = f(s)$ be the corresponding succession function (where $s$ is a parameter on the arc $l$). Let $s_0$ be the parameter value corresponding to the point at which $l$ crosses the closed path. It can be shown that the quantity $f'(s_0)$ is invariant under regular mappings. However, in problems encountered in applications one is interested not in $f'(s_0)$ itself but in whether it is positive or negative, i. e., whether the limit cycle is stable or unstable. Again, therefore, the proper tool here is invariants not of regular mappings but of topological mappings.

## 2. DIFFERENT APPROACHES TO DEFINITION OF REGIONS CHARACTERIZED BY "IDENTICAL" PHASE PORTRAITS.

In this subsection we briefly discuss certain concepts appearing in the mathematical literature and related to the concept of o r b i t a l  s t a b i l i t y* introduced in this book, touching also upon the corresponding definition of regions in which the phase portraits are in some sense "similar."

One of the pioneers of this field was Brouwer, who considered the partition of the sphere into regions characterized by "similar" phase portraits for an extremely general case, that of a continuous vector field on the sphere with finitely many singular points. (Since Brouwer assumed the field to be only continuous, and not continuously differentiable as in the present book, more than one path may pass through a nonsingular point.) If Brouwer's classification of regions is applied to our case of a continu-ously differentiable field, each Brouwer region will generally be the union of several "cells" in the sense of §17. An example is the region illustrated

---

* The concepts of "singular" and "nonsingular" paths were first introduced by Andronov and Pontryagin for a certain class of dynamic systems — the so-called "structurally stable" systems. The concept of orbitally stable and orbitally unstable paths, introduced in /46/ and presented in Chapter VII of this book, is a natural generalization of these concepts.

in Figure 341, which is a single Brouwer region but the union of two cells. Brouwer did not consider the question of defining the paths which determine the topological structure of the phase portrait.

It is also natural to compare $\omega\,(\alpha)$-orbital stability with the concept of $\omega\,(\alpha)$-regular points, defined by Birkhoff in connection with mappings of a surface into itself. This concept can be carried over quite naturally to the dynamic systems considered in this book, on the sphere or in a bounded plane region, as follows. Let $N$ be the set of all limit points of all paths on the sphere (in our case, where the number of singular paths is finite, this set is closed), and $G$ the complement of $N$ in the sphere, which is of course an open set. Let $\sigma$ be some region contained together with its boundary in $G$.

Consider the possible motions $M = f\,(t - t_0,\,M_0)$, where $M_0$ is any point of $\sigma$. Call $\sigma$ an $\omega\,(\alpha)$-regular region if for any $\varepsilon > 0$ there exists a number $T > 0\ (T < 0)$, independent of the point $M_0$, such that the point $M = f\,(\tau,\,M_0)$ lies within $U_\varepsilon\,(N)$ for all $\tau > T\ (\tau < T)$. The points of $\sigma$ will be called $\omega\,(\alpha)$-regular points.

FIGURE 341.

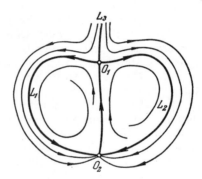

FIGURE 342.

It is not difficult to see that in our case (finitely many singular paths) all points of nonclosed orbitally stable paths are $\omega$- and $\alpha$-regular. When there are infinitely many singular paths, this may not be true. On the other hand, points of $\omega\,(\alpha)$-orbitally unstable paths may be $\omega\,(\alpha)$-regular. A simple example is the separatrix in Figure 342. The set $N$ of all limit points contains the paths $L_1$ and $L_2$ (the equilibrium state $O_1$ is a saddle point, $O_2$ is a saddle node). The separatrix $L_3$ is $\omega$-orbitally unstable, but all its points are $\omega$-regular.

Birkhoff proved an analog of Theorem 54 (§16) stating that regions filled by $\omega\,(\alpha)$-regular points are at most doubly-connected.

We now discuss some more recent work, that of Markus and Vrublevskaya, which has a direct bearing on systems of type (1) satisfying the conditions of §1.

Markus /47/ also considers the partition of a plane region into "cells" containing "identical" phase portraits. His approach, however, is slightly different from ours; he does not single out "singular" and "nonsingular" paths, but from the start examines regions in which the phase portraits are "identical." (For exact definitions, see /47/.) When the number of singular paths is finite, Markus's regions are precisely cells in the sense of Chapter VII. He then considers a closed set of paths, the complement of the union of all regions containing "identical" phase portraits (this union is an open set), but does not define individual "singular" paths. Markus's approach leads in a natural way to consideration of both finite and infinite sets of singular paths (in our terminology). Thus, for example, one must consider the case of a countable set of limit cycles converging to a closed path.*

Vrublevskaya /49/ defines a r e g u l a r   d e f o r m a t i o n and uses this concept to define "geometric equivalence" of two sets, in particular, of two paths; she also defines "kinematic equivalence" of two paths, as a special case of geometric equivalence. The family of all paths (phase portrait) of a dynamic system is the union of geometric equivalence classes. The set-theoretic union of all paths of one class is called a "geometric cell." When the number of singular paths is finite, Vrublevskaya's "geometric cells" are precisely the cells of Chapter VII.

Each of these three approaches to path classification, that presented in this book and those of Markus and Vrublevskaya, carries over to dynamic systems of order $n > 2$, and each possesses its own advantages.

**3. CASE OF INFINITELY MANY ORBITALLY UNSTABLE PATHS.** Throughout the book it is assumed that the number of singular (orbitally unstable) paths is finite. As stated in §15.9, this case is the most natural from various standpoints. One can also follow Markus in considering infinite sets of orbitally unstable paths.

We have already given an example of an infinite set of singular paths — a countable set of limit cycles whose topological limit is a certain closed path. Figure 343 shows another simple example, in which an infinite set of "alternating" saddle points and nodes converge to a point $O$ and the separatrices have topological limits $L_0$ and $L_0'$ passing through $O$.

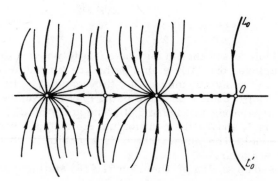

FIGURE 343.

***

* Note that when the limit cycles converge to a closed path $L_0$ (from both outside and inside $L_0$), this closed path is obviously the topological limit of the limit cycles and is orbitally stable. Thus, the topological limit of infinitely many orbitally unstable paths m a y  be  a n  orbitally  stable  path.

In both these examples, the set of orbitally unstable paths is infinite but nowhere dense on the sphere.

Confining ourselves to the case of finitely many equilibrium states, we now consider the natural question: can the set of orbitally unstable paths be dense, or do the orbitally unstable paths always form a nowhere dense set? Markus provides no answer to this question.

In the next subsection we present a geometric example, due to Maier, which in a certain sense answers the question. Maier constructs a family of curves, possessing the main properties of paths (and we shall therefore call them paths), which contains a dense set of orbitally unstable paths. Though the example is geometric and does not present a dynamic system satisfying the conditions of §1.1 and defining precisely this phase portrait, we are nevertheless heuristically convinced of the existence of such a system. Our description of Maier's example will employ the term "path" rather than curve, and also the terms "separatrix" and "elliptic region," defined usually for dynamic systems; this cannot lead to misunderstandings.

**4. MAIER'S GEOMETRIC EXAMPLE OF A DENSE SET OF ORBITALLY UNSTABLE PATHS — SEPARATRICES OF AN EQUILIBRIUM STATE.** Before proceeding to the geometric construction of a dense set of separatrices tending to a single singular point, we note that, first, the very definition of a separatrix implies that there must be a hyperbolic region on at least one of its "sides" and, second, in the neighborhood of an equilibrium state there may be an infinite set of different elliptic regions whose diameters tend to zero. The construction involves the following steps:

1. The separatrices are constructed successively; together with each separatrix we construct an elliptic region adjoining it on the left.

2. At each successive step, separatrices and elliptic regions are constructed in the unhatched regions of Figure 344a, b, c.

3. At each step, one new separatrix is constructed between every two previously constructed ones.

A few steps of the construction will be illustrated. Let $C$ be a circle about the origin.

Figure 344a shows the first step. No further elliptic regions will be constructed between the separatrix $L_1$ and the elliptic region $\sigma_1$ (i. e., in the hatched sector). In the final analysis the hatched sector will be a hyperbolic region. The next step is to construct a new separatrix and elliptic region in the unhatched sector. Let the points $A_1$ and $A_1'$ in Figure 344a be antipodal and let $A_2$ and $A_2'$ be another pair of antipodal points which bisect the semicircles $A_1A_1'$ and $A_1'A_1$.

Figure 344b illustrates the second step. As in the preceding step, no elliptic regions lie in the hatched sectors. The unhatched sectors will contain the separatrices and elliptic regions of the next step.

The third step is illustrated in Figure 344c; the points $A_3$, $A_3'$, $A_4$ and $A_4'$ bisect the arcs $A_1A_2$, $A_1'A_2'$, $A_2A_1'$ and $A_2'A_1$, respectively.

At the next step, separatrices are drawn through the points bisecting previously constructed arcs, and an elliptic region constructed on the negative side of each such separatrix (see §3 of this Appendix). Then the resulting arcs on the circumference are again bisected, and so on. After each step there is an unhatched sector between two previously constructed separatrices, in which a new separatrix and elliptic region may be constructed. The diameters of successively constructed elliptic regions tend to zero.

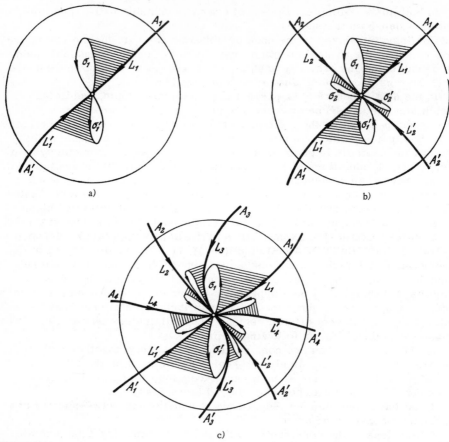

FIGURE 344.

As a result, we obtain a dense set of points on the circumference, through each of which passes a separatrix. Denote the interior of the circle by $S$ and the set of points in the closed elliptic regions by $\bar{E}$. Then the set $S \setminus \bar{E}$ is a region. It is readily seen that the union of the separatrices is dense in $S \setminus \bar{E}$. All other paths are uniquely determined by continuity considerations.

It is not hard to see that the family of curves thus obtained satisfies all necessary conditions for phase portrait of a dynamic system. Examination of the paths crossing the circle at points not on the separatrices shows that they are all orbitally stable.

## §10. BENDIXSON'S THEOREM ON THE INDEX OF A MULTIPLE EQUILIBRIUM STATE

In this final section we prove Bendixson's theorem on the relation between the number of hyperbolic and elliptic sectors of an equilibrium state and its Poincaré index. Let (I) be a dynamic system, $O$ an isolated equilibrium state, $h$ the number of hyperbolic sectors (i. e., the number of hyperbolic sectors in a sufficiently small neighborhood of $0$), $e$ the number of elliptic sectors, $I = I(0)$ the Poincaré index.

*Bendixson's theorem:*

$$I = 1 + \frac{e-h}{2}. \tag{1}$$

Bendixson proved this theorem for analytic dynamic systems (see Bendixson /33/). To simplify the proof, we shall assume that the system (I) is not necessarily analytic but satisfies the following condition:

(*) Let $L$ be a path of system (I) which tends to an equilibrium state $0$ as $t \to +\infty$ $(t \to -\infty)$ and is not a spiral, $M = M(t)$ a point on $L$ and $MT$ the tangent to $L$ at $M$; then the tangent $MT$ tends to a limit position as $t \to +\infty$ $(t \to -\infty)$.

It follows easily from condition (*) that the curve formed by adding $0$ to the path $L$ has a tangent at $0$ which is the limit position of the tangent $MT$. We shall call this the tangent to the path $L$ at the point $0$.

One would naturally like to know whether condition (*) is always satisfied for analytic systems. It was proved in §20.2 (Remark 1) that as a rule this is true, but the question is nevertheless still open in one exceptional case.

The following proof is a sharpened version of that in Lefschetz /13/ (Chapter X, §2).

Proof of Bendixson's theorem. If the equilibrium state $0$ is a center, its Poincaré index is unity by Theorem 28 (§11); but $e = h = 0$, so that (1) is true.

Now let $0$ be a topological node, that is to say, all paths passing sufficiently close to $0$ tend to $0$ (as $t \to +\infty$ or $t \to -\infty$; see §17, Lemma 8). Then, by Lemma 3 of §18, an arbitrarily small neighborhood of $0$ contains a cycle without contact with $0$ in its interior. But then (see §11, Theorem 29) $I(0) = 1$. Since in this case too $e = h = 0$, we again have (1).

We must therefore consider the case in which the equilibrium state $0$ is neither a center nor a topological node, so that the neighborhood of $0$ contains at least one elliptic or hyperbolic sector. Let $p$ denote the number of parabolic sectors in a canonical neighborhood of $0$ and $n$ the number of all sectors, $(n = h + e + p)$.

Let $L_1^{()}, L_2^{()}, \ldots, L_n^{()}$ be the semipaths comprising the boundaries of the sectors in the canonical neighborhood, enumerated in cyclic order (in the positive direction around $0$), and $\theta_i$ $(i = 1, 2, \ldots, n)$ the directions along which these semipaths reach the equilibrium state $0$ (as $t$ tends to $+\infty$ or $-\infty$). We denote the smallest nonnegative angle between rays in directions $\theta_i$ and $\theta_{i+1}$ $(i = 1, 2, \ldots, n; \theta_{n+1} = \theta_1)$ by $\alpha_k$, $\beta_e$ or $\gamma_m$, according to whether the sector bounded by semipaths $L_i^{()}$ and $L_{i+1}^{()}$ is hyperbolic, elliptic or parabolic. Obviously,

$$\sum_1^h \alpha_k + \sum_1^e \beta_e + \sum_1^p \gamma_m = 2\pi. \tag{2}$$

Note that some of the angles $\alpha$, $\beta$ or $\gamma$ may vanish.

Let $C_0$ be a circle of radius $r_0$ about $O$, having points in common with each of the semipaths $L_i^{(\prime)}$ ($i = 1, 2, \ldots, n$), and $P_i$ the "last" point of intersection of the semipath $L_i^{(\prime)}$ with the circle $C_0$. It follows from condition (*), as is readily seen, that if $r_0$ is sufficiently small the following conditions will hold:

1) each semipath $L_i^{(\prime)}$ crosses the circle $C_0$ at $P_i$ at an angle arbitrarily close to $\pi/2$;

2) the rotation of the vector field along the segment of $L_i^{(\prime)}$ with endpoints $P_i$ and $O$ is arbitrarily small.

It clearly follows from 1) that

3) any sufficiently small arc of the circle $C_0$ containing the point $P_i$ ($i = 1, 2, \ldots, n$) is an arc without contact for the system.

Take $r_0 > 0$ so small that conditions 1) and 2) are satisfied, and construct a canonical closed curve $E$ of the equilibrium state $O$ through the points $P_1, P_2, \ldots, P_n$ (Figure 345), in such a way that the saddle arcs without contact are sufficiently small arcs of $C_0$. The existence of a curve $E$ satisfying these conditions was established in §19.2.

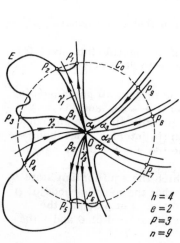

FIGURE 345.

$h = 4$
$e = 2$
$P = 3$
$n = 9$

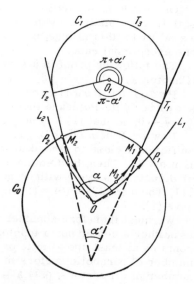

FIGURE 346.

The rotation of the vector field of system (I) along the closed curve $E$ is equal to the sum of rotations of the field along the various components of $E$: elliptic and hyperbolic arcs, parabolic arcs without contact and saddle arcs without contact (see §19). It follows from 1) and 3) that the sum of rotations of the vector field along the saddle arcs without contact may be assumed arbitrarily small (by making the saddle arcs sufficiently small). We now compute the rotation of the field along the hyperbolic arcs without contact.

Rotation of the field along a hyperbolic arc. To fix
ideas, assume that the arc $M_1M_3M_2$ in question lies in the hyperbolic
sector bounded by separatrices $L_1$ and $L_2$, and denote by $\alpha$ the angle $\alpha_h$ of
this sector. Suppose first that $0 < \alpha < \pi$. Draw rays tangent to the arc of
at its endpoints $M_1$ and $M_2$ and lying outside $C_0$. By condition 2) and the
fact that the arcs $P_1M_1$ and $P_2M_2$ are small, the angle $\alpha'$ between these rays
may be made arbitrarily close to $\alpha$ (in particular, we may assume that
$0 < \alpha' < \pi$. Let $T_1T_3T_2$ be an arc of a circle $C_1$ about a point $O_1$, tangent to
the rays at the points $T_1$ and $T_2$, concave toward $C_0$ and not cutting $C_0$
(Figure 346; the last condition holds automatically if the center $O_1$ of $C_1$
is sufficiently distant).

Let $\Gamma$ denote the simple closed curve consisting of the hyperbolic arc
$M_2M_3M_1$, the segments $M_1T_1$, $T_2M_2$ and the arc $T_1T_3T_2$ of $C_1$. By Theorem 28
(§11) the rotation of the tangent field along this curve is $2\pi$. The rotation
of this field along the straight-line segments $M_1T_1$ and $T_2M_2$ is zero, and
along the arc $T_1T_3T_2$ of $C_1$ it is obviously $\pi + \alpha'$ (Figure 346). Thus the
rotation of the tangent field along the arc $M_2M_3M_1$ is $2\pi - (\pi + \alpha') = \pi - \alpha'$.
Since this field is identical with the vector field of our dynamic system, it
follows that as the hyperbolic arc is described in the sense $M_1M_3M_2$ (induced
by the positive traversal of the canonical curve $E$) the rotation of the field
is $\alpha' - \pi$. The same obviously holds if the direction on the separatrices
$L_1$ and $L_2$ and the path $M_1M_3M_2$ is opposite to that shown in the figure.

We have assumed hitherto that $0 < \alpha' < \pi$. Now let $\alpha' = \pi$. Then the
rays $M_1T_1$ and $M_2T_2$ have opposite directions, and the rotation of the field
along the arc $M_1M_3M_2$ can be determined by drawing two auxiliary semicircles and a straight line parallel to the rays (Figure 347). Examination
of the new curve $\Gamma$ and its tangent field shows that in this case again the
rotation along the hyperbolic arc is 0, i. e., $\alpha' - \pi$.

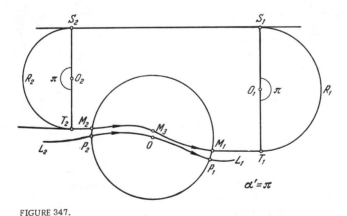

FIGURE 347.

We leave it to the reader to consider the cases $\pi \leqslant \alpha' < 2\pi$ and $\alpha = 0$
and to show that in these cases too the rotation of the field along the
hyperbolic arc is $\alpha' - \pi$. In all cases, therefore, the rotation may be
made arbitrarily close to $\alpha - \pi$.

Rotation of the field along an elliptic arc. Let $P_1SP_2$ be an elliptic arc, $L$ the path ("loop") of which it is a subarc, $L_1$ and $L_2$ the corresponding semipaths, $P_1$ and $P_2$ their last points of intersection with the circle $C_0$ (Figure 348). Let $\beta$ be the angle $\beta_k$ corresponding to our elliptic sector and $\beta'$ the angle between rays tangent to the path $L$ at $P_1$ and $P_2$. By virtue of our assumptions, $\beta'$ may be made arbitrarily close to $\beta$.

To compute the rotation of the field along $P_1SP_2$, draw a smooth simple arc $P_1TP_2$ lying entirely (except for its endpoints $P_1$ and $P_2$) inside the circle $C_0$ and inside the loop $L$, touching the path $L$ at $P_1$ and $P_2$ and having no points other than $P_1$ and $P_2$ in common with $P_1SP_2$. These conditions may be met by drawing $P_1TP_2$ inside the circle $C_0$, between the semipaths $L_1$ and $L_2$ and near them.

Let $\Gamma$ denote the curve consisting of the elliptic arc $P_1SP_2$ and the auxiliary arc $P_1TP_2$. $\Gamma$ is a smooth simple closed curve and its field of tangents coincides on the arc with the field of system (I). The rotation of the tangent field along $\Gamma$ is $2\pi$. The rotation of the tangent field along the curve $P_1TP_2$ may be determined exactly as for the case of a hyperbolic arc, and consequently is equal to $\beta' - \pi$. Thus the rotation along the elliptic arc $P_1SP_2$ is $2\pi - (\pi - \beta') = \pi + \beta'$, i.e., arbitrarily close to $\pi + \beta$.

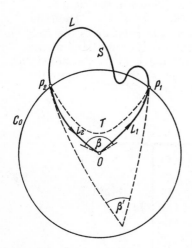

FIGURE 348.

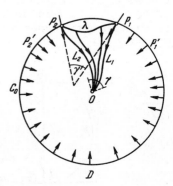

FIGURE 349.

Rotation of the field along a parabolic arc without contact. Let $\lambda$ be a parabolic arc without contact, $P_1$ and $P_2$ its endpoints, $L_1$ and $L_2$ the semipaths through these points, comprising the boundary of the corresponding parabolic sector, $\gamma$ the angle between them at the point $0$, $\gamma'$ the angle between the tangents to $L_1$ and $L_2$ at $P_1$ and $P_2$, respectively (Figure 349). As before, the ange $\gamma'$ may be made arbitrarily close to $\gamma$. To fix ideas, let the sector be $\omega$-parabolic, so that the paths passing through it tend to $0$ as $t \to +\infty$.

By condition 3) above, any sufficiently small arc of $C_0$ which contains $P_i$ is an arc without contact. Hence, by Remark 3 to Lemma 8 in §3, it follows that we may assume the arc $\lambda$ to be tangent to $C_0$ at $P_1$ and $P_2$. Then the curve $\Gamma$ consisting of $\lambda$ and the arc $P_2DP_1$ of $C_0$ (Figure 349) is a smooth simple closed curve. We construct a continuous vector field $v = v(M)$, $M \in \Gamma$, on this curve, without singularities, as follows. On the arc $\lambda$ the directions of the vectors $v$ coincide with those of the vectors of the dynamic system. Now let $P_1'$ and $P_2'$ be two points on the arc $P_2DP_1$ of $C_0$, close to $P_1$ and $P_2$, respectively. The vectors of the field on the arc $P_2'DP_1'$ of $C_0$ point to the center. Finally, the directions of the vectors on the arc $P_iP_i'$ $(i = 1, 2)$ are chosen so that, as the arc is described from $P_i$ to $P_i'$, the vector rotates uniformly in the same sense and always points into the interior of $C_0$ (this uniquely determines the vectors if $P_i'$ is sufficiently close to $P_i$ and the radius of $C_0$ is sufficiently small).

It is clear that all the vectors $v$ of this field point into the interior of $\Gamma$. But then, by the corollary to Theorem 29 (§11), the rotation of the field $v$ along $\Gamma$ is $2\pi$. A direct computation shows that the rotation of the field $v$ along the arc $P_2DP_1$ of $C_0$ is $2\pi - \gamma'$. Thus the rotation of the field $v$, hence also that of our system, along the curve $\lambda$ is $\gamma'$, i. e., arbitrarily close to $\gamma$.

Rotation of the field of the system along $E$. It now follows that the rotation of the field of system (I) along the canonical curve without contact $E$ may be made arbitrarily close to

$$\sum_1^h (\alpha_k - \pi) + \sum_1^e (\pi + \beta_l) + \sum_1^p \gamma_m.$$

By (2), this number is equal to $e\pi - h\pi + 2\pi$. Hence the index $I$ of the equilibrium state $O$ is arbitrarily close to

$$\frac{e\pi - h\pi + 2\pi}{2\pi} = 1 + \frac{e-h}{2},$$

and since the index is an integer, this implies that

$$I = 1 + \frac{e-h}{2}.$$

This completes the proof of Bendixson's theorem.

Corollary. If $O_1$ and $O_2$ are isolated equilibrium states of dynamic systems $(A_1)$ and $(A_2)$, respectively, which have the same topological structure, then their Poincaré indices are equal.

In other words, the Poincaré index is an invariant of path-preserving topological mappings. The underlying reason for this is that a mapping of this kind maps equilibrium states into equilibrium states and each sector into a sector of the same type.

In conclusion, we observe that by Bendixson's formula the numbers of elliptic and hyperbolic sectors must have the same parity, $e \equiv h \pmod 2$. However, this is easily proved directly by considering the directions on the semipaths $L_1, L_2, \ldots, L_n$ comprising the boundaries of the sectors.

## Bibliography*

1. Poincaré, H. La valeur de la science, Chap. VI. Paris. 1906.
2. Laplace, P.S. Essai philosophique sur le fondement des probabilités. Paris. 1814.
3. Bohr, N. Essays 1958 − 1962 on Atomic Physics and Human Knowledge. New York, Interscience Publishers. 1963.
4. Lyapunov, A. M. Obshchaya zadacha ob ustoichivosti dvizheniya (The General Problem of Stability of Motion). Moskva − Leningrad, Gostekhizdat. 1950. Also in: Sobranie sochinenii A. M. Lyapunova (Collected Works of A. M. Lyapunov). Moskva − Leningrad, Izdatel'stvo AN SSSR. 1956.
5. Poincaré, H. Sur les courbes définies par une équation différentielle. − Oeuvres, Vol. I. Paris, Gauthier-Villars. 1928.
6. Andronov, A. A., A. A. Vitt, and S. E. Khaikin. Teoriya kolebanii (Theory of Oscillations) (Second edition, edited by N. A. Zheleztsov). Moskva, Fizmatgiz. 1959.
7. Mandel'shtam, L. I. Voprosy elektricheskikh kolebatel'nykh sistem i radiotekhniki (Problems of Electrical Oscillating Systems and Radio-Engineering). − In: Pervaya Vsesoyuznaya konferentsiya po kolebaniyam (First All-Union Conference on Oscillations), Vol. I. GTTI. 1933.
8. Mandel'shtam, L. I., N. D. Papaleksi, A. A. Andronov, A. A. Vitt, G. S. Gorelik, and S. E. Khaikin. Novye issledovaniya v oblasti nelineinykh kolebanii (New Research in the Field of Nonlinear Oscillations). Moskva, Radioizdat. 1936.
9. Andronov, A. A. 1) Matematicheskie problemy teorii kolebanii (Mathematical Problems of the Theory of Oscillations). 2) L. I. Mandel'shtam i teoriya nelineinykh kolebanii (Mandel'shtam and the Theory of Nonlinear Oscillations). − In: Sobranie sochinenii (Collected Works). Moskva − Leningrad, Izdatel'stvo AN SSSR. 1956.
10. Birkhoff, G. D. Dynamical Systems. New York, American Mathematical Society. 1927.
11. Pontryagin, L. S. Obyknovennye differentsial'nye uravneniya (Ordinary Differential Equations). Moskva, Gostekhizdat. 1961.
12. Coddington, E. A. and N. Levinson. Theory of Ordinary Differential Equations. New York, McGraw-Hill. 1955.
13. Lefschetz, S. Differential Equations: Geometric Theory. New York, Interscience Publishers. 1957.
14. Kamke, E. Differentialgleichungen reeller Funktionen. Leipzig, Akademische Verlagsgesellschaft. 1952.
15. Hilbert, D. and S. Cohn-Vossen. Geometry and the Imagination. New York, Chelsea. 1952.

* [Abbreviations of periodical names are as in Mathematical Reviews (except for transliteration). ]

16. B e l y u s t i n a, L. N.  K dinamike simmetrichnogo poleta samoleta (The Dynamics of Symmetric Aircraft Motion). Izv. Akad. Nauk SSSR, Otd. Tekhn. Nauk, No. 11. 1956.

17. B a t a l o v a, Z. S. and L. N. B e l y u s t i n a.  Issledovanie odnoi nelineinoi sistemy na tore (Investigation of a Certain Nonlinear System on the Torus).  Izv. Vyssh. Uchebn. Zaved. Radiofizika 6. 1963.

18. M a i e r, A. G.  O traektoriyakh na orientiruemykh poverkhnostyakh (On Paths on Orientable Surfaces).  Mat. Sb. 12 (54).  1943.

19. B e n d i x s o n, I.  Sur les courbes définies par des équations différentielles. Acta Math. 24. 1901.

20. K o n s t a n t i n o v, N. N.  O nesamoperesekayushchikhsya krivykh na ploskosti (On Non-self-intersecting Curves on the Plane). Mat. Sb. 54 (96). 1961.

21. M a l ' t s e v, A. I.  Osnovy lineinoi algebry (Elements of Linear Algebra). Moskva, Gostekhizdat. 1956.

22. S t e p a n o v, V. V.  Kurs differentsial'nykh uravnenii (Course of Differential Equations) (Fourth edition).  Moskva, Gostekhizdat. 1945.

23. N e m y t s k i i, V. V. and V. V. S t e p a n o v.  Kachestvennaya teoriya differentsial'nykh uravnenii (Qualitative Theory of Differential Equations).  Moskva — Leningrad, Gostekhizdat. 1949.

24. P i c a r d, E.  Traité d'analyse, Vol. III. Paris, Gauthier-Villars. 1896.

25. P o i n c a r é, H.  Sur les propriétés des fonctions définies par les équations aux différences partielles (Thèse inaugurale, Paris 1879). — In: Oeuvres, Vol. I. Paris, Gauthier-Villars. 1928.

26. D u l a c, H.  Solutions d'un système d'équations différentielles dans le voisinage des valeurs singulières.  Bull. Soc. Math. France 40. 1912.

27. P e r r o n, O.  Über Stabilität und asymptotisches Verhalten der Integrale von Differentialgleichungssystemen.  Math. Z., 29. 1928.

28. P e r r o n, O.  Die Stabilitätsfrage bei Differentialgleichungen. Math. Z. 32. 1930.

29. P e t r o v s k i i, I. G.  Über das Verhalten der Integralkurven eines Systems gewöhnlicher Differentialgleichungen in des Nähe eines singulären Punktes.  Mat. Sb.  41. 1934.

30. P e t r o v s k i i, I. G.  O povedenii integral'nykh krivykh sistemy differentsial'nykh uravnenii v okrestnosti osoboi tochki (On the Behavior of Integral Curves of a System of Differential Equations in the Neighborhood of a Singular Point).  Mat. Sb. 41. 1934. [Apparently the same as /29/.]

31. B e l l m a n, R.  On the Boundedness of Solutions of Nonlinear Differential and Difference Equations.  Trans. Amer. Math. Soc. 62. 1947.

32. B i e b e r b a c h, L.  Differentialgleichungen.  Berlin, Springer. 1930.

33. B e n d i x s o n, I.  [See /19/.]

34. K r a s n o s e l ' s k i i, M. A.  Vektornye polya na ploskosti (Vector Fields on the Plane).  Moskva, Fizmatgiz. 1963.

35. S t e b a k o v, S. A.  Analiz staticheski ustoichivykh dinamicheskikh sistem (Analysis of Statically Stable Dynamic Systems). — Dokl. Akad. Nauk SSSR 95. 1954.

36. D u l a c, H. Recherche des cycles limites. C. R. Acad. Sci. Paris, 204. 1937.

37. D u l a c, H. Sur les cycles limites. Bull. Soc. Math. France 51. 1923.

38. S a n s o n e, G. and R. C o n t i. Soluzioni periodiche dell equazione avente due soluzione singolare. Abh. Math. Sem. Univ. Hamburg 20. 1956.

39. S a n s o n e, G. Equazioni differenziali nel campo reale, Parte seconda (Second edition). Bologna, Zanichelli. 1949.

40. M a r k o s y a n, S. Kachestvennoe issledovanie sistemy dvukh differentsial'nykh uravnenii metodom "dvukh izoklin" (Qualitative Investigation of a System of Two Differential Equations by the Method of "Two Isoclines"). Izv. Vyssh. Uchebn. Zaved. Matematika, No. 1/8. 1959.

41. N e m y t s k i i, V. V. Kachestvennoe integrirovanie sistemy differentsial'nykh uravnenii (Qualitative Integration of a System of Differential Equations). Mat. Sb. 16. 1946.

42. N e m y t s k i i, V. V. Kachestvennoe integrirovanie sistemy s pomoshch'yu universal'nykh lomanykh (Qualitative Integration of a System Using Universal Polygons). Uchen. Zap. Moskov. Gos. Univ. 100. 1946.

43. C o l l a t z, L. Numerische Behandlung von Differentialgleichungen. Berlin, Springer. 1951.

44. M i l n e, W. E. Numerical Solution of Differential Equations. New York, Wiley. 1957.

45. B e r e z i n, I. S. and N. P. Z h i d k o v. Metody vychislenii (Computing Methods), Vol. II. Moskva, Fizmatgiz. 1959.

46. L e o n t o v i c h, E. A. and A. G. M a i e r. O traektoriyakh, opredelya-yushchikh kachestvennuyu strukturu razbieniya na traektorii (On Paths Defining the Qualitative Structure of the Partition into Paths). Dokl. Akad. Nauk SSSR 14. 1937.

47. M a r k u s, L. Global Structure of Ordinary Differential Equations in the Plane. Trans. Amer. Math. Soc. 76. 1954.

48. B r u b l e v s k a y a, I. N. Nekotorye kriterii ekvivalentnosti traektorii i polutraektorii dinamicheskikh sistem (Some Criteria for Equi-valence of Paths and Semipaths of Dynamic Systems). Dokl. Akad. Nauk SSSR 97. 1954.

49. B r u b l e v s k a y a, I. N. O geometricheskoi ekvivalentnosti traektorii i polutraektorii dinamicheskikh system (On the Geometric Equivalence of Paths and Semipaths of Dynamic Systems). Mat. Sb. 42 (84). 1947.

50. L e o n t o v i c h, E. A. and A. G. M a i e r. O skheme, opredelyayushchei topologicheskuyu strukturu razbieniya na traektorii (On a Scheme Defining the Topological Structure of the Partition into Paths). Dokl. Akad. Nauk SSSR 103. 1955.

51. B r o u w e r, L. Continuous One-one Transformations of Surfaces in Themselves. Nederl. Akad. Wetensch. 11. 1908; 12. 1909; 13. 1910.

52. F r o m m e r, M. Die Integralkurven einer gewöhnlichen Differentialgleichung erster Ordnung in der Umgebung rationaler Unbestimmtheitsstellen. Math. Ann. 99. 1928.

53. A n d r e e v , A. F. Issledovanie povedeniya integral'nykh krivykh odnoi sistemy dvukh differentsial'nykh uravnenii v okrestnosti osoboi tochki (Investigation of the Behavior of Integral Curves of a Certain System of Two Differential Equations in the Neighborhood of a Singular Point). Vestnik Leningrad. Univ. 8. 1955.

54. K h a i m o v , N. B. Issledovanie uravneniya, pravaya chast' kotorogo soderzhit lineinye chleny (Investigation of an Equation Whose Right-Hand Side Contains Linear Terms). Uchen. Zap. Fiz.-Mat. Fakul'teta Stalinabad. Ped. Uchitel. Inst. 2. 1952.

55. G u b a r ' , N. A. Kharakteristika slozhnykh osobykh tochek sistemy dvukh differentsial'nykh uravnenii pri pomoshchi grubykh osobykh tochek blizkikh sistem (Classification of Multiple Singular Points of a System of Two Differential Equations Using Structurally Stable Singular Points of Close Systems). Mat. Sb. 40 (82). 1956.

56. H a u s d o r f f , F. Mengenlehre. New York, Dover. 1944.

57. K e r e k j a r t o , B. Vorlesungen über Topologie. Berlin, Springer. 1923.

58. F i l i p p o v , A. F. Elementarnoe dokazatel'stvo teoremy Zhordana (An Elementary Proof of Jordan's Theorem). Uspekhi Mat. Nauk 5. 1950.

59. F i k h t e n g o l ' t s , G. M. Kurs differentsial'nogo i integral'nogo ischisleniya (Course of Differential and Integral Calculus) (Third edition). Moskva — Leningrad, Gostekhizdat. 1951.

60. C o u r a n t , R. Differential and Integral Calculus, Vol. II (Second edition). London, Blackie. 1958.

61. P e t r o v s k i i , I. G. Lektsii po teorii obyknovennykh differentsial'nykh uravnenii (Lectures on the Theory of Ordinary Differential Equations) (Fourth edition). Moskva — Leningrad, Gostekhizdat. 1952.

62. B i r k h o f f , G. D. and P. A. S m i t h . Structure Analysis of Surface Transformations. J. Math. Pures Appl. 9. 1928.

63. M o i s e e v , N. D. Ob odnom "metode" otyskaniya predel'nykh tsiklov (On a "Method" for Determination of Limit Cycles). Zh. Eksper. Teoret. Fiz. 9. 1939.

64. B e l l y u s t i n , S. V. K teorii toka v vakuume (On the Theory of Current in a Vacuum). Zh. Eksper. Teoret. Fiz. 9. 1939.

65. B a u t i n , N. N. Ob odnom sluchae negarmonicheskikh kolebanii (On a Certain Case of Non-Harmonic Oscillations). Uchen. Zap. Gor'kov. Gos. Univ. 12. 1939.

66. B a u t i n , N. N. O periodicheskikh resheniyakh odnoi sistemy differentsial'nykh uravnenii (On Periodic Solutions of a Certain System of Differential Equations). Prikl. Mat. Mekh. 18. 1954.

67. M i n t s , R. M. O nekotorykh differentsial'nykh uravneniyakh dopuskayushchikh ponizhenie poryadka (On Certain Differential Equations Whose Order Can be Lowered). Trudy Gor'kov. Issled. Fiz.-Tekhn. Inst. i Radiofaka Gor'kov. Gos. Univ., Uchen. Zap. 35. 1957.

68. S a n s o n e , G. and R. C o n t i . Sull' equazione di T. Uno ed R. Yokomi. Ann. Mat. Pura Appl. (4) 37. 1954.

69. B a u t i n , N. N. Ob odnom differentsial'nom uravnenii imeyushchem predel'nyi tsikl (On a Certain Differential Equation Having a Limit Cycle). Zh. Tekhn. Fiz. 9. 1939.

70. **Mints, R. M.** Issledovanie traektorii sistemy trekh differentsial'nykh uravnenii v beskonechnosti (Investigation of the Paths of a System of Three Differential Equations at Infinity). — In: Sbornik pamyati A. A. Andronova (Collection of Papers in Memory of A. A. Andronov). Moskva, Izdatel'stvo AN SSSR. 1955.

71. **Buchel, W.** Die physikalischen Bedeutungen der durch die Differentialgleichung $\frac{dy}{dx} = \frac{Y(x,y)}{X(x,y)}$ definierten Kurvenschar. Mitt. Math. Ges. Hamburg 4. 1908.

72. **Sharpe, F. R.** The Topography of Certain Curves Defined by a Differential Equation. Ann. of Math. 11. 1909.

73. **Chandrasekhar, S.** An Introduction to the Study of Stellar Structure. New York, Dover. 1957.

74. **Kostitzin, V. A.** Sur le développement des populations bactériennes. C. R. Acad. Sci. Paris 242. 1956.

75. **Bellman, R.** Stability Theory of Differential Equations. New York, McGraw-Hill. 1953.

76. **Lemke, H.** Über die Differentialgleichungen, welche den Gleichgewichtszustand eines gasförmingen Himmelskörpers bestimmen, dessen Teile gegeneinander nach dem Newtonschen Gesetze gravitieren. J. Reine Angew. Math. 1942. 1912.

77. **Sommerfeld, A.** Atombau und Spektrallinien — Wellenmechanischer Ergänzungsband. Braunschweig, Vieweg. 1929.

78. **Aronovich, G. V.** and **L. N. Belyustina.** Ob ustoichivosti kolebanii gorizonta v uravnitel'noi bashne (On the Stability of Level Fluctuations in a Surge Tank). Inzhenernyi Sbornik, Inst. Mekh. AN SSSR 13. 1952.

79. **Tsin' Yuan'-syun' (Chin, Yuan-shun).** On Algebraic Limit Cycles of Degree 2 of the Differential Equation $\dfrac{dy}{dx} = \dfrac{\sum\limits_{0 \leqslant i+j \leqslant 2} a_{ij} x^i y^j}{\sum\limits_{0 \leqslant i+j \leqslant 2} b_{ij} x^i y^j}$ (Chinese, German summary). Acta Math. Sinica 8. 1958.

80. **Tun Tszin'-chzhu (Tung, Chin-chu).** Position of the Limit Cycles of the System $\dfrac{dx}{dt} = \sum\limits_{0 \leqslant i+j \leqslant 2} a_{ij} x^i y^j, \quad \dfrac{dy}{dt} = \sum\limits_{0 \leqslant i+j \leqslant 2} b_{ij} x^i y^j$ (Chinese, Russian summary). Acta Math. Sinica 8. 1958. Russian translation: Matematika 6. 1962.

81. **Sal'nikov, I. E.** K teorii periodicheskogo protekaniya gomogennykh khimicheskikh reaktsii (On the Theory of the Periodic Course of Homogeneous Chemical Reactions). Zh. Fiz. Khimii 23. 1949.

82. **Bautin, N. N.** K teorii sinkhronizatsii (On the Theory of Synchronization). Zh. Teoret. Fiz. 9. 1939.

83. **Bol'shakov, V. M., E. S. Zel'din, R. M. Mints,** and **N. A. Fufaev.** K dinamike sistem ostsillyator — rotator (On the Dynamics of Oscillator — Rotator Systems). Izv. Vyssh. Uchebn. Zaved. Radiofizika, No. 2. 1965.

84. **Gubar', N. A.** Issledovanie metodom Bendiksona topologicheskoi struktury raspolozheniya traektorii v okrestnosti osoboi tochki odnoi dinamicheskoi sistemy (Investigation by Bendixson's Method of the Topological Structure of the Configuration of Paths in the Neighborhood of a Singular Point of a Certain Dynamic System). Izv. Vyssh. Uchebn. Zaved. Radiofizika 2, No. 6. 1959.

# SUBJECT INDEX